T0249597

Escherichia coli

Escherichia coli

Pathotypes and Principles of Pathogenesis

Second Edition

Edited by

Michael S. Donnenberg

Division of Infectious Diseases
University of Maryland School of Medicine
Baltimore, Maryland, USA

AMSTERDAM • BOSTON • HEIDELBERG • LONDON
NEW YORK • OXFORD • PARIS • SAN DIEGO
SAN FRANCISCO • SINGAPORE • SYDNEY • TOKYO

Academic Press is an Imprint of Elsevier

Academic Press is an imprint of Elsevier
32 Jamestown Road, London NW1 7BY, UK
225 Wyman Street, Waltham, MA 02451, USA
525 B Street, Suite 1800, San Diego, CA 92101-4495, USA

Second edition

Notice
No responsibility is assumed by the publisher for any injury and/or damage to persons
or property as a matter of products liability, negligence or otherwise, or from any use
or operation of any methods, products, instructions or ideas contained in the material
herein. Because of rapid advances in the medical sciences, in particular, independent
verification of diagnoses and drug dosages should be made

British Library Cataloguing-in-Publication Data
A catalogue record for this book is available from the British Library

Library of Congress Cataloging-in-Publication Data
A catalog record for this book is available from the Library of Congress

ISBN: 978-0-12-397048-0

For information on all Academic Press publications
visit our website at www.store.elsevier.com

Typeset by TNQ Books and Journals

Working together
to grow libraries in
developing countries

www.elsevier.com • www.bookaid.org

Contents

Section II
Escherichia coli pathotypes   **73**

4.   Enteropathogenic *Escherichia coli*   **75**

Shahista Nisa, Karen M. Scanlon, and Michael S. Donnenberg

List of Contributors

Sowmya Balasubramanian Tufts University School of Medicine, Boston, MA, USA

Cedric N. Berger Imperial College London, London, UK

Julien R.C. Bergeron University of British Columbia, Vancouver, BC, Canada

Nadia Boisen University of Virginia School of Medicine, Charlottesville, VA, USA

Sujay Chattopadhyay University of Washington, Seattle, WA, USA

Abigail Clements Imperial College London, London, UK

Leon G. De Masi University of Maryland School of Medicine, Baltimore, MD, USA

Mauricio J. Farfan Universidad de Chile, Santiago, Chile

James M. Fleckenstein Washington University School of Medicine, St. Louis, MO, USA

Gad Frankel Imperial College London, London, UK

Victor A. Garcia-Angulo University of Texas Medical Branch, Galveston, TX, USA

David M. Gordon The Australian National University, Canberra, ACT, Australia

Ian R. Henderson University of Birmingham, Birmingham, UK

Scott J. Hultgren Washington University School of Medicine, St. Louis, MO, USA

Dakshina M. Jandhyala Tufts-New England Medical Center, Boston, MA, USA

Vasilios Kalas Washington University School of Medicine, St. Louis, MO, USA

Kwang Sik Kim Johns Hopkins University School of Medicine, Baltimore, MD, USA

Karen A. Krogfelt Statens Serum Institut Copenhagen, Denmark

John M. Leong Tufts University School of Medicine, Boston, MA, USA

Joshua A. Lieberman University of Maryland School of Medicine, Baltimore, MD, USA

Mariella Lomma Imperial College London, London, UK

Emily M. Mallick University of Massachusetts Medical School, Worcester, MA, USA

Anthony T. Maurelli Uniformed Services University, Bethesda, MD, USA

Harry L.T. Mobley University of Michigan Medical School, Ann Arbor, MI, USA

Carolyn R. Morris University of Maryland School of Medicine, Baltimore, MD, USA

James P. Nataro University of Virginia School of Medicine, Charlottesville, VA, USA

Shahista Nisa University of Maryland School of Medicine, Baltimore, MD, USA

David A. Rasko University of Maryland School of Medicine, Baltimore, MD, USA

Jason W. Sahl Translational Genomics Research Institute, Flagstaff, AZ, USA

Karen M. Scanlon University of Maryland School of Medicine, Baltimore, MD, USA

Evgeni V. Sokurenko University of Washington, Seattle, WA, USA

Rachel R. Spurbeck University of Michigan Medical School, Ann Arbor, MI, USA

Natalie C.J. Strynadka University of British Columbia, Vancouver, BC, Canada

Courntey D. Sturey University of Maryland School of Medicine, Baltimore, MD, USA

Alfredo G. Torres University of Texas Medical Branch, Galveston, TX, USA

Sivapriya Kailasan Vanaja Tufts University School of Medicine, Boston, MA, USA

Ender Volkan Washington University School of Medicine, St. Louis, MO, USA

Timothy J. Wells University of Birmingham, Birmingham, UK

Chris Whitfield University of Guelph, Guelph, ON, Canada

Lisa M. Willis University of Guelph, Guelph, ON, Canada

Liam J. Worrall University of British Columbia, Vancouver, BC, Canada

In his studies of the neonatal and infant fecal flora, Theodore Escherich (1857–1911) used the nascent techniques of bacterial isolation in pure culture, Gram staining, and fermentation reactions to identify 19 bacterial species (Shulman et al., 2007). He aptly chose the designation *Bacterium coli commune* (the common colon bacterium) for the organism that now bears his name, indeed the most common facultative anaerobe in the intestinal tract of humans and many other endothermic species. As he noted, *Escherichia coli* colonizes neonates within hours of birth, an event that probably occurs during delivery as these initial strains are usually serologically identical to those found in the mother (Bettelheim et al., 1974). We remain colonized with *E. coli* bacteria throughout life, although particular strains come and go over time. Most of these strains are non-pathogenic, coexisting in harmony with their hosts. Indeed, the relationship may be symbiotic, in that the bacteria, in addition to benefiting from the host, synthesize cofactors and contribute to colonization resistance against pathogenic organisms.

This pacific image of *E. coli* belies the fact that this species can also be regarded as the prototypical pleuripotent pathogen capable of causing a wide variety of illnesses in a broad array of species. The gastrointestinal tract, the meninges, and the kidneys are among the organs targeted by *E. coli*. Diseases resulting from *E. coli* infections include diarrhea, dysentery, pyelonephritis, and the hemolytic-uremic syndrome. Outcomes include sepsis, renal failure, and death. How is it possible for this Jekyll and Hyde species to both coexist peacefully with its host and cause devastating illness?

The answer lies in the existence of different strains of *E. coli* with variable pathogenic potential. Indeed, as early as 1897, Lesage postulated this point of view (cited in Robins-Browne, 1987) and the concept ultimately achieved general acceptance when Bray established that strains that we now term enteropathogenic *E. coli* (EPEC) cause devastating outbreaks of neonatal diarrhea (Bray, 1945). Since then, a plethora of pathogenic *E. coli* varieties or pathotypes has been described. The goal of this book, now in its second edition, is to review the current state of knowledge regarding those pathotypes which cause disease in humans, placing particular emphasis on mechanisms shared among strains.

The differences in the ability of strains to cause disease and the diverse syndromes caused by the various pathotypes can be attributed to specific genes encoding virulence factors and to the capacity of *E. coli* for genetic exchange. The core *E. coli* genome, that portion shared among all strains of the species, amounts to only about 20% of its average genome size (Rasko et al., 2008). In contrast, the total pool of genes available to be sampled by *E. coli* is much larger, at least six times that amount. Genes are constantly acquired and exchanged through plasmid transfer, bacteriophages, and perhaps by mechanisms unknown, to be tested by evolution. More subtle pathoadaptive mutations also contribute to disease. The diversity of *E. coli* and the pressures and outcomes of evolutionary forces are the focus of the first section of this book. The population structure and ecology of *E. coli* in humans, animals, and the environment is explored in the first chapter. Chapter 2 tackles the rapidly expanding universe of *E. coli* genome sequences, bringing some order to the genetic traits that define, contrast, and obscure the distinctions among pathotypes and placing these issues in the context of the radiation of *E. coli* strains from their most recent common ancestor millions of years ago. In Chapter 3 more emphasis is placed on evolutionary forces that drive the continual changes in *E. coli* genomes and the emergence of new variants capable of causing disease.

The pathogenic potential of a particular *E. coli* strain depends on the repertoire of the specific virulence genes it may possess. Particular virulence gene combinations define specific pathotypes of *E. coli*, and each pathotype has a propensity to cause a limited variety of clinical syndromes. A number of forces conspire to challenge clinicians and microbiologists to remain current in their appreciation of the diversity of *E. coli* infections. The complexity of the nomenclature is a product of the number of *E. coli* pathotypes, the similarities of their names, inconsistencies in usage in the literature, advances in our understanding of evolution and pathogenesis, and the emergence of new pathotypes. This nomenclature may be viewed as existing in a state of flux as new strains are described and the relationships among previously described pathotypes are clarified. Figure I.1 represents an attempt to illustrate these complex relationships. It is useful to view pathogenic strains as belonging to two groups: those which primarily cause gastrointestinal illness and those which primarily cause extraintestinal infections. However, there are strains with virulence potential that bridge these boundaries. Among the extraintestinal strains, it seems likely that most, if not all, strains capable of causing neonatal meningitis also can cause urinary tract infections, although the converse does not appear to be true. Among the gastrointestinal pathotypes, the situation is even more complex, especially given the overlap in attributes of EPEC and Shiga-toxin-producing *E. coli* (STEC) and the transmission of stx genes by transduction. However, a precise lexicon remains possible within the classification scheme presented. In the introduction to the first edition of this book, it was predicted that new strains would emerge with traits attributed to more than one of the pathotypes described then. Indeed, the 2011 outbreak of severe disease caused by a Shiga-toxin-producing enteroaggregative *E. coli* (Frank et al., 2011) validated this view and Figure I.1 has been updated to include such strains.

The second part of this book contains chapters detailing the molecular pathogenesis of infections due to each of the major *E. coli* pathotypes that cause human disease. These chapters provide a detailed profile of each of these

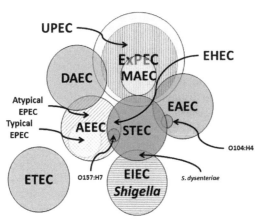

FIGURE I.1   Venn diagram illustrating the complex relationships among different pathotypes of
E. coli that cause disease in humans. Extraintestinal pathogenic *E. coli* (ExPEC, yellow) strains include
meningitis-associated *E. coli* (MAEC, bright yellow) and uropathogenic *E. coli* (UPEC, vertical
stripes) and strains from patients with pneumonia, cholecystitis, peritonitis, and other infections.
These strains share many virulence factors, and it is clear that single clones can cause both meningitis
and urinary tract infections (Russo and Johnson, 2000). It is less clear whether or not strains exist that
are capable of causing one syndrome and not the other. Among the UPEC, some strains exhibit diffuse
adherence to tissue culture cells and share with diffuse adhering *E. coli* (DAEC, orange) the same
adhesins. DAEC is a heterogeneous pathotype that has been epidemiologically linked to diarrhea.
There are reports of DAEC strains recovered from individuals with both urinary tract infections (UTIs)
and diarrhea (Germani et al., 1997). There are also reports of Shiga-toxin-producing *E. coli* (STEC,
green) strains causing UTI (Tarr et al., 1996) and other extraintestinal infections. STEC are defined
by production of Shiga toxins, usually encoded by bacteriophages. Among STEC, some strains are
also capable of attaching intimately to epithelial cells, effacing microvilli, and eliciting the formation
of adhesion pedestals composed of cytoskeletal proteins, a property that defines the attaching and
effacing *E. coli* (AEEC, diagonal stripes). Strains, which are both STEC and AEEC, are known as
enterohemorrhagic *E. coli* (EHEC). The most important serotype found within the EHEC pathotype
is O157:H7. AEEC strains that do not produce Shiga toxins are referred to as enteropathogenic *E. coli*
(EPEC). Among EPEC, many strains produce a bundle-forming pilus and attach to tissue culture
cells in a localized adherence pattern. These are referred to as typical EPEC (checkered), whereas
those which produce neither Shiga toxins nor bundle-forming pili are known as atypical EPEC. Some
strains of atypical EPEC exhibit diffuse adherence. Enteroinvasive *E. coli* (EIEC, horizontal stripes)
invade tissue culture cells with high efficiency, multiply in the cytoplasm, and spread from cell to cell.
These strains include the organisms commonly classified in the genus *Shigella,* which in fact all lie
phylogenetically within the species *E. coli*. Strains classified as *S. dysenteriae* serogroup 1 produce
Shiga toxins and therefore could be described as members of both the EIEC and STEC pathotypes.
Enteroaggregative *E. coli* (EAEC, blue) cause acute and persistent diarrhea and are defined by their
pattern of adherence. In 2011 a large outbreak of severe diarrhea was caused by EAEC belonging to
serotype O104:H4 that produced Shiga toxins, but other O104:H4 EAEC strains do not. Although
not commonly recognized as an extraintestinal pathogen, an outbreak of community acquired UTI in
Copenhagen was caused by EAEC (Olesen et al., 2012). Enterotoxigenic *E. coli* (ETEC, violet) strains
cause acute diarrhea and are defined by production of heat-labile and/or heat-stable enterotoxins.

categories of organisms. It should be recognized that additional pathogenic varieties exist that cause disease exclusively in non-human species. A remarkable
feature of this section is the number of distinct molecular pathways to human
disease that may be employed by *E. coli*. A chapter devoted to strains that are

hybrids of other pathotypes and strains for which the pathogenic potential in humans is less well established emphasizes the dynamic nature of an evolving field. Many of these pathogenic strategies are employed by other species that cause disease in animals and humans. Thus *E. coli* can serve as a model organism for the study of bacterial pathogenesis as well as intermediary metabolism.

Despite our attempts to distinguish strains of *E. coli*, there is much overlap in the mechanisms of pathogenesis for various pathotypes. Similar virulence pathways may be pursued by more than one type of strain. For example, pili of the chaperone-usher family are ubiquitous among pathogenic and non-pathogenic *E. coli* strains. Type 3 secretion systems (T3SSs) play an important role in the pathogenesis of EPEC, enterohemorrhagic *E. coli* (EHEC), and enteroinvasive *E. coli* infections. Type IV pili and the closely related type 2 secretion systems are expressed by EPEC and enterotoxigenic *E. coli* (ETEC). Hemolysins of the RTX family are produced by many strains of *E. coli* that cause extraintestinal infections, by EHEC, and occasionally by other strains associated with intestinal infections, while proteins exported by the autotransporter or type 5 pathway are ubiquitous among *E. coli*. Many strains of pathogenic *E. coli*, especially those that cause extraintestinal infections, elaborate polysaccharide capsules, and all strains make lipopolysaccharide. To allow these critical virulence factors to be explored in more detail than is possible in the second section, the final part of this book contains chapters devoted to virulence systems that are common to more than one pathotype. The explosion of information on the structure and function of T3SSs and the function of effector proteins employed by more than one pathotype are explored in two separate chapters. By design, each chapter of this book can stand alone while references among the chapters allow the reader to explore further detail on virulence mechanisms and how different pathotypes exploit similar systems.

The interrelationships among various pathogenic and non-pathogenic *E. coli* strains, the complexities of the disease pathways navigated by each pathotype, and the overlap in virulence mechanisms employed by different types reveal an intricate web of information about the organism. Yet there remains much to learn. Despite our advances in the cellular and molecular details of the interactions between these organisms and host cells, we remain ignorant of the mechanisms by which most strains of *E. coli* actually cause disease. Interactions with host factors likely dictate outcome for many infections. For some pathotypes, the virulence mechanisms that define the group remain largely mysterious. For other pathotypes that we thought we understood, further research has revealed new surprises, the significance of which has yet to be fully realized. The ever-increasing availability of genomic sequences will continue to reveal unanticipated genes that may help to unravel disease mechanisms, clarify relationships among pathotypes, and provide insight into the evolution of the species. It remains likely that additional pathotypes of *E. coli* lurk unrecognized, awaiting characterization until new assays are applied to strains isolated from patients and controls. *E. coli* has been subject to intensive scrutiny for more than a century and will continue to be regarded with interest for a long time to come.

REFERENCES

Bettelheim, K.A., Breadon, A., Faiers, M.C., O'Farrell, S.M., Shooter, R.A., 1974. The origin of O serotypes of *Escherichia coli* in babies after normal delivery. J. Hyg. (Lond) 72, 67–70.

Bray, J., 1945. Isolation of antigenically homogeneous strains of *Bact. coli neapolitanum* from summer diarrhoea of infants. J. Pathol. Bacteriol. 57, 239–247.

Frank, C., Werber, D., Cramer, J.P., Askar, M., Faber, M., an der, H.M., 2011. Epidemic profile of Shiga-toxin-producing *Escherichia coli* O104:H4 outbreak in Germany. N. Eng. J. Med. 365, 1771–1780.

Germani, Y., Bégaud, E., Duval, P., Le Bouguénec, C., 1997. An *Escherichia coli* clone carrying the adhesin-encoding *afa* operon is involved in both diarrhoea and cystitis in twins. T. R. Soc. Trop. Med. Hyg. 91, 573.

Olesen, B., Scheutz, F., Andersen, R.L., et al., 2012. Enteroaggregative *Escherichia coli* O78:H10, the cause of an outbreak of urinary tract infection. J. Clin. Microbiol. 50, 3703–3711.

Rasko, D.A., Rosovitz, M.J., Myers, G.S., et al., 2008. The pangenome structure of *Escherichia coli*: comparative genomic analysis of *E. coli* commensal and pathogenic isolates. J. Bacteriol. 190, 6881–6893.

Robins-Browne, R.M., 1987. Traditional enteropathogenic *Escherichia coli* of infantile diarrhea. Rev. Infect. Dis. 9, 28–53.

Russo, T.A., Johnson, J.R., 2000. Proposal for a new inclusive designation for extraintestinal pathogenic isolates of *Escherichia coli*: ExPEC. J. Infect. Dis. 181, 1753–1754.

Shulman, S.T., Friedmann, H.C., Sims, R.H., 2007. Theodor Escherich: the first pediatric infectious diseases physician? Clin. Infect. Dis. 45, 1025–1029.

Tarr, P.I., Fouser, L.S., Stapleton, A.E., et al., 1996. Hemolytic-uremic syndrome in a six-year-old girl after a urinary tract infection with shiga-toxin-producing *Escherichia coli* O103:H2. N. Eng. J. Med. 335, 635–638.

Escherichia coli, the organism

Chapter 1

The ecology of *Escherichia coli*

David M. Gordon

The Australian National University, Canberra, ACT, Australia

THE GENUS *ESCHERICHIA*

The genus *Esherichia* and the species *E. coli* have been recognized for over a century. In 1985, Farmer and colleagues described the first new species in the genus, *E. fergusonii* (Farmer et al., 1985). In 2003, Huys and colleagues described the second new species, *E. albertii* (Huys et al., 2003). Hyma et al. (2005) described the evolutionary relationship of *E. albertii* to *E. coli*, and its identity to the diarrheal pathogen, *Shigella boydii* serotypes 7 and 13. All of the other named species and serotypes of *Shigella* are actually members of *E. coli* (Sims and Kim, 2011). There are three other named species of *Escherichia*: *E. blattae*, *E. vulneris*, and *E. hermannii*, but strains of these species are only distantly related to other *Escherichia* and are not valid members of the genus (Walk et al., 2009).

The techniques of multi-locus sequence typing (MLST) and multi-locus sequence analysis (MLSA) have revolutionized our understanding of the evolution, ecology, epidemiology, and population genetics of bacteria (Maiden et al., 1998). There are three MLST schemes used for *E. coli* and MLST data have been collected for a great many isolates (http://mlst.ucc.ie/mlst/dbs/Ecoli, www.pasteur.fr/recherche/genopole/PF8/mlst/EColi.html, and http://www.shigatox.net/stec/cgi-bin/index). Coupled with the growth of the MLST databases, there are an increasing number of studies that isolate and characterize *E. coli* from non-clinical sources (Gordon, 1997; Souza et al., 1999; Pupo et al., 2000,b; Gordon and Cowling, 2003; Power et al., 2005; Wirth et al., 2006; Walk et al., 2007), and as a consequence our understanding of the diversity of bacteria has increased enormously. These studies have revealed substantially more genetic variation in the genus *Escherichia*, and five 'cryptic clades' of *Escherichia* were described (Walk et al., 2009). The term cryptic clades was used because, based on standard phenotypic methods, strains belonging to these novel *Escherichia* species are phenotypically indistinguishable from *E. coli*. Our current understanding of the relationships among the various *Escherichia* lineages is illustrated in Figure 1.1.

Escherichia coli. http://dx.doi.org/10.1016/B978-0-12-397048-0.00001-2

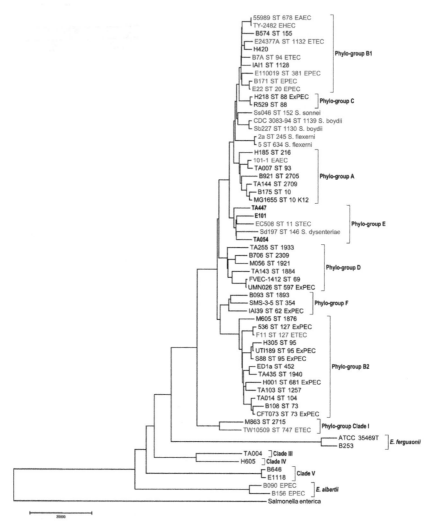

FIGURE 1.1 This phylogeny presented depicts our current understanding of the diversity to be found in the genus *Escherichia* and the phylo-group structure observed with the species *E. coli*. The phylogeny also demonstrates that both major types of *E. coli* pathogens, those causing diarrheal disease (EPEC, ETEC, STEC, EAEC, and *Shigella*; in red) and extraintestinal infection (ExPEC; in blue) are present in multiple phylogenic lineages. Strain names are presented as well as ST numbers (MLST hosted at http://mlst.ucc.ie/mlst/dbs/Ecoli) when available and the pathovar designation for the strain. Strains in black were isolated from the feces of asymptomatic hosts.

E. fergusonii has been implicated as an opportunistic extraintestinal pathogen of humans (Farmer et al., 1985; Funke et al., 1993; Savini et al., 2008), birds, and mammals (Herráez et al., 2005; Hariharan et al., 2007). It has been suggested that *E. fergusonii* is also capable of causing intestinal disease

(Bain and Green, 1999; Chaudhury et al., 1999), but the evidence for this supposition is weak. Genes associated with extraintestinal disease (e.g. *papG*, *sfa/focG*, *cnf1*, *ibeA*, *fyuA*, *iroN*, and *ompT*) are lacking in *E. fergusonii*, as are intestinal disease factors such as *astA*, *sat*, *pic*, *stx1*, and *eaeA*. By comparison, *E. albertii* is a known diarrheal pathogen of humans (Huys et al., 2003; Oaks et al., 2010; Ooka et al., 2012), an avian pathogen responsible for epidemic mortality in finches in Alaska and Scotland (Foster et al., 1998; Oaks et al., 2010), and has been implicated in the death of other bird species (Oaks et al., 2010). All strains of *E. albertii* appear to possess the intimin locus and produce cyto-lethal distending toxin B (Oaks et al., 2010). *Escherichia* strains belonging to the cryptic lineages CI–CV encode a variety of extraintestinal and intestinal virulence factors, sometimes at high frequencies (Ingle et al., 2011). However, with the exception of clade I strains, there appears to be little evidence to suggest that they are potential pathogens. Enterotoxigenic *E. coli* (ETEC) appear to have evolved on multiple occasions and Steinsland et al. (2010) identified one clonal group of ETEC strains, CG12, that they estimated to be atleast ten times older than all other ETEC clonal groups. ETEC CG12 strains are members of cryptic clade I suggesting that at least some members of this clade are potential diarrheal pathogens.

WHERE DOES *E. COLI* OCCUR?

E. coli can be isolated from plants and a wide variety of animals. While *E. coli* can be recovered from ectothermic vertebrates it is more frequently encountered in homeotherms (Gordon and Cowling, 2003). However, even among birds and mammals, factors such as body size and gut morphology are important predictors of the likelihood of isolating *E. coli* from a particular host (Gordon and Cowling, 2003). In birds and carnivorous mammals the probability of detecting *E. coli* in a host increases with the body size of the host. This outcome is likely a consequence of the relationship that exists between body mass and gut transit times. For example, in the carnivorous marsupials (Dasyuridae) gut transit times vary from about 1 hour for the 18-g *Sminthopsis crassicaudata* to 13 hours for the 1000-g *Dasyurus viverrinus*. Gut morphology also appears to play a role. Insectivorous bats have relatively short undifferentiated tube-like intestines that lack a cecum, while a similar-sized rodent has a well-differentiated intestine and possesses a cecum. *E. coli* is much less frequently isolated from bats than from similar-sized rodents. Experiments with rats fed diets differing in the concentration of crude fiber resulted, as expected, in food transit times through the gut declining with increasing fiber concentration (O'Brien and Gordon, 2011). These experiments also demonstrated that *E. coli* cell densities declined with decreasing transit times. Extrapolating the observed linear relationship between cell density and gut transit time predicts that *E. coli* should not be found in animals such as bats (O'Brien and Gordon, 2011). A variety of other factors may also influence *E. coli* cell densities in a host. For example, pregnancy coupled with excessive weight gain (Santacruz et al., 2010) or starvation in children (Monira et al., 2011) may elevate *E. coli* cell densities. Many different antibiotics have also been shown to lead to significant increases in *E. coli* cell density (Looft and Allen, 2012). However, there are little specific data on the nature of the interactions occurring between the host, *E. coli,* and other members of the gut microbiota.

Host effects, gut morphology and dynamics, and the gut microbiota are not the only factors determining the likelihood of detecting *E. coli* in a host. Background levels of contamination are also important. *E. coli* is more likely to be recovered from frogs, reptiles, or birds living in association with humans than in those living in undisturbed habitats such as national parks (Gordon and Cowling, 2003). The climate where the host lives also appears to be a factor in determining whether *E. coli* can be detected in a mammalian host. In Australia, *E. coli* is unlikely to be detected in hosts living in the desert and less likely to be detected in hosts living in the tropics, compared to hosts living in temperate regions of the country (Gordon and Cowling, 2003).

GENETIC STRUCTURE OF *E. COLI*

It has long been recognized that there is considerable genetic substructure in *E. coli* (reviewed in Chauduri and Henderson, 2012). In addition to the well-known phylo-groups A, B1, B2, and D, there is phylo-group E of which O157:H7 is the best-known member. Other workers recognize the existence of phylo-groups known as C and F (Tenaillon et al., 2010). Phylo-group C strains are closely related to phylo-group B1 strains and phylo-group F strains are related to D and B2 strains. Strains of the cryptic clade I should also be considered a phylo-group of *E. coli* (Luo et al., 2011). The relationships among the different phylo-groups of *E. coli* are depicted in Figure 1.1. Although there is a well-established and reliable PCR-based method of determining the phylo-group membership of an *E. coli* isolate (Clermont et al., 2000; Gordon et al., 2008), MLST is the only method, at present, capable of identifying strains belonging to phylo-groups C, E, and F.

Numerous studies have demonstrated that the distribution of strains belonging to the phylo-groups A, B1, B2, and D is very non-random. For example, among Australian vertebrates, strains belonging to phylo-group B1 are most frequently isolated from frogs, reptiles, birds, and carnivorous mammals such as bats and quolls, whilst B2 strains are rare in such hosts (Figure 1.2). Among humans living in developed countries such as Australia, the United States, and Europe, B1 strains are less often encountered than strains belonging to the other phylo-groups (Figure 1.3). By contrast, phylo-group A and B1 strains appear to be predominant in humans living in developing countries (Figure 1.3).

Consequently, the morphology and dynamics of the gastrointestinal tract as well as host diet appear to influence the phylo-group membership of strains present in a host. There is some experimental evidence to support these conclusions. B2 strains occurred at lower cell densities in rodents fed diets high in crude fiber as compared to the cell densities they achieved when the crude fiber concentration of the diet was low (O'Brien and Gordon, 2011). Diet effects may

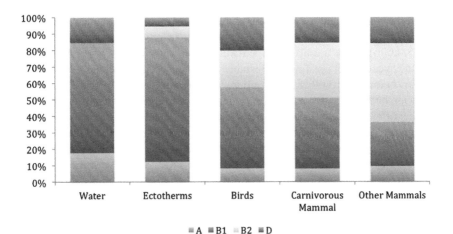

FIGURE 1.2   Relative abundance of the main *E. coli* phylo-groups (A, B1, B2, D) with respect to the source of isolation. *Unpublished data and data taken from Gordon and Cowling (2003), and Power et al. (2005).*

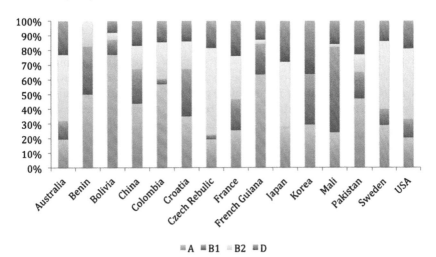

FIGURE 1.3   Relative abundance of the main *E. coli* phylo-groups (A, B1, B2, D) with respect to the source of isolation. *Data taken from Unno et al. (2009), Tenaillon et al. (2010), and Li et al. (2010).*

explain the very different relative abundances of the phylo-groups recovered from humans living in different parts of the world.

Therefore, strains belonging to the different phylo-groups appear to have different ecological niches and life-history characteristics. Phylo-group A and B1 strains appear to be generalists able to occupy a broad range of vertebrate hosts. By contrast, B2 and D strains are more commonly isolated from birds and mammals than ectotherms. It has been argued that phylo-group B2 strains are the most host adapted as, at least in humans, there are data to suggest they persist longer in a host than strains of the other phylo-groups (Nowrouzian et al., 2005, 2006). They are also competitively dominant, as hosts harboring a B2 strain have, on average, fewer detectable *E. coli* genotypes than hosts harboring strains belonging to other phylo-groups (Moreno et al., 2009). There is also a growing body of evidence to indicate that the transmission dynamics of strains of the different phylo-groups also differs. For example, phylo-group A and B1 strains are over-represented in freshwater samples, whilst B2 and D strains are rare in such samples (Power et al., 2005; Walk et al., 2007; Ratajczak et al., 2010). The persistence of *E. coli* strains in soil also varies with their phylo-group membership (Bergholz et al., 2010). A phenotype commonly linked to a strain's ability to survive stressful conditions is the red dry and rough (rdar) phenotype, where cells produce an extracellular matrix composed of curli fimbriae and a variety of polysaccharides, is more frequent in B1 strains than in strains belonging to the other phylogroups (White et al., 2011).

While it is clear that strains belonging to the different phylo-groups of *E. coli* differ in their phenotypic, ecological and life-history characteristics, there is little understanding of the underlying mechanisms leading to these differences. To a very great extent this difficulty has arisen because of the enormous genetic diversity to be found in *E. coli* (Touchon et al., 2009). A typical *E. coli* genome consists of about 4700 genes, however, only about 2000 of these genes are common to all *E. coli* strains. The balance of the genes in the genome of a strain is drawn from a gene pool that is in excess of 10 000 unique genes, after eliminating all transposable elements and prophages. To date, there has been no systematic attempt to determine whether or not any genes are over-represented in a particular phylo-group.

Strains of the various phylo-groups do vary in their phenotypic properties, such as carbon source utilization patterns (Gordon, 2004) and their ability to cause disease in a mouse model of extraintestinal infection (Johnson et al., 2006). Strains of the various phylo-groups also differ in their gene content, with phylo-group B2 strains in particular harboring a variety of traits, often at high frequency, thought to enhance the ability of a strain to cause extraintestinal disease. These same traits have been shown to be important determinants of a strain's ability to colonize the intestine (Diard et al., 2010).

The hierarchical structure of *E. coli* extends beyond the species or phylo-group level to clonal complexes and clonal lineages (MLST sequence type (ST) or clonal complex (CC)) (Figure 1.1). Researchers studying *E. coli* responsible for intestinal infections have long recognized the significance of particular clonal groups, for example, the infamous O157:H7 clonal group (ST 11). Particular clonal lineages are also often responsible for extraintestinal infections such as the phylo-group D clonal group A strains (ST 69) (Johnson et al., 2011); or the phylo-group B2 strains O1:K1:H7/NM (ST 95) (Mora et al., 2009). Other lineages are less often responsible for extraintestinal infection, but very commonly isolated from the feces of humans and other animals. The group of strains related to *E. coli* K12, belonging to ST 10, is the best known of these as it represents a very large fraction of the sequence types in the *E. coli* MLST database hosted by University College Cork.

Strains responsible for extraintestinal infection may originate from any of the *E. coli* phylo-groups, although most isolates are members of phylo-groups B2 and to a lesser extent, D. All of the available evidence indicates that the main pathotypes responsible for intestinal disease have arisen from each of the phylo-groups, often on multiple occasions (Figure 1.1). Although for the intestinal pathogens, phylo-group A, B1, D, and E strains were more likely to have provided the ancestral lineages than phylo-group B2 strains (Escobar-Páramo et al., 2004; Yang et al., 2007). It is often also the case that a particular pathogenic lineage has as its closest relative a non-pathogenic variant (Figure 1.1).

WITHIN AND AMONG HOST *E. COLI* DIVERSITY

Studies based on non-selective plating (e.g. MacConkey) of fecal *E. coli* isolates have revealed that typically one or two genotypes are recovered per host, although as many as six or more may be detected (Caugant et al., 1983; Alm et al., 2011). However, such studies seldom use sample sizes that permit the detection of genotypes occurring at a frequency of less than 5% of the total *E. coli* population. Selective plating, usually for antibiotic-resistant variants, almost always reveals the presence of additional genotypes that represent less than 1% of the total *E. coli* population (Gordon et al., 2002). *E. coli* have been isolated from different regions of the small and large intestines of domestic pigs and wild boars (Dixit et al., 2004; Schierack et al., 2009). These studies have revealed that the focus on fecal isolates underestimates the diversity of isolates present in the gut. A similar outcome has been observed for O157:H7 in cattle, where rectal swabs can result in significantly higher frequencies of positive animals than do fecal isolations (Naylor et al., 2003). Significantly, the pig studies have shown that there are isolates detected in the small intestine that are not detected in the colon or feces. The reasons for this outcome are unknown but probably reflect the fact that cell densities achieved in the small intestine are from 100 to 1000 times lower than those achieved in the colon. Further, there is experimental evidence to suggest that non-adherent *E. coli* cells do not divide (Poulsen et al., 1995). Hence non-adhering cells originating in the small intestine would be significantly outnumbered in the feces. Although there are very limited data, strains from different gut regions appear to have different characteristics. For example, bacteriocin production was less frequent among isolates from the duodenum as compared to isolates from downstream regions and the kinds of bacteriocins produced differed between strains isolated from the ileum and those taken from feces (Abraham et al., 2012). Food transit times are more rapid in the upper small intestine and nutrient concentrations high compared to the lower intestinal tract (Timm et al., 2011). Theoretical studies indicate that bacteriocin producers should be disadvantaged under these conditions compared to non-bacteriocin producers (Frank, 1994; Barnes et al., 2007).

As a first approximation it may be stated that every human harbors a different numerically dominant strain of *E. coli.* In a study of fecal isolates recovered from 228 people, it was found that only 27% of the observed clones were recovered from more than one individual (Johnson et al., 2008). However, the sharing that did occur was highly structured. Strain sharing was observed in 313 of the total potential sharing pairs, however 27% of within-household pairs shared a clone, while only 0.8% of across-household pairs did.

HOST SPECIFICITY

Does *E. coli* exhibit host-specificity? It depends who you ask. Part of the problem relates to definition. Do we confine host specificity to situations where there is a one-to-one match between host species and partner organism? Or do we think of it in terms of host preference, where the probability of detecting the partner varies among different host species. At the species level there is clear evidence that *E. coli* is most likely to be isolated from warm-blooded vertebrates that have either a cecum or a body mass greater than a kilogram (Gordon and Cowling, 2003).

Although there is some evidence to suggest that strains of the various phylo-groups exhibit some degree of host preference (Figure 1.2), it is also clear that other factors can influence phylo-group membership of *E. coli* isolated from a single host species, such as humans (Figure 1.3). Until recently there has been little evidence to indicate that the vast majority of commensal isolates of *E. coli* exhibit host specificity. However, it has been suggested that there are animal-specific strains (Escobar-Páramo et al., 2006), as well as human-specific strains (Clermont et al., 2008). Some clonal groups, ST 95 and ST 69, for example, are frequently isolated from the feces of asymptomatic humans living in Australia, yet neither of these STs has ever been detected in a native Australian mammal (unpublished data).

Among researchers primarily concerned with *E. coli* that causes extraintestinal infection, many believe that hosts such as poultry and companion animals can serve as reservoirs for strains responsible for human infections. Indeed, near identical strains can be isolated from human and animal feces (Reeves et al., 2011). Among researchers focused on the intestinal pathogens, host specificity is virtually dogma. Whereby, most of the various pathotypes (EPEC, EIEC, ETEC, DAEC, EAEC) are host specific, with the exception of classic human EHEC strains that have a ruminant reservoir. For the non-EHEC pathotypes, the available evidence does indicate that the great majority of strains capable of causing disease are indeed host specific (Nataro and Kaper, 1998; Robins-Browne and Hartland, 2002; Bardiau et al., 2010; Croxen and Finlay, 2010).

POPULATION DYNAMICS OF INTESTINAL PATHOGENS

The population dynamics of many of the intestinal *E. coli* pathogens of humans and domesticated animals have many similarities to classic infectious diseases such as measles (Hartley et al., 2005). For a pathogen to persist in a host population requires that every infection will cause at least one new infection, that is, the basic reproductive value R_0 must be >1. The basic reproductive value R_0 of a diarrheal pathogen has three components; 1, the average number of cells shed per infected host over the period of infection; 2, new hosts are infected at a rate that depends on the average number of cells ingested per host and the number of cells required to initiate an infection; and 3, the lifespan of cells in the external environment.

In developing countries, it seems reasonable to assume that most of the diarrheal pathotypes of *E. coli* can be maintained in the human population in the absence of any other host. Asymptomatic carriage coupled with low levels of sanitation and a lack of clean water, are probably sufficient to maintain these diseases even if they are human-specific. The maintenance of EHEC strains can probably be attributed to their asymptomatic carriage in ruminants and subsequent distribution via the food chain to humans, although there are some doubts concerning the ability of cattle to be the sole maintenance host (Besser et al., 2011).

The maintenance of human-specific diarrheal pathotypes in developed countries is more difficult to understand. The number of hosts shedding the pathogen will represent only a small subset of the total population. Acquired immunity and age-related resistance are the most important determinants of the number of hosts infected by a particular pathotype (Nataro and Kaper, 1998). Acquired immunity is thought to be important in determining the nature of the susceptible population (young children) in regions where ETEC is endemic, while it is age-related physiological changes that have been suggested as the dominant factor in EPEC infections being restricted to children and the young of other species. By contrast, there is no evidence to suggest that acquired immunity influences the establishment of commensal *E. coli* in the human gut. Host age effects may occur, as there are some data to suggest that strains belonging to phlyo-group B1 are more likely to be recovered from children than from older adults living in developed countries (Figure 1.4).

Although the data for commensal isolates are far from adequate, the evidence indicates that a commensal *E. coli* strain can persist in an individual host for periods of months to years (Caugant et al., 1981; Clermont et al., 2008; Reeves et al., 2011). Throughout this period commensal strains will shed cells at the rate of about 10^8 cells per day, although among-sample variation in the number of cells recovered per gram of feces may vary by more than 100-fold for the same individual (McOrist et al., 2005). By comparison, *E. coli*-induced diarrheal disease generally lasts for less than 2 weeks, although some shedding of cells may continue after symptoms have passed (Nataro and Kaper, 1998). However, the number of cells diarrheal pathotypes shed per day is generally much larger than for commensal strains.

In developed countries asymptomatic carriage appears to be rare and the genes associated with the various pathotypes are seldom detected. For example, in a survey of 489 human fecal samples collected in Melbourne, Australia,

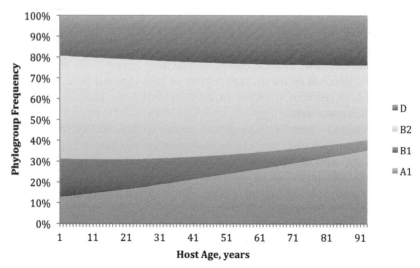

FIGURE 1.4   Change in the relative abundance of the main *E. coli* phylo-groups (A, B1, B2, D) with respect to human host age. *Data taken from Gordon et al. (2005).*

EAEC were detected in 3.1% of samples and atypical EPEC in 2.4% of samples (Robins-Browne et al., 2004). All other pathotypes were detected in less than 0.2% of the samples. Similarly, asymptomatic carriage in many mammal species also seems to be rare. Few of almost 500 *E. coli* strains isolated from native Australian mammals (Gordon and Cowling, 2003) were found to possess virulence factors associated with intestinal pathotypes. For example, *eaeA* was detected in 5% of strains, either stx_1 or stx_2 were detected in 1% of strains, and *bfpA* and *aggR* were not detected (unpublished data). A growing body of data indicates there are strains present in the intestinal tract that are not detected in fecal samples. This means that it is possible that in asymptomatic hosts, cells of most pathotypes are shed at frequencies too low to be detected. The significance of such hosts to the maintenance of the disease is unknown.

The minimum infectious dose is a significant component of a parasite's basic reproductive value. The infectious dose for EIEC and EHEC is thought to be in the order of hundreds of cells while the infectious dose for EPEC, ETEC, and EAEC pathotypes is thought to be many orders of magnitude larger (Nataro and Kaper, 1998). It is, however, unclear whether these differences in infectious dose reflect the site of infection, small (EIEC and EHEC) versus large bowel (EPEC, ETEC, EAEC) or a dichotomy in the manner in which infectious dose is measured; experimentally or epidemiologically. It is well accepted among people working with commensal *E. coli* that it is virtually impossible to establish an *E. coli* strain in a host with an existing *E. coli* population and such attempts will only be successful when using germ-free or antibiotic-treated animals. However, these outcomes are at odds with the observation that international travelers often return with antibiotic-resistant strains they did not appear to have prior to travel and that these outcomes occur in the absence of antibiotic usage while traveling (Rogers et al., 2012).

It is very likely that we have a poor understanding of the true infectious dose required to initiate an infection. Certainly most experimental studies attempting to establish an *E. coli* strain in a host use cells grown in vitro and it is likely that such cells are not in the appropriate physiological state to successfully establish in the intestine. In *Vibrio cholera* O1 Inaba El Tor there are data to indicate that passage through a host results in cells that are, for a time, hyperinfectious relative to cells that have spent some time in the external environment (Hartley et al., 2005). A hyperinfectious state has also been observed for the diarrheal pathogen *Citrobacter rodentium* (Bishop et al., 2007). Mathematical models incorporating this hyperinfectious state for *Vibrio cholera* suggest that this state plays a significant role in determining the dynamics of the early stages of a cholera outbreak (Hartley et al., 2005). It may well be true that infectious doses of many of the intestinal pathotypes are significantly lower than those estimated.

The final component of the basic reproductive value of diarrheal pathogens is the factors affecting the persistence of the pathogen in the external environment. Since *E. coli* is a characteristic component of mammalian feces, Schardinger proposed in 1892 that its presence in water could indicate the presence of fecal contamination and therefore the potential presence of enteric pathogens. The abundance of *E. coli* in feces, combined with the ease with which *E. coli* can be detected and enumerated in contaminated water, led to its development as an indicator organism for water quality monitoring. Although the indicator concept and criteria have undergone a number of modifications over the years, *E. coli* remains the preferred indicator of fecal contamination for water quality monitoring (EPA, 1986; WHO, 2008). Indeed, when there are known sources of fecal contamination, *E. coli* counts in water are a good predictor of gastrointestinal disease (Cabelli et al., 1979, 1982).

Savageau (1983) was the first to recognize that the biotic and abiotic conditions of the lower gastrointestinal tract differ markedly from those found in soil, sediments, and water, and that enteric bacteria such as *E. coli* must have evolved mechanisms to cope with both the host and external environments. Although we have a growing understanding of the molecular and physiological basis of *E. coli*'s interaction with the intestinal environment, we do not have a good understanding of the phenotypic response of a cell when it moves from a host to the external environment and vice versa. The nature of the cell's response to the shift between these environments will be a prime determinant of the cell's fate in the environment it is moving to.

For many years the assumption was that the density of *E. coli* cells in the external environment was driven by the balance between the rate at which cells entered the environment and the rate at which they died, with the average lifespan of an *E. coli* cell being about 1 day in water to 5 days in soil (Van Donsel et al., 1967; Van Donsel and Geldreich, 1971; Faust et al.,

1975; Yates, 2007; Ingle et al., 2011). However, 1980s' studies in the tropics revealed high *E. coli* densities in environments where there were no obvious sources of fecal contamination (Carrillo et al., 1985; Fujioka et al., 1988; Rivera et al., 1988). High *E. coli* counts were then observed in subtropical environments (Solo-Gabriele et al., 2000) and then in a variety of temperate environments (Alm et al., 2003; Haack et al., 2003; Whitman et al., 2003). These studies indicated that there are environments where fecal inputs are only one factor determining *E. coli* cell densities and, overall, *E. coli* rates of cell division can exceed cell death rates. Subsequently, population genetic studies revealed that similar or identical *E. coli* genotypes can be recovered at multiple locations, and at different seasons, from a range of environments (Gordon et al., 2002; Walk et al., 2007; Ratajczak et al., 2010). Indeed one study of a river system in Canada found that six genotypes (0.08% of all genotypes observed) represented 28% of the over 21 000 isolates characterized (Lyautey et al., 2010). Another study found what appear to be season-specific genotypes (Jang et al., 2011). Thus, it seems while these *E. coli* may have originally been host-adapted, they have evolved to the point that their persistence in the external environment may have little to do with fecal inputs.

An Australian study has identified an *E. coli* strain that appears to be entirely free-living. This strain (Bloom B1-001) is frequently responsible for elevated coliform counts (>10 000/100 ml), in recreational lakes and water reservoirs (Power et al., 2005). As many as 2–3 of the 'bloom' events may occur in a particular water body each year and such bloom events are observed over many years in the same water body. Near identical variants of this bloom strain have been isolated from multiple water bodies across southeastern Australia. The bloom strain can be regularly isolated from water samples at different times of the year in the absence of elevated counts. The strain has also never been detected in the feces of any vertebrate. In Lake Burley Griffin, Canberra, this strain can increase its density over 100 000-fold in less than a week. For these densities to be achieved as a result of fecal contamination would require that the lake directly receive all of the feces produced by every resident of Canberra for over a week. This would also require that every person in Canberra harbored this *E. coli* strain.

Genetic analysis of this strain (B1-001) has revealed that, based on its core genome, it is most closely related to strains of *Shigella boydii, S. sonnei,* and *S. flexerni* (unpublished data). It also has a relatively small genome of 4.6 mb, more like the genome size of *Shigella* strains than typical *E. coli*. Phenotypically it also resembles *Shigella* more so than other *E. coli*. However, the strain lacks most, if not all, virulence factors associated with *Shigella*. Thus we have two groups of strains that, phylogenetically, appear to be very closely related. Yet one represents what appear to be free-living variants of *E. coli*, while the other represents strains that are facultative intracellular parasites. Since there are no data to suggest that *Shigella* and other EIEC can be carried asymptomatically by anything other than humans in developed countries, could the maintenance 'host' of *Shigella* represent the external environment and is its persistence due to, as yet unidentified, traits also present in the 'free-living' *E. coli* strain? Are some of the same traits perhaps shared by the other diarrheal pathotypes of *E. coli*?

Although there is an increasingly sophisticated understanding of the molecular and physiological means by which *E. coli* diarrheal pathogens cause disease, there are many significant gaps in our understanding of the ecology/epidemiology of these diseases.

REFERENCES

Abraham, S., Gordon, D.M., Chin, J., et al., 2012. Molecular characterization of commensal Escherichia coli adapted to different compartments of the porcine gastrointestinal tract. Appl. Environ. Microbiol. 78, 6799–6803.

Alm, E.W., Burke, J., Spain, A., 2003. Fecal indicator bacteria are abundant in wet sand at freshwater beaches. Water Res. 37, 3978–3982.

Alm, E.W., Walk, S.T., Gordon, D.M., 2011. The niche of *Escherichia coli*. In: Walk, S.T., Feng, P.C.H. (Eds.), Population Genetics of Bacteria, ASM Press, Washington, DC, pp. 107–123.

Bain, M.S., Green, C.C., 1999. Isolation of *Escherichia fergusonii* in cases clinically suggestive of salmonellosis. Vet. Rec. 144, 511.

Bardiau, M., Szalo, M., Mainil, J.G., 2010. Initial adherence of EPEC, EHEC and VTEC to host cells. Vete. Rese. 41, 57.

Barnes, B., Sidhu, H., Gordon, D.M., 2007. Host-gastro-intestinal dynamics and the frequency of colicin production in *Escherichia coli*. Microbiology 153, 2823–2827.

Bergholz, P.W., Noar, J.D., Buckley, D.,H., 2010. Environmental patterns are imposed on the population structure of *Escherichia coli* after fecal deposition. Appl. Environ. Microbiol. 77, 211–219.

Besser, TE., Davis, M.A., Walk, S.T., 2011. Escherichia coli O157:H7 in reservoir hosts. In: Walk S.T., Feng, P.C.H. (Eds.), Population Genetics of Bacteria, ASM Press, Washington, DC, pp. 303–324.

Bishop, A.L., Wiles, S., Dougan, G., Frankel, G., 2007. Cell attachment properties and infectivity of host-adapted and environmentally adapted *Citrobacter rodentium*. Microbes. Infect. 9 (11), 1316–1324.

Cabelli, V.J., Dufour, A.P., Levin, M.A., McCabe, L.J., Haberman, P.W., 1979. Relationship of microbial indicators to health effects at marine bathing beaches. Am. J. Public Health 69, 690–696.

Cabelli, V.J., Dufour, A.P., McCabe, L.J., Levin, M.A., 1982. Swimming associated gastroenteritis and water quality. Am. J. Epidemiol. 115, 606–616.

Carrillo, M., Estrada, E., Hazen, T.C., 1985. Survival and enumeration of the fecal indicators *Bifidobacterium adolescentis* and *Escherichia coli* in a tropical rain forest watershed. Appl. Environ. Microbiol. 50, 468–476.

Caugant, D.A., Levin, B.R., Selander, R.K., 1981. Genetic diversity and temporal variation in the *E. coli* populations of a human host. Genetics 98, 467–490.

Caugant, D.A., Levin, B.R., Selander, R.K., 1983. Distribution of multilocus genotypes of *Escherichia coli* within and between host families. J. Hyg. 92, 377–384.

Chaudhuri, R.R., Henderson, I.R., 2012. The evolution of the *Escherichia coli* phylogeny. Infect. Genet. Evol. 12, 214–226.

Chaudhury, A., Nath, G., Tikoo, A., Sanyal, S.C., 1999. Enteropathogenicity and antimicrobial susceptibility of new *Escherichia* spp. J. Diarrhoeal. Dis. Res. 17, 85–87.

Clermont, O., Bonacorsi, S., Bingen, E., 2000. Rapid and simple determination of the *Escherichia coli* phylogenetic group. Appl. Environ. Microbiol. 66, 4555–4558.

Clermont, O., Lescat, M., O'Brien, C.L., Gordon, D.M., Tenaillon, O., Denamur, E., 2008. Evidence for a human-specific *Escherichia coli* clone. Environ. Microbiol. 10, 1000–1006.

Croxen, M., Finlay, B.B., 2010. Molecular mechanisms of *Escherichia coli* pathogenicity. Nat. Rev. Microbiol. 8, 26–38.

Diard, M., Garry, L., Selva, M., Mosser, T., Denamur, E., Matic, I., 2010. Pathogenicity-associated islands in extraintestinal pathogenic *Escherichia coli* are fitness elements involved in intestinal colonization. J. Bacteriol. 192, 4885–4893.

Dixit, S.M., Gordon, D.M., Wu, X.Y., Chapman, T., Kailasapathy, K., Chin, J.J., 2004. Diversity analysis of commensal porcine *Escherichia coli* – associations between genotypes and habitat in the porcine gastrointestinal tract. Microbiology 150, 1735–1740.

EPA (United States Environmental Protection Agency), 1986. Ambient Water Quality Criteria for Bacteria – 1986. USEPA, Washington, DC.

Escobar-Páramo, P., Grenet, K., Le Menac'h, A., et al., 2004. Large-scale population structure of human commensal *Escherichia coli* isolates. Appl. Environ. Microbiol. 70, 5698–5700.

Escobar-Páramo, P., Clermont, O., Blanc-Potard, A.B., Bui, H., Le Bouguénec, C., Denamur, E., 2006. A specific genetic background is required for acquisition and expression of virulence factors in *Escherichia coli*. Mol. Biol. Evol. 21, 1085–1094.

Farmer 3rd, J.J., Fanning, G.R., Davis, B.R., et al., 1985. *Escherichia fergusonii* and *Enterobacter taylorae*, two new species of Enterobacteriaceae isolated from clinical specimens. J. Clin. Microbiol. 21, 77–81.

Faust, M.A., Aotaky, A.E., Hargadon, M.T., 1975. Effect of physical parameters on the in situ survival of *Escherichia coli* MC-6 in an estuarine environment. Appl. Microbiol. 30, 800–806.

Foster, G., Ross, H.M., Pennycott, T.W., Hopkins, G.F., McLaren, I.M., 1998. Isolation of *Escherichia coli* O86:K61 producing cyto-lethal distending toxin from wild birds of the finch family. Lett. Appl. Microbiol. 26, 395–398.

Frank, S.A., 1994. Spatial polymorphism of bacteriocins and other allelopathic traits. Evol. Ecol. 8, 369–386.

Fujioka, R.S., Tenno, K., Kansako, S., 1988. Naturally occurring fecal coliforms and fecal streptococci in Hawaii's freshwater streams. Toxicity Assessment 3, 613–630.

Funke, G., Hany, A., Altwegg, M., 1993. Isolation of *Escherichia fergusonii* from four different sites in a patient with pancreatic carcinoma and cholangiosepsis. J. Clin. Microbiol. 31, 2201–2203.

Gordon, D.M., 1997. The genetic structure of *Escherichia coli* populations in feral house mice. Microbiology 143, 2039–2046.

Gordon, D.M., 2004. The influence of ecological factors on the distribution and genetic structure of *Escherichia coli*. In: Neidhardt, F. (Ed.), *Escherichia coli* and *Salmonella typhimurium*: cellular and molecular biology, American Society for Microbiology, Washington, DC. http://www.ecosal.org [Online].

Gordon, D.M., Bauer, S., Johnson, J.R., 2002. The genetic structure of *Escherichia coli* populations in primary and secondary habitats. Microbiology 148, 1513–1522.

Gordon, D.M., Cowling, A., 2003. The distribution and genetic structure of *Escherichia coli* in Australian vertebrates: Host and geographic effects. Microbiology 149, 3575–3586.

Gordon, D.M., Stern, S.E., Collignon, P.J., 2005. The influence of the age and sex of human hosts on the distribution of *Escherichia coli* ECOR groups and virulence traits. Microbiology 151, 15–23.

Gordon, D.M., Clermont, O., Tolley, H., Denamur, E., 2008. Assigning *Escherichia coli* strains to phylogenetic groups: multi-locus sequence typing versus the PCR triplex method. Environ. Microbiol. 10, 2484–2496.

Haack, S.K., Fogarty, L.R., Wright, C., 2003. *Escherichia coli* and enterococci at beaches in the Grand Traverse Bay, Lake Michigan: sources, characteristics, and environmental pathways. Environ. Sci. Tech. 37, 3275–3282.

Hariharan, H., López, A., Conboy, G., Coles, M., Muirhead, T., 2007. Isolation of *Escherichia fergusonii* from the feces and internal organs of a goat with diarrhoea. Cana. Vete. J. 48, 630–631.

Hartley, D.M., Morris Jr., J.G., Smith, D.L., 2005. Hyperinfectivity: a critical element in the ability of *V. cholerae* to cause epidemics? Public Lib. Sci. Med. 3 (1), e7.

Herráez, P., Rodríguez, A.F., Espinosa de los Monteros, A., et al., 2005. Fibrino-necrotic typhlitis caused by *Escherichia fergusonii* in ostriches (*Struthio camelus*). Avian Dis. 49, 167–169.

Huys, G., Cnockaert, M., Janda, J.M., Swings, J., 2003. *Escherichia albertii* sp. nov., a diarrhoeagenic species isolated from stool specimens of Bangladeshi children. Int. J. Syst. Evol. Microbiol. 53, 807–810.

Hyma, K.E., Lacher, D.W., Nelson, A.M., et al., 2005. Evolutionary genetics of a new pathogenic *Escherichia* species: *Escherichia albertii* and related *Shigella boydii* strains. J. Bacteriol. 187, 619–628.

Ingle, D.J., Clermont, O., Skurnik, D., Denamur, E., Walk, S.T., Gordon, D.M., 2011. Biofilm formation by and thermal niche and virulence characteristics of *Escherichia* spp. Appl. Environ. Microbiol. 77, 2695–2700.

Jang, J., Unno, T., Lee, S.W., et al., 2011. Prevalence of season-specific *Escherichia coli* strains in the Yeongsan River Basin of South Korea. Environ. Microbiol. 13, 3103–3113.

Johnson, J.R., Clermont, O., Menard, M., Kuskowski, M.A., Picard, B., Denamur, E., 2006. Experimental mouse lethality of *Escherichia coli* isolates, in relation to accessory traits, phylogenetic group, and ecological source. J. Infect. Dis. 194, 1141–1150.

Johnson, J.R., Owens, K., Gajewski, A., Clabots, C., 2008. *Escherichia coli* colonization patterns among human household members and pets, with attention to acute urinary tract infection. J. Infect. Dis. 197, 218–224.

Johnson, J.R., Menard, M.E., Lauderdale, T.L., et al., 2011. Global distribution and epidemiologic associations of *Escherichia coli* clonal group A, 1998–2007. Emerg. Infect. Dis. 17, 2001–2009.

Lyautey, E., Lu, Z., Lapen, D.R., et al., 2010. Distribution and diversity of *Escherichia coli* populations in the South Nation River drainage basin, eastern Ontario, Canada. Appl. Environ. Microbiol. 76, 1486–1496.

Li, B., Sun, J.Y., Han, L.Z., Huang, X.H., Fu, Q., Ni, Y.X., 2010. Phylogenetic groups and pathogenicity island markers in fecal *Escherichia coli* isolates from asymptomatic humans in China. Appl. Environ. Microbiol. 76, 6698–6700.

Looft, T., Allen, H.K., 2012. Collateral effects of antibiotics on mammalian gut microbiomes. Gut. Microbes. Sep. 1 3 (5) [Epub ahead of print].

Luo, C., Walk, S.T., Gordon, D.M., Feldgarden, M., Tiedje, J.M., Konstantinidis, K.T., 2011. Genome sequencing of environmental *Escherichia coli* expands understanding of the ecology and speciation of the model bacterial species. Proc. Natl. Acad. Sci. USA 108, 7200–7205.

Maiden, M.C., Bygraves, J.A., Feil, E., et al., 1998. Multilocus sequence typing: a portable approach to the identification of clones within populations of pathogenic microorganisms. Proc. Natl. Acad. Sci. USA 95, 3140–3145.

McOrist, A., Veuilett, G., Vuaran, M., Bird, A., Noakes, M., Topping, D., 2005. Population and virulence factor dynamics in fecal *Escherichia coli* from healthy adults consuming weight control diets. Can. J. Microbiol. 51, 467–475.

Monira, S., Nakamura, S., Gotoh, K., et al., 2011. Gut microbiota of healthy and malnourished children in Bangladesh. Frontiers Microbiol. 2, 228.

Mora, A., López, C., Dabhi, G., et al., 2009. Extraintestinal pathogenic *Escherichia coli* O1:K1:H7/ NM from human and avian origin: detection of clonal groups B2 ST95 and D ST59 with different host distribution. BMC Microbiol. 9, 132.

Moreno, E., Johnson, J.R., Pérez, T., Prats, G., Kuskowski, M.A., Andreu, A., 2009. Structure and urovirulence characteristics of the fecal *Escherichia coli* population among healthy women. Microbes. Infect. 11, 274–280.

Nataro, J.P., Kaper, J.B., 1998. Diarrheagenic *Escherichia coli.* Clin. Microbiol. Rev. 11, 142–201.

Naylor, S.W., Low, J.C., Besser, T.E., et al., 2003. Lymphoid follicle-dense mucosa at the terminal rectum is the principal site of colonization of enterohemorrhagic *Escherichia coli* O157:H7 in the bovine host. Infect. Immun. 71, 1505–1512.

Nowrouzian, F.L., Wold, A.E., Adlerberth, I., 2005. *Escherichia coli* strains belonging to phylogenetic group B2 have superior capacity to persist in the intestinal microflora of infants. J. Infect. Dis. 191, 1078–1083.

Nowrouzian, F.L., Adlerberth, I., Wold, A.E., 2006. Enhanced persistence in the colonic microbiota of *Escherichia coli* strains belonging to phylogenetic group B2: role of virulence factors and adherence to colonic cells. Microbes. Infect. 8, 834–840.

Oaks, J.L., Besser, T.E., Walk, S.T., et al., 2010. *Escherichia albertii* in wild and domestic birds. Emerg. Infect. Dis. 16, 638–646.

O'Brien, C.L., Gordon, D.M., 2011. Effect of diet and gut dynamics on the establishment and persistence of *Escherichia coli.* Microbiology 157, 1375–1384.

Ooka, T., Seto, K., Kawano, K., et al., 2012. Clinical significance of *Escherichia albertii.* Emerg. Infect. Dis. 18, 488–492.

Poulsen, L.K., Licht, T.R., Rang, C., Krogfelt, K.A., Molin, S., 1995. Physiological state of *Escherichia coli* BJ4 growing in the large intestines of streptomycin-treated mice. J. Bacteriol. 177, 5840–5845.

Power, M.L., Littlefield-Wyer, J., Gordon, D.M., Veal, D.A., Slade, M.B., 2005. Phenotypic and genotypic characterization of encapsulated *Escherichia coli* isolated from blooms in two Australian lakes. Environ. Microbiol. 7, 631–640.

Pupo, G.M., Lan, R., Reeves, P.R., 2000a. Multiple independent origins of *Shigella* clones of *Escherichia coli* and convergent evolution of their many characteristics. Proc. Natl. Acad. Sci. USA 97, 10567–10572.

Pupo, G.M., Lan, R., Reeves, P.R., Baverstock, P.R., 2000b. Population genetics of *Escherichia coli* in a natural population of native Australian rats. Environ. Microbiol. 2, 594–610.

Ratajczak, M., Laroche, E., Berthe, T., et al., 2010. Influence of hydrological conditions on the *Escherichia coli* population structure in the water of a creek on a rural watershed. BMC Microbiol. 10, 222.

Reeves, P.R., Liu, B., Zhou, Z., et al., 2011. Rates of mutation and host transmission for an *Escherichia coli* clone over 3 years. PLoS One 6 (10), e26907.

Rivera, S.C., Hazen, T.C., Toranzos, G.A., 1988. Isolation of fecal coliforms from pristine sites in a tropical rain forest. Appl. Environ. Microbiol. 54, 513–517.

Robins-Browne, R.M., Hartland, E.L., 2002. *Escherichia coli* as a cause of diarrhea. J. Gastroenterol. Hepatol. 17, 467–475.

Robins-Browne, R.M., Bordun, A.M., Tauschek, M., et al., 2004. *Escherichia coli* and community-acquired gastroenteritis, Melbourne, Australia. Emerg. Infect. Dis. 10, 1797–1805.

Rogers, B.A., Kennedy, K.J., Sidjabat, H.E., Jones, M., Collignon, P., Paterson, D.L., 2012. Prolonged carriage of resistant *E. coli* by returned travellers: clonality, risk factors and bacterial characteristics. Eur. J. Clin. Microbiol. Infect. Dis. 31 (9), 2413–2420.

Santacruz, A., Collado, M.C., García-Valdés, L., et al., 2010. Gut microbiota composition is associated with body weight, weight gain and biochemical parameters in pregnant women. Br. J. Nutr. 104, 83–92.

Savageau, M.A., 1983. *Escherichia coli* habitats, cell types, and molecular mechanisms of gene control. Am. Nat. 122, 732–744.

Savini, V., Catavitello, C., Talia, M., et al., 2008. Multidrug-resistant *Escherichia fergusonii*: a case of acute cystitis. J. Clin. Microbiol. 46, 1551–1552.

Schierack, P., Romer, A., Jores, J., et al., 2009. Isolation and characterization of intestinal *Escherichia coli* clones from wild boars in Germany. Appl. Environ. Microbiol. 75, 695–702.

Sims, G.E., Kim, S.H., 2011. Whole-genome phylogeny of *Escherichia coli/Shigella* group by feature frequency profiles (FFPs). Proc. Natl. Acad. Sci. USA 108, 8329–8334.

Solo-Gabriele, H.M., Wolfert, M.A., Desmarais, T.R., Palmer, C.J., 2000. Sources of *Escherichia coli* in a coastal subtropical environment. Appl. Environ. Microbiol. 66, 230–237.

Souza, V., Rocha, M., Valera, A., Eguiarte, L.E., 1999. Genetic structure of natural populations of *Escherichia coli* in wild hosts on different continents. Appl. Environ. Microbiol. 65, 3373–3385.

Steinsland, H., Lacher, D.W., Sommerfelt, H., Whittam, T.S., 2010. Ancestral lineages of human enterotoxigenic *Escherichia coli*. J. Clin. Microbiol. 48, 2916–2924.

Tenaillon, O., Skurnik, D., Picard, B., Denamur, E., 2010. The population genetics of commensal *Escherichia coli*. Nat. Rev. Microbiol. 8, 207–217.

Timm, D., Willis, H., Thomas, W., Sanders, L., Boileau, T., Slavin, J., 2011. The use of a wireless motility device (SmartPill®) for the measurement of gastrointestinal transit time after a dietary fibre intervention. Br. J. Nutr. 105, 1337–1342.

Touchon, M., Hoede, C., Tenaillon, O., et al., 2009. Organised genome dynamics in *Escherichia coli* results in highly diverse adaptive paths. PloS Genetics 5, e1000344.

Unno, T., Han, D., Jang, J., et al., 2009. Absence of *Escherichia coli* phylogenetic group B2 strains in humans and domesticated animals from Jeonnam Province, Republic of Korea. Appl. Environ. Microbiol. 75, 5659–5666.

Van Donsel, D.J., Geldreich, E.E., 1971. Relationships of salmonellae to fecal coliforms in bottom sediments. Water Res. 5, 1079–1087.

Van Donsel, D.J., Geldreich, E.E., Clarke, N.A., 1967. Seasonal variations in survival of indicator bacteria in soil and their contribution to storm-water pollution. Appl. Environ. Microbiol. 15, 1362–1370.

Walk, S.T., Alm, E.W., Calhoun, L.M., Mladonicky, J.M., Whittam, T.S., 2007. Genetic diversity and population structure of *Escherichia coli* isolated from freshwater beaches. Environ. Microbiol. 9, 2274–2288.

Walk, S.T., Alm, E.W., Gordon, D.M., et al., 2009. Cryptic lineages of the genus *Escherichia*. Appl. Environ. Microbiol. 75, 6534–6544.

White, A.P., Sibley, K.A., Sibley, C.D., et al., 2011. Intergenic sequence comparison of *Escherichia coli* isolates reveals lifestyle adaptations but not host specificity. Appl. Environ. Microbiol. 277, 7620–7632.

Whitman, R.L., Shively, D.A., Pawlik, H.M., Nevers, M.B., Byappanahalli, M.N., 2003. Occurrence of *Escherichia coli* and enterococci in Cladophora (Chlorophyta) in nearshore water and beach sand of Lake Michigan. Appl. Environ. Microbiol. 69, 4714–4719.

WHO (World Health Organization), 2008. Guidelines for Drinking-water Quality, third ed. vol. 1. WHO Press, Geneva, Switzerland.

Wirth, T., Falush, D., Lan, R., et al., 2006. Sex and virulence in *Escherichia coli*: an evolutionary perspective. Mol. Microbiol. 60, 1136–1151.

Yang, J., Nie, H., Chen, L., et al., 2007. Revisiting the molecular evolutionary history of *Shigella* spp. J. Mol. Evol. 64, 71–79.

Yates, M.V., 2007. Classical indicators in the 21st century – far and beyond the coliform. Water Environ. Res. 79, 279–286.

Chapter 2

Comparative genomics of pathogenic *Escherichia coli*

Jason W. Sahl[1], Carolyn R. Morris[2], and David A. Rasko[2]

[1]*Translational Genomics Research Institute, Flagstaff, AZ, USA*, [2]*University of Maryland School of Medicine, Baltimore, MD, USA*

INTRODUCTION

Escherichia coli is a human gut commensal isolate and deadly human pathogen (Kaper et al., 2004). *E. coli* is easily cultured from the human gut and has been the focus of scientific studies for greater than one hundred years (as will be discussed elsewhere in this book). The availability of clinical, laboratory, and commensal isolates, as well as the associated clinical/epidemiological data have provided highly characterized isolates for whole genome sequencing.

The first *Escherichia coli* genome sequenced was the laboratory-adapted isolate, K12 MG1655 (Blattner et al., 1997). The single chromosome consisted of approximately 4.6 Mb in sequence that encodes approximately 4300 genes. At the time of sequencing, 38% of all coding regions had no predicted function. The sequencing of this isolate was rapidly followed by the publication of genomes from O157:H7 isolates EDL933 (Perna et al., 2001) and Sakai (Hayashi et al., 2001). Comparisons were made between these genomes with the genome of K12 to determine the genetic variability between isolates in the same species. In 2002, the genome of the uropathogenic isolate, CFT073, was completed (Welch et al., 2002). Comparisons among the three sequenced isolates at that time demonstrated that all isolates only shared ~39% of all coding regions. At the time, the low conservation of genes and coding regions in a single species changed the existing paradigm of gene conservation. Early thoughts on genome sequencing were that 'representative isolates' could be sequenced and they would represent the species or in this case pathovar. This concept was rapidly discarded in light of the low level of conservation in this species.

In 2008, the first confirmed intestinal commensal isolate (HS) was published (Rasko et al., 2008). A pan-genome analysis, based on peptide identity and conservation, was conducted on 17 *E. coli* isolates, including eight new genomes, sequenced at that time. The results demonstrated that the conserved genomic core of *E. coli* consists of ~2200 genes. The analysis of representatives of multiple

Escherichia coli. http://dx.doi.org/10.1016/B978-0-12-397048-0.00002-4

pathovars demonstrated that few pathovar-specific genes were identified. It was hypothesized that the low level of conservation was in part due to the strain selection bias that attempted to include as broad a collection of isolates as possible. These early studies highlighted the diversity in the species and pathovars.

In 2009, Touchon et al. published a paper that performed a pan-genome analysis on a larger number of sequenced genomes ($n = 20$) (Touchon et al., 2009). The analysis demonstrated that the number of core genes in *E. coli* remained around 2000. The analysis demonstrated the size variability in the *E. coli* chromosome, from 4.6–5.3 Mb, and the variability in plasmid content between isolates. Variation in the pan-genome conserved core between these two studies could be attributed to using different bioinformatic methodologies, as well as defining the term 'core pan-genome' differently. While these variations in the absolute numbers are small, both studies highlight the fact that approximately only half of the *E. coli* genome is highly conserved in all isolates examined. At the time these studies were published this was an interesting finding that impacted the definition of species.

To date approximately 60 *E. coli/Shigella* genomes have been completed (Table 2.1) with representatives from each of the pathovars. Recent advances

TABLE 2.1 Complete *E. coli/Shigella* genomes

Species/strain name	Pathovar	Genome size (Mb)	Year completed
Escherichia coli CFT073	UPEC	5.23	2002
Escherichia coli 536	UPEC	4.94	2006
Escherichia coli UTI89	UPEC	5.18	2006
Escherichia coli IAI39	UPEC	5.13	2009
Escherichia coli UMN026	UPEC	5.36	2009
Escherichia coli NA114	UPEC	4.97	2011
Escherichia coli str. clone D	UPEC	5.04	2011
Escherichia coli S88	ExPEC - Meningitis	5.17	2009
Escherichia coli IHE3034	ExPEC - Meningitis	5.11	2010
Escherichia coli O7:K1 str. CE10	ExPEC - Meningitis	5.38	2011
Escherichia coli HS	Commensal	4.64	2008
Escherichia coli strain SE11	Commensal	5.16	2008
Escherichia coli ED1a	Commensal	5.21	2009
Escherichia coli IAI1	Commensal	4.7	2009

TABLE 2.1 Complete *E. coli/Shigella* genomes—cont'd

Species/strain name	Pathovar	Genome size (Mb)	Year completed
Escherichia coli SE15	Commensal	4.84	2009
Escherichia coli ABU 83972	Commensal	5.13	2010
Escherichia coli KO11FL	Commensal	5.03	2011
Escherichia coli str. K-12 substr. MG1655	Laboratory	4.64	1997
Escherichia coli str. K-12 substr. W3110	Laboratory	4.65	2006
Escherichia coli ATCC 8739	Laboratory	4.75	2008
Escherichia coli str. K-12 substr. DH10B	Laboratory	4.69	2008
Escherichia coli 'BL21-Gold(DE3) pLysS AG'	Laboratory	4.57	2009
Escherichia coli B str. REL606	Laboratory	4.63	2009
Escherichia coli BL21(DE3)	Laboratory	4.56	2009
Escherichia coli BW2952	Laboratory	4.58	2009
Escherichia coli DH1	Laboratory	4.63	2009
Escherichia coli W	Laboratory	5.01	2011
Escherichia coli P12b	Laboratory	4.94	2012
Escherichia coli 042	EAEC	5.36	2009
Escherichia coli 55989	EAEC	5.15	2009
Escherichia coli O104:H4 str. 2009EL-2050	STEC/EAEC	5.44	2012
Escherichia coli O104:H4 str. 2009EL-2071	STEC/EAEC	5.39	2012
Escherichia coli O104:H4 str. 2011C-3493	STEC/FAEC	5.44	2012
Escherichia coli O157:H7 str. EDL933	EHEC	5.62	2001
Escherichia coli O157:H7 str. Sakai	EHEC	5.59	2001
Escherichia coli O103:H2 str. 12009	EHEC	5.52	2009
Escherichia coli O111:H- str. 11128	EHEC	5.77	2009

Continued

TABLE 2.1 Complete *E. coli/Shigella* genomes—cont'd

Species/strain name	Pathovar	Genome size (Mb)	Year completed
Escherichia coli O157:H7 str. TW14359	EHEC	5.62	2009
Escherichia coli O26:H11 str. 11368	EHEC	5.86	2009
Escherichia coli O157:H7 str. EC4115	EHEC	5.7	2011
Escherichia coli Xuzhou21	EHEC	5.52	2012
Escherichia coli O127:H6 str. E2348/69	EPEC	5.07	2008
Escherichia coli O55:H7 str. CB9615	EPEC	5.45	2010
Escherichia coli O55:H7 str. RM12579	EPEC	5.45	2012
Escherichia coli E24377A	ETEC	5.25	2008
Escherichia coli H10407	ETEC	5.33	2010
Escherichia coli UMNK88	porcine ETEC	5.67	2011
Shigella flexneri 2a str. 301	Shigella	4.83	2002
Shigella flexneri 2a str. 2457T	Shigella	4.6	2003
Shigella boydii Sb227	Shigella	4.65	2004
Shigella dysenteriae Sd197	Shigella	4.56	2005
Shigella sonnei Ss046	Shigella	5.06	2005
Shigella flexneri 5 str. 8401	Shigella	4.57	2006
Shigella boydii CDC 3083-94	Shigella	4.87	2008
Shigella flexneri 2002017	Shigella	4.89	2009
Shigella sonnei 53G	Shigella	5.22	2011
Escherichia coli LF82	AIEC	4.77	2010
Escherichia coli O83:H1 str. NRG 857C	AIEC	4.89	2010
Escherichia coli UM146	AIEC	5.11	2010
Escherichia coli APEC O1	APEC	5.5	2007
Escherichia coli SMS-3-5	Environmental	5.22	2008

in bench-top next generation sequencing platforms have provided independent investigators with the ability to sequence isolates of *E. coli*. As of September 2012, there are >1000 registered *E. coli* sequencing projects available in Genbank (http://www.ncbi.nlm.nih.gov/). These sequencing projects may further clarify the evolutionary history of the species, the diversity of gene content within the species, and identify genes associated with pathogenesis within each pathovar. Ideally, the ongoing genomic and phylogenomic studies will identify regions of the genome that are not encoded on mobile elements, and thus may represent more stable and effective biomarkers for each pathovar. Additionally, the sequencing will provide deeper insight into genomic diversity, combined with studies of the transcriptome, will provide clues into the regulatory networks of these pathogens.

This chapter will be separated into descriptions of the genomic studies that have been completed on isolates on each of the specific pathovars.

UROPATHOGENIC *E. COLI*

Uropathogenic *E. coli* (UPEC) are thought to be innocuous in the gastrointestinal tract, but become pathogenic in the urinary tract (Chen et al., 2006; Schwartz et al., 2011). High-throughput sequencing has shed light on the diverse genomic organization and distribution of virulence factors in this pathotype. The UPEC are the prototypes of the extraintestinal pathogenic *E. coli* (ExPEC) isolates and often people refer to UPEC isolates as ExPEC isolates, but in this case we will only discuss the UPEC, as the majority of the sequencing has focused specifically on the UPEC.

The first UPEC isolate, CFT073, was sequenced, as a collaborative effort, by Welch et al., in 2002. CFT073 was isolated from the blood of a patient with pyelonephritis; the genome consists of 5.2 Mbp, 5533 protein-coding genes, and no identified virulence plasmids. This strain is considered to be the prototype UPEC isolate, but the isolation from the blood of a patient suggests that it may be particularly pathogenic in comparison to other UPEC isolates. UPEC have acquired virulence genes that allow them to survive, and possibly thrive, in the GI tract but result in a disease presentation in the urinary tract. These genes are often found on horizontally acquired pathogenicity islands (Hacker et al., 1992; Hacker and Kaper, 2000). The CFT073 sequence confirmed the presence of previously described pathogenicity associated islands (PAIs) located at tRNA genes *pheV*, *pheU*, and *asnT*. PAI-*pheV* contains genes for pili associated with pyelonephritis (*pap*), aerobactin synthesis, hemolysin (*hly*), capsule synthesis, and two autotransporters. PAI-*pheU* also contains *pap* genes as well as a siderphore receptor (Welch et al., 2002). PAI-*asnT* is similar to the high pathogenicity island of *Yersinia pestis* and contains yersiniabactin genes (Perry and Fetherston, 2011). The location and composition of the PAIs in CFT073 were surprisingly different from the known PAIs in two other well-studied UPEC isolates 536 and J96 (Swenson et al., 1996; Middendorf et al., 2001). The PAI-*pheV* of J96 also contains *hly* but additionally encodes <u>h</u>eat <u>r</u>esistance <u>h</u>emmagglutinin (*hra*) and cytotoxic necrotizing factor (*cnf-1*). J96 and 536 both encode yersiniabactin at PAI-*asnT*, but 536 has three additional PAIs at *selC*, *leuX*, and *thrW*. In 536, *pap* and *hly* are located at PAI-*selC*, a second copy of *hly* and P-related fimbriae genes (*prf*) are found at PAI-*lueX*, and S-type fimbrial adhesion (*sfa*) and iron siderphore (*iro*) genes are located at PAI-*thrW*. In addition to the PAI encoded genes, CFT073 also has several potential virulence factors located outside of PAIs (Welch et al., 2002). These include seven putative auto-transporters and 12 putative fimbriae (10 chaperone-usher and two type IV).

In 2006, two additional UPEC isolates were sequences. UTI89 was sequenced by Chen et al. (2006) at Washington University. UTI189 was isolated from the urine of a patient with cystitis and the genome consists of 5.1 Mbps and one plasmid, pUTI89, which is 114 230 bps. There are 5066 predicted protein-coding genes and four large PAIs. In contrast, Brzuszkiewicz et al. (2006) sequenced UPEC strain 536, isolated from the urine of a patient with pyelone-phritis. The 536 genome contains 4.9 Mbps, 4747 predicted coding sequences, and no plasmids. Comparison of five *E. coli* genomes revealed 432 genes that were present in UPEC strains 536 and CFT073 but not in EDL933, Sakai, or laboratory strain K12 MG1655; this analysis suggested that these genes may have a role in urovirulence. Additionally, 427 genes were identified that were present in 536 but absent in all other published *E. coli* genomes (Brzuszkiewicz et al., 2006). Many of these genes are found in PAIs. PAI-*selC* encodes *hyl* and two sets of fimbriae genes; PAI-*leuX* contains another copy of *hyl* and *prf*; PAI-*thrW* contains *sfa* and *iro* genes; PAI-*asnT* contains yersiniabactin genes; and PAI-*pheV* contains capsule genes. By creating PAI deletion mutants they found PAI-*selC*, -*leuX*, and -*asnT* all contributed to virulence in a mouse model of ascending UTI, while only deletion of PAI-*selC* and PAI-*leuX* together had an impact on virulence in a mouse model of urosepsis. This result suggested that there are many factors that contribute to UTI establishment, but *hly*, which is found in two copies on PAI-*selC* and PAI-*leuX* is associated with the later stages of urosepsis.

In 2007, Lloyd et al. described ten new genomic islands in CFT073, in addition to the three previously described PAIs, by comparative genome hybridization of 10 *E. coli* strains (Lloyd et al., 2007). In 2009, Lloyd et al. compared nine genomic island mutants in a mouse model of ascending UTI. PAI-*metV* and PAI-*aspV* mutants were out-competed by wild-type CFT073 in the bladder and genes c3405-c3409 of PAI-*metV* were found to be important for colonization of both bladder and kidneys and specific to UPEC. PAI-aspV was further dissected, and contact-dependent inhibition gene cdiA and the autotransporter protease gene *picU* were determined to be important for colonization of the bladder. The RTX family exoprotein A gene, *tosA*, was important for colonization of the kidneys. PAI-aspV additionally contains a copy of the ferric binding protein (*fbp*) operon, which is also present in genomic island cobU. Deletion of both copies of the *fbp* operon was required for attenuation. This work emphasized that no single factor was responsible for virulence, even within a single strain, and that multiple genes and genomic regions contribute to the fitness of CFT073 in the urinary tract.

Later in 2009, two more UPEC strains, UMN026 and IAI39, were sequenced by Touchon et al. (2009). UMN026 was isolated from the urine of a patient with cystitis. The genome contains 5.2 Mbps with 4918 protein-coding genes, and two plasmids; one plasmid was 122 kb with 149 protein-coding genes and the other was 34 kb with 54 protein-coding genes. IAI39 was isolated from the urine of a patient with pyelonephritis. The genome consists of 5.1 Mbps, 4906 protein-coding genes, and no plasmids. This group compared UMN026, IAI39, and the three previously published UPEC genomes with 15 other diarrheagenic or commensal *E. coli* strains (Touchon et al., 2009). When considering intrinsic extraintestinal virulence (in a mouse model of bacteremia) (Johnson et al., 2006), the authors were not able to identify any single genes that were specific to a virulent phenotype (Touchon et al., 2009). This demonstrated that the distinct pathogenic potential of UPEC, as compared to other *E. coli*, was not the result of a fixed group of virulence factors, but rather a variable collection of factors.

More recently, in 2010, Hagan et al. published the first transcriptional profile of pathogenic *E. coli* taken directly from patients with a naturally occurring infection (Hagan et al., 2010). Isolates were collected from the urine of eight women with bacteriuria, and the transcriptomes were compared by microarray to the same isolates grown statically in sterilized urine. When these data were compared to gene expression data from a mouse model (Snyder et al., 2004), there was an overall positive correlation. Iron acquisition genes and metabolic genes were most strongly correlated with colonization, toxin genes were moderately correlated, but fimbriae and adhesion genes were poorly correlated. Multiple fimbriae genes are highly expressed by CFT073 in mice, including the *fim* operon. Despite being highly expressed in mice, only two of six isolates tested from humans expressed the *fim* operon in the urine, even though seven of the eight strains were confirmed to make functional fimbria. These findings demonstrate that while the mouse model largely reflects the transcriptional profile of UPEC in humans, there are some differences, and continued study of UPEC transcriptomics may lead to more effective therapeutics (Hagan et al., 2010).

New approaches to studying global changes in gene expression, such as RNA-seq, will further our understanding of UPEC pathogenesis. While genomic analyses have demonstrated that there is variable collection of virulence factors that confer pathogenic potential to UPEC isolates, the coordinated expression of these factors is also likely to influence pathogenesis. Further study of UPEC transcriptomics may identify novel virulence factors and targets for vaccine and therapeutic development.

SHIGA-TOXIN PRODUCING *E. COLI*/ENTEROHEMORRHAGIC *E. COLI* (STEC/EHEC)

The Shiga toxin, encoded by a lambdoid prophage in *E. coli* isolates (Campbell et al., 1992), was first characterized in an epidemic case of dysentery caused by

Shigella dysenteriae (Trofa et al., 1999). Shiga toxin-producing *E. coli* (STEC) have also been described and are associated with a range of symptoms in the human host, from mild diarrhea to severe hemorrhagic colitis (see Chapter 5) (Kaper et al., 2004). Most importantly, STEC are associated with hemolytic uremic syndrome (HUS) (Corrigan et al., 2001), which differentially affects young children in developing countries. The *E. coli* outbreak in Europe in the summer of 2011 was caused by an isolate that had the Shiga-toxin phage in the context of a phylogenetic group that did not previously harbor this genetic element (Frank et al., 2011; Rasko et al., 2011). If strict molecular definitions are observed, the outbreak isolate could be classified as an STEC, but not EHEC, as the majority of the genome was most similar to an enteroaggregative *E. coli* (see below). These findings from genomic studies highlight the limitations with regard to inferred phylogeny of the current typing schema that rely on mobile genomic regions as biomarkers or small amounts of DNA for typing.

The STEC isolates can be segregated into groups that either contain the Locus of Enterocyte Effacement (LEE) (McDaniel et al., 1995) pathogenicity island (Kaper et al., 2004) and those that do not. The LEE encodes a type III secretion system that injects infectors into the host cell that results in the formation of attaching and effacing lesions (see Chapters 5, 6, 14, and 15) (Donnenberg and Kaper, 1992; McDaniel et al., 1995). LEE-positive STEC are frequently termed enterohemorrhagic *E. coli* (EHEC), due to the frequent manifestation of hemorrhagic colitis in infected hosts. LEE-positive isolates can also be classified as attaching and effacing *E. coli* (AEEC), which also includes enteropathogenic *E. coli* (EPEC, see below). Thus there is some ambiguity in how these isolates that can be phylogenetically related (Figure 2.1) are classified based on the presence or absence of specific molecular markers. As with the other pathovars, it is hoped that continued sequencing will identify stable regions of the genome that can be utilized as potential biomarkers for rapid and accurate classification of these pathogens. It is possible that in some cases, as with the EHEC/STEC division, that multiple markers will be required.

The first EHEC genome sequenced was O157:H7 strain EDL933, which was isolated from an outbreak in Michigan (Perna et al., 2001). A comparison of this genome with K12 identified ~1400 new genes associated with virulence factors, prophages, and genes associated with variable novel metabolic pathways. This publication was followed closely by a manuscript that described the sequence of O157:H7 strain Sakai (Hayashi et al., 2001), which was isolated from an outbreak in Japan (Watanabe et al., 1996). The results of this comparison identified 1632 proteins present in the Sakai strain and absent in K12. These two isolates from the same pathovar provided the data for the first intrapathotype comparisons (Kudva et al., 2002). These studies provided evidence of greater intrapathotype diversity than was previously recognized and signaled that the evolutionary processes leading to the creation of these pathogens were not likely to be simple, linear or easily understood (see Chapter 3).

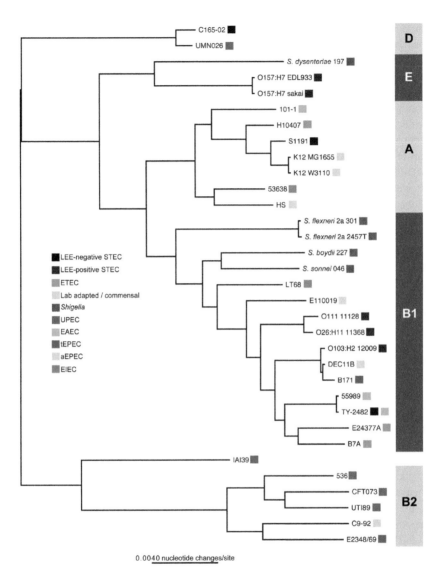

FIGURE 2.1   A phylogenetic tree inferred from 2.5 Mb of genomic sequence conserved in all isolates. The alignment was parsed from a Mugsy (Angiuoli and Salzberg, 2011) whole genome alignment of all genomes. A phylogeny was inferred on this concatenated alignment with FastTree2. Boxes to the right of each genome indicate the corresponding pathotype.

Outbreaks of O157:H7 infection, such as the spinach-associated outbreak in 2006 have occurred in developed countries, including the United States (Parker et al., 2012). An analysis of 25 O157:H7 genomes associated with three food-borne outbreak events identified 1225 single nucleotide polymorphisms (SNPs) in this clonal expansion (Eppinger et al., 2011). In fact, within the O157:H7 clade, variation in virulence (Manning et al., 2008) has been associated with specific clades in the clonal complex. Whole genome sequence data have been used to not only characterize the evolution of virulence of this important pathogen, but will help in the surveillance of this pathogen.

In addition to O157:H7 EHEC isolates, genomes have also been sequenced from other STEC/EHEC serotypes; these genomes include those of O111:H-strain 11128, O26:H11 strain 11368, and O103:H2 strain 12009 (Ogura et al., 2009). A phylogeny of 345 orthologous coding regions demonstrated that the non-O157 EHEC isolates are not contained on a single phylogenetic branch (Ogura et al., 2009). However, a phylogenetic tree based on the entire gene repertoire demonstrated that all EHEC genomes could be grouped together, most likely due to the presence of common secreted effectors and LEE-associated genes (Ogura et al., 2009).

Other LEE-negative STEC isolates have also been associated with severe diarrheal symptoms and HUS (Johnson et al., 2006; Newton et al., 2009; Rasko et al., 2011). Recently nine LEE-negative STEC isolates were sequenced and comparative analysis demonstrates the extreme phylogenetic diversity of this group (Steyert et al., 2012). This study of LEE-negative STEC has revealed that these isolates are even less similar than the LEE-positive STEC, suggesting a diverse evolutionary history. One would expect this finding considering the typing schema is based on a mobile element that can insert into most *E. coli* genomes. Additional detailed comparative analysis demonstrated the diversity of secreted effectors and *stx* insertion sites in the group. The major finding in the study was that the Shiga-toxin phage and the genomic backbone are not intimately linked, but some features of the phage define the location of insertion (Steyert et al., 2012).

The genomic examination of STEC and EHEC isolates highlights the juxtaposition of the current typing schema; while the virulence factors are important for the clinical treatment and identification of pathogens, these features do not always concur with the phylogenetic relationships of the isolates. The advent of the new sequencing technologies will allow rapid identification of whole genome sequencing into the clinical paradigm and then it will be possible to characterize isolates by their genomic content, and not phenotypic presentation; problems of associating phenotype with gene content may be avoided.

ENTEROPATHOGENIC *E. COLI* (EPEC)

Enteropathogenic *E. coli* (EPEC) are traditionally classified by the presence of the LEE and the absence of *stx* genes encoding Shiga toxins (Kaper, 1996). As with EHEC, there are subclassifications within EPEC based on virulence factor presence and absence. Isolates that contain the LEE region and the EPEC adherence factor (EAF) plasmid (Kaper et al., 2004) encoding genes for a bundle forming pilus (BFP) (Giron et al., 1991) are frequently termed typical EPEC (tEPEC), while LEE-positive, BFP-negative isolates are classified as atypical

EPEC (aEPEC) (Kaper et al., 2004). The bundle forming pilus creates a network of filaments that bind the bacteria together into what is known as a microcolony. Both tEPEC and aEPEC have been associated with diarrheal disease (Trabulsi et al., 2002), which suggests that the BFP is not absolutely required for EPEC virulence. Indeed, volunteer studies demonstrate that loss of the EAF plasmid or mutation of *bfp* genes reduces, but does not eliminate pathogenicity (Levine et al., 1985; Bieber et al., 1998). Phylogenetic analysis based on concatenated multilocus sequence typing (MLST) genes demonstrated that tEPEC are found in two primary lineages (EPEC1 and EPEC2), although several smaller clades have been identified (Orskov et al., 1990; Lacher et al., 2007). This diversity suggests that the acquisition of LEE and the EAF plasmid have occurred on multiple, independent occasions.

The genomic landscape of EPEC is as follows: The first two EPEC genomes sequenced were B171 and E110019 (Rasko et al., 2008); B171 is a tEPEC isolate from the EPEC2 lineage (Giron et al., 1991) and E110019 is an aEPEC that was the etiological agent of a large outbreak in Finland in 1990 (Viljanen et al., 1990). A comparative analysis revealed that although 200 unique genes were identified in the two EPEC genomes, compared to other sequenced genomes, few ($n = 9$) pathovar-specific genes were identified (Rasko et al., 2008). The prototype EPEC isolate E2348/69, which is part of the EPEC1 clade, was sequenced in 2009 (Iguchi et al., 2009). A comparative analysis demonstrated greater than 400 unique genes in this isolate. Furthermore, only 21 secreted effectors were identified in the genome of E2348/69 compared to approximately 50 described in the genomes of O157:H7 isolates (Tobe et al., 1999).

The diversity of EPEC has been previously determined by the typing of intimin variants (Lacher et al., 2006); the intimin gene (*eae*) is encoded by the LEE. In a study of 151 *eae*-positive isolates, 26 distinct profiles were observed using fluorescent restriction fragment length polymorphism. Phylogenetic analyses of EPEC have typically been performed from concatenated MLST markers (Lacher et al., 2006). Phylogenies inferred from whole genome sequence data have demonstrated that trees inferred from concatenated MLST markers do a poor job at recapitulating the whole genome phylogeny (Leopold et al., 2011; Sahl et al., 2011). Thus whole genome phylogeny will help better understand the phylogenetic history of EPEC isolates in the broader context of all *E. coli* and *Shigella* genomes.

The sequencing of additional EPEC genomes, as part of the GSCID project (http://gscid.igs.umaryland.edu/wp.php?wp=emerging_diarrheal_pathogens), has provided additional information on the evolution of EPEC. For example, some BFP-positive EPEC are closely related to BFP-negative EPEC phylogenetically (Hazen et al., 2012); this may be due to the loss of the EAF plasmid prior to or during lab passage. Previous studies have demonstrated that in E2348/69, the prototype EPEC isolate, the plasmid is very stable (Levine et al., 1985; Donnenberg et al., 1993). However, as more isolates are sequenced and investigated, the existing dogma is increasingly challenged. For example, based on a PCR screen analysis, BFP-positive and BFP-negative isolates would be classified as tEPEC or aEPEC, respectively, despite potentially sharing a closely related common ancestor. To incorporate phylogenetic information into a clinical assay, comparative genomics were used to identify genomic markers that distinguish between the different lineages of EPEC. A multiplex PCR reaction was designed to detect classical virulence factors as well as phylogenetic markers (Hazen et al., 2012). This methodology will define a new paradigm in which whole genome sequence data focuses better diagnostics to understand both virulence profiles and phylogenetic history.

ENTEROTOXIGENIC *E. COLI* (ETEC)

Enterotoxigenic *E. coli* is a diverse pathovar (Shaheen et al., 2003) characterized by the presence of a heat-labile (LT) and/or heat-stable (ST) enterotoxin (see Chapter 6) (So et al., 1976, 1978). ETEC is responsible for approximately 300 000 to 500 000 deaths annually, primarily in children in the developing world (WHO, 2006). In addition to the enterotoxins, ETEC possess plasmid-encoded fimbrial appendages known as colonization factors (CFs) (Gaastra and Svennerholm, 1996); ETEC CFs are structurally diverse, with greater than 30 known structural CF proteins described (Nada et al., 2011).

The first two ETEC genomes, E24377A and B7A, were sequenced in 2008 as part of an *E. coli* pan-genome analysis (Rasko et al., 2008). Comparative genome analysis of the completed genome of *E. coli* E24377A demonstrated that the isolate contained six plasmids, several of which encoded known ETEC virulence factors. The analysis also identified a limited number of features that appeared to be unique to ETEC (*n* = 9). Both of the ETEC genomes grouped into the B1 phylogroup, however it is known that ETEC isolates occupy greater phylogenetic space than just the B1 group.

The prototypical ETEC isolate is H10407, which was sequenced in 2010 (Crossman et al., 2010). The authors concluded that H10407, which falls within *E. coli* phylogroup A, was a commensal isolate that acquired virulence plasmids to become a human pathogen. Many of the virulence genes (*cexE, tibA, tia*) broadly associated with ETEC were identified in H10407. However, comparative studies have suggested that several of these virulence genes are not broadly distributed across diverse ETEC isolates and thus will not make acceptable vaccine or therapeutic targets (Turner et al., 2006). A recent transcriptomics analysis demonstrated that the transcriptional response of ETEC isolates E24377A and H10407 differed significantly in the presence of chemical signals, such as glucose and bile salts (Sahl and Rasko, 2012). This result demonstrates that the use of a single prototypical isolate is insufficient at describing either the genomic diversity or the pathogenesis of a pathovar.

As part of a large study, the multilocus sequence typing (MLST) profiles were identified from 1019 ETEC isolates (Steinsland et al., 2010). From this analysis, five isolates from the most dominant sequence types were sequenced and analyzed (Sahl et al., 2011). A whole genome analysis demonstrated that these genomes fell into phylogroups B1 and A. A comparative genomic analysis demonstrated that ETEC genomes share more chromosomal sequence with one another than they do with non-ETEC *E. coli* genomes (Sahl et al., 2011).

Decreasing sequencing costs have allowed the sequencing and comparison of a large number of clinical ETEC isolates. In a single study, 71 ETEC isolates were sequenced; 38 isolates were taken from patients with active diarrheal disease and 33 isolates were taken from patients with no diarrheal symptoms (Sahl et al., 2012). A whole genome phylogeny demonstrated that ETEC are distributed across the *E. coli* phylogeny. This result is not unanticipated because the acquisition of the virulence plasmid that harbors the LT and/or ST genes is all that is required to be typed as an ETEC. A comparative analysis of diarrheal-associated and asymptomatic isolates demonstrated that several genes were differentially conserved in each group irrespective of their phylogroup membership. In addition, an analysis was performed by which the genomic content of ETEC and non-ETEC genomes in phylogroups A and B1 were compared. The results demonstrate that greater than 100 non-plasmid-associated genes were differentially conserved in ETEC isolates. Therefore, a genomic background in ETEC appears to exist beyond the simple acquisition and expression of an ETEC virulence plasmid. Additional large-scale projects are underway with hundreds of isolates being sequenced (http://gscid.igs.umaryland.edu/wp.php?wp=comparative_genome_analysis_of_enterotoxigenic_e._coli_isolates_from_infections_of_different_clinical_severity). Genomics has changed the understanding of ETEC evolution, gene content, and pathogenesis. Genomic comparative analyses have identified putative virulence factors (Sahl and Rasko, 2012) that will help focus functional characterization studies. A global understanding of the genomic diversity of an entire pathovar will help move beyond the limitations of single-isolate, prototypical analyses. With these types of studies underway, we can begin to apply epidemiological principles to the identification of genomic regions that are associated with isolates from health, colonization, and disease presentations. The application of genomics in this way is termed genomic epidemiology.

ENTEROAGGREGATIVE *E. COLI* (EAEC)

Enteroaggregative *E. coli* (EAEC) are characterized by the phenotypic 'stacked brick' attachment to host cells. EAEC isolates cannot be defined by a single molecular marker that distinguishes all isolates of this pathovar (Kaper et al., 2004), which presents a challenge for surveillance of potential outbreak events. However, the aggregative adherence (AA) phenotype appears to be encoded by genes harbored on the pAA virulence plasmid (Vial et al., 1988). Whole genome analysis has demonstrated that the genetic composition of pAA plasmids can differ dramatically between divergent EAEC isolates (Rasko et al., 2011). The original definition of the EAEC was predicated on the lack of the ST or LT of

ETEC, but associated with diarrhea caused by *E. coli*. The identification of a group of pathogens based on the features that are 'not present' often leads to the inclusion of isolates that truly do not belong together. The phylogeny in Figure 2.1 demonstrates that the included EAEC isolates are diverse and separated in the phylogeny, highlighting this diversity.

The best-characterized regulator associated with virulence in EAEC is AggR; this regulator controls expression of the *aap* gene (encoding dispersin) (Sheikh et al., 2002), the aggregative adherence fimbriae (AAF) (Nataro et al., 1993; Bernier et al., 2002), and the *aai* type VI secretion system (Dudley et al., 2006). In addition to these factors, serine protease autotransporters of Enterobacteriaceae (SPATEs) are thought to be important in EAEC pathogenesis (Boisen et al., 2009). One study examined the virulence profiles of EAEC from children in Mali to correlate genomic content with clinical outcomes (Boisen et al., 2012). Through a classification and regression tree analysis (CART), they concluded that the SepA SPATE associated with intestinal inflammation was positively correlated ($P = 0.0006$) with diarrhea. These studies examine the virulence profiles of the pathogen with a preconceived concept of virulence in mind. In contrast, unbiased genome-wide studies have not been published.

The first EAEC draft genome sequenced was the isolate 101-1 (Rasko et al., 2008). EAEC 101-1 does not possess typical EAEC virulence factors, including the pAA plasmid. Since then, the genomes of 55989 (Touchon et al., 2009), isolated from an HIV-positive patient in Africa, and the prototypical isolate 042 (Chaudhuri et al., 2010), have been sequenced. A global phylogeny demonstrated that EAEC are phylogenetically diverse, reflecting the fact that the AA phenotype is encoded by features contained on mobile elements. Surveillance of EAEC infection is difficult because an accurate diagnosis is based on the AA phenotype and not a molecular marker.

EAEC was brought to the forefront of public interest due to an outbreak in Germany in 2011 (Frank et al., 2011) that caused ~3500 hospitalizations, 850 cases of HUS and killed 50 (Grad et al., 2012). Although outbreaks of *E. coli*-related HUS in the United States have largely been associated with the enterohemorrhagic O157:H7 lineage (Parker et al., 2012), the German outbreak isolate was serotyped as O104:H4 (Frank et al., 2011); this is the same serotype as that of the 55989 EAEC isolate, which had previously been sequenced. Panels of polymerase chain reaction (PCR) assays indicated that this strain contained Shiga-toxin genes, but also contained several virulence factors, including pAA, *aggR*, and *aaiC*, associated with EAEC (Frank et al., 2011). The rapid sequencing and public release of sequence data initiated a global crowd-source analysis of the outbreak isolate (Rohde et al., 2011). Although incidents of the acquisition of the Stx bacteriophage had been reported previously by EAEC (Morabito et al., 1998; Mellmann et al., 2008), no cases of widespread sickness and mortality had been published. A comprehensive comparative analysis of the genomes from multiple O104:H4 isolates revealed that the apparent hypervirulence of the outbreak isolate was most likely due to the independent acquisition of a plasmid encoding antibiotic resistance genes and the *stx* phage (Rasko et al., 2011). Overall, this outbreak demonstrated both the speed with which genomic data can be generated and analyzed and how comparative analyses can now be conducted in order to understand the pathogenic potential of outbreak isolates. However, one must keep in mind that a new assemblage of virulence factors in any bacterium is only one part of the infectious process, with the host factors, including immune status and existing microbiota being other parts of the infectious equation.

DIFFUSELY ADHERENT *E. COLI* (DAEC) AND ADHERENT INVASIVE *E. COLI* (AIEC)

These two relatively new pathovars are primarily described based on the genome sequence and limited functional data have been published.

Diffusely adherent *E. coli* (DAEC), implicated in diarrheal disease in children (Scaletsky et al., 2002), are characterized by their diffuse adherence to HEp-2 cell monolayers (see Chapter 11) (Kaper et al., 2004). The DAEC group is poorly defined on a molecular level, with a relatively little understanding of gene composition due to a lack of multiple whole genome sequence data and functional studies. A fimbrial adhesion called F1845 (Bilge et al., 1989) has been identified in ~75% of DAEC isolates (Kaper et al., 2004), but is not yet used widely as an accurate diagnostic marker. A phylogeny inferred from a multiple enzyme electrophoresis analysis (MLEE) demonstrated that the DAEC isolates sequenced in one study were isolated to two primary groups (Czeczulin et al., 1999). Without additional genome sequence data, the evolution of this pathovar and the global distribution of virulence factors cannot be fully determined.

Adherent and invasive *E. coli* (AIEC) are found in ileal lesions of Crohn's disease (CD) patients, where they adhere to intestinal epithelial cells and invade into and survive in epithelial cells and macrophages, thereby gaining access to a typically restricted host niche (Nash et al., 2010). The genes responsible for these phenotypes have not yet been identified, despite the sequencing of two isolates (Miquel et al., 2010; Nash et al., 2010). Our limited understanding may be due to the fact that the infection process is extremely diverse and involves many host processes or it may be due to the lack of the ability to rapidly and accurately identify AIEC isolates.

Further genome studies are required to elaborate the molecular details of these two pathovars.

SHIGELLA AND ENTEROINVASIVE *E. COLI* (EIEC)

For the purposes of this chapter we have grouped the EIEC and *Shigella* species isolates as they have a similar phenotype in the human body, as well as sharing the virulence plasmid required for colonization, infection, and survival.

Shigella is an obligate intracellular pathogen that causes intestinal disease, known as shigellosis, in humans (see Chapter 7) (Kotloff et al., 1999). *Shigella* are separated into four 'species' based on serotyping: *S. boydii, S. sonnei, S. dysenteriae,* and *S. flexneri* (Hale, 1991). Despite their name, only one clade, *S. dysenteriae,* contains isolates that produce Shiga toxin. Clinical symptoms cannot distinguish *Shigella* infection from some types of *E. coli* infection (Johnson, 2000). Furthermore, biochemical tests have been used to differentiate *Shigella* from *E. coli,* but cannot always differentiate *Shigella* from some enteroinvasive *E. coli* (EIEC) (Johnson, 2000). *Shigella* are characterized by an invasive phenotype encoded by the large (~200 kb) invasion plasmid, pINV(28). pINV contains the genetic machinery for a T3SS, which consists of approximately 50 proteins involved in the assembly and regulation of the T3SS (Parsot et al., 1995). *Shigella* are also characterized by gene loss compared to *E. coli* (Hershberg et al., 2007), which is often associated with obligate pathogens (Moran, 2002).

Molecular analyses has determined that *Shigella* is actually a phenotype of *E. coli* that has emerged on multiple independent occasions (Ochman et al., 1983). A phylogeny based on a concatenated alignment of multiple conserved genes (MLST) demonstrated that the *Shigella* phenotype has emerged at least seven times from *E. coli* (Pupo et al., 2000). Therefore, the emergence of *Shigella* is associated with plasmid acquisition and convergent evolution, including genome size reduction.

The first *Shigella* genome sequenced was *S. flexneri* 301 in 2002 (Jin et al., 2002). A comparison of this genome with K12 and O157:H7 EDL933 revealed that the *S. flexneri* genome contains a repertoire of unique insertion elements and pathogenicity islands. Despite these differences, all three genomes share an essentially co-linear 3.9 Mb backbone. The related genome, *S. flexneri* 2a 2457T, was published in 2003 (Wei et al., 2003). A comparison of this genome with the genome of K12 and O157:H7 identified ~2800 ORFs shared between the three genomes. A comparison of *S. flexneri* 2a 2457T with *S. flexneri* 2a 301 demonstrated that the two genomes differ by 45 insertion sequence loci. This large number of insertion sequences is characteristic of *Shigella* genomes.

In 2005, the genomes of *S. boydii* 227, *S. sonnei* 046, and *S. dysenteriae* 197 were all sequenced to completion (Table 2.1); all of these isolates were collected from epidemics in China in the 1950s (Yang et al., 2005). A comparative analysis of multiple *Shigella* genomes demonstrated genomic rearrangements in both the chromosome and in the invasion plasmid. Phylogenetic analyses have previously demonstrated that *S. dysenteriae* 197 is more closely related to O157:H7 EHEC than to other *Shigella* species (Touchon et al., 2009; Sahl et al., 2011). This finding has been re-confirmed when comparing additional phylogenetically diverse isolates (Figure 2.1).

A phylogeny inferred from a large number of concatenated, recombination filtered (Bruen et al., 2006), coding regions identified that *Shigella* falls into two main clusters (Ogura et al., 2009). A feature frequency profile method also demonstrated that *Shigella* groups into two clusters (Sims and Kim, 2011). If the accessory genome is included in the analysis, *Shigella* group into one monophyletic cluster (Sims and Kim, 2011; Zhang and Lin, 2012); this is likely due to the shared presence of pINV. Nine complete genomes are now publically available for *Shigella* (Table 2.1). A recent study sequenced an additional 55 genomes available in high-quality draft status. A whole genome alignment and phylogeny of 69 *Shigella* genomes demonstrated that all *Shigella* group into five monophyletic clades that contain a mix of 'species' based on serotyping (Sahl et al., 2012). A large-scale genomic analysis of 337 *E. coli*/*Shigella* genomes demonstrated the presence of three exclusive clusters of *Shigella* that contain the five monophyletic clades. Previous studies using limited conserved gene-based data had suggested that *Shigella* had evolved from *E. coli* multiple times, up to seven defining events (Ochman et al., 1983; Pupo et al., 2000), however the genomic data indicate that there have only been three radiations from *E. coli*; these three events include the separation of the *S. dysenteriae* from the enterohemorrhagic *E. coli*, and the separation of two other mixed *Shigella* species groups in phylogenetic group A and B1. While these findings do not follow the 'species' lines in *Shigella*, they will allow clarification of the *Shigella* evolutionary path and the relationship to *E. coli*.

Shigella are almost indistinguishable from several enteroinvasive *E. coli* (EIEC), which also contain the invasion plasmid, pINV, and show signs of genome reduction (Pupo et al., 2000). Only one EIEC genome, *E. coli* 53638, is available in Genbank and is frequently not included in studies of *Shigella* gene content and evolution (Pupo et al., 2000; Zhang and Lin, 2012). Additional genomics studies are required in order to characterize the genomic content and phylogenetic diversity of EIEC isolates.

FUTURE DIRECTIONS

Whole genome sequence analysis has revealed information on important evolutionary relationships between *E. coli* and *Shigella* isolates. Comparative analyses have also been used to identify conserved regions that will result in more accurate and more rapid diagnostic assays. Accurate diagnostics will likely help with food safety, as food supplies can be quickly surveyed for the presence of potential human pathogens. In addition to diagnostics, food safety, and surveillance, whole genome sequence data can be used to better understand outbreak events in real time. This was showcased by the *E. coli* outbreak in Germany, in which a global crowd-sourcing effort was conducted to analyze the isolate (Rohde et al., 2011). These analyses are likely to be more common as data sharing becomes more common.

Whole genome sequence data will also be more important in the diagnosis of human infections. If human samples can be quickly sequenced and the infecting pathogen accurately identified, then treatment therapies can be more quickly applied and modified. Furthermore, understanding the complete complement of virulence genes and antibiotic resistance profiles, can help tailor specific treatments. A complete understanding of *E. coli* and *Shigella* genomics will build the foundation for these types of clinical applications.

ACKNOWLEDGMENTS

The work included in this chapter was funded in part by federal funds from the National Institute of Allergy and Infectious Diseases, National Institutes of Health, grant numbers RO1 AI089894 and 1RC4AI092828.

REFERENCES

Angiuoli, S.V., Salzberg, S.L., 2011. Mugsy: fast multiple alignment of closely related whole genomes. Bioinformat. 27, 334–342.

Bernier, C., Gounon, P., Le Bouguenec, C., 2002. Identification of an aggregative adhesion fimbria (AAF) type III-encoding operon in enteroaggregative *Escherichia coli* as a sensitive probe for detecting the AAF-encoding operon family. Infect. Immun. 70, 4302–4311.

Bieber, D., Ramer, S.W., Wu, C.Y., et al., 1998. Type IV pili, transient bacterial aggregates, and virulence of enteropathogenic *Escherichia coli*. Science 280, 2114–2118.

Bilge, S.S., Clausen, C.R., Lau, W., Moseley, S.L., 1989. Molecular characterization of a fimbrial adhesin, F1845, mediating diffuse adherence of diarrhea-associated *Escherichia coli* to HEp-2 cells. J. Bacteriol. 171, 4281–4289.

Blattner, F.R., Plunkett 3rd, G., Bloch, C.A., et al., 1997. The complete genome sequence of *Escherichia coli* K-12. Science 277, 1453–1462.

Boisen, N., Ruiz-Perez, F., Scheutz, F., Krogfelt, K.A., Nataro, J.P., 2009. Short report: high prevalence of serine protease autotransporter cytotoxins among strains of enteroaggregative *Escherichia coli*. Am. J. Trop. Med. Hyg. 80, 294–301.

Boisen, N., Scheutz, F., Rasko, D.A., et al., 2012. Genomic characterization of enteroaggregative *Escherichia coli* from children in Mali. J. Infect. Dis. 205, 431–444.

Bruen, T.C., Philippe, H., Bryant, D., 2006. A simple and robust statistical test for detecting the presence of recombination. Genetics 172, 2665–2681.

Brzuszkiewicz, E., Bruggemann, H., Liesegang, H., et al., 2006. How to become a uropathogen: comparative genomic analysis of extraintestinal pathogenic *Escherichia coli* strains. Proc. Nat. Acad. Sci. USA 103, 12879–12884.

Campbell, A., Schneider, S.J., Song, B., 1992. Lambdoid phages as elements of bacterial genomes (integrase/phage21/*Escherichia coli* K-12/icd gene). Genetica 86, 259–267.

Chaudhuri, R.R., Sebaihia, M., Hobman, J.L., et al., 2010. Complete genome sequence and comparative metabolic profiling of the prototypical enteroaggregative *Escherichia coli* strain 042. PloS One 5, e8801.

Chen, S.L., Hung, C.S., Xu, J., et al., 2006. Identification of genes subject to positive selection in uropathogenic strains of *Escherichia coli*: a comparative genomics approach. Proc. Natl. Acad. Sci. USA 103, 5977–5982.

Corrigan Jr., J.J., Boineau, F.G., 2001. Hemolytic-uremic syndrome. Pediatr. Rev. 22, 365–369.

Crossman, L.C., Chaudhuri, R.R., Beatson, S.A., et al., 2010. A commensal gone bad: complete genome sequence of the prototypical enterotoxigenic *Escherichia coli* strain H10407. J. Bacteriol. 192, 5822–5831.

Czeczulin, J.R., Whittam, T.S., Henderson, I.R., Navarro-Garcia, F., Nataro, J.P., 1999. Phylogenetic analysis of enteroaggregative and diffusely adherent *Escherichia coli*. Infect. Immun. 67, 2692–2699.

Donnenberg, M.S., Kaper, J.B., 1992. Enteropathogenic *Escherichia coli*. Infect. Immun. 60, 3953–3961.

Donnenberg, M.S., Tacket, C.O., James, S.P., et al., 1993. Role of the eaeA gene in experimental enteropathogenic *Escherichia coli* infection. J. Clin. Invest. 92, 1412–1417.

Dudley, E.G., Thomson, N.R., Parkhill, J., Morin, N.P., Nataro, J.P., 2006. Proteomic and microarray characterization of the AggR regulon identifies a *pheU* pathogenicity island in enteroaggregative *Escherichia coli*. Mol. Microbiol. 61, 1267–1282.

Eppinger, M., Mammel, M.K., Leclerc, J.E., Ravel, J., Cebula, T.A., 2011. Genomic anatomy of *Escherichia coli* O157:H7 outbreaks. Proc. Natl. Acad. Sci. USA 108, 20142–20147.

Frank, C., Werber, D., Cramer, J.P., et al., 2011. Epidemic profile of Shiga-toxin-producing *Escherichia coli* O104:H4 outbreak in Germany. NEJM 365, 1771–1780.

Gaastra, W., Svennerholm, A.M., 1996. Colonization factors of human enterotoxigenic *Escherichia coli* (ETEC). Trends. Microbiol. 4, 444–452.

Giron, J.A., Ho, A.S., Schoolnik, G.K., 1991. An inducible bundle-forming pilus of enteropathogenic *Escherichia coli*. Science 254, 710–713.

Grad, Y.H., Lipsitch, M., Feldgarden, M., et al., 2012. Genomic epidemiology of the *Escherichia coli* O104:H4 outbreaks in Europe, 2011. Proc. Natl. Acad. Sci. USA 109, 3065–3070.

Hacker, J., Kaper, J.B., 2000. Pathogenicity islands and the evolution of microbes. Annu. Rev. Microbiol. 54, 641–679.

Hacker, J., Ott, M., Blum, G., et al., 1992. Genetics of *Escherichia coli* uropathogenicity: analysis of the O6:K15:H31 isolate 536. Zentralbl. Bakteriol. 276, 165–175.

Hagan, E.C., Lloyd, A.L., Rasko, D.A., Faerber, G.J., Mobley, H.L., 2010. *Escherichia coli* global gene expression in urine from women with urinary tract infection. PLoS Path 6, e1001187.

Hale, T.L., 1991. Genetic basis of virulence in *Shigella* species. Microbiol. Rev. 55, 206–224.

Hayashi, T., Makino, K., Ohnishi, M., et al., 2001. Complete genome sequence of enterohemorrhagic *Escherichia coli* O157:H7 and genomic comparison with a laboratory strain K-12. DNA Res. 8, 11–22.

Hazen, T.H., Sahl, J.W., Donnenberg, M.S., Scheutz, F., Rasko, D.A., 2012. Phylogenomics of the attaching and effacing *Escherichia coli* (AEEC): redefining the pathotype paradigm using whole-genome sequencing. Unpublished.

Hershberg, R., Tang, H., Petrov, D.A., 2007. Reduced selection leads to accelerated gene loss in *Shigella*. Genome. Biol. 8, R164.

Iguchi, A., Thomson, N.R., Ogura, Y., et al., 2009. Complete genome sequence and comparative genome analysis of enteropathogenic *Escherichia coli* O127:H6 strain E2348/69. J. Bacteriol. 191, 347–354.

Jin, Q., Yuan, Z., Xu, J., et al., 2002. Genome sequence of *Shigella flexneri* 2a: insights into pathogenicity through comparison with genomes of *Escherichia coli* K12 and O157. Nuc. Acid. Res. 30, 4432–4441.

Johnson, J.R., 2000. *Shigella* and *Escherichia coli* at the crossroads: machiavellian masqueraders or taxonomic treachery? J. Med. Micro. 49, 583–585.

Johnson, J.R., Clermont, O., Menard, M., Kuskowski, M.A., Picard, B., Denamur, E., 2006. Experimental mouse lethality of *Escherichia coli* isolates, in relation to accessory traits, phylogenetic group, and ecological source. J. Infect. Dis. 194, 1141–1150.

Johnson, K.E., Thorpe, C.M., Sears, C.L., 2006. The emerging clinical importance of non-O157 Shiga toxin-producing *Escherichia coli*. Clin. Infect. Dis. 43, 1587–1595.

Kaper, J.B., 1996. Defining EPEC Rev. Microbiol. Sao Paulo 27, 130–133.

Kaper, J.B., Nataro, J.P., Mobley, H.L., 2004. Pathogenic *Escherichia coli.* Nat. Rev. Microbiol. 2, 123–140.

Kotloff, K.L., Winickoff, J.P., Ivanoff, B., et al., 1999. Global burden of *Shigella* infections: implications for vaccine development and implementation of control strategies. Bull. World Health Organ. 77, 651–666.

Kudva, I.T., Evans, P.S., Perna, N.T., et al., 2002. Strains of *Escherichia coli* O157:H7 differ primarily by insertions or deletions, not single-nucleotide polymorphisms. J. Bacteriol. 184, 1873–1879.

Lacher, D.W., Steinsland, H., Blank, T.E., Donnenberg, M.S., Whittam, T.S., 2007. Molecular evolution of typical enteropathogenic *Escherichia coli*: clonal analysis by multilocus sequence typing and virulence gene allelic profiling. J. Bacteriol. 189, 342–350.

Lacher, D.W., Steinsland, H., Whittam, T.S., 2006. Allelic subtyping of the intimin locus (*eae*) of pathogenic *Escherichia coli* by fluorescent RFLP. FEMS. Microbiol. Lett. 261, 80–87.

Leopold, S.R., Sawyer, S.A., Whittam, T.S., Tarr, P.I., 2011. Obscured phylogeny and possible recombinational dormancy in *Escherichia coli.* BMC Evolut. Biol. 11, 183.

Levine, M.M., Nataro, J.P., Karch, H., et al., 1985. The diarrheal response of humans to some classic serotypes of enteropathogenic *Escherichia coli* is dependent on a plasmid encoding an entero-adhesiveness factor. J. Infect. Dis. 152, 550–559.

Lloyd, A.L., Rasko, D.A., Mobley, H.L., 2007. Defining genomic islands and uropathogen-specific genes in uropathogenic *Escherichia coli.* J. Bacteriol. 189, 3532–3546.

Manning, S.D., Motiwala, A.S., Springman, A.C., et al., 2008. Variation in virulence among clades of *Escherichia coli* O157:H7 associated with disease outbreaks. Proc. Natl. Acad. Sci. USA 105, 4868–4873.

McDaniel, T.K., Jarvis, K.G., Donnenberg, M.S., Kaper, J.B., 1995. A genetic locus of enterocyte effacement conserved among diverse enterobacterial pathogens. Proc. Natl. Acad. Sci. USA 92, 1664–1668.

Mellmann, A., Bielaszewska, M., Kock, R., et al., 2008. Analysis of collection of hemolytic uremic syndrome-associated enterohemorrhagic *Escherichia coli.* Emerg. Infect. Dis. 14, 1287–1290.

Middendorf, B., Blum-Oehler, G., Dobrindt, U., Muhldorfer, I., Salge, S., Hacker, J., 2001. The pathogenicity islands (PAIs) of the uropathogenic *Escherichia coli* strain 536: island probing of PAI II536. J. Infect. Dis. 183 (Suppl. 1), S17–S20.

Miquel, S., Peyretaillade, E., Claret, L., et al., 2010. Complete genome sequence of Crohn's disease-associated adherent-invasive *E. coli* strain LF82. PloS One 5.

Morabito, S., Karch, H., Mariani-Kurkdjian, P., et al., 1998. Enteroaggregative, Shiga toxin-producing *Escherichia coli* O111:H2 associated with an outbreak of hemolytic-uremic syndrome. J. Clin. Microbiol. 36, 840–842.

Moran, N.A., 2002. Microbial minimalism: genome reduction in bacterial pathogens. Cell 108, 583–586.

Nada, R.A., Shaheen, H.I., Khalil, S.B., et al., 2011. Discovery and phylogenetic analysis of novel members of class b enterotoxigenic *Escherichia coli* adhesive fimbriae. J. Clin. Microbiol. 49, 1403–1410.

Nash, J.H., Villegas, A., Kropinski, A.M., et al., 2010. Genome sequence of adherent-invasive *Escherichia coli* and comparative genomic analysis with other *E. coli* pathotypes. BMC Genomics 11, 667.

Nataro, J.P., Yikang, D., Giron, J.A., Savarino, S.J., Kothary, M.H., Hall, R., 1993. Aggregative adherence fimbria I expression in enteroaggregative *Escherichia coli* requires two unlinked plasmid regions. Infect. Immun. 61, 1126–1131.

Newton, H.J., Sloan, J., Bulach, D.M., et al., 2009. Shiga toxin-producing *Escherichia coli* strains negative for locus of enterocyte effacement. Emerg. Infect. Dis. 15, 372–380.

Ochman, H., Whittam, T.S., Caugant, D.A., Selander, R.K., 1983. Enzyme polymorphism and genetic population structure in *Escherichia coli* and *Shigella*. J. Gen. Microbiol. 129, 2715–2726.

Ogura, Y., Ooka, T., Iguchi, A., et al., 2009. Comparative genomics reveal the mechanism of the parallel evolution of O157 and non-O157 enterohemorrhagic *Escherichia coli*. Proc. Natl. Acad. Sci. USA 106, 17939–17944.

Orskov, F., Whittam, T.S., Cravioto, A., Orskov, I., 1990. Clonal relationships among classic enteropathogenic *Escherichia coli* (EPEC) belong to different O groups. J. Infect. Dis. 162, 76–81.

Parker, C.T., Kyle, J.L., Huynh, S., Carter, M.Q., Brandl, M.T., Mandrell, R.E., 2012. Distinct transcriptional profiles and phenotypes exhibited by *Escherichia coli* O157:H7 isolates related to the 2006 spinach-associated outbreak. Appl. Environ. Microbiol. 78, 455–463.

Parsot, C., Menard, R., Gounon, P., Sansonetti, P.J., 1995. Enhanced secretion through the *Shigella flexneri* Mxi-Spa translocon leads to assembly of extracellular proteins into macromolecular structures. Mol. Microbiol. 16, 291–300.

Perna, N.T., Plunkett 3rd, G., Burland, V., et al., 2001. Genome sequence of enterohaemorrhagic *Escherichia coli* O157:H7. Nature 409, 529–533.

Perry, R.D., Fetherston, J.D., 2011. Yersiniabactin iron uptake: mechanisms and role in *Yersinia pestis* pathogenesis. Microbes. Infect. 13, 808–817.

Pupo, G.M., Lan, R., Reeves, P.R., 2000. Multiple independent origins of *Shigella* clones of *Escherichia coli* and convergent evolution of many of their characteristics. Proc. Natl. Acad. Sci. USA 97, 10567–10572.

Rasko, D.A., Rosovitz, M.J., Myers, G.S., et al., 2008. The pangenome structure of *Escherichia coli*: comparative genomic analysis of *E. coli* commensal and pathogenic isolates. J. Bacteriol. 190, 6881–6893.

Rasko, D.A., Webster, D.R., Sahl, J.W., et al., 2011. Origins of the *E. coli* strain causing an outbreak of hemolytic-uremic syndrome in Germany. N. Engl. J. Med. 365, 725–729.

Rohde, H., Qin, J., Cui, Y., et al., 2011. Open-source genomic analysis of Shiga-toxin-producing *E. coli* O104:H4. N. Engl. J. Med. 365, 718–724.

Sahl, J.W., Morris, C.R., Emberger, J., et al., 2012. The phylogenomics of *Shigella*: a pathway to a new diagnostic assay. Unpublished.

Sahl, J.W., Rasko, D.A., 2012. Analysis of the global transcriptional profiles of enterotoxigenic *Escherichia coli* (ETEC) isolate E24377A. Infect. Immun. 80, 1232–1242.

Sahl, J.W., Steinsland, H., Rasko, D.A., 2012. Insights into enterotoxigenic *Escherichia coli* (ETEC) pathogenesis based genomic epidemiology of symptomatic and asymptomatic isolates. Unpublished.

Sahl, J.W., Steinsland, H., Redman, J.C., et al., 2011. A comparative genomic analysis of diverse clonal types of enterotoxigenic *Escherichia coli* reveals pathovar-specific conservation. Infect. Immun. 79, 950–960.

Scaletsky, I.C., Fabbricotti, S.H., Carvalho, R.L., et al., 2002. Diffusely adherent *Escherichia coli* as a cause of acute diarrhea in young children in Northeast Brazil: a case-control study. J. Clin. Microbiol. 40, 645–648.

Schwartz, D.J., Chen, S.L., Hultgren, S.J., Seed, P.C., 2011. Population dynamics and niche distribution of uropathogenic *Escherichia coli* during acute and chronic urinary tract infection. Infect. Immun. 79, 4250–4259.

Shaheen, H.I., Kamal, K.A., Wasfy, M.O., et al., 2003. Phenotypic diversity of enterotoxigenic *Escherichia coli* (ETEC) isolated from cases of travelers' diarrhea in Kenya. Internat. J. Infect. Dis. 7, 35–38.

Sheikh, J., Czeczulin, J.R., Harrington, S., et al., 2002. A novel dispersin protein in enteroaggregative *Escherichia coli.* J. Clin. Invest. 110, 1329–1337.

Sims, G.E., Kim, S.H., 2011. Whole-genome phylogeny of *Escherichia coli/Shigella* group by feature frequency profiles (FFPs). Proc. Natl. Acad. Sci. USA 108, 8329–8334.

Snyder, J.A., Haugen, B.J., Buckles, E.L., et al., 2004. Transcriptome of uropathogenic *Escherichia coli* during urinary tract infection. Infect. Immun. 72, 6373–6381.

So, M., Boyer, H.W., Betlach, M., Falkow, S., 1976. Molecular cloning of an *Escherichia coli* plasmid determinant that encodes for the production of heat-stable enterotoxin. J. Bacteriol. 128, 463–472.

So, M., Dallas, W.S., Falkow, S., 1978. Characterization of an *Escherichia coli* plasmid encoding for synthesis of heat-labile toxin: molecular cloning of the toxin determinant. Infect. Immun. 21, 405–411.

Steinsland, H., Lacher, D.W., Sommerfelt, H., Whittam, T.S., 2010. Ancestral lineages of human enterotoxigenic *Escherichia coli.* J. Clin. Microbiol. 48, 2916–2924.

Steyert, S.R., Sahl, J.W., Fraser-Liggett, C.M., Teel, L.D., Scheutz, F., Rasko, D.A., 2012. Comparative genomics and stx phage characterization of LEE-negative Shiga toxin-producing *Escherichia coli.* Front Cell Infect. Microbiol. 2, 133.

Swenson, D.L., Bukanov, N.O., Berg, D.E., Welch, R.A., 1996. Two pathogenicity islands in uropathogenic *Escherichia coli* J96: cosmid cloning and sample sequencing. Infect. Immun. 64, 3736–3743.

Tobe, T., Hayashi, T., Han, C.G., Schoolnik, G.K., Ohtsubo, E., Sasakawa, C., 1999. Complete DNA sequence and structural analysis of the enteropathogenic *Escherichia coli* adherence factor plasmid. Infect. Immun. 67, 5455–5462.

Touchon, M., Hoede, C., Tenaillon, O., et al., 2009. Organised genome dynamics in the *Escherichia coli* species results in highly diverse adaptive paths. PLoS Genet 5, e1000344.

Trabulsi, L.R., Keller, R., Tardelli Gomes, T.A., 2002. Typical and atypical enteropathogenic *Escherichia coli.* Emerg. Infect. Dis. 8, 508–513.

Trofa, A.F., Ueno-Olsen, H., Oiwa, R., Yoshikawa, M., 1999. Dr. Kiyoshi Shiga: discoverer of the dysentery bacillus. Clin. Infect. Dis. 29, 1303–1306.

Turner, S.M., Chaudhuri, R.R., Jiang, Z.D., et al., 2006. Phylogenetic comparisons reveal multiple acquisitions of the toxin genes by enterotoxigenic *Escherichia coli* strains of different evolutionary lineages. J. Clin. Microbiol. 44, 4528–4536.

Vial, P.A., Robins-Browne, R., Lior, H., et al., 1988. Characterization of enteroadherent-aggregative *Escherichia coli,* a putative agent of diarrheal disease. J. Infect. Dis. 158, 70–79.

Viljanen, M.K., Peltola, T., Junnila, S.Y., et al., 1990. Outbreak of diarrhoea due to *Escherichia coli* O111:B4 in schoolchildren and adults: association of Vi antigen-like reactivity. Lancet 336, 831–834.

Watanabe, H., Wada, A., Inagaki, Y., Itoh, K.-i., Tamura, K., 1996. Outbreaks of enterohaemorrhagic *Escherichia coli* O157:H7 infection by two different genotype strains in Japan, 1996. Lancet 348, 831–832.

Wei, J., Goldberg, M.B., Burland, V., et al., 2003. Complete genome sequence and comparative genomics of *Shigella flexneri* serotype 2a strain 2457T. Infect. Immun. 71, 2775–2786.

Welch, R.A., Burland, V., Plunkett 3rd, G., et al., 2002. Extensive mosaic structure revealed by the complete genome sequence of uropathogenic *Escherichia coli.* Proc. Natl. Acad. Sci. USA 99, 17020–17024.

WHO, 2006. Future directions for research on enterotoxigenic *Escherichia coli* vaccines for developing countries. Wkly. Epidemiol. Rec. 81, 97–104.

Yang, F., Yang, J., Zhang, X., et al., 2005. Genome dynamics and diversity of *Shigella* species, the etiologic agents of bacillary dysentery. Nucleic Acids Res. 33, 6445–6458.

Zhang, Y., Lin, K., 2012. A phylogenomic analysis of *Escherichia coli/Shigella* group: implications of genomic features associated with pathogenicity and ecological adaptation. BMC Evolut. Biol. 12, 174.

Chapter 3

Evolution of pathogenic *Escherichia coli*

Sujay Chattopadhyay and Evgeni V. Sokurenko
University of Washington, Seattle, WA, USA

INTRODUCTION

As with evolution in general, bacterial evolution happens through the action of selection and drift on random genetic variations, affecting their frequency in nature, in space and in time. The evolution of virulence is viewed from the host's perspective, where evolution of microbial genomes results in the ability of the microorganism to cause clinically manifested damage of the host. The ability to cause disease reflects the microbial fitness (i.e. ability to survive and reproduce) during the infection itself. To understand the driving forces and mechanisms behind the evolution of microbial virulence, one might compare the genomic content of organisms that are able to cause the disease to those that are unable to do so. There are relatively few (<200) bacterial species that are isolated as the cause of human infections, among which less than a dozen cause the vast majority of the infections. It is possible to compare these species to the species that do not act as common pathogens. While this approach is valid, the high level of genomic diversity between even closely related species makes such a strategy difficult. Frequent horizontal gene transfer between prokaryotic species results in a high level of genome mosaicism, which adds to the genome plasticity along with intragenomic variations, like point mutations, gene deletion/amplification, or genomic rearrangements. It is easier to compare organisms from the same bacterial species that differ significantly in their ability to cause disease. *Escherichia coli* offers an ideal example of such within-species virulence diversity.

WITHIN-SPECIES DIVERSITY OF PATHOGENIC *E. COLI*

There is a huge diversity of phenotypic traits across *E. coli* strains, both quantitatively or qualitatively, which allow these strains to differ in their appearance, behavior, metabolism, as well as in their ability to cause disease in humans. At one end, commensal strains, the vast majority of *E. coli* population, have adapted

Escherichia coli. http://dx.doi.org/10.1016/B978-0-12-397048-0.00003-6

to colonize the host without causing disease. Colonization of pathogenic counterparts, on the other hand, can result in clinically significant pathologies. For example, the recommended regular dose of probiotic MutaFlor™, purported to promote health, contains billions of live bacteria of *E. coli* strain Nissle-1917, while the ingestion of only 100 bacteria of *E. coli* O157 strain or even fewer of *Shigella flexneri* can produce fatal disease (Todd et al., 2008; Allen et al., 2010). Also, the spectrum of disease caused by *E. coli* varies significantly, from diarrhea to meningitis, from asymptomatic bacteriuria to lethal urosepsis. This spectrum is caused by several main pathotypes of *E. coli* – enterotoxigenic (ETEC), enteropathogenic (EPEC), enterohemorrhagic (EHEC), enteroinvasive (EIEC/*Shigella*), enteroaggregative (EAEC) and extraintestinal-pathogenic (ExPEC, among which uropathogenic, UPEC, is the most common), as detailed elsewhere in this volume.

Despite being from the same species, different strains of *E. coli* are also highly diverse genetically. In fact, a given strain of *E. coli* shares only a minority of its genes with every other strain of the species (Rasko et al., 2008). This diversity is especially obvious from the analysis of clonally unrelated strains, i.e. those with different genotype profile based on the multi-locus sequence typing (MLST). MLST profile (or sequence type, ST) is defined in *E. coli* by comparing sequence identity of 400–500 bp long regions of seven housekeeping genes that are spread across the bacterial chromosome (Wirth et al., 2006).

Here, to assess the level of clonal diversity of *E. coli*, we selected 22 clonally unrelated *E. coli* strains, ranging from commensal to different major pathotypes (Figure 3.1),with fully assembled annotated genome sequences publically available at the time of preparation of this chapter. The strains' genome size ranged from 4116 to 5379 genes. However, based on 95% nucleotide identity and length preservation, a total of 16 148 genes were found in it least one of the 22 strains examined, comprising a minimal estimate of the *E. coli* pangenome. Out of these genes, 8573 genes were mosaic in nature, i.e. found in multiple (but not all!) strains that, on average, comprised 49–67% of individual genomes. Only 1996 genes were found in every strain and could be defined as core genes, which constituted from 37 to 49% of the genomes of individual strains. Finally, the rest of the genes either were found only in a single strain so far or were highly diverse orthologs (with less than 95% sequence identity) found in two or more strains.

The nucleotide-level difference found between the clonally unrelated *E. coli* was even greater. Even if only the shared genes more than 95% identical in sequence are considered, strain-to-strain polymorphisms affect over 100 000 nucleotides (2% of all gene sequences across the genome) on average. Out of these, about 20 000 nucleotide changes resulted in allelic amino acid replacements, with an average coded protein variant differing in 4–5 amino acids from the variant coded by the same gene but in a different strain.

Furthermore, strains from different pathotypes tend not only to be from different STs, but to belong to separate major phylogenetic clades that form so-called ECOR groups of *E. coli* (Figure 3.1) (Ochman and Selander, 1984).

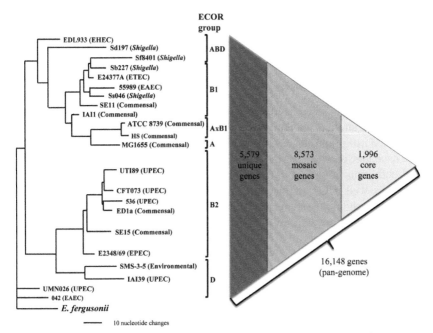

FIGURE 3.1   Phylogenetic tree (phylogram) of 22 *E. coli* isolates with fully assembled genomes. The tree is based on 7-loci Multi-Locus Sequence Type (MLST) profiles, using *E. fergusonii* as an outgroup. Core genes are present in all 22 isolates, mosaics are present in multiple but not all isolates, and unique genes are present in one isolate only.

Thus, the genomic differences between strains from different pathotypes are even more pronounced than on average strain-to-strain, making the search for pathotype-specific traits difficult.

Interestingly, however, at least in some cases, clonally related strains with the same ST could be drastically different in their virulence (Weissman et al., 2012). For example, the probiotic strain Nissle-1917 that was isolated from the feces of a healthy individual almost a century ago has the same ST profile as model pathogenic *E. coli* strain CFT073 isolated from a patient with urosepsis in the 1980s (Vejborg et al., 2010). The vast majority of genes are shared between Nissle and CFT073, and only about 17% of genes are different. On the nucleotide level, there are about 100-fold fewer mutational differences than found between clonally unrelated strains. Thus, comparing clonally related strains might facilitate the search for the genetic basis of differential virulence and the evolutionary mechanisms of virulence acquisition.

GENETIC MECHANISMS OF VIRULENCE EVOLUTION

To gather insights on the microevolution of virulence in *E. coli*, the first step is to understand the genetic mechanisms the species adopt to achieve their within-species diversity. The major genetic events that underlie the evolution of virulence are: (a) horizontal transfer of novel genes, and (b) pathoadaptive mutation – amplification, inactivation, and variation – of existing genes (Figure 3.2).

Horizontal gene transfer (HGT)

This process, also called lateral gene transfer, involves transfer of genetic material from one strain to another, i.e. in contrast to vertical parent–progeny transfer of DNA during reproduction. According to a strict definition, HGT involves gene exchange between organisms from different species, but a more commonly used definition is applied to the gene transfer between organisms of the same species as well. HGT implies addition to the genome of novel genes, not an exchange by homologous recombination of different copies of the same gene, even though the mechanism of the gene movement between different strains could be the same.

HGT genomic elements

Genomic regions of HGT origin are typically carried by three major genetic elements in the *E. coli*: plasmids, prophages, and chromosomal islands.

Plasmids

Plasmids are usually circular and self-replicating molecules that co-exist with chromosomes. These extra-chromosomal elements harbor at least one, if not multiple, essential virulence determinants in all strains of diarrheagenic *E. coli* and *Shigella* spp. (Mellmann et al., 2009). One pioneering study demonstrated the pathogenic properties of plasmids in *E. coli* that causes diarrhea in piglets (Smith and Linggood, 1971). Subsequent studies showed that a major class of virulence factors encoded by the plasmids are genes conferring resistance to antimicrobial agents (e.g. cephalosporins, fluoroquinolones, aminoglycosides), toxic

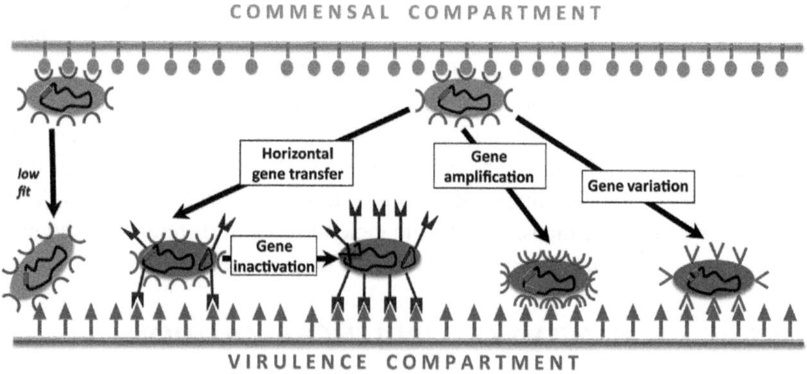

FIGURE 3.2 Different genetic mechanisms of evolution of virulence. Improved adhesion to the tissue receptors is assumed to increase virulence.

heavy metals (e.g. cadmium, mercury, silver), and other survival factors against lethal doses of antimicrobials (Mayer et al., 1995; Bennett, 2008; Hawkey, 2008). In addition, plasmid genes up-regulate important virulence and fitness genes in chromosomes through extensive cross-talk between plasmid and chromosome, as evidenced in many enteropathogenic *E. coli* (EPEC) strains (Schmidt, 2010). For example, the products of *per* genes in EPEC plasmids regulate expression of the pathogenicity locus of enterocyte effacement (LEE) genes (Mellies et al., 1999). Complex synergistic activities accelerate plasmid spread. For example, dense, structured populations as in biofilms increase the possibility of plasmid transfer by conjugation (see below) (Reisner et al., 2006). By the same token, the conjugation apparatus and the release of DNA stimulate formation and maintenance of biofilms (Molin and Tolker-Nielsen, 2003). Other examples of plasmid-borne virulence factors include Bundle-forming pili of EPEC (Donnenberg et al., 1992); EhxA in EHEC (Burland et al., 1998); pCoo in ETEC (Froehlich et al., 2005); plasmid-encoded toxin (Pet) in EAEC (Eslava et al., 1998); IcsA (VirG) in EIEC/*Shigella* (Buchrieser et al., 2000); TraT in UPEC (Timmis et al., 1985), etc.

Prophages

Temperate phages upon DNA injection into the host bacteria do not immediately enter into the lytic cycle but can instead integrate into the bacterial genome as a prophage. In fact, most of the mosaic gene elements in the *E. coli* genome are of a prophage nature (Canchaya et al., 2003). Most prophage genes are usually silent during bacterial growth and reproduction and are functional only when the prophage is activated (induced), i.e. enters the lytic pathway to produce active phages and lyse the host cell (usually under stress conditions, like UV light, antibiotic exposure, etc.), e.g. Shiga toxins Stx1 and Stx2 of EHEC (Dobrindt, 2005). However, some prophages commonly carry genes that are expressed and can add new phenotypic traits to their hosts that are important for success in colonization and competition within the habitat (Brussow et al., 2004). In EHEC and EPEC genomes, for example, a diversity of the prophages encode characteristic virulence factors (Ohnishi et al., 2002). A few examples are several type III secretion system (T3SS) effector proteins of EHEC and EPEC such as Cif (Marches et al., 2003), EspF$_U$ (Campellone et al., 2004), EspJ (Dahan et al., 2005), EspK (Vlisidou et al., 2006), NleA (Gruenheid et al., 2004), TccP (Garmendia et al., 2004) (see Chapters 4, 5, and 15). Other examples include cytolethal distending toxins in EPEC (Asakura et al., 2007); type II heat-labile enterotoxin in ETEC (Jobling and Holmes, 2012); GtrAB in EIEC/*Shigella* (Chaudhuri et al., 2010). Interestingly, until recently, there were no known major virulence determinants encoded by prophages in UPEC (Lavigne and Blanc-Potard, 2008).

Chromosomal islands

Chromosomal islands characterize a highly diverse group of DNA elements, with a broad range in size and abundance across the bacterial chromosomes

(Dobrindt et al., 2004). Originally, actually, chromosomal islands were described as pathogenicity-associated islands (PAI), coined to define large unstable regions (10–200 kb) harboring virulence determinants on uropathogenic *E. coli* chromosomes and differing in GC content from the rest of the chromosome (Hacker et al., 1990). Examples of PAI-coded virulence factors are LEE island genes in EPEC and EHEC (McDaniel et al., 1995); Tia adhesin in ETEC (Fleckenstein et al., 2000); Protein involved in colonization (Pic) in EAEC (Henderson et al., 1999); SigA in EIEC/*Shigella* (Al-Hasani et al., 2000); Hemolysin in UPEC (O'Hanley et al., 1991). Besides pathogenicity-related functions, chromosomal islands encode genes representing a wide spectrum of traits such as mercury resistance and siderophore synthesis (Larbig et al., 2002), symbiosis (Sullivan et al., 2002), sucrose and aromatic compound metabolism (Gaillard et al., 2006), etc. The genomic islands are predicted using either sequence-based methods or comparative genomic analyses (Gal-Mor and Finlay, 2006). While sequence-based approaches rely on abnormal sequence composition (e.g. bias in GC content, dinucleotide frequency, codon usage) or on features associated with mobile genetic elements (e.g. presence of direct repeats, insertion sequence elements, tRNA, integrases, transposases, etc.), comparative analyses of multiple genomes are based on detection of genomic regions with anomalous phylogenetic patterns.

Mechanisms of HGT

There are three frequent genetic mechanisms that make the transfer of DNA happen: transduction, conjugation, and transformation. These diverse forms of transfer make HGT a very important process in expanding the potential of genetic adaptation in a bacterial species (Ochman et al., 2000).

Transduction

Transduction is defined as the movement of genetic material with the help of bacteriophages, the viruses that can inject DNA into the organisms. Lytic (virulent) phages that usually immediately multiply and lyze the infected bacterium, sometimes carry DNA from previously infected organisms, accidently packed into the capsule. Thus, if such DNA is injected instead of the viral genome, it can get incorporated in the infected microbial genome. However, while lytic phage transduction is commonly observed and utilized in lab experiments (e.g. P phage), it is unclear to what extent this process occurs in nature, because in these cases there are no traces of the viral DNA being tagged to the transduced DNA. Theoretically, any region of the bacterial genome could be transferred in this way, including plasmids and chromosomal islands and, thus, such a mechanism is termed 'generalized transduction' (Berg et al., 1983). It is a different story with the temperate phages that integrate into the bacterial chromosome as a prophage (see above). Its excision from the chromosome during the lytic cycle could be accompanied by accidental packaging (and then transfer) of non-phage DNA, but commonly these are the genomic regions flanking the prophage insertion site and, thus, this process is called 'specialized transduction' (Berg, et al., 1983). Still, any region of the genome could potentially be packaged in the activated prophage.

Conjugation

Conjugation involves direct transfer of DNA from donor to recipient microbial cell (Sorensen et al., 2005). *E. coli* and other Gram-negative bacteria have conjugation machinery that involves formation of a sex pilus, which is in contrast to the conjugation in Gram-positive bacteria that involves direct cell–cell contact adhesins (Grohmann et al., 2003). The pili and transfer function is encoded by conjugative plasmids that can be used as a vehicle by other (transmissible) plasmids or chromosomal regions (Zatyka and Thomas, 1998). Integration of the conjugative plasmid into the bacterial chromosome (e.g. integration of F plasmid episome into the chromosome via homologous recombination) can lead to the conjugative transfer of the flanking chromosomal regions (Holloway, 1993; Ambrozic et al., 1998), which could be, for example, chromosomal islands.

Virulence-associated genes can also reside in transposable elements that frequently change location between chromosome and plasmid and, in this way, can move horizontally. These elements can be insertion sequences (IS elements), i.e. small mobile fragments with short terminal inverted repeat sequences. These entities can also be transposons, with or without flanking insertion sequences termed composite and non-composite transposons respectively. Many transposons include antibiotic-resistance genes, e.g. Tn5 carrying kanamycin, bleomycin, and streptomycin resistance genes (Reznikoff, 2008). Also, integrons represent multiple classes of genetic elements triggering the spread and acquisition of gene cassettes by conjugation (Mazel, 2006). They specify an integrase, attachment sites, and transcriptional elements that ensure proper expression of exogenous antibiotic-resistance determinants in recipient genomes. Examples are MDR efflux pumps, carbapenemases, etc. (Giedraitiene et al., 2011).

Transformation

Transformation is the direct uptake of naked DNA, often followed by recombination in the genomic DNA of recipient strain. This HGT process incorporates large discrete DNA segments as chromosomal islands (Chen et al., 2005). While the phenomenon was known and used for decades in experiments with laboratory strains of *E. coli*, natural transformation of wild-type *E. coli* strains has been described relatively recently (Baur et al., 1996). Potentially any type of DNA region could be transferred in this way. Another mechanism that is somewhat similar to the transformation, which does not involve a specialized genetic vehicle but where the DNA is not floating 'naked', is outer membrane vesicle-mediated transfer in *E. coli* (Yaron et al., 2000). However, unlike the transduction and conjugation, transformation mechanisms are not well understood in *E. coli*.

Pathoadaptive mutations

Pathoadaptive mutation is an alternative mechanism for the virulence evolution that involves genetic alteration of the existing genome components rather than acquisition of novel genes. The major mechanisms are gene amplification, inactivation, and variation (Figure 3.2).

Gene amplification

Amplification of gene copies can lead to increased production of the corresponding protein and, if that confers an advantage for the pathogen, be selected during the infection. It has been proposed recently that transient amplification of genetic regions is a common phenomenon in *E. coli* (and other bacteria), allowing for fast adaptation in environments and increased rates of adaptive evolution (Kugelberg et al., 2006). Thus, gene amplification might be a more significant phenomenon than appreciated previously. Examples of virulence factors with multiple copies are Acriflavin resistance protein D (AcrD) in EPEC (Fukiya et al., 2004); Iha adhesin in EHEC (Mellmann et al., 2009); P-fimbrial genes in some UPEC (Kao et al., 1997).

Gene duplication is another critical event commonly found in many enteric pathogens including *E. coli* (Ohno, 1970; Himmelreich et al., 1996; Tomb et al., 1997), providing means for novel functional evolution. An important example is the duplication of Shiga toxin 2 encoding genes (stx_2) found in some EHEC O157:H7 outbreak strains (Muniesa et al., 2003), and in vitro experiments showed correlation of the occurrence of duplicated genes with increased production of Stx2 (Bielaszewska et al., 2006). Studies showed that evolution of orthologs faces considerable selective constraints (Hughes and Hughes, 1993), and the level of functional constraints reduces immediately following a duplication event resulting in a significantly higher rate of evolution between paralogs (Lynch and Conery, 2000; Jordan et al., 2002; Kondrashov et al., 2002).Thus, duplication allows the emergence of proteins with novel structure and functions. For example, the *cspA* (the major cold-shock protein) family homologs in *E. coli* resulted from a number of gene duplication events allowing emergence of specific groups of genes that adaptively respond to different environmental stresses such as cold-shock stress (*cspA*, *cspB*, *cspG*), nutritional deprivation (*cspD*), etc. (Yamanaka et al., 1998).

Gene inactivation (anti-virulence factors)

HGT and gene amplification are so-called the 'gain-of-function' mechanisms for the evolution of microbial virulence. Another and quite common mechanism represents 'loss-of-function' that includes gene or coded protein inactivation via point substitution, frameshift mutation, insertion sequence acquisition, and gene deletion. This mechanism is often crucial to enhance the fitness of a pathogen in a novel environment.

It is necessary to distinguish between gene loss due to pathoadaption and due to reductive evolution. The model of reductive evolution is based on the lack of universal need of some genes across different niches. So, genes that are essential in one habitat but are not required in a novel habitat could be lost based on the use-it-or-lose-it principle, especially if the habitat shift is permanent. The latter trajectory of evolution develops when certain bacteria change their lifestyle to an obligate intracellular pathogen, and permanently shed off genes non-essential within the host (Moran and Plague, 2004). *Mycobacterium leprae* (Cole et al., 2001), *Coxiella burnetii* (Seshadri et al., 2003), the Rickettsiae (Andersson et al., 1998) are some examples of reductive genome evolution. Moreover, the restricted intracellular lifestyle within a host reduces the possibility of HGT events thereby maintaining a reduced genome size (Moran and Plague, 2004). However, loss of gene by such a mechanism is not driven by positive selection pressures for it, and therefore is not adaptive per se. The unnecessary genes are either not expressed or their expression does not provide fitness advantage or significant disadvantage. Thus, their loss rather happens due to the lack of negative selection against it and is driven by genetic drift and not positive selection.

Alternatively, loss of certain genes could be highly adaptive, especially for pathogens. Indeed, genes that are essential for the fitness in one habitat might be not only useless but also *detrimental* for the fitness in another habitat (Maurelli, 2007). This will force the emergence of new variants of the bacterium through elimination of genes that, for example, down-regulate those genes, increased expression of which is important for the fitness in the new niche. Another example would be elimination of genes, the product of which either directly inhibits the function of critical traits or provides liability in the new habitat, e.g. those coding for the surface structure recognized well by innate or adaptive immune mechanisms.

Genes, the function of which is detrimental to virulence, are sometimes termed 'antivirulence factors'. A great example of an antivirulence gene is *cadA* in *Shigella* spp. The gene *cadA* encodes lysine decarboxylase (LDC) and is expressed in >90% of *E. coli* isolates (Maurelli, 2007). However, all strains of *Shigella* as well as the enteroinvasive *E. coli* (EIEC that cause dysentery similar to shigellosis, see Chapter 7) lack expression of LDC (Silva et al., 1980). Experiments in an animal model showed that if *Shigella* was forced to produce LDC via transformed *cadA*, it remained invasive but its enterotoxic activity was significantly diminished compared to the wild-type level (Maurelli et al., 1998). The product of lysine decarboxylation is cadaverine, which inhibits the function of plasmid-encoded enterotoxins in *Shigella*. Another experiment showed the role of cadaverine in preventing the ability of *Shigella* to stimulate transepithelial migration of polymorphonuclear neutrophils (McCormick et al., 1999). Since expression of LDC followed by production of cadaverine attenuates virulence phenotypes, *cadA* in *Shigella* satisfies the necessary features of an antivirulence gene (Maurelli, 2007). Another example of an antivirulence gene, again in

Shigella, is *ompT*, a prophage-mediated gene encoding a surface protease. This protease degrades IcsA/VirG, a virulence protein in *Shigella* that is required for keratoconjunctivitis in guinea pigs and is involved in intra- and intercellular motility (Nakata et al., 1993). The gene *ompT* is lacking in all lineages of *Shigella*, also fulfilling the criteria to be an antivirulence gene. Multiple independent evolutionary origins of *Shigella* strains indicate convergent evolution via the loss of antivirulence factors, including inactivation/deletion of LDC by different mechanisms, which is a strong indication for the action of positive selection (Pupo et al., 2000).

Even a clean single gene deletion, however, could potentially result in toxic effects by disrupting, for example, biogenesis pathways and build-up of intermediate components (Holt et al., 2008). That is possibly why multiple gene or entire gene operons/clusters tend to be lost, resulting in so-called 'black holes' in the genomes (Maurelli, et al., 1998). Instead of gene deletion, 'loss of function' can also be achieved by a mutation introducing premature stop codons or by frameshift mutations that disrupt the reading frame. However, such mechanisms result in expression of truncated and/or misfolded proteins that could be toxic for the bacterial cell (Kuo and Ochman, 2010). In contrast, loss-of-function due to mutation in the active site of the protein might provide a better way to achieve the adaptive effect. Mutations in the active protein site have been reported, for example, in the major, type 1 fimbrial adhesin of *E. coli* species, FimH, where inactivating mutations are found in the mannose-binding pocket of the adhesive protein of some EHEC isolates (Shaikh et al., 2007). Loss of the type 1 fimbrial function has been also seen for *Shigella* and is likely to provide some kind of pathogenicity-adaptive effect in enteric pathogens (Snellings et al., 1997).

Gene variation

Point mutations that do not inactivate but modify the function of coded protein are another important player of the pathoadaptive mechanism of *E. coli* evolution. An example of pathoadaptive point mutation is the evolution of *fimH* gene encoding the type 1 fimbrial adhesin in *E. coli* (see Chapter 12). FimH is expressed by >90% of *E. coli* (Johnson and Stell, 2000), and uropathogenic isolates express some specific variants of FimH owing to accumulation of point mutations (Weissman et al., 2006). These variants significantly enhance binding to mannosylated glycoproteins on uroepithelial cell surfaces, thereby increasing bacterial tropism to uroepithelium (Sokurenko et al., 1998; Hommais et al., 2003). Point mutations leading to functional modification that is pathoadaptive in nature have been shown in at least two other types of *E. coli* adhesins – Dr family (Korotkova et al., 2007) and class 5 fimbrial adhesins in enterotoxigenic isolates (Tchesnokova et al., 2010; Chattopadhyay et al., 2012).

Non-coding mutations

Nucleotide point mutations may also have adaptive effects by affecting the rate of gene transcription or translation. Such mutations could be comparable to the gain- or loss-of-function mutation of the coding genes themselves. For example, increased/decreased gene expression could also be achieved by mutations in the non-coding regulatory regions (e.g. promoters, operators, etc.), while protein expression could be affected by mutation in ribosome-binding sites (Corvec et al., 2007; Mammeri et al., 2008; Smet et al., 2008). Another type of mutation, the importance of which has been recently recognized, is silent (synonymous) mutations in the coding genes (Kudla et al., 2009). These mutations do not affect the protein structure but can have a significant effect on the gene translation rate by introducing or eliminating rare codons or by modifying the tertiary structure (and, thus, for example stability) of the messenger RNA. Therefore, the silent mutations might not be that 'silent'!

EVOLUTIONARILY ADAPTED AND PRE-ADAPTED VIRULENCE FACTORS

While we understand relatively well the genetic mechanisms of virulence evolution, the exact nature of the virulence factors is not fully understood. Some of the examples of virulence factors that are critical for the ability of *E. coli* to cause specific types of infections are mentioned here. Others are discussed in more detail in various chapters of this book. However, in defining whether or not specific genes and/or phenotypic traits are virulence factors, we primarily rely on two rather imperfect approaches: (a) experimental, i.e. animal or cell models of the infection; and (b) epidemiological, i.e. genes/traits more common among pathogens. The experimental models are very important and led us to many major discoveries, but they could also be misleading. The pathogen's ability to survive and cause damage within a host compartment is often finely tuned to very specific conditions there that could be difficult to reproduce under the model conditions. The epidemiological analysis also does not provide a straightforward answer on the exact nature of the virulence factors. While by definition the genome of a given clinical isolate encodes all virulence factors necessary for the infection to take place, the main issue is to understand which of the genes/traits are directly relevant to the isolate's ability to cause the infection. As mentioned above, comparing content of multiple isolates from patients and healthy individuals (or environment) helps to narrow down the important genes. However, many if not most of the genes or traits associated specifically with pathogenic *E. coli* isolates might not be relevant to the pathogenicity per se. Furthermore, this approach ignores important host factors that influence the outcome of infection.

Even for well-defined, proven virulence traits, uncertainty could remain whether such traits have evolved to serve during the infection or only accidentally fit to do so. This brings us to a key question about virulence factors from evolutionary perspectives – did they evolve or not to function during the infection in humans? Any phenotypic trait could be defined as an evolutionarily adapted or pre-adapted trait (Van Derlinden et al., 2008). The adapted trait is the one that was acquired/evolved and maintained to perform a specific function in a specific environment. By using a kitchenware analogy, a vegetable knife is adapted (designed) to chop vegetables, a salad fork to pick them up on the plate and a soupspoon to fish them out in the bowl. In contrast, pre-adapted traits are those that might fit to perform a specific function though they are not designed to do so. By the same kitchenware analogy (but not so peaceful), the vegetable knife can easily be adapted as a deadly weapon in most hands. The salad fork can also be adapted as a weapon (but in rather skillful hands), while it is difficult to make use of a soupspoon for that purpose in virtually any hands.

Thus, virulence traits should be defined as adapted when they were acquired/ evolved specifically to increase the ability of a pathogen to survive in the course of infection. In contrast, pre-adapted virulence traits are those that also increase the pathogenic fitness but are originally adapted for a function that is unrelated to the virulence, e.g. evolved as a trait for commensal colonization of humans or non-human hosts. Defining whether or not a virulence factor is adapted or pre-adapted might seem a purely intellectual exercise from practical perspectives. Indeed, why should a patient, physician, or even medical researcher care whether or not the pathogen had *evolved* to impose the misery of infection or just be able to do it? As described below, however, to define the virulence trait as adapted or pre-adapted is essential for answering the most important question from evolutionary perspectives: what is the driving force behind the emergence and maintenance of the pathogenic *E. coli*? This, in turn, could help us to pinpoint what the critical virulence factors in a pathogen are, whatever nature they are.

WHY DID *E. COLI* EVOLVE TO BE PATHOGENIC?

It is likely that some *E. coli* strains just lack specific traits or their combination that would allow them to cause infection in humans, i.e. to sustain themselves long enough in protected compartments and to induce clinically manifested tissue damage. So, they can be defined as non-pathogens that can exist only asymptomatically to the human habitats (Figure 3.3). These strains are confined to either non-human hosts only or commensal colonization of healthy humans.

It appears that, from evolutionary perspectives, the non-pathogenic *E. coli* lineages are original to the species and gave rise to different pathotypes of *E. coli* and not vice versa. First, non-pathogens constitute the bulk of *E. coli* organisms in nature. Second, the clonal diversity of non-clinical *E. coli* is significantly higher than of any type of clinical isolate (Kohler and Dobrindt, 2011), implying that the latter are expanded subsets of the former and, thus, derived (evolved) from them. Thirdly, many traits that can be defined as virulence factors in pathogens (either adapted or pre-adapted) are recently acquired from evolutionary perspectives. Pathogenic strains have horizontally transferred genes on plasmids, prophages, or chromosomal islands that are missing in most of the non-clinical strains. While theoretically non-pathogens could have mostly lost these highly mobile elements, phylogenetic analysis of their distribution argues

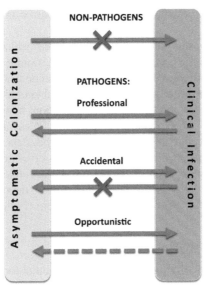

FIGURE 3.3   Different evolutionary types of pathogenic *E. coli* from the perspective of population dynamics.

against this scenario (Chattopadhyay et al., 2012). Then, the derived nature of pathogens is obvious from the fact that pathogenic isolates (like *Shigella*) tend to lose or inactivate certain core genes that are evolutionarily original to the species (Maurelli, 2007). Also, pathoadaptive point mutations are of a derived nature based on the phylogenetic analysis, with the non-pathogens inclined to have an original (ancestral) variant of the gene (Sokurenko et al., 2004; Weissman et al., 2007).

A good illustration of the non-pathogen-to-pathogen evolution is provided by *Shigella*. *Shigella* were originally grouped into four separate species (*S. boydii*, *S. dysenteriae*, *S. flexneri*, *S. sonnei*), but later it was determined that *E. coli* and *Shigella* should be considered as a single species (Jin et al., 2002). *Shigella*/EIEC genomes are colinear to *E. coli* with >90% homologous genes (Jin, et al., 2002). While *Shigella* strains belong to either B1 or ABD phylogenetic ECOR groups of *E. coli* (Figure 3.1), a more detailed phylogenetic analysis showed their independent emergence from separate *E. coli* lineages (Pupo et al., 1997, 2000). Genetic studies revealed that it is horizontal transfer of clustered virulence genes with low G+C content compared to the rest of the plasmid that occurred multiple times, each of which gave birth to a new *Shigella* clone (Pupo et al., 2000). Analysis of the region encompassing *cadA*, the antivirulence gene, from four *Shigella* species demonstrated distinct genetic rearrangements in each of the regions, again indicating independent origin of different *Shigella* lineages (Day et al., 2001). Interestingly, however, *Shigella* lineages emerged only within one major phylogenetic group of *E. coli*, suggesting that the acquisition of the virulence genes might have preferably occurred in strains of a similar genomic background.

To answer the question about the driving force behind the emergence of pathogenic lineages of *E. coli*, it is necessary to examine whether the ability of *E. coli* to cause disease contributes to its continuous survival in nature. It appears that any of the pathogenic *E. coli* are capable of asymptomatic colonization of humans or non-human hosts. By definition, however, all pathogenic *E. coli* can enter protected compartments and cause clinical infection. The key difference between different pathogens is how much they depend on the conditions asymptomatic for human habitats to maintain themselves in nature. Considering the dual habitats of pathogenic *E. coli*, from the ecological and evolutionary perspectives *E. coli* pathotypes could be subdivided into three general categories – professional, accidental, and opportunistic pathogens (Figure 3.3).

Professional pathogens

In these *E. coli* their ability to cause disease is essential for continuous circulation in nature, with pathotypes like *Shigella* and enteroinvasive *E. coli* (EIEC) being the great examples. While they can colonize humans asymptomatically, the asymptomatic carriers are predominantly individuals who recovered from symptomatic infection by the same strain (Zdziarski et al., 2010). The professional pathogens can also effectively transmit during the active infection to form new infections, i.e. from patient to patient (Stewart et al., 2003). Thus, virulence factors of *Shigella* and EIEC are likely to be maintained specifically to cause the infection or to be transmitted as a pathogen, i.e. to be of the adapted nature. This relates to both mechanisms of virulence evolution, horizontal gene transfer, and pathoadaptive mutations (Figure 3.3). Among the former category is a large virulence plasmid that encodes genes for expression of characteristic *Shigella* virulence such as invasion, intracellular replication, intercellular spread, and induction of an inflammatory response (Pupo et al., 2000). Among pathoadaptive mutations the leading role is played by the loss-of-function mutations, with the classic example of deletion of *cadA* and *ompT* antivirulence genes that interfere with the function or secretion of the enterotoxin (Day et al., 2001). Also, among different *E. coli* pathotypes, *Shigella*/EIEC carry the largest amount of point mutational changes (variations) in the core genes, a bulk of which could be pathoadaptive in nature (Chattopadhyay et al., 2009).

Other *E. coli* pathotypes that could represent professional pathogens are ETEC and EPEC, where diarrhea might be a prerequisite for establishing asymptomatic intestinal colonization in new human hosts. These new hosts are primarily individuals that did not encounter the pathogen before – either children or newcomers to the community ('travelers') (von Sonnenburg et al., 2000). However, ETEC and EPEC appear to circulate as a commensal within the community having built-up immunity against them. It is also possible that domesticated animals are primary reservoirs for some of these pathotypes (see Chapter 1).

Thus, epidemiological data supporting the essential role of virulence in the circulation of ETEC and EPEC are not as strong as for the strictly human-adapted *Shigella*/EIEC.

Accidental pathogens

In these *E. coli* pathotypes, the virulence habitat is an ecological accident and evolutionary dead-end (Figure 3.3). Causing the clinical infection in humans does not usually promote their transmission to another person or into the original host environment. EHEC strains are a good example of an accidental human pathogen by being a zoonotic pathogen (Caprioli et al., 2005). Upon ingestion of contaminated food, this pathotype causes diarrhea, sometimes accompanied by a characteristic and devastating hemolytic-uremic syndrome (Levine et al., 1987). The most notorious lineage of EHEC is comprised of strains with the O157:H7 serotype, though lately several other serotypes have become prominent as well (Riley et al., 1983). Individuals may shed EHEC for 30 days or more following infection and child-to-child transmission at home or day care centers might contribute to the outbreaks (Swerdlow and Griffin, 1997). However, such outbreaks die out relatively quickly despite potential person–person transmission and the low infectious dose. Thus, EHEC clearly cannot rely on human–human transmission to be sustained in nature. At the same time, EHEC are commonly and continuously isolated from domesticated animals (cows), which they seemingly colonize asymptomatically (Caprioli et al., 2005). Because there is a sufficient ecological separation between humans and domesticated animals, there is no significant transmission from the former to the latter even if bacteria are being shed in considerable numbers during the clinical infection. Although, due to low infectious dose, they are spread from human to human frequently leading to outbreaks, such bursts do not sustain for prolonged periods (Bell et al., 1994). Thus, EHEC lineages could not evolve specifically to cause infection in humans, and the outcome is just accidental for both the bacteria and the human host.

Because of the accidental nature of the pathogen, most EHEC virulence factors should be considered to be of the pre-adapted nature, i.e. did not evolve for virulence per se, at least not in humans. One of the critical virulence factors of EHEC is Shiga-like toxin encoded by prophage (Sandvig, 2001). The toxin is apparently expressed and secreted even without the phage activation and has powerful toxicity against human cells (Bielaszewska et al., 2006). However, it likely evolved as a weapon against protozoan predators of bacteria rather than to induce human tissue damage (Steinberg and Levin, 2007). It has been proposed that Shiga-toxin in EHEC is over-expressed or released in an overly large amount during the human infection due to activation of the prophage (Mellmann et al., 2009). The latter could be the result of stress imposed on bacteria in human intestine as it tries to establish colonization there. The nature of the stress is unclear but it appears to be absent during the colonization of bovine intestine. EHEC represent a classical example of virulence, being the result of failure to establish commensal relationships with the host. Because bacteria are being lysed during the phage activation, the toxin over-expression would unlikely evolve as an adapted trait for bacteria and, thus, can be considered as a pre-adapted virulence factor.

Examples of pathoadaptive mutations in EHEC could be mutations inactivating the type 1 fimbriae, the most ubiquitous mannose-binding adhesive trait in *E. coli* (Shaikh et al., 2007). Inactivating mutations are found both in the adhesin protein FimH and the fimbrial expression switch, suggesting it occurred repeatedly and, thus, under the action of positive selection (Leopold et al., 2009). However, these mutations are found in both clinical and non-clinical isolates of O157 serotype, consistent with their pre-adapted nature in the latter. The role of the fimbrial loss in either the human infection or bovine colonization is unclear yet.

Besides EHEC, some other infections caused by *E. coli* strains are ecological dead-ends, e.g. meningitis or bloodstream infections, rendering them as accidental pathogens. Therefore, their putative virulence factors, such as cerebrospinal barrier (CSB)-binding, sialic-acid specific S-fimbriae of meningitis-causing *E. coli* strains (Ott et al., 1986), could also be defined as pre-adapted virulence traits.

Opportunistic pathogens

This type of pathogen represents the most diverse and, from evolutionary perspectives, ambiguous group of *E. coli*. Unlike professional pathogens, these *E. coli* appear to be able to continuously circulate only as human asymptomatic colonizers (commensals). They cause clinical infections only occasionally, mostly in individuals with at least somewhat or transiently compromised defense barriers. Unlike the accidental pathogens, however, the bacteria might be readily able to return to their original environment (human intestine) while being shed during the infection, sometimes in high numbers. Thus, the infectious process does not impose an evolutionary dead-end scenario for the pathogen. Examples of such *E. coli* are primarily the ExPEC pathotype that, in developed countries, causes the vast majority of *E. coli* infections (Johnson and Russo, 2002). Just the urinary tract infections alone (primarily bladder infection – cystitis), where *E. coli* is the main pathogen, affects over 7 million women each year in the USA (Foxman, 2002). The infecting *E. coli* usually come from the intestinal microbiota, introduction of which into the bladder is facilitated by sexual intercourse, diaphragm use, or bladder catheterization (Brown and Foxman, 2000). Bladder infection commonly lasts 5–7 days, during which bacteria are shed with urine, often more than 10^5 CFU per ml. In rare cases, bacteria from the bladder ascend into kidney, resulting in a more serious clinical infection – pyelonephritis.

While the urinary tract infections are common, they affect primarily women and, in terms of frequency and bacterial load, are minute in comparison to intestinal colonization by *E. coli* in both men and women. This is true even for the clonal lineages of *E. coli* that primarily cause urinary infections (mostly of B2 phylogenetic group – see Figure 3.2) (Moreno et al., 2008). Thus, it is unlikely that the primary, intestinal colonization of ExPEC depends on the ability to cause the opportunistic infection. Therefore, the urovirulence factors that allow the infections to be initiated are most likely to be of the pre-adapted nature. Indeed, horizontally transferred virulence factors of uropathogenic *E. coli* such as di-galactose-specific P fimbriae were proposed to be important for the intestinal colonization of *E. coli* in humans with certain blood types (Herias et al., 1995).

By the same token, it has been shown that in the course of urinary tract infections, bacteria can adapt there by mutating, among other traits, FimH adhesin, LPS, flagella (Weissman et al., 2003). Thus, at least some of the virulence traits are selected for in the virulence habitat and, therefore, are of an adapted nature. In fact, some of the clearest examples of pathoadaptive mutations in the course of infection come from the UPEC. The dual nature of the virulence factors in the opportunistic pathogens (i.e. being both pre-adapted and adapted) makes this category complex from evolutionary model perspectives described in detail below.

EVOLUTIONARY MODELS, SOURCE-SINKS, AND PARADOXES

Concepts of virulence evolution

The diversity of trajectories that are taken by evolution of *E. coli* pathotypes provides illustrations of all three major models of virulence evolution: (i) virulence is increasing fitness of (i.e. adaptive for) the pathogen, (ii) virulence is a coincidental by-product of commensal evolution, and (iii) virulence is a result of a shortsighted positive selection (Levin and Bull, 1994).

According to the first model, i.e. virulence is adaptive for the pathogens, there are two main sub-models: (a) benefit-of-virulence; and (b) cost-of-virulence (Sokurenko et al., 1999). Under the benefit-of-virulence model, disease is the primary mode of the host–bacteria interaction and transmission, even when asymptomatic colonization is a more common outcome of it. In contrast, under the cost-of-virulence model, the primary form of the pathogen's existence is asymptomatic colonization, but disease is a relatively commonly associated outcome of it at the early stages of encountering the new host. Most likely, the professional pathogens like *Shigella*/EIEC, EPEC and ETEC could ostensibly fit this model.

According to the second model, virulence is the coincidental outcome of adaptation to a commensal or environmental habitat. In other words, virulence had evolved under conditions different from the disease and, thus, pathogenicity is a pure chance and not an evolutionary pre-determined property. In this case, the virulence factors are pre-adapted. Both accidental and opportunistic pathogens fit this model.

According to the third model, virulence results from short-sighted evolution during the course of infection itself, when selection improves fitness of the pathogen, no matter how transient the infection is. Because fitness is the measure of improved survival resulting from reduced elimination and/or increased reproduction rates in the habitat, the selection processes result in bacterial organisms that cause more long-lasting and, sometimes, more damaging (clinically severe) infection. Again, both accidental and opportunistic pathogens are most likely to be subjected to such selective pressures. By being pre-adapted for the ability to cause clinical infection, they are unlikely to be as optimally adapted to the pathogenic lifestyle as are professional pathogens, thus providing ample room for the action of positive selection.

Source-sink dynamics

Because evolution of accidental and opportunistic pathogens fits two different models of virulence evolution, the models are not mutually exclusive. In fact, combination of the coincidental and short-sighted virulence leads to a so-called 'source-sink' dynamics of virulence evolution (Sokurenko et al., 2006). Source-sink has been originally developed as an ecological model, where a species exists in two distinct types of habitats (Pulliam 1988). The 'source' is a continuously inhabited, stable, or reservoir habitat supporting the long-term existence of the species. In contrast, 'sink' is an occasionally invaded unstable habitat that can support the species survival only for a short period of time. The evolutionary source-sink model implies that species continuously undergo genetic adaption to the sink environment upon invasion from the source, resulting in increased fitness in the former (Sokurenko et al., 2006). However, while improving survival in the sink, the adaptive evolution there not only fails to make the sink sustainable in the long term, but also reduces the fitness-of-sink-adapted organisms in the source. This happens due to a trade-off effect of the sink-adaptive changes in the source habitat.

When applied to the pathogens, the commensal colonization reservoir could be defined as the source habitat and the infection as the sink habitat. Adaptation to the sink is likely to happen only via mutations, because transferrable genetic elements are not readily available in the course of infections (that are usually mono-infections, i.e. caused by a single strain). Clearly, loss-of-function pathoadaptive mutations that result in inactivation of the commensal traits are maladaptive back in the reservoir habitat. Mutations modifying gene function could also have the trade-off effect with the original gene function. Thus, patho-adapted bacteria will be under negative selection upon entering the original habitat.

The trade-off effect could be primarily noted with opportunistic pathogens that can readily go back from the sink to the source. Indeed, pathoadaptive mutations in, for example the FimH adhesin of uropathogenic *E. coli*, bear a signature of being only recently evolved – the likely sign of their instability in a long-term natural circulation (Weissman et al., 2007).

Virulence of evolution paradox

The concepts of virulence evolution of the accidental or opportunistic *E. coli* pathogens lead to an interesting paradox of the evolutionary mechanisms involved.

As discussed above, horizontally transferred virulence factors in these pathogens are most likely of pre-adapted nature, i.e. evolved to play a role in the commensal lifestyle. Thus, the traits that comprise the major genetic differences between pathogens and non-pathogens and that are considered to provide 'quantum leaps' in the evolution of virulence, are NOT directly relevant to the pathogenicity per se.

At the same time, virulence factors that were acquired by gene mutation could have been selected directly during the infection. Thus, the relatively minor genetic differences, as small as point mutations, that presumably fine-tune virulence could be much more relevant to the pathogenicity than the major genetic events.

Thus, 'Devil is in the details'!

Furthermore, the mobile genetic components that constitute the bulk of the horizontally transferred gene pool could be mostly of a 'selfish gene' nature, more like genetic parasitic elements (Asadulghani et al., 2009). Indeed, the majority of them are clearly of bacteriophage origin. Somewhat similar ecology could be attributed to the plasmids and chromosomal islands as well, even when they provide a direct contribution to the fitness of the bacterial organism in specific environments by encoding some fitness-improving traits. Such fitness improvement could be only transient from evolutionary perspectives and benefit primarily the spread of the mobile element in the overall bacterial population rather than the harboring bacterial host (Touchon et al., 2009). Thus, the maintenance of horizontally transferred genes in nature could be driven not by their adaptive significance for the bacterial organism, but for the genetic elements themselves. This issue of genetic conflict is only lately becoming a focus of the evolutionary biology of microorganisms (van Elsas et al., 2011).

POPULATION GENOMICS AND VARIOME OF MICROBIAL PATHOGENS

The evolution of any species, including *E. coli*, works on the gene pool of a species that is represented by the combination of all genes and their variants that occur in organisms belonging to the species. It is continuously enriched by the genes being transferred, mutated, sifted, and shuffled under various selective pressures and by random genetic drift. Structure and dynamics of the gene pool can be assessed to a certain extent by comparing sequences of one or several genetic loci across different populations of the same species. Such studies have defined the field of population genetics that, to date, successfully characterized the genotypic heterogeneity level and the action of various types of selection processes in a large number of species (Kassen and Rainey, 2004; Mes, 2008). Until recently, however, population genetics studies have mostly employed a very limited number of loci due to sequencing and labor expenses (Butlin, 2010).

With the present flood of new data unleashed by the arrival of affordable, rapid, high-quality sequencing technologies, the focus is rapidly shifting from the mechanics of generating sequence data to the problems of analyzing it. This creates an urgent need for better ways to compare and visualize genomic data (Field et al., 2006). By analogy with the application of the Human Variome Database (Ring et al., 2006), an important aim of microbial pathogenomics would be development of a Microbial Variome Database (Chattopadhyay et al., 2013). Such a database would constitute a species-specific genomic resource. The information about all the changes in sequence, origin of isolation of the strains where such changes accumulated and their potential adaptive values will bring bacterial evolutionary genomics to a new level, both quantitatively and qualitatively, offering broad applicability of population genomics tools to experimental research, clinical diagnostics, epidemiology, and environmental control of pathogens. Similar to the case with the human genome, such a database would be extremely helpful to associate genetic variation in bacteria (such as gene presence/absence or mutation) with the bacterial ability to cause disease and, thus, greatly contribute to understanding virulence evolution. It would provide genome-wide information on potential targets for vaccines, antibiotics, and other therapeutics development on one hand, while on the other it would offer a global surveillance system to enable rapid determination of newly emerging pathogenic clones and genetic mechanisms behind the emergence.

REFERENCES

Al-Hasani, K., Henderson, I.R., Sakellaris, H., et al., 2000. The sigA gene which is borne on the she pathogenicity island of *Shigella flexneri* 2a encodes an exported cytopathic protease involved in intestinal fluid accumulation. Infect. Immun. 68 (5), 2457–2463.

Allen, S.J., Martinez, E.G., Gregorio, G.V., Dans, L.F., 2010. Probiotics for treating acute infectious diarrhoea. [Meta-Analysis Review]. Cochrane Database Syst. Rev. (11) CD003048.

Ambrozic, J., Ostroversnik, A., Starcic, M., Kuhar, I., Grabnar, M., Zgur-Bertok, D., 1998. *Escherichia coli* CoIV plasmid pRK100: genetic organization, stability and conjugal transfer. Microbiology 144 (Pt 2), 343–352.

Andersson, S.G., Zomorodipour, A., Andersson, J.O., et al., 1998. The genome sequence of *Rickettsia prowazekii* and the origin of mitochondria. Nature 396 (6707), 133–140.

Asadulghani, M., Ogura, Y., Ooka, T., et al., 2009. The defective prophage pool of *Escherichia coli* O157 prophage–prophage interactions potentiate horizontal transfer of virulence determinants. PLoS Pathog. 5 (5), e1000408.

Asakura, M., Hinenoya, A., Alam, M.S., et al., 2007. An inducible lambdoid prophage encoding cytolethal distending toxin (Cdt-I) and a type III effector protein in enteropathogenic *Escherichia coli*. Proc. Natl. Acad. Sci. USA 104 (36), 14483–14488.

Baur, B., Hanselmann, K., Schlimme, W., Jenni, B., 1996. Genetic transformation in freshwater: *Escherichia coli* is able to develop natural competence. Appl. Environ. Microbiol. 62 (10), 3673–3678.

Bell, B.P., Goldoft, M., Griffin, P.M., et al., 1994. A multistate outbreak of *Escherichia coli* O157:H7-associated bloody diarrhea and hemolytic uremic syndrome from hamburgers. The Washington experience. JAMA 272 (17), 1349–1353.

Bennett, P.M., 2008. Plasmid encoded antibiotic resistance: acquisition and transfer of antibiotic resistance genes in bacteria. Br. J. Pharmacol. 153 (Suppl. 1), S347–357.

Berg, C.M., Grullon, C.A., Wang, A., Whalen, W.A., Berg, D.E., 1983. Transductional instability of Tn5-induced mutations: generalized and specialized transduction of Tn5 by bacteriophage P1. Genetics 105 (2), 259–263.

Bielaszewska, M., Friedrich, A.W., Aldick, T., Schurk-Bulgrin, R., Karch, H., 2006. Shiga toxin activatable by intestinal mucus in *Escherichia coli* isolated from humans: predictor for a severe clinical outcome. Clin. Infect. Dis. 43 (9), 1160–1167.

Brown, P.D., Foxman, B., 2000. Pathogenesis of urinary tract infection: the role of sexual behavior and sexual transmission. Curr. Infect. Dis. Rep. 2 (6), 513–517.

Brussow, H., Canchaya, C., Hardt, W.D., 2004. Phages and the evolution of bacterial pathogens: from genomic rearrangements to lysogenic conversion. Microbiol. Mol. Biol. Rev. 68 (3), 560–602 table of contents.

Buchrieser, C., Glaser, P., Rusniok, C., et al., 2000. The virulence plasmid pWR100 and the repertoire of proteins secreted by the type III secretion apparatus of *Shigella flexneri*. Mol. Microbiol. 38 (4), 760–771.

Burland, V., Shao, Y., Perna, N.T., Plunkett, G., Sofia, H.J., Blattner, F.R., 1998. The complete DNA sequence and analysis of the large virulence plasmid of *Escherichia coli* O157:H7. Nucleic Acids Res. 26 (18), 4196–4204.

Butlin, R.K., 2010. Population genomics and speciation. Genetica. 138 (4), 409–418.

Campellone, K.G., Robbins, D., Leong, J.M., 2004. EspFU is a translocated EHEC effector that interacts with Tir and N-WASP and promotes Nck-independent actin assembly. Dev. Cell 7 (2), 217–228.

Canchaya, C., Proux, C., Fournous, G., Bruttin, A., Brussow, H., 2003. Prophage genomics. Microbiol. Mol. Biol. Rev. 67 (2), 238–276 table of contents.

Caprioli, A., Morabito, S., Brugere, H., Oswald, E., 2005. Enterohaemorrhagic *Escherichia coli*: emerging issues on virulence and modes of transmission. Vet. Res. 36 (3), 289–311.

Chattopadhyay, S., Weissman, S.J., Minin, V.N., Russo, T.A., Dykhuizen, D.E., Sokurenko, E.V., 2009. High frequency of hotspot mutations in core genes of *Escherichia coli* due to short-term positive selection. Proc. Natl. Acad. Sci. USA 106 (30), 12412–12417.

Chattopadhyay, S., Paul, S., Kisiela, D.I., Linardopoulou, E.V., Sokurenko, E.V., 2012. Convergent molecular evolution of genomic cores in *Salmonella enterica* and *Escherichia coli*. J. Bacteriol. 194 (18), 5002–5011.

Chattopadhyay, S., Tchesnokova, V., McVeigh, A., et al., 2012. Adaptive evolution of class 5 fimbrial genes in enterotoxigenic *Escherichia coli* and its functional consequences. J. Biol. Chem. 287 (9), 6150–6158.

Chattopadhyay, S., Taub, F., Paul, S., Weissman, S.J., Sokurenko, E.V., 2013. Microbial Variome Database: point mutations, adaptive or not, in bacterial core genomes. Mol. Biol. Evol. 30 (6), 1465–1470.

Chaudhuri, R.R., Sebaihia, M., Hobman, J.L., et al., 2010. Complete genome sequence and comparative metabolic profiling of the prototypical enteroaggregative *Escherichia coli* strain 042. PLoS One 5 (1), e8801.

Chen, I., Christie, P.J., Dubnau, D., 2005. The ins and outs of DNA transfer in bacteria. Science 310 (5753), 1456–1460.

Cole, S.T., Eiglmeier, K., Parkhill, J., et al., 2001. Massive gene decay in the leprosy bacillus. Nature 409 (6823), 1007–1011.

Corvec, S., Prodhomme, A., Giraudeau, C., Dauvergne, S., Reynaud, A., Caroff, N., 2007. Most *Escherichia coli* strains overproducing chromosomal AmpC beta-lactamase belong to phylogenetic group A. J. Antimicrob. Chemother. 60 (4), 872–876.

Dahan, S., Wiles, S., La Ragione, R.M., et al., 2005. EspJ is a prophage-carried type III effector protein of attaching and effacing pathogens that modulates infection dynamics. Infect. Immun. 73 (2), 679–686.

Day Jr., W.A., Fernandez, R.E., Maurelli, A.T., 2001. Pathoadaptive mutations that enhance virulence: genetic organization of the cadA regions of *Shigella* spp. Infect. Immun. 69 (12), 7471–7480.

Dobrindt, U., 2005. (Patho-)Genomics of *Escherichia coli*. Int. J. Med. Microbiol. 295 (6-7), 357–371.

Dobrindt, U., Hochhut, B., Hentschel, U., Hacker, J., 2004. Genomic islands in pathogenic and environmental microorganisms. Nat. Rev. Microbiol. 2 (5), 414–424.

Donnenberg, M.S., Giron, J.A., Nataro, J.P., Kaper, J.B., 1992. A plasmid-encoded type IV fimbrial gene of enteropathogenic *Escherichia coli* associated with localized adherence. Mol. Microbiol. 6 (22), 3427–3437.

Eslava, C., Navarro-Garcia, F., Czeczulin, J.R., Henderson, I.R., Cravioto, A., Nataro, J.P., 1998. Pet, an autotransporter enterotoxin from enteroaggregative *Escherichia coli*. Infect. Immun. 66 (7), 3155–3163.

Field, D., Wilson, G., van der Gast, C., 2006. How do we compare hundreds of bacterial genomes? Curr. Opin. Microbiol. 9 (5), 499–504.

Fleckenstein, J.M., Lindler, L.E., Elsinghorst, E.A., Dale, J.B., 2000. Identification of a gene within a pathogenicity island of enterotoxigenic *Escherichia coli* H10407 required for maximal secretion of the heat-labile enterotoxin. Infect. Immun. 68 (5), 2766–2774.

Foxman, B., 2002. Epidemiology of urinary tract infections: incidence, morbidity, and economic costs. Am. J. Med. 113 (Suppl. 1A), 5S–13S.

Froehlich, B., Parkhill, J., Sanders, M., Quail, M.A., Scott, J.R., 2005. The pCoo plasmid of enterotoxigenic *Escherichia coli* is a mosaic cointegrate. J. Bacteriol. 187 (18), 6509–6516.

Fukiya, S., Mizoguchi, H., Tobe, T., Mori, H., 2004. Extensive genomic diversity in pathogenic *Escherichia coli* and *Shigella* strains revealed by comparative genomic hybridization microarray. J. Bacteriol. 186 (12), 3911–3921.

Gaillard, M., Vallaeys, T., Vorholter, F.J., et al., 2006. The clc element of *Pseudomonas* sp. strain B13, a genomic island with various catabolic properties. J. Bacteriol. 188 (5), 1999–2013.

Gal-Mor, O., Finlay, B.B., 2006. Pathogenicity islands: a molecular toolbox for bacterial virulence. Cell Microbiol. 8 (11), 1707–1719.

Garmendia, J., Phillips, A.D., Carlier, M.F., et al., 2004. TccP is an enterohaemorrhagic *Escherichia coli* O157:H7 type III effector protein that couples Tir to the actin-cytoskeleton. Cell Microbiol. 6 (12), 1167–1183.

Giedraitiene, A., Vitkauskiene, A., Naginiene, R., Pavilonis, A., 2011. Antibiotic resistance mechanisms of clinically important bacteria. Medicina (Kaunas) 47 (3), 137–146.

Grohmann, E., Muth, G., Espinosa, M., 2003. Conjugative plasmid transfer in gram-positive bacteria. Microbiol. Mol. Biol. Rev. 67 (2), 277–301 table of contents.

Gruenheid, S., Sekirov, I., Thomas, N.A., et al., 2004. Identification and characterization of NleA, a non-LEE-encoded type III translocated virulence factor of enterohaemorrhagic *Escherichia coli* O157:H7. Mol. Microbiol. 51 (5), 1233–1249.

Hacker, J., Bender, L., Ott, M., et al., 1990. Deletions of chromosomal regions coding for fimbriae and hemolysins occur in vitro and in vivo in various extraintestinal *Escherichia coli* isolates. Microb. Pathog. 8 (3), 213–225.

Hawkey, P.M., 2008. The growing burden of antimicrobial resistance. J. Antimicrob. Chemother. 62 (Suppl. 1), i1–9.

Henderson, I.R., Czeczulin, J., Eslava, C., Noriega, F., Nataro, J.P., 1999. Characterization of pic, a secreted protease of *Shigella flexneri* and enteroaggregative *Escherichia coli*. Infect. Immun. 67 (11), 5587–5596.

Herias, M.V., Midtvedt, T., Hanson, L.A., Wold, A.E., 1995. Role of *Escherichia coli* P fimbriae in intestinal colonization in gnotobiotic rats. Infect. Immun. 63 (12), 4781–4789.

Himmelreich, R., Hilbert, H., Plagens, H., Pirkl, E., Li, B.C., Herrmann, R., 1996. Complete sequence analysis of the genome of the bacterium *Mycoplasma pneumoniae*. Nucleic Acids Res. 24 (22), 4420–4449.

Holloway, B.W., 1993. Genetics for all bacteria. Annu. Rev. Microbiol. 47, 659–684.

Holt, K.E., Parkhill, J., Mazzoni, C.J., et al., 2008. High-throughput sequencing provides insights into genome variation and evolution in *Salmonella typhi*. Nat. Genet. 40 (8), 987–993.

Hommais, F., Gouriou, S., Amorin, C., et al., 2003. The FimH A27V mutation is pathoadaptive for urovirulence in *Escherichia coli* B2 phylogenetic group isolates. Infect. Immun. 71 (6), 3619–3622.

Hughes, M.K., Hughes, A.L., 1993. Evolution of duplicate genes in a tetraploid animal. Xenopus laevis. Mol. Biol. Evol. 10 (6), 1360–1369.

Jin, Q., Yuan, Z., Xu, J., et al., 2002. Genome sequence of *Shigella flexneri* 2a: insights into pathogenicity through comparison with genomes of *Escherichia coli* K12 and O157. Nucleic Acids Res. 30 (20), 4432–4441.

Jobling, M.G., Holmes, R.K., 2012. Type II heat-labile enterotoxins from 50 diverse *Escherichia coli* isolates belong almost exclusively to the LT-IIc family and may be prophage encoded. PLoS One 7 (1), e29898.

Johnson, J.R., Russo, T.A., 2002. Extraintestinal pathogenic *Escherichia coli*: the other bad *E coli*. J. Lab. Clin. Med. 139 (3), 155–162.

Johnson, J.R., Stell, A.L., 2000. Extended virulence genotypes of *Escherichia coli* strains from patients with urosepsis in relation to phylogeny and host compromise. J. Infect. Dis. 181 (1), 261–272.

Jordan, I.K., Rogozin, I.B., Wolf, Y.I., Koonin, E.V., 2002. Microevolutionary genomics of bacteria. Theor. Popul. Biol. 61 (4), 435–447.

Kao, J.S., Stucker, D.M., Warren, J.W., Mobley, H.L., 1997. Pathogenicity island sequences of pyelonephritogenic *Escherichia coli* CFT073 are associated with virulent uropathogenic strains. Infect. Immun. 65 (7), 2812–2820.

Kassen, R., Rainey, P.B., 2004. The ecology and genetics of microbial diversity. Annu. Rev. Microbiol. 58, 207–231.

Kohler, C.D., Dobrindt, U., 2011. What defines extraintestinal pathogenic *Escherichia coli?* Int. J. Med. Microbiol. 301 (8), 642–647.

Kondrashov, F.A., Rogozin, I.B., Wolf, Y.I., Koonin, E.V., 2002. Selection in the evolution of gene duplications. Genome. Biol. 3 (2) RESEARCH0008.

Korotkova, N., Chattopadhyay, S., Tabata, T.A., et al., 2007. Selection for functional diversity drives accumulation of point mutations in Dr adhesins of *Escherichia coli*. Mol. Microbiol. 64 (1), 180–194.

Kudla, G., Murray, A.W., Tollervey, D., Plotkin, J.B., 2009. Coding-sequence determinants of gene expression in *Escherichia coli*. Science 324 (5924), 255–258.

Kugelberg, E., Kofoid, E., Reams, A.B., Andersson, D.I., Roth, J.R., 2006. Multiple pathways of selected gene amplification during adaptive mutation. Proc. Natl. Acad. Sci. USA 103 (46), 17319–17324.

Kuo, C.H., Ochman, H., 2010. The extinction dynamics of bacterial pseudogenes. PLoS Genet. 6 (8).

Larbig, K.D., Christmann, A., Johann, A., et al., 2002. Gene islands integrated into tRNA(Gly) genes confer genome diversity on a *Pseudomonas aeruginosa* clone. J. Bacteriol. 184 (23), 6665–6680.

Lavigne, J.P., Blanc-Potard, A.B., 2008. Molecular evolution of *Salmonella enterica* serovar Typhimurium and pathogenic *Escherichia coli*: from pathogenesis to therapeutics. Infect. Genet. Evol. 8 (2), 217–226.

Leopold, S.R., Magrini, V., Holt, N.J., et al., 2009. A precise reconstruction of the emergence and constrained radiations of *Escherichia coli* O157 portrayed by backbone concatenomic analysis. Proc. Natl. Acad. Sci. USA 106 (21), 8713–8718.

Levin, B.R., Bull, J.J., 1994. Short-sighted evolution and the virulence of pathogenic microorganisms. Trends Microbiol. 2 (3), 76–81.

Levine, M.M., Xu, J.G., Kaper, J.B., et al., 1987. A DNA probe to identify enterohemorrhagic *Escherichia coli* of O157:H7 and other serotypes that cause hemorrhagic colitis and hemolytic uremic syndrome. J. Infect. Dis. 156 (1), 175–182.

Lynch, M., Conery, J.S., 2000. The evolutionary fate and consequences of duplicate genes. Science 290 (5494), 1151–1155.

Mammeri, H., Eb, F., Berkani, A., Nordmann, P., 2008. Molecular characterization of AmpC-producing *Escherichia coli* clinical isolates recovered in a French hospital. J. Antimicrob. Chemother. 61 (3), 498–503.

Marches, O., Ledger, T.N., Boury, M., et al., 2003. Enteropathogenic and enterohaemorrhagic *Escherichia coli* deliver a novel effector called Cif, which blocks cell cycle G2/M transition. Mol. Microbiol. 50 (5), 1553–1567.

Maurelli, A.T., 2007. Black holes, antivirulence genes, and gene inactivation in the evolution of bacterial pathogens. FEMS Microbiol. Lett. 267 (1), 1–8.

Maurelli, A.T., Fernandez, R.E., Bloch, C.A., Rode, C.K., Fasano, A., 1998. "Black holes" and bacterial pathogenicity: a large genomic deletion that enhances the virulence of *Shigella* spp. and enteroinvasive *Escherichia coli*. Proc. Natl. Acad. Sci. USA 95 (7), 3943–3948.

Mayer, H.B., Wanke, C.A., Du, B., Hammer, S.M., Terwilliger, E.F., 1995. HIV-1 Tat modulates invasion by a bacterial enteric pathogen into a human intestinal cell line. AIDS 9 (11), 1237–1242.

Mazel, D., 2006. Integrons: agents of bacterial evolution. Nat. Rev. Microbiol. 4 (8), 608–620.

McCormick, B.A., Fernandez, M.I., Siber, A.M., Maurelli, A.T., 1999. Inhibition of *Shigella flexneri*-induced transepithelial migration of polymorphonuclear leucocytes by cadaverine. Cell Microbiol. 1 (2), 143–155.

McDaniel, T.K., Jarvis, K.G., Donnenberg, M.S., Kaper, J.B., 1995. A genetic locus of enterocyte effacement conserved among diverse enterobacterial pathogens. Proc. Natl. Acad. Sci. USA 92 (5), 1664–1668.

Mellies, J.L., Elliott, S.J., Sperandio, V., Donnenberg, M.S., Kaper, J.B., 1999. The Per regulon of enteropathogenic *Escherichia coli*: identification of a regulatory cascade and a novel transcriptional activator, the locus of enterocyte effacement (LEE)-encoded regulator (Ler). Mol. Microbiol. 33 (2), 296–306.

Mellmann, A., Bielaszewska, M., Karch, H., 2009. Intrahost genome alterations in enterohemorrhagic *Escherichia coli*. Gastroenterology 136 (6), 1925–1938.

Mes, T.H., 2008. Microbial diversity – insights from population genetics. Environ. Microbiol. 10 (1), 251–264.

Molin, S., Tolker-Nielsen, T., 2003. Gene transfer occurs with enhanced efficiency in biofilms and induces enhanced stabilisation of the biofilm structure. Curr. Opin. Biotechnol. 14 (3), 255–261.

Moran, N.A., Plague, G.R., 2004. Genomic changes following host restriction in bacteria. Curr. Opin. Genet. Dev. 14 (6), 627–633.

Moreno, E., Andreu, A., Pigrau, C., Kuskowski, M.A., Johnson, J.R., Prats, G., 2008. Relationship between *Escherichia coli* strains causing acute cystitis in women and the fecal *E. coli* population of the host. J. Clin. Microbiol. 46 (8), 2529–2534.

Muniesa, M., de Simon, M., Prats, G., Ferrer, D., Panella, H., Jofre, J., 2003. Shiga toxin 2-converting bacteriophages associated with clonal variability in *Escherichia coli* O157:H7 strains of human origin isolated from a single outbreak. Infect. Immun. 71 (8), 4554–4562.

Nakata, N., Tobe, T., Fukuda, I., et al., 1993. The absence of a surface protease, OmpT, determines the intercellular spreading ability of *Shigella*: the relationship between the ompT and kcpA loci. Mol. Microbiol. 9 (3), 459–468.

O'Hanley, P., Lalonde, G., Ji, G., 1991. Alpha-hemolysin contributes to the pathogenicity of piliated digalactoside-binding *Escherichia coli* in the kidney: efficacy of an alpha-hemolysin vaccine in preventing renal injury in the BALB/c mouse model of pyelonephritis. Infect. Immun. 59 (3), 1153–1161.

Ochman, H., Lawrence, J.G., Groisman, E.A., 2000. Lateral gene transfer and the nature of bacterial innovation. Nature 405 (6784), 299–304.

Ochman, H., Selander, R.K., 1984. Standard reference strains of *Escherichia coli* from natural populations. J. Bacteriol. 157 (2), 690–693.

Ohnishi, M., Terajima, J., Kurokawa, K., et al., 2002. Genomic diversity of enterohemorrhagic *Escherichia coli* O157 revealed by whole genome PCR scanning. Proc. Natl. Acad. Sci. USA 99 (26), 17043–17048.

Ohno, S., 1970. Evolution by Gene Duplication. London: George Alien & Unwin Ltd. Springer-Verlag, Berlin, Heidelberg and New York.

Ott, M., Hacker, J., Schmoll, T., Jarchau, T., Korhonen, T.K., Goebel, W., 1986. Analysis of the genetic determinants coding for the S-fimbrial adhesin (sfa) in different *Escherichia coli* strains causing meningitis or urinary tract infections. Infect. Immun. 54 (3), 646–653.

Pulliam, H.R., 1988. Sources, sinks, and population regulation. Am. Nat. 652–661.

Pupo, G.M., Karaolis, D.K., Lan, R., Reeves, P.R., 1997. Evolutionary relationships among pathogenic and nonpathogenic *Escherichia coli* strains inferred from multilocus enzyme electrophoresis and mdh sequence studies. Infect. Immun. 65 (7), 2685–2692.

Pupo, G.M., Lan, R., Reeves, P.R., 2000. Multiple independent origins of *Shigella* clones of *Escherichia coli* and convergent evolution of many of their characteristics. Proc. Natl. Acad. Sci. USA 97 (19), 10567–10572.

Rasko, D.A., Rosovitz, M.J., Myers, G.S., et al., 2008. The pangenome structure of *Escherichia coli*: comparative genomic analysis of *E. coli* commensal and pathogenic isolates. J. Bacteriol. 190 (20), 6881–6893.

Reisner, A., Holler, B.M., Molin, S., Zechner, E.L., 2006. Synergistic effects in mixed *Escherichia coli* biofilms: conjugative plasmid transfer drives biofilm expansion. J. Bacteriol. 188 (10), 3582–3588.

Reznikoff, W.S., 2008. Transposon Tn5. Annu. Rev. Genet. 42, 269–286.

Riley, L.W., Remis, R.S., Helgerson, S.D., et al., 1983. Hemorrhagic colitis associated with a rare *Escherichia coli* serotype. N. Engl. J. Med. 308 (12), 681–685.

Ring, H.Z., Kwok, P.Y., Cotton, R.G., 2006. Human Variome Project: an international collaboration to catalogue human genetic variation. Pharmacogenomics 7 (7), 969–972.

Sandvig, K., 2001. Shiga toxins. Toxicon. 39 (11), 1629–1635.

Schmidt, M.A., 2010. LEEways: tales of EPEC, ATEC and EHEC. Cell Microbiol. 12 (11), 1544–1552.

Seshadri, R., Paulsen, I.T., Eisen, J.A., et al., 2003. Complete genome sequence of the Q-fever pathogen *Coxiella burnetii.* Proc. Natl. Acad. Sci. USA 100 (9), 5455–5460.

Shaikh, N., Holt, N.J., Johnson, J.R., Tarr, P.I., 2007. Fim operon variation in the emergence of enterohemorrhagic *Escherichia coli*: an evolutionary and functional analysis. FEMS Microbiol. Lett. 273 (1), 58–63.

Silva, R.M., Toledo, M.R., Trabulsi, L.R., 1980. Biochemical and cultural characteristics of invasive *Escherichia coli.* J. Clin. Microbiol. 11 (5), 441–444.

Smet, A., Martel, A., Persoons, D., et al., 2008. Diversity of extended-spectrum beta-lactamases and class C beta-lactamases among cloacal *Escherichia coli* isolates in Belgian broiler farms. Antimicrob. Agents Chemother. 52 (4), 1238–1243.

Smith, H.W., Linggood, M.A., 1971. Observations on the pathogenic properties of the K88, Hly and Ent plasmids of *Escherichia coli* with particular reference to porcine diarrhoea. J. Med. Microbiol. 4 (4), 467–485.

Snellings, N.J., Tall, B.D., Venkatesan, M.M., 1997. Characterization of *Shigella* type 1 fimbriae: expression, FimA sequence, and phase variation. Infect. Immun. 65 (6), 2462–2467.

Sokurenko, E.V., Chesnokova, V., Dykhuizen, D.E., et al., 1998. Pathogenic adaptation of *Escherichia coli* by natural variation of the FimH adhesin. Proc. Natl. Acad. Sci. USA 95 (15), 8922–8926.

Sokurenko, E.V., Feldgarden, M., Trintchina, E., et al., 2004. Selection footprint in the FimH adhesin shows pathoadaptive niche differentiation in *Escherichia coli.* Mol. Biol. Evol. 21 (7), 1373–1383.

Sokurenko, E.V., Gomulkiewicz, R., Dykhuizen, D.E., 2006. Source-sink dynamics of virulence evolution. Nat. Rev. Microbiol. 4 (7), 548–555.

Sokurenko, E.V., Hasty, D.L., Dykhuizen, D.E., 1999. Pathoadaptive mutations: gene loss and variation in bacterial pathogens. Trends Microbiol. 7 (5), 191–195.

Sorensen, S.J., Bailey, M., Hansen, L.H., Kroer, N., Wuertz, S., 2005. Studying plasmid horizontal transfer in situ: a critical review. Nat. Rev. Microbiol. 3 (9), 700–710.

Steinberg, K.M., Levin, B.R., 2007. Grazing protozoa and the evolution of the *Escherichia coli* O157:H7 Shiga toxin-encoding prophage. Proc. Biol. Sci. 274 (1621), 1921–1929.

Stewart, G.R., Robertson, B.D., Young, D.B., 2003. Tuberculosis: a problem with persistence. Nat. Rev. Microbiol. 1 (2), 97–105.

Sullivan, J.T., Trzebiatowski, J.R., Cruickshank, R.W., et al., 2002. Comparative sequence analysis of the symbiosis island of *Mesorhizobium loti* strain R7A. J. Bacteriol. 184 (11), 3086–3095.

Swerdlow, D.L., Griffin, P.M., 1997. Duration of fecal shedding of *Escherichia coli* O157:H7 among children in day-care centers. The Lancet 349 (9054), 745–746.

Tchesnokova, V., McVeigh, A.L., Kidd, B., et al., 2010. Shear-enhanced binding of intestinal colonization factor antigen I of enterotoxigenic *Escherichia coli.* Mol. Microbiol. 76 (2), 489–502.

Timmis, K.N., Boulnois, G.J., Bitter-Suermann, D., Cabello, F.C., 1985. Surface components of *Escherichia coli* that mediate resistance to the bactericidal activities of serum and phagocytes. Curr. Top Microbiol. Immunol. 118, 197–218.

Todd, E.C., Greig, J.D., Bartleson, C.A., Michaels, B.S., 2008. Outbreaks where food workers have been implicated in the spread of foodborne disease. Part 4. Infective doses and pathogen carriage. J. Food. Prot. 71 (11), 2339–2373.

Tomb, J.F., White, O., Kerlavage, A.R., et al., 1997. The complete genome sequence of the gastric pathogen *Helicobacter pylori.* Nature 388 (6642), 539–547.

Touchon, M., Hoede, C., Tenaillon, O., et al., 2009. Organised genome dynamics in the *Escherichia coli* species results in highly diverse adaptive paths. PLoS Genet. 5 (1), e1000344.

Van Derlinden, E., Bernaerts, K., Van Impe, J.F., 2008. Dynamics of *Escherichia coli* at elevated temperatures: effect of temperature history and medium. J. Appl. Microbiol. 104 (2), 438–453.

van Elsas, J.D., Semenov, A.V., Costa, R., Trevors, J.T., 2011. Survival of *Escherichia coli* in the environment: fundamental and public health aspects. ISME J. 5 (2), 173–183.

Vejborg, R.M., Friis, C., Hancock, V., Schembri, M.A., Klemm, P., 2010. A virulent parent with probiotic progeny: comparative genomics of *Escherichia coli* strains CFT073, Nissle 1917 and ABU 83972. Mol. Genet. Genomics. 283 (5), 469–484.

Vlisidou, I., Marches, O., Dziva, F., Mundy, R., Frankel, G., Stevens, M.P., 2006. Identification and characterization of EspK, a type III secreted effector protein of enterohaemorrhagic *Escherichia coli* O157:H7. FEMS Microbiol. Lett. 263 (1), 32–40.

von Sonnenburg, F., Tornieporth, N., Waiyaki, P., et al., 2000. Risk and aetiology of diarrhoea at various tourist destinations. Lancet 356 (9224), 133–134.

Weissman, S.J., Beskhlebnaya, V., Chesnokova, V., et al., 2007. Differential stability and trade-off effects of pathoadaptive mutations in the *Escherichia coli* FimH adhesin. Infect. Immun. 75 (7), 3548–3555.

Weissman, S.J., Chattopadhyay, S., Aprikian, P., et al., 2006. Clonal analysis reveals high rate of structural mutations in fimbrial adhesins of extraintestinal pathogenic *Escherichia coli*. Mol. Microbiol. 59 (3), 975–988.

Weissman, S.J., Johnson, J.R., Tchesnokova, V., et al., 2012. High-resolution two-locus clonal typing of extraintestinal pathogenic *Escherichia coli*. Appl. Environ. Microbiol. 78 (5), 1353–1360.

Weissman, S.J., Moseley, S.L., Dykhuizen, D.E., Sokurenko, E.V., 2003. Enterobacterial adhesins and the case for studying SNPs in bacteria. Trends Microbiol. 11 (3), 115–117.

Wirth, T., Falush, D., Lan, R., et al., 2006. Sex and virulence in *Escherichia coli*: an evolutionary perspective. Mol. Microbiol. 60 (5), 1136–1151.

Yamanaka, K., Fang, L., Inouye, M., 1998. The CspA family in *Escherichia coli*: multiple gene duplication for stress adaptation. Mol. Microbiol. 27 (2), 247–255.

Yaron, S., Kolling, G.L., Simon, L., Matthews, K.R., 2000. Vesicle-mediated transfer of virulence genes from *Escherichia coli* O157:H7 to other enteric bacteria. Appl. Environ. Microbiol. 66 (10), 4414–4420.

Zatyka, M., Thomas, C.M., 1998. Control of genes for conjugative transfer of plasmids and other mobile elements. FEMS Microbiol. Rev. 21 (4), 291–319.

Zdziarski, J., Brzuszkiewicz, E., Wullt, B., et al., 2010. Host imprints on bacterial genomes – rapid, divergent evolution in individual patients. PLoS Pathog. 6 (8), e1001078.

Escherichia coli pathotypes

Chapter 4

Enteropathogenic
Escherichia coli

Shahista Nisa, Karen M. Scanlon, and Michael S. Donnenberg

University of Maryland School of Medicine, Baltimore, MD, USA

BACKGROUND

Definition and classification

Enteropathogenic *E. coli* (EPEC) strains are diarrhea-causing, non-Shiga-toxin-producing *E. coli*. In healthy adults, EPEC-induced diarrhea may be initiated by a dose of 10^8 to 10^{10} CFU, though this infectious dose is likely lower for children (Donnenberg et al., 1998). Upon intimate attachment, these strains efface the host cell microvilli, generating a characteristic histopathology known as the attaching and effacing (A/E) effect. In these ways, EPEC is distinguished from other diarrheagenic pathotypes, as enteroaggregative *E. coli* (EAEC), enteroinvasive *E. coli* (EIEC), and enterotoxigenic *E. coli* (ETEC) do not display an A/E effect, while enterohemorrhagic *E. coli* (EHEC) express Shiga toxins (Nataro and Kaper, 1998). The genetic elements responsible for the production of A/E pathology are harbored on a large pathogenicity island known as the locus of enterocyte effacement (LEE) (McDaniel and Kaper, 1997). In addition, EPEC strains are further classified as typical or atypical by the presence or absence of an EPEC adherence factor (EAF) plasmid encoding the BFP.

History

In the 1930s, serological methods were used to characterize *E. coli* infections among both infants and calves, but the significance of *E. coli* in diarrheal disease was not fully appreciated until 1945 (Taylor, 1961). Bray associated specific *E. coli*, later recognized as serogroup O111, with an outbreak of infantile diarrhea, and this was quickly succeeded by a number of publications which further established the importance of O111 and extended the implicated serogroups to include O55, O26, and O119 (Bray, 1945). In the 1950s, with the advent of an internationally accepted serological testing method, it was soon

Escherichia coli. http://dx.doi.org/10.1016/B978-0-12-397048-0.00004-8

reported that EPEC had a worldwide distribution. For decades thereafter, EPEC was described solely based on the O and H serotypes prevalent amongst children under 2 years with diarrhea (Robins-Browne, 1987). WHO recognizes 12 EPEC serotypes, including typical and atypical EPEC strains, along with other diarrheagenic *E. coli*, such as EAEC and EHEC (Campos et al., 1994; Rodrigues et al., 1996; Scotland et al., 1996; Do Valle et al., 1997; Trabulsi et al., 2002). However, it is now appreciated that assessments of genotype and phenotype, and not serotype, are more accurate methods for identifying EPEC. The use of these descriptors has shed new light on the epidemiological and evolutionary features of this pathogen, in addition to bringing about a change in the classification of certain strains.

Epidemiology and global impact

EPEC is a major cause of human diarrhea in the developing world; with rare outbreaks in developed countries typically associated with nurseries and day-care centers (Donnenberg and Kaper, 1992; Nataro and Kaper, 1998). Adult infections are uncommon and typically the result of common source outbreaks or concordant with other conditions (Nataro and Kaper, 1998). Previously, rates of community-acquired EPEC infection were thought to be highest within the first 6 months after birth (Chatkaeomorakot et al., 1987; Cravioto et al., 1988; Gomes et al., 1991), but recent studies of children 5 years and younger indicate that prevalence increases with age (Behiry et al., 2011; Ochoa and Contreras, 2011). The disparity observed among publications describing age-associated incidence of infection may be due to changes in EPEC diagnosis, promotion of breastfeeding, or evolution of the pathogen. Disease severity, however, remains inversely proportionate to age (Toledo et al., 1983; Cravioto et al., 1990). In addition to age, it has been proposed that bacterial load may be a factor in EPEC-induced diarrhea (Barletta et al., 2011). The described incidence of infection differs greatly among studies, with EPEC-associated diarrhea ranging from 3–50% (Gomes et al., 1991; Alikhani et al., 2006; Nweze, 2010). Emergent reports, in which EPEC was characterized genetically and not by serotype, indicate a high prevalence of atypical EPEC (aEPEC), particularly in developed regions (Afset et al., 2003; Robins-Browne et al., 2004; Liebchen et al., 2011) and associated prolonged diarrhea with these strains (Afset et al., 2004; Nguyen et al., 2006). While humans are the sole reservoir for typical EPEC (tEPEC), animals may act as reservoirs for aEPEC; this chapter will focus on those typical and atypical strains affecting humans.

MOLECULAR PATHOGENESIS

Regulation

Over the years, multiple proven and potential virulence factors that help facilitate EPEC colonization of the intestinal epithelium have been identified.

The expression, secretion, and translocation of effector proteins that enable the bacteria to subvert host cellular pathways and cause disease are tightly regulated. These virulence factors are regulated by elements encoded on the EAF plasmid as well as the LEE pathogenicity island.

The EAF plasmid

The large (50–70 MDa) EAF plasmid harbored by tEPEC strains is required for localized adherence (LA; Figure 4.1A) (Baldini et al., 1983; Nataro et al., 1985; McConnell et al., 1989). Curing the EAF plasmid from EPEC abolished LA and autoaggregation phenotypes (Figure 4.1B) (Baldini et al., 1983; Knutton et al., 1987a; Chart et al., 1988) and led to attentuated virulence in colostrum-deprived piglets (Baldini et al., 1983) and healthy volunteers (Levine et al., 1985). However, plasmid-cured EPEC retained the ability to form A/E lesions (Figure 4.1C) (Baldini et al., 1983; Knutton et al., 1987a,b; Tzipori et al., 1989) and transfer of the EAF plasmid to non-EPEC strains endowed them with the ability to carry out LA (Baldini et al., 1983; Knutton et al., 1987b).

Sequence analysis of the EAF plasmid from E2348/69, pMAR7, a modified pMAR2 plasmid that carries an ampicillin resistance marker, revealed it to be a

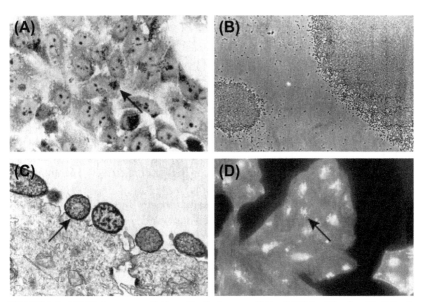

FIGURE 4.1   EPEC phenotypes: (A) Localized adherence (LA) of EPEC to HEp-2 epithelial cells in tissue culture. Arrow indicates a representative microcolony of EPEC. (B) Autoaggregation of EPEC in DMEM tissue culture medium. Portions of two aggregates are shown. (C) Attaching and effacing (A/E) of T84 cells by EPEC. The arrow indicates a site of intimate adherence between a bacterium and host cell (courtesy of Barry McNamara). (D) Fluorescent-actin staining (FAS) of EPEC adhering to HeLa cells. FITC-labeled phalloidin was used to identify highly concentrated actin filaments localizing beneath EPEC.

self-transmissible plasmid due to the presence of the highly conserved *tra* locus (Brinkley et al., 2006). The complete *tra* region is only found in representatives of the EPEC1 clonal group (Tobe et al., 1999; Brinkley et al., 2006), confirming previous findings that some EAF plasmids can be mobilized to other *E. coli* strains by conjugation (Baldini et al., 1983; Laporta et al., 1986; McConnell et al., 1989) while others cannot be transmitted as they lack the required transfer genes (Hales et al., 1992; Tobe et al., 1999).

Two loci important for pathogenicity on the EAF plasmid are the type IV BFP gene cluster (*bfp*) and the *per* locus encoding for transcriptional activators. The EAF plasmid also regulates the expression of the *eae* gene, located in the chromosomal pathogenicity island LEE (Jerse and Kaper, 1991) (see below).

Bundle-forming pili

BFP is a member of the type IV-B class of fimbriae that are produced by a number of Gram-negative pathogenic bacteria (e.g. ETEC, *Salmonella enerica* serovar Typhi and *Vibrio cholerae*). BFPs form ropelike bundles extending from the cell surface of typical EPEC strains (Girón et al., 1991). These pili are important virulence factors and are required for LA to epithelial cells (Donnenberg et al., 1992a), autoaggregation and virulence in volunteers (Bieber et al., 1998; Knutton et al., 1999).

The *bfp* operon consists of 14 genes, all of which are required for BFP biogenesis (Ramer et al., 1996; Sohel et al., 1996; Stone et al., 1996; Anantha et al., 2000; Blank and Donnenberg, 2001; Schreiber et al., 2002; Hwang et al., 2003). Expression of this gene cluster in a laboratory strain of *E. coli* via a plasmid confers both LA and BFP biogenesis (Stone et al., 1996). BFP is composed of repeating subunits of bundlin, pilin monomers encoded by *bfpA*, the first gene in the *bfp* operon. *bfpA* encodes pre-bundlin, which is processed by the pre-pilin peptidase, encoded by *bfpP*, producing mature bundlin (Donnenberg et al., 1992a,b; Zhang et al., 1994). Mutation of the *bfpA* gene leads to loss of LA and autoaggregation (Donnenberg et al., 1992a; Bieber et al., 1998; Anantha et al., 2000).

Structural analysis of bundlin revealed that while EPEC and *Vibrio cholerae* show distinct differences in their monomer subunits and are unable to complement one another, they assemble into filaments with similar helical organization (Ramboarina et al., 2005). The genes in the *bfp* operon are regulated by environmental and host factors together with a plasmid encoded regulator, Per (Puente et al., 1996; Rosenshine et al., 1996a; Martínez-Laguna et al., 1999). For more detailed information on BFP and other type IV pili, see Chapter 13.

per *(plasmid-encoded regulator)*

The *per* operon consists of three ORFs, designated *perA*, *perB*, and *perC*, which are also known as *bfpT*, *bfpV*, and *bfpW*, respectively. *perA*, the first gene, codes for a protein that belongs to the AraC family of transcriptional activators. The C-terminal region of PerA contains a DNA-binding helix-turn-helix motif which is highly conserved between strains (Okeke et al., 2001). PerA activates the expression of *bfpA* as *perA* inactivation leads to the loss of *bfpA* expression and thus LA (Tobe et al., 1996; Okeke et al., 2001; Ibarra et al., 2003). Volunteer studies demonstrated the contribution of *perA* and *bfpA* to diarrhea, with reduced virulence displayed by mutant strains (Bieber et al., 1998). PerA also auto activates its own promoter, binding to AT-rich regions upstream of the gene (Martínez-Laguna et al., 1999). This region is significantly homologous to the PerA binding site upstream of *bfpA* (Bustamante et al., 1998). PerA is closely related to VirF and Rns, the activator of the virulence gene regulatory cascade in *Shigella flexneri* and the activator of virulence genes in ETEC, respectively (Dorman, 1992). However, while VirF and Rns could be functionally substituted (Porter et al., 1998; Munson et al., 2001), complementation of *virF* and *rns* mutant strains was not observed with PerA and likewise a *perA* mutant could not be complemented with either VirF or Rns (Porter et al., 2004).

Apart from activating the EAF plasmid encoded genes, the *per* regulon has been shown to activate the expression of chromosomally encoded genes present at the LEE, including the *eae* gene (Gómez-Duarte and Kaper, 1995). By ensuring adequate expression of PerC, PerA indirectly activates Ler, the LEE transcriptional activator (see below) (Mellies et al., 1999; Bustamante et al., 2011).

The LEE pathogenicity island and the type 3 secretion/translocation system

The LEE is a 35.6 kb pathogenicity island. Introduction of this region into a non-pathogenic *E. coli* strain confers the ability to induce the A/E effect (Figure 4.1C) (McDaniel and Kaper, 1997). The EPEC LEE contains 41 ORFs (Elliott et al., 1998); encoding for an adhesin, the type 3 secretion system (T3SS), effector proteins, type 3 specific chaperones and transcriptional regulators. The LEE contains five polycistronic operons, LEE1 to LEE5. The principal function of the LEE is to form the T3SS apparatus, enabling EPEC to transport proteins across three membranes: the cytoplasmic and outer membranes of EPEC and the cytoplasmic membrane of the host cell to which the EPEC are attached (Jarvis et al., 1995). T3SSs are composed of more than 20 proteins that form a complex apparatus spanning both bacterial membranes and culminating in a needle that extends from the outer membrane (see Chapter 14 for details). The EPEC T3SS also includes the translocator proteins EspA, EspB, and EspD. Mutations of *espA*, *espB*, or *espD* result in strains that are unable to translocate effector proteins into host cells (Kenny et al., 1997; Knutton et al., 1998; Taylor et al., 1998; Wolff et al., 1998; Kenny and Jepson, 2000) and therefore unable to cause an A/E effect and reorganization of the host cell cytoskeleton (Donnenberg et al., 1990a,b, 1993a,b,c; Foubister et al., 1994a,b; Kenny et al., 1996; Abe et al., 1997, 1998; Kenny and Finlay, 1997; Lai et al., 1997; Nougayrède et al., 1999). EspB is required for virulence in volunteers (Tacket et al., 2000), while in a rabbit model, EspA, EspB, and EspD have each been shown to be required for virulence (Abe et al., 1998). EspA is a component of a prominent filamentous structure that connects EPEC to host cells (Knutton et al., 1998). It forms a sheath around the needle complex which is encoded by *escF* (Sekiya et al., 2001; Wilson et al., 2001). EspB and EspD are both localized to the host cell membrane after attachment and together, these proteins form the translocation pore (Knutton et al., 1998; Wolff et al., 1998; Wachter et al., 1999; Ide et al., 2001; Luo and Donnenberg, 2011). In addition, EspB has been shown to be an effector protein secreted by the T3SS. EspB binding to α-catenin (Kodama et al., 2002) and myosin (Iizumi et al., 2007) within the host cell regulates actin cytoskeletal rearrangement and thus promotes morphological changes (Taylor et al., 1999).

Recently, other proteins have been identified as being crucial for the T3SS. EscA, a small protein encoded by Orf15, localizes to the periplasm, associates with the inner membrane, interacts with outer membrane secretin EscC and is required for the structural integrity of the T3SS complex (Sal-Man et al., 2012a). EscI has also been shown to be essential for T3S and may function as an inner rod protein and act as an early secreted effector (Sal-Man et al., 2012b). EscI interacts with EscU, which is an essential inner membrane ring of the T3SS that self-cleaves to form a secretin-competent state for accommodating effectors (Zarivach et al., 2008; Thomassin et al., 2011). The EscF needle protein associates with multiple proteins: EscD, an inner membrane structural protein, EscJ, an inner ring protein, and EscC. EscC, EscD, and EscJ are required for the formation of the T3SS apparatus (Ogino et al., 2006). EscI, EscU, EscF, and the chaperone CesT also interact with EscP (Orf16) (Monjaras et al., 2012). While EscP is not an essential protein of the T3SS, it appears to be required for regulating needle length and participating in substrate specificity (Monjaras et al., 2012). EscN is a type III ATPase (Andrade et al., 2007) that provides a potential energy source for the unfolding of T3S effectors (Song et al., 2004; Zarivach et al., 2007).

Delivery of effectors by the T3SS leads to the interaction between the products of two highly characterized genes in the LEE, *eae* and *tir*, encoding intimin and Tir, respectively. Intimin is a 94-kDa outer membrane adhesin (Jerse and Kaper, 1991) required for the intimate attachment of EPEC to epithelial cells at the site of A/E lesions (Figure 4.1C) (Jerse et al., 1990), while Tir is the translocated intimin receptor which binds intimin at the host cell surface (Kenny et al., 1997). While intimin is not strictly required for EPEC protein secretion or translocation (Rosenshine et al., 1992; Foubister et al., 1994b; Haigh et al., 1995; Kenny and Finlay, 1995, 1997; Wolff et al., 1998), the absence of intimin results in a significant decrease in EspB translocation (Wolff et al., 1998) and intiminis required for full virulence in human and rabbit infection (Donnenberg et al., 1992b; Marchès et al., 2000). There are 29 subtypes of intimin described to date. These subtypes are differentiated by a variable C-terminal region comprising the last 280 amino acids that is required for intimin–Tir binding and controlling specific host and tissue tropism (Fitzhenry et al., 2002; Ito et al., 2007; Mundy et al., 2007). Intimin β is the most prevalent subtype found amongst isolates of EPEC from humans and animals (Ramachandran et al., 2003; Nakazato et al., 2004; Blanco et al., 2005). Mutation of the *eae* gene leads to loss of A/E and host cell invasion (Francis et al., 1991), in addition to preventing the promotion of pseudopod formation (Rosenshine et al., 1996b) and EPEC attachment to human mucosal explants (Hicks et al., 1998). Unlike T3SS mutants, *eae* mutants retain the ability to redistribute host cell actin but have a blunted capacity to concentrate host cell cytoskeletal elements beneath the attachment site (Donnenberg et al., 1990a; Rosenshine et al., 1992). Mutant *eae* strains of EPEC can complement strains carrying mutations in *espA*, *espB* or other T3SS genes, allowing them to establish A/E (Rosenshine et al., 1992; Donnenberg et al., 1993b; Kenny et al., 1996). In addition, pre-infection with an *eae* mutant allows a laboratory *E. coli* strain carrying the cloned intimin gene (Rosenshine et al., 1996a) or beads coated with intimin (Liu et al., 1999) to bind to epithelial cells. Such *trans* complementation is possible because the *eae* mutant remains capable of translocating Tir into the host cell membrane. Apart from binding Tir, intimin may also bind to host intimin receptors (Hir) such as nucleolin via its variable C-terminus (Mundy et al., 2007).

Tir is a multifunctional protein which, after injection into the host cell via the T3SS, becomes integrated into the host cell membrane in a protease-sensitive conformation (Rosenshine et al., 1996a; Kenny et al., 1997). This protein exhibits a dimeric hairpin topology, such that an extracellular loop from each monomer binds an intimin molecule (De Grado et al., 1999; Hartland et al., 1999; Kenny, 1999) with the N-terminal and C-terminal domains projecting into the host cell cytoplasm (Figure 4.2B). Tir serves as the focus for the actin reorganization that occurs at the site of EPEC pedestals (Figure 4.1D). Some, but not all EPEC strains express Tir variants that have a tyrosine at position 474 (Y474). Upon translocation of this form of Tir into the host cell, it is phosphorylated at Y474 (Kenny et al., 1997) by host cell kinases (Nakazato et al., 2004; Swimm et al., 2004; Blanco et al., 2005). This event is essential for actin remodeling in such strains (Kenny, 1999), as mutation of Tir results in the loss of the A/E phenotype and reduced virulence in a rabbit model (Kenny et al., 1997; Marchès et al., 2000). As discussed below, there are other pathways leading to indistinguishable actin remodeling.

Apart from Tir, other translocated LEE encoded effector molecules include EspF, EspG, EspH, EspZ, and Map (Pallen et al., 2005). More recently, a repertoire of non-LEE encoded effectors (Nles) that are secreted by the T3SS have been discovered. For more detailed information on the type 3 effectors, see Chapter 15.

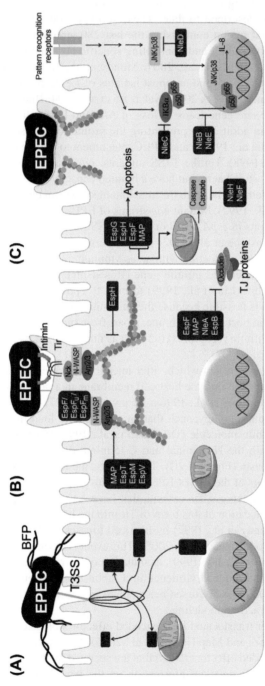

FIGURE 4.2   A model of EPEC attachment and host cell subversion: (A) Typical EPEC uses bundle-forming pili (BFPs) to attach to intestinal epithelial cells and translocates effector proteins (purple) via a type 3 secretion system (T3SS). (B) Effector proteins serve as receptors (Tir) for intimin, subvert the host cell cytoskeleton (EspF; EspF$_u$, EspF$_m$, MAP, EspT, EspM, EspV), interacting with host proteins such as N-WASP and the Arp2/3 complex to cause actin rearrangement and the formation of a cup-shaped pedestal, and disrupting barrier function (EspF, MAP, NleA, EspB) by inducing the redistribution of tight junction (TJ) proteins. (C) In addition to altering the regulation of ion transport (not shown), EPEC effector proteins also modulate host survival responses; both promoting (EspF, EspG, EspH, MAP) and antagonizing (NleF, NleH) apoptosis and inhibiting innate immune responses (NleB, NleC, NleD, NleE) activated by pathogen-associated molecular patterns.

Ler (LEE-encoded regulator)

Ler is a 15-kDa protein encoded by the first gene in the LEE1 operon (Mellies et al., 1999). Ler regulates the expression of LEE as well as some non-LEE-encoded genes (Elliott et al., 2000). This protein induces expression by counteracting the repression mediated by nucleoid-associated protein H-NS (Bustamante et al., 2001). H-NS binds to AT-rich sequences and negatively modulates expression of a number of genes (Atlung and Ingmer, 1997). Expression of Ler is influenced by many factors in response to environmental signals (Mellies et al., 2007) including quorum sensing (Sperandio et al., 1999). Quorum sensing activates quorum-sensing *E. coli* regulator A (QseA), which activates transcription of Ler independent of *per* (Sircili et al., 2004). Ler expression is also dependent on integration host factor (IHF). IHF positively regulates *ler* by binding upstream of the *ler* promoter (Friedberg et al., 1999). GlrA (Global Regulator of LEE-Activator) and GrlR (Repressor) encoded within the LEE positively and negatively regulate Ler expression respectively (Deng et al., 2004). PerC, encoded in the EAF plasmid, has also been shown to induce the expression of Ler (Lio and Syu, 2004), independent of activation via GlrA (Bustamante et al., 2011). Ler is also a negative auto-regulator (Berdichevsky et al., 2005). It binds to the LEE1 regulatory region to reduce Ler, enabling the activation of the LEE2, LEE3, and other promoters.

Other virulence factors

Lymphostatin/Efa1

The *lifA* gene, encoding a 366-kDa protein lymphostatin, is located on the IE6 pathogenicity island. This pathogenicity island also encodes several non-LEE-encoded effectors (Iguchi et al., 2009). To date, lymphostatin is the largest T3S effector protein identified (Deng et al., 2012), both in EPEC and EHEC. The prevalence of this locus in EHEC strains isolated from patients with hemolytic uremia and outbreak strains suggests it may play a role in virulence (Afset et al., 2006; Wickham et al., 2006). A portion of lymphostatin displays sequence similarity to the glycosyltransferase region of large clostridial cytotoxins, which is required for glycosylation of the Rho family GTPases (Klapproth et al., 2000). A *lifA* mutant of *Citrobacter rodentium* was deficient in decreasing epithelial barrier function. This change was dependent on the glycosyltransferase region and was associated with a decreased activation of GTPase Cdc42 (Babbin et al., 2009). Multiple functions have been described for this extremely large protein. *C. rodentium* lymphostatin is also involved in bacterial colonization, crypt cell proliferation, and epithelial cell regeneration (Klapproth et al., 2005). In EPEC, lymphostatin inhibits the transcription of many lymphokines, including interleukin 2, interleukin 4, and interferon-γ, and inhibits lymphocyte proliferation (Klapproth et al., 1995, 1996, 2000; Malstrom and James, 1998). Similar activity is produced by other A/E pathogens, including EHEC O157:H7 strains and *C. rodentium*. The *lifA* gene is also known as Efa1 for EHEC factor for adherence as an Efa1 mutant in EHEC has significantly lower adherence to Chinese hamster ovary cells and is deficient in human red blood cell agglutination and autoaggregation (Nicholls et al., 2000). A similar phenotype is seen in EPEC, where the *efa1* mutant is considerably less adherent to epithelial cells compared to its parent strain (Badea et al., 2003).

EspC

EspC is another large (110 kDa) secreted protein encoded on a pathogenicity island of EPEC (Vidal and Navarro-Garcia, 2008). EspC is a member of the SPATE (serine protease autotransporters of the enterobacteriaceae) family of autotransporter proteins encoding its own transport mechanism (see Chapter 16) (Henderson and Nataro, 2001). Such proteins possess: (a) an N-terminal signal sequence that promotes secretion through the inner membrane via the sec apparatus; (b) a C-terminal domain that forms a beta barrel pore in the outer membrane and exports; and (c) a central 'passenger' domain of the protein to the bacterial cell surface (Stein et al., 1996a). Although secreted via the type 5 mechanism, EspC requires the T3SS for translocation into host cells (Vidal and Navarro-Garcia, 2008). EspC causes cytotoxicity via cytoskeletal damage, which is dependent on the internalization of the serine protease motif (Navarro-Garcia et al., 2004). EspC is also an enterotoxin that induces a change in short-circuit current in rat jejunal tissue mounted on Ussing chambers (Mellies et al., 2001). The regulation of EspC is mediated by the global regulator Ler (Mellies et al., 1999).

Other toxins

EAEC strains produce an enterotoxin known as enteroaggregative heat-stable enterotoxin 1 (EAST1). A survey of diarrheagenic *E. coli* strains reported that 14 of 65 EPEC strains tested (22%) hybridized with an EAST1 probe (Savarino et al., 1996). The E2348/69 strain contains two copies of the *east1* gene, one in the chromosome and another in the EAF plasmid.

Another toxin characterized in an EPEC strain is the cytolethal distending toxin (CDT) (Scott and Kaper, 1994). CDT is composed of three polypeptides *cdtA*, *cdtB*, and *cdtC*, where CDT-B is the catalytic subunit. The translocation of CDT-B is mediated by CDT-A and CDT-C subunits. Upon reaching the nucleus, CDT causes DNA damage (Ohara et al., 2004), i.e. chromatin disruption, leading to G_2/M-phase growth arrest of the target cell and ultimately cell death (Lara-Tejero and Galan, 2000). There are other sporadic reports of production of CDT by EPEC. A study of CDT-producing *E. coli* in Bangladeshi children found that CDT-positive EPEC strains were isolated from more children with diarrhea than healthy controls; however, this difference did not reach statistical significance (Albert et al., 1995, 1996). Another study described the possible association of CDT-producing EPEC with diarrheal patients in the Republic of Korea (Kim et al., 2009). The E2348/69 strain does not encode CDT.

Other fimbriae and pili

Some EPEC strains elaborate pili or fimbriae other than or in addition to BFP (Bradley and Thompson, 1992; Girón et al., 1993). Atypical EPEC strains lacking BFP have been shown to express other fimbrial genes that could potentially aid in adherence to epithelial cells (Hernandes et al., 2011). The *E. coli* common pilus (ECP) has been implicated in the LA of aEPEC (Scaletsky et al., 2010b). The expression of the long, fine fimbriae; rigid bent fimbriae; and short, fine fimbriae is ablated in a *ler* mutant of the EPEC strain E2348/69 (Elliott et al., 2000).

Flagella have also been shown to be involved in EPEC adherence to epithelial cells. A mutation in the flagella gene, *fliC*, severely decreased the ability of EPEC to adhere to and form microcolonies on host cells. Adherence was also blocked by purified EPEC flagella and anti-flagellum antibodies (Girón et al., 2002). This Δ*fliC* phenotype was restored upon growth in preconditioned tissue culture media, suggesting that a signal from the mammalian cells influences flagella production.

Type I pili of EPEC have been found to be antigenic in volunteer studies (Karch et al., 1987). While type I pili do not have a role in adherence on epithelial cells in vitro (Elliott and Kaper, 1997), they have been associated with biofilm development at low temperatures in aEPEC O55:H7 (Weiss-Muszkat et al., 2010).

Adherence and invasion

Localized adherence

Typical EPEC strains adhere to HeLa, HEp-2, and other cell lines, and to organ cultures in vitro, in a distinctive pattern of three-dimensional clusters or microcolonies (Clausen and Christie, 1982; Scaletsky et al., 1984; Knutton et al., 1987a,b; Nataro et al., 1985, 1987; Yamamoto et al., 1992; Hicks et al., 1998). The LA pattern is a characteristic specific to EPEC strains and therefore has been used widely as a diagnostic tool (Figure 4.1A). A similar adherence pattern had been observed on tissue biopsies from EPEC-infected infants (Rothbaum et al., 1982, 1983). LA is due to the production of BFP by the bacterium. BFP binds to specific receptors on the host cell surface. Different investigators have found that LA can be inhibited by various sugar moieties, including galactose (Vanmaele and Armstrong, 1997), N-acetylgalactosamine (Scaletsky et al., 1988), N-acetyllactosamine (Vanmaele et al., 1999), and fucosylated oligosaccharides and gangliosides from milk (Jagannatha et al., 1991; Costa-Carvalho et al., 1994). These saccharides could be moieties of host cell glycolipids or glycoproteins that serve as EPEC receptors (Vanmaele et al., 1995; Manjarrez-Hernandez et al.,

1997). A binding affinity between bundlin, the major structural subunit of BFP and N-acetyllactosamine has been established (Hyland et al., 2008). Although this interaction is weak, it may have a significant effect on LA due to the potentially large numbers of ligands and receptors involved. More recent work indicates that N-acetyllactosamine may trigger BFP retraction in addition to acting as a competitive inhibitor of EPEC binding to the host receptors (Hyland et al., 2006). EPEC also binds to phosphatidylethanol-amine, a component of cell membranes (Foster et al., 1999; Barnett et al., 2000; Khursigara et al., 2001).

Other surface components contributing to EPEC adherence include intimin, the EspA filament and other fimbriae and/or pili. Single, double, and triple mutants of the *bfp*, *espA*, and *eae* genes revealed that while BFP and EspA both play a role in adherence, BFP plays a more dominant role; atypical EPEC strains that do not express BFP may use EspA for adherence. While a triple *bfpA espAeae* mutant is not adherent to epithelial cells (Cleary et al., 2004). EPEC strains lacking BFP remain able to adhere to pediatric small intestine tissue explants, although they form smaller colonies that lack the three-dimensional structure (Hicks et al., 1998). Initial adherence in such strains has been attributed to the production of *E. coli* common pilus (ECP) (Scaletsky et al., 2010a,b) and in some strains to the afimbrial adhesion, locus for diffuse adherence (LDA) (Torres et al., 2007). *E. coli* strains fitting the definition of EPEC but having non-LA pattern of adhesion, i.e. diffused adherence, have been noted (Knutton et al., 1991; Rodrigues et al., 1996; Beinke et al., 1998; Pelayo et al., 1999; Scaletsky et al., 1999). This phenotype has been attributed to the omicron subtype of intimin where the adherence was seen due to an invasive process (Hernandes et al., 2008).

Autoaggregation

When grown in tissue culture medium at 37°C, tEPEC strains aggregate into large clusters that may contain hundreds or thousands of bacteria (Figure 4.1B) (Vuopio-Varkila and Schoolnik, 1991). These autoaggregates are readily visible under a low-power microscope or even to the naked eye. They are unstable, dispersing quickly when moved to sub-optimal conditions (Bieber et al., 1998), with microcolonies in tissue culture forming and then dispersing over a period of 6 hours (Knutton et al., 1999). Like LA, autoaggregation requires BFP. Retraction of the pilus fiber causes bundlin pilin subunits to dissociate and thus the observed dispersion (Humphries et al., 2010). The BfpF protein drives pilus retraction and thus a *bfpF* mutant fails to disperse over time (Knutton et al., 1999).

Attaching and effacing

EPEC has the ability to strikingly alter the surface of the cells to which they attach. The characteristic phenotype observed during EPEC infection is the formation of A/E lesions (Moon et al., 1983), in which the brush border microvilli are sloughed off the apical surface of enterocytes (effacement) and transiently replaced by elongated microvillus-like processes (Frankel et al., 1996) (Figure 4.1C). Effacement of microvilli is followed by actin polymerization (Figure 4.1D) and ultimately the formation of prominent cup-like pedestals and elongated (up to 10 μm) pseudopod structures (Rosenshine et al., 1996a,b). EPEC utilize these pedestals to attach in very close (intimate) apposition to the host cell membrane. The pedestal structures are dynamic, being able to change length, shape, and position over time, and move the attached EPEC along the host cell surface (Sanger et al., 1996). A/E lesions are observed in model EPEC infections with cultured cells and mucosal explants (Knutton et al., 1987a,b; Hicks et al., 1998) as well as in intestinal biopsies from EPEC-infected infants or animals (Rothbaum et al., 1983; Peeters et al., 1988), but not on formalin-fixed epithelial cells (Knutton et al., 1997). The ability to form A/E lesions is shared among EPEC, EHEC, and strains of *Escherichia albertii* and *C.rodentium* (Tzipori et al., 1986; Schauer and Falkow, 1993; Huys et al., 2003). All attaching and effacing pathogens carry the LEE, which harbors the genes encoding intimin, *eae*, the adhesin required for attaching and effacing and the T3SS (Elliott et al., 1998).

Invasion

The ability of EPEC to be internalized by epithelial cells has been noted both in tissue culture (Andrade et al., 1989; Donnenberg et al., 1989, 1990b) and in small intestinal biopsies from EPEC-infected infants (Fagundes-Neto et al., 1995). Several critical EPEC virulence genes, including the *bfp* operon and the LEE, were first identified via the characterization of mutants lacking the ability to invade HEp-2 cells (Donnenberg et al., 1990a,b). The ability of EPEC to invade epithelial cells is in marked contrast to their capacity to evade phagocytic engulfment. Invasion requires the same virulence factors as A/E, therefore, it has been suggested that invasion is a byproduct of the cytoskeletal rearrangements that occur during the A/E process (Rosenshine et al., 1996b). MAP, or mitochondrial associated protein, is located immediately upstream of the operon encoding intimin and Tir. It is targeted to the host cell mitochondrial membrane where it disrupts the membrane potential. Map also mediates filopodia formation via Cdc42 (Kenny and Jepson, 2000; Kenny et al., 2002). This latter function of MAP has been shown to be involved in EPEC invasion (Jepson et al., 2003). An invasion phenotype has also been associated with other T3S effectors. EspT causes membrane ruffling and the generation of lamellipodia, via Rac1 and Cdc42 activation, leading to EPEC invasion (Bulgin et al., 2009). EspF and one of its host binding proteins, sorting nexin 9 (SNX9), which regulates vesicle trafficking and endocytosis, has also been shown to aid in the epithelial cell invasion of EPEC (Weflen et al., 2010). Despite its invasive potential, EPEC remains classified as an extracellular pathogen; this is likely due to the rarity of severe inflammation and bacteremia associated with EPEC disease.

Avoidance of host responses

Subversion by the T3SS

Like other bacterial pathogens, EPEC has multiple activators of host pattern recognition receptors including lipopolysaccharide, flagellin, and unmethylated CpG nucleotides and must contend with the host innate immune response to establish infection. Histologic examination of EPEC-infected tissue confirms the induction of an inflammatory response by these bacteria (Ulshen and Rollo, 1980; Higgins et al., 1999a,b; Marchès et al., 2000). However, some reports noted that the response is rather mild (Rothbaum et al., 1982). In cultured epithelial cells, EPEC activates NF-κB which induces the transcription of interleukin 8 (IL-8) (Figure 4.2C) (Savkovic et al., 1996, 1997). NF-κB functions as a homo- or heterodimer composed of various subtypes, e.g. RelA (p65), RelB, c-Rel, p50, and p52. This transcription factor is inactive while sequestered in the cytoplasm in an IκB-bound complex. Upon stimulation, IκB is phosphorylated by IκB kinase (IKK) and targeted for degradation, releasing NF-κB and allowing the subunits to translocate to the nucleus where they regulate gene expression (Karin and Ben-Neriah, 2000; Chen, 2005). EPEC also induces the activation of the MAP kinase cascade. c-Jun N-terminal kinases (JNKs) are serine/threonine kinases which belong to the MAP kinase family. Upon activation, JNKs phosphorylate c-Jun, an AP-1 transcription factor. Despite measurable activation of these pro-inflammatory cascades, it appears that multiple EPEC T3S effectors also negatively regulate the host cell response,with T3SS mutants inducing greater levels of activation (Sham et al., 2011). Most of these effectors are interdependent, multifunctional and redundant (Dean and Kenny, 2009). More details on T3SS effectors can be found in Chapter 15.

The E2348/69 strain contains two *nleH* genes, both of which have been shown to inhibit pro-inflammatory cytokine expression and promote colonization (Royan et al., 2010). Both NleH1 and NleH2 reduce the abundance of nuclear ribosomal protein S3 (RPS3), a non-Rel subunit of p65 homodimer and p65–p50 heterodimer (Gao et al., 2009), however, only NleH1 inhibited the NF-κB activity independent of IκBα phosphorylation and degradation (Pham et al., 2012). NleC and NleD are zinc metalloproteases that specifically cleave NF-κBp65 and JNK respectively (Baruch et al., 2011). This cleavage prevents transcription of NF-κB- and AP-1-induced genes leading to a marked reduction in IL-8 secretion (Yen et al., 2010; Baruch et al., 2011). NleC has also been found to bind and cleave host acetyltransferase p300, decreasing the inflammatory response. NleB and NleE act to inhibit p65 translocation to the nucleus (Newton et al., 2010). NleB inhibits the TNF-α pathway, while NleE is an S-adenosyl-L-methionine-dependent methyltransferase that prevents the activation of IKKβ (Nadler et al., 2010) by inactivating its kinase activity (Zhang et al., 2012). Recently, Tir has been shown to inhibit NF-κB activation by inducing TNF-α receptor associated factor (TRAF) protein degradation (Ruchaud-Sparagano et al., 2011).

Apoptosis is another part of the host innate immune response induced by bacterial infection; this action aids in the elimination of the infecting pathogen and promotes a dendritic-cell-mediated immune response. EPEC subverts this response at multiple stages throughout the process. UV-induced apoptosis is JNK-dependent and may be inhibited by transient expression of NleD. In addition, infection with an *nleD* mutant strain of EPEC induces a greater percentage of apoptosis compared with wild-type. These results indicate that NleD inhibits JNK-dependent apoptosis, likely via JNK cleavage (Baruch et al., 2011). EspZ, a LEE-encoded effector, promotes the survival of epithelial cells by enhancing the phosphorylation of the focal adhesion kinase (FAK) and thereby may stabilize the epithelium upon infection (Shames et al., 2010). NleH inhibits apoptosis by preventing pro-caspase-3 cleavage through direct binding to Bax inhibitor-1 (BI-1) protein (Hemrajani et al., 2010). The NleF effector, highly conserved in EPEC and EPEC, binds the active site of caspase-9 to inhibit its activity (Blasche et al., 2013). For additional information, see 'Apoptosis', below.

Damage

Cytoskeletal rearrangement

Numerous signaling and actin-associated/binding proteins including α-actinin, ezrin, and myosin light chain II (MLC) are recruited to the site of bacterial intimate adhesion (Manjarrez-Hernandez et al., 1996; Cantarelli et al., 2001; Goosney et al., 2001). These proteins are involved in pedestal formation, which leads to loss of absorptive surface (effacing). EPEC achieves this effect by subverting fundamental host cell functions to build focal adhesion-like structures that anchor the bacteria to the host cytoskeleton. Upon EPEC infection and insertion of Tir into the host cell membrane, Tir becomes clustered at the site of bacterial attachment, where it serves as the receptor for intimin. F-actin and other host cytoskeletal proteins accumulate at this site and generate a pedestal (Figure 4.2B) (Knutton et al., 1987a,b; da Silva et al., 1989; Finlay et al., 1992). The localized F-actin accumulation can be demonstrated using fluorescent probes, which is the basis of the diagnostic fluorescent-actin staining (FAS) test (Figure 4.1D). Recent advances have begun to decipher the detailed molecular architecture of the pedestals and how EPEC remodels the host cytoskeleton at the surface of the cell. These studies have revealed that different strains of EPEC use different pathways to achieve the same end. For the E2348/69 strain, Tir is phosphorylated on Y474 by redundant kinases from the Src family, including Fyn and Tec/Abl family kinases (Phillips et al., 2004; Bommarius et al., 2007). This phosphorylation is essential for actin remodeling and pedestal formation (Rosenshine et al., 1996a,b; Kenny, 1999) but it is not required for the membrane insertion of Tir or intimin binding (Kenny et al., 1997; Gauthier et al., 2000). After phosphorylation, the residues flanking Y474 directly bind to the host SH3/SH2 adaptor protein Nck (Gruenheid et al., 2001; Campellone et al., 2002). Nck recruits the neural-Wiskott-Aldrich syndrome protein (N-WASP), which activates the actin-related proteins 2 and 3 (Arp2/3) complex and leads to the formation of the actin pedestals (Figure 4.2B) (Lommel et al., 2001). However, some EPEC strains lacking Tir Y474 are still capable of promoting actin polymerization. These Nck-independent modes of recruiting N-WASP are attributed to phosphorylation of Y454 (Campellone and Leong, 2005), and to the TccP (Tir-cytoskeleton coupling protein) (Whale et al., 2006) effectors one and two (also known as $EspF_U$ and $EspF_M$). These effectors bind the GTPase binding domain of N-WASP (Campellone et al., 2004). More recently, phosphorylation of Y474 has been shown to recruit redundant kinases. The polyproline region of EspF interacts with the SH3 domains of these kinases, while the Y474 site interacts with the SH2 domains (Bommarius et al., 2007). This activity constitutes a positive feedback loop in which the recruited kinases phosphorylate other Tir molecules.

Recent studies have highlighted the importance of other T3S effectors in remodeling the host cytoskeleton by modulating Rho GTPases. EspF is a 206-amino-acid protein, the C-terminus of which contains three identical repeats of a proline-rich sequence that resembles those recognized by eukaryotic signaling proteins containing SH3 domains (Donnenberg et al., 1997). The C-terminus of each repeat also contains an N-WASP binding motif (Alto et al., 2007). EspF activates the Arp2/3 complex and induces actin polymerization in vitro (Weflen et al., 2009). Map and EspM display guanine exchange factor (GEF) activity (Huang et al., 2009; Arbeloa et al., 2010). Map and EspT activate Cdc42 activity, which induces filopodia formation (Kenny et al., 2002; Berger et al., 2009), while EspM activates RhoA, inducing actin stress fibers (Alto et al., 2006; Arbeloa et al., 2008). Other T3S effectors involved in pedestal dynamics are EspH and EspV (Tu et al., 2003; Arbeloa et al., 2011). The mechanism by which EspV modulates the cytoskeleton is unknown. EspH competitively binds to tandem DH-PH domains of endogenous Dbl-homology and pleckstrin-homology RhoGEFs and prevents Rho activation (Dong et al., 2010). Thus, EPEC strains secrete effectors that inactivate endogenous RhoGEFs, in addition to secreting their own GEF to hijack the host signaling pathway. EspH also stimulates actin polymerization by recruiting N-WASP and the Arp2/3 complex independently of the tyrosine phosphorylation of Tir but via the C-terminus of Tir and the WH1 domain of N-WASP (Wong et al., 2012a,b).

Disruption of the epithelial barrier

The apical junction complex, which consists of tight junctions (TJ) and adherent junctions connecting neighboring cells, is critical for separation of intestinal tissues from luminal contents (Harhaj and Antonetti, 2004). The TJ is comprised of three different transmembrane proteins; occludin, claudins, and junctional adhesion molecule (JAM). These proteins are attached to actin filaments and MLC via the zonula occludens (ZO) family of proteins, which are peripheral membrane proteins associated with the cytoplasmic surface of TJ and adherent junctions. Adherent junctions are composed of the transmembrane protein E-cadherin and cytoplasmic proteins from the catenin family (Kuphal and Bosserhoff, 2012).

EPEC infection of polarized cultured intestinal cells (Caco-2 or T84) increases the permeability of the monolayer, as reflected by a decrease in trans-epithelial electrical resistance (TER) and a flux of markers across the mono-layer via the paracellular pathway. These permeability changes are mediated by the modification of the TJ structure (Canil et al., 1993; Spitz et al., 1995; Philpott et al., 1996). EPEC infection activates the MLC kinase which phos-phorylates MLC, inducing the contraction of the actinomysin ring and increasing paracellular permeability (Manjarrez-Hernandez et al., 1996). EPEC also induces the redistribution of occludin, ZO-1, clearing of claudin-1 (Simonovic et al., 2000; McNamara et al., 2001; Muza-Moons et al., 2004; Shifflett et al., 2005) and relocalization and activation of ezrin, a protein that links TJ pro-teins to the actin cytoskeleton (Simonovic et al., 2001). This redistribution decreases protein interactions at the TJ and disrupts barrier function. Redis-tribution of these junctional proteins is dependent on the EspB translocation protein and on the effector EspF (Simonovic et al., 2001). EspB is also required for the phosphorylation of protein kinase C which increases association with E-cadherins resulting in redistribution of β-catenin and increased paracel-lular permeability (Malladi et al., 2004). EPEC infection causes β_1-integrins to migrate from the basolateral side to the apical side, where they interact with inti-min and enhance barrier defects (Muza-Moons et al., 2003). Another T3S effec-tor, EspG, alters paracellular permeability but does not alter the transepithelial electrical resistance (Matsuzawa et al., 2005). EspG also causes localized deple-tion of microtubules triggering disruption in the network (Shaw et al., 2005).

Alteration of water and ion secretion

EPEC infection can disrupt the tightly regulated flow of ions across the intesti-nal epithelium. Such changes can be measured in intestinal monolayers in vitro using a patch-clamp technique (Stein et al., 1996b). Caco-2 cells infected with EPEC induce a rapid and transient increase in short-circuit current (Isc) that is dependent on a functional T3SS and intimate adherence. This response was partially attributed to chloride ion (Cl^-) secretion (Collington et al., 1998a,b). Another study showed that EPEC altered bicarbonate transport and decreased Isc in T84 cells in a Cl^--independent manner (Hecht and Koutsouris, 1999). The difference in Cl^- transport observed between these two studies can potentially be explained by the net absorptive and secretory nature of the respective cell lines used. There are a number of other factors that can influence ion transport upon infection. By activating NF-κB, EPEC infection significantly increases the expression of the galanin receptor in vitro and in vivo. Galanin leads to Cl^- secretion by acting as a secretagog (Savkovic et al., 1997; Benya et al., 1999; Hecht et al., 1999).

EPEC infection also affects sodium ion transport in Caco-2 cells by reducing the expression of sodium hydrogen exchanger 3 (NHE3) in an EspF-dependent manner (Hodges et al., 2008). NHE3 contributes to sodium ion (Na^+) absorption, verified by the production of watery stool by NHE3-deficient mice (Gawenis et al., 2002). EPEC also reduces Cl^-/HCO_3^- exchange both in vivo and in vitro (Gill et al., 2007). This effect is mediated by the T3S effectors EspG and EspG2, which disrupt the microtubular network, altering the distribution of the membrane targeting Cl^-/HCO_3^- exchanger DRA, and resulting in accumulation of Cl^- in the lumen and loss of water. Other EPEC virulence factors including EspF, Map, Tir, and intimin have been shown to alter function of SGLT-1, a cotransporter that is responsible for the majority of fluid uptake in the small intestine (Dean et al., 2006). In a murine infection, *C. rodentium* causes the mislocalization of aquaporins from the cell membrane to the cytoplasm, which correlates to diarrhea-like phenotype and which is partially dependent on the EspF and EspG effector proteins (Guttman et al., 2007).

Apoptosis

Apoptosis, the programmed death of cells in response to internal or external stimuli, is an important defense mechanism launched by multicellular organisms where infected cells are induced to die for the benefit of the remaining cells (Elmore, 2007). EPEC is able to manipulate the host cell death pathway by two distinct mechanisms; intrinsic (mitochondrial- and ER-mediated pathway) and extrinsic (receptor-mediated pathway) (Figure 4.2C) (Elmore, 2007).The intrinsic pathway involves the activation of the Bcl-2 family proteins Bak and Bax, which leads to mitochondrial outer membrane permeabilization and release of cytochrome C. The cytosolic cytochrome C interacts with apoptosis activating factor 1 (Apaf-1) and procaspase-9 to activate downstream caspases-3, -6, and -7 (Li et al., 1997). This pathway is induced by translocation of the T3S effector EspF. The N-terminus of EspF (residues 1 to 20) contains the bacterial secretion signal (Charpentier and Oswald, 2004) and superimposed onto this region is the organelle-targeting domain (Nougayrède and Donnenberg, 2004; Nagai et al., 2005; Dean et al., 2010a,b). The mitochondrial targeting of EspF induces apoptosis (Crane et al., 2001; Nougayrède and Donnenberg, 2004) via disruption of the mitochondrial membrane potential, causing the release of cytochrome C and subsequent activation of caspase 9 (Nougayrède and Donnenberg, 2004). EspF also binds to and depletes Abcf2, an anti-apoptotic factor (Nougayrède et al., 2007). Another T3S effector, EspG, has been shown to activate the host cysteine protease calpain inducing rapid cell death and necrosis, a process modulated by the presence of Tir (Dean et al., 2010a). A second EspG homolog, EspG2, encoded on the EspC pathogenicity island (Mellies et al., 2001), also homologous to VirA, the *Shigella flexneri* effector protein, activates calpain (Elliott et al., 2001; Bergounioux et al., 2012). Recently, EspH has been shown to induce caspase-3 activation and this activity is counteracted by translocation of bacterial RhoGEFs EspT and EspM which override the cellular Rho GTPase signaling (Wong et al., 2012a,b).

Apart from secreting effector proteins that cause apoptosis, EPEC also secretes effectors that are anti-apoptotic (see subversion by the T3SS). The balance between these processes remains a relatively unexplored subject.

CLINICAL MANIFESTATIONS

Transmission

EPEC strains are transmitted via the fecal–oral route through direct contact, via contaminated food and water sources, or fomites acting as vehicles. Hospitals may also serve as a source of infection (Blake et al., 1993). In addition, one case study described the concurrent infection of a child and pet dog with the same human EPEC clone, indicating the possibility of household pets acting as reservoirs (Rodrigues et al., 2004). Outbreaks in nurseries and daycare are no longer common, but in those cases where disease had occurred, the source of infection was rarely identified. However, two outbreaks in Chongqing, China, affecting a total of 69 neonates, were sourced back to one mother, and were determined to have spread via medical staff's hands (Wu and Peng, 1992). In 2000, 45 Japanese students (ages 12–15 years) became infected with aEPEC whilst on a fieldwork program held at a farm. A contaminated water source was determined to be the vehicle for transmission, with a marked difference between rates of attack for those students who had drank the tap water (86.1%) and those who had washed their hands with the water (26.3%). This outbreak is an example in which EPEC may have been transmitted from animals via a water source to humans, as the provided water was from a brook which also served as a water source for farm animals (Yatsuyanagi et al., 2003). Conversely, an outbreak of diarrheal illness affecting >100 people in the US was linked to a single buffet serving uncooked and cooked food. Environmental health inspectors had observed extensive contact between uncooked seafood and eggs, and fresh fruits and vegetables (Hedberg et al., 1997). Person-to-person transmission was highlighted by a Finnish school aEPEC outbreak, where spread to associated households was observed in one third of those who had responded to the survey questionnaire (Viljanen et al., 1990).

Clinical features

EPEC infections often last between 5–15 days, but prolonged illness can occur (Lacroix et al., 1984; Hill et al., 1991). Diarrhea is the most common clinical presentation of EPEC, with recent studies suggesting an association between atypical strains and persistent diarrhea (Levine et al., 1978; Afset et al., 2004; Nguyen et al., 2006). Diarrhea may contain mucus but rarely blood. Greater than 90% of patients present with diarrhea, with additional features observed such as abdominal pain (80–90%), fever (20–60%), myalgia (20–30%), vomiting (10–20%), and nausea (30–60%) (Hedberg et al., 1997; Yatsuyanagi et al., 2003). In infants, dehydration is also common and severe cases may result in weight loss, malnutrition, and death (Levine, 1987; Donnenberg, 1995; Nataro and Kaper, 1998; Fagundes-Neto and Andrade, 1999).

Complications

Children with EPEC are more likely to fail to respond to rehydration therapy (Fagundes-Neto and Scaletsky, 2000). Food intolerance is also prevalent amongst infants <6 months (Fagundes-Neto et al., 1996). These factors greatly impact recovery from the chronic state and create a condition where mortality rates may be as high 50% (Rothbaum et al., 1983).

Diagnosis

Recent approaches towards diagnosing EPEC have been multifaceted; with most current studies now undertaking both phenotypic and molecular analyses to elucidate the pathotype of an infecting pathogen. In the past 15 years, there has been an exponential increase in the number of publications working towards the generation of an ideal set of genes with which to characterize EPEC strains. However, the increase in the genomic libraries of these strains has further increased our understanding of their genetic complexity. Hence, it appears that a dual approach incorporating both phenotyping and genotyping will remain the gold standard for some time. Furthermore, in many countries clinical laboratories rarely perform tests required for EPEC identification, which remains the purview of reference and research labs.

Traditionally EPEC strains were identified by their serotype. This method has proven to be erroneous, as not all described EPEC serotypes display localized adherence, or lack genes encoding Shiga toxins. Also, strains belonging to non-EPEC serotypes, upon more extensive characterization, have been found to display the current classifying features (see above). Presently, phenotypic features used to identify EPEC strains are more focused on the defining actions of the pathogen. The assessment of LA is a standard phenotypic diagnostic tool. Complications arise however, when this method is utilized to diagnose a strain that is atypical, as these strains do not exhibit typical LA (Vieira et al., 2001; Dulguer et al., 2003). Since EPEC induce the generation of actin-rich pedestals in host cells, fluorescent-actin staining (FAS) as a means of pedestal detection and a diagnostic technique has been included in a number of outbreak investigations and epidemiological studies of EPEC prevalence amongst food sources, livestock, and domestic animals (Knutton et al., 1991; Morelli et al., 1994; Scotland et al., 1996; Saridakis et al., 1997; Rosa et al., 1998; Nakazato et al., 2004; Carneiro et al., 2006). However, this method is limited by its inability to differentiate EPEC from EHEC.

The use of an EAF probe to detect and diagnose EPEC infection has diminished. This detection method required the infecting strain to possess the EAF plasmid and for that plasmid not to be altered in a way which might affect probe hybridization. Multiplex PCR approaches are now frequently utilized. These multi-gene detections are focused on the identification of specific virulence genes associated with EPEC classification, and those capable of distinguishing between closely related pathotypes. The *eae* or *escV* genes are typically used to identify EHEC, tEPEC, and aEPEC. Shiga toxin-producing strains are then excluded based on the presence of *stx* genes. To discriminate typical from atypical EPEC, multiplex PCR detect genes of the bundle forming pili (*bfp* genes) (Aranda et al., 2004; Fujioka et al., 2012). In addition to this method being the most comprehensive, the ability to amplify directly from fecal samples means that this method is also relatively quick (Barletta et al., 2011; Wiemer et al., 2011). Quantitative PCR has recently been proposed as a tool to determine symptomatic from asymptomatic infection as the former is associated with a higher pathogen burden (Barletta et al., 2011).

Treatment

EPEC infections are usually self-limiting. Paramount attention should be paid to correct for fluid and electrolyte loss and nutritional imbalances associated with diarrhea; rehydration therapy alone constitutes the primary treatment. The severity of the disease usually dictates the treatment, as oral rehydration may be sufficient for milder cases, while more severe cases will require the inclusion of parenteral rehydration or total parenteral nutrition (Donnenberg, 1995; Nataro and Kaper, 1998).

Whilst antibiotic treatment for EPEC is not the primary therapeutic, EPEC has been successfully cleared with antibiotics (Thorén et al., 1980; Hill et al., 1991). However, the emergence of antibiotic- and multi-drug-resistant strains has further complicated this means of treatment for severe cases. Resistance is more prevalent in typical EPEC than atypical, largely due to coding sequences residing on plasmids (Laporta et al., 1986; Scaletsky et al., 2010b). EPEC is most often resistant to ampicillin, streptomycin, tetracycline, and triple sulfa or its components (Laporta et al., 1986; Lim et al., 1992; Guerra et al., 2006; Garcia et al., 2011).

There are many existing and emerging therapies that directly target EPEC colonization. Mixed results have been obtained in studies in which EPEC-infected infants were treated by passive immunization with anti-EPEC bovine immunoglobulin concentrate (Mietens et al., 1979; Casswall et al., 2000). However, work on anti-EPEC antibody development continues. A group in Brazil detailed the development of IgY antibodies against the O-antigens of EPEC O111 and EHEC O111 and O157. While these antibodies were reactive against their respective strain and inhibited the growth of the pathogen, they remain to be tested clinically (Amaral et al., 2008). Extracts from medicinal plants may avoid the problems associated with antibiotic resistance but may be disadvantaged by their broad specificity. A bifunctional diaryl heptaniod isolated from *Alpinia officinarum*, displays bactericidal activity against EPEC and suppresses EPEC LPS-induced inflammation. However, it is predicted that the antimicrobial function is derived from its ability to interact with the A subunit of *E. coli* gyrase (Subramanian et al., 2009). Hence, the non-specificity of this potential drug may present some challenges in its use. Alkaloids from *Holarrhena antidysenterica* (AHA) seeds are antibacterial and antidiarrheal. AHA has been shown to inhibit EPEC in both disk diffusion and agar well diffusion assays, and to inhibit bacterial attachment to host epithelial cells (Kavitha and Niranjali, 2009). In addition, these compounds were shown to reduce EPEC-induced apoptosis (Kavitha and Niranjali, 2009). However, no assessment of AHA alkaloids on commensal strains has been performed. A novel approach towards biocontrol and treatment of both EHEC and EPEC was described with the identification of two coliphages (MVBS and MVSS) that demonstrate specificity against pathogenic strains of *E. coli* (Viscardi et al., 2008). A number of characteristics are specific to EPEC and related strains; these include auto-aggregation, intimate attachment and translocation of effector proteins. One approach targeting the T3SS with a linear polyketide compound inhibits secretion of EspB, EspF, and MAP by EPEC. This study also detailed improved survival using the *C. rodentium* murine model of EPEC infection (Kimura et al., 2011). Benefiting from the emergent field of miRNA, one study identified three miRNA that target tight junction proteins and indirectly modulate the junctional complexes. The levels of these miRNA were altered by the probiotic *E. coli* strain Nissle 1917 to enhance barrier integrity and may potentially be exploited for drug therapy in the future (Veltman et al., 2012).

Other indirect methods of treating EPEC infections include the use of probiotic bacterial strains and factors produced by them. Both the whole cells and the isolated surface-associated proteins (SAP) of *Lactobacillus fermentum* demonstrate inhibition of enteropathogenic bacterial attachment (Varma et al., 2010). The surface layer adhesive protein (SLAP) of *L. plantarum* also decreases EPEC adhesion, in addition to up-regulating the expression of TJ proteins (Liu et al., 2011). However, few clinical studies have been performed. One study using *Saccharomyces boulardii* proved this strategy to be of social and economic benefit, demonstrating slight improvement to disease progression. Children treated with *S. boulardii* and oral rehydration solution displayed a reduction in mean duration of diarrhea of 1.6 days compared with children who received oral rehydration solution alone (Htwe et al., 2008).

Immune response

Whilst volunteer studies have provided some information on the immune response to EPEC, much of our knowledge has been acquired from the murine model of A/E infection, which utilizes the murine EPEC homolog *C. rodentium*.

Human studies have supplied us with details on proteins that are antigenic and proteins that are required for the establishment of a strong humoral immune response. Volunteers challenged with E2348/69 generate antibodies specific to bundlin, intimin, and O127 LPS (Donnenberg et al., 1998; Fernandes et al., 2007). In a separate study, anti-LPS titer was demonstrated to be greatly decreased when the *eae* gene was deleted (Donnenberg et al., 1993a). Diagnostic tests on hospitalized infants validated bundlin as immunogenic and also described the presence of antibodies against EspB (de Souza Campos Fernandes et al., 2003).

An innate immune response to EPEC is initiated by various bacterial products. Flagellin alone has been described as an inducer of IL-8 secretion and gastroenteritis, mediated by the activation of the NF-κB pathway (Schüller et al., 2009). In addition, LPS, bundlin, EspA, and EspC have been shown to induce an inflammatory host response (Edwards et al., 2011). IL-8 secretion in response to EPEC infection is concordant with the previously observed recruitment of neutrophils and macrophages (Ulshen and Rollo, 1980). In mice infected with *C. rodentium*, localization of these cells is due to activation of the CXCR2 chemokine receptor (Spehlmann et al., 2009). In the Peyer's patch, M cells sample antigens from the intestinal lumen and present them to specialized antigen-presenting cells within the intraepithelial spaces of the intestine (Neutra et al., 1996) and may transport some pathogens across the intestinal barrier. Inhibition of EPEC transcytosis by M cells is T3SS-dependent and is likely due to EspF (Martinez-Argudo et al., 2007; Tahoun et al., 2011).

The adaptive response to *C. rodentium* is typically T_H1-mediated and involves the production of protective IgG antibodies (Higgins et al., 1999a,b; Bry and Brenner, 2004). Phagocytosis of *C. rodentium* by macrophages is significantly impaired by loss of the IgG receptor, Fcγ, highlighting the importance of this antibody isotype to A/E infection (Masuda et al., 2008). However, this mode of clearance may be inhibited by EPEC, as the effector EspH can inhibit Fcγ receptor-induced phagocytosis (Dong et al., 2010). A balance between a pro- and anti-immune response is acquired in a T3SS manner, where effector proteins translocated into the host cell cytoplasm modulate the host response and affect the activation of inflammatory mediators (see above) (Sharma et al., 2006).

Control and prevention

EPEC is transmitted via a fecal–oral route and hence the primary means of prevention include improved sanitation conditions and hygiene. However, there are other means by which disease onset may be prevented.

Vaccines

Along with EPEC, rotavirus infection represents a major cause of infant diarrhea. A rotavirus vaccine is now being introduced into global vaccination programs, and a push for the development of vaccines against ETEC and *Shigella* is underway. Natural progression would put forward EPEC as the next prioritized pathogen for vaccine development in pediatric disease. Early efforts using killed bacteria exhibited some efficacy (Kubinyi et al., 1974). Expression of an EPEC antigen by a bacterial vector and the elicitation of a sustained humoral response was described using bundlin as an antigenic component (Schriefer et al., 1999). Vaccines against O-antigens have also been described, with rabbits immunized with capsulated EHEC O111ac:H⁻ generating antibodies that reacted against all O111 *E. coli* tested, including EPEC (Santos et al., 2010). More recent projects have focused on the immunogenicity of intimin. Mice immunized with *L. casei* expressing the C-terminal fragment the β-intimin required for Tir binding generated specific antibodies that reacted against native intimin on the surface of EPEC and inhibited EPEC attachment to epithelial cells (Ferreira et al., 2008). In addition, C3H/HePas mice (a strain highly susceptible to *C. rodentium* infection) displayed significantly reduced mortality when immunized with the same recombinant *L. casei* prior to *C. rodentium* infection (Ferreira et al., 2011). The delivery of intimin protein via a *Vibrio cholerae* vector in a rabbit model further corroborates the use of this antigen. Rabbits immunized with this strain and infected with rabbit EPEC exhibited milder diarrhea, reduced weight loss, and reduced colonization by the pathogen (Keller et al., 2010). In addition, a chitosan nanoparticle packed with porcine IL-4 and IL-6 has been identified as a promising adjuvant, increasing the concentration and specificity of antibodies in the sera of mice vaccinated against *E. coli*, and this was shown to be protective against EPEC (Zhang et al., 2007).

Protective effects of breastfeeding

Numerous studies have reported the protective effects of breast-feeding. IgA, IgM, and IgG are found in mothers' colostrum, with secretory IgA most associated with EPEC prevention (Araujo et al., 2005). IgA specific to bundlin, intimin, EspA, EspB, EspC, EspF, Tir, and other EPEC proteins have been described (Cravioto et al., 1991; Adu-Bobie et al., 1998; Loureiro et al., 1998; Manjarrez-Hernandez et al., 2000; Parissi-Crivelli et al., 2000; Sanches et al., 2000). Antibody levels are highest in colostrum and decrease during lactation (Araujo et al., 2005). Severe morbidity and mortality are most often associated with non-breastfed or partially breastfed infants, and a decreasing odds ratio of dehydration is observed (Creek et al., 2010; Barletta et al., 2011). Other components of milk, notably oligosaccharides, are also capable of inhibiting or modifying EPEC adherence (Cravioto et al., 1991; Jagannatha et al., 1991; Idota and Kawakami, 1995). These findings support the importance of breast-feeding as an essential prophylactic measure in EPEC-endemic areas.

REFERENCES

Abe, A., Heczko, U., Hegele, R.G., Finlay, B.B., 1998. Two enteropathogenic *Escherichia coli* type III secreted proteins, EspA and EspB, are virulence factors. J. Exp. Med. 188, 1907–1916.

Abe, P., Kenny, B., Stein, M., Finlay, B.B., 1997. Characterization of two virulence proteins secreted by rabbit enteropathogenic *Escherichia coli*, EspA and EspB, whose maximal expression is sensitive to host body temperature. Infect. Immun. 65, 3547–3555.

Adu-Bobie, J., Trabulsi, L.R., Carneiro-Sampaio, M.M.S., Dougan, G., Frankel, G., 1998. Identification of immunodominant regions within the C-terminal cell binding domain of intimin α and intimin β from enteropathogenic *Escherichia coli*. Infect. Immun. 66, 5643–5649.

Afset, J.E., Bruant, G., Brousseau, R., et al., 2006. Identification of virulence genes linked with diarrhea due to atypical enteropathogenic *Escherichia coli* by DNA microarray analysis and PCR. J. Clin. Microbiol. 44, 3703–3711.

Afset, J.E., Bergh, K., Bevanger, L., 2003. High prevalence of atypical enteropathogenic *Escherichia coli* (EPEC) in Norwegian children with diarrhoea. J. Med. Microbiol. 52, 1015–1019.

Afset, J.E., Bevanger, L., Romundstad, P., Bergh, K., 2004. Association of atypical enteropathogenic *Escherichia coli* (EPEC) with prolonged diarrhoea. J. Med. Microbiol. 53, 1137–1144.

Albert, M.J., Faruque, S.M., Faruque, A.S.G., et al., 1995. Controlled study of *Escherichia coli* diarrheal infections in Bangladeshi children. J. Clin. Microbiol. 33, 973–977.

Albert, M.J., Faruque, S.M., Faruque, A.S.G., et al., 1996. Controlled study of cytolethal distending toxin-producing *Escherichia coli* infections in Bangladeshi children. J. Clin. Microbiol. 34, 717–719.

Alikhani, M.Y., Mirsalehian, A., Aslani, M.M., 2006. Detection of typical and atypical enteropathogenic *Escherichia coli* (EPEC) in Iranian children with and without diarrhoea. J. Med. Microbiol. 55, 1159–1163.

Alto, N.M., Shao, F., Lazar, C.S., et al., 2006. Identification of a bacterial type III effector family with G protein mimicry functions. Cell 124, 133–145.

Alto, N.M., Weflen, A.W., Rardin, M.J., et al., 2007. The type III effector EspF coordinates membrane trafficking by the spatiotemporal activation of two eukaryotic signaling pathways. J. Cell. Biol. 178, 1265–1278.

Amaral, J.A., De Franco, M.T., Zapata-Quintanilla, L., Carbonare, S.B., 2008. In vitro reactivity and growth inhibition of EPEC serotype O111 and STEC serotypes O111 and O157 by homologous and heterologous chicken egg yolk antibody. Vet. Res. Commun. 32, 281–290.

Andrade, A., Pardo, J.P., Espinosa, N., Perez-Hernández, G., González-Pedrajo, B., 2007. Enzymatic characterization of the enteropathogenic *Escherichia coli* type III secretion ATPase EscN. Arch. Biochem. Biophys. 468, 121–127.

Andrade, J.R., Da Veiga, V.F., De Santa Rosa, M.R., Suassuna, I., 1989. An endocytic process in HEp-2 cells induced by enteropathogenic *Escherichia coli*. J. Med. Microbiol. 28, 49–57.

Anantha, R.P., Stone, K.D., Donnenberg, M.S., 2000. Effects of *bfp* mutations on biogenesis of functional enteropathogenic *Escherichia coli* type IV pili. J. Bacteriol. 182, 2498–2506.

Aranda, K.R., Fagundes-Neto, U., Scaletsky, I.C., 2004. Evaluation of multiplex PCRs for diagnosis of infection with diarrheagenic *Escherichia coli* and *Shigella spp.* J. Clin. Microbiol. 42, 5849–5853.

Araujo, E.D., Carbonare, S.B., de Araujo, M.C., Palmeira, P., Amaral, J.A., Sales, V.S., 2005. Total and specific IgA in colostrum and milk of mothers of Natal-Rio Grande do Norte, Brasil Acta Cir. Bras. 20 (Suppl. 1), 178–184.

Arbeloa, A., Bulgin, R.R., MacKenzie, G., et al., 2008. Subversion of actin dynamics by EspM effectors of attaching and effacing bacterial pathogens. Cell Microbiol. 10, 1429–1441.

Arbeloa, A., Garnett, J., Lillington, J., et al., 2010. EspM2 is a RhoA guanine nucleotide exchange factor. Cell Microbiol. 12, 654–664.

Arbeloa, A., Oates, C.V., Marchés, O., Hartland, E.L., Frankel, G., 2011. Enteropathogenic and enterohemorrhagic *Escherichia coli* type III secretion effector EspV induces radical morphological changes in eukaryotic cells. Infect. Immun. 79, 1067–1076.

Atlung, T., Ingmer, H., 1997. H-NS: a modulator of environmentally regulated gene expression. Mol. Microbiol. 24, 7–17.

Babbin, B.A., Sasaki, M., Gerner-Schmidt, K.W., Nusrat, A., Klapproth, J.M., 2009. The bacterial virulence factor lymphostatin compromises intestinal epithelial barrier function by modulating rho GTPases. Am. J. Pathol. 174, 1347–1357.

Badea, L., Doughty, S., Nicholls, L., Sloan, J., Robins-Browne, R.M., Hartland, E.L., 2003. Contribution of Efa1/LifA to the adherence of enteropathogenic *Escherichia coli* to epithelial cells. Microb. Pathog. 34, 205–215.

Baldini, M.M., Kaper, J.B., Levine, M.M., Candy, D.C., Moon, H.W., 1983. Plasmid-mediated adhesion in enteropathogenic *Escherichia coli*. J. Pediatr. Gastroenterol. Nutr. 2, 534–538.

Barletta, F., Ochoa, T.J., Mercado, E., et al., 2011. Quantitative real-time polymerase chain reaction for enteropathogenic *Escherichia coli*: a tool for investigation of asymptomatic versus symptomatic infections. Clin. Infect. Dis. 53, 1223–1229.

Barnett, F.D., Abul-Milh, M., Huesca, M., Lingwood, C.A., 2000. Enterohemorrhagic *Escherichia coli* induces apoptosis which augments bacterial binding and phosphatidylethanolamine exposure on the plasma membrane outer leaflet. Infect. Immun. 68, 3108–3115.

Baruch, K., Gur-Arie, L., Nadler, C., et al., 2011. Metalloprotease type III effectors that specifically cleave JNK and NF-kappaB. EMBO J. 30, 221–231.

Behiry, I.K., Abada, E.A., Ahmed, E.A., Labeeb, R.S., 2011. Enteropathogenic *Escherichia coli* associated with diarrhea in children in Cairo. Egypt Scientific World J. 11, 2613–2619.

Beinke, C., Laarmann, S., Wachter, C., Karch, H., Greune, L., Schmidt, M.A., 1998. Diffusely adhering *Escherichia coli* strains induce attaching and effacing phenotypes and secrete homologs of Esp proteins. Infect. Immun. 66, 528–539.

Benya, R.V., Marrero, J.A., Ostrovskiy, D.A., Koutsouris, A., Hecht, G., 1999. Human colonic epithelial cells express galanin-1 receptors, which when activated cause Cl- secretion. Am. J. Physiol. 276, G64–G72.

Berdichevsky, T., Friedberg, D., Nadler, C., Rokney, A., Oppenheim, A., Rosenshine, I., 2005. Ler is a negative autoregulator of the LEE1 operon in enteropathogenic *Escherichia coli*. J. Bacteriol. 187, 349–357.

Berger, C.N., Crepin, V.F., Jepson, M.A., Arbeloa, A., Frankel, G., 2009. The mechanisms used by enteropathogenic *Escherichia coli* to control filopodia dynamics. Cell Microbiol. 11, 309–322.

Bergounioux, J., Elisee, R., Prunier, A.L., et al., 2012. Calpain activation by the *Shigella flexneri* effector VirA regulates key steps in the formation and life of the bacterium's epithelial niche. Cell Host. Microbe. 11, 240–252.

Bieber, D., Ramer, S.W., Wu, C.Y., et al., 1998. Type IV pili, transient bacterial aggregates, and virulence of enteropathogenic *Escherichia coli*. Science 280, 2114–2118.

Blake, P.A., Ramos, S., Macdonald, K.L., et al., 1993. Pathogen-specific risk factors and protective factors for acute diarrheal disease in urban Brazilian infants. J. Infect. Dis. 167, 627–632.

Blanco, M., Schumacher, S., Tasara, T., et al., 2005. Serotypes, intimin variants and other virulence factors of eae positive *Escherichia coli* strains isolated from healthy cattle in Switzerland. Identification of a new intimin variant gene (eae-η2). BMC Microbiol. 5, 23.

Blank, T.E., Donnenberg, M.S., 2001. Novel topology of BfpE, a cytoplasmic membrane protein required for type IV fimbrial biogenesis in enteropathogenic *Escherichia coli*. J. Bacteriol. 183, 4435–4450.

Blasche, S., Mörtl, M., Steuber, H., et al., 2013. The *E. coli* effector protein NleF is a caspase inhibitor. PLoS ONE 8, e58937.

Bommarius, B., Maxwell, D., Swimm, A., et al., 2007. Enteropathogenic *Escherichia coli* Tir is an SH2/3 ligand that recruits and activates tyrosine kinases required for pedestal formation. Mol. Microbiol. 63, 1748–1768.

Bradley, D.E., Thompson, C.R., 1992. Synthesis of unusual thick pili by *Escherichia coli* of EPEC serogroup O119. FEMS Microbiol. Lett. 94, 31–36.

Bray, J., 1945. Isolation of antigenically homogeneous strains of *Bact. coli neapolitanum* from summer diarrhoea of infants. J. Pathol. Bacteriol. 57, 239–247.

Brinkley, C., Burland, V., Keller, R., et al., 2006. Nucleotide sequence analysis of the enteropathogenic *Escherichia coli* adherence factor plasmid pMAR7. Infect. Immun. 74, 5408–5413.

Bry, L., Brenner, M.B., 2004. Critical role of T cell-dependent serum antibody, but not the gut-associated lymphoid tissue, for surviving acute mucosal infection with *Citrobacter rodentium*, an attaching and effacing pathogen. J. Immunol. 172, 433–441.

Bulgin, R.R., Arbeloa, A., Chung, J.C., Frankel, G., 2009. EspT triggers formation of lamellipodia and membrane ruffles through activation of Rac-1 and Cdc42. Cell Microbiol. 11, 217–229.

Bustamante, V.H., Calva, E., Puente, J.L., 1998. Analysis of *cis*-acting elements required for *bfpA* expression in enteropathogenic *Escherichia coli*. J. Bacteriol. 180, 3013–3016.

Bustamante, V.H., Santana, F.J., Calva, E., Puente, J.L., 2001. Transcriptional regulation of type III secretion genes in enteropathogenic *Escherichia coli*: Ler antagonizes H-NS-dependent repression. Mol. Microbiol. 39, 664–678.

Bustamante, V.H., Villalba, M.I., Garcia-Angulo, V.A., et al., 2011. PerC and GrlA independently regulate Ler expression in enteropathogenic *Escherichia coli*. Mol. Microbiol. 82, 398–415.

Campellone, K.G., Giese, A., Tipper, D.J., Leong, J.M., 2002. A tyrosine-phosphorylated 12-amino-acid sequence of enteropathogenic *Escherichia coli* Tir binds the host adaptor protein Nck and is required for Nck localization to actin pedestals. Mol. Microbiol. 43, 1227–1241.

Campellone, K.G., Leong, J.M., 2005. Nck-independent actin assembly is mediated by two phosphorylated tyrosines within enteropathogenic *Escherichia coli* Tir. Mol. Microbiol. 56, 416–432.

Campellone, K.G., Robbins, D., Leong, J.M., 2004. EspF$_U$ is a translocated EHEC effector that interacts with Tir and N-WASP and promotes Nck-independent actin assembly. Dev. Cell 7, 217–228.

Campos, L.C., Whittam, T.S., Gomes, T.A.T., Andrade, J.R.C., Trabulsi, L.R., 1994. *Escherichia coli* serogroup O111 includes several clones of diarrheagenic strains with different virulence properties. Infect. Immun. 62, 3282–3288.

Canil, C., Rosenshine, I., Ruschkowski, S., Donnenberg, M.S., Kaper, J.B., Finlay, B.B., 1993. Enteropathogenic *Escherichia coli* decreases the transepithelial electrical resistance of polarized epithelial monolayers. Infect. Immun. 61, 2755–2762.

Cantarelli, V.V., Takahashi, A., Yanagihara, I., et al., 2001. Talin, a host cell protein, interacts directly with the translocated intimin receptor, Tir, of enteropathogenic *Escherichia coli*, and is essential for pedestal formation. Cell Microbiol. 3, 745–751.

Carneiro, L.A.M., Lins, M.C., Garcia, F.R.A., et al., 2006. Phenotypic and genotypic characterisation of *Escherichia coli* strains serogrouped as enteropathogenic *E. coli* (EPEC) isolated from pasteurised milk. Int. J. Food Microbiol. 108, 15–21.

Casswall, T.H., Sarker, S.A., Faruque, S.M., et al., 2000. Treatment of enterotoxigenic and enteropathogenic *Escherichia coli*-induced diarrhoea in children with bovine immunoglobulin milk concentrate from hyperimmunized cows: a double-blind, placebo-controlled, clinical trial. Scand. J. Gastroenterol. 35, 711–718.

Charpentier, X., Oswald, E., 2004. Identification of the secretion and translocation domain of the enteropathogenic and enterohemorrhagic *Escherichia coli* effector Cif, using TEM-1 beta-lactamase as a new fluorescence-based reporter. J. Bacteriol. 186, 5486–5495.

Chart, H., Scotland, S.M., Willshaw, G.A., Rowe, B., 1988. HEp-2 adhesion and the expression of a 94 kDa outer-membrane protein by strains of *Escherichia coli* belonging to enteropathogenic serogroups. J. Gen. Microbiol. 134, 1315–1321.

Chatkaeomorakot, A., Echeverria, P., Taylor, D.N., et al., 1987. HeLa cell-adherent *Escherichia coli* in children with diarrhea in Thailand. J. Infect. Dis. 156, 669–672.

Chen, Z.J., 2005. Ubiquitin signalling in the NF-κB pathway. Nat. Cell Biol. 7, 758–765.

Clausen, C.R., Christie, D.L., 1982. Chronic diarrhea in infants caused by adherent enteropathogenic *Escherichia coli*. J. Pediatr. 100, 358–361.

Cleary, J., Lai, L.-C., Donnenberg, M.S., Frankel, G., Knutton, S., 2004. Enteropathogenic *E. coli* (EPEC) adhesion to intestinal epithelial cells: role of bundle-forming pili (BFP), EspA filaments and intimin. Microbiology 150, 527–538.

Collington, G.K., Booth, I.W., Donnenberg, M.S., Kaper, J.B., Knutton, S., 1998a. Enteropathogenic *Escherichia coli* virulence genes encoding secreted signalling proteins are essential for modulation of Caco-2 cell electrolyte transport. Infect. Immun. 66, 6049–6053.

Collington, G.K., Booth, I.W., Knutton, S., 1998b. Rapid modulation of electrolyte transport in Caco-2 cell monolayers by enteropathogenic *Escherichia coli* (EPEC) infection. Gut. 42, 200–207.

Costa-Carvalho, B.T., Bertipaglia, A., Solé, D., Naspitz, C.K., Scaletsky, I.C.A., 1994. Detection of immunoglobulin (IgG and IgA) anti-outer-membrane proteins of enteropathogenic *Escherichia coli* (EPEC) in saliva, colostrum, breast milk, serum, cord blood and amniotic fluid. Study of inhibition of localized adherence of EPEC to HeLa cells. Acta Paediatr. 83, 870–873.

Crane, J.K., McNamara, B.P., Donnenberg, M.S., 2001. Role of EspF in host cell death induced by enteropathogenic *Escherichia coli*. Cell Microbiol. 3, 197–211.

Cravioto, A., Reyes, R., Ortega, R., Fernández, G., Hernández, R., López, D., 1988. Prospective study of diarrhoeal disease in a cohort of rural Mexican children: incidence and isolated pathogens during the first two years of life. Epidemiol. Infect. 101, 123–134.

Cravioto, A., Reyes, R.E., Trujillo, F., et al., 1990. Risk of diarrhea during the first year of life associated with initial and subsequent colonization by specific enteropathogens. Am. J. Epidemiol. 131, 886–904.

Cravioto, A., Tello, A., Villafán, H., Ruiz, J., Del Vedovo, S., Neeser, J.-R., 1991. Inhibition of localized adhesion of enteropathogenic *Escherichia coli* to HEp-2 cells by immunoglobulin and oligosaccharide fractions of human colostrum and breast milk. J. Infect. Dis. 163, 1247–1255.

Creek, T.L., Kim, A., Lu, L., et al., 2010. Hospitalization and mortality among primarily nonbreastfed children during a large outbreak of diarrhea and malnutrition in Botswana, 2006. J. Acquir. Immune Defic. Syndr. 53, 14–19.

da Silva, M.L., Mortara, R.A., Barros, H.C., de Souza, W., Trabulsi, L.R., 1989. Aggregation of membrane-associated actin filaments following localized adherence of enteropathogenic *Escherichia coli* to HeLa cells. J. Cell Sci. 93, 439–446.

De Grado, M., Abe, A., Gauthier, A., Steele-Mortimer, O., DeVinney, R., Finlay, B.B., 1999. Identification of the intimin-binding domain of Tir of enteropathogenic *Escherichia coli*. Cell Microbiol. 1, 7–17.

de Souza Campos Fernandes, R.C., Quintana Flores, V.M., Sousa de Macedo, Z., Medina-Acosta, E., 2003. Coproantibodies to the enteropathogenic *Escherichia coli* vaccine candidates BfpA and EspB in breastfed and artificially fed children. Vaccine 21, 1725–1731.

Dean, P., Kenny, B., 2009. The effector repertoire of enteropathogenic *E. coli*: ganging up on the host cell. Curr. Opin. Microbiol. 12, 101–109.

Dean, P., Maresca, M., Schuller, S., Phillips, A.D., Kenny, B., 2006. Potent diarrheagenic mechanism mediated by the cooperative action of three enteropathogenic *Escherichia coli*-injected effector proteins. Proc. Natl. Acad. Sci. USA 103, 1876–1881.

Dean, P., Muhlen, S., Quitard, S., Kenny, B., 2010a. The bacterial effectors EspG and EspG2 induce a destructive calpain activity that is kept in check by the co-delivered Tir effector. Cell Microbiol. 12, 1308–1321.

Dean, P., Scott, J.A., Knox, A.A., Quitard, S., Watkins, N.J., Kenny, B., 2010b. The enteropathogenic *E. coli* effector EspF targets and disrupts the nucleolus by a process regulated by mitochondrial dysfunction. PLoS Pathog. 6, e1000961.

Deng, W., Puente, J.L., Gruenheid, S., et al., 2004. Dissecting virulence: Systematic and functional analyses of a pathogenicity island. Proc. Natl. Acad. Sci. USA 101, 3597–3602.

Deng, W., Yu, H.B., de Hoog, C.L., et al., 2012. Quantitative proteomic analysis of type III secretome of enteropathogenic *Escherichia coli* reveals an expanded effector repertoire for attaching/effacing bacterial pathogens. Mol. Cell Proteomics. 11, 692–709.

Do Valle, G.R.F., Gomes, T.A.T., Irino, K., Trabulsi, L.R., 1997. The traditional enteropathogenic *Escherichia coli* (EPEC) serogroup O125 comprises serotypes which are mainly associated with the category of enteroaggregative *E-coli*. FEMS Microbiol. Lett. 152, 95–100.

Dong, N., Liu, L., Shao, F., 2010. A bacterial effector targets host DH-PH domain RhoGEFs and antagonizes macrophage phagocytosis. EMBO J. 29, 1363–1376.

Donnenberg, M.S., 1995. Enteropathogenic *Escherichia coli*. In: Blaser, M.J., Smith, P.D., Ravdin, J.I., Greenberg, H.B., Guerrant, R.L. (Eds.), Infections of the gastrointestinal tract, Raven Press, Ltd, New York, pp. 709–726.

Donnenberg, M.S., Calderwood, S.B., Donohue-Rolfe, A., Keusch, G.T., Kaper, J.B., 1990a. Construction and analysis of Tn*phoA* mutants of enteropathogenic *Escherichia coli* unable to invade HEp-2 cells. Infect. Immun. 58, 1565–1571.

Donnenberg, M.S., Donohue-Rolfe, A., Keusch, G.T., 1989. Epithelial cell invasion: an overlooked property of enteropathogenic *Escherichia coli* (EPEC) associated with the EPEC adherence factor. J. Infect. Dis. 160, 452–459.

Donnenberg, M.S., Donohue-Rolfe, A., Keusch, G.T., 1990b. A comparison of HEp-2 cell invasion by enteropathogenic and enteroinvasive *Escherichia coli*. FEMS. Microbiol. Lett. 57, 83–86.

Donnenberg, M.S., Girón, J.A., Nataro, J.P., Kaper, J.B., 1992a. A plasmid-encoded type IV fimbrial gene of enteropathogenic *Escherichia coli* associated with localized adherence. Mol. Microbiol. 6, 3427–3437.

Donnenberg, M.S., Kaper, J.B., 1992. Minireview: Enteropathogenic *Escherichia coli*. Infect. Immun. 60, 3953–3961.

Donnenberg, M.S., Lai, L.C., Taylor, K.A., 1997. The locus of enterocyte effacement pathogenicity island of enteropathogenic *Escherichia coli* encodes secretion functions and remnants of transposons at its extreme right end. Gene 184, 107–114.

Donnenberg, M.S., Tacket, C.O., James, S.P., et al., 1993a. The role of the *eaeA* gene in experimental enteropathogenic *Escherichia coli* infection. J. Clin. Invest. 92, 1412–1417.

Donnenberg, M.S., Tacket, C.O., Losonsky, G., et al., 1998. Effect of prior experimental human enteropathogenic *Escherichia coli* infection on illness following homologous and heterologous rechallenge. Infect. Immun. 66, 52–58.

Donnenberg, M.S., Tacket, C.O., Losonsky, G., Nataro, J.P., Kaper, J.B., Levine, M.M., 1992b. The role of the *eae* gene in experimental human enteropathogenic *Escherichia coli* (EPEC) infection. Clin. Res. 40, 214A.

Donnenberg, M.S., Tzipori, S., McKee, M., O'Brien, A.D., Alroy, J., Kaper, J.B., 1993b. The role of the *eae* gene of enterohemorrhagic *Escherichia coli* in intimate attachment in vitro and in a porcine model. J. Clin. Invest. 92, 1418–1424.

Donnenberg, M.S., Yu, J., Kaper, J.B., 1993c. A second chromosomal gene necessary for intimate attachment of enteropathogenic *Escherichia coli* to epithelial cells. J. Bacteriol. 175, 4670–4680.

Dorman, C.J., 1992. The VirF protein from *Shigella flexneri* is a member of the AraC transcription factor superfamily and is highly homologous to Rnsa positive regulator of virulence genes in enterotoxigenic *Escherichia coli*. Mol. Microbiol. 6, 1575.

Dulguer, M.V., Fabbricotti, S.H., Bando, S.Y., Moreira-Filho, C.A., Fagundes-Neto, U., Scaletsky, I.C., 2003. Atypical enteropathogenic *Escherichia coli* strains: phenotypic and genetic profiling reveals a strong association between enteroaggregative *E. coli* heat-stable enterotoxin and diarrhea. J. Infect. Dis. 188, 1685–1694.

Edwards, L.A., Bajaj-Elliott, M., Klein, N.J., Murch, S.H., Phillips, A.D., 2011. Bacterial-epithelial contact is a key determinant of host innate immune responses to enteropathogenic and enteroaggregative *Escherichia. coli*. PLoS ONE 6, e27030.

Elliott, S.J., Kaper, J.B., 1997. Role of type 1 fimbriae in EPEC infections. Microb. Pathog. 23, 113–118.

Elliott, S.J., Krejany, E.O., Mellies, J.L., Robins-Browne, R.M., Sasakawa, C., Kaper, J.B., 2001. EspG, a novel type III secreted protein from enteropathogenic *E.coli* with similarities to VirA of *Shigella*. Infect. Immun. 69, 4027–4033.

Elliott, S.J., Sperandio, V., Giron, J.A., et al., 2000. The locus of enterocyte effacement (LEE)-encoded regulator controls expression of both LEE- and non-LEE-encoded virulence factors in enteropathogenic and enterohemorrhagic *Escherichia coli*. Infect. Immun. 68, 6115–6126.

Elliott, S.J., Wainwright, L.A., McDaniel, T.K., et al., 1998. The complete sequence of the locus of enterocyte effacement (LEE) of enteropathogenic *E. coli* E2348/69. Mol. Microbiol. 28, 1–4.

Elmore, S., 2007. Apoptosis: a review of programmed cell death. Toxicol. Pathol. 35, 495–516.

Fagundes-Neto, U., Andrade, J.A.B., 1999. Acute diarrhea and malnutrition: lethality risk in hospitalized infants. J. Am. Coll. Nutr. 18, 303–308.

Fagundes-Neto, U., Freymuller, E., Gatti, M.S.V., Schmitz, L.G., Scaletsky, I., 1995. Enteropathogenic *Escherichia coli* O111ab:H2 penetrates the small bowel epithelium in an infant with acute diarrhoea. Acta Paediatr. 84, 453–455.

Fagundes-Neto, U., Scaletsky, I.C.A., 2000. The gut at war: the consequences of enteropathogenic *Escherichia coli* infection as a factor of diarrhea and malnutrition. Sao Paulo Med. J. 118, 21–29.

Fagundes-Neto, U., Schmitz, L.G., Scaletsky, I., 1996. Acute diarrhea due to enteropathogenic *Escherichia coli*: Epidemiological and clinical features in Brasília. Brazil International J. Infect. Dis. 1, 65–69.

Fernandes, P.J., Guo, Q., Donnenberg, M.S., 2007. Functional consequences of sequence variation in bundlin, the enteropathogenic *Escherichia coli* type IV pilin protein. Infect. Immun. 75, 4687–4696.

Ferreira, P.C.D., Campos, I.B., Abe, C.M., et al., 2008. Immunization of mice with *Lactobacillus casei* expressing intimin fragments produces antibodies able to inhibit the adhesion of enteropathogenic *Escherichia coli* to cultivated epithelial cells. FEMS Immunol. Med. Microbiol. 54, 245–254.

Ferreira, P.C.D., da Silva, J.B., Piazza, R.M.F., Eckmann, L., Ho, P.L., Oliveira, M.L.S., 2011. Immunization of mice with *Lactobacillus casei* expressing a beta-intimin fragment reduces intestinal colonization by *Citrobacter rodentium*. Clin. Vaccine Immunol. 18, 1823–1833.

Finlay, B.B., Rosenshine, I., Donnenberg, M.S., Kaper, J.B., 1992. Cytoskeletal composition of attaching and effacing lesions associated with enteropathogenic *Escherichia coli* adherence to HeLa cells. Infect. Immun. 60, 2541–2543.

Fitzhenry, R.J., Pickard, D.J., Hartland, E.L., et al., 2002. Intimin type influences the site of human intestinal mucosal colonisation by enterohaemorrhagic *Escherichia coli* O157:H7. Gut. 50, 180–185.

Foster, D.B., Philpott, D., Abul-Milh, M., Huesca, M., Sherman, P.M., Lingwood, C.A., 1999. Phosphatidylethanolamine recognition promotes enteropathogenic *E. coli* and enterohemorrhagic *E. coli* host cell attachment. Microb. Pathog. 27, 289–301.

Foubister, V., Rosenshine, I., Donnenberg, M.S., Finlay, B.B., 1994a. The *eaeB* gene of enteropathogenic *Escherichia coli* is necessary for signal transduction in epithelial cells. Infect. Immun. 62, 3038–3040.

Foubister, V., Rosenshine, I., Finlay, B.B., 1994b. A diarrheal pathogen, enteropathogenic *Escherichia coli* (EPEC), triggers a flux of inositol phosphates in infected epithelial cells. J. Exp. Med. 179, 993–998.

Francis, C.L., Jerse, A.E., Kaper, J.B., Falkow, S., 1991. Characterization of interactions of enteropathogenic *Escherichia coli* O127:H6 with mammalian cells in vitro. J. Infect. Dis. 164, 693–703.

Frankel, G., Phillips, A.D., Novakova, M., et al., 1996. Intimin from enteropathogenic *Escherichia coli* restores murine virulence to a *Citrobacter rodentiumeaeA* mutant: induction of an immunoglobulin A response to intimin and EspB. Infect. Immun. 64, 5315–5325.

Friedberg, D., Umanski, T., Fang, Y.A., Rosenshine, I., 1999. Hierarchy in the expression of the locus of enterocyte effacement genes of enteropathogenic *Escherichia coli*. Mol. Microbiol. 34, 941–952.

Fujioka, M., Otomo, Y., Ahsan, C.R., 2012. A novel single-step multiplex polymerase chain reaction assay for the detection of diarrheagenic *Escherichia coli*. J. Microbiol. Methods

Gao, X., Wan, F., Mateo, K., et al., 2009. Bacterial effector binding to ribosomal protein s3 subverts NF-kappaB function. PLoS Pathog. 5, e1000708.

Garcia, P.G., Silva, V.L., Diniz, C.G., 2011. Occurrence and antimicrobial drug susceptibility patterns of commensal and diarrheagenic *Escherichia coli* in fecal microbiota from children with and without acute diarrhea. J. Microbiol. 49, 46–52.

Gauthier, A., De Grado, M., Finlay, B.B., 2000. Mechanical fractionation reveals structural requirements for enteropathogenic *Escherichia coli* Tir insertion into host membranes. Infect. Immun. 68, 4344–4348.

Gawenis, L.R., Stien, X., Shull, G.E., et al., 2002. Intestinal NaCl transport in NHE2 and NHE3 knockout mice. Am. J. Physiol. Gastrointest Liver Physiol. 282, G776–G784.

Gill, R.K., Borthakur, A., Hodges, K., et al., 2007. Mechanism underlying inhibition of intestinal apical Cl⁻/OH⁻ exchange following infection with enteropathogenic *Escherichia coli*. J. Clin. Invest. 117, 428–437.

Girón, J.A., Ho, A.S.Y., Schoolnik, G.K., 1991. An inducible bundle-forming pilus of enteropathogenic *Escherichia coli*. Science 254, 710–713.

Girón, J.A., Ho, A.S.Y., Schoolnik, G.K., 1993. Characterization of fimbriae produced by enteropathogenic *Escherichia coli*. J. Bacteriol. 175, 7391–7403.

Girón, J.A., Torres, A.G., Freer, E., Kaper, J.B., 2002. The flagella of enteropathogenic *Escherichia coli* mediate adherence to epithelial cells. Mol. Microbiol. 44, 361–379.

Gomes, T.A.T., Rassi, V., Macdonald, K.L., et al., 1991. Enteropathogens associated with acute diarrheal disease in urban infants in São Paulo. Brazil. J. Infect. Dis. 164, 331–337.

Gómez-Duarte, O.G., Kaper, J.B., 1995. A plasmid-encoded regulatory region activates chromosomal eaeA expression in enteropathogenic *Escherichia coli*. Infect. Immun. 63, 1767–1776.

Goosney, D.L., DeVinney, R., Finlay, B.B., 2001. Recruitment of cytoskeletal and signaling proteins to enteropathogenic and enterohemorrhagic *Escherichia coli* pedestals. Infect. Immun. 69, 3315–3322.

Gruenheid, S., DeVinney, R., Bladt, F., et al., 2001. Enteropathogenic *E. coli* Tir binds Nck to initiate actin pedestal formation in host cells. Nat. Cell Biol. 3, 856–859.

Guerra, B., Junker, E., Schroeter, A., Helmuth, R., Guth, B.E.C., Beutin, L., 2006. Phenotypic and genotypic characterization of antimicrobial resistance in *Escherichia coli* O111 isolates. J. Antimicrob. Chemother. 57, 1210–1214.

Guttman, J.A., Samji, F.N., Li, Y., Deng, W., Lin, A., Finlay, B.B., 2007. Aquaporins contribute to diarrhoea caused by attaching and effacing bacterial pathogens. Cell Microbiol. 9, 131–141.

Haigh, R., Baldwin, T., Knutton, S., Williams, P.H., 1995. Carbon dioxide regulated secretion of the EaeB protein of enteropathogenic *Escherichia coli*. FEMS Microbiol. Lett. 129, 63–67.

Hales, B.A., Hart, C.A., Batt, R.M., Saunders, J.R., 1992. The large plasmids found in enterohemorrhagic and enteropathogenic *Escherichia coli* constitute a related series of transfer-defective Inc F-IIA replicons. Plasmid 28, 183–193.

Harhaj, N.S., Antonetti, D.A., 2004. Regulation of tight junctions and loss of barrier function in pathophysiology. Int. J. Biochem. Cell Biol. 36, 1206–1237.

Hartland, E.L., Batchelor, M., Delahay, R.M., et al., 1999. Binding of intimin from enteropathogenic *Escherichia coli* to Tir and to host cells. Mol. Microbiol. 32, 151–158.

Hecht, G., Koutsouris, A., 1999. Enteropathogenic *E. coli* attenuates secretagogue-induced net intestinal ion transport but not Cl- secretion. Am. J. Physiol. Gastrointes. Liver Physiol. 276, G781–G788.

Hecht, G., Marrero, J.A., Danilkovich, A., et al., 1999. Pathogenic *Escherichia coli* increase Cl- secretion from intestinal epithelia by upregulating galanin-1 receptor expression. J. Clin. Invest. 104, 253–262.

Hedberg, C.W., Savarino, S.J., Besser, J.M., et al., 1997. An outbreak of foodborne illness caused by *Escherichia coli* O39:NM, an agent not fitting into the existing scheme for classifying diarrheogenic *E. coli*. J. Infect. Dis. 176, 1625–1628.

Hemrajani, C., Berger, C.N., Robinson, K.S., Marches, O., Mousnier, A., Frankel, G., 2010. NleH effectors interact with Bax inhibitor-1 to block apoptosis during enteropathogenic *Escherichia coli* infection. Proc. Natl. Acad. Sci. USA 107, 3129–3134.

Henderson, I.R., Nataro, J.P., 2001. Virulence functions of autotransporter proteins. Infect. Immun. 69, 1231–1243.

Hernandes, R.T., Silva, R.M., Carneiro, S.M., et al., 2008. The localized adherence pattern of an atypical enteropathogenic *Escherichia coli* is mediated by intimin omicron and unexpectedly promotes HeLa cell invasion. Cell Microbiol. 10, 415–425.

Hernandes, R.T., Velsko, I., Sampaio, S.C., et al., 2011. Fimbrial adhesins produced by atypical enteropathogenic *Escherichia coli* strains. Appl. Environ. Microbio. 77, 8391–8399.

Hicks, S., Frankel, G., Kaper, J.B., Dougan, G., Phillips, A.D., 1998. Role of intimin and bundle-forming pili in enteropathogenic *Escherichia coli* adhesion to pediatric intestinal tissue in vitro. Infect. Immun. 66, 1570–1578.

Higgins, L.M., Frankel, G., Connerton, I., Gonçalves, N.S., Dougan, G., MacDonald, T.T., 1999a. Role of bacterial intimin in colonic hyperplasia and inflammation. Science 285, 588–591.

Higgins, L.M., Frankel, G., Douce, G., Dougan, G., MacDonald, T.T., 1999b. *Citrobacter rodentium* infection in mice elicits a mucosal Th1 cytokine response and lesions similar to those in murine inflammatory bowel disease. Infect. Immun. 67, 3031–3039.

Hill, S.M., Phillips, A.D., Walker-Smith, J.A., 1991. Enteropathogenic *Escherichia coli* and life threatening chronic diarrhoea. Gut. 32, 154–158.

Hodges, K., Alto, N.M., Ramaswamy, K., Dudeja, P.K., Hecht, G., 2008. The enteropathogenic *Escherichia coli* effector protein EspF decreases sodium hydrogen exchanger 3 activity. Cell Microbiol. 10, 1735–1745.

Htwe, K., Yee, K.S., Tin, M., Vandenplas, Y., 2008. Effect of *Saccharomyces boulardii* in the treatment of acute watery diarrhea in myanmar children: a randomized controlled study. Am. J. Trop. Med. Hyg. 78, 214–216.

Huang, Z., Sutton, S.E., Wallenfang, A.J., et al., 2009. Structural insights into host GTPase isoform selection by a family of bacterial GEF mimics. Nat. Struct. Mol. Biol. 16, 853–860.

Humphries, R.M., Griener, T.P., Vogt, S.L., et al., 2010. N-acetyllactosamine-induced retraction of bundle-forming pili regulates virulence-associated gene expression in enteropathogenic *Escherichia coli*. Mol. Microbiol. 76, 1111–1126.

Huys, G., Cnockaert, M., Janda, J.M., Swings, J., 2003. *Escherichia albertii* sp. nov., a diarrhoeagenic species isolated from stool specimens of Bangladeshi children. Int. J. Syst. Evol. Microbiol. 53, 807–810.

Hwang, J., Bieber, D., Ramer, S.W., Wu, C.Y., Schoolnik, G.K., 2003. Structural and topographical studies of the type IV bundle-forming pilus assembly complex of enteropathogenic *Escherichia coli*. J. Bacteriol. 185, 6695–6701.

Hyland, R.M., Griener, T.P., Mulvey, G.L., et al., 2006. Basis for N-acetyllactosamine-mediated inhibition of enteropathogenic *Escherichia coli* localized adherence. J. Med. Microbiol. 55, 669–675.

Hyland, R.M., Sun, J., Griener, T.P., et al., 2008. The bundlin pilin protein of enteropathogenic *Escherichia coli* is an N-acetyllactosamine-specific lectin. Cell Microbiol. 10, 177–187.

Ibarra, J.A., Villalba, M.I., Puente, J.L., 2003. Identification of the DNA binding sites of PerA, the transcriptional activator of the *bfp* and *per* operons in enteropathogenic *Escherichia coli*. J. Bacteriol. 185, 2835–2847.

Ide, T., Laarmann, S., Greune, L., Schillers, H., Oberleithner, H., Schmidt, M.A., 2001. Characterization of translocation pores inserted into plasma membranes by type III-secreted Esp proteins of enteropathogenic *Escherichia coli*. Cell Microbiol. 3, 669–679.

Idota, T., Kawakami, H., 1995. Inhibitory effects of milk gangliosides on the adhesion of *Escherichia coli* to human intestinal carcinoma cells. Biosci. Biotechnol. Biochem. 59, 69–72.

Iguchi, A., Thomson, N.R., Ogura, Y., et al., 2009. Complete genome sequence and comparative genome analysis of enteropathogenic *Escherichia coli* O127:H6 strain E2348/69. J. Bacteriol. 191, 347–354.

Iizumi, Y., Sagara, H., Kabe, Y., et al., 2007. The enteropathogenic *E. coli* effector EspB facilitates microvillus effacing and antiphagocytosis by inhibiting myosin function. Cell Host. Microbe. 2, 383–392.

Ito, K., Iida, M., Yamazaki, M., et al., 2007. Intimin types determined by heteroduplex mobility assay of intimin gene (*eae*)-positive *Escherichia coli* strains. J. Clin. Microbiol. 45, 1038–1041.

Jagannatha, H.M., Sharma, U.K., Ramaseshan, T., Surolia, A., Balganesh, T.S., 1991. Identification of carbohydrate structures as receptors for localised adherent enteropathogenic *Escherichia coli*. Microb. Pathog. 11, 259–268.

Jarvis, K.G., Girón, J.A., Jerse, A.E., McDaniel, T.K., Donnenberg, M.S., Kaper, J.B., 1995. Entero-pathogenic *Escherichia coli* contains a putative type III secretion system necessary for the export of proteins involved in attaching and effacing lesion formation. Proc. Natl. Acad. Sci. USA 92, 7996–8000.

Jepson, M.A., Pellegrin, S., Peto, L., et al., 2003. Synergistic roles for the Map and Tir effector molecules in mediating uptake of enteropathogenic *Escherichia coli* (EPEC) into non-phagocytic cells. Cell Microbiol. 5, 773–783.

Jerse, A.E., Kaper, J.B., 1991. The *eae* gene of enteropathogenic *Escherichia coli* encodes a 94-Kilodalton membrane protein, the expression of which is influenced by the EAF plasmid. Infect. Immun. 59, 4302–4309.

Jerse, A.E., Yu, J., Tall, B.D., Kaper, J.B., 1990. A genetic locus of enteropathogenic *Escherichia coli* necessary for the production of attaching and effacing lesions on tissue culture cells. Proc. Natl. Acad. Sci. USA 87, 7839–7843.

Karch, H., Heesemann, J., Laufs, R., Kroll, H.P., Kaper, J.B., Levine, M.M., 1987. Serological response to type 1-like somatic fimbriae in diarrheal infection due to classical enteropathogenic *Escherichia coli*. Microb. Pathog. 2, 425–434.

Karin, M., Ben-Neriah, Y., 2000. Phosphorylation meets ubiquitination: the control of NF-κB activity. Annu. Rev. Immunol. 18, 621–663.

Kavitha, D., Niranjali, S., 2009. Inhibition of enteropathogenic *Escherichia coli* adhesion on host epithelial cells by *Holarrhena antidysenterica* (L.) WALL. Phytotherapy. Research 23, 1229–1236.

Keller, R., Hilton, T.D., Rios, H., Boedeker, E.C., Kaper, J.B., 2010. Development of a live oral attaching and effacing *Escherichia coli* vaccine candidate using *Vibrio cholerae* CVD 103-HgR as antigen vector. Microb. Pathog. 48, 1–8.

Kenny, B., 1999. Phosphorylation of tyrosine 474 of the enteropathogenic *Escherichia coli* (EPEC) Tir receptor molecule is essential for actin nucleating activity and is preceded by additional host modifications. Mol. Microbiol. 31, 1229–1241.

Kenny, B., DeVinney, R., Stein, M., Reinscheid, D.J., Frey, E.A., Finlay, B.B., 1997. Enteropathogenic *E. coli* (EPEC) transfers its receptor for intimate adherence into mammalian cells. Cell 91, 511–520.

Kenny, B., Ellis, S., Leard, A.D., Warawa, J., Mellor, H., Jepson, M.A., 2002. Co-ordinate regulation of distinct host cell signalling pathways by multifunctional enteropathogenic *Escherichia coli* effector molecules. Mol. Microbiol. 44, 1095–1107.

Kenny, B., Finlay, B.B., 1995. Protein secretion by enteropathogenic *Escherichia coli* is essential for transducing signals to epithelial cells. Proc. Natl. Acad. Sci. USA 92, 7991–7995.

Kenny, B., Finlay, B.B., 1997. Intimin-dependent binding of enteropathogenic *Escherichia coli* to host cells triggers novel signaling events, including tyrosine phosphorylation of phospholipase C-gamma1. Infect. Immun. 65, 2528–2536.

Kenny, B., Jepson, M., 2000. Targeting of an enteropathogenic *Escherichia coli* (EPEC) effector protein to host mitochondria. Cell Microbiol. 2, 579–590.

Kenny, B., Lai, L.-C., Finlay, B.B., Donnenberg, M.S., 1996. EspA, a protein secreted by enteropathogenic *Escherichia coli* (EPEC), is required to induce signals in epithelial cells. Mol. Microbiol. 20, 313–323.

Khursigara, C., Abul-Milh, M., Lau, B., Girón, J.A., Lingwood, C.A., Foster, D.E., 2001. Enteropathogenic *Escherichia coli* virulence factor bundle-forming pilus has a binding specificity for phosphatidylethanolamine. Infect. Immun. 69, 6573–6579.

Kim, J.H., Kim, J.C., Choo, Y.A., et al., 2009. Detection of cytolethal distending toxin and other virulence characteristics of enteropathogenic *Escherichia coli* isolates from diarrheal patients in Republic of Korea. J. Microbiol. Biotechnol. 19, 525–529.

Kimura, K., Iwatsuki, M., Nagai, T., et al., 2011. A small-molecule inhibitor of the bacterial type III secretion system protects against in vivo infection with *Citrobacter rodentium*. J. Antibiot. 64, 197–203.

Klapproth, J.M., Donnenberg, M.S., Abraham, J.M., James, S.P., 1996. Products of enteropathogenic *E. coli* inhibit lymphokine production by gastrointestinal lymphocytes. Am. J. Physiol. Gastrointest. Liver Physiol. 271, G841–G848.

Klapproth, J.M., Donnenberg, M.S., Abraham, J.M., Mobley, H.L.T., James, S.P., 1995. Products of enteropathogenic *Escherichia coli* inhibit lymphocyte activation and lymphokine production. Infect. Immun. 63, 2248–2254.

Klapproth, J.M., Sasaki, M., Sherman, M., et al., 2005. *Citrobacter rodentium lifA/efa1* is essential for colonic colonization and crypt cell hyperplasia in vivo. Infect. Immun. 73, 1441–1451.

Klapproth, J.M., Scaletsky, I.C.A., McNamara, B.P., et al., 2000. A large toxin from pathogenic *Escherichia coli* strains that inhibits lymphocyte activation. Infect. Immun. 68, 2148–2155.

Knutton, S., Adu-Bobie, J., Bain, C., Phillips, A.D., Dougan, G., Frankel, G., 1997. Down regulation of intimin expression during attaching and effacing enteropathogenic *Escherichia coli* adhesion. Infect. Immun. 65, 1644–1652.

Knutton, S., Baldini, M.M., Kaper, J.B., McNeish, A.S., 1987a. Role of plasmid-encoded adherence factors in adhesion of enteropathogenic *Escherichia coli* to HEp-2 cells. Infect. Immun. 55, 78–85.

Knutton, S., Lloyd, D.R., McNeish, A.S., 1987b. Adhesion of enteropathogenic *Escherichia coli* to human intestinal enterocytes and cultured human intestinal mucosa. Infect. Immun. 55, 69–77.

Knutton, S., Phillips, A.D., Smith, H.R., et al., 1991. Screening for enteropathogenic *Escherichia coli* in infants with diarrhea by the fluorescent-actin staining test. Infect. Immun. 59, 365–371.

Knutton, S., Rosenshine, I., Pallen, M.J., et al., 1998. A novel EspA-associated surface organelle of enteropathogenic *Escherichia coli* involved in protein translocation into epithelial cells. EMBO J. 17, 2166–2176.

Knutton, S., Shaw, R.K., Anantha, R.P., Donnenberg, M.S., Zorgani, A.A., 1999. The type IV bundle-forming pilus of enteropathogenic *Escherichia coli* undergoes dramatic alterations in structure associated with bacterial adherence, aggregation and dispersal. Mol. Microbiol. 33, 499–509.

Kodama, T., Akeda, Y., Kono, G., et al., 2002. The EspB protein of enterohaemorrhagic *Escherichia coli* interacts directly with alpha-catenin. Cell Microbiol. 4, 213–222.

Kubinyi, L., Kiss, I., Lendvai, K.G., 1974. Epidemiological-statistical evaluation of oral vaccination against infantile *Escherichia coli* enteritis. Acta Microbiol. Acad. Sci. Hung. 21, 187–191.

Kuphal, S., Bosserhoff, A.K., 2012. E-cadherin cell-cell communication in melanogenesis and during development of malignant melanoma. Arch. Biochem. Biophys. 524, 43–47.

Lacroix, J., Delage, G., Gosselin, F., Chicoine, L., 1984. Severe protracted diarrhea due to multiresistant adherent *Escherichia coli*. Am. J. Dis. Child. 138, 693–696.

Lai, L.C., Wainwright, L.A., Stone, K.D., Donnenberg, M.S., 1997. A third secreted protein that is encoded by the enteropathogenic *Escherichia coli* pathogenicity island is required for transduction of signals and for attaching and effacing activities in host cells. Infect. Immun. 65, 2211–2217.

Laporta, M.Z., Silva, M.L.M., Scaletsky, I.C.A., Trabulsi, L.R., 1986. Plasmids coding for drug resistance and localized adherence to HeLa cells in enteropathogenic *Escherichia coli* O55:H- and O55:H6. Infect. Immun. 51, 715–717.

Lara-Tejero, M., Galan, J.E., 2000. A bacterial toxin that controls cell cycle progression as a deoxyribonuclease I-like protein. Science 290, 354–357.

Levine, M.M., 1987. *Escherichia coli* that cause diarrhea: enterotoxigenic, enteropathogenic, enteroinvasive, enterohemorrhagic, and enteroadherent. J. Infect. Dis. 155, 377–389.

Levine, M.M., Bergquist, E.J., Nalin, D.R., et al., 1978. *Escherichia coli* strains that cause diarrhoea but do not produce heat-labile or heat-stable enterotoxins and are non-invasive. Lancet 1, 1119–1122.

Levine, M.M., Nataro, J.P., Karch, H., et al., 1985. The diarrheal response of humans to some classic serotypes of enteropathogenic *Escherichia coli* is dependent on a plasmid encoding an enteroadhesiveness factor. J. Infect. Dis. 152, 550–559.

Li, P., Nijhawan, D., Budihardjo, I., et al., 1997. Cytochrome c and dATP-dependent formation of Apaf-1/caspase-9 complex initiates an apoptotic protease cascade. Cell 91, 479–489.

Liebchen, A., Benz, I., Mellmann, A., et al., 2011. Characterization of *Escherichia coli* strains isolated from patients with diarrhea in São Paulo, Brazil: identification of intermediate virulence factor profiles by multiplex PCR. J. Clin. Microbiol. 49, 2274–2278.

Lim, Y.S., Ngan, C.C.L., Tay, L., 1992. Enteropathogenic *Escherichia coli* as a cause of diarrhoea among children in Singapore. J. Trop. Med. Hyg. 95, 339–342.

Lio, J.C., Syu, W.J., 2004. Identification of a negative regulator for the pathogenicity island of enterohemorrhagic *Escherichia coli* O157:H7. J. Biomed. Sci. 11, 855–863.

Liu, H., Magoun, L., Luperchio, S., Schauer, D.B., Leong, J.M., 1999. The Tir-binding region of enterohaemorrhagic *Escherichia coli* intimin is sufficient to trigger actin condensation after bacterial-induced host cell signalling. Mol. Microbiol. 34, 67–81.

Liu, Z., Shen, T., Zhang, P., Ma, Y., Qin, H., 2011. *Lactobacillus plantarum* surface layer adhesive protein protects intestinal epithelial cells against tight junction injury induced by enteropathogenic *Escherichia coli*. Mol. Biol. Rep. 38, 3471–3480.

Lommel, S., Benesch, S., Rottner, K., Franz, T., Wehland, J., Kuhn, R., 2001. Actin pedestal formation by enteropathogenic *Escherichia coli* and intracellular motility of *Shigella flexneri* are abolished in N-WASP- defective cells. EMBO Rep. 2, 850–857.

Loureiro, I., Frankel, G., Adu-Bobie, J., Dougan, G., Trabulsi, L.R., Carneiro-Sampaio, M.M., 1998. Human colostrum contains IgA antibodies reactive to enteropathogenic *Escherichia coli* virulence-associated proteins: intimin, BfpA, EspA, and EspB. J. Pediatr. Gastroenterol. Nutr. 27, 166–171.

Luo, W., Donnenberg, M.S., 2011. Interactions and predicted host membrane topology of enteropathogenic *Escherichia coli* translocator protein EspB. J. Bacteriol. 193, 2972–2980.

Malladi, V., Puthenedam, M., Williams, P.H., Balakrishnan, A., 2004. Enteropathogenic *Escherichia coli* outer membrane proteins induce iNOS by activation of NF-κB and MAP kinases. Inflammation 28, 345–353.

Malstrom, C., James, S., 1998. Inhibition of murine splenic and mucosal lymphocyte function by enteric bacterial products. Infect. Immun. 66, 3120–3127.

Manjarrez-Hernandez, A., Gavilanes-Parra, S., Chavez-Berrocal, M.E., Molina-Lopez, J., Cravioto, A., 1997. Binding of diarrheagenic *Escherichia coli* to 32- to 33-kilodalton human intestinal brush border proteins. Infect. Immun. 65, 4494–4501.

Manjarrez-Hernandez, H.A., Baldwin, T.J., Williams, P.H., Haigh, R., Knutton, S., Aitken, A., 1996. Phosphorylation of myosin light chain at distinct sites and its association with the cytoskeleton during enteropathogenic *Escherichia coli* infection. Infect. Immun. 64, 2368–2370.

Manjarrez-Hernandez, H.A., Gavilanes-Parra, S., Chavez-Berrocal, E., Navarro-Ocana, A., Cravioto, A., 2000. Antigen detection in enteropathogenic *Escherichia coli* using secretory immunoglobulin A antibodies isolated from human breast milk. Infect. Immun. 68, 5030–5036.

Marchès, O., Nougayrède, J.P., Boullier, S., et al., 2000. Role of Tir and intimin in the virulence of rabbit enteropathogenic *Escherichia coli* serotype O103:H2. Infect. Immun. 68, 2171–2182.

Martinez-Argudo, I., Sands, C., Jepson, M.A., 2007. Translocation of enteropathogenic *Escherichia coli* across an in vitro M cell model is regulated by its type III secretion system. Cell Microbiol. 9, 1538–1546.

Martínez-Laguna, Y., Calva, E., Puente, J.L., 1999. Autoactivation and environmental regulation of *bfpT* expression, the gene coding for the transcriptional activator of *bfpA* in enteropathogenic *Escherichia coli*. Mol. Microbiol. 33, 153–166.

Masuda, A., Yoshida, M., Shiomi, H., et al., 2008. Fcγ receptor regulation of *Citrobacter rodentium* infection. Infect. Immun. 76, 1728–1737.

Matsuzawa, T., Kuwae, A., Abe, A., 2005. Enteropathogenic *Escherichia coli* type III effectors EspG and EspG2 alter epithelial paracellular permeability. Infect. Immun. 73, 6283–6289.

McConnell, M.M., Chart, H., Scotland, S.M., Smith, H.R., Willshaw, G.A., Rowe, B., 1989. Properties of adherence factor plasmids of enteropathogenic *Escherichia coli* and the effect of host strain on expression of adherence to HEp-2 cells. J. Gen. Microbiol. 135, 1123–1134.

McDaniel, T.K., Kaper, J.B., 1997. A cloned pathogenicity island from enteropathogenic *Escherichia coli* confers the attaching and effacing phenotype on K-12 *E. coli*. Mol Microbiol. 23, 399–407.

McNamara, B.P., Koutsouris, A., O'Connell, C.B., Nougayrède, J.P., Donnenberg, M.S., Hecht, G., 2001. Translocated EspF protein from enteropathogenic *Escherichia coli* disrupts host intestinal barrier function. J. Clin. Invest. 107, 621–629.

Mellies, J.L., Barron, A.M., Carmona, A.M., 2007. Enteropathogenic and enterohemorrhagic *Escherichia coli* virulence gene regulation. Infect. Immun. 75, 4199–4210.

Mellies, J.L., Elliott, S.J., Sperandio, V., Donnenberg, M.S., Kaper, J.B., 1999. The Per regulon of enteropathogenic *Escherichia coli*: identification of a regulatory cascade and a novel transcriptional activator, the locus of enterocyte effacement (LEE)-encoded regulator (Ler). Mol. Microbiol. 33, 296–306.

Mellies, J.L., Navarro-Garcia, F., Okeke, I., Frederickson, J., Nataro, J.P., Kaper, J.B., 2001. *espC* pathogenicity island of enteropathogenic *Escherichia coli* encodes an enterotoxin. Infect. Immun. 69, 315–324.

Mietens, C., Keinhorst, H., Hilpert, H., Gerber, H., Amster, H., Pahud, J.J., 1979. Treatment of infantile *E. coli* gastroenteritis with specific bovine anti-*E. coli* milk immunoglobulins. European. J. Pediatr. 132, 239–252.

Monjaras, F.J., Garcia-Gomez, E., Espinosa, N., Minamino, T., Namba, K., Gonzalez-Pedrajo, B., 2012. Role of EscP (Orf16) in injectisome biogenesis and regulation of type III protein secretion in enteropathogenic *Escherichia coli*. J. Bacteriol. 194, 6029–6045.

Moon, H.W., Whipp, S.C., Argenzio, R.A., Levine, M.M., Giannella, R.A., 1983. Attaching and effacing activities of rabbit and human enteropathogenic *Escherichia coli* in pig and rabbit intestines. Infect. Immun. 41, 1340–1351.

Morelli, R., Baldassarri, L., Falbo, V., Donelli, G., Caprioli, A., 1994. Detection of enteroadherent *Escherichia coli* associated with diarrhoea in Italy. J. Med. Microbiol. 41, 399–404.

Mundy, R., Schuller, S., Girard, F., Fairbrother, J.M., Phillips, A.D., Frankel, G., 2007. Functional studies of intimin *in vivo* and *ex vivo*: implications for host specificity and tissue tropism. Microbiology 153, 959–967.

Munson, G.P., Holcomb, L.G., Scott, J.R., 2001. Novel group of virulence activators within the AraC family that are not restricted to upstream binding sites. Infect. Immun. 69, 186–193.

Muza-Moons, M.M., Koutsouris, A., Hecht, G., 2003. Disruption of cell polarity by enteropathogenic *Escherichia coli* enables basolateral membrane proteins to migrate apically and to potentiate physiological consequences. Infect. Immun. 71, 7069–7078.

Muza-Moons, M.M., Schneeberger, E.E., Hecht, G.A., 2004. Enteropathogenic *Escherichia coli* infection leads to appearance of aberrant tight junctions strands in the lateral membrane of intestinal epithelial cells. Cell Microbiol. 6, 783–793.

Nadler, C., Baruch, K., Kobi, S., et al., 2010. The type III secretion effector NleE inhibits NF-κB activation. PLoS Pathog. 6, e1000743.

Nagai, T., Abe, A., Sasakawa, C., 2005. Targeting of enteropathogenic *Escherichia coli* EspF to host mitochondria is essential for bacterial pathogenesis: critical role of the 16th leucine residue in EspF. J. Biol. Chem. 280, 2998–3011.

Nakazato, G., Gyles, C., Ziebell, K., et al., 2004. Attaching and effacing *Escherichia coli* isolated from dogs in Brazil: characteristics and serotypic relationship to human enteropathogenic *E. coli* (EPEC). Vet. Microbiol. 101, 269–277.

Nataro, J.P., Kaper, J.B., Robins-Browne, R., Prado, V., Vial, P., Levine, M.M., 1987. Patterns of adherence of diarrheagenic *Escherichia coli* to HEp-2 cells. Pediatr. Infect. Dis. J. 6, 829–831.

Nataro, J.P., Kaper, J.B., 1998. Diarrheagenic *Escherichia coli*. Clin. Microbiol. Rev. 11, 142–201.

Nataro, J.P., Scaletsky, I.C.A., Kaper, J.B., Levine, M.M., Trabulsi, L.R., 1985. Plasmid-mediated factors conferring diffuse and localized adherence of enteropathogenic *Escherichia coli*. Infect. Immun. 48, 378–383.

Navarro-Garcia, F., Canizalez-Roman, A., Sui, B.Q., Nataro, J.P., Azamar, Y., 2004. The serine protease motif of EspC from enteropathogenic *Escherichia coli* produces epithelial damage by a mechanism different from that of Pet toxin from enteroaggregative *E. coli*. Infect. Immun. 72, 3609–3621.

Neutra, M.R., Pringault, E., Kraehenbuhl, J.P., 1996. Antigen sampling across epithelial barriers and induction of mucosal immune responses. Annu. Rev. Immunol. 14, 275–300.

Newton, H.J., Pearson, J.S., Badea, L., et al., 2010. The type III effectors NleE and NleB from enteropathogenic *E. coli* and OspZ from *Shigella* block nuclear translocation of NF-kappaB p65. PLoS Pathog. 6, e1000898.

Nguyen, R.N., Taylor, L.S., Tauschek, M., Robins-Browne, R.M., 2006. Atypical enteropathogenic *Escherichia coli* infection and prolonged diarrhea in children. Emerg. Infect. Dis. 12, 597–603.

Nicholls, L., Grant, T.H., Robins-Browne, R.M., 2000. Identification of a novel genetic locus that is required for *in vitro* adhesion of a clinical isolate of enterohaemorrhagic *Escherichia coli* to epithelial cells. Mol. Microbiol. 35, 275–288.

Nougayrède, J.P., Donnenberg, M.S., 2004. Enteropathogenic *Escherichia coli* EspF is targeted to mitochondria and is required to initiate the mitochondrial death pathway. Cell Microbiol. 6, 1097–1111.

Nougayrède, J.P., Foster, G.H., Donnenberg, M.S., 2007. Enteropathogenic *Escherichia coli* effector EspF interacts with host protein Abcf2. Cell Microbiol. 9, 680–693.

Nougayrède, J.P., Marchès, O., Boury, M., et al., 1999. The long-term cytoskeletal rearrangement induced by rabbit enteropathogenic *Escherichia coli* is Esp dependent but intimin independent. Mol. Microbiol. 31, 19–30.

Nweze, E.I., 2010. Aetiology of diarrhoea and virulence properties of diarrhoeagenic *Escherichia coli* among patients and healthy subjects in southeast Nigeria. J. Health Popul. Nutr. 28, 245–252.

Ochoa, T.J., Contreras, C.A., 2011. Enteropathogenic *Escherichia coli* infection in children. Curr. Opin. Infect. Dis. 24, 478–483.

Ogino, T., Ohno, R., Sekiya, K., et al., 2006. Assembly of the type III secretion apparatus of enteropathogenic *Escherichia coli*. J. Bacteriol. 188, 2801–2811.

Ohara, M., Oswald, E., Sugai, M., 2004. Cytolethal distending toxin: a bacterial bullet targeted to nucleus. J. Biochem. 136, 409–413.

Okeke, I.N., Borneman, J.A., Shin, S., Mellies, J.L., Quinn, L.E., Kaper, J.B., 2001. Comparative sequence analysis of the plasmid-encoded regulator of enteropathogenic *Escherichia coli* Strains. Infect. Immun. 69, 5553–5564.

Pallen, M.J., Beatson, S.A., Bailey, C.M., 2005. Bioinformatics analysis of the locus for enterocyte effacement provides novel insights into type-III secretion. BMC Microbiol. 5, 9.

Parissi-Crivelli, A., Parissi-Crivelli, J.M., Girón, J.A., 2000. Recognition of enteropathogenic *Escherichia coli* virulence determinants by human colostrum and serum antibodies. J. Clin. Microbiol. 38, 2696–2700.

Peeters, J.E., Geeroms, R., Orskov, F., 1988. Biotype, serotype, and pathogenicity of attaching and effacing enteropathogenic *Escherichia coli* strains isolated from diarrheic commercial rabbits. Infect. Immun. 56, 1442–1448.

Pelayo, J.S., Scaletsky, I.C.A., Pedroso, M.Z., et al., 1999. Virulence properties of atypical EPEC strains. J. Med. Microbiol. 48, 41–49.

Pham, T.H., Gao, X., Tsai, K., Olsen, R., Wan, F., Hardwidge, P.R., 2012. Functional differences and interactions between the *Escherichia coli* type III secretion system effectors NleH1 and NleH2. Infect. Immun. 80, 2133–2140.

Phillips, N., Hayward, R.D., Koronakis, V., 2004. Phosphorylation of the enteropathogenic *E. coli* receptor by the Src-family kinase c-Fyn triggers actin pedestal formation. Nat. Cell Biol. 6, 618–625.

Philpott, D.J., McKay, D.M., Sherman, P.M., Perdue, M.H., 1996. Infection of T84 cells with enteropathogenic *Escherichia coli* alters barrier and transport functions. Am. J. Physiol Gastrointest. Liver Physiol. 270, G634–G645.

Porter, M.E., Mitchell, P., Roe, A.J., Free, A., Smith, D.G., Gally, D.L., 2004. Direct and indirect transcriptional activation of virulence genes by an AraC-like protein, PerA from enteropathogenic *Escherichia coli*. Mol. Microbiol. 54, 1117–1133.

Porter, M.E., Smith, S.G., Dorman, C.J., 1998. Two highly related regulatory proteins, *Shigella flexneri* VirF and enterotoxigenic *Escherichia coli* Rns, have common and distinct regulatory properties. FEMS Microbiol. Lett. 162, 303–309.

Puente, J.L., Bieber, D., Ramer, S.W., Murray, W., Schoolnik, G.K., 1996. The bundle-forming pili of enteropathogenic *Escherichia coli*: Transcriptional regulation by environmental signals. Mol. Microbiol. 20, 87–100.

Ramachandran, V., Brett, K., Hornitzky, M.A., et al., 2003. Distribution of intimin subtypes among *Escherichia coli* isolates from ruminant and human sources. J. Clin. Microbiol. 41, 5022–5032.

Ramboarina, S., Fernandes, P.J., Daniell, S., et al., 2005. Structure of the bundle-forming pilus from enteropathogenic *Escherichia coli*. J. Biol. Chem. 280, 40252–40260.

Ramer, S.W., Bieber, D., Schoolnik, G.K., 1996. BfpB, an outer membrane lipoprotein required for the biogenesis of bundle-forming pili in enteropathogenic *Escherichia coli*. J. Bacteriol. 178, 6555–6563.

Robins-Browne, R.M., 1987. Traditional enteropathogenic *Escherichia coli* of infantile diarrhea. Rev. Infect. Dis. 9, 28–53.

Robins-Browne, R.M., Bordun, A.M., Tauschek, M., et al., 2004. *Escherichia coli* and community-acquired gastroenteritis, Melbourne, Australia. Emerg. Infect. Dis. 10, 1797–1805.

Rodrigues, J., Scaletsky, I.C.A., Campos, L.C., Gomes, T.A.T., Whittam, T.S., Trabulsi, L.R., 1996. Clonal structure and virulence factors in strains of *Escherichia coli* of the classic serogroup O55. Infect. Immun. 64, 2680–2686.

Rodrigues, J., Thomazini, C.M., Lopes, C.A.M., Dantas, L.O., 2004. Concurrent infection in a dog and colonization in a child with a human enteropathogenic *Escherichia coli* clone. J. Clin. Microbiol. 42, 1388–1389.

Rosa, A.C.P., Mariano, A.T., Pereira, A.M.S., Tibana, A., Gomes, T.A.T., Andrade, J.R.C., 1998. Enteropathogenicity markers in *Escherichia coli* isolated from infants with acute diarrhoea and healthy controls in Rio de Janeiro, Brazil. J. Med. Microbiol. 47, 781–790.

Rosenshine, I., Donnenberg, M.S., Kaper, J.B., Finlay, B.B., 1992. Signal exchange between entero-pathogenic *Escherichia coli* (EPEC) and epithelial cells: EPEC induce tyrosine phosphoryla-tion of host cell protein to initiate cytoskeletal rearrangement and bacterial uptake. EMBO J. 11, 3551–3560.

Rosenshine, I., Ruschkowski, S., Finlay, B.B., 1996a. Expression of attaching effacing activity by enteropathogenic *Escherichia coli* depends on growth phase, temperature, and protein synthesis upon contact with epithelial cells. Infect. Immun. 64, 966–973.

Rosenshine, I., Ruschkowski, S., Stein, M., Reinscheid, D.J., Mills, S.D., Finlay, B.B., 1996b. A pathogenic bacterium triggers epithelial signals to form a functional bacterial receptor that mediates actin pseudopod formation. EMBO J. 15, 2613–2624.

Rothbaum, R., McAdams, A.J., Giannella, R., Partin, J.C., 1982. A clinicopathological study of enterocyte-adherent *Escherichia coli*: a cause of protracted diarrhea in infants. Gastroenterol-ogy 83, 441–454.

Rothbaum, R.J., Partin, J.C., Saalfield, K., McAdams, A.J., 1983. An ultrastructural study of entero-pathogenic *Escherichia coli* infection in human infants. Ultrastruct. Pathol. 4, 291–304.

Royan, S.V., Jones, R.M., Koutsouris, A., et al., 2010. Enteropathogenic *E. coli* non-LEE encoded effectors NleH1 and NleH2 attenuate NF-κB activation. Mol. Microbiol. 78, 1232–1245.

Ruchaud-Sparagano, M.H., Muhlen, S., Dean, P., Kenny, B., 2011. The enteropathogenic *E. coli* (EPEC) Tir effector inhibits NF-kappaB activity by targeting TNFalpha receptor-associated factors. PLoS Pathog. 7, e1002414.

Sal-Man, N., Biemans-Oldehinkel, E., Sharon, D., et al., 2012a. EscA is a crucial component of the type III secretion system of enteropathogenic *Escherichia coli*. J. Bacteriol. 194, 2819–2828.

Sal-Man, N., Deng, W., Finlay, B.B., 2012b. EscI: a crucial component of the type III secretion system forms the inner rod structure in enteropathogenic *Escherichia coli*. Biochemical. J. 442, 119–125.

Sanches, M.I., Keller, R., Hartland, E.L., et al., 2000. Human colostrum and serum contain antibod-ies reactive to the intimin- binding region of the enteropathogenic *Escherichia coli* translocated intimin receptor. J. Pediatr. Gastroenterol. Nutr. 30, 73–77.

Sanger, J.M., Chang, R., Ashton, F., Kaper, J.B., Sanger, J.W., 1996. Novel form of actin-based motil-ity transports bacteria on the surface of infected cells. Cell Motil. Cytoskeleton. 34, 279–287.

Santos, M.F., New, R.R.C., Andrade, G.R., et al., 2010. Lipopolysaccharide as an antigen target for the formulation of a universal vaccine against *Escherichia coli* O111 strains. Clin. Vaccine Immunol. 17, 1772–1780.

Saridakis, H.O., El Gared, S.A., Vidotto, M.C., Guth, B.E.C., 1997. Virulence properties of *Escherichia coli* strains belonging to enteropathogenic (EPEC) serogroups isolated from calves with diarrhea. Vet. Microbiol. 54, 145–153.

Savarino, S.J., McVeigh, A., Watson, J., et al., 1996. Enteroaggregative *Escherichia coli* heat-stable enterotoxin is not restricted to enteroaggregative *Escherichia coli*. J. Infect. Dis. 173, 1019–1022.

Savkovic, S.D., Koutsouris, A., Hecht, G., 1996. Attachment of a noninvasive enteric pathogen, enteropathogenic *Escherichia coli*, to cultured human intestinal epithelial monolayers induces transmigration of neutrophils. Infect. Immun. 64, 4480–4487.

Savkovic, S.D., Koutsouris, A., Hecht, G., 1997. Activation of NF-kappaB in intestinal epithelial cells by enteropathogenic *Escherichia coli*. Am. J. Physiol. 273, C1160–C1167.

Scaletsky, I., Souza, T., Aranda, K., Okeke, I., 2010b. Genetic elements associated with antimicrobial resistance in enteropathogenic *Escherichia coli* (EPEC) from Brazil. BMC Microbiol. 10, 25.

Scaletsky, I.C., Aranda, K.R., Souza, T.B., Silva, N.P., 2010a. Adherence factors in atypical enteropathogenic *Escherichia coli* strains expressing the localized adherence-like pattern in HEp-2 cells. J. Clin. Microbiol. 48, 302–306.

Scaletsky, I.C.A., Milani, S.R., Trabulsi, L.R., Travassos, L.R., 1988. Isolation and characterization of the localized adherence factor of enteropathogenic *Escherichia coli*. Infect. Immun. 56, 2979–2983.

Scaletsky, I.C.A., Pedroso, M.Z., Oliva, C.A.G., Carvalho, R.L.B., Morais, M.B., Fagundes-Neto, U., 1999. A localized adherence-like pattern as a second pattern of adherence of classic enteropathogenic *Escherichia coli* to HEp-2 cells that is associated with infantile diarrhea. Infect. Immun. 67, 3410–3415.

Scaletsky, I.C.A., Silva, M.L.M., Trabulsi, L.R., 1984. Distinctive patterns of adherence of enteropathogenic *Escherichia coli* to HeLa cells. Infect. Immun. 45, 534–536.

Schauer, D.B., Falkow, S., 1993. Attaching and effacing locus of a *Citrobacter freundii* biotype that causes transmissible murine colonic hyperplasia. Infect. Immun. 61, 2486–2492.

Schreiber, W., Stone, K.D., Strong, M.A., DeTolla Jr., L.J., Hoppert, M., Donnenberg, M.S., 2002. BfpU, a soluble protein essential for type IV pilus biogenesis in enteropathogenic *Escherichia coli*. Microbiology 148, 2507–2518.

Schriefer, A., Maltez, J.R., Silva, N., Stoeckle, M.Y., Barral-Netto, M., Riley, L.W., 1999. Expression of a pilin subunit BfpA of the bundle-forming pilus of enteropathogenic *Escherichia coli* in an *aroA* live salmonella vaccine strain. Vaccine 17, 770–778.

Schüller, S., Lucas, M., Kaper, J.B., Girón, J.A., Phillips, A.D., 2009. The ex vivo response of human intestinal mucosa to enteropathogenic *Escherichia coli* infection. Cell Microbiol. 11, 521–530.

Scotland, S.M., Smith, H.R., Cheasty, T., et al., 1996. Use of gene probes and adhesion tests to characterise *Escherichia coli* belonging to enteropathogenic serogroups isolated in the United Kingdom. J. Med. Microbiol. 44, 438–443.

Scott, D.A., Kaper, J.B., 1994. Cloning and sequencing of the genes encoding *Escherichia coli* cytolethal distending toxin. Infect. Immun. 62, 244–251.

Sekiya, K., Ohishi, M., Ogino, T., Tamano, K., Sasakawa, C., Abe, A., 2001. Supermolecular structure of the enteropathogenic *Escherichia coli* type III secretion system and its direct interaction with the EspA-sheath-like structure. Proc. Natl. Acad. Sci. USA 98, 11638–11643.

Sham, H.P., Shames, S.R., Croxen, M.A., et al., 2011. Attaching and effacing bacterial effector NleC suppresses epithelial inflammatory responses by inhibiting NF-kappaB and p38 mitogen-activated protein kinase activation. Infect. Immun. 79, 3552–3562.

Shames, S.R., Deng, W., Guttman, J.A., et al., 2010. The pathogenic *E. coli* type III effector EspZ interacts with host CD98 and facilitates host cell prosurvival signalling. Cell Microbiol. 12, 1322–1339.

Sharma, R., Tesfay, S., Tomson, F.L., Kanteti, R.P., Viswanathan, V.K., Hecht, G., 2006. Balance of bacterial pro- and anti-inflammatory mediators dictates net effect of enteropathogenic *Escherichia coli* on intestinal epithelial cells. Am. J. Physiol. G. Liver Physiol. 290, G685–G694.

Shaw, R.K., Smollett, K., Cleary, J., et al., 2005. Enteropathogenic *Escherichia coli* type III effectors EspG and EspG2 disrupt the microtubule network of intestinal epithelial cells. Infect. Immun. 73, 4385–4390.

Shifflett, D.E., Clayburgh, D.R., Koutsouris, A., Turner, J.R., Hecht, G.A., 2005. Enteropathogenic *E. coli* disrupts tight junction barrier function and structure *in vivo*. Lab Invest. 85, 1308–1324.

Simonovic, I., Arpin, M., Koutsouris, A., Falk-Krzesinski, H.J., Hecht, G., 2001. Enteropathogenic *Escherichia coli* activates ezrin, which participates in disruption of tight junction barrier function. Infect. Immun. 69, 5679–5688.

Simonovic, I., Rosenberg, J., Koutsouris, A., Hecht, G., 2000. Enteropathogenic *Escherichia coli* dephosphorylates and dissociates occludin from intestinal epithelial tight juctions. Cell Microbiol. 2, 305–315.

Sircili, M.P., Walters, M., Trabulsi, L.R., Sperandio, V., 2004. Modulation of enteropathogenic *Escherichia coli* virulence by quorum sensing. Infect. Immun. 72, 2329–2337.

Sohel, I., Puente, J.L., Ramer, S.W., Bieber, D., Wu, C.-Y., Schoolnik, G.K., 1996. Enteropathogenic *Escherichia coli*: identification of a gene cluster coding for bundle-forming pilus morphogenesis. J. Bacteriol. 178, 2613–2628.

Song, Y.C., Jin, S., Louie, H., et al., 2004. FlaC, a protein of *Campylobacter jejuni* TGH9011 (ATCC43431) secreted through the flagellar apparatus, binds epithelial cells and influences cell invasion. Mol. Microbiol. 53, 541–553.

Spehlmann, M.E., Dann, S.M., Hruz, P., Hanson, E., McCole, D.F., Eckmann, L., 2009. CXCR2-dependent mucosal neutrophil influx protects against colitis-associated diarrhea caused by an attaching/effacing lesion-forming bacterial pathogen. J. Immunol. 183, 3332–3343.

Sperandio, V., Mellies, J.L., Nguyen, W., Shin, S., Kaper, J.B., 1999. Quorum sensing controls expression of the type III secretion gene transcription and protein secretion in enterohemorrhagic and enteropathogenic *Escherichia coli*. Proc. Natl. Acad. Sci. USA 96, 15196–15201.

Spitz, J., Yuhan, R., Koutsouris, A., Blatt, C., Alverdy, J., Hecht, G., 1995. Enteropathogenic *Escherichia coli* adherence to intestinal epithelial monolayers diminishes barrier function. Am. J. Physiol. Gastrointes. Liver Physiol. 268, G374–G379.

Stein, M., Kenny, B., Stein, M.A., Finlay, B.B., 1996a. Characterization of EspC, a 110-kilodalton protein secreted by enteropathogenic *Escherichia coli* which is homologous to members of the IgA protease-like family of secreted proteins. J. Bacteriol. 178, 6546–6554.

Stein, M.A., Mathers, D.A., Yan, H., Baimbridge, K.G., Finlay, B.B., 1996b. Enteropathogenic *Escherichia coli* (EPEC) markedly decreases the resting membrane potential of Caco-2 and HeLa human epithelial cells. Infect. Immun. 64, 4820–4825.

Stone, K.D., Zhang, H.-Z., Carlson, L.K., Donnenberg, M.S., 1996. A cluster of fourteen genes from enteropathogenic *Escherichia coli* is sufficient for biogenesis of a type IV pilus. Mol. Microbiol. 20, 325–337.

Subramanian, K., Selvakkumar, C., Vinaykumar, K.S., et al., 2009. Tackling multiple antibiotic resistance in enteropathogenic *Escherichia coli* (EPEC) clinical isolates: a diarylheptanoid from *Alpinia officinarum* shows promising antibacterial and immunomodulatory activity against EPEC and its lipopolysaccharide-induced inflammation. Int. J. Antimicrob. Agents. 33, 244–250.

Swimm, A., Bommarius, B., Li, Y., et al., 2004. Enteropathogenic *Escherichia coli* use redundant tyrosine kinases to form actin pedestals. Mol. Biol. Cell 15, 3520–3529.

Tacket, C.O., Sztein, M.B., Losonsky, G., et al., 2000. Role of EspB in experimental human enteropathogenic *Escherichia coli* infection. Infect. Immun. 68, 3689–3695.

Tahoun, A., Siszler, G., Spears, K., et al., 2011. Comparative analysis of EspF variants in inhibition of *Escherichia coli* phagocytosis by macrophages and inhibition of *E. coli* translocation through human- and bovine-derived M cells. Infect. Immun. 79, 4716–4729.

Taylor, J., 1961. Host specificity and enteropathogenicity of *Escherichia coli*. J. Appl. Microbiol. 24, 316–325.

Taylor, K.A., Luther, P.W., Donnenberg, M.S., 1999. Expression of the EspB protein of enteropathogenic *Escherichia coli* within HeLa cells affects stress fibers and cellular morphology. Infect. Immun. 67, 120–125.

Taylor, K.A., O'Connell, C.B., Luther, P.W., Donnenberg, M.S., 1998. The EspB protein of enteropathogenic *Escherichia coli* is targeted to the cytoplasm of infected HeLa cells. Infect. Immun. 66, 5501–5507.

Thomassin, J.L., He, X., Thomas, N.A., 2011. Role of EscU auto-cleavage in promoting type III effector translocation into host cells by enteropathogenic *Escherichia coli*. BMC Microbiol. 11, 205.

Thorén, A., Wolde-Mariam, T., Stintzing, G., Wadström, T., Habte, D., 1980. Antibiotics in the treatment of gastroenteritis caused by enteropathogenic *Escherichia coli*. J. Infect. Dis. 141, 27–31.

Tobe, T., Hayashi, T., Han, C.G., Schoolnik, G.K., Ohtsubo, E., Sasakawa, C., 1999. Complete DNA sequence and structural analysis of the enteropathogenic *Escherichia coli* adherence factor plasmid. Infect. Immun. 67, 5455–5462.

Tobe, T., Schoolnik, G.K., Sohel, I., Bustamante, V.H., Puente, J.L., 1996. Cloning and characterization of *bfpTVW*, genes required for the transcriptional activation of *bfpA* in enteropathogenic *Escherichia coli*. Mol. Microbiol. 21, 963–975.

Toledo, M.R., Alvariza, M., Murahovschi, J., Ramos, S.R., Trabulsi, L.R., 1983. Enteropathogenic *Escherichia coli* serotypes and endemic diarrhea in infants. Infect. Immun. 39, 586–589.

Torres, A.G., Tutt, C.B., Duval, L., et al., 2007. Bile salts induce expression of the afimbrial LDA adhesin of atypical enteropathogenic *Escherichia coli*. Cell Microbiol. 9, 1039–1049.

Trabulsi, L.R., Keller, R., Gomes, T.A.T., 2002. Typical and atypical enteropathogenic *Escherichia coli*. Emerg. Infect. Dis. 8, 508–513.

Tu, X., Nisan, I., Yona, C., Hanski, E., Rosenshine, I., 2003. EspH, a new cytoskeleton-modulating effector of enterohaemorrhagic and enteropathogenic *Escherichia coli*. Mol. Microbiol. 47, 595–606.

Tzipori, S., Gibson, R., Montanaro, J., 1989. Nature and distribution of mucosal lesions associated with enteropathogenic and enterohemorrhagic *Escherichia coli* in piglets and the role of plasmid-mediated factors. Infect. Immun. 57, 1142–1150.

Tzipori, S., Wachsmuth, I.K., Chapman, C., et al., 1986. The pathogenesis of hemorrhagic colitis caused by *Escherichia coli* O157:H7 in gnotobiotic pigs. J. Infect. Dis. 154, 712–716.

Ulshen, M.H., Rollo, J.L., 1980. Pathogenesis of *Escherichia coli* gastroenteritis in man - another mechanism. N. Engl. J. Med. 302, 99–101.

Vanmaele, R.P., Armstrong, G.D., 1997. Effect of carbon source on localized adherence of enteropathogenic *Escherichia coli*. Infect. Immun. 65, 1408–1413.

Vanmaele, R.P., Finlayson, M.C., Armstrong, G.D., 1995. Effect of enteropathogenic *Escherichia coli* on adherent properties of Chinese hamster ovary cells. Infect. Immun. 63, 191–198.

Vanmaele, R.P., Heerze, L.D., Armstrong, G.D., 1999. Role of lactosyl glycan sequences in inhibiting enteropathogenic *Escherichia coli* attachment. Infect. Immun. 67, 3302–3307.

Varma, P., Dinesh, K.R., Menon, K.K., Biswas, R., 2010. *Lactobacillus Fermentum* isolated from human colonic mucosal biopsy inhibits the growth and adhesion of enteric and foodborne pathogens. J. Food Sci. 75, M546–M551.

Veltman, K., Hummel, S., Cichon, C., Sonnenborn, U., Schmidt, M.A., 2012. Identification of specific miRNAs targeting proteins of the apical junctional complex that simulate the probiotic effect of *E. coli* Nissle 1917 on T84 epithelial cells. Int. J. Biochem. Cell Biol. 44, 341–349.

Vidal, J.E., Navarro-Garcia, F., 2008. EspC translocation into epithelial cells by enteropathogenic *Escherichia coli* requires a concerted participation of type V and III secretion systems. Cell Microbiol. 10, 1975–1986.

Vieira, M.A., Andrade, J.R., Trabulsi, L.R., et al., 2001. Phenotypic and genotypic characteristics of *Escherichia coli* strains of non-enteropathogenic *E. coli* (EPEC) serogroups that carry *eae* and lack the EPEC adherence factor and Shiga toxin DNA probe sequences. J. Infect. Dis. 183, 762–772.

Viljanen, M.K., Peltola, T., Junnila, S.Y.T., et al., 1990. Outbreak of diarrhoea due to *Escherichia coli* O111:B4 in schoolchildren and adults: association of Vi antigen-like reactivity. Lancet 336, 831–834.

Viscardi, M., Perugini, A.G., Auriemma, C., et al., 2008. Isolation and characterisation of two novel coliphages with high potential to control antibiotic-resistant pathogenic *Escherichia coli* (EHEC and EPEC). Int. J. Antimicrob. Agents. 31, 152–157.

Vuopio-Varkila, J., Schoolnik, G.K., 1991. Localized adherence by enteropathogenic *Escherichia coli* is an inducible phenotype associated with the expression of new outer membrane proteins. J. Exp. Med. 174, 1167–1177.

Wachter, C., Beinke, C., Mattes, M., Schmidt, M.A., 1999. Insertion of EspD into epithelial target cell membranes by infecting enteropathogenic *Escherichia coli*. Mol. Microbio. 31, 1695–1707.

Weflen, A.W., Alto, N.M., Hecht, G.A., 2009. Tight junctions and enteropathogenic *E. coli*. Ann. NY Acad. Sci. 1165, 169–174.

Weflen, A.W., Alto, N.M., Viswanathan, V.K., Hecht, G., 2010. *E. coli* secreted protein F promotes EPEC invasion of intestinal epithelial cells via an SNX9-dependent mechanism. Cell Microbiol. 12, 919–929.

Weiss-Muszkat, M., Shakh, D., Zhou, Y., et al., 2010. Biofilm formation by and multicellular behavior of *Escherichia coli* O55:H7, an atypical enteropathogenic strain. Appl. Environ. Microbiol. 76, 1545–1554.

Whale, A.D., Garmendia, J., Gomes, T.A., Frankel, G., 2006. A novel category of enteropathogenic *Escherichia coli* simultaneously utilizes the Nck and TccP pathways to induce actin remodelling. Cell Microbiol. 8, 999–1008.

Wickham, M.E., Lupp, C., Mascarenhas, M., et al., 2006. Bacterial genetic determinants of non-O157 STEC outbreaks and hemolytic-uremic syndrome after infection. J. Infect. Dis. 194, 819–827.

Wiemer, D., Loderstaedt, U., von Wulffen, H., et al., 2011. Real-time multiplex PCR for simultaneous detection of *Campylobacter jejuni, Salmonella, Shigella* and *Yersinia* species in fecal samples. Int. J. Med. Microbiol. 301, 577–584.

Wilson, R.K., Shaw, R.K., Daniell, S., Knutton, S., Frankel, G., 2001. Role of EscF, a putative needle complex protein, in the type III protein translocation system of enteropathogenic *Escherichia coli*. Cell Microbiol. 3, 753–762.

Wolff, C., Nisan, I., Hanski, E., Frankel, G., Rosenshine, I., 1998. Protein translocation into host epithelial cells by infecting enteropathogenic *Escherichia coli*. Mol. Microbiol. 28, 143–155.

Wong, A.R., Clements, A., Raymond, B., Crepin, V.F., Frankel, G., 2012a. The interplay between the *Escherichia coli* Rho guanine nucleotide exchange factor effectors and the mammalian. RhoGEF. inhibitor. EspH. MBio. 3 [doi: 10.1128/mBio.00250-11].

Wong, A.R., Raymond, B., Collins, J.W., Crepin, V.F., Frankel, G., 2012b. The enteropathogenic *E. coli* effector EspH promotes actin pedestal formation and elongation via WASP-interacting protein (WIP). Cell Microbiol. 14, 1051–1070.

Wu, S.-X., Peng, R.-Q., 1992. Studies on an outbreak of neonatal diarrhea caused by EPEC O127:H6 with plasmid analysis restriction analysis and outer membrane protein determination. Acta Paediatr. Scand. 81, 217–221.

Yamamoto, T., Koyama, Y., Matsumoto, M., et al., 1992. Localized, aggregative, and diffuse adherence to HeLa cells, plastic, and human small intestines by *Escherichia coli* isolated from patients with diarrhea. J. Infect. Dis. 166, 1295–1310.

Yatsuyanagi, J., Saito, S., Miyajima, Y., Amano, K., Enomoto, K., 2003. Characterization of atypical enteropathogenic *Escherichia coli* strains harboring the *astA* gene that were associated with a waterborne outbreak of diarrhea in Japan. J. Clin. Microbiol. 41, 2033–2039.

Yen, H., Ooka, T., Iguchi, A., Hayashi, T., Sugimoto, N., Tobe, T., 2010. NleC, a type III secretion protease, compromises NF-kappaB activation by targeting p65/RelA. PLoS Pathog. 6, e1001231.

Zarivach, R., Deng, W., Vuckovic, M., et al., 2008. Structural analysis of the essential self-cleaving type III secretion proteins EscU and SpaS. Nature 453, 124–127.

Zarivach, R., Vuckovic, M., Deng, W., Finlay, B.B., Strynadka, N.C., 2007. Structural analysis of a prototypical ATPase from the type III secretion system. Nat. Struct. Mol. Biol. 14, 131–137.

Zhang, H., Cheng, C., Zheng, M., et al., 2007. Enhancement of immunity to an *Escherichia coli* vaccine in mice orally inoculated with a fusion gene encoding porcine interleukin 4 and 6. Vaccine 25, 7094–7101.

Zhang, H.-Z., Lory, S., Donnenberg, M.S., 1994. A plasmid-encoded prepilin peptidase gene from enteropathogenic *Escherichia coli*. J. Bacteriol. 176, 6885–6891.

Zhang, L., Ding, X., Cui, J., et al., 2012. Cysteine methylation disrupts ubiquitin-chain sensing in NF-κB activation. Nature 481, 204–208.

Enterohemorrhagic and other Shigatoxin-producing *Escherichia coli*

Sivapriya Kailasan Vanaja[1], Dakshina M. Jandhyala[2], Emily M. Mallick[3], John M. Leong[1], and Sowmya Balasubramanian[1]

[1]Tufts University School of Medicine, Boston, MA, USA, [2]Division of Geographic Medicine and Infectious Disease Tufts-New England Medical Center, Boston, MA, USA, [3]University of Massachusetts Medical School, Worcester, MA, USA

BACKGROUND

Definition and classification

Shiga toxin-producing *Escherichia coli*, known as STEC, produce a typical AB_5 toxin, with a single A (enzymatically active) subunit and five B (binding) subunits and characterized by its profound and irreversible cytopathic effect on Vero cells (Paton and Paton, 1998). Enterohemorrhagic *E. coli*, known as EHEC, is a subset of STEC that harbors additional virulence factors, such as a type III secretion system (T3SS) that translocates bacterial effectors into mammalian cells, and a large virulence plasmid (Levine, 1987). In addition to the characteristic virulence factor armament, EHEC are also defined by their ability to induce characteristic attaching and effacing (AE) lesions in the host cells (see below) and to cause manifestations such as hemorrhagic colitis and hemolytic uremic syndrome (HUS) in humans (Croxen and Finlay, 2010). Similar to the related pathogen enteropathogenic *E. coli* (EPEC; Chapter 4), the T3SS is encoded on the LEE (locus of enterocyte effacement) pathogenicity island, and is responsible for the AE lesions induced by EHEC (Elliott et al., 1998; Croxen and Finlay, 2010). *E. coli* O157:H7 is the predominant serotype of EHEC that causes human infections globally (Tarr et al., 2005; Pennington, 2010). The lack of β-glucuronidase activity (GUD⁻) and the inability to ferment sorbitol (SOR⁻) differentiate STEC O157:H7 from other *E. coli* strains (Paton and Paton, 1998).

Escherichia coli. http://dx.doi.org/10.1016/B978-0-12-397048-0.00005-X

STEC are characterized by enormous strain diversity. They are classified into four groups, EHEC1, EHEC2, STEC1, and STEC2 (Whittam, 1998), based on multi-locus enzyme electrophoresis. STEC O157:H7 and its evolutionary ancestor, O55:H7, constitute EHEC1. EHEC2 is the most common group of non-O157 Stx-producing strains and is comprised of serotypes such as O111:H8, O111:H-, O26:H11, and O111:H11. STEC1 is highly diverse; the common serotypes in this group are O113:H21, OX3:H21, and O91:H21. Compared with the other groups, little is known about the virulence of STEC2, which is composed of serotypes O103:H2, O103:H6, and O45:H2 (Whittam, 1998). Serotype O157:H7 and six 'non-*E. coli* O157' serogroups O26, O111, O103, O121, O45, and O145 are highly significant contributors to human disease in the US (Brooks et al., 2005).

History

In 1977, almost 100 years after Stx was originally identified in culture extracts of *Shigella dysenteriae*, two groups independently demonstrated that certain *E. coli* strains produce cytotoxins that can kill Vero cells and are neutralized by anti-Stx serum (Konowalchuk et al., 1977; O'Brien and LaVeck, 1983). These cytotoxins were termed Shiga-like toxins, owing to their striking similarity to Stx of *Shigella*. In 1982, following two hemorrhagic colitis outbreaks in Oregon and Michigan in the US (Centers for Disease Control, 1982), STEC came to the attention of scientists and the public alike as a prominent human pathogen. During these food-borne disease outbreaks, serotype *E. coli* O157:H7 was isolated from both human patients and from frozen ground beef patties. Shortly afterwards, in a seminal study, Karmali and colleagues reported the presence of cytotoxins and cytotoxin-producing *E. coli* in the stools of children affected with HUS (Karmali et al., 1983). Subsequent studies led to the identification of Stx and *E. coli* O157:H7 as the causative agents of post-diarrheal HUS. The following years were marked by several outbreaks of STEC, predominantly *E. coli* O157:H7, in the US, Europe, and Japan.

Evolution

STEC evolution is characterized by the acquisition of numerous virulence factors through horizontal gene transfer effected predominantly by bacteriophage transduction, allowing further rapid gene gain or loss through duplication or deletion (Kaper et al., 2004). *E. coli* O157:H7 is believed to have diverged from a common ancestor with *E. coli* K-12 about 4.5 million years ago (Reid et al., 2000). Interestingly, an atypical EPEC serotype, O55:H7, is considered the most recent precursor of *E. coli* O157:H7 (Whittam, 1998; Reid et al., 2000).

The emergence of *E. coli* O157:H7 has been analyzed extensively by several groups, leading to the formulation of a step-wise evolutionary model for this pathogen. *E. coli* O157:H7 strains are highly clonal in nature and are distinguished from other *E. coli* strains by their SOR⁻ and GUD⁻ phenotype. The O55:H7 SOR⁺ GUD⁺ strains evolved from ancestral EPEC-like strains through the acquisition of the LEE pathogenicity island (Figure 5.1). This was followed by the acquisition of Stx2 phage through transduction of a toxin-converting bacteriophage. The next step was marked by the gain of the EHEC virulence plasmid and an antigenic shift from O55 to O157. In further steps, *E. coli* O157:H7 acquired Stx1 phage and lost

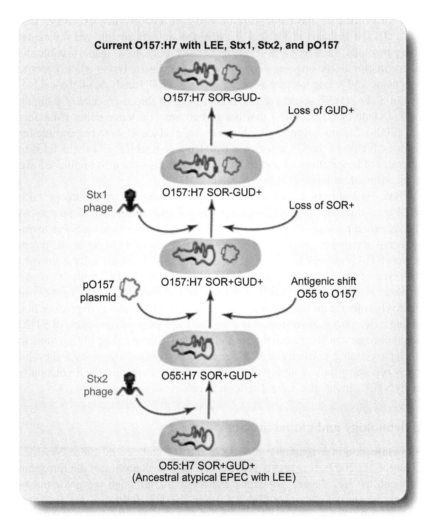

FIGURE 5.1   Step-wise model of *E. coli* O157:H7 evolution. In the first step, anatypical EPEC-like O55:H7 SOR⁺ GUD⁺ strain carrying the LEE pathogenicity island acquired Stx2 phage through transduction of a toxin-converting bacteriophage. In the next step EHEC virulence plasmid was gained and an antigenic shift occurred from O55 to O157. Subsequently, O157:H7 acquired Stx1 phage and lost the SOR fermentation and GUD activities, resulting in the current O157:H7 GUD⁻ SOR⁻ strain containing Stx1, Stx2, and the EHEC plasmid (Modified from Kaper et al., 2004).

SOR fermentation and GUD activities, resulting in the present day GUD⁻ SOR⁻ strain containing Stx1, Stx2, and the EHEC plasmid (Reid et al., 2000; Wick et al., 2005).

E. coli O157:H7 is believed to be continuously evolving by mutating existing genes and by acquiring novel virulence factors, mostly by means of gene transfer through bacteriophage transduction. Several T3SS effectors in EHEC are phage-encoded, and recent studies have found that phage-encoded regulators also influence the expression of virulence factors such as T3SS (Tree et al., 2011; Flockhart et al., 2012). Advances in sequencing and molecular typing methods, such as multi-locus sequence typing, have made feasible the documentation of the ongoing evolution of this organism. Using whole genome sequencing and single nucleotide polymorphism (SNP) analysis of clinical isolates of STEC O157, several recent studies suggest the emergence of a highly virulent clade of STEC O157 that has gained new virulence genes (Manning et al., 2008). Strains belonging to this new clade (clade 8) were responsible for recent outbreaks in the US associated with leafy vegetables, all of which led to high rates of hospitalization and HUS. Clade 8 strains were also implicated in a recent outbreak in Japan (Yokoyama et al., 2011).

Thus, the ability of *E. coli* strains to acquire virulence factors through horizontal gene transfer continues to generate novel and unusual STEC pathotypes with increased pathogenicity. A clear example is the highly virulent Shiga toxin-producing enteroaggregative *E. coli* O104:H4, the causative agent of the recent German STEC outbreak (see Chapter 11). This strain is believed to have emerged as a result of transduction of enteroaggregative *E. coli* (EAEC) with Stx phage (Rasko et al., 2011). This hybrid strain therefore apparently has the capacity to effectively colonize the human gastrointestinal tract and the ability to produce Stx, making it exceptionally virulent. It is clear that the continuous evolution of STEC can produce new public health threats. A thorough understanding of the extent to which the genomic plasticity of STEC contributes to its pathogenicity and evolution is necessary for our ability to rapidly respond to emerging and potentially deadly STEC strains. For more details on *E. coli* evolution, see Chapter 3.

Epidemiology and global impact

STEC infection is a notifiable disease in the US and several other countries (Chang et al., 2009). It is highly prevalent in the US, countries of the European continent, the UK, Japan, Argentina, and Australia, although the predominant serotype varies with location (Table 5.1). In the US, following the first outbreak in 1982, the number and severity of STEC outbreaks has dramatically increased. According to the CDC (Centers for Disease Control and Prevention, 2012), STEC causes approximately 265,000 cases of illness, including 3,600 hospitalizations, and 30 deaths in the US each year (Scallan et al., 2011). *E. coli* O157:H7 is the predominant STEC serotype responsible for human disease in the US (Tarr et al., 2005), causing more than 96,000 cases of illness

TABLE 5.1 STEC serogroup distribution across the world

Country/Continent	Prevalent serogroups	Predominant serogroup O157 vs non-O157
United States	O157, O26, O111, O103, O121, O45, O145	O157
Canada	O157, O55, O125, O26, O126, O128, O18	O157
South America	O1, O2, O15, O25, O26, O49, O92, O11	Non-O157
United Kingdom	O157	O157
Continental Europe	O157, O26, O111, O104, O103, O128, O91, O113, O2, O9, O145	Both
Australia	O157, O111, O26	Both
Japan	O157, O26, O111	O157

and 17 outbreaks annually. Between 1982 and 2002 there were 350 reported out-breaks of *E. coli* O157:H7 in the US. Of those affected, 17% were hospitalized, 4% developed HUS, and 0.5% suffered lethal infection (Scallan et al., 2011). In 2011 alone, STEC O157 caused three of the four multistate STEC outbreaks in the US. While the transmission route has been primarily (52%) foodborne, other routes of transmission such as waterborne, person-to-person, and animal contact have also caused outbreaks. Ground beef has been the major food transmission vehicle (causing 41% of outbreaks), followed by produce (21%) (Rangel et al., 2005). However, while adoption of stringent control measures has resulted in a decrease in ground beef-associated outbreaks in the US (Rangel et al., 2005; CDC, 2007) fresh produce, particularly green leafy vegetables such as spinach and lettuce, have been increasingly linked to STEC outbreaks in recent years (see below).

Although the *E. coli* O157:H7 serotype remains most commonly associated with STEC disease in the US, a few recent outbreaks have brought non-*E. coli* O157 serotypes of STEC into focus (Brooks et al., 2005; Bosilevac and Koohmaraie, 2011; Bradley et al., 2012). Indeed, the burden of illness resulting from non-*E. coli* O157 serotypes may not be fully recognized because of diagnostic limitations and inadequate surveillance. Non-*E. coli* O157 strains have been estimated to cause approximately 160,000 cases of illness annually in the US (Scallan et al., 2011). Major non-*E. coli* O157 STEC serotypes isolated in the

US between 1983–2002 include O26 (22%), O111 (16%), O103 (12%), O121 (8%), O45 (7%), and O145 (5%) (Brooks et al., 2005).

In Canada, as in the US, *E. coli* O157:H7 is the predominant serotype, causing 93% of STEC infections (Pennington, 2010). Other major serotypes isolated from humans in Canada include O55, O125, O26, O126, O128, and O18 (Table 5.1). The incidence of STEC in Canada ranges from 3.0 to 5.3 cases per 100 000, peaking in 1989 and declining thereafter (CCDR, 1997). Ground beef has been the most common source of outbreaks, followed by raw milk, apple juice, and contaminated water.

In contrast to the North American countries, STEC epidemiology in South America is marked by a strikingly low incidence of the *E. coli* O157:H7 serotype (Lopez et al., 1995, 1998; Aidar-Ugrinovich et al., 2007). For example, *E. coli* O157:H7 was isolated from 0–18% of STEC cases in Argentina, 9% in Chile, and none in Uruguay. Common serotypes isolated from HUS and bloody diarrhea cases include *E. coli* with serotypes O1:NM, O2:NM, O15:H(–), O25:H(–), O26:H(–), O49:H10, O92:H3, and O11:NM (Table 5.1). South America, specifically Argentina, has an exceptionally high HUS frequency (22 cases per 100,000 children less than 5 years old) (Lopez et al., 1998).

STEC infections are a very important public health problem in the UK, causing 1% of all food poisoning cases in England and Wales and 3% in Scotland (Money et al., 2012). In the UK, O157:H7 is the most common STEC serotype; other serotypes of STEC are rarely isolated from human cases. *E. coli* O157:H7 was isolated first in the UK from an outbreak of HUS in 1983 (Taylor et al., 1986). In contrast to the UK, both STEC O157 and non-*E.coli* O157 serotypes are equally important in Continental Europe, especially in Germany where STEC O157 causes only 30.5% of STEC infections (Eurosurveillance Editorial, 2012). However, compared to the US and the UK, STEC outbreaks are observed less frequently in Continental Europe. In 2010, a total of 4000 cases were reported in the European Union (EU), with Germany accounting for 56.8% and the Netherlands accounting for 22.0% (Eurosurveillance Editorial, 2012). Common non-*E. coli* O157 serotypes isolated from the EU include serotypes O26, O111, O103, O128, O91, O113, O2, O9, and O145 (Pennington, 2010) (Table 5.1). It is important to note that in the EU in 2011, an unusual serotype of STEC, O104:H4, caused one of the deadliest known STEC outbreaks ever recorded (~3.3% mortality rate). This outbreak originated in Germany in May 2011 and later spread to most of the countries in Europe. Approximately 4200 cases were identified globally. This outbreak was marked by a strikingly high rate of HUS (22%) and hemorrhagic colitis (78%) (Buchholz et al., 2011; Rasko et al., 2011; Wu et al., 2011).

The overall yearly rate of STEC infections in Australia is markedly lower than in North American and European countries: only 0.4 cases per 100,000 between 2000–2010, of which 0.12 was STEC O157 serotype (Vally et al., 2012). Thus, STEC O157 and non-*E.coli* O157 serotypes have been isolated at similar rates from Australian STEC patients (Table 5.1). The predominant non-*E.coli* O157 serotypes are O111 (13.7%) and O26 (11.1%) (Vally et al., 2012). New Zealand, Australia's nearest neighbor has a higher rate, e.g. 3.3 cases per 100,000 in 2009 (Leotta et al., 2008; Vally et al., 2012).

Among Asian countries, STEC disease is most prevalent in Japan (0.74 cases per 100,000 in 1999–2004), where multiple outbreaks occur every year (Sakuma et al., 2006). Although other serogroups, such as O26 and O111, have been reported to cause outbreaks, serogroup O157 is the predominant serogroup in Japan (Table 5.1). In 1996, multiple STEC O157 outbreaks throughout Japan affected 11 826 people and caused 12 deaths. Indeed, one of the largest reported *E. coli* O157:H7 outbreaks in history occurred that year in Sakai City, affecting approximately 8000 people, although with very low mortality (i.e. two deaths) (Izumiya et al., 1997). Until recently, only a few studies on STEC epidemiology have been done in other Asian countries. Human cases of STEC have been reported in India, Thailand, Vietnam, and Bangladesh. In 2006, the prevalence of STEC in Dhaka, Bangladesh was estimated to be 0.5% among hospitalized patients. Interestingly, no STEC O157 strains were isolated from these patients; the common serotypes isolated included O32:H25, O2:H45, O76:H19, ONT:H25, and ONT:H19 (Islam et al., 2007, 2008). Because of the low incidence of severe disease and complications such as HUS, the STEC strains in these developing countries are believed to be less virulent.

Economic impact

STEC outbreaks inflict a severe global economic burden both on the public and on the food industry due to costs associated with illness, product recalls, and compensations. The Economic Research Service of the USDA estimated the annual cost of illness due to STEC in the US at $1 billion in 2000. As of 2009, illness due to STEC O157 alone was estimated to cost $478 million in the US (Frenzen et al., 2005). Additionally, the total cost of each outbreak is amplified by legal costs and losses due to product recalls. For instance, a 1996 outbreak associated with an apple juice outbreak in the US resulted in $1.5 million in federal fines, $6.5 million in losses due to a product recall, and $12 million in claim settlements (Pennington, 2010). In Canada, the annual economic impact of STEC O157 illness alone is CAD$21 million, and more than CAD$82 million due to product recalls and other associated costs (George Morris Study Excerpt, 2007). Illness from STEC illnesses costs €9.1 million annually in the Netherlands (Tariq et al., 2011), and the 2011 German STEC outbreak accrued a total cost of $2.84 billion throughout the EU. Additionally, EU farmers claimed to have lost $600 million per week during the outbreak period (Kirk, 2011).

MOLECULAR PATHOGENESIS

The molecular pathogenesis of EHEC has been the subject of intense study due not only to its global clinical significance but also its diversity of pathogenic strategies that serve as models for other bacterial infections. As outlined below,

EHEC interacts with the intestinal epithelium in a highly characteristic fashion, manipulates many host cell functions by injecting dozens of bacterial effectors, regulates gene expression by sensing products of both nearby bacteria and mammalian cells, and causes life-threatening disease by the local production of a potent cytotoxin that is absorbed systemically. One potential limitation of our understanding of the pathogenesis of infection by EHEC is that the majority of the findings discussed in this chapter are derived from studies of a few *E. coli* O157:H7 strains. Investigation of how broadly these pathogenic features are shared by other EHEC may be a fruitful area of future investigation.

Entry

E. coli O157:H7 and other STECs are primarily transmitted to humans through consumption of contaminated foods (Bell et al., 1994; Itoh et al., 1998; Cody et al., 1999; Guh et al., 2010; Buchholz et al., 2011). The infectious dose of EHEC is extremely low (<100 organisms), indicating that it can pass through the stomach with high efficiency (Tilden et al., 1996; Nataro and Kaper, 1998; Doyle and Pariza, 2001; Yoon and Hovde, 2008). A critical aspect of successful host entry by *E. coli* is its ability to withstand the highly acidic (pH 1.5–3.0) gastric environment (Waterman and Small, 1996; Large et al., 2005). EHEC possess at least three acid resistance (AR) systems (AR1, AR2, and AR3) that regulate an acid resistance response (Smith et al., 1992; Stim and Bennett, 1993; Lin et al., 1995; Hersh et al., 1996; Castanie-Cornet et al., 1999) (for review see Foster (2004)). In addition, the chaperone HdeA, RNA polymerase-associated-protein SspA, small non-coding DsrA RNA, and DNA-binding protein Dps have been shown to have a role in acid resistance (Lease et al., 2004; Hansen et al., 2005; Kern et al., 2007; Calhoun and Kwon, 2011; Hong et al., 2012).

Adherence

Initial attachment

Upon transit through the stomach, EHEC must adhere to the luminal surface of the large intestine in order to effectively colonize the host and compete with normal microbiota. EHEC encodes a multitude of surface factors including fimbrial and non-fimbrial adhesins (Table 5.2; for review, see Torres et al. (2005) and Bardiau et al. (2010)) that may mediate the initial interaction with the host mucosa and promote colonization and disease (La Ragione et al., 2000; Perna et al., 2001; Cergole-Novella et al., 2007).

The EHEC-specific long polar fimbriae (LPF) (Torres et al., 2002a, 2004) was recently shown to be involved in intestinal colonization of infant rabbits (Lloyd et al., 2012). Additionally, bacteria deficient in production of hemorrhagic coli pilus (HCP), a type IV pilus that EHEC shares with most other *E. coli* (see Chapter 13), display a significant reduction in adherence

TABLE 5.2 STEC adherence factors

Potential adhesins	Function
EspA	Filamentous appendage that promotes type III effector delivery and mediates interaction between host cells and AE pathogens (Ebel et al., 1998; Knutton et al., 1998; Daniell et al., 2001).
Long polar fimbriae (LPF)	Required for adherence and microcolony formation in eukaryotic cells (Torres et al., 2002; Torres et al., 2004). Has a role in pig, sheep (Jordan et al., 2004) and infant rabbit colonization (Lloyd et al., 2012) and influencing intestinal tissue tropism (Fitzhenry et al., 2006). Binds to ECM proteins (Farfan et al., 2011).
F9 Fimbriae (Type I pilus homolog)	Potential role in colonization in young calves and in adherence to cells in culture (Low et al., 2006).
Curli fimbriae	Associated with increased virulence of EHEC in mice (Uhlich et al., 2002). Promotes attachment of EHEC (O111:H-7-57C+, O157:H7 5-11C+ and O103:H2 7-52C+) to certain abiotic surfaces (Cookson et al., 2002; Pawar et al., 2005).
OmpA	Promotes adherence to HeLa and polarized Caco-2 cells (Torres and Kaper, 2003; Torres et al., 2006).
IrgA homolog adhesion (Iha)	Confers the adherence phenotype on laboratory *E. coli*; possess siderophore-receptor activity (Tarr et al., 2000; Rashid et al., 2006).
EHEC factor for adherence (Efa)	Important for adhesion of non-O157 EHEC to CHO cells *in vitro,* hemagglutination, autoaggregation and colonization of bovine intestines (Nicholls et al., 2000; Stevens et al., 2002; Deacon et al., 2010).
Calcium binding Antigen 43 homolog (Cah)	Binds to calcium and promotes autoaggregation and biofilm formation (Torres et al., 2002).
Hemorrhagic coli pilus (HCP) (Type IV pili)	Binds to laminin and fibronectin, promotes twitching motility, biofilm formation and hemagglutination (Xicohtencatl-Cortes et al., 2007, 2009). Activates release of pro- and anti-inflammatory cytokines from polarized, cultured intestinal cells (Ledesma et al., 2010).
Escherichia coli common pilus (ECP)	Important for adherence to cultured epithelial cells and virulence of *E. coli* O157 in humans (Rendon et al., 2007).
ToxB	Homolog of Efa1, encoded on plasmid pO157 and promotes adherence to cultured epithelial cells (Tatsuno et al., 2001). Affects production and secretion of some virulence factors required for the development of A/E lesions (Stevens et al., 2004).

Continued

TABLE 5.2 STEC adherence factors—cont'd

Potential adhesins	Function
Porcine attaching-effacing associated protein (Paa)	Required for the A/E activity by many A/E strains, including O157:H7 and promotes initial bacterial adherence (An et al., 1999; Batisson et al., 2003).
STEC autoagglutinatingadhesin (Saa)	Plasmid encoded and isolated from LEE-negative STEC (Paton et al., 2001). Saa-harboring STEC exhibit differential binding properties to epithelial cells (Toma et al., 2008). Saa positive STEC frequently isolated from bovine rather than human origin (Jenkins et al., 2003).
Flagella	O157:H7 flagella agglutinates rabbit red blood cells, promotes adherence to mucins I and II and persistence in chicken in the absence of intimin (Best et al., 2005; Erdem et al., 2007).
AdfO	Secreted from T2SS and encoded on the cryptic prophage CP-933O, AdfO promotes for adherence of EHEC to cultured HeLa cells (Ho et al., 2008).
YodA	AdfO facilitates YodA secretion from pO157-encoded T2SS and promotes adherence of EHEC to HeLa cells and colonization of the infant rabbit intestine (Ho et al., 2008).
StcE	A zinc metalloprotease secreted by the T2SS on the plasmid pO157, cleaves serpin C1 esterase inhibitor (Lathem et al., 2002, 2004). Reduces the mucous glycocalyx cell surface barrier and promotes a closer interaction between the bacterium and host cell (Grys et al., 2005, 2006).

to bovine gut explants (Xicohtencatl-Cortes et al., 2007). Moreover, HCP generates an antibody response in patients with HUS, suggesting that these pili are produced during infection (Xicohtencatl-Cortes et al., 2007). EspA, which forms a filamentous appendage from EHEC's T3SS, has been suggested to be an adhesin promoting initial interaction between host cells and bacteria (Ebel et al., 1998). YodA and StcE, two effectors secreted through a type II secretion system, promote EHEC adherence to cultured monolayers, as well as during colonization of infant rabbits (Grys et al., 2005; Ho et al., 2008). Non-fimbrial EHEC adhesins include OmpA, and have been shown to facilitate bacterial binding to cultured monolayers (Torres and Kaper, 2003). Finally, H7 flagella have been hypothesized to promote adherence to mucus and extracellular matrix proteins within the host intestinal tract (Erdem et al., 2007).

The best-characterized adhesin of EHEC, and an absolutely essential virulence factor, is the ~94-kDa outer-membrane protein intimin, encoded by the *eae* gene (Jerse et al., 1990; Donnenberg and Kaper, 1991; Beebakhee et al., 1992; Yu and Kaper, 1992). Intimin is an integral outer membrane protein with three domains: a flexible N-terminal region thought to be located in the periplasm, a central region that forms a β-barrel in the outer membrane, and a surface-exposed C-terminal receptor-binding domain (Ross and Miller, 2007; Yi et al., 2010). The latter domain exhibits considerable allelic variation (Frankel et al., 1994), with at least ten different intimin subtypes (Adu-Bobie et al., 1998; Oswald et al., 2000).

Intimin was first identified by its essential role in the formation of the characteristic attaching and effacing (AE) lesions on the surface of intestinal epithelial cells (Jerse et al., 1990; Donnenberg and Kaper, 1991). AE lesions are characterized by effacement of microvilli, intimate bacterial attachment to epithelial cells, and the formation of localized actin assembly, i.e. filamentous (F–) actin 'pedestals', beneath bound bacteria (Staley et al., 1969; Ulshen and Rollo, 1980; Moon et al., 1983). Intimin promotes AE lesion formation by binding to Tir (translocated intimin receptor), a type III-secreted effector that localizes in the host plasma membrane after translocation into mammalian cells (see below) (Kenny et al., 1997b; Hartland et al., 1999). Thus, intimin is required for tight attachment to cultured cells, as well as for colonization and virulence during mammalian infection (Donnenberg et al., 1993b; McKee et al., 1995; Tzipori et al., 1995; Dean-Nystrom et al., 1998).

Intimin may possess Tir-independent functions as well. Intimin of the related EPEC has been shown to contribute to the disruption of epithelial barrier function of polarized monolayers in a Tir-independent manner (Dean and Kenny, 2004), a function that may be related to Tir-independent host cell attachment. Indeed, intimin promotes binding to cultured cells in the absence of Tir, albeit with much lower efficiency than in the presence of Tir binding. (Frankel et al., 1994; McKee and O'Brien, 1995; Frankel et al., 1998a, b; Hartland et al., 1999). Several host cell receptors, including β1-chain integrins (Frankel et al., 1996; Sinclair et al., 2006) and nucleolin (Sinclair and O'Brien, 2002; Sinclair et al., 2006) have been shown to have the capacity to bind intimin. Interestingly, Stx2 (see below) has been shown to enhance adherence of EHEC to intestinal epithelial cells and intestinal colonization in mice by increasing the surface localization of nucleolin (Robinson et al., 2006; Liu et al., 2010). Whereas all alleles of intimin bind to Tir, alleles differ in promoting epithelial colonization in some experimental models, consistent with a Tir-independent colonization function (Tzipori et al., 1995; Phillips and Frankel, 2000; Fitzhenry et al., 2002; Girard et al., 2005; Mundy et al., 2007). Finally, intimin of the murine pathogen Citrobacter rodentium, which also generates AE lesions (see below), plays a Tir-independent role in colonization of streptomycin-pretreated mice (Mallick et al., 2012a).

Formation of attaching and effacing lesions

The best-characterized interactions between EHEC and gut epithelia involve the formation of AE lesions, described above, which are characteristic of EHEC and the other AE pathogens EPEC and *C. rodentium*. AE lesion formation requires Tir, which is translocated to the host cell by a T3SS, a multi-protein export apparatus that spans the inner and outer membranes and facilitates direct injection of bacterial effectors from the bacterial cytoplasm into the host cell (see Chapter 14; Garmendia et al., 2005). Tir, intimin, and the entire T3SS are encoded on a ~35 kb chromosomal pathogenicity island called the Locus of Enterocyte Effacement (LEE) (McDaniel et al., 1995).

Tir, like intimin, is absolutely required for AE lesion formation and colonization in many animal models (Marches et al., 2000; Deng et al., 2003; Ritchie et al., 2003). Once injected into the host cell, Tir integrates into the host cell membrane in a hairpin loop conformation, with its central extracellular domain serving as the receptor for the C-terminal extracellular domain of intimin (Batchelor et al., 2000; Luo et al., 2000; Liu et al., 2002). Intimin-binding by Tir induces Tir clustering (Touze et al., 2004) and permits the cytoplasmic domains of Tir to coordinate a downstream signaling cascade resulting in the formation of F-actin pedestals (Campellone et al., 2006; Hayward et al., 2006; Brady et al., 2007).

A cytoplasmic activity of Tir critical for F-actin pedestal formation is recognition by mammalian adaptor proteins IRTKS (Insulin Receptor Tyrosine Kinase Substrate) and/or IRSp53 (Insulin Receptor Substrate Protein 53 kDa) (Vingadassalom et al., 2009; Weiss et al., 2009; de Groot et al., 2011). These homologous proteins encode an I-BAR (membrane-binding) and an SH3 (polyproline-binding) domain, and can stimulate low-level actin assembly. However, in EHEC strain O157:H7, robust actin pedestal formation requires a second-type III-secreted effector, EspF$_U$/TccP, which is encoded on the cryptic prophage CP-933U/Sp14 (Campellone et al., 2004; Garmendia et al., 2004), and which is recruited to sites of bacterial binding by the IRTKS/IRSp53 SH3 domain. In turn, EspF$_U$/TccP binds directly to the GTPase binding (GBD) region of N-WASP, resulting in its activation (Cheng et al., 2008; Sallee et al., 2008; Campellone, 2010) and the subsequent activation of the actin nucleator Arp 2/3 (reviewed in Goley and Welch (2006) and Stradal and Scita (2006)), finally resulting in actin pedestal formation.

Mutants of EHEC or other AE pathogens that entirely lack intimin or Tir are incapable of generating AE lesions and unable to colonize a range of mammalian hosts (Donnenberg et al., 1993a; Tzipori et al., 1995; Marches et al., 2000; Deng et al., 2003; Ritchie et al., 2003). Given that intimin- or Tir-deficient mutants are almost entirely defective for mammalian cell attachment, their inability to colonize animals may reflect a binding defect rather than a pedestal formation defect per se. Instead, the potential role of actin assembly in the pathogenesis of disease requires the analysis of mutants of AE pathogens that express intimin and Tir and are competent for attachment but are specifically defective for actin assembly. In fact, a mutant with this phenotype, i.e. an EHEC *espF$_U$/tccP* mutant, is defective for colonization of infant rabbits at late time points and forms unusually small bacterial aggregates in intestines of gnotobiotic piglets (Ritchie et al., 2008). Additionally, *C. rodentium* mutants similarly specifically defective in generating actin pedestals have diminished colonization ability during coinfection with wild-type bacteria (Crepin et al., 2010), or were found to be deficient in colonizing the colonic mucosa and causing lethal disease in an Stx-producing *C. rodentium* model (E. Mallick, unpublished observation). These findings indicate that pedestal formation by EHEC promotes colonization and stx-mediated disease.

Regulation of gene expression

Expression of major EHEC virulence factor genes, such as *stx*, the large pO157 plasmid, and the LEE, is regulated by multiple genes and environmental signals. (A detailed description of the highly complex pathways of EHEC regulation is beyond the scope of this chapter and readers are referred to recent reviews (Mellies et al., 2007; Croxen and Finlay, 2010)). Regulation of the LEE, which is comprised of 41 open reading frames arranged in five major polycistronic operons, *LEE1* through *LEE5*, has been the subject of considerable investigation. An important element of LEE regulation involves the global nucleoprotein-like regulator H-NS, which acts as a major suppressor of LEE expression (Bustamante et al., 2001). Ler, the major *LEE*-encoded regulator, is central to the control of LEE expression, and disrupts H-NS-mediated silencing of LEE (Mellies et al., 2007), leading to increased transcription of operons *LEE2* through *LEE5*. The additional LEE-encoded regulators include GrlA (global regulator of LEE activator) and GrlR (global regulator of LEE repressor). GrlA binds to Ler promoter and acts as an H-NS antagonist thereby promoting the expression of Ler (Jimenez et al., 2010). GrlR interacts with GrlA, interfering with the GrlA-mediated induction of Ler leading to an inhibition of LEE expression (Iyoda et al., 2006).

LEE expression is also controlled by regulators encoded elsewhere on the EHEC genome. Some of these, such as Pch (Iyoda and Watanabe, 2004), IHF (Yona-Nadler et al., 2003), RcsCDB and GrvA (Tobe et al., 2005) induce LEE expression by increasing Ler production. Negative regulators of LEE, such as GadE, GadF (Tatsuno et al., 2003; Kailasan Vanaja et al., 2009), EtrA and EivF (two regulatory proteins of *E. coli* type III secretion system 2, ETT2) (Zhang et al., 2004), Hha (Sharma and Zuerner, 2004), CadA (Vazquez-Juarez et al., 2008), and SdiA (Kanamaru et al., 2000; Sharma et al., 2010) regulate LEE expression mainly by inhibiting Ler. Finally, the sRNA chaperone Hfq regulates the LEE in a strain-specific manner, indicating that the LEE is subject to complex regulation by sRNA (Hansen and Kaper, 2009).

During intestinal infection, EHEC is exposed to extracellular factors, such as EHEC-secreted components, products of the existing microbiota, and host-derived signals. Several extracellular factors that regulate LEE expression by targeting some of the above pathways have been investigated. Growth of EPEC and EHEC in Dulbecco's modified Eagle's medium (DMEM) at host body temperature (37°C), pH 7.0, and physiological osmolarity induces LEE expression, an observation that has been historically utilized to facilitate investigation of the T3SS (Kenny and Finlay, 1995; Kenny et al., 1997a). Interestingly, expression of EHEC LEE is regulated by a quorum-sensing system that recognizes both the bacterial autoinducer AI-3 and host-derived catecholamines. This inter-kingdom communication involves multiple two-component systems that positively and negatively regulate both the LEE and non-LEE-encoded type III effectors (Kendall and Sperandio, 2007; Reading et al., 2007; Hughes and Sperandio, 2008). Other extracellular factors critical to LEE expression include iron, sodium bicarbonate, ammonium chloride, calcium, glucose, and short-chain fatty acids, which are end-products of dietary carbohydrate fermentation whose concentrations in the intestine vary depending on the microbiota and nutrient conditions (Abe et al., 1997; Ide et al., 2003; Herold et al., 2009; Delcenserie et al., 2012).

Although clearly the LEE encodes a central virulence attribute of EHEC O157:H7, many other loci contribute to the virulence of this bacterium. Several of the regulatory pathways that control LEE expression also function as regulators of other virulence factors. For example, Ler increases expression of long polar fimbriae and the secreted protease StcE encoded on the pO157 virulence plasmid (Lathem et al., 2002; Torres et al., 2007). The quorum-sensing system that recognizes the bacterial autoinducer AI-3 and host catecholamines also regulates Shiga toxin and flagellae production in a complex manner (Sperandio et al., 2002; Jeon and Itoh, 2007).

Disruption of host defense

Immune responses may promote clearance of the bacterium, or could facilitate host damage, and an important component of the pathogenesis of EHEC and other AE pathogens is the ability to disrupt bacterial recognition by the innate immune system. It has been suggested that EHEC/EPEC infection blocks phagocytosis, one of the first steps in the development of an antimicrobial immune response. T3SS proteins such as EspB, EspF, EspH, and EspJ are implicated in this antiphagocytic activity of EHEC/EPEC (Wong et al., 2011). EspB binds to the actin binding domains of several myosin family members and may block the closure of the phagocytic cup (Diakonova et al., 2002; Iizumi et al., 2007). EspF also appears to inhibit phagocytosis by binding to actin (Alto et al., 2007). Additionally, EspH blocks phagocytosis by binding and inhibiting DH-PH domain-containing Rho GTPases, which control cytoskeletal remodeling during phagocytosis (Dong et al., 2010). In contrast, the mechanism of EspJ-mediated antiphagocytic activity remains unknown even though it has been shown to inhibit two phagocytic pathways mediated by FCγR and CR3 (Marches et al., 2008).

EHEC/EPEC further disrupts mounting of an effective proinflammatory response by interfering with innate immune signaling pathways. A common target in the host for this purpose is the nuclear factor kappa B (NF-κB), a key protein in expression, and production of proinflammatory cytokines. Normally, the NF-κB-inhibitory protein IκB sequesters NF-κB in an inactive state, and upon activation of pattern recognition receptors such as TLRs by pathogen-associated molecular patterns (PAMPs), active NF-κB is released from IκB and translocates from the cytosol to the nucleus to activate transcription of target genes, such as those encoding proinflammatory cytokines and chemokines (Kawai and Akira, 2007). These proinflammatory mediators potentially promote clearance of the infection.

Some of the most detailed studies addressing disruption of inflammatory signaling in the host cell involve the action of non-LEE encoded (Nle) type-III-secreted effectors such as NleE, NleB, NleC, NleD, and NleH. Partly because type III translocation by *in vitro* cultivated EPEC is more efficient than that by EHEC, work on type III effectors shared by the two pathogens has been performed largely in EPEC. Given the high sequence homology of EPEC and EHEC (and *C. rodentium*) effector alleles, it seems likely that the mechanisms of action are conserved among all the AE pathogens (see also Table 5.3). NleE appears to inhibit TNFα-induced NF-κB activation by inhibiting its nuclear translocation (Zurawski et al., 2008; Newton et al., 2010; Vossenkamper et al., 2010). Recent studies indicate that NleE modifies host ubiquitin chain sensory proteins, TAB2/3, to disrupt host NF-κB signaling (Zhang et al., 2012). NleE also prevents degradation of IκB by inhibiting activation of IKKbeta resulting in further NF-κB suppression (Nadler et al., 2010). NleB, has been shown to enhance this particular activity of NleE (Nadler et al., 2010). Furthermore, NleC and NleD have been shown to cleave the p65 subunit of NF-κB and c-Jun N-terminal kinase (JNK), preventing inflammatory responses (Yen et al., 2010). Additionally, EHEC homologs NleH1 and NleH2 also can bind to NF-κB subunit RPS3 (Wan et al., 2011). All these findings demonstrate that EHEC/EPEC has evolved a strategy to interfere with the induction of a proinflammatory response by simultaneously blocking multiple steps in the inflammatory signaling cascade through the concerted activity of several T3SS effector proteins.

Host damage

During infection, AE pathogens attach to the luminal surfaces of the host intestinal epithelia where they efface localized regions of microvilli, form actin pedestals, and translocate effector proteins into the host cell. Many of these effector proteins have been shown to subvert various cellular processes and dramatically alter the function of intestinal cells, influencing colonization and likely contributing significantly to the development of diarrhea (for review see Frankel et al. (1998a) and Wong et al. (2011)).

TABLE 5.3 Effector proteins

Effectors	Function or activity
LEE-encoded	
Tir	Translocated intimin receptor; required for AE lesion formation, colonization, and virulence (Kenny et al., 1997; Marches et al., 2000; Deng et al., 2003; Ritchie et al., 2003); binds IRTKS/IRSp53 (Vingadassalom et al., 2009; Weiss et al., 2009).
Map	Promotes colonization (Ritchie and Waldor, 2005); disrupts tight junctions (Ma et al., 2006); induces transient filopodia (Kenny et al., 2002); targets mitochondria (Kenny and Jepson, 2000; Dean and Kenny, 2004); WxxxE family Cdc42 guanine-nucleotide exchange factor (GEF) (Huang et al., 2009); targets Na+/H+ exchanger regulatory factor 1 (NHERF1) (Simpson et al., 2006).
EspF	Promotes colonization (Nagai et al., 2005; Ritchie and Waldor, 2005); disrupts tight junctions (McNamara et al., 2001; Elliott et al., 2002; Guttman et al., 2006; Peralta-Ramirez et al., 2008); disrupts microvilli (Dean et al., 2006); targets mitochondria and promotes apoptosis (Crane et al., 2001; Nougayrede and Donnenberg, 2004; Nagai et al., 2005; Dean et al., 2010); anti-phagocytic (Quitard et al., 2006; Martinez-Argudo et al., 2007; Marches et al., 2008; Tahoun et al., 2011); targets transporters Abcf2 (Nougayrede et al., 2007) and SGLT-1 (Dean et al., 2006) and sodium hydrogen exchanger 3 (NHE3) (Hodges et al., 2008); binds 14-3-3 and cytokeratin 18 (Viswanathan et al., 2004); nucleates sorting nexin 9 (SNX9) and N-WASP to induce membrane remodeling (Alto et al., 2007).
EspG	Disrupts microtubules (Tomson et al., 2005); promotes actin stress fibers (Tomson et al., 2005); alters epithelial paracellular permeability (Matsuzawa et al., 2005); promotes colonization of small-intestine in rabbits (Ritchie and Waldor, 2005); targets ADP-ribosylation factor GTPase (ARF), p21-activated kinases (PAKs) (Selyunin and Alto, 2011; Selyunin et al., 2011), and GM130 (Clements et al., 2011) to disrupt Golgi function
EspH	Modulates actin pedestal formation (Tu et al., 2003); promotes brush border remodeling (Shaw et al., 2005); recruits N-WASP and Arp2/3 (Wong et al., 2012); inactivates RhoGEFs (Wong et al., 2012).
EspZ (SepZ)	Interacts with CD98 and induces FAK phosphorylation and cell detachment *in vitro* (Shames et al., 2010); interacts with inner mitochondrial membranetranslocase TIM17b to protect against rapid cell death (Kanack et al., 2005; Shames et al., 2011; Roxas et al., 2012).
Non-LEE encoded	
EspFu/ TccP	Required for AE lesion formation (Campellone et al., 2004; Garmendia et al., 2004); promotes colonization (Vlisidou et al., 2006); binds to host proteins IRTKS/IRSp53 (Vingadassalom et al., 2009; Weiss et al., 2009), F-BAR proteins (Campellone et al., 2012) and N-WASP (Garmendia et al., 2006; Cheng et al., 2008; Sallee et al., 2008).

TABLE 5.3 Effector proteins—cont'd

Effectors	Function or activity
TccP2	EspF$_U$/TccP homolog (Ogura et al., 2007; Ooka et al., 2007; Whale et al., 2007).
EspJ	Anti-opsonophagocytic (Dahan et al., 2005; Marches et al., 2008).
EspK	Colonization factor (Calves) (Vlisidou et al., 2006).
EspL2	Interacts with Annexin A2, induces microcolony formation and F-actin aggregation (Miyahara et al., 2009; Tobe, 2010).
EspM1/M2	WxxxE family RhoA GEF (Arbeloa et al., 2009, 2010); induces stress fibers, modulates pedestal formation (Arbeloa et al., 2008); alters polarized epithelial architecture (Simovitch et al., 2010).
EspI (NleA)	Colonization factor (Lee et al., 2008); binds Sec24 and inhibits COPII-dependent vesicular transport (Kim et al., 2007; Thanabalasuriar et al., 2012); disrupts tight junctions (Thanabalasuriar et al., 2010a, b); binds PDZ domains (Lee et al., 2008).
NleB	Colonization factor (Cattle) (Misyurina et al., 2010); blocks nuclear translocation of NF-κB p65 subunit (Nadler et al., 2010; Newton et al., 2010).
NleC	Zinc-metalloprotease that represses IL-8 secretion by cleaving NF-kB and acetyltransferase p300 (Yen et al., 2010; Baruch et al., 2011; Muhlen et al., 2011; Pearson et al., 2011; Sham et al., 2011; Shames et al., 2011).
NleD	Zinc-metalloprotease that inactivates JNK and diminishes IL-8 secretion (Baruch et al., 2011).
NleE	Inhibits NF-kB activation by blocking IkB degradation (Nadler et al., 2010; Newton et al., 2010; Vossenkamper et al., 2010).
NleF NleG/NleI	Colonization factor (Gnotobiotic piglets) (Echtenkamp et al., 2008) U-Box-E3 ubiquitin ligase (Li et al., 2006; Wu et al., 2010).
NleH1/H2	Colonization factor (Royan et al., 2010); binds human ribosomal protein S3 (RPS3), a cofactor of NF-kB transcriptional complex and disrupts innate immune response (Hemrajani et al., 2008; Gao et al., 2009; Pham et al., 2012); interacts with Bax inhibitor-1 to block apoptosis (Hemrajani et al., 2010); decreases IKK-β-activity (Royan et al., 2010).
NleL	E3 ubiquitin ligase; modulates pedestal formation (Lin et al., 2011; Piscatelli et al., 2011).
Cif	Cycle inhibiting factor; binds NEDD8 and blocks CRL ubiquitin ligase activity (Morikawa et al., 2010); induces cell cycle arrest (Marches et al., 2003; Taieb et al., 2006; Cui et al., 2010); induces delayed apoptosis (Samba-Louaka et al., 2009).

Translocated effectors are responsible for much of the damage to host cells. The precise number of effectors has not been definitively determined, but certainly numbers in the dozens (Tobe et al., 2006). The set of effectors includes those that (i) disrupt actin cytoskeleton, resulting in the formation of pedestals (see above), the disruption of tight junctions and/or microvilli; (ii) interfere with vesicle trafficking pathways; (iii) induce mitochondrial dysfunction; or (iv) trigger apoptosis. Given that many effectors are described in some detail in Chapters 4 and 15 of this volume, in this narrative we give a relatively brief overview of some of the critical effectors. In addition, Table 5.3 describes the *in vitro* and *in vivo* functions of many of the EHEC effector proteins, as well as references to primary literature.

Alteration of host vesicle trafficking pathways

EHEC and other AE pathogens manipulate host vesicle trafficking pathways. The effector EspG binds several eukaryotic proteins, including ADP-ribosylating GTPase (ARF), which regulates Golgi membrane trafficking, the Golgi matrix protein GM130, and p21-activated kinases (PAKs), which in turn regulate a wide variety of cellular processes (Clements et al., 2011; Selyunin et al., 2011b). Consistent with the hypothesis that EspG functions as a scaffold to generate a novel signaling complex on the Golgi membrane, EspG localizes to Golgi, inhibits endomembrane trafficking, and induces Golgi fragmentation. NleA inhibits vesicle trafficking between the endoplasmic reticulum and the Golgi and interacts with the Sec24 subunit of the COPII vesicle complex, which controls membrane fusion events (Kim et al., 2007; Lee et al., 2008a; Thanabalasuriar et al., 2012). Finally, overexpression of EspF (*E. coli* secreted protein F) in mammalian cells induces membrane tabulation (Alto et al., 2007) and like EspG, appears to function as a scaffold to generate a novel complex by targeting more than one eukaryotic target, in this case SNX9, which binds to membranes and regulates endocytosis, and the actin nucleation promoting factor N-WASP (Weflen et al., 2010).

Disruption of epithelial barrier and absorptive function

Epithelial tight junctions (TJs) serve to separate the luminal and adluminal environments as well as the apical membrane and basolateral membrane proteins. AE pathogens target tight junctions, leading to compromised barrier function that has been assayed by increased permeability and decreased transepithelial resistance of polarized monolayers (Canil et al., 1993; Dean and Kenny, 2004; Shifflett et al., 2005; Guttman et al., 2006a, b). The functional compromise of TJs correlates with morphological changes, for example the redistribution of TJ-associated proteins such as ZO-1, claudin, and occludin (Simonovic et al., 2000; Roxas et al., 2010; Zhang et al., 2010). Changes in epithelial barrier function observed during infection of monolayers have also been documented during animal infection by *C. rodentium* (Guttman et al., 2006a; Ma et al., 2006;

Flynn and Buret, 2008). The mechanism of TJ disruption is incompletely understood. The effector EspF, described above, is required for this step (McNamara et al., 2001; Dean and Kenny, 2004; Guttman et al., 2006a; Peralta-Ramirez et al., 2008), and its function can be supplied by the related effector EspF$_U$/TccP, which plays a critical role in pedestal formation and which, like EspF, is capable of binding to the actin nucleation promoting factor N-WASP (Viswanathan et al., 2004a). Map (Mitochondrial-associated protein), which activates Rho GTPases, is also required for tight junction disruption (McNamara et al., 2001; Dean and Kenny, 2004; Viswanathan et al., 2004a; Simpson et al., 2006; Huang et al., 2009). Finally, intimin (but not Tir) plays a role in this activity (Dean and Kenny, 2004), although the precise function of intimin in this process is not known.

The mechanisms of effacement, i.e. the loss of absorptive microvilli of host cells, are not well understood, but intimin, at least three type III effectors, i.e. Map, EspF, and Tir, and one or more host signaling pathways dependent on the host Ca^{++}-dependent protease calpain are required (Potter et al., 2003; Dean et al., 2006; Lai et al., 2011). As enterocyte microvilli vastly increase cell surface area and thus the absorptive capacity of the intestine, the effacement of intestinal epithelium may contribute to the diarrhea characteristic of EHEC infection.

Induction of mitochondrial dysfunction and apoptosis

In addition to their involvement in disruption of other host processes, EspF and Map also have mitochondria-associated functions. EspF localizes to mitochondria and disrupts the mitochondrial membrane potential, leading to mitochondrial swelling and damage (Kenny and Jepson, 2000; Kenny, 2002) and triggering cell apoptosis (Crane et al., 2001; Nougayrede and Donnenberg, 2004; Nagai et al., 2005). Map also is targeted to mitochondria, causing mitochondrial dysfunction both in cultured cells and in colonoctyes during murine infection by *C. rodentium* (Kenny and Jepson, 2000; Ma et al., 2006; Papatheodorou et al., 2006). Apoptosis can apparently be also triggered by mitochondrial-independent means, as another effector, Cif, impairs the ubiquitin proteosome system, leading to disruption of the cell-cycle (Marches et al., 2003) and delayed apoptosis (Samba-Louaka et al., 2009; Jubelin et al., 2010).

Shiga toxins and development of HUS

Infection with Shiga toxin-producing *Escherichia* coli (STEC) can result in a microangiopathic sequela termed the hemolytic uremic syndrome (HUS). Defined by a triad of hemolytic anemia, thrombocytopenia, and renal failure, HUS is the most frequent cause of renal failure in children (Siegler, 2003). As mentioned above, Shiga toxin, so-named because it was first identified in *Shigella dysenteriae*, is the critical STEC virulence factor associated with the development of HUS during STEC infection (Paton & Paton, 1998).

Shiga toxins are classified as belonging to one of two types (or groups), Stx1 or Stx2, based on their serological neutralization profile; within the two types there are several subtypes (Scheutz et al., 2012). All Shiga toxins are proteins with an AB_5 quaternary structure. The enzymatic Shiga toxin A-subunit (StxA) has N-glycosidase activity, and the five B-subunits (StxB) function to bind glycolipid cell surface receptors (reviewed in Thorpe et al. (2002)). In addition, these different toxin types share similar operon structures, in which *stxA* is immediately upstream of *stxB*. Despite these similarities, Shiga toxins of a given type or subtype often have different epidemiological profiles. For example, although data suggest that O157:H7 *E. coli* strains that produce Stx2 but not Stx1 are more likely to be associated with HUS and severe disease in humans (Griffin and Tauxe, 1991; Boerlin et al., 1999; Thorpe et al., 2002; Beutin et al., 2008), not all Stx2 subtypes are associated with HUS in humans.

The genes encoding Shiga toxins generally reside in the late region of either functional or cryptic lambdoid prophages (Unkmeir and Schmidt, 2000; Schmidt, 2001), and induction of the phages have been shown to result in a marked up-regulation of toxin production in both Stx1- and Stx2-expressing STEC (Wagner et al., 2001, 2002). Phage induction can be triggered by activation of the SOS response, a bacterial stress response which is activated by DNA damage (Little and Mount, 1982). Certain antibiotics have been shown to activate this response and promote transcription of *stx1* and *stx2* genes, production of toxin, and mortality in laboratory mice (Walterspiel et al., 1992; Kimmitt et al., 2000; Zhang et al., 2000). Furthermore, during infection, antibiotic-induced bacterial cell lysis may increase the level of free luminal Stx available for systemic absorption (Kimmitt et al., 2000; Tarr et al., 2005). Consistent with these notions, antibiotic administration has been associated with an increased risk of subsequent HUS in some studies (Bell et al., 1997; Wong et al., 2000). It has therefore been suggested that administration of antibiotics, particularly those known to activate the SOS response, should be avoided for treating STEC infection. However, recent *in vitro* studies suggest that certain antibiotics that do not induce Stx production in STEC may be worthy of further study (Bielaszewska et al., 2012; Corogeanu et al., 2012).

The first step in Stx-mediated intoxication is binding of the B-subunit pentamer to a glycolipid receptor on the host cell surface (Figure 5.2, Step 1). The best-studied Stx receptor is globotriaosylceramide (Gb_3) (Jacewicz et al., 1986; Waddell et al., 1988; Hoffmann et al., 2010). Gb_3 bound toxin is endocytosed (Step 2), and trafficked retrograde through the Golgi apparatus to the endoplasmic reticulum (ER; Step 3; Sandvig et al., 1992; Arab and Lingwood, 1998; Girod et al., 1999; Sandvig and van Deurs, 2000; Tam and Lingwood, 2007). During this process, the A-subunit is proteolytically cleaved, possibly by furin located in the endosome and/or the trans-Golgi network (Garred et al., 1995a, b; Tam and Lingwood, 2007). However, it should be noted that the A-subunit is also sensitive to cleavage by cytosolic calpain (Garred et al., 1997). The resulting A1 fragment remains linked to the A2 fragment via an intra A-subunit disulfide bond (Garred et al., 1997), and investigations into Stx1 suggest the holotoxin remains intact during translocation

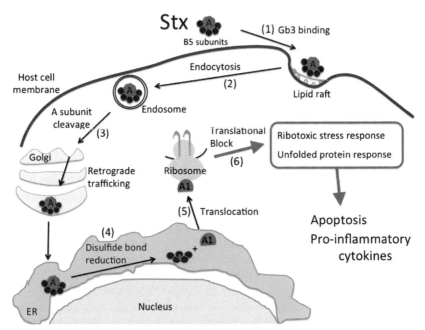

FIGURE 5.2   Generalized model of cellular intoxication by Stxs. (1) Stx holotoxin via its five B-subunits recognizes and binds to host cell Gb₃ receptors localized in lipid rafts on the surface of host cells. (2) Shiga toxin undergoes endocytosis and (3) is transported retrograde through the Golgi apparatus to the lumen of the endoplasmic reticulum (ER). During this time the StxA-subunit may undergo proteolytic cleavage, possibly by furin present in the endosome or the trans Golgi network (Garred et al., 1995a,b). (4) The proteolytically cleaved StxA1 fragment is detached from the StxA2 fragment and associated StxB-subunits through reduction of an intramolecular disulfide bond. (5) The StxA1 fragment is then transported to the cytosol (Tam and Lingwood, 2007) where it depurinates the alpha sarcin loop of the 28S ribosomal RNA. (6) This damage to the ribosome results in protein synthesis inhibition, the ribotoxic stress response, and the unfolded protein response all three of which may contribute to pathology through increased inflammation and/or apoptosis.

to the ER (Tam and Lingwood, 2007). Once in the ER, it is presumed that the disulfide bond is reduced (Step 4) based on the observation that toxins such as ricin and cholera toxin have been shown to undergo disulfide bond reduction in this compartment (Majoul et al., 1997; Tsai et al., 2001; Bellisola et al., 2004; Spooner et al., 2004). Furthermore, fluorescently labeled A- and B-subunits of Stx1 do not separate during retrograde translocation to the ER, suggesting that Stx disulfide bond reduction does not occur during transport to this compartment (Tam and Lingwood, 2007). The free A1 fragment, which possesses the *N*-glycosidase activity, is then translocated to the cytosol (Step 5; LaPointe et al., 2005; Yu and Haslam, 2005; Tam and Lingwood, 2007), where its *N*-glycosidase activity results in the cleavage of a single adenine residue (A4324) located on the α-sarcin/ricin loop of the 28S ribosomal RNA (Obrig et al., 1987; Endo et al., 1988). The depurination of the ribosome results in cessation of protein synthesis (Step 6; Endo et al., 1988; Saxena et al., 1989).

Despite being an inhibitor of protein synthesis, *in vitro* studies have demonstrated that intoxication by Shiga toxin results in a paradoxical increase in cytokine expression (Sakiri et al., 1998; Thorpe et al., 1999; Foster et al., 2000; Thorpe et al., 2001). Shiga toxins have been shown to regulate cytokine expression through the ribotoxic stress response (RSR) and the phosphatidylinositol 3-kinase/Akt/mammalian target of rapamycin signaling (PI3/AKT/mTOR) pathway (Colpoys et al., 2005; Cherla et al., 2006, 2009; Jandhyala et al., 2008, 2012). Cytokines induced by Shiga toxins include interleukin-8 (IL-8), tumor necrosis factor-α (TNF-α), interleukin-6 (IL-6), and interleukin-1β (IL-1β) (van Setten et al., 1996; Thorpe et al., 1999; Cherla et al., 2006, 2009; Jandhyala et al., 2008). In addition to promoting cytokine expression, *in vitro* studies have also demonstrated that Shiga toxins induce apoptotic signaling via the RSR and the unfolded protein response (UPR) (Smith et al., 2003; Lee et al., 2008b).

As STEC are enteric pathogens and generally non-invasive, in order to cause systemic disease such as HUS, Shiga toxins have to pass from the intestinal lumen, where they are produced, to the systemic circulation, where the toxin can access the microvasculature of the kidney and the central nervous system. Intestinal infection of mice with a genetically engineered Stx-producing *C. rodentium*, which like EHEC generates AE lesions on intestinal epithelial cells, suggests that colonization of the mucosal surface may promote systemic intoxication (E. Mallick and J. Leong, unpublished observations). *In vitro* data have suggested that Shiga toxins may cross the epithelial barrier via a transcellular route that does not result in the death of the enterocyte (Acheson et al., 1996; Philpott et al., 1997; Malyukova et al., 2009). Alternatively, Shiga toxins may enter the systemic circulation via a paracellular route, consistent with *in vitro* studies suggesting that luminal infiltration of polymorphonuclear leukocytes (PMNs) may promote the basolateral uptake of Stxs (Hurley et al., 2001). Indeed, histological analyses of patients with STEC-associated hemorrhagic colitis show large infiltrates of neutrophils in the lamina propria and crypts of the large intestine (Griffin et al., 1990).

There is currently a lack of consensus as to the role Shiga toxins play in promoting intestinal damage. Murine infection with Stx-producing *C. rodentium* results in toxin-dependent damage to the intestinal epithelium (Mallick et al., 2012b) although it has not been determined whether Stx damages the epithelium directly or indirectly, e.g. by compromising blood supply to the mucosa. Several studies have suggested that, with the exception of the enteric endothelium and Paneth cells, the human intestine is devoid of the Stx receptor Gb_3 (Holgersson et al., 1991; Berin et al., 2002; Schuller et al., 2004, 2007; Miyamoto et al., 2006). In contrast, Zumbrum et al. suggested that human colonic epithelium expresses the Stx receptor Gb_3, albeit at low levels (Zumbrun et al., 2010). Whether or not Gb_3 is expressed in the human colonic epithelium, Stx has been detected in epithelium from the illeocecal valves of patients infected with STEC, suggesting that Stx has the potential to enter these cells (Malyukova et al., 2009). In addition, Stx was found in expelled cells in the lumen, although whether these cells were damaged directly by Stx was not determined. It has been proposed that susceptibility to intoxication by Stx requires that Gb_3 be localized to lipid rafts (Falguieres et al., 2001; Kovbasnjuk et al., 2001; Hanashima et al., 2008). It is therefore conceivable that Gb_3-positive and/or -negative intestinal epithelial cells take up Stx but are not intoxicated by them.

Once in the systemic circulation, Stx can reach target organs such as the brain and kidney, where it can induce life-threatening disease. Data are inconclusive as to whether Stx circulates freely in the serum and/or is carried by blood cells to target organs. While small amounts of free Stx have been detected in HUS patients, Stx has been shown to bind *in vitro* to a variety of blood components including PMNs, platelets, monocytes, and red blood cells (RBCs) (Bitzan et al., 1994; van Setten et al., 1996; Te Loo et al., 2001b; Ghosh et al., 2004; Stahl et al., 2006), and Stx-positive PMNs have also been detected in patients with HUS (Brigotti et al., 2011).

HUS is a microangiopathic disorder featuring deposition of thrombi rich in fibrin and platelets in the renal microvasculature (reviewed in Blackall and Marques (2004)). Glomerular capillaries are occluded by these thrombi resulting in ischemic damage to the endothelium. Endothelial expression of IL-8, fractaline, and monocyte chemotactic protein (MCP-1) promotes endothelial adherence by leukocytes in the presence of Stx (Zoja et al., 2002; Geelen et al., 2008; Zanchi et al., 2008). Furthermore, PMNs are capable of transferring toxin to endothelial cells (Te Loo et al., 2000; Brigotti et al., 2010). Stx has been found bound to platelets (Stahl et al., 2006), PMN–platelet complexes, and monocyte–platelet complexes in HUS patients (Stahl et al., 2009). The toxin also binds to fibrinogen and promotes platelet aggregation and adherence to endothelial cells (Karpman et al., 2001), and thus could contribute to the fibrin-rich clots associated with HUS. Finally, Stx can induce the expression of tissue factor on monocytes, and trigger the release of monocyte-derived microparticles, which may contribute to thrombosis by increasing the amount of circulatory tissue factor (Murata et al., 2006; Stahl et al., 2009).

Renal vascular damage is a prominent feature of HUS, and kidney failure is one of the most serious sequelae during EHEC infection (reviewed in Obrig and Karpman (2012)). A plausible hypothesis is that renal pathology results from pro-thrombotic damage to Gb_3-expressing microvascular endothelium of the kidney by circulating Stx, in particular Stx2 (Obrig and Karpman, 2012). Indeed, Stx has been detected in kidney tissue from patients with HUS (Uchida et al., 1999; Chaisri et al., 2001). A possible reason the kidney is specifically targeted may be an elevated Gb_3 density in that tissue, because cultured human renal microvascular endothelial cells express 50-fold more Gb_3 than human umbilical vein endothelial cells (Obrig et al., 1993). Human glomerular podocytes, mesangial, as well as several other kidney cell types also express Gb_3 (reviewed in Obrig and Karpman (2012)). Therefore in addition to the endothelium, other tissues in the kidney may be subject to damage by Stx.

A second clinically important aspect of HUS is the potential development of neurological symptoms. These can result in epileptic seizures, alterations of consciousness, and paresis (Nathanson et al., 2010). Stx2 is able to induce hindlimb paralysis when administered to mice by intraperitoneal injection, suggesting that

Stx is directly responsible for some aspects of neurologic disease in EHEC infection (Obata et al., 2008). Consistent with this notion, Stx2 was shown *in vivo* to act on Gb_3-expressing neurons, and treatment of murine brain slices *ex vivo* resulted in increased Ca^{2+} transients by astrocytes in response to electric field stimulation.

Complement activation contributes to the vascular damage associated with over 50% of cases of 'atypical' HUS, i.e. HUS not associated with diarrheal (e.g. STEC) disease (Besbas et al., 2006; Loirat and Fremeaux-Bacchi, 2011) and may also play a role in Shiga toxin- (diarrhea-)associated HUS ('D+HUS') (reviewed in Noris et al. (2012)). C3 deposits on human microvascular endothelial cells that have been treated with Stx1 and perfused with human serum (Morigi et al., 2011). When these cells are perfused with whole blood, Stx1 pretreatment is associated with an increase in cell surface thrombi, an event that is blocked by the complement inhibitor sCR1. Treatment of human serum with Stx2 results in the formation of sC5b-9, indicating that Stx2 is also able to activate complement (Orth et al., 2009). A complement component 3a receptor antagonist diminishes glomerular fibrin deposition and platelet clumps when administered to mice prior to challenge with Stx2 and LPS (Morigi et al., 2011) and mice deficient in factor B of the alternative complement pathway did not suffer renal impairment upon Stx2/LPS challenge. Children with documented or suspected D+HUS had perturbations in serum levels of complement components, e.g. decreased levels of C3 or increased levels of C3b, Ba, Bb, and/or sC5b-9 (Cameron and Vick, 1973; Kaplan et al., 1973; Monnens et al., 1974, 1980; Robson et al., 1992; Thurman et al., 2009; Lapeyraque et al., 2011). One study demonstrated that increased plasma levels of C3a and sC5b-9 during the onset of D+HUS subsequently normalized upon recovery (Stahl et al., 2011). Thus, although the signaling mechanism(s) by which Stxs promote complement activation has not been elucidated, these data suggest that complement activation is associated with D+HUS.

As mentioned above, Stx can induce pro-inflammatory signaling via the RSR, PI3/AKT/mTOR pathways, and apoptotic signaling via the RSR and UPR pathways *in vitro*. HUS patients manifest elevated urinary and/or serum levels of MCP-1, IL-6, IL-8, and TNF-α (Karpman et al., 1995; Inward et al., 1997; van Setten et al., 1998). PMN counts are elevated during HUS, and together with increased IL-8 levels, are prognostic indicators of severe disease (Fitzpatrick et al., 1992; Robson et al., 1992; Fernandez et al., 2007). Furthermore, apoptotic glomerular and renal tubular cells are present in the kidneys of HUS patients (Karpman et al., 1998; Kaneko et al., 2001; Te Loo et al., 2001a). Thus, it is reasonable to hypothesize that the RSR, PI3/ AKT/mTOR, and UPR signaling pathways contribute at least in part to the pathology associated with HUS. However, future studies are required to better characterize the mechanism by which HUS develops, and whether other bacterial factors such as LPS, which is an integral component in one murine Stx model (Keepers et al., 2006), act in concert with Stx to promote severe disease.

Subtilase-like toxin (SubAB)

Although STEC encoding LEE are most often associated with major STEC outbreaks, LEE-negative STEC are also a significant contributor to Shiga toxin-mediated disease including HUS (Bonnet et al., 1998; Paton et al., 1999; Newton et al., 2009; Galli et al., 2010). In 1998 a STEC strain, 98NK2, responsible for an outbreak of D+HUS, was found to lack several virulence factors (including LEE) thought to be important to the onset of systemic disease (HUS) (Paton et al., 1999). However, this strain was found to express a new cytotoxin, Subtilase-like toxin AB (SubAB) (Paton et al., 2004). Most often the gene encoding SubAB has been associated with strains that lack *eae* and has been found in both Stx-encoding and non-Stx encoding *E. coli* (Wolfson et al., 2009; Buvens et al., 2010; Galli et al., 2010; Tozzoli et al., 2010). Also most studies of HUS disease by *subAB*-encoding *E. coli* have been performed in the context of STEC.

SubAB is a novel AB5 toxin that damages cells by proteolytically cleaving the ER-chaperone and master ER-regulator BiP, also termed GRP78 (Paton et al., 2006). This activity by SubAB has been shown to result in activation of the unfolded protein response in human, African green monkey, and mouse-derived cell lines, as well as after injection into mice (Paton et al., 2006; Wolfson et al., 2008). A role for SubAB in promoting thrombotic sequelae such as HUS is supported by studies showing that SubAB treatment of U937 and HUVECs results in increased tissue factor gene expression as well as increased tissue factor procoagulant activity (Wang et al., 2010). In addition, it has been demonstrated that intraperitoneal administration of SubAB into mice causes TMA with damage occurring to the kidney, liver, spleen, and brain (Wang et al., 2007), hemorrhaging in the small intestine (Furukawa et al., 2011), and severe leukocytosis (Wang et al., 2011). These data support a role for SubAB as an ancillary virulence determinant that may promote severe disease in humans by LEE-negative STEC. However, since HUS disease by *subAB* encoding *E. coli* has been addressed only in Stx-producing strains, a specific role for SubAB in virulence has not been established.

Exit

Following a typical incubation period of 3–5 days, individuals with EHEC infection characteristically develop diarrhea, at first watery, but if the illness progresses, bloody, reflecting hemorrhagic colitis (Griffin and Tauxe, 1991). It has been hypothesized that diarrhea could be a result of multiple mechanisms resulting in loss of absorptive surface (Dean et al., 2006), increased intestinal permeability, and active ion secretion (Ewe, 1988; Schmitz et al., 2000). Tight junction disruption, e.g. by the type III effectors Map and EspF, potentially allows lumen contents to enter the adlumenal compartment of the epithelium, triggering inflammation and resulting in diarrhea (Spitz et al., 1995; Elliott et al., 2002; Guttman and Finlay, 2008).

CLINICAL MANIFESTATIONS

Sources of STEC

STEC infections are predominantly zoonotic in origin and food-borne in nature with ruminants serving as the primary animal reservoirs. Cattle are the major reservoir hosts of STEC and they exist as a part of the intestinal microbiota of cattle largely without causing clinical symptoms (Borczyk et al., 1987; Chapman et al., 1993; Laegreid et al., 1999). STEC colonizes the bovine lymphoid follicle-dense mucosa of the recto–anal junction (Naylor et al., 2003; Fox et al., 2008). Prevalence of STEC in cattle ranges from 0.2% to 48.8% in the US and Canada (Pennington, 2010). Although earlier surveys in cattle indicated a low (e.g. 1.8%) prevalence of STEC in feces, possibly due to the poor sensitivity of isolation methods (Herriott et al., 1998), more recent studies have indicated a markedly higher prevalence of STEC in cattle (Heuvelink et al., 1998; Jackson et al., 1998). For example, one study indicated that 87% of beef cattle had at least one *E. coli* O157-positive pre-evisceration sample (Elder et al., 2000). Currently, STEC is considered ubiquitous in cattle farms and the shedding rate in cattle farms is estimated to be greater than 10%, sometimes approaching 100% (Hancock et al., 2001; Besser et al., 2007).

Prevalence of STEC also varies within farms. Calves with a functional rumen appear to carry STEC at a higher rate compared to heifers and other types of cattle (Pennington, 2010). Furthermore, individual animals also show tremendous variation with some animals designated as 'supershedders' excreting far more bacteria than others. A study by Chase-Topping et al. (2008) in Scotland demonstrated that 80% of the *E. coli* O157 transmission originates from 20% of the high-shedding animals in a farm. The type of cattle and animal stress due to movement and weaning are considered risk factors for high colonization and shedding (Chase-Topping et al., 2007). Interestingly, there is also a seasonal effect in the prevalence of STEC in cattle with peak prevalence in summer and early fall (Hancock et al., 2001). This corresponds with the peak in outbreaks involving ground beef in summer (Rangel et al., 2005). In addition to cattle, other domestic animals such as sheep, goats, and pigs also serve as reservoirs of STEC. Wild animals such as deer can also harbor and transmit STEC (Oliveira et al., 2008; Oporto et al., 2008).

Not all of the over 100 different serotypes of STEC that have been isolated from cattle are equally pathogenic to humans. Based on the Stx-bacteriophage insertion sites, STEC O157 strains identified in the US are classified into either bovine-biased genotypes, which are primarily found in cattle and rarely isolated from human cases, or clinical genotypes, isolated from both human cases and cattle (Besser et al., 2007). It is believed that bovine-biased genotypes may have evolved towards reduced human virulence. Supporting this hypothesis, a recent study demonstrated that infection of piglets and Dutch belted rabbits with clinical genotypes of *E. coli* O157 resulted in markedly severe disease compared to infection by bovine-biased genotypes (Shringi et al., 2012). Similarly, in Europe where *E. coli* strains are classified based on phage types (PT), strains of PT8, PT2, PT28/21 predominate in human isolates, whereas PT21 is a common PT isolated from cattle but found less frequently in clinical cases (Mora et al., 2004).

Transmission

Transmission of STEC from animal reservoirs to humans occurs predominantly through a variety of food vehicles (Table 5.4). One important factor that distinguishes STEC, particularly *E. coli* O157:H7, from many other food-borne pathogens is its very low infective dose. *E. coli* O157:H7 has an estimated infective dose of 10–100 organisms (Keene et al., 1994; Tuttle et al., 1999; Strachan

TABLE 5.4 Major multistate STEC outbreaks in the United States between 2006–2012

Year	Serotype	Vehicle/Source of infection	Number infected	Number of states affected
2012	O26	Raw clover sprouts	25 (as of 03/08/12)	8
2011	O157:H7	Romaine lettuce	60	10
	O104:H4	Travel to Germany	6	–
	O157:H7	Bologna	14	5
	O157:H7	In-shell hazel nuts	8	3
2010	O157:H7	Cheese	38	5
	O145	Shredded romaine lettuce	26	5
	O157:H7	Beef	21	16
2009	O157:H7	Beef	26	8
	O157:H7	Beef	23	9
	O157:H7	Prepackaged cookie dough	72	30
2008	O157:H7	Beef	49	7
2007	O157:H7	Pizza	21	10
	O157:H7	Ground beef patties	40	8
2006	O157:H7	Lettuce	71	5
	O157:H7	Fresh spinach	199	26

et al., 2001). Consequently, low doses of bacteria present in food products are sufficient to elicit serious infections. Contamination of meat with feces or intestinal contents of cattle during slaughter followed by consumption of undercooked meat is considered the most common route of transmission. Hamburger patties made from ground beef have caused a strikingly large number of outbreaks in the US (Rangel et al., 2005). Specifically, between 1982 and 2002, 41% of all food-borne *E. coli* O157:H7 outbreaks were associated with contaminated ground beef. Other common food vehicles of STEC include unpasteurized milk and other dairy products, fruit juices, and leafy vegetables. The significance of vegetables as a source of STEC infection has been increasing in the recent past as several outbreaks in the US and Europe were mediated by vegetables such as spinach, lettuce, alfalfa, and bean sprouts (Table 5.4) (Centers for Disease Control and Prevention, 2012). Contamination of irrigation water with animal feces is believed to be the source of tainted vegetables in such outbreaks. Interestingly, *E. coli* O157:H7 utilize multiple mechanisms to colonize and persist in plants, e.g. survival in the substomatal cavity and intercellular spaces of leafy greens (Itoh et al., 1998; Solomon et al., 2002), in some cases utilizing the T3SS (Shaw et al., 2008; Xicohtencatl-Cortes et al., 2009b). Finally, certain unconventional food vehicles have also been identified in recent multistate STEC outbreaks in the US including prepackaged raw cookie dough (2009), Lebanon bologna (2011), and in-shell hazel nuts (2011) (Centers for Disease Control and Prevention, 2012).

Animal contact, contaminated water, and environment are other important agents of transmission. In the 90 outbreaks that occurred between 1982 and 2006 in the US, Canada, UK, Ireland, Denmark, Norway, Finland, and Japan, 7.8% were transmitted by animal contact, 6.7% by water, and 2.2% by other environmental sources (Snedeker et al., 2009). Transmission from the environment is strikingly predominant in Scotland where 54% of outbreaks between 1994 and 2003 were environmental (Strachan et al., 2006). Occupational infections have also been reported in laboratory personnel (Spina et al., 2005). Additionally, it has been estimated that approximately 20% of the cases within an outbreak are caused by secondary person-to-person transmissions, and perhaps a greater proportion during outbreaks in day care and institutional facilities (Parry and Salmon, 1998; Snedeker et al., 2009).

Clinical features and complications

STEC disease displays a spectrum of manifestations varying from asymptomatic carriage to diarrhea to lethal HUS. The average incubation period for STEC infection is 3.7 days (Bell et al., 1994). The disease starts as non-bloody diarrhea, which lasts for 1–3 days followed in 90% of the cases by bloody diarrhea due to severe hemorrhagic colitis. Most patients remain afebrile and the abdominal pain is more severe compared to other bacterial gastroenteritis (Tarr et al., 2005). Abdominal tenderness and painful defecation are also common symptoms of STEC disease. Interestingly, STEC patients do not develop bacteremia. Thus, patients with acute and painful bloody diarrhea without fever are considered to have a possible STEC infection.

In most cases, diarrheal disease is self-limiting, but some individuals develop severe complications, the most lethal of which is HUS, defined as a triad of microangiopathic anemia, thrombocytopenia, and acute renal failure (Scheiring et al., 2008; Pennington, 2010). HUS associated with Stx accounts for 90% of all the HUS cases (Ray and Liu, 2001). The onset of HUS occurs typically between 5–13 days of the infection and generally starts as thrombocytopenia followed by hemolysis, anemia, and decrease in packed cell volume, and finally elevated serum creatinine level, which is often used as a guide for the clinical management. HUS causes renal failure in 15% of STEC-infected children younger than 10 years old (Ostroff et al., 1989; Bell et al., 1997), and can develop even in the absence of preceding diarrhea. In addition to young age, administration of anti-motility drugs or antibiotics and elevated white blood cell count are correlated with an increased risk of HUS (Pavia et al., 1990; Bell et al., 1997; Wong et al., 2000). The hematologic manifestations of HUS usually disappear within 1–2 weeks (Tarr et al., 2005).

The characteristic histopathologic lesions of HUS include microvascular thrombi and swollen endothelial cells, reflecting the thrombotic nature of this disorder. The renal disease is marked by hematuria, proteinuria, oliguria, and in severe cases renal failure (Tarr et al., 2005; Pennington, 2010). Increased activity of plasminogen activator inhibitor 1 (PAI-1) results in inhibition of fibrinolysis in HUS patients leading to increased intravascular generation of fibrin and thrombin, which are manifested as high plasma levels of D-dimer and thrombin fragment 1+2, respectively (Nevard et al., 1997). Consequently, high concentrations of D-dimer and fragment 1+2 when presented with diarrhea are believed to be risk factors for the development of HUS.

While most patients recover from acute renal disease, they retain a long-term risk of renal failure even many years after recovery. Furthermore, in certain cases, STEC toxicity affects other organ systems such as the pulmonary, cardiovascular, and central nervous system (CNS). Fluid overload, pleural effusions, and adult respiratory distress are frequent pulmonary complications. Adult HUS patients are more prone to congestive heart failure whereas other cardiac disorders are common in children. Additionally, complications such as intestinal perforations and necrosis, acidosis, pancreatitis, and glucose intolerance can develop during HUS (Tarr et al., 2005). Importantly, CNS complications, which develop in 25% of the HUS cases, are the most perilous and are the main precedent of HUS lethality (Eriksson et al., 2001). Ischemia-hypoxia resulting from cerebral microvascular thrombi, along with direct CNS toxicity of Stx, can lead to irritability, lethargy, and confusion. Approximately, 10% of the patients develop stroke, seizures, and coma (Taylor et al., 1986; Brandt et al., 1994).

Diagnosis

Rapid detection of STEC infection is crucial for effective management, as well as for prompt detection of outbreaks and concomitant institution of control measures. The ability to culture STEC from feces is considered to be the most definitive diagnostic procedure and is generally combined with tests to detect Stx in fecal extracts or cultures. As mentioned above, unique biochemical characteristics of *E. coli* O157:H7, such as the inability to ferment sorbitol and failure to produce β-D-glucuronidase, are utilized for detection in clinical as well as in food samples. For example, *E. coli* O157 grows as colorless (sorbitol-negative) colonies on Sorbitol MacConkey (SMAC) agar, in contrast to pink colonies produced by most fecal *E. coli* strains. Supplementation of SMAC agar with additional compounds that inhibit the growth of other bacteria or enrich *E. coli* O157 strains is often utilized (Chapman et al., 1991; Zadik et al., 1993). Commercial media containing chromogenic substrates for β-D-glucuronidase are also available for identification of *E. coli* O157 strains. A combination of two or more methods is often used to confirm the presence of *E. coli* O157:H7 (Paton and Paton, 1998), and positive colonies are typically further confirmed by slide or tube agglutination tests with O157 and H7 antisera.

A limitation of the culture detection methods described above is that they are specific to the *E. coli* O157:H7 serotype. Many non-*E. coli* O157 STEC serotypes associated with serious outbreaks lack distinguishing biochemical characteristics. However, most STEC strains produce EHEC-Hly, a hemolysin that can be used to assist in identification (Paton and Paton, 1998). In addition, assays to detect Stx can be used to identify all STEC strains. For example, Vero cells have a high plasma membrane concentration of Gb_3 and Gb_4 receptors and are therefore exquisitely sensitive to all variants of Stx, a property that has been used to detect Stx in fecal extracts and cultures (Konowalchuk et al., 1977). ELISAs using monoclonal or polyclonal antibodies to Stx are easier and less expensive than cell culture methods to detect STEC, but somewhat less sensitive, an important limitation for diagnosis at late stages of disease when the number of STEC in feces may be extremely low. Clinicians often resort to anti-Stx, anti-O Ag, or anti-H Ag serology for diagnosis (Yamada et al., 1993; De Boer and Heuvelink, 2000).

More recently, molecular techniques for the detection of STEC have gained popularity. These techniques primarily depend on the detection of Stx genes using PCR or RT-PCR, which are highly sensitive and specific (Nguyen et al., 2004). These detection techniques are particularly useful in the analysis of microbiologically complex samples or those containing non-viable bacteria (Paton and Paton, 1998). Multiplex PCR for simultaneous detection of *stx*, *eae*, and EHEC-*hly* genes is also frequently used for detection of STEC, especially EHEC strains (Fratamico et al., 1995; Paton and Paton, 2002).

Treatment

Treatment of STEC disease does not typically involve specific regimens, but rather involves supportive therapy for management of symptoms and mitigation of sequelae. The main goal of therapy is to restrain the severity and duration of gastrointestinal symptoms and hinder the development of systemic complications such as HUS. Currently, antibiotics are not recommended for treatment of STEC disease because as mentioned above, several epidemiological studies have concluded that antibiotic administration either does not provide significant benefit to EHEC-infected individuals or results in increased risk of HUS (Bell et al., 1997; Wong et al., 2000). Anti-motility agents and narcotics are also contraindicated during the diarrheal phase of STEC disease as these agents have been shown to be associated with increased risk of HUS development (Cimolai et al., 1992; Bell et al., 1997). Similarly, non-steroidal anti-inflammatory drugs can diminish renal blood flow and are therefore avoided (Murray and Brater, 1993).

One of the most serious consequences of STEC infection is renal failure, and standard rehydration protocols are considered insufficient for STEC disease management. Instead, optimal nephroprotection during the diarrhea phase of the disease is achieved by aggressive intravenous rehydration and maintenance, and patients are often admitted to the hospital for careful monitoring during administration of high sodium infusions (Ake et al., 2005; Tarr et al., 2005). After development of HUS, more intense monitoring and management strategies are required. The therapeutic focus at this stage is to maintain renal perfusion while avoiding fluid overload. HUS patients are highly prone to rapid development of anemia, which is treated with erythrocyte transfusion. In cases of clinically significant central volume overload, dialysis is preferred to diuretics. Other indications for dialysis in HUS include hyperkalemia, high serum urea concentration, persistent acidosis, oliguria or anuria, and hypertension irresponsive to treatments (Tarr et al., 2005; Pennington, 2010).

Development of therapeutic strategies that focus on hindering the progress of STEC disease has recently gained momentum. Stx is a potent activator of complement, resulting in complement hyperactivation observed in STEC-HUS in children (Noris et al., 2012). During the 2011 German STEC O104:H4 outbreak, eculizumab, a monoclonal antibody that modulates the hemolytic cascade by inhibiting cleavage of complement component C5, was widely used in patients with severe symptoms (Loos et al., 2012). The FDA has approved eculizumab for treating atypical HUS, and its utility for treating STEC-HUS deserves further careful investigation (Gruppo and Rother, 2009; Schrezenmeier and Hochsmann, 2012).

Immune response

An innate immune response may exacerbate STEC disease by inducing an inflammatory response that results in increased Stx toxicity. For example, pretreatment of human vascular endothelial cells with TNF-α or IL-1β induced expression of Gb$_3$ receptors, rendering cells more sensitive to Stx challenge (van Setten et al., 1997). TNF-α and IL-1β up-regulate the expression of several leukocyte adhesion molecules, such as VCAM-1, E-selectin, and ICAM-1, potentially enhancing inflammatory damage. LPS and flagellin are believed to act synergistically with Stx to induce proinflammatory cytokines, and LPS greatly potentiates Stx in one murine Stx injection model (Keepers et al., 2006; Jandhyala et al., 2010). In addition, Stx is a potent inducer of proinflammatory cytokines such as TNF-α, IL-6, and IL-1β (Harrison et al., 2004), and depending on the dose, administration of IL-1β enhances or diminishes Stx toxicity in mice (Palermo et al., 2000; Lentz et al., 2010).

IL-1β levels are elevated in HUS patients, raising the possibility that this cytokine contributes to human disease (Taneike et al., 2002). A recent study unraveled a novel signaling pathway through which EHEC (as well as other Gram-negative bacteria) induce IL-1β (Rathinam et al., 2012). Unlike other cytokines, the production of IL-1β requires two signals received in a stepwise fashion. Signal 1, sensed by toll-like receptors (TLRs), such as the LPS-recognition receptor TLR4, results in the production and accumulation of the proform of the cytokine, proIL-1β (Yu and Finlay, 2008). Signal 2, induced in response to distinct microbial and endogenous signals in the cytosol, results in the assembly and activation of inflammasome complexes, which are multiprotein platforms containing a pattern recognition receptor (PRR), the adapter protein ASC, and the autolytic protease caspase-1 (Schroder and Tschopp, 2010). Activated inflammasomes process the constitutively expressed pro-caspase-1 to active caspase-1, which in turn catalyzes the cleavage of proIL-1β into active IL-1β (Franchi et al., 2010; Schroder and Tschopp, 2010). Inflammasome-associated PRRs include Nod-like receptors (NLRs) such as NLRP3, which is activated by a wide variety of microbial (e.g. pore-forming toxins) or endogenous products (Broz and Monack, 2011). EHEC infection is sensed by TLR4, which leads not only to the production of proIL-1β, but also to the activation of a downstream signaling cascade mediated by an adaptor protein TRIF. The TLR4–TRIF pathway then induces the production of the proinflammatory type I interferons, which in turn induce the expression of caspase-11, an inflammatory caspase, that functions in concert with the assembled NLRP3 inflammasome to activate caspase-1, leading to IL-1β production (Rathinam et al., 2012). Thus, not only does EHEC proinflammatory signaling likely contribute to tissue damage during EHEC infection, this pathogen has been instrumental in identifying a TRIF-mediated control point that is critical for regulating innate immune responses to Gram-negative pathogens.

Whereas the innate immune response may promote disease, an adaptive response may provide a degree of protection against STEC infection. For example, studies in animal models have demonstrated a protective role for Stx antibodies (Gupta et al., 2011). Clinical studies have shown that STEC patients develop serum antibodies against Stx, LPS, T3SS proteins, and plasmid-encoded enterohemolysin (Verweyen et al., 1999; Jenkins et al., 2000; Fernandez-Brando et al., 2011). However, STEC does not appear to induce a sufficient antibody response in humans to confer protective immunity (Orth et al., 2008). Indeed, antibody response to Stx is not universal among STEC-infected individuals (Karmali et al., 1994), and the apparent low amount of Stx antibodies may reflect of the small amount of this highly potent toxin produced during human infection. It is also possible that the altered regulation of innate immune response by EHEC type III effectors interferes with the development of an effective adaptive immune response.

Control and prevention

Control and prevention of STEC infection is of tremendous importance because of the lack of specific treatments for this potent pathogen. Attempts to develop human vaccines against STEC based on Stx, T3SS proteins, and STEC LPS antigens have been unsuccessful, and current preventive measures focus primarily on breaking the transmission cycle by controlling the contamination of food. Most of the past STEC outbreaks were associated with a failure to adhere to the proper control measures. The 1994 declaration of *E. coli* O157:H7 as an adulterant in ground beef in the US resulted in a zero tolerance policy and magnified the legal liability of the meat industry (Dodd and Powell, 2009). Since then there have been important advances in pre- and post-harvest control measures in the meat industry, regulated by the USDA Food Safety and Inspection Service (FSIS), that have contributed to a decline in beef-associated outbreaks in the US (Koohmaraie et al., 2007; CDC 2007). Recently, a policy update from the FSIS classified six new non-O157 STEC serogroups as adulterants in beef that require testing. These strains, known as 'Big Six *E. coli*', include the serogroups O26, O45, O103, O111, O121, and O145 (Kalchayanand et al., 2012).

Pre-harvest intervention both diminishes beef carcass contamination, which is closely associated with STEC fecal shedding and hide prevalence, and also reduces the environmental prevalence of STEC, thereby diminishing the risk of contamination of other food products such as leafy vegetables (Berry and Wells, 2010) (Table 5.5). Vaccines that prevent STEC colonization of cattle are an emerging pre-harvest control measure. Current cattle vaccines licensed for use in the US include Epitopix, which targets siderophore receptors and porin proteins and has been shown to reduce fecal shedding of *E. coli* O157:H7 (Thornton et al., 2009). Vaccination of feedlot cattle with T3SS proteins, which are essential for cattle colonization, reduced environmental contamination with STEC and the frequency of hides that were positive for *E. coli* O157:H7. This vaccine is commercially available in Canada (Econiche™, Bioniche Life Sciences Inc) and is currently being licensed for use in the US (Smith et al., 2009). Other experimental vaccines targeting intimin, EspA, and a combination of T3SS proteins are being explored (Berry and Wells, 2010).

TABLE 5.5 Common control and preventive strategies against STEC

Level of control	Target	Control measures
Pre-harvest	Cattle	Anti-colonization vaccines
		Probiotics
		Chlorate treated water
		Neomycin sulphate
	Pen surface/manure	Dry bedding
		Carbonate treatment
		Alkali treatment
		Plant essential oils
Post-harvest	Slaughterhouse	Hide wash
		Steam pasteurization
		Organic acid washes
		Hot water washes
	Processing plants/FSIS	Routine sampling and testing for contamination
		Prevention of contaminated sample from entering food supply
	Restaurants	HACCP – Raise cooking temperature
	Vegetables	Sodium hypochlorite
		Peroxyacetic acid
		Acidic electrolysed water
		Aqueous chlorine dioxide
		Irradiation
Consumer	Humans	Anti-Stx/T3SS/LPS vaccines (under research)
		Cooking ground beef thoroughly to an internal temperature of 160°F
		Drinking only pasteurized milk, apple juice and cider
		Hygienic food handling to prevent cross contamination

Pre-harvest control measures also include probiotics, bacteriophages, and treatment with chlorate, neomycin sulfate and other dietary supplements. Probiotics may prevent colonization by competing with STEC for environmental niches, by producing antibacterial compounds, or by promoting a healthy immune system (Berry and Wells, 2010). *Lactobacillus acidophilus* alone or in combination with other probiotics has been shown to reduce *E. coli* O157:H7 shedding in cattle and prevalence on hides (Brashears et al., 2003). Bacteriophages that directly lyse STEC have been shown to eliminate *E. coli* O157:H7 *in vitro*, but results in animal models have been variable (Raya et al., 2006; Rozema et al., 2009). The USDA is currently reviewing treatment of cattle with chlorate, which is toxic to *E. coli* O157:H7 after its reduction in the intestine to chlorite (Callaway et al., 2002). Finally, measures that control STEC persistence in manure (e.g. treatment with carbonate or alkali) or cattle pen surfaces (e.g. providing dry bedding) may also reduce *E. coli* O157:H7 prevalence (Berry and Wells, 2010).

Post-harvest measures focus on reducing STEC contamination of cattle carcasses through hide washes, steam pasteurization, organic acid or hot water washes, or combinations of these treatments (Berry and Wells, 2010) (Table 5.5). Furthermore, individual beef processing plants and the FSIS routinely collect and test ground beef and beef trim samples for *E. coli* O157:H7 contamination (Koohmaraie et al., 2007). In addition, FDA has recently approved gamma irradiation of ground beef to eliminate contaminating bacteria such as EHEC (Hijaz and Smith, 2010). Further, outbreaks associated with ground beef from fast food restaurants in the US have been diminished by raising cooking temperatures, and those associated with dairy products and fruit juices have been diminished by pasteurization (Dodd and Powell, 2009; Berry and Wells, 2010). Consumers are advised to kill contaminating bacteria by cooking ground beef to an internal temperature of 160°F and to avoid cross contamination between meat and other food products through stringent hygienic practices in the kitchen (Kassenborg et al., 2004).

The recent increased frequency of vegetable-associated STEC outbreaks has highlighted the need to develop effective treatments to eliminate STEC contamination of vegetables, and compared to efforts to prevent beef contamination, development of post-harvest control measures for other food sources of STEC is still in its infancy. In this regard, the efficacy of several antimicrobial agents such as sodium hypochlorite, peroxyacetic acid, acidic electrolyzed water, aqueous chlorine dioxide, and irradiation is currently under investigation.

CONCLUSION

STEC are zoonotic pathogens that cause food-borne disease outbreaks associated with ground beef and fresh produce worldwide. The lethal complications associated with STEC disease and the severe economic impact on the food industry cause STEC to be regarded as an exceptionally hazardous pathogen.

Indeed, several recent STEC outbreaks are associated with the emergence of highly virulent clades and unusual serotypes. EHEC is a subset of STEC that carries specific virulence factors and can cause an infection that can progress to severe disease. Factors that cause tissue damage include T3SS effectors that manipulate important epithelial cell processes and Stx, a potent inhibitor of translation that is essential to systemic damage in STEC disease. Given that specific treatment measures are still lacking for STEC disease, new strategies to prevent and treat infection are urgently needed.

ACKNOWLEDGMENTS

We thank Dr. Marcia S. Osburne for critical reading of the manuscript. This work was supported by NIH RO1 HS1301 and R21 HS1337 to J.M.L., NIH R21 AI088336 to D.M.J., and NERCE Post-Doctoral Fellowship Award NIH/NIAID U54 AI057159 to S.K.V.

REFERENCES

Abe, A., Kenny, B., Stein, M., et al., 1997. Characterization of two virulence proteins secreted by rabbit enteropathogenic *Escherichia coli*, EspA and EspB, whose maximal expression is sensitive to host body temperature. Infect. Immun. 65 (9), 3547–3555.

Acheson, D.W., Moore, R., De Breucker, S., et al., 1996. Translocation of Shiga toxin across polarized intestinal cells in tissue culture. Infect. Immun. 64 (8), 3294–3300.

Adu-Bobie, J., Frankel, G., Bain, C., et al., 1998. Detection of intimins alpha, beta, gamma, and delta, four intimin derivatives expressed by attaching and effacing microbial pathogens. J. Clin. Microbiol. 36 (3), 662–668.

Aidar-Ugrinovich, L., Blanco, J., Blanco, M., et al., 2007. Serotypes, virulence genes, and intimin types of Shiga toxin-producing *Escherichia coli* (STEC) and enteropathogenic *E. coli* (EPEC) isolated from calves in Sao Paulo, Brazil. Int. J. Food Microbiol. 115 (3), 297–306.

Ake, J.A., Jelacic, S., Ciol, M.A., et al., 2005. Relative nephroprotection during *Escherichia coli* O157:H7 infections: association with intravenous volume expansion. Pediatrics 115 (6), e673–680.

Alto, N.M., Weflen, A.W., Rardin, M.J., et al., 2007. The type III effector EspF coordinates membrane trafficking by the spatiotemporal activation of two eukaryotic signaling pathways. J. Cell Biol. 178 (7), 1265–1278.

An, H., Fairbrother, J.M., Desautels, C., et al., 1999. Distribution of a novel locus called Paa (porcine attaching and effacing associated) among enteric *Escherichia coli*. Adv. Exp. Med. Biol. 473, 179–184.

Arab, S., Lingwood, C.A., 1998. Intracellular targeting of the endoplasmic reticulum/nuclear envelope by retrograde transport may determine cell hypersensitivity to verotoxin via globotri-aosyl ceramide fatty acid isoform traffic. J. Cell Physiol. 177 (4), 646–660.

Arbeloa, A., Blanco, M., Moreira, F.C., et al., 2009. Distribution of espM and espT among enteropathogenic and enterohaemorrhagic *Escherichia coli*. J. Med. Microbiol. 58 (Pt 8), 988–995.

Arbeloa, A., Bulgin, R.R., MacKenzie, G., et al., 2008. Subversion of actin dynamics by EspM effectors of attaching and effacing bacterial pathogens. Cell Microbiol. 10 (7), 1429–1441.

Arbeloa, A., Garnett, J., Lillington, J., et al., 2010. EspM2 is a RhoA guanine nucleotide exchange factor. Cell Microbiol. 12 (5), 654–664.

Bardiau, M., Szalo, M., Mainil, J.G., 2010. Initial adherence of EPEC, EHEC and VTEC to host cells. Vet. Res. 41 (5), 57.

Baruch, K., Gur-Arie, L., Nadler, C., et al., 2011. Metalloprotease type III effectors that specifically cleave JNK and NF-kappaB. EMBO J. 30 (1), 221–231.

Batchelor, M., Prasannan, S., Daniell, S., et al., 2000. Structural basis for recognition of the translocated intimin receptor (Tir) by intimin from enteropathogenic *Escherichia coli*. EMBO J. 19 (11), 2452–2464.

Batisson, I., Guimond, M.P., Girard, F., et al., 2003. Characterization of the novel factor paa involved in the early steps of the adhesion mechanism of attaching and effacing *Escherichia coli*. Infect. Immun. 71 (8), 4516–4525.

Beebakhee, G., Louie, M., De Azavedo, J., et al., 1992. Cloning and nucleotide sequence of the eae gene homologue from enterohemorrhagic *Escherichia coli* serotype O157:H7. FEMS Microbiol. Lett. 70 (1), 63–68.

Bell, B.P., Goldoft, M., Griffin, P.M., et al., 1994. A multistate outbreak of *Escherichia coli* O157:H7-associated bloody diarrhea and hemolytic uremic syndrome from hamburgers. The Washington experience JAMA 272 (17), 1349–1353.

Bell, B.P., Griffin, P.M., Lozano, P., et al., 1997. Predictors of hemolytic uremic syndrome in children during a large outbreak of *Escherichia coli* O157:H7 infections. Pediatrics 100 (1), E12.

Bellisola, G., Fracasso, G., Ippoliti, R., et al., 2004. Reductive activation of ricin and ricin A-chain immunotoxins by protein disulfide isomerase and thioredoxin reductase. Biochem. Pharmacol. 67 (9), 1721–1731.

Berin, M.C., Darfeuille-Michaud, A., Egan, L.J., et al., 2002. Role of EHEC O157:H7 virulence factors in the activation of intestinal epithelial cell NF-kappaB and MAP kinase pathways and the upregulated expression of interleukin 8. Cell Microbiol. 4 (10), 635–648.

Berry, E.D., Wells, J.E., 2010. *Escherichia coli* O157:H7 recent advances in research on occurrence, transmission, and control in cattle and the production environment. Adv. Food Nutr. Res. 60, 67–117.

Besbas, N., Karpman, D., Landau, D., et al., 2006. A classification of hemolytic uremic syndrome and thrombotic thrombocytopenic purpura and related disorders. Kidney Int. 70 (3), 423–431.

Besser, T.E., Shaikh, N., Holt, N.J., et al., 2007. Greater diversity of Shiga toxin-encoding bacterio-phage insertion sites among *Escherichia coli* O157:H7 isolates from cattle than in those from humans. Appl. Environ. Microbiol. 73 (3), 671–679.

Best, A., La Ragione, R.M., Sayers, A.R., et al., 2005. Role for flagella but not intimin in the persistent infection of the gastrointestinal tissues of specific-pathogen-free chicks by shiga toxin-negative *Escherichia coli* O157:H7. Infect. Immun. 73 (3), 1836–1846.

Beutin, L., Kruger, U., Krause, G., et al., 2008. Evaluation of major types of Shiga toxin 2E-producing *Escherichia coli* bacteria present in food, pigs, and the environment as potential pathogens for humans. Appl. Environ. Microbiol. 74 (15), 4806–4816.

Bielaszewska, M., Idelevich, E.A., Zhang, W., et al., 2012. Effects of antibiotics on Shiga toxin 2 production and bacteriophage induction by epidemic *Escherichia coli* O104:H4 strain. Antimi-crob. Agents Chemother. 56 (6), 3277–3282.

Bitzan, M., Richardson, S., Huang, C., et al., 1994. Evidence that verotoxins (Shiga-like toxins) from *Escherichia coli* bind to P blood group antigens of human erythrocytes *in vitro*. Infect. Immun. 62 (8), 3337–3347.

Blackall, D.P., Marques, M.B., 2004. Hemolytic uremic syndrome revisited: Shiga toxin, factor H, and fibrin generation. Am. J. Clin. Pathol. 121 (Suppl.), S81–S88.

Boerlin, P., McEwen, S.A., Boerlin-Petzold, F., et al., 1999. Associations between virulence factors of Shiga toxin-producing *Escherichia coli* and disease in humans. J. Clin. Microbiol. 37 (3), 497–503.

Bonnet, R., Souweine, B., Gauthier, G., et al., 1998. Non-O157:H7 Stx2-producing *Escherichia coli* strains associated with sporadic cases of hemolytic-uremic syndrome in adults. J. Clin. Microbiol. 36 (6), 1777–1780.

Borczyk, A.A., Karmali, M.A., Lior, H., et al., 1987. Bovine reservoir for verotoxin-producing *Escherichia coli* O157:H7. Lancet 1 (8524), 98.

Bosilevac, J.M., Koohmaraie, M., 2011. Prevalence and characterization of non-O157 shiga toxin-producing *Escherichia coli* isolates from commercial ground beef in the United States. Appl. Environ. Microbiol. 77 (6), 2103–2112.

Bradley, K.K., Williams, J.M., Burnsed, L.J., et al., 2012. Epidemiology of a large restaurant-associated outbreak of Shiga toxin-producing *Escherichia coli* O111:NM. Epidemiol. Infect. 140 (9), 1644–1654.

Brady, M.J., Campellone, K.G., Ghildiyal, M., et al., 2007. Enterohaemorrhagic and entero-pathogenic *Escherichia coli* Tir proteins trigger a common Nck-independent actin assembly pathway. Cell Microbiol. 9 (9), 2242–2253.

Brandt, J.R., Fouser, L.S., Watkins, S.L., et al., 1994. *Escherichia coli* O157:H7-associated hemolytic-uremic syndrome after ingestion of contaminated hamburgers. J. Pediatr. 125 (4), 519–526.

Brashears, M.M., Galyean, M.L., Loneragan, G.H., et al., 2003. Prevalence of *Escherichia coli* O157:H7 and performance by beef feedlot cattle given Lactobacillus direct-fed microbials. J. Food Prot. 66 (5), 748–754.

Brigotti, M., Tazzari, P.L., Ravanelli, E., et al., 2010. Endothelial damage induced by Shiga toxins delivered by neutrophils during transmigration. J. Leukoc. Biol. 88 (1), 201–210.

Brigotti, M., Tazzari, P.L., Ravanelli, E., et al., 2011. Clinical relevance of shiga toxin concentrations in the blood of patients with hemolytic uremic syndrome. Pediatr. Infect. Dis. J. 30 (6), 486–490.

Brooks, J.T., Sowers, E.G., Wells, J.G., et al., 2005. Non-O157 Shiga toxin-producing *Escherichia coli* infections in the United States, 1983–2002. J. Infect. Dis. 192 (8), 1422–1429.

Broz, P., Monack, D.M., 2011. Molecular mechanisms of inflammasome activation during microbial infections. Immunol. Rev. 243 (1), 174–190.

Buchholz, U., Bernard, H., Werber, D., et al., 2011. German outbreak of *Escherichia coli* O104:H4 associated with sprouts. N. Engl. J. Med. 365 (19), 1763–1770.

Bustamante, V.H., Santana, F.J., Calva, E., et al., 2001. Transcriptional regulation of type III secretion genes in enteropathogenic *Escherichia coli*: Ler antagonizes H-NS-dependent repression. Mol. Microbiol. 39 (3), 664–678.

Buvens, G., Lauwers, S., Pierard, D., et al., 2010. Prevalence of subtilase cytotoxin in verocyto-toxin-producing *Escherichia coli* isolated from humans and raw meats in Belgium. Eur. J. Clin. Microbiol. Infect. Dis. 29 (11), 1395–1399.

Calhoun, L.N., Kwon, Y.M., 2011. Structure, function and regulation of the DNA-binding protein Dps and its role in acid and oxidative stress resistance in *Escherichia coli*: a review. J. Appl. Microbiol. 110 (2), 375–386.

Callaway, T.R., Anderson, R.C., Genovese, K.J., et al., 2002. Sodium chlorate supplementation reduces *E. coli* O157:H7 populations in cattle. J. Anim. Sci. 80 (6), 1683–1689.

Cameron, J.S., Vick, R., 1973. Letter: Plasma-C3 in haemolytic-uraemic syndrome and thrombotic thrombocytopenic purpura. Lancet 2 (7835), 975.

Campellone, K.G., 2010. Cytoskeleton-modulating effectors of enteropathogenic and enterohaemor-rhagic *Escherichia coli*: Tir, EspFu and actin pedestal assembly. FEBS J. 277 (11), 2390–2402.

Campellone, K.G., Brady, M.J., Alamares, J.G., et al., 2006. Enterohaemorrhagic *Escherichia coli* Tir requires a C-terminal 12-residue peptide to initiate EspF$_U$-mediated actin assembly and harbours N-terminal sequences that influence pedestal length. Cell Microbiol. 8 (9), 1488–1503.

Campellone, K.G., Robbins, D., Leong, J.M., et al., 2004. EspFU is a translocated EHEC effector that interacts with Tir and N-WASP and promotes Nck-independent actin assembly. Dev. Cell 7 (2), 217–228.

Campellone, K.G., Siripala, A.D., Leong, J.M., et al., 2012. Membrane-deforming proteins play distinct roles in actin pedestal biogenesis by enterohemorrhagic *Escherichia coli*. J. Biol. Chem. 287 (24), 20613–20624.

Canil, C., Rosenshine, I., Ruschkowski, S., et al., 1993. Enteropathogenic *Escherichia coli* decreases the transepithelial electrical resistance of polarized epithelial monolayers. Infect, Immun 61 (7), 2755 2762.

Castanie-Cornet, M.P., Penfound, T.A., Smith, D., et al., 1999. Control of acid resistance in *Escherichia coli*. J. Bacteriol. 181 (11), 3525–3535.

CCDR, 1997. Notifiable diseases annual summary. 1995. Can. Commun. Dis. Rep. 23 (Suppl. 9), 1–104.

CDC, 2007. Preliminary FoodNet data on the incidence of infection with pathogens transmitted commonly through food – 10 states, 2006. MMWR Morb. Mortal. Wkly. Rep. 56 (14), 336–339.

CDC, 1982. Isolation of E. coli O157:H7 from sporadic cases of hemorrhagic colitis – United States. MMWR Morb Mortal Wkly Rep, 31 (43), 580, 585.

CDC, 2010. Surveillance for foodborne disease outbreaks- United States, 2007. MMWR Morb Mortal Wkly Rep, 59 (31),, 973–979.

CDC, 2012. Multi state food borne outbreak. http://www.cdc.gov/outbreaknet/outbreaks.html#ecoli.

Cergole-Novella, M.C., Nishimura, L.S., Dos Santos, L.F., et al., 2007. Distribution of virulence profiles related to new toxins and putative adhesins in Shiga toxin-producing *Escherichia coli* isolated from diverse sources in Brazil. FEMS Microbiol. Lett. 274 (2), 329–334.

Chaisri, U., Nagata, M., Kurazono, H., et al., 2001. Localization of Shiga toxins of enterohaemor-rhagic *Escherichia coli* in kidneys of paediatric and geriatric patients with fatal haemolytic uraemic syndrome. Microb. Pathog. 31 (2), 59–67.

Chang, M., Groseclose, S.L., Zaidi, A.A., et al., 2009. An ecological analysis of sociodemo-graphic factors associated with the incidence of salmonellosis, shigellosis, and E. coli O157:H7 infections in US counties. Epidemiol. Infect. 137 (6), 810–820.

Chapman, P.A., Siddons, C.A., Wright, D.J., et al., 1993. Cattle as a possible source of verocyto-toxin-producing *Escherichia coli* O157 infections in man. Epidemiol. Infect. 111 (3), 439–447.

Chapman, P.A., Siddons, C.A., Zadik, P.M., et al., 1991. An improved selective medium for the isolation of *Escherichia coli* O157. J. Med. Microbiol. 35 (2), 107–110.

Chase-Topping, M., Gally, D., Low, C., et al., 2008. Super-shedding and the link between human infection and livestock carriage of *Escherichia coli* O157. Nat. Rev. Microbiol. 6 (12), 904–912.

Chase-Topping, M.E., McKendrick, I.J., Pearce, M.C., et al., 2007. Risk factors for the presence of high-level shedders of *Escherichia coli* O157 on Scottish farms. J. Clin. Microbiol. 45 (5), 1594–1603.

Cheng, H.C., Skehan, B.M., Campellone, K.G., et al., 2008. Structural mechanism of WASP activa-tion by the enterohaemorrhagic E. coli effector EspF(U). Nature 454 (7207), 1009–1013.

Cherla, R.P., Lee, S.Y., Mees, P.L., et al., 2006. Shiga toxin 1-induced cytokine production is mediated by MAP kinase pathways and translation initiation factor eIF4E in the macrophage-like THP-1 cell line. J. Leukoc. Biol. 79 (2), 397–407.

Cherla, R.P., Lee, S.Y., Mulder, R.A., et al., 2009. Shiga toxin 1-induced proinflammatory cytokine production is regulated by the phosphatidylinositol 3-kinase/Akt/mammalian target of rapamycin signaling pathway. Infect. Immun. 77 (9), 3919–3931.

Cimolai, N., Morrison, B.J., Carter, J.E., 1992. Risk factors for the central nervous system manifes-tations of gastroenteritis-associated hemolytic-uremic syndrome. Pediatrics 90 (4), 616–621.

Clements, A., Smollett, K., Lee, S.F., et al., 2011. EspG of enteropathogenic and enterohemorrhagic *E. coli* binds the Golgi matrix protein GM130 and disrupts the Golgi structure and function. Cell Microbiol. 13 (9), 1429–1439.

Cody, S.H., Glynn, M.K., Farrar, J.A., et al., 1999. An outbreak of *Escherichia coli* O157:H7 infection from unpasteurized commercial apple juice. Ann. Intern. Med. 130 (3), 202–209.

Colpoys, W.E., Cochran, B.H., Carducci, T.M., et al., 2005. Shiga toxins activate translational regulation pathways in intestinal epithelial cells. Cell Signal 17 (7), 891–899.

Cookson, A.L., Cooley, W.A., Woodward, M.J., 2002. The role of type 1 and curli fimbriae of Shiga toxin-producing *Escherichia coli* in adherence to abiotic surfaces. Int. J. Med. Microbiol. 292 (3-4), 195–205.

Corogeanu, D., Willmes, R.Wolke, M., et al., 2012. Therapeutic concentrations of antibiotics inhibit Shiga toxin release from enterohemorrhagic *E. coli* O104:H4 from the 2011 German outbreak. BMC Microbiol. 12 (1), 160.

Crane, J.K., McNamara, B.P., Donnenberg, M.S., 2001. Role of EspF in host cell death induced by enteropathogenic *Escherichia coli*. Cell Microbiol. 3 (4), 197–211.

Crepin, V.F., Girard, F., Schuller, S., et al., 2010. Dissecting the role of the Tir:Nck and Tir:IRTKS/IRSp53 signalling pathways *in vivo*. Mol. Microbiol. 75 (2), 308–323.

Croxen, M.A., Finlay, B.B., 2010. Molecular mechanisms of *Escherichia coli* pathogenicity. Nat. Rev. Microbiol. 8 (1), 26–38.

Cui, J., Yao, Q., Li, S., et al., 2010. Glutamine deamidation and dysfunction of ubiquitin/NEDD8 induced by a bacterial effector family. Science 329 (5996), 1215–1218.

Dahan, S., Wiles, S., La Ragione, R.M., et al., 2005. EspJ is a prophage-carried type III effector protein of attaching and effacing pathogens that modulates infection dynamics. Infect. Immun. 73 (2), 679–686.

Daniell, S.J., Delahay, R.M., Shaw, R.K., et al., 2001. Coiled-coil domain of enteropathogenic *Escherichia coli* type III secreted protein EspD is involved in EspA filament-mediated cell attachment and hemolysis. Infect. Immun. 69 (6), 4055–4064.

De Boer, E., Heuvelink, A.E., 2000. Methods for the detection and isolation of Shiga toxin-producing *Escherichia coli*. Symp. Ser. Soc. Appl. Microbiol. 29, 133S–143S.

de Groot, J.C., Schluter, K., Carius, Y., et al., 2011. Structural basis for complex formation between human IRSp53 and the translocated intimin receptor Tir of enterohemorrhagic *E. coli*. Structure 19 (9), 1294–1306.

Deacon, V., Dziva, F., van Diemen, P.M., et al., 2010. Efa-1/LifA mediates intestinal colonization of calves by enterohaemorrhagic *Escherichia coli* O26: H- in a manner independent of glycosyltransferase and cysteine protease motifs or effects on type III secretion. Microbiology 156 (Pt 8), 2527–2536.

Dean, P., Kenny, B., 2004. Intestinal barrier dysfunction by enteropathogenic *Escherichia coli* is mediated by two effector molecules and a bacterial surface protein. Mol. Microbiol. 54 (3), 665–675.

Dean, P., Maresca, M., Schuller, S., et al., 2006. Potent diarrheagenic mechanism mediated by the cooperative action of three enteropathogenic *Escherichia coli*-injected effector proteins. Proc. Natl. Acad. Sci. USA 103 (6), 1876–1881.

Dean, P., Scott, J.A., Knox, A.A., et al., 2010. The enteropathogenic *E. coli* effector EspF targets and disrupts the nucleolus by a process regulated by mitochondrial dysfunction. PLoS Pathog. 6 (6), e1000961.

Dean-Nystrom, E.A., Bosworth, B.T., Moon, H.W., et al., 1998. *Escherichia coli* O157:H7 requires intimin for enteropathogenicity in calves. Infect. Immun. 66 (9), 4560–4563.

Delcenserie, V., LaPointe, G., Charaslertrangsi, T., et al., 2012. Glucose decreases virulence gene expression of *Escherichia coli* O157:H7. J. Food Prot. 75 (4), 748–752.

Deng, W., Vallance, B.A., Li, Y., et al., 2003. Citrobacter rodentium translocated intimin receptor (Tir) is an essential virulence factor needed for actin condensation, intestinal colonization and colonic hyperplasia in mice. Mol. Microbiol. 48 (1), 95–115.

Diakonova, M., Bokoch, G., Swanson, J.A., et al., 2002. Dynamics of cytoskeletal proteins during Fcgamma receptor-mediated phagocytosis in macrophages. Mol. Biol. Cell 13 (2), 402–411.

Dodd, C., Powell, D., 2009. Regulatory management and communication of risk associated with *Escherichia coli* O157:H7 in ground beef. Foodborne Pathog. Dis. 6 (6), 743–747.

Dong, N., Liu, L., Shao, F., 2010. A bacterial effector targets host DH-PH domain RhoGEFs and antagonizes macrophage phagocytosis. EMBO J. 29 (8), 1363–1376.

Donnenberg, M.S., Kaper, J.B., 1991. Construction of an *eae* deletion mutant of enteropathogenic *Escherichia coli* by using a positive-selection suicide vector. Infect. and Immuni. 59 (12), 4310–4317.

Donnenberg, M.S., Tacket, C.O., James, S.P., et al., 1993a. Role of the *eaeA* gene in experimental enteropathogenic *Escherichia coli* infection. J. Clin. Invest. 92 (3), 1412–1417.

Donnenberg, M.S., Tzipori, S., McKee, M.L., et al., 1993b. The role of the *eae* gene of enterohemorrhagic *Escherichia coli* in intimate attachment *in vitro* and in a porcine model. J. Clin. Invest. 92 (3), 1418–1424.

Doyle, M.E., Pariza, M.W., 2001. Foodborne microbial pathogens and the food research institute. Adv. Appl. Microbiol. 49, 143–161.

Ebel, F., Podzadel, T., Rohde, M., et al., 1998. Initial binding of Shiga toxin-producing *Escherichia coli* to host cells and subsequent induction of actin rearrangements depend on filamentous EspA-containing surface appendages. Mol. Microbiol. 30 (1), 147–161.

Echtenkamp, F., Deng, W., Wickham, M.E., et al., 2008. Characterization of the NleF effector protein from attaching and effacing bacterial pathogens. FEMS Microbiol. Lett. 281 (1), 98–107.

Elder, R.O., Keen, J.E., Siragusa, G.R., et al., 2000. Correlation of enterohemorrhagic *Escherichia coli* O157 prevalence in feces, hides, and carcasses of beef cattle during processing. Proc. Natl. Acad. Sci. USA 97 (7), 2999–3003.

Elliott, S.J., O'Connell, C.B., Koutsouris, A., et al., 2002. A gene from the locus of enterocyte effacement that is required for enteropathogenic *Escherichia coli* to increase tight-junction permeability encodes a chaperone for EspF. Infect. Immun. 70 (5), 2271–2277.

Elliott, S.J., Wainwright, L.A., McDaniel, T.K., et al., 1998. The complete sequence of the locus of enterocyte effacement (LEE) from enteropathogenic *Escherichia coli* E2348/69. Mol. Microbiol. 28 (1), 1–4.

Endo, Y., Tsurugi, K., Yutsudo, T., et al., 1988. Site of action of a Vero toxin (VT2) from *Escherichia coli* O157:H7 and of Shiga toxin on eukaryotic ribosomes. RNA N-glycosidase activity of the toxins. Eur. J. Biochem. 171 (1-2), 45–50.

Erdem, A.L., Avelino, F., Xicohtencatl-Cortes, J., et al., 2007. Host protein binding and adhesive properties of H6 and H7 flagella of attaching and effacing *Escherichia coli*. J. Bacteriol. 189 (20), 7426–7435.

Eriksson, K.J., Boyd, S.G., Tasker, R.C., 2001. Acute neurology and neurophysiology of haemolytic-uraemic syndrome. Arch. Dis. Child 84 (5), 434–435.

Eurosurveillance Editorial, 2012. The European Union summary report on trends and sources of zoonoses, zoonotic agents and food-borne outbreaks in 2010. Euro. Surveill. 17 (10).

Ewe, K., 1988. Intestinal transport in constipation and diarrhoea. Pharmacology 36 (Suppl. 1), 73–84.

Falguieres, T., Mallard, F., Baron, C., et al., 2001. Targeting of Shiga toxin B-subunit to retrograde transport route in association with detergent-resistant membranes. Mol. Biol. Cell 12 (8), 2453–2468.

Farfan, M.J., Cantero, L., Vidal, R., et al., 2011. Long polar fimbriae of enterohemorrhagic *Escherichia coli* O157:H7 bind to extracellular matrix proteins. Infect. Immun. 79 (9), 3744–3750.

Fernandez, G.C., Gomez, S.A., Ramos, M.V., et al., 2007. The functional state of neutrophils correlates with the severity of renal dysfunction in children with hemolytic uremic syndrome. Pediatr. Res. 61 (1), 123–128.

Fernandez-Brando, R.J., Bentancor, L.V., Mejias, M.P., et al., 2011. Antibody response to Shiga toxins in Argentinean children with enteropathic hemolytic uremic syndrome at acute and long-term follow-up periods. PLoS One 6 (4), e19136.

Fitzhenry, R., Dahan, S., Torres, A.G., et al., 2006. Long polar fimbriae and tissue tropism in *Escherichia coli* O157:H7. Microbes Infect. 8 (7), 1741–1749.

Fitzhenry, R.J., Pickard, D.J., Hartland, E.L., et al., 2002. Intimin type influences the site of human intestinal mucosal colonisation by enterohaemorrhagic *Escherichia coli* O157:H7. Gut 50 (2), 180–185.

Fitzpatrick, M.M., Shah, V., Trompeter, R.S., et al., 1992. Interleukin-8 and polymorphoneutrophil leucocyte activation in hemolytic uremic syndrome of childhood. Kidney Int. 42 (4), 951–956.

Flockhart, A.F., Tree, J.J., Xu, X., et al., 2012. Identification of a novel prophage regulator in *Escherichia coli* controlling the expression of type III secretion. Mol. Microbiol. 83 (1), 208–223.

Flynn, A.N., Buret, A.G., 2008. Tight junctional disruption and apoptosis in an *in vitro* model of *Citrobacter rodentium* infection. Microb. Pathog. 45 (2), 98–104.

Foster, G.H., Armstrong, C.S., Sakiri, R., et al., 2000. Shiga toxin-induced tumor necrosis factor alpha expression: requirement for toxin enzymatic activity and monocyte protein kinase C and protein tyrosine kinases. Infect. Immun. 68 (9), 5183–5189.

Foster, J.W., 2004. *Escherichia coli* acid resistance: tales of an amateur acidophile. Nat. Rev. Microbiol. 2 (11), 898–907.

Fox, J.T., Shi, X., Nagaraja, T.G., 2008. *Escherichia coli* O157 in the rectoanal mucosal region of cattle. Foodborne Pathog. Dis. 5 (1), 69–77.

Franchi, L., Munoz-Planillo, R., Reimer, T., et al., 2010. Inflammasomes as microbial sensors. Eur. J. Immunol. 40 (3), 611–615.

Frankel, G., Candy, D.C., Everest, P., et al., 1994. Characterization of the C-terminal domains of intimin-like proteins of enteropathogenic and enterohemorrhagic *Escherichia coli, Citrobacter freundii*, and *Hafnia alvei*. Infect. Immun. 62 (5), 1835–1842.

Frankel, G., Lider, O., Hershkoviz, R., et al., 1996. The cell-binding domain of intimin from entero-pathogenic *Escherichia coli* binds to beta1 integrins. J. Biol. Chem. 271 (34), 20359–20364.

Frankel, G., Philips, A.D., Novakova, M., et al., 1998a. Generation of *Escherichia coli* intimin derivatives with differing biological activities using site-directed mutagenesis of the intimin C- terminus domain. Mol. Microbiol. 29 (2), 559–570.

Frankel, G., Phillips, A.D., Rosenshine, I., et al., 1998b. Enteropathogenic and enterohaemorrhagic *Escherichia coli*: more subversive elements. Mol. Microbiol. 30 (5), 911–921.

Fratamico, P.M., Sackitey, S.K., Wiedmann, M., et al., 1995. Detection of *Escherichia coli* O157:H7 by multiplex PCR. J. Clin. Microbiol. 33 (8), 2188–2191.

Frenzen, P.D., Drake, A., Angulo, F.J., 2005. Economic cost of illness due to *Escherichia coli* O157 infections in the United States. J. Food Prot. 68 (12), 2623–2630.

Furukawa, T., Yahiro, K., Tsuji, A.B., et al., 2011. Fatal hemorrhage induced by subtilase cytotoxin from Shiga-toxigenic *Escherichia coli*. Microb. Pathog. 50 (3-4), 159–167.

Galli, L., Miliwebsky, E., Irino, K., et al., 2010. Virulence profile comparison between LEE-negative Shiga toxin-producing *Escherichia coli* (STEC) strains isolated from cattle and humans. Vet. Microbiol. 143 (2-4), 307–313.

Gao, X., Wan, F., Mateo, K., et al., 2009. Bacterial effector binding to ribosomal protein s3 subverts NF-kappaB function. PLoS Pathog. 5 (12), e1000708.

Garmendia, J., Carlier, M.F., Egile, C., et al., 2006. Characterization of TccP-mediated N-WASP activation during enterohaemorrhagic *Escherichia coli* infection. Cell Microbiol. 8 (9), 1444–1455.

Garmendia, J., Frankel, G., Crepin, V.F., 2005. Enteropathogenic and enterohemorrhagic *Escherichia coli* infections: translocation, translocation, translocation. Infect. Immun. 73 (5), 2573–2585.

Garmendia, J., Phillips, A.D., Carlier, M.F., et al., 2004. TccP is an enterohaemorrhagic *Escherichia coli* O157:H7 type III effector protein that couples Tir to the actin-cytoskeleton. Cell Microbiol. 6 (12), 1167–1183.

Garred, O., Dubinina, E., Holm, P.K., et al., 1995a. Role of processing and intracellular transport for optimal toxicity of Shiga toxin and toxin mutants. Exp. Cell Res. 218 (1), 39–49.

Garred, O., Dubinina, E., Polesskaya, A., et al., 1997. Role of the disulfide bond in Shiga toxin A-chain for toxin entry into cells. J. Biol. Chem. 272 (17), 11414–11419.

Garred, O., van Deurs, B., Sandvig, K., 1995b. Furin-induced cleavage and activation of Shiga toxin. J. Biol. Chem. 270 (18), 10817–10821.

Geelen, J., Valsecchi, F., van der Velden, T., et al., 2008. Shiga-toxin-induced firm adhesion of human leukocytes to endothelium is in part mediated by heparan sulfate. Nephrol. Dial Transplant 23 (10), 3091–3095.

George Morris Study Excerpt, 2007. Cost of E. coli O157:H7 illness in Canada. www.fightecoli. com/docs/MorrisReportExcerpt-en.pdf.

Ghosh, S.A., Polanowska-Grabowska, R.K., Fujii, J., et al., 2004. Shiga toxin binds to activated platelets. J. Thromb. Haemost. 2 (3), 499–506.

Girard, F., Batisson, I., Frankel, G.M., et al., 2005. Interaction of enteropathogenic and Shiga toxin-producing *Escherichia coli* and porcine intestinal mucosa: role of intimin and Tir in adherence. Infect. Immun. 73 (9), 6005–6016.

Girod, A., Storrie, B., Simpson, J.C., et al., 1999. Evidence for a COP-I-independent transport route from the Golgi complex to the endoplasmic reticulum. Nat. Cell Biol. 1 (7), 423–430.

Goley, E.D., Welch, M.D., 2006. The ARP2/3 complex: an actin nucleator comes of age. Nat. Rev. Mol. Cell Biol. 7 (10), 713–726.

Griffin, P.M., Olmstead, L.C., Petras, R.E., 1990. *Escherichia coli* O157:H7-associated colitis. A clinical and histological study of 11 cases. Gastroenterology 99 (1), 142–149.

Griffin, P.M., Tauxe, R.V., 1991. The epidemiology of infections caused by *Escherichia coli* O157:H7, other enterohemorrhagic E. coli, and the associated hemolytic uremic syndrome. Epidemiol. Rev. 13, 60–98.

Gruppo, R.A., Rother, R.P., 2009. Eculizumab for congenital atypical hemolytic-uremic syndrome. N. Engl. J. Med. 360 (5), 544–546.

Grys, T.E., Siegel, M.B., Lathem, W.W., et al., 2005. The StcE protease contributes to intimate adherence of enterohemorrhagic *Escherichia coli* O157:H7 to host cells. Infect. Immun. 73 (3), 1295 1303.

Grys, T.E., Walters, L.L., Welch, R.A., 2006. Characterization of the StcE protease activity of *Escherichia coli* O157:H7. J. Bacteriol. 188 (13), 4646–4653.

Guh, A., Phan, Q., Nelson, R., et al., 2010. Outbreak of *Escherichia coli* O157 associated with raw milk, Connecticut, 2008. Clin. Infect. Dis. 51 (12), 1411–1417.

Gupta, P., Singh, M.K., Singh, Y., et al., 2011. Recombinant Shiga toxin B subunit elicits protection against Shiga toxin via mixed Th type immune response in mice. Vaccine 29 (45), 8094–8100.

Guttman, J.A., Finlay, B.B., 2008. Subcellular alterations that lead to diarrhea during bacterial pathogenesis. Trends Microbiol. 16 (11), 535–542.

Guttman, J.A., Li, Y., Wickham, M.E., et al., 2006a. Attaching and effacing pathogen-induced tight junction disruption *in vivo*. Cell Microbiol. 8 (4), 634–645.

Guttman, J.A., Samji, F.N., Li, Y., et al., 2006b. Evidence that tight junctions are disrupted due to intimate bacterial contact and not inflammation during attaching and effacing pathogen infection *in vivo*. Infect. Immun. 74 (11), 6075–6084.

Hanashima, T., Miyake, M., Yahiro, K., et al., 2008. Effect of Gb$_3$ in lipid rafts in resistance to Shiga-like toxin of mutant Vero cells. Microb. Pathog. 45 (2), 124–133.

Hancock, D., Besser, T., Lejeune, J., et al., 2001. The control of VTEC in the animal reservoir. Int. J. Food Microbiol. 66 (1–2), 71–78.

Hansen, A.M., Kaper, J.B., 2009. Hfq affects the expression of the LEE pathogenicity island in enterohaemorrhagic *Escherichia coli*. Mol. Microbiol. 73 (3), 446–465.

Hansen, A.M., Qiu, Y., Yeh, N., et al., 2005. SspA is required for acid resistance in stationary phase by downregulation of H-NS in *Escherichia coli*. Mol. Microbiol. 56 (3), 719–734.

Harrison, L.M., van Haaften, W.C., Tesh, V.L., 2004. Regulation of proinflammatory cytokine expression by Shiga toxin 1 and/or lipopolysaccharides in the human monocytic cell line THP-1. Infect. Immun. 72 (5), 2618–2627.

Hartland, E.L., Batchelor, M., Delahay, R.M., et al., 1999. Binding of intimin from enteropathogenic *Escherichia coli* to Tir and to host cells. Mol. Microbiol. 32 (1), 151–158.

Hayward, R.D., Leong, J.M., Koronakis, V., et al., 2006. Exploiting pathogenic *Escherichia coli* to model transmembrane receptor signalling. Nat. Rev. Microbiol. 4 (5), 358–370.

Hemrajani, C., Berger, C.N., Robinson, K.S., et al., 2010. NleH effectors interact with Bax inhibitor-1 to block apoptosis during enteropathogenic *Escherichia coli* infection. Proc. Natl. Acad. Sci. USA 107 (7), 3129–3134.

Hemrajani, C., Marches, O., Wiles, S., et al., 2008. Role of NleH, a type III secreted effector from attaching and effacing pathogens, in colonization of the bovine, ovine, and murine gut. Infect. Immun. 76 (11), 4804–4813.

Herold, S., Paton, J.C., Srimanote, P., et al., 2009. Differential effects of short-chain fatty acids and iron on expression of iha in Shiga-toxigenic *Escherichia coli*. Microbiology 155 (Pt 11), 3554–3563.

Herriott, D.E., Hancock, D.D., Ebel, E.D., et al., 1998. Association of herd management factors with colonization of dairy cattle by Shiga toxin-positive *Escherichia coli* O157. J. Food Prot. 61 (7), 802–807.

Hersh, B.M., Farooq, F.T., Barstad, D.N., et al., 1996. A glutamate-dependent acid resistance gene in *Escherichia coli*. J. Bacteriol. 178 (13), 3978–3981.

Heuvelink, A.E., van den Biggelaar, F.L., Zwartkruis-Nahuis, J., et al., 1998. Occurrence of vero-cytotoxin-producing *Escherichia coli* O157 on Dutch dairy farms. J. Clin. Microbiol. 36 (12), 3480–3487.

Hijaz, F.M., Smith, J.S., 2010. Levels of 2-dodecylcyclobutanone in ground beef patties irradiated by low-energy X-ray and gamma rays. J. Food Sci. 75 (9), T156–160.

Ho, T.D., Davis, B.M., Ritchie, J.M., et al., 2008. Type 2 secretion promotes enterohemorrhagic *Escherichia coli* adherence and intestinal colonization. Infect. Immun. 76 (5), 1858–1865.

Hodges, K., Alto, N.M., Ramaswamy, K., et al., 2008. The enteropathogenic *Escherichia coli* effector protein EspF decreases sodium hydrogen exchanger 3 activity. Cell Microbiol. 10 (8), 1735–1745.

Hoffmann, P., Hulsewig, M., Duvar, S., et al., 2010. On the structural diversity of Shiga toxin glycosphingolipid receptors in lymphoid and myeloid cells determined by nanoelectrospray ionization tandem mass spectrometry. Rapid. Commun. Mass. Spectrom. 24 (15), 2295–2304.

Holgersson, J., Jovall, P.A., Breimer, M.E., 1991. Glycosphingolipids of human large intestine: detailed structural characterization with special reference to blood group compounds and bacterial receptor structures. J. Biochem. 110 (1), 120–131.

Hong, W., Wu, Y.E., Fu, X., et al., 2012. Chaperone-dependent mechanisms for acid resistance in enteric bacteria. Trends Microbiol. 20 (7), 328–335.

Huang, Z., Sutton, S.E., Wallenfang, A.J., et al., 2009. Structural insights into host GTPase isoform selection by a family of bacterial GEF mimics. Nat. Struct. Mol. Biol. 16 (8), 853–860.

Hughes, D.T., Sperandio, V., 2008. Inter-kingdom signalling: communication between bacteria and their hosts. Nat. Rev. Microbiol. 6 (2), 111–120.

Hurley, B.P., Thorpe, C.M., Acheson, D.W., 2001. Shiga toxin translocation across intestinal epithelial cells is enhanced by neutrophil transmigration. Infect. Immun. 69 (10), 6148–6155.

Ide, T., Michgehl, S., Knappstein, S., et al., 2003. Differential modulation by Ca2+ of type III secretion of diffusely adhering enteropathogenic *Escherichia coli*. Infect. Immun. 71 (4), 1725–1732.

Iizumi, Y., Sagara, H., Kabe, Y., et al., 2007. The enteropathogenic *E. coli* effector EspB facilitates microvillus effacing and antiphagocytosis by inhibiting myosin function. Cell Host Microbe 2 (6), 383–392.

Inward, C.D., Varagunam, M., Adu, D., et al., 1997. Cytokines in haemolytic uraemic syndrome associated with verocytotoxin-producing *Escherichia coli* infection. Arch. Dis. Child 77 (2), 145–147.

Islam, M.A., Heuvelink, A.E., de Boer, E., et al., 2007. Shiga toxin-producing *Escherichia coli* isolated from patients with diarrhoea in Bangladesh. J. Med. Microbiol. 56 (Pt 3), 380–385.

Islam, M.A., Mondol, A.S., de Boer, E., et al., 2008. Prevalence and genetic characterization of shiga toxin-producing *Escherichia coli* isolates from slaughtered animals in Bangladesh. Appl. Environ. Microbiol. 74 (17), 5414–5421.

Itoh, Y., Sugita-Konishi, Y., Kasuga, F., et al., 1998. Enterohemorrhagic *Escherichia coli* O157:H7 present in radish sprouts. Appl. Environ. Microbiol. 64 (4), 1532–1535.

Iyoda, S., Koizumi, N., Satou, H., et al., 2006. The GrlR-GrlA regulatory system coordinately controls the expression of flagellar and LEE-encoded type III protein secretion systems in enterohemorrhagic *Escherichia coli*. J. Bacteriol. 188 (16), 5682–5692.

Iyoda, S., Watanabe, H., 2004. Positive effects of multiple pch genes on expression of the locus of enterocyte effacement genes and adherence of enterohaemorrhagic *Escherichia coli* O157: H7 to HEp-2 cells. Microbiology 150 (Pt 7), 2357–2571.

Izumiya, H., Terajima, J., Wada, A., et al., 1997. Molecular typing of enterohemorrhagic *Escherichia coli* O157:H7 isolates in Japan by using pulsed-field gel electrophoresis. J. Clin. Microbiol. 35 (7), 1675–1680.

Jacewicz, M., Clausen, H., Nudelman, E., et al., 1986. Pathogenesis of shigella diarrhea. XI. Isolation of a shigella toxin-binding glycolipid from rabbit jejunum and HeLa cells and its identification as globotriaosylceramide. J. Exp. Med. 163 (6), 1391–1404.

Jackson, S.G., Goodbrand, R.B., Johnson, R.P., et al., 1998. *Escherichia coli* O157:H7 diarrhoea associated with well water and infected cattle on an Ontario farm. Epidemiol. Infect. 120 (1), 17–20.

Jandhyala, D.M., Ahluwalia, A., Obrig, T., et al., 2008. ZAK: a MAP3Kinase that transduces Shiga toxin- and ricin-induced proinflammatory cytokine expression. Cell Microbiol. 10 (7), 1468–1477.

Jandhyala, D.M., Rogers, T.J., Kane, A., et al., 2010. Shiga toxin 2 and flagellin from shigatoxigenic *Escherichia coli* superinduce interleukin-8 through synergistic effects on host stress-activated protein kinase activation. Infect. Immun. 78 (7), 2984–2994.

Jandhyala, D.M., Thorpe, C.M., Magun, B., 2012. Ricin and Shiga toxins: effects on host cell signal transduction. Curr. Top. Microbiol. Immunol. 357, 41–65.

Jenkins, C., Chart, H., Smith, H.R., et al., 2000. Antibody response of patients infected with verocytotoxin-producing *Escherichia coli* to protein antigens encoded on the LEE locus. J. Med. Microbiol. 49 (1), 97–101.

Jenkins, C., Perry, N.T., Cheasty, T., et al., 2003. Distribution of the saa gene in strains of Shiga toxin-producing *Escherichia coli* of human and bovine origins. J. Clin. Microbiol. 41 (4), 1775–1778.

Jeon, B., Itoh, K., 2007. Production of shiga toxin by a luxS mutant of *Escherichia coli* O157:H7 *in vivo* and *in vitro*. Microbiol. Immunol. 51 (4), 391–396.

Jerse, A.E., Yu, J., Tall, B.D., Kaper, J.B., 1990. A genetic locus of enteropathogenic *Escherichia coli* necessary for the production of attaching and effacing lesions on tissue culture cells. Proc. Nat. Acad. Sci., USA 87, 7839–7843.

Jimenez, R., Cruz-Migoni, S.B., Huerta-Saquero, A., et al., 2010. Molecular characterization of GrlA, a specific positive regulator of ler expression in enteropathogenic *Escherichia coli*. J. Bacteriol. 192 (18), 4627–4642.

Jordan, D.M., Cornick, N., Torres, A.G., et al., 2004. Long polar fimbriae contribute to colonization by *Escherichia coli* O157:H7 *in vivo*. Infect. Immun. 72 (10), 6168–6171.

Jubelin, G., Taieb, F., Duda, D.M., et al., 2010. Pathogenic bacteria target NEDD8-conjugated cullins to hijack host-cell signaling pathways. PLoS Pathog. 6 (9), e1001128.

Kailasan Vanaja, S., Bergholz, T.M., Whittam, T.S., 2009. Characterization of the *Escherichia coli* O157:H7 Sakai GadE regulon. J. Bacteriol. 191 (6), 1868–1877.

Kalchayanand, N., Arthur, T.M., Bosilevac, J.M., et al., 2012. Evaluation of commonly used antimicrobial interventions for fresh beef inoculated with Shiga toxin-producing *Escherichia coli* serotypes O26, O45, O103, O111, O121, O145, and O157:H7. J. Food Prot. 75 (7), 1207–1212.

Kanack, K.J., Crawford, J.A., Tatsuno, I., et al., 2005. SepZ/EspZ is secreted and translocated into HeLa cells by the enteropathogenic *Escherichia coli* type III secretion system. Infect. Immun. 73 (7), 4327–4337.

Kanamaru, K., Kanamaru, K., Tatsuno, I., et al., 2000. SdiA, an *Escherichia coli* homologue of quorum-sensing regulators, controls the expression of virulence factors in enterohaemorrhagic *Escherichia coli* O157:H7. Mol. Microbiol. 38 (4), 805–816.

Kaneko, K., Kiyokawa, N., Ohtomo, Y., et al., 2001. Apoptosis of renal tubular cells in Shiga-toxin-mediated hemolytic uremic syndrome. Nephron 87 (2), 182–185.

Kaper, J.B., Nataro, J.P., Mobley, H.L., 2004. Pathogenic *Escherichia coli*. Nat. Rev. Microbiol. 2 (2), 123–140.

Kaplan, B.S., Thomson, P.D., MacNab, G.M., 1973. Letter: Serum-complement levels in haemolytic-uraemic syndrome. Lancet 2 (7844), 1505–1506.

Karmali, M.A., Petric, M., Winkler, M., et al., 1994. Enzyme-linked immunosorbent assay for detection of immunoglobulin G antibodies to *Escherichia coli* Vero cytotoxin 1. J. Clin. Microbiol. 32 (6), 1457–1463.

Karmali, M.A., Steele, B.T., Petric, M., et al., 1983. Sporadic cases of haemolytic-uraemic syndrome associated with faecal cytotoxin and cytotoxin-producing *Escherichia coli* in stools. Lancet 1 (8325), 619–620.

Karpman, D., Andreasson, A., Thysell, H., et al., 1995. Cytokines in childhood hemolytic uremic syndrome and thrombotic thrombocytopenic purpura. Pediatr. Nephrol. 9 (6), 694–699.

Karpman, D., Hakansson, A., Perez, M.T., et al., 1998. Apoptosis of renal cortical cells in the hemolytic-uremic syndrome: *in vivo* and *in vitro* studies. Infect. Immun. 66 (2), 636–644.

Karpman, D., Papadopoulou, D., Nilsson, K., et al., 2001. Platelet activation by Shiga toxin and circulatory factors as a pathogenetic mechanism in the hemolytic uremic syndrome. Blood 97 (10), 3100–3108.

Kassenborg, H.D., Hedberg, C.W., Hoekstra, M., et al., 2004. Farm visits and undercooked hamburgers as major risk factors for sporadic *Escherichia coli* O157:H7 infection: data from a case-control study in 5 FoodNet sites. Clin. Infect. Dis. 38 (Suppl. 3), S271–278.

Kawai, T., Akira, S., 2007. Signaling to NF-kappaB by Toll-like receptors. Trends Mol. Med. 13 (11), 460–469.

Keene, W.E., McAnulty, J.M., Hoesly, F.C., et al., 1994. A swimming-associated outbreak of hemorrhagic colitis caused by *Escherichia coli* O157:H7 and Shigella sonnei. N. Engl. J. Med. 331 (9), 579–584.

Keepers, T.R., Psotka, M.A., Gross, L.K., et al., 2006. A murine model of HUS: Shiga toxin with lipopolysaccharide mimics the renal damage and physiologic response of human disease. J. Am. Soc. Nephrol. 17 (12), 3404–3414.

Kendall, M.M., Sperandio, V., 2007. Quorum sensing by enteric pathogens. Curr. Opin. Gastroenterol 23 (1), 10–15.

Kenny, B., 2002. Mechanism of action of EPEC type III effector molecules. Int. J. Med. Microbiol. 291 (6-7), 469–477.

Kenny, B., Abe, A., Stein, M., et al., 1997a. Enteropathogenic *Escherichia coli* protein secretion is induced in response to conditions similar to those in the gastrointestinal tract. Infect. Immun. 65 (7), 2606–2612.

Kenny, B., DeVinney, R., Stein, M., et al., 1997b. Enteropathogenic *E. coli* (EPEC) transfers its receptor for intimate adherence into mammalian cells. Cell 91 (4), 511–520.

Kenny, B., Ellis, S., Leard, A.D., et al., 2002. Co-ordinate regulation of distinct host cell signalling pathways by multifunctional enteropathogenic *Escherichia coli* effector molecules. Mol. Microbiol. 44 (4), 1095–1107.

Kenny, B., Finlay, B.B., 1995. Protein secretion by enteropathogenic *Escherichia coli* is essential for transducing signals to epithelial cells. Proc. Natl. Acad. Sci. USA 92 (17), 7991–7995.

Kenny, B., Jepson, M., 2000. Targeting of an enteropathogenic *Escherichia coli* (EPEC) effector protein to host mitochondria. Cell Microbiol. 2 (6), 579–590.

Kern, R., Malki, A., Abdallah, J., et al., 2007. *Escherichia coli* HdeB is an acid stress chaperone. J. Bacteriol. 189 (2), 603–610.

Kim, J., Thanabalasuriar, A., Chaworth-Musters, T., et al., 2007. The bacterial virulence factor NleA inhibits cellular protein secretion by disrupting mammalian COPII function. Cell Host Microbe 2 (3), 160–171.

Kimmitt, P.T., Harwood, C.R., Barer, M.R., 2000. Toxin gene expression by shiga toxin-producing *Escherichia coli*: the role of antibiotics and the bacterial SOS response. Emerg. Infect. Dis. 6 (5), 458–465.

Kirk, M.D., 2011. The German *Escherichia coli* outbreak – could it happen here? Med. J. Aust. Australia 195, 325–326.

Knutton, S., Rosenshine, I., Pallen, M.J., et al., 1998. A novel EspA-associated surface organelle of enteropathogenic *Escherichia coli* involved in protein translocation into epithelial cells. EMBO J. 17 (8), 2166–2176.

Konowalchuk, J., Speirs, J.I., Stavric, S., 1977. Vero response to a cytotoxin of *Escherichia coli*. Infect. Immun. 18 (3), 775–779.

Koohmaraie, M., Arthur, T.M., Bosilevac, J.M., et al., 2007. Interventions to reduce/eliminate *Escherichia coli* O157:H7 in ground beef. Meat Sci. 77 (1), 90–96.

Kovbasnjuk, O., Edidin, M., Donowitz, M., 2001. Role of lipid rafts in Shiga toxin 1 interaction with the apical surface of Caco-2 cells. J. Cell Sci. 114 (Pt 22), 4025–4031.

La Ragione, R.M., Cooley, W.A., Woodward, M.J., 2000. The role of fimbriae and flagella in the adherence of avian strains of *Escherichia coli* O78:K80 to tissue culture cells and tracheal and gut explants. J. Med. Microbiol. 49 (4), 327–338.

Laegreid, W.W., Elder, R.O., Keen, J.E., 1999. Prevalence of *Escherichia coli* O157:H7 in range beef calves at weaning. Epidemiol. Infect. 123 (2), 291–298.

Lai, Y., Riley, K., Cai, A., et al., 2011. Calpain mediates epithelial cell microvillar effacement by enterohemorrhagic *Escherichia coli*. Front Microbiol. 2, 222.

Lapeyraque, A.L., Malina, M., Fremeaux-Bacchi, V., et al., 2011. Eculizumab in severe Shiga-toxin-associated HUS. N. Engl. J. Med. 364 (26), 2561–2563.

LaPointe, P., Wei, X., Gariepy, J., 2005. A role for the protease-sensitive loop region of Shiga-like toxin 1 in the retrotranslocation of its A1 domain from the endoplasmic reticulum lumen. J. Biol. Chem. 280 (24), 23310–23318.

Large, T.M., Walk, S.T., Whittam, T.S., 2005. Variation in acid resistance among shiga toxin-producing clones of pathogenic *Escherichia coli*. Appl. Environ. Microbiol. 71 (5), 2493–2500.

Lathem, W.W., Bergsbaken, T., Welch, R.A., et al., 2004. Potentiation of C1 esterase inhibitor by StcE, a metalloprotease secreted by *Escherichia coli* O157:H7. J. Exp. Med. 199 (8), 1077–1087.

Lathem, W.W., Grys, T.E., Witowski, S.E., et al., 2002. StcE, a metalloprotease secreted by *Escherichia coli* O157:H7, specifically cleaves C1 esterase inhibitor. Mol. Microbiol. 45 (2), 277–288.

Lease, R.A., Smith, D., McDonough, K., et al., 2004. The small noncoding DsrA RNA is an acid resistance regulator in *Escherichia coli*. J. Bacteriol. 186 (18), 6179–6185.

Ledesma, M.A., Ochoa, S.A., Cruz, A., et al., 2010. The hemorrhagic coli pilus (HCP) of *Escherichia coli* O157:H7 is an inducer of proinflammatory cytokine secretion in intestinal epithelial cells. PLoS One 5 (8), e12127.

Lee, S.F., Kelly, M., McAlister, A., et al., 2008a. A C-terminal class I PDZ binding motif of EspI/ NleA modulates the virulence of attaching and effacing *Escherichia coli* and Citrobacter rodentium. Cell Microbiol. 10 (2), 499–513.

Lee, S.Y., Lee, M.S., Cherla, R.P., et al., 2008b. Shiga toxin 1 induces apoptosis through the endoplasmic reticulum stress response in human monocytic cells. Cell Microbiol. 10 (3), 770–780.

Lentz, E.K., Cherla, R.P., Jaspers, V., et al., 2010. Role of tumor necrosis factor alpha in disease using a mouse model of Shiga toxin-mediated renal damage. Infect. Immun. 78 (9), 3689–3699.

Leotta, G.A., Miliwebsky, E.S., Chinen, I., et al., 2008. Characterisation of Shiga toxin-producing *Escherichia coli* O157 strains isolated from humans in Argentina, Australia and New Zealand. BMC Microbiol. 8, 46.

Levine, M.M., 1987. *Escherichia coli* that cause diarrhea: enterotoxigenic, enteropathogenic, enteroinvasive, enterohemorrhagic, and enteroadherent. J. Infect. Dis. 155 (3), 377–389.

Li, M., Rosenshine, I., Yu, H.B., et al., 2006. Identification and characterization of NleI, a new non-LEE-encoded effector of enteropathogenic *Escherichia coli* (EPEC). Microbes Infect. 8 (14–15), 2890–2898.

Lin, D.Y., Diao, J., Zhou, D., et al., 2011. Biochemical and structural studies of a HECT-like ubiquitin ligase from *Escherichia coli* O157:H7. J. Biol. Chem. 286 (1), 441–449.

Lin, J., Lee, I.S., Frey, J., et al., 1995. Comparative analysis of extreme acid survival in Salmonella typhimurium, Shigella flexneri, and *Escherichia coli*. J. Bacteriol. 177 (14), 4097–4104.

Little, J.W., Mount, D.W., 1982. The SOS regulatory system of *Escherichia coli*. Cell 29 (1), 11–22.

Liu, B., Yin, X., Feng, Y., et al., 2010. Verotoxin 2 enhances adherence of enterohemorrhagic *Escherichia coli* O157:H7 to intestinal epithelial cells and expression of beta1-integrin by IPEC-J2 cells. Appl. Environ. Microbiol. 76 (13), 4461–4468.

Liu, H., Radhakrishnan, P., Magoun, L., et al., 2002. Point mutants of EHEC intimin that diminish Tir recognition and actin pedestal formation highlight a putative Tir binding pocket. Mol. Microbiol. 45 (6), 1557–1573.

Lloyd, S.J., Ritchie, J.M., Rojas-Lopez, M., et al., 2012. A double, long polar fimbria mutant of *Escherichia coli* O157:H7 expresses Curli and exhibits reduced *in vivo* colonization. Infect. Immun. 80 (3), 914–920.

Loirat, C., Fremeaux-Bacchi, V., 2011. Atypical hemolytic uremic syndrome. Orphanet J. Rare Dis. 6, 60.

Loos, S., Ahlenstiel, T., Kranz, B., et al., 2012. An outbreak of shiga toxin-producing *Escherichia coli* O104:H4 hemolytic uremic syndrome in Germany: presentation and short-term outcome in children. Clin. Infect. Dis. 55 (6), 753–759.

Lopez, E.L., Contrini, M.M., Devoto, S., et al., 1995. Incomplete hemolytic-uremic syndrome in Argentinean children with bloody diarrhea. J. Pediatr. 127 (3), 364–367.

Lopez, E.L., Contrini, M.M., De Rosa, M.F., 1998. Epidemiology of shiga toxin producing *Escherichia coli* in South America. *Escherichia coli* O157:H7 and other shiga toxin-producing *E. coli* strains. In: Kaper, J.B., O'Brien, A.D. (Eds.), American Society for Microbiology, Washington, D.C.

Low, A.S., Dziva, F., Torres, A.G., et al., 2006. Cloning, expression, and characterization of fimbrial operon F9 from enterohemorrhagic *Escherichia coli* O157:H7. Infect. Immun. 74 (4), 2233–2244.

Luo, Y., Frey, E.A., Pfuetzner, R.A., et al., 2000. Crystal structure of enteropathogenic *Escherichia coli* intimin-receptor complex. Nature 405 (6790), 1073–1077.

Ma, C., Wickham, M.E., Guttman, J.A., et al., 2006. Citrobacter rodentium infection causes both mitochondrial dysfunction and intestinal epithelial barrier disruption *in vivo:* role of mitochondrial associated protein (Map). Cell Microbiol. 8 (10), 1669–1686.

Majoul, I., Ferrari, D., Soling, H.D., 1997. Reduction of protein disulfide bonds in an oxidizing environment. The disulfide bridge of cholera toxin A-subunit is reduced in the endoplasmic reticulum. FEBS Lett. 401 (2–3), 104–108.

Mallick, E.M., Brady, M.J., Luperchio, S.A., et al., 2012a. Allele- and Tir-independent functions of intimin in diverse animal infection models. Front Microbiol. 3, 11.

Mallick, E.M., McBee, M.E., Vanguri, V.K., et al., 2012b. A novel murine infection model for Shiga toxin-producing *Escherichia coli*. J. Clin. Invest. 122 (11), In press.

Malyukova, I., Murray, K.F., Zhu, C., et al., 2009. Macropinocytosis in Shiga toxin 1 uptake by human intestinal epithelial cells and transcellular transcytosis. Am. J. Physiol. Gastrointest. Liver Physiol. 296 (1), G78–92.

Manning, S.D., Motiwala, A.S., Springman, A.C., et al., 2008. Variation in virulence among clades of *Escherichia coli* O157:H7 associated with disease outbreaks. Proc. Natl. Acad. Sci. USA 105 (12), 4868–4873.

Marches, O., Covarelli, V., Dahan, S., et al., 2008. EspJ of enteropathogenic and enterohaemorrhagic *Escherichia coli* inhibits opsono-phagocytosis. Cell Microbiol. 10 (5), 1104–1115.

Marches, O., Ledger, T.N., Boury, M., et al., 2003. Enteropathogenic and enterohaemorrhagic *Escherichia coli* deliver a novel effector called Cif, which blocks cell cycle G2/M transition. Mol. Microbiol. 50 (5), 1553–1567.

Marches, O., Nougayrede, J.P., Boullier, S., et al., 2000. Role of tir and intimin in the virulence of rabbit enteropathogenic *Escherichia coli* serotype O103:H2. Infect. Immun. 68 (4), 2171–2182.

Martinez-Argudo, I., Sands, C., Jepson, M.A, 2007. Translocation of enteropathogenic *Escherichia coli* across an in vitro M cell model is regulated by its type III secretion system. Cell Microbiol. 9 (6), 1538–1546.

Matsuzawa, T., Kuwae, A., Abe, A., 2005. Enteropathogenic *Escherichia coli* type III effectors EspG and EspG2 alter epithelial paracellular permeability. Infect. Immun. 73 (10), 6283–6289.

McDaniel, T.K., Jarvis, K.G., Donnenberg, M.S., et al., 1995. A genetic locus of enterocyte effacement conserved among diverse enterobacterial pathogens. Proc. Natl. Acad. Sci. USA 92 (5), 1664–1668.

McKee, M.L., O'Brien, A.D., 1995. Investigation of enterohemorrhagic *Escherichia coli* O157:H7 adherence characteristics and invasion potential reveals a new attachment pattern shared by intestinal *E. coli*. Infect. Immun. 63 (5), 2070–2074.

McKee, M.L., Melton-Celsa, A.R., Moxley, R.A., et al., 1995. Enterohemorrhagic *Escherichia coli* O157:H7 requires intimin to colonize the gnotobiotic pig intestine and to adhere to HEp-2 cells. Infect. Immun. 63 (9), 3739–3744.

McNamara, B.P., Koutsouris, A., O'Connell, C.B., et al., 2001. Translocated EspF protein from enteropathogenic *Escherichia coli* disrupts host intestinal barrier function. J. Clin. Invest. 107 (5), 621–629.

Mellies, J.L., Barron, A.M., Carmona, A.M., et al., 2007. Enteropathogenic and enterohemorrhagic *Escherichia coli* virulence gene regulation. Infect. Immun. 75 (9), 4199–4210.

Misyurina, O., Asper, D.J., Deng, W., et al., 2010. The role of Tir, EspA, and NleB in the colonization of cattle by Shiga toxin producing *Escherichia coli* O26:H11. Can. J. Microbiol. 56 (9), 739–747.

Miyahara, A., Nakanishi, N., Ooka, T., et al., 2009. Enterohemorrhagic *Escherichia coli* effector EspL2 induces actin microfilament aggregation through annexin 2 activation. Cell Microbiol. 11 (2), 337–350.

Miyamoto, Y., Iimura, M., Kaper, J.B., et al., 2006. Role of Shiga toxin versus H7 flagellin in enterohaemorrhagic *Escherichia coli* signalling of human colon epithelium *in vivo*. Cell Microbiol. 8 (5), 869–879.

Money, P., Kelly, A.F., Gould, S.W., et al., 2012. Cattle, weather and water: mapping *Escherichia coli* O157:H7 infections in humans in England and Scotland. Environ. Microbiol. 12 (10), 2633–2644.

Monnens, L., Hendrickx, G., van Wieringen, P., et al., 1974. Letter: Serum-complement levels in haemolytic-uraemic syndrome. Lancet 2 (7875), 294.

Monnens, L., Molenaar, J., Lambert, P.H., et al., 1980. The complement system in hemolytic-uremic syndrome in childhood. Clin. Nephrol. 13 (4), 168–171.

Moon, H.W., Whipp, S.C., Argenzio, R.A., et al., 1983. Attaching and effacing activities of rabbit and human enteropathogenic *Escherichia coli* in pig and rabbit intestines. Infect. Immun. 41, 1340–1351.

Mora, A., Blanco, M., Blanco, J.E., et al., 2004. Phage types and genotypes of shiga toxin-producing *Escherichia coli* O157:H7 isolates from humans and animals in Spain: identification and characterization of two predominating phage types (PT2 and PT8). J. Clin. Microbiol. 42 (9), 4007–4015.

Morigi, M., Galbusera, M., Gastoldi, S., et al., 2011. Alternative pathway activation of complement by Shiga toxin promotes exuberant C3a formation that triggers microvascular thrombosis. J. Immun. 187 (1), 172–180.

Morikawa, H., Kim, M., Mimuro, H., et al., 2010. The bacterial effector Cif interferes with SCF ubiquitin ligase function by inhibiting deneddylation of Cullin1. Biochem. Biophys. Res. Commun. 401 (2), 268–274.

Muhlen, S., Ruchaud-Sparagano, M.H., Kenny, B., 2011. Proteasome-independent degradation of canonical NFkappaB complex components by the NleC protein of pathogenic *Escherichia coli*. J. Biol. Chem. 286 (7), 5100–5107.

Mundy, R., Schuller, S., Girard, F., et al., 2007. Functional studies of intimin *in vivo* and *ex vivo*: implications for host specificity and tissue tropism. Microbiology 153 (Pt 4), 959–967.

Murata, K., Higuchi, T., Takada, K., et al., 2006. Verotoxin-1 stimulation of macrophage-like THP-1 cells up-regulates tissue factor expression through activation of c-Yes tyrosine kinase: possible signal transduction in tissue factor up-regulation. Biochim. Biophys. Acta 1762 (9), 835–843.

Murray, M.D., Brater, D.C., 1993. Renal toxicity of the nonsteroidal anti-inflammatory drugs. Annu. Rev. Pharmacol. Toxicol. 33, 435–465.

Nadler, C., Baruch, K., Kobi, S., et al., 2010. The type III secretion effector NleE inhibits NF-kappaB activation. PLoS Pathog. 6 (1), e1000743.

Nagai, T., Abe, A., Sasakawa, C., 2005. Targeting of enteropathogenic *Escherichia coli* EspF to host mitochondria is essential for bacterial pathogenesis: critical role of the 16th leucine residue in EspF. J. Biol. Chem. 280 (4), 2998–3011.

Nataro, J.P., Kaper, J.B., 1998. Diarrheagenic *Escherichia coli*. Clin. Microbiol. Rev. 11 (1), 142–201.

Nathanson, S., Kwon, T., Elmaleh, M., et al., 2010. Acute neurological involvement in diarrhea-associated hemolytic uremic syndrome. Clin. J. Am. Soc. Nephrol. 5 (7), 1218–1228.

Naylor, S.W., Low, J.C., Besser, T.E., et al., 2003. Lymphoid follicle-dense mucosa at the terminal rectum is the principal site of colonization of enterohemorrhagic *Escherichia coli* O157:H7 in the bovine host. Infect. Immun. 71 (3), 1505–1512.

Nevard, C.H., Jurd, K.M., Lane, D.A., et al., 1997. Activation of coagulation and fibrinolysis in childhood diarrhoea-associated haemolytic uraemic syndrome. Thromb. Haemost. 78 (6), 1450–1455.

Newton, H.J., Pearson, J.S., Badea, L., et al., 2010. The type III effectors NleE and NleB from enteropathogenic *E. coli* and OspZ from Shigella block nuclear translocation of NF-kappaB p65. PLoS Pathog. 6 (5), e1000898.

Newton, H.J., Sloan, J., Bulach, D.M., et al., 2009. Shiga toxin-producing *Escherichia coli* strains negative for locus of enterocyte effacement. Emerg. Infect. Dis. 15 (3), 372–380.

Nguyen, L.T., Gillespie, B.E., Nam, H.M., et al., 2004. Detection of *Escherichia coli* O157:H7 and Listeria monocytogenes in beef products by real-time polymerase chain reaction. Foodborne Pathog. Dis. 1 (4), 231–240.

Nicholls, L., Grant, T.H., Robins-Browne, R.M., 2000. Identification of a novel genetic locus that is required for *in vitro* adhesion of a clinical isolate of enterohaemorrhagic *Escherichia coli* to epithelial cells. Mol. Microbiol. 35 (2), 275–288.

Noris, M., Mescia, F., Remuzzi, G., 2012. STEC-HUS, atypical HUS and TTP are all diseases of complement activation. Nat. Rev. Nephrol. 8 (11), 622–633.

Nougayrede, J.P., Donnenberg, M.S., 2004. Enteropathogenic *Escherichia coli* EspF is targeted to mitochondria and is required to initiate the mitochondrial death pathway. Cell Microbiol. 6 (11), 1097–1111.

Nougayrede, J.P., Foster, G.H., Donnenberg, M.S., 2007. Enteropathogenic *Escherichia coli* effector EspF interacts with host protein Abcf2. Cell Microbiol. 9 (3), 680–693.

O'Brien, A.D., LaVeck, G.D., 1983. Purification and characterization of a Shigella dysenteriae 1-like toxin produced by *Escherichia coli*. Infect. Immun. 40 (2), 675–683.

Obata, F., Tohyama, K., Bonev, A.D., et al., 2008. Shiga toxin 2 affects the central nervous system through receptor globotriaosylceramide localized to neurons. J. Infect. Dis. 198 (9), 1398–1406.

Obrig, T.G., Karpman, D., 2012. Shiga toxin pathogenesis: kidney complications and renal failure. Curr. Top. Microbiol. Immunol. 357, 105–136.

Obrig, T.G., Louise, C.B., Lingwood, C.A., et al., 1993. Endothelial heterogeneity in Shiga toxin receptors and responses. J. Biol. Chem. 268 (21), 15484–15488.

Obrig, T.G., Moran, T.P., Brown, J.E., 1987. The mode of action of Shiga toxin on peptide elongation of eukaryotic protein synthesis. Biochem. J. 244 (2), 287–294.

Ogura, Y., Ooka, T., Whale, A., et al., 2007. TccP2 of O157:H7 and non-O157 enterohemorrhagic *Escherichia coli* (EHEC): challenging the dogma of EHEC-induced actin polymerization. Infect. Immun. 75 (2), 604–612.

Oliveira, M.G., Brito, J.R., Gomes, T.A., et al., 2008. Diversity of virulence profiles of Shiga toxin-producing *Escherichia coli* serotypes in food-producing animals in Brazil. Int. J. Food Microbiol. 127 (1–2), 139–146.

Ooka, T., Vieira, M.A., Ogura, Y., et al., 2007. Characterization of tccP2 carried by atypical enteropathogenic *Escherichia coli*. FEMS Microbiol. Lett. 271 (1), 126–135.

Oporto, B., Esteban, J.I., Aduriz, G., et al., 2008. *Escherichia coli* O157:H7 and non-O157 Shiga toxin-producing *E. coli* in healthy cattle, sheep and swine herds in Northern Spain. Zoonoses Public Health 55 (2), 73–81.

Orth, D., Grif, K., Zimmerhackl, L.B., et al., 2008. Prevention and treatment of enterohemorrhagic *Escherichia coli* infections in humans. Expert Rev. Anti. Infect. Ther. 6 (1), 101–108.

Orth, D., Khan, A.B., Naim, A., et al., 2009. Shiga toxin activates complement and binds factor H: evidence for an active role of complement in hemolytic uremic syndrome. J. Immunol. 182 (10), 6394–6400.

Ostroff, S.M., Kobayashi, J.M., Lewis, J.H., 1989. Infections with *Escherichia coli* O157:H7 in Washington State. The first year of statewide disease surveillance. JAMA 262 (3), 355–359.

Oswald, E., Schmidt, H., Morabito, S., et al., 2000. Typing of intimin genes in human and animal enterohemorrhagic and enteropathogenic *Escherichia coli*: characterization of a new intimin variant. Infect. Immun. 68 (1), 64–71.

Palermo, M., Alves-Rosa, F., Rubel, C., et al., 2000. Pretreatment of mice with lipopolysaccharide (LPS) or IL-1beta exerts dose-dependent opposite effects on Shiga toxin-2 lethality. Clin. Exp. Immunol. 119 (1), 77–83.

Papatheodorou, P., Domanska, G., Oxle, M., et al., 2006. The enteropathogenic *Escherichia coli* (EPEC) Map effector is imported into the mitochondrial matrix by the TOM/Hsp70 system and alters organelle morphology. Cell Microbiol. 8 (4), 677–689.

Paton, A.W., Beddoe, T., Thorpe, C.M., et al., 2006. AB5 subtilase cytotoxin inactivates the endoplasmic reticulum chaperone BiP. Nature 443 (7111), 548–552.

Paton, A.W., Paton, J.C., 2002. Direct detection and characterization of Shiga toxigenic *Escherichia coli* by multiplex PCR for stx1, stx2, eae, ehxA, and saa. J. Clin. Microbiol. 40 (1), 271–274.

Paton, A.W., Srimanote, P., Talbot, U.M., et al., 2004. A new family of potent AB(5) cytotoxins produced by Shiga toxigenic *Escherichia coli*. J. Exp. Med. 200 (1), 35–46.

Paton, A.W., Srimanote, P., Woodrow, M.C., et al., 2001. Characterization of Saa, a novel auto-agglutinating adhesin produced by locus of enterocyte effacement-negative Shiga-toxigenic *Escherichia coli* strains that are virulent for humans. Infect. Immun. 69 (11), 6999–7009.

Paton, A.W., Woodrow, M.C., Doyle, R.M., et al., 1999. Molecular characterization of a Shiga toxigenic *Escherichia coli* O113:H21 strain lacking eae responsible for a cluster of cases of hemolytic-uremic syndrome. J. Clin. Microbiol. 37 (10), 3357–3361.

Paton, J.C., Paton, A.W., 1998. Pathogenesis and diagnosis of Shiga toxin-producing *Escherichia coli* infections. Clin. Microbiol. Rev. 11 (3), 450–479.

Parry, S.M., Salmon, R.L., 1998. Sporadic STEC O157 infection: secondary household transmission in Wales. Emerg. Infect. Dis. 4 (4), 657–661.

Pavia, A.T., Nichols, C.R., Green, D.P., et al., 1990. Hemolytic-uremic syndrome during an outbreak of *Escherichia coli* O157:H7 infections in institutions for mentally retarded persons: clinical and epidemiologic observations. J. Pediatr. 116 (4), 544–551.

Pawar, D.M., Rossman, M.L., Chen, J., 2005. Role of curli fimbriae in mediating the cells of entero-haemorrhagic *Escherichia coli* to attach to abiotic surfaces. J. Appl. Microbiol. 99 (2), 418–425.

Pearson, J.S., Riedmaier, P., Marches, O., et al., 2011. A type III effector protease NleC from enteropathogenic *Escherichia coli* targets NF-kappaB for degradation. Mol. Microbiol. 80 (1), 219–230.

Pennington, H., 2010. *Escherichia coli* O157. Lancet 376 (9750), 1428–1435.

Peralta-Ramirez, J., Hernandez, J.M., Manning-Cela, R., et al., 2008. EspF Interacts with nucleation-promoting factors to recruit junctional proteins into pedestals for pedestal maturation and disruption of paracellular permeability. Infect. Immun. 76 (9), 3854–3868.

Perna, N.T., Plunkett 3rd, G., Burland, V., et al., 2001. Genome sequence of enterohaemorrhagic *Escherichia coli* O157:H7. Nature 409 (6819), 529–533.

Pham, T.H., Gao, X., Tsai, K., et al., 2012. Functional differences and interactions between the *Escherichia coli* type III secretion system effectors NleH1 and NleH2. Infect. Immun. 80 (6), 2133–2140.

Phillips, A.D., Frankel, G., 2000. Intimin-mediated tissue specificity in enteropathogenic *Escherichia coli* interaction with human intestinal organ cultures. J. Infect. Dis. 181 (4), 1496–1500.

Philpott, D.J., Ackerley, C.A., Kiliaan, A.J., et al., 1997. Translocation of verotoxin-1 across T84 monolayers: mechanism of bacterial toxin penetration of epithelium. Am. J. Physiol. 273 (6 Pt 1), G1349–G1358.

Piscatelli, H., Kotkar, S.A., McBee, M.E., et al., 2011. The EHEC type III effector NleL is an E3 ubiquitin ligase that modulates pedestal formation. PLoS One 6 (4), e19331.

Potter, D.A., Srirangam, A., Fiacco, K.A., et al., 2003. Calpain regulates enterocyte brush border actin assembly and pathogenic *Escherichia coli*-mediated effacement. J. Biol. Chem. 278 (32), 30403–30412.

Quitard, S., Dean, P., Maresca, M., et al., 2006. The enteropathogenic *Escherichia coli* EspF effector molecule inhibits PI-3 kinase-mediated uptake independently of mitochondrial targeting. Cell Microbiol. 8 (6), 972–981.

Rangel, J.M., Sparling, P.H., Crowe, C., et al., 2005. Epidemiology of *Escherichia coli* O157:H7 outbreaks, United States, 1982–2002. Emerg. Infect. Dis. 11 (4), 603–609.

Rashid, R.A., Tarr, P.I., Moseley, S.L., 2006. Expression of the *Escherichia coli* IrgA homolog adhesin is regulated by the ferric uptake regulation protein. Microb. Pathog. 41 (6), 207–217.

Rasko, D.A., Webster, D.R., Sahl, J.W., et al., 2011. Origins of the *E. coli* strain causing an outbreak of hemolytic-uremic syndrome in Germany. N. Engl. J. Med. 365 (8), 709–717.

Rathinam, V.A., Vanaja, S.K., Waggoner, L., et al., 2012. TRIF licenses caspase-11-dependent NLRP3 inflammasome activation by gram-negative bacteria. Cell 150 (3), 606–619.

Ray, P.E., Liu, X.H., 2001. Pathogenesis of Shiga toxin-induced hemolytic uremic syndrome. Pediatr. Nephrol. 16 (10), 823–839.

Raya, R.R., Varey, P., Oot, R.A., et al., 2006. Isolation and characterization of a new T-even bacteriophage, CEV1, and determination of its potential to reduce *Escherichia coli* O157:H7 levels in sheep. Appl. Environ. Microbiol. 72 (9), 6405–6410.

Reading, N.C., Torres, A.G., Kendall, M.M., et al., 2007. A novel two-component signaling system that activates transcription of an enterohemorrhagic *Escherichia coli* effector involved in remodeling of host actin. J. Bacteriol. 189 (6), 2468–2476.

Reid, S.D., Herbelin, C.J., Bumbaugh, A.C., et al., 2000. Parallel evolution of virulence in pathogenic *Escherichia coli*. Nature 406 (6791), 64–67.

Rendon, M.A., Saldana, Z., Erdem, A.L., et al., 2007. Commensal and pathogenic *Escherichia coli* use a common pilus adherence factor for epithelial cell colonization. Proc. Natl. Acad. Sci. USA 104 (25), 10637–10642.

Ritchie, J.M., Brady, M.J., Riley, K.N., et al., 2008. EspFU, a type III-translocated effector of actin assembly, fosters epithelial association and late-stage intestinal colonization by *E. coli* O157:H7. Cell Microbiol. 10 (4), 836–847.

Ritchie, J.M., Thorpe, C.M., Rogers, A.B., et al., 2003. Critical roles for stx2, eae, and tir in entero-hemorrhagic *Escherichia coli*-induced diarrhea and intestinal inflammation in infant rabbits. Infect. Immun. 71 (12), 7129–7139.

Ritchie, J.M., Waldor, M.K., 2005. The locus of enterocyte effacement-encoded effector proteins all promote enterohemorrhagic *Escherichia coli* pathogenicity in infant rabbits. Infect. Immun. 73 (3), 1466–1474.

Robinson, C.M., Sinclair, J.F., Smith, M.J., et al., 2006. Shiga toxin of enterohemorrhagic *Escherichia coli* type O157:H7 promotes intestinal colonization. Pro. Nat. Acad. Sci. USA 103 (25), 9667–9672.

Robson, W.L., Leung, A.K., Fick, G.H., et al., 1992. Hypocomplementemia and leukocytosis in diarrhea-associated hemolytic uremic syndrome. Nephron 62 (3), 296–299.

Ross, N.T., Miller, B.L., 2007. Characterization of the binding surface of the translocated intimin receptor, an essential protein for EPEC and EHEC cell adhesion. Protein Sci. 16 (12), 2677–2683.

Roxas, J.L., Koutsouris, A., Bellmeyer, A., et al., 2010. Enterohemorrhagic *E. coli* alters murine intestinal epithelial tight junction protein expression and barrier function in a Shiga toxin independent manner. Lab. Invest. 90 (8), 1152–1168.

Roxas, J.L., Wilbur, J.S., Zhang, X., et al., 2012. The enteropathogenic *Escherichia coli* secreted protein EspZ inhibits intrinsic apoptosis of host intestinal epithelial cells. Infect. Immun. 80 (11), 3850–3870.

Royan, S.V., Jones, R.M., Koutsouris, A., et al., 2010. Enteropathogenic *E. coli* non-LEE encoded effectors NleH1 and NleH2 attenuate NF-kappaB activation. Mol. Microbiol. 78 (5), 1232–1245.

Rozema, E.A., Stephens, T.P., Bach, S.J., et al., 2009. Oral and rectal administration of bacterio-phages for control of *Escherichia coli* O157:H7 in feedlot cattle. J. Food Prot. 72 (2), 241–250.

Sakiri, R., Ramegowda, B., Tesh, V.L., 1998. Shiga toxin type 1 activates tumor necrosis factor-alpha gene transcription and nuclear translocation of the transcriptional activators nuclear factor-kappaB and activator protein-1. Blood 92 (2), 558–566.

Sakuma, M., Urashima, M., Okabe, N., 2006. Verocytotoxin-producing *Escherichia coli*, Japan, 1999–2004. Emerg. Infect. Dis. 12 (2), 323–325.

Sallee, N.A., Rivera, G.M., Dueber, J.E., et al., 2008. The pathogen protein EspF(U) hijacks actin polymerization using mimicry and multivalency. Nature 454 (7207), 1005–1008.

Samba-Louaka, A., Nougayrede, J.P., Watrin, C., et al., 2009. The enteropathogenic *Escherichia coli* effector Cif induces delayed apoptosis in epithelial cells. Infect. Immun. 77 (12), 5471–5477.

Sandvig, K., Garred, O., Prydz, K., et al., 1992. Retrograde transport of endocytosed Shiga toxin to the endoplasmic reticulum. Nature 358 (6386), 510–512.

Sandvig, K., van Deurs, B., 2000. Entry of ricin and Shiga toxin into cells: molecular mechanisms and medical perspectives. EMBO J. 19 (22), 5943–5950.

Saxena, S.K., O'Brien, A.D., Ackerman, E.J., 1989. Shiga toxin, Shiga-like toxin II variant, and ricin are all single-site RNA N-glycosidases of 28 S RNA when microinjected into Xenopus oocytes. J. Biol. Chem. 264 (1), 596–601.

Scallan, E., Hoekstra, R.M., Angulo, F.J., et al., 2011. Foodborne illness acquired in the United States – major pathogens. Emerg. Infect. Dis. 17 (1), 7–15.

Scheiring, J., Andreoli, S.P., Zimmerhackl, L.B., 2008. Treatment and outcome of Shiga-toxin-associated hemolytic uremic syndrome (HUS). Pediatr. Nephrol. 23 (10), 1749–1760.

Scheutz, F., Teel, L.D., Beutin, L., et al., 2012. A multi-center evaluation of a sequence-based protocol to subtype Shiga toxins and standardize Stx nomenclature. J. Clin. Microbiol. 50 (9), 2951–2963.

Schmidt, H., 2001. Shiga-toxin-converting bacteriophages. Res. Microbiol. 152 (8), 687–695.

Schmitz, H., Barmeyer, C., Gitter, A.H., et al., 2000. Epithelial barrier and transport function of the colon in ulcerative colitis. Annu. NY Acad. Sci. 915, 312–326.

Schrezenmeier, H., Hochsmann, B., 2012. Drugs that inhibit complement. Transfus. Apher. Sci. 46 (1), 87–92.

Schroder, K., Tschopp, J., 2010. The inflammasomes. Cell 140 (6), 821–832.

Schuller, S., Frankel, G., Phillips, A.D., 2004. Interaction of Shiga toxin from *Escherichia coli* with human intestinal epithelial cell lines and explants: Stx2 induces epithelial damage in organ culture. Cell Microbiol. 6 (3), 289–301.

Schuller, S., Heuschkel, R., Torrente, F., et al., 2007. Shiga toxin binding in normal and inflamed human intestinal mucosa. Microbes Infect. 9 (1), 35–39.

Selyunin, A.S., Alto, N.M., 2011a. Activation of PAK by a bacterial type III effector EspG reveals alternative mechanisms of GTPase pathway regulation. Small GTPases 2 (4), 217–221.

Selyunin, A.S., Sutton, S.E., Weigele, B.A., et al., 2011b. The assembly of a GTPase-kinase signalling complex by a bacterial catalytic scaffold. Nature 469 (7328), 107–111.

Sham, H.P., Shames, S.R., Croxen, M.A., et al., 2011. Attaching and effacing bacterial effector NleC suppresses epithelial inflammatory responses by inhibiting NF-kappaB and p38 mitogen-activated protein kinase activation. Infect. Immun. 79 (9), 3552–3562.

Shames, S.R., Bhavsar, A.P., Croxen, M.A., et al., 2011a. The pathogenic *Escherichia coli* type III secreted protease NleC degrades the host acetyltransferase p300. Cell Microbiol. 13 (10), 1542–1557.

Shames, S.R., Croxen, M.A., Deng, W., et al., 2011b. The type III system-secreted effector EspZ localizes to host mitochondria and interacts with the translocase of inner mitochondrial membrane 17b. Infect. Immun. 79 (12), 4784–4790.

Shames, S.R., Deng, W., Guttman, J.A., et al., 2010. The pathogenic *E. coli* type III effector EspZ interacts with host CD98 and facilitates host cell prosurvival signalling. Cell Microbiol. 12 (9), 1322–1339.

Sharma, V.K., Bearson, S.M., Bearson, B.L., 2010. Evaluation of the effects of sdiA, a luxR homologue, on adherence and motility of *Escherichia coli* O157: H7. Microbiology 156 (Pt 5), 1303–1312.

Sharma, V.K., Zuerner, R.L., 2004. Role of hha and ler in transcriptional regulation of the esp operon of enterohemorrhagic *Escherichia coli* O157:H7. J. Bacteriol. 186 (21), 7290–7301.

Shaw, R.K., Berger, C.N., Feys, B., et al., 2008. Enterohemorrhagic *Escherichia coli* exploits EspA filaments for attachment to salad leaves. Appl. Environ. Microbiol. 74 (9), 2908–2914.

Shaw, R.K., Cleary, J., Murphy, M.S., et al., 2005. Interaction of enteropathogenic *Escherichia coli* with human intestinal mucosa: role of effector proteins in brush border remodeling and formation of attaching and effacing lesions. Infect. Immun. 73 (2), 1243–1251.

Shifflett, D.E., Clayburgh, D.R., Koutsouris, A., et al., 2005. Enteropathogenic *E. coli* disrupts tight junction barrier function and structure *in vivo*. Lab. Invest. 85 (10), 1308–1324.

Shringi, S., Garcia, A., Lahmers, K.K., et al., 2012. Differential virulence of clinical and bovine-biased enterohemorrhagic *Escherichia coli* O157:H7 genotypes in piglet and Dutch belted rabbit models. Infect. Immun. 80 (1), 369–380.

Siegler, R.L., 2003. Postdiarrheal Shiga toxin-mediated hemolytic uremic syndrome. JAMA 290 (10), 1379–1381.

Simonovic, I., Rosenberg, J., Koutsouris, A., et al., 2000. Enteropathogenic *Escherichia coli* dephosphorylates and dissociates occludin from intestinal epithelial tight junctions. Cell Microbiol. 2 (4), 305–315.

Simovitch, M., Sason, H., Cohen, S., et al., 2010. EspM inhibits pedestal formation by entero-haemorrhagic *Escherichia coli* and enteropathogenic *E. coli* and disrupts the architecture of a polarized epithelial monolayer. Cell Microbiol. 12 (4), 489–505.

Simpson, N., Shaw, R., Crepin, V.F., et al., 2006. The enteropathogenic *Escherichia coli* type III secretion system effector Map binds EBP50/NHERF1: implication for cell signalling and diarrhoea. Mol. Microbiol. 60 (2), 349–363.

Sinclair, J.F., Dean-Nystrom, E.A., O'Brien, A.D., 2006. The established intimin receptor Tir and the putative eucaryotic intimin receptors nucleolin and beta1 integrin localize at or near the site of enterohemorrhagic *Escherichia coli* O157:H7 adherence to enterocytes *in vivo*. Infect. Immun. 74 (2), 1255–1265.

Sinclair, J.F., O'Brien, A.D., 2002. Cell surface-localized nucleolin is a eukaryotic receptor for the adhesin intimin-gamma of enterohemorrhagic *Escherichia coli* O157:H7. J. Biol. Chem. 277 (4), 2876–2885.

Smith, D.K., Kassam, T., Singh, B., et al., 1992. *Escherichia coli* has two homologous glutamate decarboxylase genes that map to distinct loci. J. Bacteriol. 174 (18), 5820–5826.

Smith, D.R., Moxley, R.A., Peterson, R.E., et al., 2009. A two-dose regimen of a vaccine against type III secreted proteins reduced *Escherichia coli* O157:H7 colonization of the terminal rectum in beef cattle in commercial feedlots. Foodborne Pathog. Dis. 6 (2), 155–161.

Smith, W.E., Kane, A.V., Campbell, S.T., et al., 2003. Shiga toxin 1 triggers a ribotoxic stress response leading to p38 and JNK activation and induction of apoptosis in intestinal epithelial cells. Infect. Immun. 71 (3), 1497–1504.

Snedeker, K.G., Shaw, D.J., Locking, M.E., et al., 2009. Primary and secondary cases in *Escherichia coli* O157 outbreaks: a statistical analysis. BMC Infect. Dis. 9, 144.

Solomon, E.B., Yaron, S., Matthews, K.R., 2002. Transmission of *Escherichia coli* O157:H7 from contaminated manure and irrigation water to lettuce plant tissue and its subsequent internalization. Appl. Environ. Microbiol. 68 (1), 397–400.

Sperandio, V., Torres, A.G., Kaper, J.B., 2002. Quorum sensing *Escherichia coli* regulators B and C (QseBC): a novel two-component regulatory system involved in the regulation of flagella and motility by quorum sensing in *E. coli*. Mol. Microbiol. 43 (3), 809–821.

Spina, N., Zansky, S., Dumas, N., et al., 2005. Four laboratory-associated cases of infection with *Escherichia coli* O157:H7. J. Clin. Microbiol. 43 (6), 2938–2939.

Spitz, J., Yuhan, R., Koutsouris, A., et al., 1995. Enteropathogenic *Escherichia coli* adherence to intestinal epithelial monolayers diminishes barrier function. Am. J. Physiol. 268 (2 Pt 1), G374–379.

Spooner, R.A., Watson, P.D., Marsden, C.J., et al., 2004. Protein disulphide-isomerase reduces ricin to its A and B chains in the endoplasmic reticulum. Biochem. J. 383 (Pt 2), 285–293.

Stahl, A.L., Sartz, L., Karpman, D., 2011. Complement activation on platelet-leukocyte complexes and microparticles in enterohemorrhagic *Escherichia coli*-induced hemolytic uremic syndrome. Blood 117 (20), 5503–5513.

Stahl, A.L., Sartz, L., Nelsson, A., et al., 2009. Shiga toxin and lipopolysaccharide induce platelet-leukocyte aggregates and tissue factor release, a thrombotic mechanism in hemolytic uremic syndrome. PLoS One 4 (9), e6990.

Stahl, A.L., Svensson, M., Morgelin, M., et al., 2006. Lipopolysaccharide from enterohemorrhagic *Escherichia coli* binds to platelets through TLR4 and CD62 and is detected on circulating platelets in patients with hemolytic uremic syndrome. Blood 108 (1), 167–176.

Staley, T.E., Jones, E.W., Corley, L.D., 1969. Attachment and penetration of *Escherichia coli* into intestinal epithelium of the ileum in newborn pigs. Amer. J. Path. 56, 371–392.

Stevens, M.P., Roe, A.J., Vlisidou, I., et al., 2004. Mutation of toxB and a truncated version of the efa-1 gene in *Escherichia coli* O157:H7 influences the expression and secretion of locus of enterocyte effacement-encoded proteins but not intestinal colonization in calves or sheep. Infect. Immun. 72 (9), 5402–5411.

Stevens, M.P., van Diemen, P.M., Frankel, G., et al., 2002. Efa1 influences colonization of the bovine intestine by shiga toxin-producing *Escherichia coli* serotypes O5 and O111. Infect. Immun. 70 (9), 5158–5166.

Stim, K.P., Bennett, G.N., 1993. Nucleotide sequence of the adi gene, which encodes the biodegradative acid-induced arginine decarboxylase of *Escherichia coli*. J. Bacteriol. 175 (5), 1221–1234.

Strachan, N.J., Dunn, G.M., Locking, M.E., et al., 2006. *Escherichia coli* O157: burger bug or environmental pathogen? Int. J. Food Microbiol. 112 (2), 129–137.

Strachan, N.J., Fenlon, D.R., Ogden, I.D., 2001. Modelling the vector pathway and infection of humans in an environmental outbreak of *Escherichia coli* O157. FEMS Microbiol. Lett. 203 (1), 69–73.

Stradal, T.E., Scita, G., 2006. Protein complexes regulating Arp2/3-mediated actin assembly. Curr. Opin. Cell Biol. 18 (1), 4–10.

Tahoun, A., Siszler, G., Spears, K., et al., 2011. Comparative analysis of EspF variants in inhibition of *Escherichia coli* phagocytosis by macrophages and inhibition of *E. coli* translocation through human- and bovine-derived M cells. Infect. Immun. 79 (11), 4716–4729.

Taieb, F., Nougayrede, J.P., Watrin, C., et al., 2006. *Escherichia coli* cyclomodulin Cif induces G2 arrest of the host cell cycle without activation of the DNA-damage checkpoint-signalling pathway. Cell Microbiol. 8 (12), 1910–1921.

Tam, P.J., Lingwood, C.A., 2007. Membrane cytosolic translocation of verotoxin A1 subunit in target cells. Microbiology 153 (Pt 8), 2700–2710.

Taneike, I., Zhang, H.M., Wakisaka-Saito, N., et al., 2002. Enterohemolysin operon of Shiga toxin-producing *Escherichia coli*: a virulence function of inflammatory cytokine production from human monocytes. FEBS Lett. 524 (1-3), 219–224.

Tariq, L., Haagsma, J., Havelaar, A., 2011. Cost of illness and disease burden in The Netherlands due to infections with Shiga toxin-producing *Escherichia coli* O157. J. Food Prot. 74 (4), 545–552.

Tarr, P.I., Bilge, S.S., Vary, J.C., et al., 2000. Iha: a novel *Escherichia coli* O157:H7 adherence-conferring molecule encoded on a recently acquired chromosomal island of conserved structure. Infect. Immun. 68 (3), 1400–1407.

Tarr, P.I., Gordon, C.A., Chandler, W.L., 2005. Shiga-toxin-producing *Escherichia coli* and haemolytic uraemic syndrome. Lancet 365 (9464), 1073–1086.

Tatsuno, I., Horie, M., Abe, H., et al., 2001. toxB gene on pO157 of enterohemorrhagic *Escherichia coli* O157:H7 is required for full epithelial cell adherence phenotype. Infect. Immun. 69 (11), 6660–6669.

Tatsuno, I., Nagano, K., Taguchi, K., et al., 2003. Increased adherence to Caco-2 cells caused by disruption of the yhiE and yhiF genes in enterohemorrhagic *Escherichia coli* O157:H7. Infect. Immun. 71 (5), 2598–2606.

Taylor, C.M., White, R.H., Winterborn, M.H., et al., 1986. Haemolytic-uraemic syndrome: clinical experience of an outbreak in the West Midlands. Br. Med. J. Clin. Res. 292, 1513–1516, (6534).

Te Loo, D.M., Monnens, L.A., van Der Velden, T.J., et al., 2000. Binding and transfer of verocytotoxin by polymorphonuclear leukocytes in hemolytic uremic syndrome. Blood 95 (11), 3396–3402.

Te Loo, D.M., Monnens, L.A., van den Heuvel, L.P., et al., 2001a. Detection of apoptosis in kidney biopsies of patients with D+ hemolytic uremic syndrome. Pediatr. Res. 49 (3), 413–416.

Te Loo, D.M., van Hinsbergh, V.W., van den Heuvel, L.P., et al., 2001b. Detection of verocytotoxin bound to circulating polymorphonuclear leukocytes of patients with hemolytic uremic syndrome. J. Am. Soc. Nephrol. 12 (4), 800–806.

Thanabalasuriar, A., Bergeron, J., Gillingham, A., et al., 2012. Sec24 interaction is essential for localization and virulence-associated function of the bacterial effector protein NleA. Cell Microbiol. 14 (8), 1206–1218.

Thanabalasuriar, A., Koutsouris, A., Hecht, G., et al., 2010a. The bacterial virulence factor NleA's involvement in intestinal tight junction disruption during enteropathogenic *E. coli* infection is independent of its putative PDZ binding domain. Gut. Microbes 1 (2), 114–118.

Thanabalasuriar, A., Koutsouris, A., Weflen, A., et al., 2010b. The bacterial virulence factor NleA is required for the disruption of intestinal tight junctions by enteropathogenic *Escherichia coli*. Cell Microbiol. 12 (1), 31–41.

Thornton, A.B., Thomson, D.U., Loneragan, G.H.., et al., 2009. Effects of a siderophore receptor and porin proteins-based vaccination on fecal shedding of *Escherichia coli* O157:H7 in experimentally inoculated cattle. J. Food Prot. 72 (4), 866–869.

Thorpe, C.M., Hurley, B.P., Lincicome, L.L., et al., 1999. Shiga toxins stimulate secretion of interleukin-8 from intestinal epithelial cells. Infect. Immun. 67 (11), 5985–5993.

Thorpe, C.M., Ritchie, J.M., Acheson, D.W.K., 2002. Enterohemorrhagic and other Shiga toxin-producing *Escherichia coli*. *Escherichia coli* Virulence Mechanisms of a Versatile Pathogen. In: Donnenberg, M.S. (Ed.), Elsevier Science, San Diego, pp. 119–154.

Thorpe, C.M., Smith, W.E., Hurley, B.P., et al., 2001. Shiga toxins induce, superinduce, and stabilize a variety of C-X-C chemokine mRNAs in intestinal epithelial cells, resulting in increased chemokine expression. Infect. Immun. 69 (10), 6140–6147.

Thurman, J.M., Marians, R., Emlen, W., et al., 2009. Alternative pathway of complement in children with diarrhea-associated hemolytic uremic syndrome. Clin. J. Am. Soc. Nephrol. 4 (12), 1920–1924.

Tilden Jr., J., Young, W., McNamara, A.M., et al., 1996. A new route of transmission for *Escherichia coli*: infection from dry fermented salami. Am. J. Public Health 86 (8), 1142–1145.

Tobe, T., 2010. Cytoskeleton-modulating effectors of enteropathogenic and enterohemorrhagic *Escherichia coli*: role of EspL2 in adherence and an alternative pathway for modulating cytoskeleton through Annexin A2 function. FEBS J. 277 (11), 2403–2408.

Tobe, T., Ando, H., Ishikawa, H., et al., 2005. Dual regulatory pathways integrating the RcsC-RcsD-RcsB signalling system control enterohaemorrhagic *Escherichia coli* pathogenicity. Mol. Microbiol. 58 (1), 320–333.

Tobe, T., Beatson, S.A., Taniguchi, H., et al., 2006. An extensive repertoire of type III secretion effectors in *Escherichia coli* O157 and the role of lambdoid phages in their dissemination. Proc. Natl. Acad. Sci. USA 103 (40), 14941–14946.

Toma, C., Nakasone, N., Miliwebsky, E., et al., 2008. Differential adherence of Shiga toxin-producing *Escherichia coli* harboring saa to epithelial cells. Int. J. Med. Microbiol. 298 (7–8), 571–578.

Tomson, F.L., Viswanathan, V.K., Kanack, K.J., et al., 2005. Enteropathogenic *Escherichia coli* EspG disrupts microtubules and in conjunction with Orf3 enhances perturbation of the tight junction barrier. Mol. Microbiol. 56 (2), 447–464.

Torres, A.G., Giron, J.A., Perna, N.T., et al., 2002a. Identification and characterization of lpfABCC'DE, a fimbrial operon of enterohemorrhagic *Escherichia coli* O157:H7. Infect. Immun. 70 (10), 5416–5427.

Torres, A.G., Kanack, K.J., Tutt, C.B., et al., 2004. Characterization of the second long polar (LP) fimbriae of *Escherichia coli* O157:H7 and distribution of LP fimbriae in other pathogenic *E. coli* strains. FEMS Microbiol. Lett. 238 (2), 333–344.

Torres, A.G., Kaper, J.B., 2003. Multiple elements controlling adherence of enterohemorrhagic *Escherichia coli* O157:H7 to HeLa cells. Infect. Immun. 71 (9), 4985–4995.

Torres, A.G., Li, Y., Tutt, C.B., et al., 2006. Outer membrane protein A of *Escherichia coli* O157:H7 stimulates dendritic cell activation. Infect. Immun. 74 (5), 2676–2685.

Torres, A.G., Lopez-Sanchez, G.N., Milflores-Flores, L., et al., 2007. Ler and H-NS, regulators controlling expression of the long polar fimbriae of *Escherichia coli* O157:H7. J. Bacteriol. 189 (16), 5916–5928.

Torres, A.G., Perna, N.T., Burland, V., et al., 2002b. Characterization of Cah, a calcium-binding and heat-extractable autotransporter protein of enterohaemorrhagic *Escherichia coli*. Mol. Microbiol. 45 (4), 951–966.

Torres, A.G., Zhou, X., Kaper, J.B., 2005. Adherence of diarrheagenic *Escherichia coli* strains to epithelial cells. Infect. Immun. 73 (1), 18–29.

Touze, T., Hayward, R.D., Eswaran, J., et al., 2004. Self-association of EPEC intimin mediated by the beta-barrel-containing anchor domain: a role in clustering of the Tir receptor. Mol. Microbiol. 51 (1), 73–87.

Tozzoli, R., Caprioli, A., Cappannella, S., et al., 2010. Production of the subtilase AB5 cytotoxin by Shiga toxin-negative *Escherichia coli*. J. Clin. Microbiol. 48 (1), 178–183.

Tree, J.J., Roe, A.J., Flockhart, A., et al., 2011. Transcriptional regulators of the GAD acid stress island are carried by effector protein-encoding prophages and indirectly control type III secretion in enterohemorrhagic *Escherichia coli* O157:H7. Mol. Microbiol. 80 (5), 1349–1365.

Tsai, B., Rodighiero, C., Lencer, W.I., et al., 2001. Protein disulfide isomerase acts as a redox-dependent chaperone to unfold cholera toxin. Cell 104 (6), 937–948.

Tu, X., Nisan, I., Yona, C., et al., 2003. EspH, a new cytoskeleton-modulating effector of entero-haemorrhagic and enteropathogenic *Escherichia coli*. Mol. Microbiol. 47 (3), 595–606.

Tuttle, J., Gomez, T., Doyle, M.P., et al., 1999. Lessons from a large outbreak of *Escherichia coli* O157:H7 infections: insights into the infectious dose and method of widespread contamination of hamburger patties. Epidemiol. Infect. 122 (2), 185–192.

Tzipori, S., Gunzer, F., Donnenberg, M.S., et al., 1995. The role of the *eaeA* gene in diarrhea and neurological complications in a gnotobiotic piglet model of enterohemorrhagic *Escherichia coli* infection. Infect. Immun. 63 (9), 3621–3627.

Uchida, H., Kiyokawa, N., Horie, H., et al., 1999. The detection of Shiga toxins in the kidney of a patient with hemolytic uremic syndrome. Pediatr. Res. 45 (1), 133–137.

Uhlich, G.A., Keen, J.E., Elder, R.O., 2002. Variations in the csgD promoter of *Escherichia coli* O157:H7 associated with increased virulence in mice and increased invasion of HEp-2 cells. Infect. Immun. 70 (1), 395–399.

Ulshen, M.H., Rollo, J.L., 1980. Pathogenesis of *Escherichia coli* gastroenteritis in man–another mechanism. N. Engl. J. Med. 302 (2), 99–101.

Unkmeir, A., Schmidt, H., 2000. Structural analysis of phage-borne stx genes and their flanking sequences in shiga toxin-producing *Escherichia coli* and Shigella dysenteriae type 1 strains. Infect. Immun. 68 (9), 4856–4864.

Vally, H., Hall, G., Dyda, A., et al., 2012. Epidemiology of Shiga toxin producing *Escherichia coli* in Australia, 2000–2010. BMC Public Health 12 (1), 63.

van Setten, P.A., Monnens, L.A., Verstraten, R.G., et al., 1996. Effects of verocytotoxin-1 on nonad-herent human monocytes: binding characteristics, protein synthesis, and induction of cytokine release. Blood 88 (1), 174–183.

van Setten, P.A., van Hinsbergh, V.W., van der Velden, T.J., et al., 1997. Effects of TNF alpha on verocytotoxin cytotoxicity in purified human glomerular microvascular endothelial cells. Kidney Int. 51 (4), 1245–1256.

van Setten, P.A., van Hinsbergh, V.W., van den Heuvel, L.P., et al., 1998. Monocyte chemoattractant protein-1 and interleukin-8 levels in urine and serum of patients with hemolytic uremic syndrome. Pediatr. Res. 43 (6), 759–767.

Vazquez-Juarez, R.C., Kuriakose, J.A., Rasko, D.A., et al., 2008. CadA negatively regulates *Escherichia coli* O157:H7 adherence and intestinal colonization. Infect. Immun. 76 (11), 5072–5081.

Verweyen, H.M., Karch, H., Allerberger, F., et al., 1999. Enterohemorrhagic *Escherichia coli* (EHEC) in pediatric hemolytic-uremic syndrome: a prospective study in Germany and Austria. Infection 27 (6), 341–347.

Vingadassalom, D., Kazlauskas, A., Skehan, B., et al., 2009. Insulin receptor tyrosine kinase substrate links the *E. coli* O157:H7 actin assembly effectors Tir and EspF(U) during pedestal formation. Proc. Natl. Acad. Sci. USA 106 (16), 6754–6759.

Viswanathan, V.K., Koutsouris, A., Lukic, S., et al., 2004a. Comparative analysis of EspF from enteropathogenic and enterohemorrhagic *Escherichia coli* in alteration of epithelial barrier function. Infect. Immun. 72 (6), 3218–3227.

Viswanathan, V.K., Lukic, S., Koutsouris, A., et al., 2004b. Cytokeratin 18 interacts with the enteropathogenic *Escherichia coli* secreted protein F (EspF) and is redistributed after infection. Cell Microbiol. 6 (10), 987–997.

Vlisidou, I., Dziva, F., La Ragione, R.M., et al., 2006a. Role of intimin-tir interactions and the tir-cytoskeleton coupling protein in the colonization of calves and lambs by *Escherichia coli* O157:H7. Infect. Immun. 74 (1), 758–764.

Vlisidou, I., Marches, O., Dziva, F., et al., 2006b. Identification and characterization of EspK, a type III secreted effector protein of enterohaemorrhagic *Escherichia coli* O157:H7. FEMS Microbiol. Lett. 263 (1), 32–40.

Vossenkamper, A., Marches, O., Fairclough, P.D., et al., 2010. Inhibition of NF-kappaB signaling in human dendritic cells by the enteropathogenic *Escherichia coli* effector protein NleE. J. Immunol. 185 (7), 4118–4127.

Waddell, T., Head, S., Petric, M., et al., 1988. Globotriosyl ceramide is specifically recognized by the *Escherichia coli* verocytotoxin 2. Biochem. Biophys. Res. Commun. 152 (2), 674–679.

Wagner, P.L., Livny, J., Neely, M.N., et al., 2002. Bacteriophage control of Shiga toxin 1 production and release by *Escherichia coli*. Mol. Microbiol. 44 (4), 957–970.

Wagner, P.L., Neely, M.N., Zhang, X., et al., 2001. Role for a phage promoter in Shiga toxin 2 expression from a pathogenic *Escherichia coli* strain. J. Bacteriol. 183 (6), 2081–2085.

Walterspiel, J.N., Ashkenazi, S., Morrow, A.L., et al., 1992. Effect of subinhibitory concentrations of antibiotics on extracellular Shiga-like toxin I. Infection 20 (1), 25–29.

Wan, F., Weaver, A., Gao, X., et al., 2011. IKKbeta phosphorylation regulates RPS3 nuclear translocation and NF-kappaB function during infection with *Escherichia coli* strain O157:H7. Nat. Immunol. 12, 335–343.

Wang, H., Paton, A.W., McColl, S.R., et al., 2011. In vivo leukocyte changes induced by *Escherichia coli* subtilase cytotoxin. Infect. Immun. 79 (4), 1671–1679.

Wang, H., Paton, J.C., Paton, A.W., 2007. Pathologic changes in mice induced by subtilase cytotoxin, a potent new *Escherichia coli* AB5 toxin that targets the endoplasmic reticulum. J. Infect. Dis. 196 (7), 1093–1101.

Wang, H., Paton, J.C., Thorpe, C.M., et al., 2010. Tissue factor-dependent procoagulant activity of subtilase cytotoxin, a potent AB5 toxin produced by shiga toxigenic *Escherichia coli*. J. Infect. Dis. 202 (9), 1415–1423.

Waterman, S.R., Small, P.L., 1996. Characterization of the acid resistance phenotype and rpoS alleles of shiga-like toxin-producing *Escherichia coli*. Infect. Immun. 64 (7), 2808–2811.

Weflen, A.W., Alto, N.M., Viswanathan, V.K., et al., 2010. *E. coli* secreted protein F promotes EPEC invasion of intestinal epithelial cells via an SNX9-dependent mechanism. Cell Microbiol. 12 (7), 919–929.

Weiss, S.M., Ladwein, M., Schmidt, D., et al., 2009. IRSp53 links the enterohemorrhagic *E. coli* effectors Tir and EspFU for actin pedestal formation. Cell Host Microbe. 5 (3), 244–258.

Whale, A.D., Hernandes, R.T., Ooka, T., et al., 2007. TccP2-mediated subversion of actin dynamics by EPEC 2 –a distinct evolutionary lineage of enteropathogenic *Escherichia coli*. Microbiology 153 (Pt 6), 1743–1755.

Whittam, T.S., 1998. Evolution of *Escherichia coli* O157:H7 and other Shiga toxin-producing *E. coli* strains. *Escherichia coli* O157:H7 and Other Shiga Toxin-producing *E. coli* Strains. In: Kaper, J.B., O'Brien, A.D. (Eds.), American Society for Microbiology, Washington, D.C, pp. 195–212.

Wick, L.M., Qi, W., Lacher, D.W., et al., 2005. Evolution of genomic content in the stepwise emergence of *Escherichia coli* O157:H7. J. Bacteriol. 187 (5), 1783–1791.

Wolfson, J.J., Jandhyala, D.M., Gorczyca, L.A., et al., 2009. Prevalence of the operon encoding subtilase cytotoxin in non-O157 Shiga toxin-producing *Escherichia coli* isolated from humans in the United States. J. Clin. Microbiol. 47 (9), 3058–3059.

Wolfson, J.J., May, K.L., Thorpe, C.M., et al., 2008. Subtilase cytotoxin activates PERK, IRE1 and ATF6 endoplasmic reticulum stress-signalling pathways. Cell Microbiol. 10 (9), 1775–1786.

Wong, A.R., Clements, A., Raymond, B., et al., 2012a. The interplay between the *Escherichia coli* Rho guanine nucleotide exchange factor effectors and the mammalian RhoGEF inhibitor EspH. MBio 3 (1).

Wong, A.R., Pearson, J.S., Bright, M.D., et al., 2011. Enteropathogenic and enterohaemorrhagic *Escherichia coli*: even more subversive elements. Mol. Microbiol. 80 (6), 1420–1438.

Wong, A.R., Raymond, B., Collins, J.W., et al., 2012b. The enteropathogenic *E. coli* effector EspH promotes actin pedestal formation and elongation via WASP-interacting protein (WIP). Cell Microbiol. 14 (7), 1051–1070.

Wong, C.S., Jelacic, S., Habeeb, R.L., et al., 2000. The risk of the hemolytic-uremic syndrome after antibiotic treatment of *Escherichia coli* O157:H7 infections. N. Engl. J. Med. 342 (26), 1930–1936.

Wu, B., Skarina, T., Yee, A., et al., 2010. NleG Type 3 effectors from enterohaemorrhagic *Escherichia coli* are U-Box E3 ubiquitin ligases. PLoS Pathog. 6 (6), e1000960.

Wu, C.J., Hsueh, P.R., Ko, W.C., 2011. A new health threat in Europe: Shiga toxin-producing *Escherichia coli* O104:H4 infections. J. Microbiol. Immunol. Infect. 44 (5), 390–393.

Xicohtencatl-Cortes, J., Monteiro-Neto, V., Ledesma, M.A., et al., 2007. Intestinal adherence associated with type IV pili of enterohemorrhagic *Escherichia coli* O157:H7. J. Clin. Invest. 117 (11), 3519–3529.

Xicohtencatl-Cortes, J., Monteiro-Neto, V., Saldana, Z., et al., 2009a. The type 4 pili of enterohemorrhagic *Escherichia coli* O157:H7 are multipurpose structures with pathogenic attributes. J. Bacteriol. 191 (1), 411–421.

Xicohtencatl-Cortes, J., Sanchez Chacon, E., Saldana, Z., et al., 2009b. Interaction of *Escherichia coli* O157:H7 with leafy green produce. J. Food Prot. 72 (7), 1531–1537.

Yamada, S., Matsushita, S., Kai, A., et al., 1993. Detection of verocytotoxin from stool and serological testing of patients with diarrhea caused by *Escherichia coli* O157: H7. Microbiol. Immunol. 37 (2), 111–118.

Yen, H., Ooka, T., Iguchi, A., et al., 2010. NleC, a type III secretion protease, compromises NF-kappaB activation by targeting p65/RelA. PLoS Pathog. 6 (12), e1001231.

Yi, Y., Ma, Y., Gao, F., et al., 2010. Crystal structure of EHEC intimin: insights into the complementarity between EPEC and EHEC. PLoS One 5 (12), e15285.

Yokoyama, E., Etoh, Y., Ichihara, S., et al., 2011. Emergence of enterohemorrhagic *Escherichia coli* serovar O157 strains in clade 8 with highly similar pulsed-field gel electrophoresis patterns. J. Food Prot. 74 (8), 1324–1327.

Yona-Nadler, C., Umanski, T., Aizawa, S., et al., 2003. Integration host factor (IHF) mediates repression of flagella in enteropathogenic and enterohaemorrhagic *Escherichia coli*. Microbiology 149 (Pt 4), 877–884.

Yoon, J.W., Hovde, C.J., 2008. All blood, no stool: enterohemorrhagic *Escherichia coli* O157:H7 infection. J. Vet. Sci. 9 (3), 219–231.

Yu, H.B., Finlay, B.B., 2008. The caspase-1 inflammasome: a pilot of innate immune responses. Cell Host Microbe 4 (3), 198–208.

Yu, J., Kaper, J.B., 1992. Cloning and characterization of the *eae* gene of enterohaemorrhagic *Escherichia coli* O157:H7. Mol. Microbiol. 6 (3), 411–417.

Yu, M., Haslam, D.B., 2005. Shiga toxin is transported from the endoplasmic reticulum following interaction with the luminal chaperone HEDJ/ERdj3. Infect. Immun. 73 (4), 2524–2532.

Zadik, P.M., Chapman, P.A., Siddons, C.A., 1993. Use of tellurite for the selection of verocytotoxigenic *Escherichia coli* O157. J. Med. Microbiol. 39 (2), 155–158.

Zanchi, C., Zoja, C., Morigi, M., et al., 2008. Fractalkine and CX3CR1 mediate leukocyte capture by endothelium in response to Shiga toxin. J. Immunol. 181 (2), 1460–1469.

Zhang, L., Chaudhuri, R.R., Constantinidou, C., et al., 2004. Regulators encoded in the *Escherichia coli* type III secretion system 2 gene cluster influence expression of genes within the locus for enterocyte effacement in enterohemorrhagic *E. coli* O157:H7. Infect. Immun. 72 (12), 7282–7293.

Zhang, L., Ding, X., Cui, J., et al., 2012. Cysteine methylation disrupts ubiquitin-chain sensing in NF-kappaB activation. Nature 481 (7380), 204–208.

Zhang, Q., Li, Q., Wang, C., et al., 2010. Enteropathogenic *Escherichia coli* changes distribution of occludin and ZO-1 in tight junction membrane microdomains in vivo. Microb. Pathog. 48 (1), 28–34.

Zhang, X., McDaniel, A.D., Wolf, L.E., et al., 2000. Quinolone antibiotics induce Shiga toxin-encoding bacteriophages, toxin production, and death in mice. J. Infect. Dis. 181 (2), 664–670.

Zoja, C., Angioletti, S., Donadelli, R., et al., 2002. Shiga toxin-2 triggers endothelial leukocyte adhesion and transmigration via NF-kappaB dependent up-regulation of IL-8 and MCP-1. Kidney Int. 62 (3), 846–856.

Zumbrun, S.D., Hanson, L., Sinclair, J.F., et al., 2010. Human intestinal tissue and cultured colonic cells contain globotriaosylceramide synthase mRNA and the alternate Shiga toxin receptor globotetraosylceramide. Infect. Immun. 78 (11), 4488–4499.

Zurawski, D.V., Mumy, K.L., Badea, L., et al., 2008. The NleE/OspZ family of effector proteins is required for polymorphonuclear transepithelial migration, a characteristic shared by enteropathogenic *Escherichia coli* and Shigella flexneri infections. Infect. Immun. 76 (1), 369–379.

Chapter 6

Enterotoxigenic *Escherichia coli*

James M. Fleckenstein

Washington University School of Medicine, St. Louis, MO, USA

BACKGROUND

Definition and/or classification

The enterotoxigenic *E. coli* (ETEC) comprise a diverse pathotype of diarrheagenic organisms that share the ability to produce and effectively deliver heat-labile (LT) and/or heat-stable (ST) enterotoxins to target receptors in the small intestine. While a number of additional virulence traits including both fimbrial and non-fimbrial adhesins have been described, ETEC by definition universally produce one of these toxins.

History

Enterotoxigenic *E. coli* were discovered in the course of clinical investigation of patients with *Vibrio cholerae* culture-negative stools presenting with clinical cholera characterized by acute onset of watery diarrhea and severe dehydration (Sack, 2011). In the 1950s De, working in Calcutta, first described patients with clinical cholera whose stool yielded pure cultures of *E. coli* (then referred to as *Bacterium coli*) (De et al., 1956). Strains from these patients caused fluid accumulation similar to *V. cholerae* in the rabbit ileal loop model that De had used previously to identify the cholera toxin (De, 1959), suggesting that these organisms also produced an enterotoxin. Subsequent study by a team of cholera investigators from Johns Hopkins University working at the Calcutta School of Tropical Medicine and the Infectious Disease Hospital in Calcutta in the late 1960s led to definitive identification of enterotoxin-producing *E. coli* from patients with diarrheal disease clinically indistinguishable from severe cholera (Carpenter et al., 1965; Lindenbaum et al., 1965; Gorbach et al., 1971). Similar to De, these investigators recognized that many patients presenting with cholera-like syndromes had pure cultures of *E. coli* in their stools, and that these organisms produced a heat-labile filterable enterotoxin (Sack et al., 1971). Analysis of patients presenting with severe acute watery diarrheal syndromes

Escherichia coli. http://dx.doi.org/10.1016/B978-0-12-397048-0.00006-1

in Bangladesh (Evans and Evans, 1973; Nalin, 1975; Ryder et al., 1976; Sack et al., 1977) later corroborated these early studies.

Soon after the initial discovery of enterotoxin-producing *E. coli* in patients with severe cholera-like syndromes came reports of travelers' diarrhea linked to *E. coli*. While travelers' diarrhea was clearly a previously known entity, Rowe and Taylor first described an outbreak of diarrhea related to particular *E. coli* serotype O148:H28 in British soldiers deployed to the United Kingdom of Aden (now Yemen) (Rowe and Taylor, 1969), and interestingly, ETEC of the same serotype was shortly thereafter identified in soldiers deployed to Vietnam (Rowe et al., 1970; DuPont et al., 1971). One of these isolates, B7A, was used in early volunteer challenge studies which firmly established the pathogenic nature of ETEC (Levine et al., 1979). Following these initial reports, came the publication of a number of studies establishing the importance of ETEC as the principal etiologic agent of travelers' diarrhea (Shore et al., 1974; Merson et al., 1976).

Evolution

Enterotoxigenic *E. coli* strains exhibit both phenotypic and genetic diversity. This likely relates to the fact that genes for both ST and LT are encoded on plasmids (Gyles et al., 1974). Several lines of evidence support the idea that ETEC have arisen through independent acquisition of these essential toxin genes by a genetically diverse population of *E. coli*. First, while some serotypes appear more commonly in collections of ETEC than others, enterotoxigenic *E. coli* are represented by multiple O and H serotypes (Wolf, 1997). Some phylogenetic comparisons based on multi-locus sequence typing (MLST) would likewise suggest that the chromosomal background of ETEC strains is not highly conserved (Turner et al., 2006). Nevertheless, in the most extensive phylogenetic analysis of ETEC to date, performed on over 1000 ETEC isolates, Steinsland et al. determined that strains segregated into distinct clonal groups represented the majority of ETEC (Steinsland et al., 2010). Their analysis suggested that the population of currently circulating ETEC strains likely emerged on several occasions from distinct established globally distributed lineages. One implication of these later studies is that while the evolution of ETEC has certainly been complex, the existence of distinct pathogen lineages may facilitate the identification of conserved antigens useful in vaccine development.

Epidemiology and global impact

Diarrheal illnesses are a leading cause of death in developing countries where more than a fifth of all deaths in children under the age of 5 years can be attributed to infectious diarrhea (Kosek et al., 2003). Collectively, diarrheal pathogens are estimated to cause between 1 and 2 million deaths annually (Kosek et al., 2003; Boschi-Pinto et al., 2008). ETEC contributes significantly to mortality associated with diarrheal illness, accounting for hundreds of thousands of deaths each year. Moreover, these illnesses appear to contribute substantially to overall morbidity and may relate to delayed growth in infected children (Petri et al., 2008).

Traveler's diarrhea

The diarrheal attack rate among travelers is appreciable with roughly one third to one half (Merson et al., 1976; Sack, 1990) of them becoming ill, often within days of arriving at their destination. Since shortly after ETEC were identified as a causative agent of diarrheal illness, they have been linked to diarrhea in travelers. In virtually every series ETEC is the predominant pathogen accounting for more than half of the cases where an etiologic agent is identified (Black, 1990; Sack, 1990). Largely, this is a reflection of the fact that many travelers under study arrive from regions with adequate sanitation at destinations where sanitation is substandard. Without immunity established by repeated exposure, travelers to developing countries are at high risk for acquisition of ETEC infection as these organisms are ubiquitously distributed in regions where sanitation is poor (Subekti et al., 2003).

Non-governmental organization (NGO) personnel military deployment and diarrhea

Soldiers, NGO personnel, and Peace Corp volunteers deployed to regions where sanitation is often quite poor often fall victim to infectious diarrhea during their deployment. Data from this subset of travelers provide valuable insight into the global epidemiology of infectious diarrheal pathogens as these individuals are often deployed for extended durations to areas not frequented by tourists where local microbiologic data are otherwise lacking (Haberberger et al., 1991; Hyams et al., 1991; Bourgeois et al., 1993). Of the many studies performed in this population, most have also found ETEC to be the predominant pathogen (Riddle et al., 2006).

Seasonality in developing countries

In Bangladesh, there is considerable seasonal variation in ETEC with biannual peaks of illness in the warmer spring and early fall periods (Qadri et al., 2005, 2007). Similarly, diarrheal illness in Egypt has been shown to peak in warmer months of the year (Abu-Elyazeed et al., 1999), and in studies of travelers with diarrhea in Mexico, the rate of ETEC infections increased by 7% for every degree centigrade increase in ambient temperature (Paredes-Paredes et al., 2011).

Emergence of ETEC in developed countries

Interestingly, despite the clear preponderance of ETEC infections in developing regions of the world, a spate of large-scale ETEC outbreaks in the US (Rosenberg et al., 1977; CDC, 1994; Dalton et al., 1999; Beatty et al., 2004;

Devasia et al., 2006) attests to the fact that even in industrialized countries these organisms can cause serious harm. While some outbreaks can be traced to importation of food from beyond our borders (MacDonald et al., 1985), in many cases the ultimate source of ETEC is unclear (Beatty et al., 2004). Because screening for ETEC is not routinely performed in clinical microbiology laboratories, these organisms usually escape recognition in sporadic cases and are only identified during the course of investigation of large clusters of patients presenting with diarrheal disease (Jain et al., 2008). On systematic screening of patients with diarrhea, domestically acquired ETEC infections have been identified in studies from Sweden (Svenungsson et al., 2000) and in Minnesota (Beatty et al., 2004), suggesting that these organisms are responsible for some small percentage of sporadic diarrheal illness in industrialized countries including the US. One of the larger recorded outbreaks occurred in the suburban Chicago area when a single delicatessen catered multiple events, disseminating ETEC-laden food, ultimately affecting thousands of people (New Tork Times, 1998; Beatty et al., 2004).

MOLECULAR PATHOGENESIS

Regulation

cAMP receptor protein (CRP) modulation of gene expression in ETEC

Like other pathogens, ETEC respond to their environment by modulating virulence gene expression, often as a result of sensing small molecules such as glucose. Interestingly, some genes such as those encoding LT are optimally expressed in the presence of glucose (catabolite activation), while other genes such as those involved in elaboration of some pili, including CFA/I, are repressed by glucose (catabolite repression). As in other important pathogens (McDonough and Rodriguez, 2012), cyclic AMP appears to play a central role in modulating the expression of these and other ETEC virulence molecules. High levels of glucose inhibit bacterial adenylate cyclase and reduce levels of cAMP and consequently its interaction with the cAMP receptor protein (CRP). In the case of heat-labile toxin, the CRP–cAMP complex represses expression of the *eltAB* genes encoding LT; *eltAB* genes are therefore de-repressed or stimulated in the presence of glucose as cAMP intracellular concentrations fall and there is no CRP–cAMP complex to bind to the *eltA* promoter region and inhibit initiation of transcription by RNA polymerase (Bodero and Munson, 2009) (Figure 6.1).

Recent studies in non-pathogenic strains of *E. coli* demonstrate that more than 100 genes are modulated in response to CRP (Zheng et al., 2004). Similarly, ETEC have adapted this central regulatory mechanism to modulate expression of a number of other virulence genes including *tibA* (Espert et al., 2011), and the heat-stable toxin gene, *estA* (Bodero and Munson, 2009) (Table 6.1).

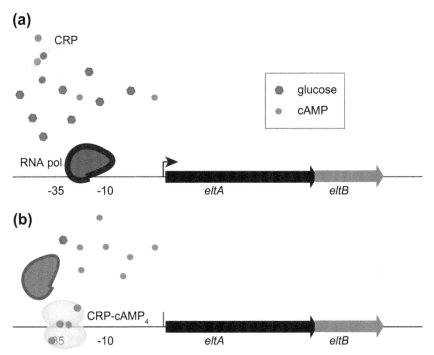

FIGURE 6.1 Modulation of LT gene transcription by CRP provides an example of catabolite activation in ETEC. (a) At high glucose concentrations, cAMP levels in the cell are low and the homodimeric CRP molecule (shown with four potential cAMP binding sites) is inactive allowing RNA polymerase to interact with the promoter region and initiate transcription of *eltA*. (b) At low glucose concentrations, cAMP levels in the bacteria increase, activating CRP and permitting the CRP–cAMP complex to bind to a CRP binding site (operator centered at −31.5 upstream from *eltA*) within the *eltA* promoter region. This prevents transcription by preventing the RNA polymerase from forming an open complex at the promoter. *(Adapted from Bodero and Munson (2009).)*

Interestingly, intestinal epithelial cells possess high-affinity cAMP transporters that efflux cAMP into the surrounding milieu (Li et al., 2007) as cAMP levels in the host cell increase. Therefore it is intriguing to hypothesize that ETEC actually sense cAMP generated by target epithelial cells in response to successful delivery of LT, ultimately governing the interaction of ETEC and target epithelium.

Rns and CfaD

Two other transcriptional regulators, Rns (Caron et al., 1989) and CfaD, have been shown to modulate expression of virulence genes in ETEC. Both belong to the AraC/XylS family of transcriptional regulators and their amino acid sequences are 95% identical (Pilonieta et al., 2007). CfaD, located on the large 94.8 kbp virulence plasmid of ETEC strain H10407, has been shown to activate production of CFA/I (Jordi et al., 1992), while Rns is known to activate

TABLE 6.1 Modulation of virulence determinants by CRP

Gene(s)/ operon	cAMP-CRP modulation	Reference(s)
eltAB	–	Bodero and Munson, 2009
estA	+	Bodero and Munson, 2009
fliC	+	Chilcott and Hughes, 2000
CS1-3	+	Evans et al., 1991
CFA/I	+	Karjalainen et al., 1991
tibDBCA	+	Espert et al., 2011
tolC	+*	Hantke et al., 2011; Zheng et al., 2004
fimA	–	Muller et al., 2009
luxS	–	Wang et al., 2005

*Indirect via activation of marRAB (Zheng et al., 2004).

a number of other genes, including those encoding several other colonization factors (Pilonieta et al., 2007) (Table 6.2).

Not surprisingly, ETEC appear to be programmed to respond to other small molecules that may be found in the small intestine such as those contained in bile. Recent array studies indicate that bile can induce production of both ST and LT (Sahl and Rasko, 2012). Therefore, expression of virulence factors by ETEC likely reflects the integration of multiple signals as these organisms make their way through different environments culminating in epithelial attachment and toxin delivery to the host cell.

Adherence and invasion

Colonization factors

A principal focus of ETEC pathogenesis has been the study of factors which promote colonization of the small intestine. Here, colonization of the intestinal mucosa is thought to be required for efficient delivery of ST and LT (Zafriri et al., 1987; Ofek et al., 1990). Fimbrial structures known as colonization factors (CFs) were among the earliest virulence factors to be identified in ETEC (Evans et al., 1975). Early ETEC volunteer challenge studies demonstrated that a strain cured of a plasmid encoding CFA/I failed to cause diarrheal illness (Figure 6.2a) (Evans et al., 1978; Satterwhite et al., 1978), and recipients of the CFA/I-negative strain had significant reductions in fecal shedding. However, the reduced pathogenicity observed in these studies likely was not simply due to absence of CFA/I as other important determinants are now known to be encoded on this plasmid (Figure 6.2b).

TABLE 6.2 ETEC colonization factors

Colonization factor	Structure	GenBank ref	Reference(s)
CFA/I group			
CFA/I	Fimbriae	M55661.1	Jordi et al., 1992
CS1	Fimbriae	AY536429.1	Froehlich et al., 2005; Levine et al., 1984; Marron and Smyth, 1995; Perez-Casal et al., 1990
CS2	Fimbriae	Z47800.1	Cravioto et al., 1982; Froehlich et al., 1995
CS4	Fimbriae	AF296132.1	Thomas et al., 1985
CS14 (PCFO166)	Fimbriae	AAQ20106	McConnell et al., 1989
CS17	Fimbriae	AAS89777.1	McConnell et al., 1990
CS19	Fimbriae	AAQ19775.1	Grewal et al., 1997
CS5 group			
CS5	Thin fimbriae	X63411.1	Clark et al., 1992; Heuzenroeder et al., 1989; Thomas et al., 1985
CS7	Helical, thin fimbriae	AAK09045.1	Hibberd et al., 1990
CS13 (PCFO9)	Thin fibrils		Heuzenroeder et al., 1990
CS20	Fimbriae	AF438156.1	Valvatne et al., 1996
CS18 (PCFO20)	Fimbriae	AF335469.1	Honarvar et al., 2003; Viboud et al., 1996
Type IV bundle forming pili			
CS21 (longus)		EF595770.1	Giron et al., 1994
CS8 (CFA/III)	Long straight pili	AB049751.1	Honda et al., 1984; Kolappan et al., 2012; Taniguchi et al., 2001
CS15 (antigen 8786)	No structure	X64623.1	Aubel et al., 1991
Unique			
CS3	Thin wiry fibrils	CBL93336.1	Levine et al., 1984
CS6	Non-fimbrial	U04846	Wolf et al., 1997

Continued

TABLE 6.2 ETEC colonization factors—cont'd

Colonization factor	Structure	GenBank ref	Reference(s)
CS10 (antigen 2230)	Non-fimbrial		Darfeuille-Michaud et al., 1986
CS11 (PCFO148)	Thin curly fibrils ~3 nm		Knutton et al., 1987
CS12 (PCFO159)	Fimbriae	AAK09047.1	Tacket et al., 1987
CS22	Thin fibrils	AF145205.1	Pichel et al., 2000
CS23	Non-fimbrial	JQ434477	Del Canto et al., 2012

Additional uncharacterized antigens with sequence homology to CS20 (Nada et al., 2011).

The molecular and structural biology of CFA/I pilus biogenesis has been described in some detail (Jordi et al., 1992; Li et al., 2009). This pilus belongs to the chaperone-usher family of adhesive organelles (see Chapter 12). The plasmid-encoded CFA/I operon encompasses genes for CfaA a putative chaperone, CfaB the major fimbrial structural subunit, CfaC a putative outer membrane usher protein, and CfaE, the minor pilin tip adhesin subunit (Baker et al., 2009). The assembled CFA/I pili are approximately 1 μm in length with approximately 1000 CfaB subunits polymerized into the shaft to present single CfaE subunits as the tip adhesin. Intriguingly, in a spring-like fashion (Mu et al., 2008), CFA/I can assume both tightly wound helical structures or more relaxed open states in response to shear stress, perhaps permitting enhanced adhesion in response to intestinal flow (Andersson et al., 2012).

Plasmid CF loci encode a wide variety of proteinaceous structures that assume the shape of fimbriae, fibrils, helical, or afimbrial surface molecules (Wolf, 1997; Qadri, et al., 2005). Since the discovery of CFA/I, new CFs continue to be identified, with more than 25 antigenically distinct factors described to date (Qadri et al., 2005). On average the fimbrial structures described to date are on the order of 1–2 μm in length. Distinct from these structures is a type IV pilus referred to as longus that extends to more than 20 μm from the bacterial surface (Giron et al., 1994). LngA, the major structural pilin subunit of longus, also known as CS20, is recognized during ETEC infections (Qadri et al., 2000) and shares significant N-terminal homology with other type-4 pilins including those which comprise the toxin co-regulated pilus (TCP) of *V. cholerae* and the bundle forming pili of enteropathogenic *E. coli* (see Chapter 13) (Giron et al., 1997).

Interestingly, intestinal colonization and adherence are complex phenotypes that involve multiple genes in addition to the classic colonization factors. Studies implicating LT in promoting epithelial cell adherence (Johnson et al.,

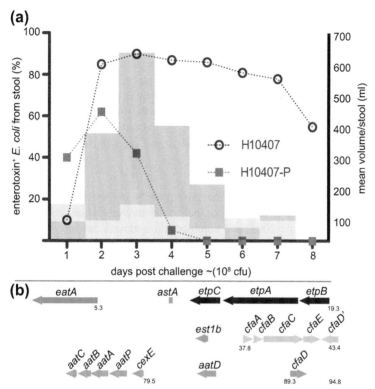

FIGURE 6.2 Importance of a large virulence plasmid in ETEC pathogenesis. (a) Summary of early clinical challenge studies (Evans et al., 1978; Satterwhite et al., 1978) with wild-type ETEC strain H10407, and H10407-P, a spontaneous deletion mutant lacking the large virulence plasmid previously known to encode CFA/I genes. Bar graph shows mean stool volume for H10407 (light blue) and H10407-P (gray); line graph shows recovery of ETEC from stool following challenge. (b) Organization of virulence genes on the large 94.8 kb virulence plasmid of ETEC H10407. In addition to CFA/I genes (yellow) are two putative toxin genes (orange), *est1b* encoding ST1b and *astA* encoding EAST1. Also encoded on this plasmid are the EtpBAC two-partner secretion system (dark blue), the *eatA* autotransporter gene (green), *cexE* and its presumed secretion system (light blue) and the *cfaD* regulator gene.

2009) and intestinal colonization (Allen et al., 2006), suggest that enterotoxins offer advantages to the bacteria that surpass simple dissemination into the environment.

Type 1 fimbriae

Genes encoding type 1 fimbriae of *E. coli* are highly conserved (see Chapter 12) and ETEC (Levine et al., 1983; Knutton et al., 1984a, b;), similar to uropathogenic *E. coli* (UPEC) are known to make these structures. However, relative to the extensive study of the role of type 1 fimbriae in UPEC pathogenesis (see Chapter 9), comparatively little is known about their involvement in

ETEC virulence. Despite some early enthusiasm for targeting type 1 pili in combination ETEC vaccines (Levine, 1981), initial efforts to protect against ETEC infections by vaccination with type 1 pili were disappointing (Levine et al., 1982). Nevertheless, tremendous advances in our understanding of the biogenesis of type 1 pili (Waksman and Hultgren, 2009), the very complex nature of their role in UPEC infections (Mulvey et al., 2000; Schilling et al., 2001), and recognized diversity in FimH tip adhesin (Sokurenko et al., 1997) structures could stimulate additional examination of a role for these highly conserved fimbriae in ETEC pathogenesis and vaccine development.

E. coli *common pili (ECP)*

The genomes of most *E. coli* encode a variety of potential pilus systems. *E. coli* common pili (ECP), originally discovered in enterohemorrhagic *E. coli* (Rendon et al., 2007) and commensal organisms, are also expressed by other *E. coli* pathotypes, including ETEC (Blackburn et al., 2009). Although at present, the contribution of these structures to the pathogenesis of ETEC is uncertain, they could act in concert with other adhesins as has been noted for EPEC (Saldana et al., 2009).

Outer membrane proteins

Two outer membrane proteins that have been identified in some strains of ETEC promote invasion of intestinal epithelial cell lines in vitro. These include Tia, a 25 kDa (Fleckenstein et al., 1996, 2002) heparin sulfate binding outer membrane protein encoded on a pathogenicity island of ETEC H10407 (Fleckenstein et al., 2000), and TibA (Elsinghorst and Weitz, 1994), a glycosylated autotransporter protein (Lindenthal and Elsinghorst, 1999). Glycosylation of TibA likely requires TibC, a hepatosyltransferase (Moormann et al., 2002). The relevance of epithelial cell invasion per se in the pathogenesis of ETEC remains uncertain, but it is likely that the in vitro invasion phenotype reflects intimate interactions between the bacteria and the host.

Novel antigens

A modified Tn*phoA* mutagenesis strategy to identify secreted or surface-expressed antigens from ETEC led to the identification of a two-partner secretion locus (Fleckenstein et al., 2006) encoded on the large 94.8 kb virulence plasmid of ETEC H10407 that also encodes CFA/I (see Figure 6.2b). Three contiguous genes, *etpB, etpA,* and *etpC* encode an outer membrane transport pore (EtpB), a secreted glycoprotein adhesin (EtpA), and a putative transglycosylase (EtpC). Interestingly, EtpA appears to promote ETEC adhesion in a unique fashion by bridging highly conserved flagellin residues exposed only at the tips of ETEC flagella, with receptors on the mucosal surface (Roy et al., 2009).

A similar strategy was also used to identify EatA (Patel et al., 2004), a secreted serine protease autotransporter (SPATE) molecule, which is also encoded on the large virulence plasmid of ETEC H10407, downstream from the *etpBAC* locus (see Figure 6.2b). Recent studies of EatA have demonstrated that this molecule modulates ETEC adhesion by degrading the EtpA adhesin molecule, in the process promoting effective delivery of the heat-labile toxin (Roy et al., 2011).

Interestingly, searches of potential CfaD binding sites led to the discovery of a novel secreted protein, CexE (Pilonieta et al., 2007), also encoded on the large 94.8 kb virulence plasmid of ETEC H10407, along with a potential secretion locus (see Figure 6.2b). While the function of CexE is uncertain, it is recognized by host antibodies during the course of ETEC infection (Roy et al., 2010).

Damage

The principal effector molecules that define ETEC are the enterotoxins that are secreted by ETEC. These include heat-labile toxin (LT), structurally and functionally similar to cholera toxin, and the small peptide heat-stable toxins (Figure 6.3). Strains may produce any or all of these toxins.

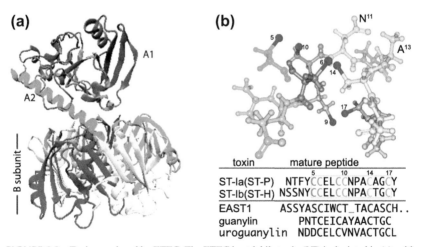

toxin	mature peptide
ST-Ia(ST-P)	NTFYCCELCCNPACAGCY
ST-Ib(ST-H)	NSSNYCCELCCNPACTGCY
EAST1	ASSYASCIWCT_TACASCH..
guanylin	PNTCEICAYAACTGC
uroguanylin	NDDCELCVNVACTGCL

FIGURE 6.3   Toxins produced by ETEC. The ETEC heat-labile toxin (LT) is depicted in (a), with A1 domain shown in blue, and the A2 domain in green, anchoring the A-subunit in the B pentamer. (b) *E. coli* heat-stable toxins. The predicted structure of the core active toxin domain of ST-P (amino acid residues C5–C17) is shown with sulfur atoms involved in disulfide bond formation in red. Alignments of ST-1a, ST-1b, EAST1 and the native human peptide guanylin are shown below the molecule. LT structure obtained from protein data bank (pdb) entry 1LTB, and rendered in VMD (Humphrey et al., 1996); ST-P was rendered in Protein Workshop (Moreland et al., 2005) from pdb entry 1etn.

Heat-labile toxin (LT)

The heat-labile toxin is a heteroheximeric molecule composed of a B-subunit pentamer and a single A-subunit. Toxin activity resides in the A-subunit, while the B-subunit is largely responsible for host cell binding. The A-subunit is comprised of two principal domains, the A1 domain, responsible for the toxicity, and A2 which non-covalently anchors the A-subunit in the center of the B pentamer (Sixma et al., 1991). Thus the overall structure (Figure 6.3a) has been compared to a 'ring (the B-pentamer) on a finger' (the A-subunit) (van Heyningen, 1991).

The B-subunit engages GM-1 gangliosides centered in lipid rafts on the surface of eukaryotic cells to trigger internalization of the toxin into the host cell following proteolytic 'nicking' of A1 and A2 into separate polypeptides (of approximately 22 and 5 kDa, respectively) by serine proteases located at the apical surface of epithelial cells (Lencer et al., 1997). The subunits remain joined by a single disulfide bond following cleavage (Lencer et al., 1997). Following uptake of the A1 subunit into the host cell cytoplasm, it acts in concert with cellular ADP ribosylating factors (ARFs) to transfer ADP-ribose from NAD to the alpha subunit of the heterotrimeric GTPase, (Gsα) (Spangler, 1992). Formation of a ternary complex of ADP-ribosylated Gsα GTP and adenylate cyclase catalyses cAMP production from ATP. The resulting increases in intracellular concentrations of cAMP activate protein kinase A, which subsequently phosphorylates the cystic fibrosis transmembrane conductance regulator (CFTR) activating the chloride channel (Cheng et al., 1991). This promotes efflux of chloride into the intestinal lumen paralleled by net decreases in Na^+ and water absorption resulting in watery diarrhea characteristic of ETEC.

LT secretion requires a functional type II secretion system

Secretion of LT from ETEC requires the presence of a functional type II secretion system (see Chapter 13), similar to the secretion apparatus for CT in *Vibrio cholerae* (Tauschek et al., 2002). In the protypical ETEC strain H10407, LT secretion appears to be further modulated by one or more genes encoded by a pathogenicity island including *leoA* (Fleckenstein et al., 2000), a GTPase (Brown and Hardwidge, 2007). Interestingly, much of the LT secreted by ETEC in vitro remains associated with lipopolysaccharide of outer membrane vesicles produced by the bacteria (Horstman and Kuehn, 2000).

Heat-stable toxins

ETEC may secrete one or more plasmid-encoded heat-stable toxins (ST), small cysteine-rich peptides that engage an extracellular domain of guanylyl cyclase C (GC-C) on the apical surface of intestinal epithelial cells (Potter, 2011). This binding triggers activation of the intracellular portion of GC-C which converts GTP to the intracellular messenger cGMP. Elevations in cGMP lead to activation of the membrane-bound (French et al., 1995; Vaandrager et al., 1998) cGMP-dependent protein kinase (cGK II or PKG) that in turn phosphorylates CFTR (Pfeifer et al., 1996; Vaandrager et al., 2000).

Several different variants of ST have been described in human strains. STa (STI) peptides come in two varieties ST-Ia(ST-P) and ST-Ib(ST-H) (Figure 6.3b), both of which are plasmid-encoded, typically on transposition elements. These peptides, as well as a similar native peptides, uroguanylin and guanylin, both activate GC-C, also known as the STa receptor or STaR (Chao et al., 1994; Giannella and Mann, 2003). Both ST1 molecules are produced as pro-peptides that undergo processing and export into the periplasm. Export of ST1b, and presumably ST1a, through the outer membrane of *E. coli* depends on the TolC (Yamanaka et al., 1998) type 1 secretion system (see Chapter 16). In contrast, STb peptides have also been identified in human strains of ETEC, but predominately are found in porcine strains, and do not bind GC-C or activate cGMP, and unlike the STa (Sack et al., 1975; Levine et al., 1977) peptides STb toxins are not as clearly linked to human illness (Weikel et al., 1986).

EAST1, or EnteroAggregative heat-Stable Toxin (Savarino et al., 1993), has predicted structural similarity to STI molecules (Figure 6.3b), and is encoded by *ast* genes residing on mobile elements (McVeigh et al., 2000) of plasmids in some ETEC strains (Savarino et al., 1996; Yamamoto and Echeverria, 1996). While the functional significance of these genes has yet to be determined, they do appear to encode a molecule with the capacity to stimulate cGMP production in vitro. Some strains, such as the prototypical ETEC strain H10407, encode LT, ST1a, ST1b, and EAST1 (Fleckenstein et al., 2010) suggesting substantial functional redundancy in the production of enterotoxins.

Candidate virulence molecules of unknown function

Subtractive DNA hybridization studies (Chen et al., 2006), the completion of multiple ETEC genomes (Rasko et al., 2008; Crossman et al., 2010; Sahl et al., 2011), and proteomic studies (Roy et al., 2010) have identified a variety of other molecules that may be important in the pathogenesis of ETEC and provide candidate targets for vaccine development. Studies currently underway are attempting to examine the utility of some of these molecules (Harris et al., 2011), and to examine their function.

Summary of ETEC pathogenesis

Our view of the molecular pathogenesis of ETEC has evolved significantly in the past decade. These organisms are significantly more complex than previously appreciated. The emerging picture of ETEC pathogenesis (depicted in Figure 6.4) suggests that these pathogens orchestrate the deployment of a sophisticated array of virulence factors in a series of complex interactions with the host.

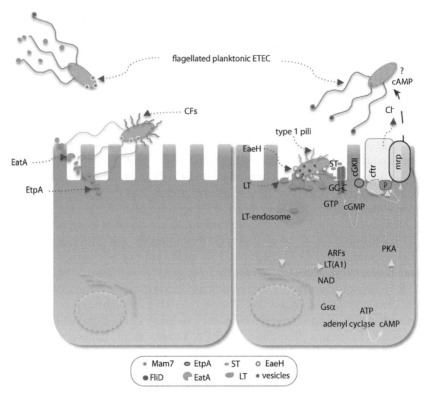

FIGURE 6.4 Summary of steps in molecular pathogenesis of ETEC. Panel on the left shows motile ETEC expressing the EtpA extracellular adhesin (green circles) which binds to the ends of flagella in the absence of the flagellar cap structure (FliD, red), and anchors ETEC via its flagella to the host cell surface. EtpA-mediated binding is modulated by EatA, a secreted auto-transporter protease. Subsequent steps in ETEC adhesion to intestinal epithelial cells include binding via fimbrial antigens including colonization factors (CFs), and type 1 fimbriae, as well as a fimbrial outer membrane proteins including EaeH, which is activated on ETEC-host cell contact. Efficient delivery of LT and ST requires intimate interaction of bacteria with the epithelial cells such that LT can bind to GM-1 gangliosides and ST can engage guanylate cyclase C (GC-C). Binding of LT to the surface triggers toxin internalization and retrograde processing via the trans-Golgi network into the endoplasmic reticulum (ER) (Wernick et al., 2010). The enzymatically active LT A1 subunit is ultimately transported into the cytosol where in the presence of ADP ribosylating factors (ARFs) it transfers ADP-ribose from NAD to the heterotrimeric G protein, $GS\alpha$, which in turn leads to constitutive activation of adenylate cyclase. The resulting rapid increases in cAMP activate protein kinase A (PKA) which phosphorylates CFTR activating the chloride channel. ST peptides (ST-P and ST-H) both engage the extracellular portion of guanylate cyclase C (GC-C) homodimers on the epithelial surface resulting in signal transduction to intracellular catalytic domain where GTP is converted to cGMP. The increases in cGMP activate membrane-associated cGMP-dependent protein kinase (cGKII or PKG) which also phosphorylates CFTR. Intracellular increases in both cyclic nucleotides result in efflux of cAMP (through MRP channels) and cGMP (through MRP5). ETEC may be programmed to respond to extracellular increases in cAMP by modulating cAMP-CRP-dependent gene transcription.

CLINICAL MANIFESTATIONS

Transmission

In both developing and industrialized countries ETEC infections are transmitted by contaminated food or water. Both epidemiologic and volunteer studies suggest that the inoculum of ETEC required to establish illness in healthy subjects is relatively high and that person-to-person transmission is uncommon (Levine et al., 1980; Roels et al., 1998). While, 10^8 CFU of ETEC H10407, the strain used in the majority of volunteer studies, typically leads to moderate to severe diarrhea in most volunteers (Porter et al., 2011), the actual dose required to establish illness could be several logs lower and the infective dose in developing countries could be substantially different (Harro et al., 2011).

Clinical features

Diarrheal illness caused by ETEC ranges from mild watery diarrhea to severe life-threatening cholera-like illness in both children and adults (Sack et al., 1971; Finkelstein et al., 1976; Vicente et al., 2005). Diarrhea caused by ETEC cannot be distinguished from that caused by *Vibrio cholerae* on clinical grounds alone (Sack et al., 1977). Other symptoms including abdominal cramping are common, while fever and vomiting occur in roughly 10% or less of cases on average (Roels et al., 1998; Porter et al., 2011). Diarrhea caused by ETEC can be somewhat protracted and last as long as 1 to 2 weeks (Roels et al., 1998; Bolin et al., 2006). Illness lasting more than 4 days, where vomiting is not a predominant symptom, should prompt a consideration of ETEC as the offending pathogen (Roels et al., 1998).

Diarrhea caused by ETEC is classically described as watery without the appearance of gross blood or mucus in the stool. While classically, ETEC infections have been unceremoniously relegated to causing 'non-inflammatory' diarrhea, this view may not be entirely accurate. Indeed, symptomatic ETEC infections elicit sizable fecal lactoferrin responses (Mercado et al., 2011), as well as interleukin-8 that are comparable to those observed with *Salmonella* infections (Greenberg et al., 2002). In studies of ETEC in travelers with diarrhea, roughly one quarter had fecal leukocytes, and nearly one third were found to have fecal occult blood (Bouckenooghe et al., 2000).

Host susceptibility

Children with A or AB blood group antigens appear to be somewhat more susceptible to developing symptomatic ETEC infections than those of blood group O (Qadri et al., 2007), suggesting a genetic predisposition to development of symptomatic ETEC infections. Likewise, children expressing Lewis blood group a (Le^{a+b-}), thought to be a receptor for CFA/I (Jansson et al., 2006), are more susceptible to diarrhea caused by ETEC expressing CFA/I (Ahmed et al., 2009). Given the genomic diversity of ETEC, and the complex interactions between these organsims and the host, it is likely that there are many such human polymorphisms that might predispose to development (or prevention) of symptomatic ETEC infections. These would include mutations in innate immune effector molecules that control ETEC, as exemplified by the identification of single-nucleotide polymorphisms (SNPs) in the human lactoferrin gene among travelers with symptomatic ETEC infection (Mohamed et al., 2007), as well as SNPs in CD14, a lipopolysaccharide receptor that is induced in travelers' diarrhea (Mohamed et al., 2011). Interestingly, polymorphisms associated with higher levels of interleukin (IL-10), an anti-inflammatory cytokine, have been associated with an increased risk of symptomatic ETEC infection (Flores et al., 2008).

Complications

The most common serious complication of ETEC diarrhea is profound dehydration. Particularly among young children, ETEC and rotavirus contribute disproportionately to episodes of life-threating dehydration in developing countries (Black et al., 1981). Similarly, among older children and adults in low- and middle-income countries, two groups of pathogens, ETEC and *Vibrio cholerae,* are most frequently associated with severe illness requiring hospitalization (Fischer Walker et al., 2010).

Another more pernicious but important effect of diarrheal illness attributed to ETEC is childhood stunting (Ricci et al., 2006; Mondal et al., 2009, 2012). The relationship between infectious diarrheal illnesses and malnutrition is complex, with multiple studies implicating diarrhea caused by ETEC and other pathogens as a cause and a consequence of malnutrition (Guerrant et al., 1992; Mondal et al., 2012). Repeated bouts of diarrhea in childhood have been associated with significant growth retardation or stunting (Checkley et al., 2008). Moreover, malnutrition appears to place children at substantially increased risk for development of diarrhea caused by ETEC. In addition, children with diarrheal illness due to ETEC tend to have more severe diarrheal illness when they are malnourished (Black et al., 1984; Mondal et al., 2012).

A small percentage of patients with traveler's diarrhea will experience more chronic symptoms lasting months to years. The pathogenesis of these presumably post-infectious sequelae, referred to as post-infectious irritable bowel syndrome, is uncertain. Nevertheless characteristic low-grade inflammatory changes in the mucosa could suggest ongoing untoward response to the prior infection (Connor, 2005).

Diagnosis

Identification of ETEC in clinical specimens relies on the identification of heat-labile and/or heat-stable toxins in fecal isolates of *E. coli*. While a number of functional or physiologic assays have been used in the past to identify toxin-producing bacteria, these have been largely supplanted by immunologic or molecular-based methods. PCR, DNA probes for genes encoding ST and LT, ELISA assays for both LT (Sack et al., 1980) and ST (Thompson et al., 1986) are used to detect toxins in clinical research settings, with multiplex PCR for LT, STa, STh emerging as the favored assay to identify ETEC (Sjoling et al., 2007). Unfortunately, none of the current assays is performed routinely in clinical laboratories, nor are they suitable for laboratories in developing countries where resources may be limited. In the US, ETEC infections are typically diagnosed during outbreaks when resources of state public health or CDC laboratories are marshaled to address unexplained cases of diarrhea. Novel rapid point-of-care technologies that can be easily applied to diagnosing infections in developing countries remain an unmet need (Qadri et al., 2005; Hauck et al., 2010).

Treatment

Travelers' diarrhea (TD) caused by ETEC is often treated with antibiotics with or without adjunctive symptomatic treatment with anti-motility agents such as loperamide (DuPont, 2007). Antibiotics currently in use in adults with TD include fluoroquinolones, azithromycin, and more recently rifaximin. Azithromycin is the only antimicrobial currently advised for use in children. In general, when active against the causative strains, all of these drugs appear to shorten the duration of diarrheal illness by approximately 48 hours. Rifaximin, a non-absorbable rifamycin antibiotic, appears equivalent to fluoroquinolones and azithromycin for treatment of sensitive ETEC strains, without systemic side effects. Emerging resistance to all three antibiotics over time will remain a concern (Ouyang-Latimer et al., 2011). Most studies have shown a modest (Dupont et al., 2007; Ericsson et al., 2007; Butler, 2008) or no (Taylor et al., 1991) benefit to adding loperamide to antimicrobial therapy in adults with travelers' diarrhea, while loperamide use is not advisable in young children due to concerns over potential CNS depression.

In severe cholera-like illness, particularly affecting young children in developing countries, treatment of ETEC infections must focus first on rehydration. Those with evidence of severe dehydration require initial management with intravenous fluids followed by oral rehydration therapy (ORT), while milder forms of illness can be treated with ORT alone. Unlike adult travelers, antimicrobial administration for ETEC diarrheal illness in children has not been well-studied.

Immune response

Innate immunity and ETEC

ETEC infections stimulate production of pro-inflammatory cytokines, including interleukin-8, a potent chemoattractant for polymorphonuclear leukocytes.

Interestingly, increased fecal levels of IL-8 have been associated with more rapid resolution of ETEC infections (Long et al., 2010), suggesting this element of innate immunity is relevant in clearing organisms from the small intestine. Recent studies have demonstrated that LT plays a role in promoting intestinal colonization by ETEC (Allen et al., 2006). One possible avenue by which LT might affect colonization is by interfering with innate immune effectors at the mucosal surface. Interestingly, increases in cAMP inhibit activation of a number of cytokines (Hajishengallis et al., 2004) including TNFα and IL-8 (Huang et al., 2004), as well as intestinal antimicrobial peptides including beta-defensin-1 and cathelicidin (O'Neil et al., 1999; Chakraborty et al., 2008) by interfering with NF-κB-mediated (Parry and Mackman, 1997) modulation of pathogen-associated molecular pattern (PAMP) responses.

Adaptive immune responses to ETEC

Multiple studies have documented striking age-associated declines in the incidence of symptomatic ETEC infections strongly suggesting that naturally occurring infections provide substantial protection against subsequent episodes of disease (Abu-Elyazeed et al., 1999; Qadri et al., 2007). Most studies of immune responses to ETEC have centered on responses to the CFs and LT. However, the precise nature of the protective antigens following ETEC infections is not really clear. While some studies have suggested that infection with strains expressing a particular CF is protective against subsequent infection with strains expressing homologous CFs (Qadri et al., 2007), other studies have failed to demonstrate a clear association with CF antigens and protection (Steinsland et al., 2003, 2004), and have implicated other as yet unspecified antigens as a possible source of protection (Steinsland et al., 2003). Similarly, there is no robust serologic correlate of protection from studies performed to date (Rao et al., 2005). Clearly, as shown by recent immuno-proteomic studies, the immune response to ETEC is extraordinarily complex, and involves recognition of multiple antigens in addition to LT and the CFs (Roy et al., 2010), including more recently discovered antigens such as the EtpA adhesin and the passenger domain of the EatA autotransporter.

CONTROL AND PREVENTION

ETEC vaccine development

Despite four decades of ETEC vaccine development effort, there is at present no vaccine against ETEC that has demonstrated sustained, broad-based protection in any of the target populations at risk, most importantly children in developing countries (Svennerholm and Tobias, 2008). Vaccinology efforts to date have focused on a limited number of antigens, primarily the CFs, LT, and more recently ST (Taxt et al., 2010) toxoids. This strategy attempts to achieve broad coverage by incorporating the most widely distributed CFs, and inducing toxin-neutralizing immunity. Unfortunately, antigenic heterogeneity and plasticity inherent in *E. coli* genomes (Rasko et al., 2008) combined within complete protection afforded by LT and poor inherent immunogenicity of ST peptides have hampered progress toward a broadly protective vaccine. This strategy has been further confounded by lack of identifiable CF antigens in a significant proportion of strains (Peruski et al., 1999), and both geographic and temporal variation in prevalent CF antigens (Isidean et al., 2011). In addition, epidemiologic studies which affirm a significant protective effect following natural ETEC infections, suggest that LT and other unexplored antigens rather than CFs (Steinsland et al., 2003) may stimulate protective immune responses. An oral vaccine consisting of recombinant cholera toxin B subunit combined with a mixture of killed ETEC expressing a variety of CF antigens induced immune responses to CFs, afforded protection against severe diarrheal illness from ETEC in travelers (Sack et al., 2007), but failed to protect children in endemic areas (Walker et al., 2007), despite immunogenicity (Savarino et al., 2002). Similarly, a compilation of three live-attenuated strains expressing six different CF antigens (Turner et al., 2011) and LTB elicited substantial immune responses to target antigens in volunteers (Harro et al., 2011b), but offered only modest protection in clinical challenge studies (NCT01060748). Remarkably, transcutaneous immunization (TCI) with heat-labile toxin has been shown to elicit significant immune responses to LT in volunteer studies (Frech et al., 2008), and to co-immunogens in experimental animal studies (Yu et al., 2002). Although the protective efficacy of LT patch-based TCI in subsequent clinical trials was disappointing (McKenzie et al., 2007; Svennerholm and Lundgren, 2012), this strategy could prove useful in subsequent iterations of ETEC vaccines that incorporate additional antigens. Collectively, these studies indicate that anti-CF and/or anti-LT immunity may not offer sufficiently robust protection, and that additional antigens coupled with technology to generate safe and effective ST toxoids (Taxt et al., 2010) may need to be incorporated in future ETEC vaccine development strategies.

CONCLUSIONS

Although ETEC were discovered now more than 40 years ago (Sack, 2011), our current understanding of the pathogenesis of these important pathogens remains insufficient, and ideal approaches to vaccine development are still not completely defined. Despite the inherent plasticity of these pathogens recent discoveries of relatively conserved novel virulence factors should afford additional avenues for investigation and approaches to development of an effective vaccine.

REFERENCES

Abu-Elyazeed, R., Wierzba, T.F., Mourad, A.S., et al., 1999. Epidemiology of enterotoxigenic *Escherichia coli* diarrhea in a pediatric cohort in a periurban area of lower Egypt. J. Infect. Dis. 179 (2), 382–389.

Ahmed, T., Lundgren, A., Arifuzzaman, M., Qadri, F., Teneberg, S., Svennerholm, A.M., 2009. Children with the Le(a+b-) blood group have increased susceptibility to diarrhea caused by enterotoxigenic *Escherichia coli* expressing colonization factor I group fimbriae. Infect. Immun. 77 (5), 2059–2064.

Allen, K.P., Randolph, M.M., Fleckenstein, J.M., 2006. Importance of heat-labile enterotoxin in colonization of the adult mouse small intestine by human enterotoxigenic *Escherichia coli* strains. Infect. Immun. 74 (2), 869–875.

Andersson, M., Bjornham, O., Svantesson, M., Badahdah, A., Uhlin, B.E., Bullitt, E., 2012. A structural basis for sustained bacterial adhesion: biomechanical properties of CFA/I pili. J. Mol. Biol. 415 (5), 918–928.

Aubel, D., Darfeuille-Michaud, A., Joly, B., 1991. New adhesive factor (antigen 8786) on a human enterotoxigenic *Escherichia coli* O117:H4 strain isolated in Africa. Infect. Immun. 59 (4), 1290–1299.

Baker, K.K., Levine, M.M., Morison, J., Phillips, A., Barry, E.M., 2009. CfaE tip mutations in enterotoxigenic *Escherichia coli* CFA/I fimbriae define critical human intestinal binding sites. Cell. Microbiol. 11 (5), 742–754.

Beatty, M.E., Bopp, C.A., Wells, J.G., Greene, K.D., Puhr, N.D., Mintz, E.D., 2004. Enterotoxin-producing *Escherichia coli* O169:H41, United States. Emerg. Infect. Dis. 10 (3), 518–521.

Black, R.E., 1990. Epidemiology of travelers' diarrhea and relative importance of various pathogens. Rev. Infect. Dis. 12 (Suppl. 1), S73–79.

Black, R.E., Merson, M.H., Eusof, A., Huq, I., Pollard, R., 1984. Nutritional status, body size and severity of diarrhoea associated with rotavirus or enterotoxigenic *Escherichia coli*. J. Trop. Med. Hyg. 87 (2), 83–89.

Black, R.E., Merson, M.H., Huq, I., Alim, A.R., Yunus, M., 1981. Incidence and severity of rotavirus and *Escherichia coli* diarrhoea in rural Bangladesh. Implications for vaccine development. Lancet 1 (8212), 141–143.

Blackburn, D., Husband, A., Saldana, Z., et al., 2009. Distribution of the *Escherichia coli* common pilus among diverse strains of human enterotoxigenic *E. coli*. J. Clin. Microbiol. 47 (6), 1781–1784.

Bodero, M.D., Munson, G.P., 2009. Cyclic AMP receptor protein-dependent repression of heat-labile enterotoxin. Infect. Immun. 77 (2), 791–798.

Bolin, I., Wiklund, G., Qadri, F., et al., 2006. Enterotoxigenic *Escherichia coli* with STh and STp genotypes is associated with diarrhea both in children in areas of endemicity and in travelers. J. Clin. Microbiol. 44 (11), 3872–3877.

Boschi-Pinto, C., Velebit, L., Shibuya, K., 2008. Estimating child mortality due to diarrhoea in developing countries. Bull. World Health Organ. 86 (9), 710–717.

Bouckenooghe, A.R., Dupont, H.L., Jiang, Z.D., et al., 2000. Markers of enteric inflammation in enteroaggregative *Escherichia coli* diarrhea in travelers. Am. J. Trop. Med. Hyg. 62 (6), 711–713.

Bourgeois, A.L., Gardiner, C.H., Thornton, S.A., et al., 1993. Etiology of acute diarrhea among United States military personnel deployed to South America and west Africa. Am. J. Trop. Med. Hyg. 48 (2), 243–248.

Brown, E.A., Hardwidge, P.R., 2007. Biochemical characterization of the enterotoxigenic *Escherichia coli* LeoA protein. Microbiology 153 (Pt 11), 3776–3784. doi:153/11/3776 [pii].

Butler, T., 2008. Loperamide for the treatment of traveler's diarrhea: broad or narrow usefulness? Clin. Infect. Dis. 47 (8), 1015–1016.

Caron, J., Coffield, L.M., Scott, J.R., 1989. A plasmid-encoded regulatory gene, rns, required for expression of the CS1 and CS2 adhesins of enterotoxigenic *Escherichia coli*. Proc. Natl. Acad. Sci. USA 86 (3), 963–967.

Carpenter, C.C., Barua, D., Wallace, C.K., et al., 1965. Clinical and physiological observations during an epidemic outbreak of non-vibrio cholera-like disease in Calcutta. Bulle. World Health Organ. 33 (5), 665–671.

CDC, 1994. Foodborne outbreaks of enterotoxigenic *Escherichia coli* – Rhode Island and New Hampshire, 1993. MMWR Morb. Mortal. Wkly. Rep. 43 (5) 81, 87–89.

Chakraborty, K., Ghosh, S., Koley, H., et al., 2008. Bacterial exotoxins downregulate cathelicidin (hCAP-18/LL-37) and human beta-defensin 1 (HBD-1) expression in the intestinal epithelial cells. Cell Microbiol. 10 (12), 2520–2537.

Chao, A.C., de Sauvage, F.J., Dong, Y.J., Wagner, J.A., Goeddel, D.V., Gardner, P., 1994. Activation of intestinal CFTR Cl- channel by heat-stable enterotoxin and guanylin via cAMP-dependent protein kinase. EMBO J. 13 (5), 1065–1072.

Checkley, W., Buckley, G., Gilman, R.H., et al., 2008. Multi-country analysis of the effects of diarrhoea on childhood stunting. Int. J. Epidemiol. 37 (4), 816–830.

Chen, Q., Savarino, S.J., Venkatesan, M.M., 2006. Subtractive hybridization and optical mapping of the enterotoxigenic *Escherichia coli* H10407 chromosome: isolation of unique sequences and demonstration of significant similarity to the chromosome of *E. coli* K-12. Microbiology 152 (Pt 4), 1041–1054.

Cheng, S.H., Rich, D.P., Marshall, J., Gregory, R.J., Welsh, M.J., Smith, A.E., 1991. Phosphorylation of the R domain by cAMP-dependent protein kinase regulates the CFTR chloride channel. Cell 66 (5), 1027–1036.

Chilcott, G.S., Hughes, K.T., 2000. Coupling of flagellar gene expression to flagellar assembly in Salmonella enterica serovar typhimurium and *Escherichia coli*. Microbiol. Mol. Biol. Rev. MMBR 64 (4), 694–708.

Clark, C.A., Heuzenroeder, M.W., Manning, P.A., 1992. Colonization factor antigen CFA/IV (PCF8775) of human enterotoxigenic *Escherichia coli*: nucleotide sequence of the CS5 determinant. Infect. Immun. 60 (3), 1254–1257.

Connor, B.A., 2005. Sequelae of traveler's diarrhea: focus on postinfectious irritable bowel syndrome. Clin. Infect. Dis. 41 (Suppl. 8), S577–586.

Cravioto, A., Scotland, S.M., Rowe, B., 1982. Hemagglutination activity and colonization factor antigens I and II in enterotoxigenic and non-enterotoxigenic strains of *Escherichia coli* isolated from humans. Infect. Immun. 36 (1), 189–197.

Crossman, L.C., Chaudhuri, R.R., Beatson, S.A., et al., 2010. A commensal gone bad: complete genome sequence of the prototypical enterotoxigenic *Escherichia coli* strain H10407. J. Bacteriol. 192 (21), 5822–5831.

Dalton, C.B., Mintz, E.D., Wells, J.G., Bopp, C.A., Tauxe, R.V., 1999. Outbreaks of enterotoxigenic *Escherichia coli* infection in American adults: a clinical and epidemiologic profile. Epidemiol. Infect. 123 (1), 9–16.

Darfeuille-Michaud, A., Forestier, C., Joly, B., Cluzel, R., 1986. Identification of a nonfimbrial adhesive factor of an enterotoxigenic *Escherichia coli* strain. Infect. Immun. 52 (2), 468–475.

De, S., 1959. Enterotoxigenicity of bacteria-free culture filtrate of *Vibrio cholerae*. Nature 183, 1533–1534.

De, S.N., Bhattacharya, K., Sarkar, J.K., 1956. A study of the pathogenicity of strains of Bacterium coli from acute and chronic enteritis. J. Pathol. Bacteriol. 71 (1), 201–209.

Del Canto, F., Botkin, D.J., Valenzuela, P., et al., 2012. Identification of the coli surface antigen 23 (CS23), a novel adhesin of enterotoxigenic *Escherichia coli*. Infect. Immun. 80 (8), 2791–2801.

Devasia, R.A., Jones, T.F., Ward, J., et al., 2006. Endemically acquired foodborne outbreak of enterotoxin-producing *Escherichia coli* serotype O169:H41. Am. J. Med. 119 (2), 168.e7–168.e10.

DuPont, H.L., 2007. Therapy for and prevention of traveler's diarrhea. Clin. Infect. Dis. 45 (Suppl. 1), S78–S84.

DuPont, H.L., Formal, S.B., Hornick, R.B., et al., 1971. Pathogenesis of *Escherichia coli* diarrhea. N. Engl. J. Med. 285 (1), 1–9.

Dupont, H.L., Jiang, Z.D., Belkind-Gerson, J., et al., 2007. Treatment of travelers' diarrhea: randomized trial comparing rifaximin, rifaximin plus loperamide, and loperamide alone. Clin. Gastroenterol. Hepatol. 5 (4), 451–456.

Elsinghorst, E.A., Weitz, J.A., 1994. Epithelial cell invasion and adherence directed by the enterotoxigenic *Escherichia coli* tib locus is associated with a 104-kilodalton outer membrane protein. Infect. Immun. 62 (8), 3463–3471.

Ericsson, C.D., DuPont, H.L., Okhuysen, P.C., Jiang, Z.D., DuPont, M.W., 2007. Loperamide plus azithromycin more effectively treats travelers' diarrhea in Mexico than azithromycin alone. J. Travel. Med. 14 (5), 312–319.

Espert, S.M., Elsinghorst, E.A., Munson, G.P., 2011. The tib adherence locus of enterotoxigenic *Escherichia coli* is regulated by cyclic AMP receptor protein. J. Bacteriol. 193 (6), 1369–1376.

Evans, D., Evans, D., Karjalainen, T., Lee, C., 1991. Production of colonization factor antigen II of enterotoxigenic *Escherichia coli* is subject to catabolite repression. Curr. Microbiol. 23, 71–74.

Evans, D.G., Satterwhite, T.K., Evans Jr., D.J., DuPont, H.L., 1978. Differences in serological responses and excretion patterns of volunteers challenged with enterotoxigenic *Escherichia coli* with and without the colonization factor antigen. Infect. Immun. 19 (3), 883–888.

Evans, D.G., Silver, R.P., Evans Jr., D.J., Chase, D.G., Gorbach, S.L., 1975. Plasmid-controlled colonization factor associated with virulence in *Escherichia coli* enterotoxigenic for humans. Infect. Immun. 12 (3), 656–667.

Evans Jr., D.J., Evans, D.G., 1973. Three characteristics associated with enterotoxigenic *Escherichia coli* isolated from man. Infect. Immun. 8 (3), 322–328.

Finkelstein, R.A., Vasil, M.L., Jones, J.R., Anderson, R.A., Barnard, T., 1976. Clinical cholera caused by enterotoxigenic *Escherichia coli*. J. Clin. Microbiol. 3 (3), 382–384.

Fischer Walker, C.L., Sack, D., Black, R.E., 2010. Etiology of diarrhea in older children, adolescents and adults: a systematic review. PLoS Negl. Trop. Dis. 4 (8), e768.

Fleckenstein, J.M., Hardwidge, P.R., Munson, G.P., Rasko, D.A., Sommerfelt, H., Steinsland, H., 2010. Molecular mechanisms of enterotoxigenic *Escherichia coli* infection. Microbes. Infect. 12 (2), 89–98.

Fleckenstein, J.M., Holland, J.T., Hasty, D.L., 2002. Interaction of an outer membrane protein of enterotoxigenic *Escherichia coli* with cell surface heparan sulfate proteoglycans. Infect. Immun. 70 (3), 1530–1537.

Fleckenstein, J.M., Kopecko, D.J., Warren, R.L., Elsinghorst, E.A., 1996. Molecular characterization of the tia invasion locus from enterotoxigenic *Escherichia coli*. Infect. Immun. 64 (6), 2256–2265.

Fleckenstein, J.M., Lindler, L.E., Elsinghorst, E.A., Dale, J.B., 2000. Identification of a gene within a pathogenicity island of enterotoxigenic *Escherichia coli* H10407 required for maximal secretion of the heat-labile enterotoxin. Infect. Immun. 68 (5), 2766–2774.

Fleckenstein, J.M., Roy, K., Fischer, J.F., Burkitt, M., 2006. Identification of a two-partner secretion locus of enterotoxigenic *Escherichia coli*. Infect. Immun. 74 (4), 2245–2258.

Flores, J., DuPont, H.L., Lee, S.A., et al., 2008. Influence of host interleukin-10 polymorphisms on development of traveler's diarrhea due to heat-labile enterotoxin-producing *Escherichia coli* in travelers from the United States who are visiting Mexico. Clin. Vaccine Immunol. CVI 15 (8), 1194–1198.

Frech, S.A., Dupont, H.L., Bourgeois, A.L., et al., 2008. Use of a patch containing heat-labile toxin from *Escherichia coli* against travellers' diarrhoea: a phase II, randomised, double-blind, placebo-controlled field trial. Lancet 371 (9629), 2019–2025.

French, P.J., Bijman, J., Edixhoven, M., et al., 1995. Isotype-specific activation of cystic fibrosis transmembrane conductance regulator-chloride channels by cGMP-dependent protein kinase II. J. Biol. Chem. 270 (44), 26626–26631.

Froehlich, B., Parkhill, J., Sanders, M., Quail, M.A., Scott, J.R., 2005. The pCoo plasmid of enterotoxigenic *Escherichia coli* is a mosaic cointegrate. J. Bacteriol. 187 (18), 6509–6516.

Froehlich, B.J., Karakashian, A., Sakellaris, H., Scott, J.R., 1995. Genes for CS2 pili of enterotoxigenic *Escherichia coli* and their interchangeability with those for CS1 pili. Infect. Immun. 63 (12), 4849–4856.

Giannella, R.A., Mann, E.A., 2003. *E. coli* heat-stable enterotoxin and guanylyl cyclase C: new functions and unsuspected actions. Trans. Am. Clin. Climatol. Assoc. 114, 67–85 discussion 85–66.

Giron, J.A., Gomez-Duarte, O.G., Jarvis, K.G., Kaper, J.B., 1997. Longus pilus of enterotoxigenic *Escherichia coli* and its relatedness to other type-4 pili – a minireview. Gene. 192 (1), 39–43.

Giron, J.A., Levine, M.M., Kaper, J.B., 1994. Longus: a long pilus ultrastructure produced by human enterotoxigenic *Escherichia coli*. Mol. Microbiol. 12 (1), 71–82.

Gorbach, S.L., Banwell, J.G., Chatterjee, B.D., Jacobs, B., Sack, R.B., 1971. Acute undifferentiated human diarrhea in the tropics. I. Alterations in intestinal micrflora. J. Clin. Invest. 50 (4), 881–889.

Greenberg, D.E., Jiang, Z.D., Steffen, R., Verenker, M.P., DuPont, H.L., 2002. Markers of inflammation in bacterial diarrhea among travelers, with a focus on enteroaggregative *Escherichia coli* pathogenicity. J. Infect. Dis. 185 (7), 944–949.

Grewal, H.M., Valvatne, H., Bhan, M.K., van Dijk, L., Gaastra, W., Sommerfelt, H., 1997. A new putative fimbrial colonization factor, CS19, of human enterotoxigenic *Escherichia coli*. Infect. Immun. 65 (2), 507–513.

Guerrant, R.L., Schorling, J.B., McAuliffe, J.F., de Souza, M.A., 1992. Diarrhea as a cause and an effect of malnutrition: diarrhea prevents catch-up growth and malnutrition increases diarrhea frequency and duration. Am. J. Trop. Med. Hyg. 47 (1 Pt 2), 28–35.

Gyles, C., So, M., Falkow, S., 1974. The enterotoxin plasmids of *Escherichia coli*. J. Infect. Dis. 130 (1), 40–49.

Haberberger Jr., R.L., Mikhail, I.A., Burans, J.P., et al., 1991. Travelers' diarrhea among United States military personnel during joint American–Egyptian armed forces exercises in Cairo, Egypt. Mil. Med. 156 (1), 27–30.

Hajishengallis, G., Nawar, H., Tapping, R.I., Russell, M.W., Connell, T.D., 2004. The type II heat-labile enterotoxins LT-IIa and LT-IIb and their respective B pentamers differentially induce and regulate cytokine production in human monocytic cells. Infect. Immun. 72 (11), 6351–6358.

Hantke, K., Winkler, K., Schultz, J.E., 2011. *Escherichia coli* exports cyclic AMP via TolC. J. Bacteriol. 193 (5), 1086–1089.

Harris, J.A., Roy, K., Woo-Rasberry, V., et al., 2011. Directed evaluation of enterotoxigenic *Escherichia coli* autotransporter proteins as putative vaccine candidates. PLoS Negl. Trop. Dis. 5 (12), e1428.

Harro, C., Chakraborty, S., Feller, A., et al., 2011a. Refinement of a human challenge model for evaluation of enterotoxigenic *Escherichia coli* vaccines. Clin. Vaccine Immunol. CVI 18 (10), 1719–1727.

Harro, C., Sack, D., Bourgeois, A.L., et al., 2011b. A combination vaccine consisting of three live attenuated enterotoxigenic *Escherichia coli* strains expressing a range of colonization factors and LTB is well tolerated and immunogenic in a placebo-controlled double-blind Phase I trial in healthy adults. Clin. Vaccine Immunol. CVI 18 (12), 2118–2127.

Hauck, T.S., Giri, S., Gao, Y., Chan, W.C., 2010. Nanotechnology diagnostics for infectious diseases prevalent in developing countries. Adv. Drug Deliv. Rev. 62 (4-5), 438–448.

Heuzenroeder, M.W., Elliot, T.R., Thomas, C.J., Halter, R., Manning, P.A., 1990. A new fimbrial type (PCFO9) on enterotoxigenic *Escherichia coli* 09:H- LT+ isolated from a case of infant diarrhea in central Australia. FEMS Microbiol. Lett. 54 (1–3), 55–60.

Heuzenroeder, M.W., Neal, B.L., Thomas, C.J., Halter, R., Manning, P.A., 1989. Characterization and molecular cloning of the PCF8775 CS5 antigen from an enterotoxigenic *Escherichia coli* 0115:H40 isolated in Central Australia. Mol. Microbiol. 3 (3), 303–310.

Hibberd, M.L., McConnell, M.M., Field, A.M., Rowe, B., 1990. The fimbriae of human enterotoxigenic *Escherichia coli* strain 334 are related to CS5 fimbriae. J. Gen. Microbiol. 136 (12), 2449–2456.

Honarvar, S., Choi, B.K., Schifferli, D.M., 2003. Phase variation of the 987P-like CS18 fimbriae of human enterotoxigenic *Escherichia coli* is regulated by site-specific recombinases. Mol. Microbiol. 48 (1), 157–171.

Honda, T., Arita, M., Miwatani, T., 1984. Characterization of new hydrophobic pili of human enterotoxigenic *Escherichia coli*: a possible new colonization factor. Infect. Immun. 43 (3), 959–965.

Horstman, A.L., Kuehn, M.J., 2000. Enterotoxigenic *Escherichia coli* secretes active heat-labile enterotoxin via outer membrane vesicles. J. Biol. Chem. 275 (17), 12489–12496.

Huang, D.B., DuPont, H.L., Jiang, Z.D., Carlin, L., Okhuysen, P.C., 2004. Interleukin-8 response in an intestinal HCT-8 cell line infected with enteroaggregative and enterotoxigenic *Escherichia coli*. Clin. Diagn. Lab. Immunol. 11 (3), 548–551.

Humphrey, W., Dalke, A., Schulten, K., 1996. VMD: visual molecular dynamics. J. Mol. Graph. 14 (1) 33–38, 27–38.

Hyams, K.C., Bourgeois, A.L., Merrell, B.R., et al., 1991. Diarrheal disease during Operation Desert Shield. N. Engl. J. Med. 325 (20), 1423–1428.

Isidean, S.D., Riddle, M.S., Savarino, S.J., Porter, C.K., 2011. A systematic review of ETEC epidemiology focusing on colonization factor and toxin expression. Vaccine 29 (37), 6167–6178.

Jain, S., Chen, L., Dechet, A., et al., 2008. An outbreak of enterotoxigenic *Escherichia coli* associated with sushi restaurants in Nevada, 2004. Clin. Infect. Dis. 47 (1), 1–7.

Jansson, L., Tobias, J., Lebens, M., Svennerholm, A.M., Teneberg, S., 2006. The major subunit, CfaB, of colonization factor antigen i from enterotoxigenic *Escherichia coli* is a glycosphingolipid binding protein. Infect. Immun. 74 (6), 3488–3497.

Johnson, A.M., Kaushik, R.S., Francis, D.H., Fleckenstein, J.M., Hardwidge, P.R., 2009. Heat-labile enterotoxin promotes *Escherichia coli* adherence to intestinal epithelial cells. J. Bacteriol. 191 (1), 178–186.

Jordi, B.J., Willshaw, G.A., van der Zeijst, B.A., Gaastra, W., 1992. The complete nucleotide sequence of region 1 of the CFA/I fimbrial operon of human enterotoxigenic *Escherichia coli*. DNA Sequence 2 (4), 257–263.

Karjalainen, T., Evans, D., Evans, D., Graham, D., Lee, C., 1991. Catabolite repression of the colonization factor antigen I (CFA/I) operon of *Escherichia coli*. Curr. Microbiol. 23 (6), 307–313.

Knutton, S., Lloyd, D.R., Candy, D.C., McNeish, A.S., 1984a. In vitro adhesion of enterotoxigenic *Escherichia coli* to human intestinal epithelial cells from mucosal biopsies. Infect. Immun. 44 (2), 514–518.

Knutton, S., Lloyd, D.R., Candy, D.C., McNeish, A.S., 1984b. Ultrastructural study of adhesion of enterotoxigenic *Escherichia coli* to erythrocytes and human intestinal epithelial cells. Infect. Immun. 44 (2), 519–527.

Knutton, S., Lloyd, D.R., McNeish, A.S., 1987. Identification of a new fimbrial structure in enterotoxigenic *Escherichia coli* (ETEC) serotype O148:H28 which adheres to human intestinal mucosa: a potentially new human ETEC colonization factor. Infect. Immun. 55 (1), 86–92.

Kolappan, S., Roos, J., Yuen, A., Pierce, O.M., Craig, L., 2012. Structural characterization of CFA/III and longus type IVB pili from enterotoxigenic *Escherichia coli*. J. Bacteriol. 194 (10), 2725–2735.

Kosek, M., Bern, C., Guerrant, R.L., 2003. The global burden of diarrhoeal disease, as estimated from studies published between 1992 and 2000. Bull. World Health Organ. 81 (3), 197–204.

Lencer, W.I., Constable, C., Moe, S., et al., 1997. Proteolytic activation of cholera toxin and *Escherichia coli* labile toxin by entry into host epithelial cells. Signal transduction by a protease-resistant toxin variant. J. Biol. Chem. 272 (24), 15562–15568.

Levine, M.M., 1981. Adhesion of enterotoxigenic *Escherichia coli* in humans and animals. Ciba. Found. Symp. 80, 142–160.

Levine, M.M., Black, R.E., Brinton Jr., C.C., et al., 1982. Reactogenicity, immunogenicity and efficacy studies of *Escherichia coli* type 1 somatic pili parenteral vaccine in man. Scand. J. Infect. Dis. Suppl. 33, 83–95.

Levine, M.M., Caplan, E.S., Waterman, D., Cash, R.A., Hornick, R.B., Snyder, M.J., 1977. Diarrhea caused by *Escherichia coli* that produce only heat-stable enterotoxin. Infect. Immun. 17 (1), 78–82.

Levine, M.M., Nalin, D.R., Hoover, D.L., Bergquist, E.J., Hornick, R.B., Young, C.R., 1979. Immunity to enterotoxigenic *Escherichia coli*. Infect. Immun. 23 (3), 729–736.

Levine, M.M., Rennels, M.B., Cisneros, L., Hughes, T.P., Nalin, D.R., Young, C.R., 1980. Lack of person-to-person transmission of enterotoxigenic *Escherichia coli* despite close contact. Am. J. Epidemiol. 111 (3), 347–355.

Levine, M.M., Ristaino, P., Marley, G., et al., 1984. Coli surface antigens 1 and 3 of colonization factor antigen II-positive enterotoxigenic *Escherichia coli*: morphology, purification, and immune responses in humans. Infect. Immun. 44 (2), 409–420.

Levine, M.M., Ristaino, P., Sack, R.B., Kaper, J.B., Orskov, F., Orskov, I., 1983. Colonization factor antigens I and II and type 1 somatic pili in enterotoxigenic *Escherichia coli*: relation to enterotoxin type. Infect. Immun. 39 (2), 889–897.

Li, C., Krishnamurthy, P.C., Penmatsa, H., et al., 2007. Spatiotemporal coupling of cAMP transporter to CFTR chloride channel function in the gut epithelia. Cell 131 (5), 940–951.

Li, Y.F., Poole, S., Nishio, K., et al., 2009. Structure of CFA/I fimbriae from enterotoxigenic *Escherichia coli*. Proc. Natl. Acad. Sci. USA 106 (26), 10793–10798.

Lindenbaum, J., Greenough 3rd, W.B., Benenson, A.S., Oseasohn, R., Rizvi, S., Saad, A., 1965. Non-Vibrio cholera. Lancet 14, 1081–1083.

Lindenthal, C., Elsinghorst, E.A., 1999. Identification of a glycoprotein produced by enterotoxigenic *Escherichia coli*. Infect. Immun. 67 (8), 4084–4091.

Long, K.Z., Rosado, J.L., Santos, J.I., et al., 2010. Associations between mucosal innate and adaptive immune responses and resolution of diarrheal pathogen infections. Infect. Immun. 78 (3), 1221–1228.

MacDonald, K.L., Eidson, M., Strohmeyer, C., et al., 1985. A multistate outbreak of gastrointestinal illness caused by enterotoxigenic *Escherichia coli* in imported semisoft cheese. J. Infect. Dis. 151 (4), 716–720.

Marron, M.B., Smyth, C.J., 1995. Molecular analysis of the cso operon of enterotoxigenic *Escherichia coli* reveals that CsoA is the adhesin of CS1 fimbriae and that the accessory genes are interchangeable with those of the cfa operon. Microbiology 141 (Pt 11), 2849–2859.

McConnell, M.M., Chart, H., Field, A.M., Hibberd, M., Rowe, B., 1989. Characterization of a putative colonization factor (PCFO166) of enterotoxigenic *Escherichia coli* of serogroup O166. J. Gen. Microbiol. 135 (5), 1135–1144.

McConnell, M.M., Hibberd, M., Field, A.M., Chart, H., Rowe, B., 1990. Characterization of a new putative colonization factor (CS17) from a human enterotoxigenic *Escherichia coli* of serotype O114:H21 which produces only heat-labile enterotoxin. J. Infect. Dis. 161 (2), 343–347.

McDonough, K.A., Rodriguez, A., 2012. The myriad roles of cyclic AMP in microbial pathogens: from signal to sword. Nat. Rev. Microbiol. 10 (1), 27–38.

McKenzie, R., Bourgeois, A.L., Frech, S.A., et al., 2007. Transcutaneous immunization with the heat-labile toxin (LT) of enterotoxigenic *Escherichia coli* (ETEC): protective efficacy in a double-blind, placebo-controlled challenge study. Vaccine 25 (18), 3684–3691.

McVeigh, A., Fasano, A., Scott, D.A., et al., 2000. IS1414, an *Escherichia coli* insertion sequence with a heat-stable enterotoxin gene embedded in a transposase-like gene. Infect. Immun. 68 (10), 5710–5715.

Mercado, E.H., Ochoa, T.J., Ecker, L., et al., 2011. Fecal leukocytes in children infected with diarrheagenic *Escherichia coli*. J. Clin. Microbiol. 49 (4), 1376–1381.

Merson, M.H., Morris, G.K., Sack, D.A., et al., 1976. Travelers' diarrhea in Mexico. A prospective study of physicians and family members attending a congress. N. Engl. J. Med. 294 (24), 1299–1305.

Mohamed, J.A., DuPont, H.L., Flores, J., et al., 2011. Single nucleotide polymorphisms in the promoter of the gene encoding the lipopolysaccharide receptor CD14 are associated with bacterial diarrhea in US and Canadian travelers to Mexico. Clin. Infect. Dis. 52 (11), 1332–1341.

Mohamed, J.A., DuPont, H.L., Jiang, Z.D., et al., 2007. A novel single-nucleotide polymorphism in the lactoferrin gene is associated with susceptibility to diarrhea in North American travelers to Mexico. Clin. Infect. Dis. 44 (7), 945–952.

Mondal, D., Haque, R., Sack, R.B., Kirkpatrick, B.D., Petri Jr., W.A., 2009. Attribution of malnutrition to cause-specific diarrheal illness: evidence from a prospective study of preschool children in Mirpur, Dhaka, Bangladesh. Am. J. Trop. Med. Hyg. 80 (5), 824–826.

Mondal, D., Minak, J., Alam, M., et al., 2012. Contribution of enteric infection, altered intestinal barrier function, and maternal malnutrition to infant malnutrition in Bangladesh. Clin. Infect. Dis. 54 (2), 185–192.

Moormann, C., Benz, I., Schmidt, M.A., 2002. Functional substitution of the TibC protein of enterotoxigenic *Escherichia coli* strains for the autotransporter adhesin heptosyltransferase of the AIDA system. Infect. Immun. 70 (5), 2264–2270.

Moreland, J.L., Gramada, A., Buzko, O.V., Zhang, Q., Bourne, P.E., 2005. The Molecular Biology Toolkit (MBT): a modular platform for developing molecular visualization applications. BMC Bioinformatics 6, 21.

Mu, X.Q., Savarino, S.J., Bullitt, E., 2008. The three-dimensional structure of CFA/I adhesion pili: traveler's diarrhea bacteria hang on by a spring. J. Mol. Biol. 376 (3), 614–620.

Muller, C.M., Aberg, A., Straseviciene, J., Emody, L., Uhlin, B.E., Balsalobre, C., 2009. Type 1 fimbriae, a colonization factor of uropathogenic *Escherichia coli*, are controlled by the metabolic sensor CRP-cAMP. PLoS pathogens 5 (2), e1000303.

Mulvey, M.A., Schilling, J.D., Martinez, J.J., Hultgren, S.J., 2000. Bad bugs and beleaguered bladders: interplay between uropathogenic *Escherichia coli* and innate host defenses. Proc. Natl. Acad. Sci. USA 97 (16), 8829–8835.

Nada, R.A., Shaheen, H.I., Khalil, S.B., et al., 2011. Discovery and phylogenetic analysis of novel members of class b enterotoxigenic *Escherichia coli* adhesive fimbriae. J. Clin. Microbiol. 49 (4), 1403–1410.

Nalin, D.R., McLaughlin, J.C., Rahaman, M., Yunus, M., Curlin, G., 1975. Enterotoxigenic *Escherichia coli* and idiopathic diarrhoea in Bangladesh. Lancet 2 (7945), 1116–1119.

New York Times, 1998. Bacteria Sicken 8 in Georgia And 4,000 in Chicago Area. The New York Times. Retrieved from: http://www.nytimes.com/1998/06/24/us/bacteria-sicken-8-in-georgia-and-4000-in-chicago-area.html.

O'Neil, D.A., Porter, E.M., Elewaut, D., et al., 1999. Expression and regulation of the human beta-defensins hBD-1 and hBD-2 in intestinal epithelium. J. Immunol. 163 (12), 6718–6724.

Ofek, I., Zafriri, D., Goldhar, J., Eisenstein, B.I., 1990. Inability of toxin inhibitors to neutralize enhanced toxicity caused by bacteria adherent to tissue culture cells. Infect. Immun. 58 (11), 3737–3742.

Ouyang-Latimer, J., Jafri, S., VanTassel, A., et al., 2011. In vitro antimicrobial susceptibility of bacterial enteropathogens isolated from international travelers to Mexico, Guatemala, and India from 2006 to 2008. Antimicrob. Agents. Chemother. 55 (2), 874–878.

Paredes-Paredes, M., Okhuysen, P.C., Flores, J., et al., 2011. Seasonality of diarrheagenic *Escherichia coli* pathotypes in the US students acquiring diarrhea in Mexico. J. Travel. Med. 18 (2), 121–125.

Parry, G.C., Mackman, N., 1997. Role of cyclic AMP response element-binding protein in cyclic AMP inhibition of NF-kappaB-mediated transcription. J. Immunol. 159 (11), 5450–5456.

Patel, S.K., Dotson, J., Allen, K.P., Fleckenstein, J.M., 2004. Identification and molecular characterization of EatA, an autotransporter protein of enterotoxigenic *Escherichia coli*. Infect. Immun. 72 (3), 1786–1794.

Perez-Casal, J., Swartley, J.S., Scott, J.R., 1990. Gene encoding the major subunit of CS1 pili of human enterotoxigenic *Escherichia coli*. Infect. Immun. 58 (11), 3594–3600.

Peruski Jr., L.F., Kay, B.A., El-Yazeed, R.A., et al., 1999. Phenotypic diversity of enterotoxigenic *Escherichia coli* strains from a community-based study of pediatric diarrhea in periurban Egypt. J. Clin. Microbio. 37 (9), 2974–2978.

Petri Jr., W.A., Miller, M., Binder, H.J., Levine, M.M., Dillingham, R., Guerrant, R.L., 2008. Enteric infections, diarrhea, and their impact on function and development. J. Clin. Invest. 118 (4), 1277–1290.

Pfeifer, A., Aszodi, A., Seidler, U., Ruth, P., Hofmann, F., Fassler, R., 1996. Intestinal secretory defects and dwarfism in mice lacking cGMP-dependent protein kinase II. Science 274 (5295), 2082–2086.

Pichel, M., Binsztein, N., Viboud, G., 2000. CS22, a novel human enterotoxigenic *Escherichia coli* adhesin, is related to CS15. Infect. Immun. 68 (6), 3280–3285.

Pilonieta, M.C., Bodero, M.D., Munson, G.P., 2007. CfaD-dependent expression of a novel extracytoplasmic protein from enterotoxigenic *Escherichia coli*. J. Bacteriol. 189 (14), 5060–5067.

Porter, C.K., Riddle, M.S., Tribble, D.R., et al., 2011. A systematic review of experimental infections with enterotoxigenic *Escherichia coli* (ETEC). Vaccine 29 (35), 5869–5885.

Potter, L.R., 2011. Guanylyl cyclase structure, function and regulation. Cell Signal 23 (12), 1921–1926.

Qadri, F., Giron, J.A., Helander, A., et al., 2000. Human antibody response to longus type IV pilus and study of its prevalence among enterotoxigenic *Escherichia coli* in Bangladesh by using monoclonal antibodies. J. Infect. Dis. 181 (6), 2071–2074.

Qadri, F., Saha, A., Ahmed, T., Al Tarique, A., Begum, Y.A., Svennerholm, A.M., 2007. Disease burden due to enterotoxigenic *Escherichia coli* in the first 2 years of life in an urban community in Bangladesh. Infect. Immun. 75 (8), 3961–3968.

Qadri, F., Svennerholm, A.M., Faruque, A.S., Sack, R.B., 2005. Enterotoxigenic *Escherichia coli* in developing countries: epidemiology, microbiology, clinical features, treatment, and prevention. Clin. Microbiol. Rev. 18 (3), 465–483.

Rao, M.R., Wierzba, T.F., Savarino, S.J., et al., 2005. Serologic correlates of protection against enterotoxigenic *Escherichia coli* diarrhea. J. Infect. Dis. 191 (4), 562–570.

Rasko, D.A., Rosovitz, M.J., Myers, G.S., et al., 2008. The pangenome structure of *Escherichia coli*: comparative genomic analysis of *E. coli* commensal and pathogenic isolates. J. Bacteriol. 190 (20), 6881–6893.

Rendon, M.A., Saldana, Z., Erdem, A.L., et al., 2007. Commensal and pathogenic *Escherichia coli* use a common pilus adherence factor for epithelial cell colonization. Proc. Natl. Acad. Sci. USA 104 (25), 10637–10642.

Ricci, K.A., Girosi, F., Tarr, et al., 2006. Reducing stunting among children: the potential contribution of diagnostics. Nature 444 (Suppl. 1), 29–38.

Riddle, M.S., Sanders, J.W., Putnam, S.D., Tribble, D.R., 2006. Incidence, etiology, and impact of diarrhea among long-term travelers (US military and similar populations): a systematic review. Am. J. Trop. Med. Hyg. 74 (5), 891–900.

Roels, T.H., Proctor, M.E., Robinson, L.C., Hulbert, K., Bopp, C.A., Davis, J.P., 1998. Clinical features of infections due to *Escherichia coli* producing heat-stable toxin during an outbreak in Wisconsin: a rarely suspected cause of diarrhea in the United States. Clin. Infect. Dis. 26 (4), 898–902.

Rosenberg, M.L., Koplan, J.P., Wachsmuth, I.K., et al., 1977. Epidemic diarrhea at Crater Lake from enterotoxigenic *Escherichia coli*. A large waterborne outbreak. Ann. Intern. Med. 86 (6), 714–718.

Rowe, B., Taylor, J., 1969. The bacteriology of traveller's diarrhoea. J. Clin. Pathol. 22 (6), 744–745.

Rowe, B., Taylor, J., Bettelheim, K.A., 1970. An investigation of traveller's diarrhoea. Lancet 1 (7636), 1–5.

Roy, K., Bartels, S., Qadri, F., Fleckenstein, J.M., 2010. Enterotoxigenic *Escherichia coli* elicits immune responses to multiple surface proteins. Infect. Immun. 78 (7), 3027–3035.

Roy, K., Hilliard, G.M., Hamilton, D.J., Luo, J., Ostmann, M.M., Fleckenstein, J.M., 2009. Enterotoxigenic *Escherichia coli* EtpA mediates adhesion between flagella and host cells. Nature 457 (7229), 594–598.

Roy, K., Kansal, R., Bartels, S.R., Hamilton, D.J., Shaaban, S., Fleckenstein, J.M., 2011. Adhesin degradation accelerates delivery of heat-labile toxin by enterotoxigenic *Escherichia coli*. J. Biol. Chem. 286 (34), 29771–29779.

Ryder, R.W., Sack, D.A., Kapikian, A.Z., et al., 1976. Enterotoxigenic *Escherichia coli* and Reovirus-like agent in rural Bangladesh. Lancet 1 (7961), 659–663.

Sack, R.B., 1990. Travelers' diarrhea: microbiologic bases for prevention and treatment. Rev. Infect. Dis. 12 (Suppl. 1), S59–63.

Sack, R.B., 2011. The discovery of cholera-like enterotoxins produced by *Escherichia coli* causing secretory diarrhoea in humans. Indian J. Med. Res. 133 (2), 171–180.

Sack, D.A., Huda, S., Neogi, P.K., Daniel, R.R., Spira, W.M., 1980. Microtiter ganglioside enzyme-linked immunosorbent assay for vibrio and *Escherichia coli* heat-labile enterotoxins and antitoxin. J. Clin. Microbiol. 11 (1), 35–40.

Sack, D.A., McLaughlin, J.C., Sack, R.B., Orskov, F., Orskov, I., 1977. Enterotoxigenic *Escherichia coli* isolated from patients at a hospital in Dacca. J. Infect. Dis. 135 (2), 275–280.

Sack, D.A., Merson, M.H., Wells, J.G., Sack, R.B., Morris, G.K., 1975. Diarrhoea associated with heat-stable enterotoxin-producing strains of *Escherichia coli*. Lancet 2 (7928), 239–241.

Sack, D.A., Shimko, J., Torres, O., et al., 2007. Randomised, double-blind, safety and efficacy of a killed oral vaccine for enterotoxigenic *E. coli* diarrhoea of travellers to Guatemala and Mexico. Vaccine 25 (22), 4392–4400.

Sack, R.B., Gorbach, S.L., Banwell, J.G., Jacobs, B., Chatterjee, B.D., Mitra, R.C., 1971. Enterotoxigenic *Escherichia coli* isolated from patients with severe cholera-like disease. J. Infect. Dis. 123 (4), 378–385.

Sahl, J.W., Rasko, D.A., 2012. Analysis of global transcriptional profiles of enterotoxigenic *Escherichia coli* isolate E24377A. Infect. Immun. 80 (3), 1232–1242.

Sahl, J.W., Steinsland, H., Redman, J.C., et al., 2011. A comparative genomic analysis of diverse clonal types of enterotoxigenic *Escherichia coli* reveals pathovar-specific conservation. Infect. Immun. 79 (2), 950–960.

Saldana, Z., Erdem, A.L., Schuller, S., et al., 2009. The *Escherichia coli* common pilus and the bundle-forming pilus act in concert during the formation of localized adherence by enteropathogenic *E. coli*. J. Bacteriol. 191 (11), 3451–3461.

Satterwhite, T.K., Evans, D.G., DuPont, H.L., Evans Jr., D.J., 1978. Role of *Escherichia coli* colonisation factor antigen in acute diarrhoea. Lancet 2 (8082), 181–184.

Savarino, S.J., Fasano, A., Watson, J., et al., 1993. Enteroaggregative *Escherichia coli* heat-stable enterotoxin 1 represents another subfamily of *E. coli* heat-stable toxin. Proc. Natl. Acad. Sci. USA 90 (7), 3093–3097.

Savarino, S.J., Hall, E.R., Bassily, S., et al., 2002. Introductory evaluation of an oral, killed whole cell enterotoxigenic *Escherichia coli* plus cholera toxin B subunit vaccine in Egyptian infants. Pediatr. Infect. Dis. J. 21 (4), 322–330.

Savarino, S.J., McVeigh, A., Watson, J., et al., 1996. Enteroaggregative *Escherichia coli* heat-stable enterotoxin is not restricted to enteroaggregative *E. coli*. J. Infect. Dis. 173 (4), 1019–1022.

Schilling, J.D., Mulvey, M.A., Hultgren, S.J., 2001. Structure and function of *Escherichia coli* type 1 pili: new insight into the pathogenesis of urinary tract infections. J. Infect. Dis. 183 (Suppl. 1), S36–S40.

Shore, E.G., Dean, A.G., Holik, K.J., Davis, B.R., 1974. Enterotoxin-producing *Escherichia coli* and diarrheal disease in adult travelers: a prospective study. J. Infect. Dis. 129 (5), 577–582.

Sixma, T.K., Pronk, S.E., Kalk, K.H., et al., 1991. Crystal structure of a cholera toxin-related heat-labile enterotoxin from *E. coli*. Nature 351 (6325), 371–377.

Sjoling, A., Wiklund, G., Savarino, S.J., Cohen, D.I., Svennerholm, A.M., 2007. Comparative analyses of phenotypic and genotypic methods for detection of enterotoxigenic *Escherichia coli* toxins and colonization factors. J. Clin. Microbiol. 45 (10), 3295–3301.

Sokurenko, E.V., Chesnokova, V., Doyle, R.J., Hasty, D.L., 1997. Diversity of the *Escherichia coli* type 1 fimbrial lectin. Differential binding to mannosides and uroepithelial cells. J. Biol. Chem. 272 (28), 17880–17886.

Spangler, B.D., 1992. Structure and function of cholera toxin and the related *Escherichia coli* heat-labile enterotoxin. Microbiol. Rev. 56 (4), 622–647.

Steinsland, H., Lacher, D.W., Sommerfelt, H., Whittam, T.S., 2010. Ancestral lineages of human enterotoxigenic *Escherichia coli*. J. Clin. Microbiol. 48 (8), 2916–2924.

Steinsland, H., Valentiner-Branth, P., Aaby, P., Molbak, K., Sommerfelt, H., 2004. Clonal related-ness of enterotoxigenic *Escherichia coli* strains isolated from a cohort of young children in Guinea-Bissau. J. Clin. Microbiol. 42 (7), 3100–3107.

Steinsland, H., Valentiner-Branth, P., Gjessing, H.K., Aaby, P., Molbak, K., Sommerfelt, H., 2003. Protection from natural infections with enterotoxigenic *Escherichia coli*: longitudinal study. Lancet 362 (9380), 286–291.

Subekti, D.S., Lesmana, M., Tjaniadi, P., et al., 2003. Prevalence of enterotoxigenic *Escherichia coli* (ETEC) in hospitalized acute diarrhea patients in Denpasar, Bali, Indonesia. Diagn. Microbiol. Infect. Dis. 47 (2), 399–405.

Svennerholm, A.M., Lundgren, A., 2012. Recent progress toward an enterotoxigenic *Escherichia coli* vaccine. Expert Rev. Vaccines 11 (4), 495–507.

Svennerholm, A.M., Tobias, J., 2008. Vaccines against enterotoxigenic *Escherichia coli*. Expert Rev. Vaccines 7 (6), 795–804.

Svenungsson, B., Lagergren, A., Ekwall, E., et al., 2000. Enteropathogens in adult patients with diarrhea and healthy control subjects: a 1-year prospective study in a Swedish clinic for infectious diseases. Clin. Infect. Dis. 30 (5), 770–778.

Tacket, C.O., Maneval, D.R., Levine, M.M., 1987. Purification, morphology, and genetics of a new fimbrial putative colonization factor of enterotoxigenic *Escherichia coli* O159:H4. Infect. Immun. 55 (5), 1063–1069.

Taniguchi, T., Akeda, Y., Haba, A., et al., 2001. Gene cluster for assembly of pilus colonization fac-tor antigen III of enterotoxigenic *Escherichia coli*. Infect. Immun. 69 (9), 5864–5873.

Tauschek, M., Gorrell, R.J., Strugnell, R.A., Robins-Browne, R.M., 2002. Identification of a protein secretory pathway for the secretion of heat-labile enterotoxin by an enterotoxigenic strain of *Escherichia coli*. Proc. Natl. Acad. Sci. USA 99 (10), 7066–7071.

Taxt, A., Aasland, R., Sommerfelt, H., Nataro, J., Puntervoll, P., 2010. Heat-stable enterotoxin of enterotoxigenic *Escherichia coli* as a vaccine target. Infect. Immun. 78 (5), 1824–1831.

Taylor, D.N., Sanchez, J.L., Candler, W., Thornton, S., McQueen, C., Echeverria, P., 1991. Treat-ment of travelers' diarrhea: ciprofloxacin plus loperamide compared with ciprofloxacin alone. A placebo-controlled, randomized trial. Ann. Intern. Med. 114 (9), 731–734.

Thomas, L.V., McConnell, M.M., Rowe, B., Field, A.M., 1985. The possession of three novel coli surface antigens by enterotoxigenic *Escherichia coli* strains positive for the putative coloniza-tion factor PCF8775. J. Gen. Microbiol. 131 (9), 2319–2326.

Thompson, M.R., Jordan, R.L., Luttrell, M.A., et al., 1986. Blinded, two-laboratory comparative analysis of *Escherichia coli* heat-stable enterotoxin production by using monoclonal antibody enzyme-linked immunosorbent assay, radioimmunoassay, suckling mouse assay, and gene probes. J. Clin. Microbiol. 24 (5), 753–758.

Turner, A.K., Stephens, J.C., Beavis, J.C., et al., 2011. Generation and characterization of a live attenuated enterotoxigenic *Escherichia coli* combination vaccine expressing six colonization factors and heat-labile toxin subunit B. Clin. Vaccine Immunol. CVI 18 (12), 2128–2135.

Turner, S.M., Chaudhuri, R.R., Jiang, Z.D., et al., 2006. Phylogenetic comparisons reveal multiple acquisitions of the toxin genes by enterotoxigenic *Escherichia coli* strains of different evolutionary lineages. J. Clin. Microbiol. 44 (12), 4528–4536.

Vaandrager, A.B., Bot, A.G., Ruth, P., Pfeifer, A., Hofmann, F., De Jonge, H.R., 2000. Differential role of cyclic GMP-dependent protein kinase II in ion transport in murine small intestine and colon. Gastroenterolog 118 (1), 108–114.

Vaandrager, A.B., Smolenski, A., Tilly, B.C., et al., 1998. Membrane targeting of cGMP-dependent protein kinase is required for cystic fibrosis transmembrane conductance regulator Cl- channel activation. Proc. Natl. Acad. Sci. USA 95 (4), 1466–1471.

Valvatne, H., Sommerfelt, H., Gaastra, W., Bhan, M.K., Grewal, H.M., 1996. Identification and characterization of CS20, a new putative colonization factor of enterotoxigenic *Escherichia coli*. Infect. Immun. 64 (7), 2635–2642.

van Heyningen, S., 1991. Enterotoxins. The ring on a finger. Nature 351 (6325), 351.

Viboud, G.I., Jonson, G., Dean-Nystrom, E., Svennerholm, A.M., 1996. The structural gene encoding human enterotoxigenic *Escherichia coli* PCFO20 is homologous to that for porcine 987P. Infect. Immun. 64 (4), 1233–1239.

Vicente, A.C., Teixeira, L.F., Iniguez-Rojas, L., et al., 2005. Outbreaks of cholera-like diarrhoea caused by enterotoxigenic *Escherichia coli* in the Brazilian Amazon Rainforest. Trans. R. Soc. Trop. Med. Hyg. 99 (9), 669–674.

Waksman, G., Hultgren, S.J., 2009. Structural biology of the chaperone-usher pathway of pilus biogenesis. Nat. Rev. Microbiol. 7 (11), 765–774.

Walker, R.I., Steele, D., Aguado, T., 2007. Analysis of strategies to successfully vaccinate infants in developing countries against enterotoxigenic *E. coli* (ETEC) disease. Vaccine 25 (14), 2545–2566.

Wang, L., Hashimoto, Y., Tsao, C.Y., Valdes, J.J., Bentley, W.E., 2005. Cyclic AMP (cAMP) and cAMP receptor protein influence both synthesis and uptake of extracellular autoinducer 2 in *Escherichia coli*. J. Bacteriol. 187 (6), 2066–2076.

Weikel, C.S., Tiemens, K.M., Moseley, S.L., Huq, I.M., Guerrant, R.L., 1986. Species specificity and lack of production of STb enterotoxin by *Escherichia coli* strains isolated from humans with diarrheal illness. Infect. Immun. 52 (1), 323–325.

Wernick, N.L., Chinnapen, D.J., Cho, J.A., Lencer, W.I., 2010. Cholera toxin: an intracellular journey into the cytosol by way of the endoplasmic reticulum. Toxins 2 (3), 310–325.

Wolf, M.K., 1997. Occurrence, distribution, and associations of O and H serogroups, colonization factor antigens, and toxins of enterotoxigenic *Escherichia coli*. Clin. Microbiol. Rev. 10 (4), 569–584.

Wolf, M.K., de Haan, L.A., Cassels, F.J., et al., 1997. The CS6 colonization factor of human enterotoxigenic *Escherichia coli* contains two heterologous major subunits. FEMS Microbiol. Lett. 148 (1), 35–42.

Yamamoto, T., Echeverria, P., 1996. Detection of the enteroaggregative *Escherichia coli* heat-stable enterotoxin 1 gene sequences in enterotoxigenic *E. coli* strains pathogenic for humans. Infect. Immun. 64 (4), 1441–1445.

Yamanaka, H., Nomura, T., Fujii, Y., Okamoto, K., 1998. Need for TolC, an *Escherichia coli* outer membrane protein, in the secretion of heat-stable enterotoxin I across the outer membrane. Microb. Pathog. 25 (3), 111–120.

Yu, J., Cassels, F., Scharton-Kersten, T., et al., 2002. Transcutaneous immunization using colonization factor and heat-labile enterotoxin induces correlates of protective immunity for enterotoxigenic *Escherichia coli*. Infect. Immun. 70 (3), 1056–1068.

Zafriri, D., Oron, Y., Eisenstein, B.I., Ofek, I., 1987. Growth advantage and enhanced toxicity of *Escherichia coli* adherent to tissue culture cells due to restricted diffusion of products secreted by the cells. J. Clin. Invest. 79 (4), 1210–1216.

Zheng, D., Constantinidou, C., Hobman, J.L., Minchin, S.D., 2004. Identification of the CRP regulon using in vitro and in vivo transcriptional profiling. Nucleic Acids Res. 32 (19), 5874–5893.

Chapter 7

Shigella and enteroinvasive *Escherichia coli*: Paradigms for pathogen evolution and host–parasite interactions

Anthony T. Maurelli

Uniformed Services University, Bethesda, MD, USA

BACKGROUND

Bacillary dysentery or shigellosis is caused by members of the *Shigella* genus and a group of pathogenic strains of *Escherichia coli* known as enteroinvasive *E. coli* (EIEC). Dysentery was the term used by Hippocrates to describe an illness characterized by frequent passage of stools containing blood and mucus accompanied by painful abdominal cramps. This disease has had a tremendous impact on human society over the centuries with perhaps one of the greatest effects being its powerful influence in military operations. Long, protracted military campaigns and sieges almost always spawned epidemics of dysentery and caused large numbers of military and civilian casualties. The Napoleonic campaigns and the American Civil War were 19th century examples of the devastation caused by dysentery. This capacity of dysentery to factor in the outcome of nation-shaping events continues in the modern era. Soldiers were stricken with the disease in essentially all major conflicts of the 20th century, including both World Wars, Korea, Vietnam, and the first Gulf War. Indeed, this ancient disease goes hand in hand with political and socio-economic strife as dysentery continues to claim the lives of hundreds of thousands of people living in the unsanitary conditions brought on by war and poverty. The low infectious dose required to cause disease coupled with oral transmission of the bacteria via fecally contaminated food and water, accounts for the spread of dysentery caused by *Shigella* spp. in the wake of many natural (earthquakes, floods, famine) as well as man-made (war) disasters. For example, refugees of a recent conflict in Zaire suffered high morbidity and mortality rates brought on by bacillary dysentery (Kim et al., 2009). Even apart from these special circumstances, shigellosis remains an important disease in developed countries

Escherichia coli. http://dx.doi.org/10.1016/B978-0-12-397048-0.00007-3
2013 Published by Elsevier Inc.

as well as in underdeveloped countries. The global burden of shigellosis has been estimated at over 163 million cases per year with nearly 1 million deaths, the majority occurring in children less than 5 years of age (Kotloff et al., 1999).

The appropriateness of a chapter on *Shigella* in a book devoted to pathogenic *E. coli* derives from the fact that EIEC cause dysentery that is clinically indistinguishable from that caused by members of the *Shigella* genus. In addition, an abundance of studies starting with the DNA–DNA hybridization work of Brenner et al. (1969) to multilocus enzyme electrophoresis (Pupo et al., 1997) and whole genome sequencing (Touchon et al., 2009; Sims and Kim, 2011) unequivocally establish that *Shigella* spp. are clones of *E. coli*. In this chapter, we will consider these pathogens together. Since the vast majority of the research on the pathogenesis of bacillary dysentery has been done on *Shigella*, we will focus on these organisms. No single review can be completely comprehensive. Therefore, the reader is encouraged to refer to several excellent recent reviews for additional information (Parsot, 2005; Schroeder and Hilbi, 2008).

Classification and biochemical characteristics

The Japanese microbiologist Shiga isolated an organism (*Shigella dysenteriae*) from the dysenteric stool of a stricken individual in 1898. Three more *Shigella* species (*S. boydii, S. flexneri*, and *S. sonnei*) were subsequently identified and grouped by serotype and metabolic activities. The four species of the genus *Shigella* are grouped serologically (41 serotypes) based on their somatic O-antigens: *S. dysenteriae* (group A), *S. flexneri* (group B), *S. boydii* (group C), and *S. sonnei* (group D). As members of the family *Enterobacteriaceae*, they are closely related to the salmonellae (Ochman et al., 1983). *Shigella* are non-motile, Gram-negative rods. Some important biochemical characteristics that distinguish these bacteria from other enterics are their inability to utilize citric acid as a sole carbon source and their inability to ferment lactose, although some strains of *S. sonnei* may ferment lactose slowly. They are oxidase-negative, do not produce H_2S (except for *S. flexneri* 6 and *S. boydii* serotypes 13 and 14), and do not produce gas from glucose. *Shigella* spp. are inhibited by potassium cyanide and do not synthesize lysine decarboxylase (Ewing, 1986).

Based on the cumulative sequence evidence from 16S rRNA genes, multiple housekeeping genes, and whole genome comparisons, the pathogenic *Shigella* can be considered a subset of non-pathogenic *E. coli* (Cilia et al., 1996; Pupo et al., 1997; Shu et al., 2000; Jin et al., 2002). Although phylogenetically it would be more accurate to treat *Shigella* as pathotypes of *E. coli*, the members of the genus *Shigella* continue to be divided for historical and medical purposes into the four species or subgroups described above (Strockbine and Maurelli, 2005).

Enteroinvasive *E. coli* (EIEC) isolates that cause a diarrheal illness identical to *Shigella* dysentery were identified in the 1970s (DuPont et al., 1971; Tulloch et al., 1973). The pathogenic and biochemical properties that EIEC share with *Shigella* pose a problem for distinguishing these pathogens. For example, unlike normal flora *E. coli*, EIEC are non-motile and 70% of isolates are unable to ferment lactose (Silva et al., 1980; van den Beld and Reubsaet, 2012). These are also features of *Shigella*. More striking is the observation that strains of EIEC are almost universally negative for lysine decarboxylase (LDC) activity whereas almost 90% of normal flora *E. coli* are positive. In this respect, EIEC also resemble *Shigella,* which uniformly lack LDC activity. Some serotypes of EIEC even share identical O-antigens with *Shigella* (Sansonetti et al., 1985). By contrast, EIEC resemble *E. coli* in their ability to ferment xylose and to produce gas from glucose, both traits for which *Shigella* are negative (Silva et al., 1980).

It is now generally accepted that the present day strains of *Shigella* arose multiple times from as many as seven independent ancestral strains of *E. coli* (Pupo et al., 2000; Yang et al., 2005). EIEC probably evolved later than *Shigella* and from different *E. coli* ancestors (Lan et al., 2004). However, the common seminal event in the evolution of both groups of pathogens was the acquisition of a large plasmid that encodes the genes necessary for invasion of mammalian cells (see below) (Lan et al., 2001).

Evolution of *Shigella* species and EIEC

The EIEC were recognized as a heterogeneous group of pathogens that resembled *Shigella* in their pathogenic potential (and in certain metabolic traits) but were more closely related to *E. coli*. Early genetic studies demonstrated that essentially all the virulence factors required for expression of the invasive phenotype are encoded on a large plasmid present in all *Shigella* and EIEC isolates examined (Sansonetti et al., 1981, 1982b; Harris et al., 1982). Therefore, from the pathogenesis perspective, EIEC are closely related to *Shigella*. These observations led many investigators to propose, incorrectly, that the EIEC represented a missing link in the evolution from *E. coli* to *Shigella*.

Several studies have established that, like the EIEC, the four species of *Shigella* are so closely related to *E. coli* that they should all be included in a single species. The chromosomes of these organisms are largely co-linear and are more than 90% homologous (Brenner et al., 1969). Therefore, *Shigella*, like EIEC, are a group of pathogenic *E. coli*. In fact, several investigators using different approaches have established that the four species of *Shigella* evolved from separate *E. coli* strains (Pupo et al., 1997, 2000; Rolland et al., 1998). Seven different *Shigella* lineages have been identified through sequence analysis of multiple chromosomal loci. Bacteria expressing the *Shigella* phenotype were generated through horizontal transfer of the virulence plasmid from an unknown donor bacterium to commensal *E. coli* (Pupo et al., 2000). Thus, horizontal transfer of the *Shigella* virulence plasmid to commensal *E. coli* has occurred multiple times, each time giving rise to new *Shigella* clones. These findings suggest that traits unique to and shared by *Shigella* species and EIEC are the result of convergent evolution either through gain-of-function mutations (e.g. horizontal transfer of the virulence plasmid) or loss-of-function mutations (deletion of genes encoding traits expressed by ancestral *E. coli* strains). These insights provide a possible answer to lingering questions of the relationship of EIEC to *Shigella* and the evolution of these pathogens. The heterogeneous characteristics of EIEC are consistent with a polyphyletic history and reflect the non-clonal origins of these isolates. Moreover, these findings strongly suggest that the EIEC are not *Shigella* ancestors, or 'missing links.' Rather, the EIEC clones may be recently evolved pathogens that have not yet completely adapted to a pathogenic lifestyle and therefore have not fully developed the *Shigella* phenotype (Pupo et al., 2000). This proposal suggests that expression of the complete *Shigella* virulence phenotype requires not only the virulence plasmid but also additional modifications to the *E. coli* genome for optimal transmissibility, fitness in host tissues, and virulence.

Several recent studies have provided evidence of a new pathway of evolution, termed antagonistic pleiotrophy, that fine-tunes pathogen genomes for maximal fitness and virulence in host tissues. This pathway and the selective forces that drive its function can be observed in the transformation of commensal *E. coli* to virulent *Shigella* and EIEC pathogens. Following acquisition of the virulence plasmid and expression of its virulence genes, commensal *E. coli* gained access to new host tissues. However, such a newly evolved pathogen, which expresses the full complement of ancestral traits as well as virulence factors, may not be optimally suited for this new pathogenic lifestyle. Selective pressures encountered in the new environment (within colonocytes) are very different to those in the ancestral niche (lumen of the colon). In fact, some ancestral traits may interfere with the expression or function of factors required for survival within host tissues. Loci encoding these interfering factors are designated antivirulence genes. Thus, the newly evolved pathogen must inactivate ancestral antivirulence genes for optimal fitness and virulence. These modifications to the new pathogen's genome are termed pathoadaptive mutations (Sokurenko et al., 1999; Maurelli, 2007).

The convergent evolution of the seven *Shigella* lineages and EIEC presents a unique opportunity for the identification and characterization of antivirulence genes and pathoadaptive mutations in addition to the study of pathogen evolution. Ancestral traits that interfere with virulence are lost from the newly evolved pathogen genome early on as the increased fitness of the adapted clones reinforces these beneficial mutations in the newly or recently evolved pathogen population. Therefore, some traits that are absent in all *Shigella* and EIEC, but commonly expressed in *E. coli*, are strong indicators of pathoadaptive mutations that have arisen by convergent evolution. Thus, evidence supporting this new pathogen evolution pathway was first provided in *Shigella*. Lysine decarboxylase (LDC) activity, which is encoded by the *cadA* gene, is expressed in over 85% of *E. coli* isolates. In contrast, no isolates of *Shigella* or EIEC express LDC activity (Silva et al., 1980). Lack of LDC activity in *Shigella* and EIEC is consistent with a pathoadaptive mutation. Experimental evidence of the antivirulence nature of the *cadA* gene was provided by the demonstration that the product of LDC activity, cadaverine, blocks the action of the virulence plasmid-encoded *Shigella* enterotoxin (Maurelli et al., 1998) and several other virulence phenotypes (McCormick et al., 1999). Sequence analysis of the *cadA* region of four *Shigella* lineages revealed novel genetic arrangements that are distinct in each strain examined (Day et al., 2001). Thus, the inactivation of *cadA* was accomplished by different mechanisms in each *Shigella* strain. Studies in EIEC showed that these strains have mutations in *cadC*, a transcriptional activator, which abolishes LDC activity (Casalino et al., 2003). These studies demonstrate that each newly evolved *Shigella* and EIEC clone increased virulence through the loss of an ancestral trait, LDC. Additional anti-virulence loci that have been lost or inactivated in *Shigella* spp. and EIEC include *nadA/nadB* (Prunier et al., 2007) and *speG* (Barbagallo et al., 2011).

MOLECULAR PATHOGENESIS

Hallmarks of virulence

Shigella spp. and EIEC cause disease by overt invasion of epithelial cells in the large intestine. The clinical symptoms of dysentery can be directly attributed to the following hallmarks of virulence: induction of diarrhea, invasion of intestinal epithelial cells, multiplication inside these cells, spread from cell to cell, and stimulation of a strong host inflammatory response (Figure 7.1).

    Shigella colonize the small intestine only transiently and cause little tissue damage (Rout et al., 1975). Production of enterotoxins by *Shigella* and EIEC in the small bowel probably results in the diarrhea that generally precedes onset of dysentery (Fasano et al., 1995; Nataro et al., 1995). It is believed that jejunal secretions elicited by these toxins facilitate passage of the bacteria through the small intestine and into the colon where they colonize and invade the epithelium.

    Formal and his colleagues demonstrated the essential role of epithelial cell invasion in *Shigella* pathogenesis in a landmark study that employed both tissue culture assays to measure invasion and animal models (LaBree et al., 1964). They showed that spontaneous colonial variants of *S. flexneri* 2a that are unable to invade mammalian cells in tissue culture do not cause disease in monkeys. Further, they showed the presence of wild-type bacteria within epithelial cells of the large intestine in experimentally infected animals.

    Gene transfer studies using *E. coli* K-12 donors and *S. flexneri* 2a recipients established the third hallmark of *Shigella* virulence. A *S. flexneri* 2a recipient that inherits the *xyl-rha* region of the *E. coli* K-12 chromosome retains the ability to invade epithelial cells but has a reduced ability to multiply within these cells (Falkow et al., 1963). This hybrid strain fails to cause a fatal infection in the opium-treated guinea pig model and is unable to cause disease when fed to rhesus monkeys (Formal et al., 1965). The high frequency of recombinants in these conjugation experiments also served to confirm the close genetic related-ness of *E. coli* and *Shigella*.

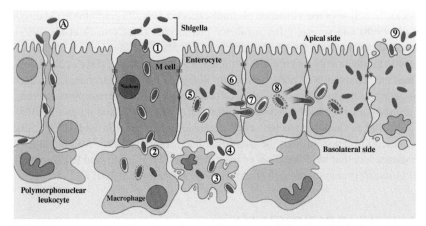

FIGURE 7.1 Stages in the invasion and intercellular spread of *Shigella*. Initially, bacteria trans-cytose through microfold (M) cells of the intestinal epithelium (1), after which they are phagocy-tosed by resident macrophages (2). The bacteria escape the macrophage phagosome and ultimately induce pyroptosis and death of the host cell to gain access to the basolateral side (3). Bacteria are then free to invade enterocytes via a T3SS-dependent process (4). Once inside the epithelial cell, the bacteria lyse the single-membrane phagosome to avoid killing and replicate inside the host cell cytosol (5). Bacteria polymerize host cell actin (6) creating an actin 'tail' that promotes cell-to-cell spread (7). Once inside a second epithelial cell, *Shigella* escapes the new double-membrane phagosome (8). Bacterial replication and spread continues throughout the colonic mucosa. Finally, infected epithelial cells undergo necrosis and these dying cells release bacteria into the intestinal lumen, where they may disseminate to the next host via the fecal–oral route (9). To maximize inva-sion, *Shigella*-infected epithelial cells induce a massive influx of polymorphonuclear leukocytes (PMNs) through the actions of chemoattractants IL-8 and hepoxilin A3 (HXA3). The transepithelial migration of these PMNs disrupts the intestinal lining, allowing the bacteria to transverse between the epithelial cells to ultimately reach the basolateral membrane for subsequent invasion (A). Figure illustrated by Kym Bliven.

While it is necessary for *Shigella* to be able to multiply within the host epi-thelial cell after invasion, intracellular multiplication is not sufficient to cause disease. The bacterium must also be able to spread through the epithelial lining of the colon by cell-to-cell spread in a manner that does not require the bacte-rium to leave the intracellular environment and be re-exposed to the intestinal lumen (Figure 7.2). Mutants of *Shigella* that are competent for invasion and multiplication but are unable to spread between cells in this fashion have been isolated. These mutants established intracellular spread as the fourth hallmark of *Shigella* virulence and will be discussed in a later section.

Along with the ability to colonize and cause disease, an intrinsic part of any bacterium's pathogenicity is the regulatory circuitry for controlling expression of virulence genes. Virulence in *Shigella* spp. and EIEC is regulated by tem-perature. After growth at 37°C, virulent strains of *Shigella* are able to invade mammalian cells but when cultivated at 30°C, they are non-invasive. This non-invasive phenotype is reversible by shifting the growth temperature to 37°C where the bacteria re-express their virulence factors (Maurelli et al., 1984).

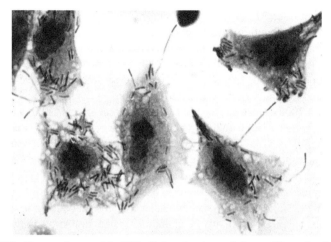

FIGURE 7.2   Tissue culture model of *Shigella* invasion, intracellular motility, and multiplication. A semi-confluent monolayer of mouse L2 fibroblasts was infected with *S. flexneri* 2a for 30 minutes. The monolayer was washed and tissue culture medium with gentamicin was added to kill extracellular bacteria. The monolayer was fixed and stained 2 hours later. Bacteria can be seen as darkly staining rods within the cytoplasm of the eukaryotic cells. Intracellular movement of the bacteria and contact with the plasma membrane causes the formation of 'fireworks,' cytoplasmic protrusions that contain bacteria at the ends.

Similarly, the invasive phenotype of EIEC is also temperature regulated (Small and Falkow, 1988). Temperature regulation of virulence gene expression is a characteristic of other human pathogens such as pathogenic *E. coli, Salmonella typhimurium, Bordetella pertussis, Yersinia* spp., and *Listeria monocytogenes*. Regulation of gene expression in response to environmental temperature is a useful strategy for bacteria. This strategy permits *Shigella* and EIEC to trigger gene expression by sensing the ambient temperature of the mammalian host (i.e. 37°C) thereby economizing energy that would otherwise be expended on the synthesis of virulence products when the bacteria are outside the host. The system also permits the bacteria to coordinately regulate expression of multiple unlinked genes that are required for the full virulence phenotype.

Cell biology

Shigellosis, or bacillary dysentery, is an acute inflammatory disease of the colonic mucosa that results from invasion and cell-to-cell spread of the bacteria. Infection occurs via the oral route by ingestion of the bacteria in contaminated food or water. *Shigella* and EIEC display a remarkable ability to survive the acidity of the stomach (Gorden and Small, 1993; Jennison and Verma, 2007). After passing through the stomach, the bacteria transit through the small intestine to reach the colon where infection is established. Studies in several different animal models demonstrate that *Shigella* gain access to the colonic subepithelium and establish an infection through M cells. These specialized cells do not produce microvilli and are not covered with mucus or glycocalyx. They sample intestinal antigens which are transcytosed unmodified to underlying lymphoid follicles. In the sub-epithelium, the bacteria are rapidly ingested by resident macrophages. However, rather than being killed, *Shigella* escape the phagosome and induce caspase-1 activation and pyroptosis of the infected macrophage (Zychlinsky et al., 1992; Suzuki et al., 2005). Death of the phagocyte has dual effects as the bacteria are released unharmed along with the pro-inflammatory cytokine IL-1β that initiates an inflammatory cascade (Zychlinsky and Sansonetti, 1997). Following release from the macrophage, *Shigella* penetrate colonic epithelial cells through the basolateral surface via interaction with α5-β1 integrins and CD44 (Watarai et al., 1996). As in the macrophage, internalized bacteria rapidly escape the endosomal vacuole (Ray et al., 2010) and spread to adjacent cells. In contrast to the fate of the macrophage, *Shigella* actively block induction of apoptosis in epithelial cells (Faherty and Maurelli, 2009). Infection of the colonocytes induces the release of still more pro-inflammatory cytokines, IL-6 and IL-8 (Sansonetti et al., 1999). The high levels of pro-inflammatory cytokines produced by epithelial cells and macrophages in response to *Shigella* infection signal the massive recruitment of polymorphonuclear neutrophils (PMNs), the only cells that are able to kill *Shigella*. Thus a paradox of *Shigella*–host interaction exists: how can the shigellae be successful pathogens (able to infect new hosts) when the organisms are read-ily killed and eventually cleared by the massive innate host immune response? One possible answer has recently been proposed.

The severe signs and symptoms of shigellosis are due to extensive tissue destruction by the intense inflammatory response. The major effectors of this damage are PMNs. Inhibition of PMN recruitment to the site of infection using anti-IL-8 antibodies significantly reduces the severity of symptoms (Sansonetti et al., 1999). PMN degranulation mediates tissue destruction and destabilizes tight junctions between colonocytes (Perdomo et al., 1994b). *Shigella* present in the intestinal lumen exploit this window of opportunity to gain access to the subepithelium where the pathogens can begin another round of infection by invasion of colonocytes through the basolateral surface (Perdomo et al., 1994a). Intracellular bacteria are likely protected from PMN killing. Thus, the initial focus of infection is amplified through successive rounds of infection made possible by the activity of PMNs. Amplification of the infection increases the number of *Shigella* in the infected host and increases the likelihood of suc-cessful transmission to another host. Consistent with this model, recruitment of PMNs across the epithelium to the apical side (which faces the intestinal lumen) requires virulent, invasive *Shigella* (McCormick et al., 1998). Recruitment of PMNs further requires that the bacteria interact with the basolateral membrane of the epithelial cells, where IL-8 is secreted. Bacteria present on the apical side do not efficiently invade colonocytes nor do they recruit PMNs. *Shigella* residing within cells (or in the subepithelium) mediate PMN recruitment across the epithelium to the apical surface in model monolayers in part by inducing the secretion of the potent PMN chemoattractant, hepoxilin A_3 (Mumy et al., 2008). Secretion of hepoxilin A_3 from the apical surface of the epithelium creates a chemotactic gradient that guides PMNs from the submucosa to the apical surface of the epithelium. Collectively, these observations suggest that *Shigella* plays dangerous games with effectors of the host innate immune response by first inducing, then exploiting, the intense inflammatory response to ensure its own survival and evolutionary success (passage to another host).

Virulence genes

Plasmid-encoded

Given the complexity of the interactions between host and pathogen, it is not surprising that the virulence of *Shigella* and EIEC is multigenic, involving both chromosomal and plasmid-encoded genes. Another landmark study on the pathogenicity of *Shigella* was the demonstration of the indispensable role for a large plasmid in invasion. A 180 kilobase pair (kb) plasmid in *S. sonnei* and a 220 kb plasmid in *S. flexneri* were first shown to be required for invasion (Sansonetti et al., 1981, 1982b). Other *Shigella* spp. as well as strains of EIEC also contain large plasmids that are functionally interchangeable and share significant degrees of DNA homology with the plasmid described in *S. flexneri* (Harris et al., 1982; Sansonetti et al., 1982a). These plasmids can be classified into two distinct but closely related forms based on sequence analysis (Lan et al., 2001). Analysis of the complete DNA sequence of the 220 kb plasmid in *S. flexneri* 5a reveals a complex history (Buchrieser et al., 2000; Venkatesan et al., 2001). This plasmid is essentially a genetic mosaic composed of blocks of genes from different origins as well as the remnants of four different ancestral plasmids. For instance, the genes required for the invasive phenotype have a G+C content of 31–37%. This value is significantly lower than the composition of loci required for intercellular spread, and the composition of these loci is significantly different from loci encoding a surface protease. These genes are interspersed with close to 100 insertion sequences comprising one-third of the virulence plasmid. Evidence exists of active transposition, rearrangements, and deletions in the plasmid mediated by homologous recombination between these elements. These events likely led to the size and organizational differences observed between the virulence plasmids of different *Shigella* species and lineages.

A 37 kb region of the invasion plasmid of *S. flexneri* 2a contains all of the genes necessary to permit the bacteria to penetrate into tissue culture cells. This DNA segment was identified as the minimal region of the virulence plasmid necessary to allow a plasmid-cured derivative of *S. flexneri* (and *E. coli* K-12) to invade tissue culture cells (Maurelli et al., 1985). The 34 virulence genes encoded in this region are organized into three large operons (*ipa*, *mxi*, and *spa*) that are transcribed in opposite directions. The essential role of these genes in expression of the invasive *Shigella* phenotype and virulence has been demonstrated using mutation analyses. The organization of these loci in large transcription units suggests that the genes are co-regulated and encode factors that function together in a multicomponent virulence (invasion) system. The *mxi* and *spa* operons encode subunits of a type III secretion system (T3SS; see Chapter 14) and the *ipa* operon encodes proteins (effectors; see Chapter 15) secreted by this system. T3SS are widely used by animal and plant pathogens (enterohemorrhagic and enteropathogenic *E. coli*, *Salmonella* spp., *Yersinia* spp., *Pseudomonas* spp., *Erwinia caratovara*, and *Xanthomonas campestris*, to name a few) for delivery of virulence factors to the surface or interior of host cells (Galan and Wolf-Watz, 2006). These secreted effectors of virulence allow the pathogens to reprogram host cells to serve the pathogen's particular needs. The T3SS delivery machines are conserved structurally, functionally, and genetically and share several core characteristics.

T3SS genes

The products of the *ipa* genes are actively secreted into the extracellular medium even though they contain no signal sequence for recognition by the usual general secretory pathway of Gram-negative bacteria. Ipa secretion is mediated by a T3SS composed of gene products from the *mxi/spa* loci (Schroeder and Hilbi, 2008). A characteristic feature of T3SS is a contact-dependent secretion mechanism. Delivery of effectors to their targets (surface and cytoplasmic) occurs only after outer membrane elements of the T3SS complex contact cholesterol in the host cell membrane (Hayward et al., 2005). This feature is observed in the *Shigella* T3SS as the Ipa proteins are deposited on the host cell surface after contact with host cells (Menard et al., 1994a; Watarai et al., 1995).

The structural elements of the *Shigella* T3SS are encoded in two adjacent operons on the virulence plasmid. The *mxi* (*m*embrane *ex*pression of *in*vasion plasmid antigens) genes encode several lipoproteins (MxiJ and MxiM), a transmembrane protein (MxiA), and proteins containing signal sequences (MxiD, MxiJ, and MxiM) (Schroeder and Hilbi, 2008). MxiH is the major subunit of the secretion needle (Blocker et al., 2001). The *spa* (*s*urface *p*resentation of Ipa *a*ntigens) genes encode additional components of the *Shigella* T3SS. The cytoplasmic protein Spa32 controls the switch for T3SS substrate specificity and thus regulates needle length (Magdalena et al., 2002; Tamano et al., 2002). Assembly of the T3SS needle and subsequent effector secretion requires Spa47 which is associated with the cytoplasmic face of the secretion apparatus and has sequence similarities with ATPases of the flagellar assembly machinery of other bacteria (Vankatesan et al., 1992; Tamano et al., 2000). Spa47 likely is the energy-generating component of the secretion apparatus.

Invasion plasmid antigens

The genes comprising the *ipaBCDA* (*i*nvasion *p*lasmid *a*ntigens) cluster encode the immunodominant antigens of *Shigella* and EIEC that are detected with sera from convalescent patients and experimentally challenged monkeys (Hale et al.,

1985; Oaks et al., 1986). *ipaBCD* are absolutely required for invasion of mammalian cells and carry out post-invasion roles as secreted effectors (Menard et al., 1993). An *ipaA* mutant is still invasive but has a 10-fold reduced ability compared to wild-type (Tran et al., 1997). This operon also encodes a molecular chaperone, IpgC (invasion plasmid gene), which binds IpaB and IpaC in the bacterial cytosol prior to secretion and prevents their interaction and proteolytic degradation (Menard et al., 1994b). Following secretion, IpaB and IpaC form a complex that binds to molecules such as α5-β1 integrins and CD44 that are present on the basolateral surface of host cells (Watarai et al., 1996; Skoudy et al., 2000). After binding, the Ipa complex inserts into the host cell membrane and causes a massive recruitment of host cell actin and other cytoskeletal factors beneath the site of attachment. This exploitation of the host cell cytoskeleton is observed with IpaBC-coated beads or whole *Shigella* bacteria and leads to the formation of localized membrane projections that encompass and internalize either beads or bacteria in roughly 5 minutes (Menard et al., 1996). Purified IpaC induces cytoskeletal reorganization, including formation of filopodia and lamellipodial extensions on permeabilized cells (Tran et al., 1999). Inside the host cell, IpaA interacts with vinculin to promote F-actin depolymerizaton. Through this action, IpaA may promote formation of a pseudo-focal adhesion complex that organizes the actin assembly complex beneath the bacterium and directs optimal uptake of the pathogen (Tran et al., 1997; Bourdet-Sicard et al., 1999).

IpaB serves several other roles in *Shigella* virulence that may be accomplished by separate domains of the protein. IpaB is required for escape of *Shigella* from both the phagosomal (following uptake by macrophages) and endosomal (following uptake by epithelial cells) compartments (High et al., 1992). This activity requires formation of a hydrophobic N-terminal alpha-helix and is linked to the ability of IpaB to insert into host cell membranes and lyse erythrocytes. In addition, IpaB binds to pro-caspase 1 in vitro and in vivo and triggers hydrolysis of the cysteine protease (Zychlinsky et al., 1994). In macrophages, this activity leads to secretion of IL-1β (which initiates inflammation) and induction of pyroptosis of the phagocyte.

The needle tip protein, IpaD, forms a pentameric ring at the end of the T3SS needle and prevents secretion of the effector proteins (Espina et al., 2006). IpaD also acts as an environmental sensor that triggers recruitment of IpaB to the needle tip upon exposure to bile salts (Olive et al., 2007). Upon contact with the cholesterol-rich plasma membrane of the mammalian cell, IpaC is recruited to the needle tip where it associates with IpaB to form a translocon pore within the host cell membrane (Epler et al., 2009). The IpaB/IpaC translocon is the conduit through which effectors pass directly from the bacterium into the host cell cytoplasm.

IpgB1 plays a critical role in invasion. It acts as a mimic of the host small GTPase RhoG at the host plasma membrane, activates Rac1, and stimulates membrane ruffling through the ELMO-Dock180 pathway (Ohya et al., 2005; Handa et al., 2007). IpgB2, a homolog of IpgB1, mimics the GTP-bound form of RhoA and triggers formation of actin stress fibers (Alto et al., 2006). IpgD displays phosphatidylinositol 4-phosphatase activity and induces morphological changes in the infected cell including membrane blebbing and ruffling at the entry site (Niebuhr et al., 2002).

Cell-to-cell spread and autophagy

The *icsA* gene (also known as *virG*), which is encoded outside the 37 kb invasion region on the *Shigella* virulence plasmid, is an essential virulence determinant. *icsA* encodes a 120 kDa autotransporter protein that is required for motility of *Shigella* inside cells and spread of the bacteria to adjacent cells (Makino et al., 1986; Bernardini et al., 1989). *icsA* mutants do not form plaques in confluent tissue culture cell monolayers (Figure 7.3) and are significantly attenuated in the macaque monkey challenge model (Sansonetti et al., 1991). Because of this latter property, *icsA* mutants have been the basis of many efforts to develop live, attenuated *Shigella* vaccines (Alexander et al., 1996; Barnoy et al., 2011).

The IcsA protein is unusual in that it is expressed asymmetrically on the bacterial surface, being found predominantly at one pole (Goldberg et al., 1993). Unipolar localization of IcsA imparts directionality of movement to the organism. IcsA mediates motility by binding host cytoplasmic N-WASP and the Arp2/Arp3 complex which catalyze polymerization of actin at that end of the bacterium (Egile et al., 1999; Suzuki et al., 2002). Polymerization of actin monomers at one end of the bacterium provides the force that drives the organism through the host cell cytoplasm and into adjacent cells (Monack and Theriot, 2001).

Correct unipolar localization of IcsA in *Shigella* is essential for virulence and is dependent on synthesis of a complete lipopolysaccharide (LPS) (Sandlin et al., 1996). LPS mutants of *Shigella* produce IcsA but fail to confine IcsA to the pole. As a result IcsA is found uniformly over the cell surface in these mutants. The protein is still capable of polymerizing actin but actin polymerizes around the entire cell rather than localizing to a single pole. Consequently, movement is restricted as the bacterium becomes encased in a shell of actin. A plasmid-encoded protease, SopA/IcsP, has been shown to cleave IcsA and is proposed to play a role in unipolar localization of IcsA (Egile et al., 1997; Shere et al., 1997). However *E. coli* and plasmid-cured derivatives of *S. flexneri* transformed with a cloned *icsA* gene localize IcsA normally (Sandlin and Maurelli, 1999; Monack and Theriot, 2001). These results suggest that IcsA localization does not require any other virulence plasmid-encoded gene and that motifs within IcsA itself contain the information that directs the protein to the pole.

IcsA is also targeted by autophagy, a conserved process in eukaryotes with diverse functions, from nutrient recycling to degradation of defective proteins and organelles to recognition and elimination of intracellular pathogens (Ogawa et al., 2005; Ogawa and Sasakawa, 2006). The autophagy protein Atg5 binds IcsA thereby inducing the autophagic degradation of *Shigella*. However IcsB, an effector encoded by the virulence plasmid and secreted by the T3SS, impairs the

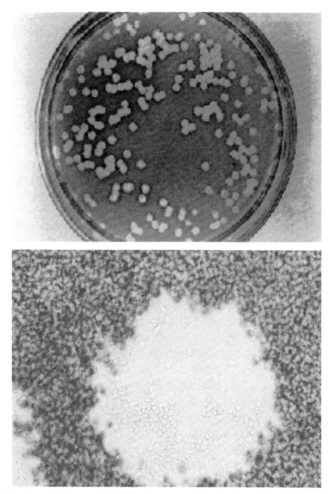

FIGURE 7.3   Intercellular spread of *Shigella* measured in the plaque assay. Bacteria are allowed to invade a confluent monolayer of cells in tissue culture for 90 min. The monolayer is washed and an overlay of 0.5% agarose-containing tissue culture medium with gentamicin is added and the monolayers incubated for 48–72 h. The gentamicin kills extracellular bacteria. Plaques are formed by intracellular bacteria that spread from cell to cell via protrusions that protect them from the antibiotic. Top panel shows plaques formed by *S. flexneri* 2a in a monolayer of L2 fibroblasts after 48 h. Bottom panel shows the appearance of a plaque at higher magnification. The monolayer is stained with neutral red, which stains living cells. Dead (unstained) cells can be seen in the center of the plaque.

interaction between Atg5 and IcsA through competitive binding to IcsA (Ogawa et al., 2005). After 4 hours post-infection, 40% of the intracellular *icsB* mutants are associated with autophagosome markers whereas only 8% of the wild-type bacteria are recognized by the autophagy machinery (Ogawa et al., 2005). IcsB and Atg5 bind to the same internal region of IcsA (amino acids 320–433) but

IcsB has a higher affinity for IcsA. This result indicates that IcsB caps IcsA to protect *Shigella* from autophagy degradation. In addition, interaction between IcsB and cholesterol is required for autophagy evasion but the ability of IcsB to bind cholesterol and IcsA involves distinct domains (Kayath et al., 2010).

Other secreted effectors

Analysis of the culture supernatants from a *Shigella* mutant which displays constitutive activation of the T3SS, revealed roughly 14 additional proteins that are secreted (Buchrieser et al., 2000). Among these proteins are VirA, IpaH7.8 and IpaH9.8, and a family of proteins known as Osp (outer *Shigella* proteins). VirA is a 44.7 kDa protein required for optimal *Shigella* invasion and intercellular spread (Uchiya et al., 1995). It was first reported to possess microtubulin cleaving activity (Yoshida et al., 2006) although this property of VirA has been challenged (Germane et al., 2008). A recent report showed that VirA activates cellular calpain which leads to degradation of p53 and inhibition of the p53/NF-κB pathway. These effects prevent induction of apoptosis but ultimately lead to necrotic cell death (Bergounioux et al., 2012). Members of the IpaH family are encoded by alleles that are located on both the chromosome and the virulence plasmid. They arose by gene duplication which was likely mediated by adjacent insertion sequences (Hartman et al., 1990). The presence of multiple copies of *ipaH* (which share a 3' constant region) on the plasmid and chromosome of both *Shigella* and EIEC has led to its widespread use as a probe to detect these organisms (Venkatesan et al., 1989). *ipaH7.8* is implicated in virulence since mutants are impaired in phagosomal escape (Fernandez-Prada et al., 2000). The E3 ubiquitin ligase activity of IpaH9.8 acts to diminish the host inflammatory response by blocking the NF-κB pathway (Rohde et al., 2007; Asida et al., 2010).

The *osp* genes, which encode secreted effector molecules, have a %G+C content very similar to that of genes in the entry region suggesting that they were acquired at the same time as genes encoding the T3SS and Ipas (Buchrieser et al., 2000). Several *osp*s have paralogs in the virulence plasmid that likely arose through gene duplication. Many of the Osp proteins have been implicated in modulation of the host inflammatory response via a variety of signaling pathways. These include OspB (Zurawski et al., 2009), OspC1 (Zurawski et al., 2006), OspF (Zurawski et al., 2009; Reiterer et al., 2011), OspG (Kim et al., 2005), OspI (Sanada et al., 2012) and OspZ (Zurawski et al., 2008; Zhang et al., 2012). More details about the secreted effector proteins of *Shigella* can be found in Chapter 15.

One of the *Shigella* enterotoxins is encoded by *ospD3/senA*. This 60 kDa T3SS effector protein induces fluid secretion in an animal model and alters the electrical potential difference across a monolayer of cultured cell lines by an unknown mechanism (Nataro et al., 1995). OspE is a secreted effector that binds integrin-linked kinase and stabilizes adherence of infected cells, thus prolonging and enhancing bacterial colonization of the epithelium (Miura et al., 2006; Kim et al., 2009).

Chromosomally encoded

In contrast to the genes of the virulence plasmid that are responsible for invasion of mammalian tissues, most of the chromosomal loci associated with *Shigella* virulence are involved in regulation or survival within the host. Mutations that alter O-antigen and LPS core synthesis or assembly lead to a 'rough' phenotype and render *Shigella* avirulent. As mentioned earlier, synthesis of a complete LPS, which is crucial for correct unipolar localization of IcsA, requires chromosomal loci such as *rfa* and *rfb*. In the case of *S. sonnei* and *S. dysenteriae* 1, plasmid-borne genes encode the enzymes necessary for synthesis of the LPS O-side chain (Brahmbhatt et al., 1992). LPS has also been implicated as an important indirect chemoattractant since the molecule induces IL-8 expression in epithelial cells to recruit PMNs to the intestinal lumen thereby promoting tissue injury and amplification of the *Shigella* infectious cycle.

The *stx* locus encodes Shiga toxin in *S. dysenteriae* 1. Shiga toxin inhibits protein synthesis by cleaving the N-glycosidic bond at adenine 4324 in the 28S ribosomal RNA of mammalian cells (Melton-Celsa et al., 2012). Animal studies suggest that Shiga toxin is responsible for the more severe vascular damage in the colonic tissue observed during *S. dysenteriae* 1 infection. Shiga toxin is also produced, at even higher levels, by enterohemorrhagic *E. coli* and is associated with glomerular damage and kidney failure. The resulting hemolytic uremic syndrome (HUS), linked to Shiga toxin-producing pathogens, may lead to death (Butler, 2012). There is no evidence for the presence of an *stx* allele in the other species of *Shigella* nor in EIEC. Thus production of Shiga toxin may account for the generally more severe infections caused by *S. dysenteriae* 1.

Two pathogenicity islands have been identified in the chromosome of *S. flexneri* and a third has been described in *S. boydii*. The *Shigella* pathogenicity island 1 (SHI-1), which is only present in *S. flexneri* 2a, contains the *set* gene that encodes a second enterotoxin, ShET-1 (Fasano et al., 1995; Rajakumar et al., 1997). The *set* gene is contained within the open reading frame of another gene, *she*, which encodes a protein with putative hemagglutinin and mucinase activity. The SHI-2 pathogenicity island is broadly distributed in the shigellae and contains the *iuc* locus, which encodes the genes for synthesis of aerobactin and its outer membrane receptor (Moss et al., 1999; Vokes et al., 1999). Aerobactin is a hydroxamate siderophore that *S. flexneri* uses to scavenge iron. When the *iuc* locus is inactivated in *S. flexneri*, the aerobactin deficient mutants retain their capacity to invade host cells but are altered in virulence as measured in animal models. These results suggest that aerobactin synthesis is important for bacterial growth within the mammalian host (Nassif et al., 1987). SHI-3 of *S. boydii* contains an aerobactin locus that is 97% identical at the DNA level to that of SHI-2 (Purdy and Payne, 2001). This island shares other genes found on SHI-2 but its location on the genome is different. SHI-2 is inserted downstream of *selC* while SHI-3 is located between *lysU* and *pheU*. Although EIEC synthesize aerobactin, the *iuc* genes are not contained on a SHI-2-like pathogenicity island near *selC*. The chromosomal location of the *iuc* genes in EIEC remains to be determined.

Virulence gene regulation

Overcoming multiple host barriers often requires coordinated expression of multiple, sometimes unlinked, genes encoding complex virulence systems such as the *Shigella* T3SS. Furthermore, it is beneficial to the invading pathogen to express virulence factors required for access to and fitness in host tissues only when these tissues are encountered. Like other bacterial pathogens, *Shigella* and EIEC have evolved regulatory cascades that respond to temperature as an environmental stimulus. This signal coordinates virulence gene expression during transit from the environment into the human gut. Temperature regulation of *Shigella* and EIEC virulence genes operates at the level of gene transcription.

Expression of the *ipa, mxi,* and *spa* operons is induced 50 to 100-fold when the bacteria are shifted from 30°C to 37°C (Hromockyj and Maurelli, 1989). Genes governing this thermal regulation include *virR/hns*, located on the chromosome, and two virulence plasmid genes, *virF* and *virB*. The histone-like protein (H-NS), encoded by *virR/hns*, is a repressor of virulence gene expression. In *hns* mutants, the *ipa, mxi,* and *spa* operons are expressed at the normally repressive temperature (30°C) (Maurelli and Sansonetti, 1988). The *virR/hns* locus is allelic with regulatory loci in other enteric bacteria and, like *virR/hns*, these alleles act as repressors of their respective regulons. The ability of H-NS to bind curved DNA, which is commonly found in promoter regions, accounts for its involvement in gene regulation in response to diverse environmental stimuli such as osmolarity, pH and temperature (Dorman, 2004). H-NS binds to the *virF* and *virB* promoters, as well as to promoters controlled by VirB, and prevents transcription of these genes at 30°C (Belboin and Dorman, 2003).

The product of the *virF* locus is a key element in temperature regulation of the *Shigella* virulence regulon. A helix-turn-helix motif in the carboxyl terminal portion of VirF is characteristic of members of the AraC family of transcriptional activators. VirF binds to sequences upstream of *virB*, which is consistent with its predicted role as a DNA-binding protein (Tobe et al., 1993). Transcription of *virB* is dependent on growth temperature and VirF (Adler et al., 1989; Tobe et al., 1991, 1993). VirB resembles DNA-binding proteins involved in plasmid partitioning and shares homology with ParB of bacteriophage P1 and SopB of plasmid F.

Current models suggest that upon a shift to 37°C, H-NS is displaced from the *virF* operator allowing VirF expression. H-NS is also displaced from the *virB* operator which allows VirF to bind and induce the expression of *virB*, which in turn induces the expression of the *ipa, mxi,* and *spa* operons. These promoters are sensitive to the level of VirB protein in the cell. They require a threshold level of VirB to be reached to displace H-NS before becoming active (Beloin and Dorman, 2003). VirB does not appear to recruit RNA polymerase but rather acts as an anti-repressor of H-NS, freeing promoters for polymerase entry (Turner and Dorman, 2007). Thus, binding of VirF may act as an antagonist to binding by H-NS and thereby provides a mechanism for responding to temperature.

In addition to temperature regulation of genes required for invasion, *Shigella* relies on post-invasion signals following contact with host cells to cue expression of virulence genes required for intracellular growth and survival (Demers et al., 1998). Growth at 37°C is not sufficient for expression of *ipaH* and *virA*. Expression of these genes and at least four additional T3SS secreted effectors is induced after entry into host cells (Kane et al., 2002). Expression of 13 "post-invasion" genes is regulated by the transcriptional activator MxiE along with the co-activator IpgC (Kane et al., 2002; Mavris et al., 2002; Le Gall et al., 2005). Included among these genes are the *osp* genes which dampen the host inflammatory response (see Table 7.1). Thus, this second level of virulence gene regulation ensures expression of factors where they are most effective and maximizes pathogen fitness in host tissues while reprograming the host cell to serve the pathogen's needs.

CLINICAL MANIFESTATIONS OF DISEASE

Infectious dose and transmission

An important aspect of *Shigella* pathogenesis is the extremely low ID_{50}, i.e. the experimentally determined oral dose required to cause disease in 50% of volunteers challenged with a virulent strain of the organism. The ID_{50} for *S. flexneri*, *S. sonnei*, and *S. dysenteriae* is approximately 5,000 organisms. Volunteers become ill when doses as low as 200 organisms are given (DuPont et al., 1989). The low infectious dose of *Shigella* underlies the high communicability of bacillary dysentery and gives the disease great explosive potential for person-to-person spread as well as foodborne and water borne outbreaks of diarrhea. In contrast, at least 10^8 EIEC must be ingested to produce disease (DuPont et al., 1971). The reason for the significantly higher infectious dose for EIEC remains unknown.

Person-to-person transmission of *Shigella* is by the fecal-oral route with most cases of shigellosis being caused by the ingestion of fecally-contaminated food or water. The highest incidence of shigellosis occurs during the warmer months of the year. *Shigella* can spread from infected carriers by several routes including food, fingers, feces, and flies. The latter usually transmit the bacteria from fecal matter to foods. Many types of raw and cooked foods have been implicated as sources and outbreak settings have included restaurants, social gatherings, airlines and cruise ships. The major factor for contamination is the poor personal hygiene of food handlers. Improper storage of contaminated foods is the second most common factor that accounts for foodborne outbreaks due to *Shigella* (Nygren et al., 2012).

Epidemiology

Shigella and EIEC are frank pathogens capable of causing disease in otherwise healthy individuals. Certain populations, however, may be predisposed to infection and disease due to the nature of transmission of the organisms. The annual number of cases of shigellosis worldwide has been estimated to exceed

TABLE 7.1 *Shigella* **virulence-associated loci encoded on the virulence plasmid**

Locus	Protein/activity	Role in virulence
Secreted effectors		
ipaA	Vinculin activation	Invasion; binds vinculin; promotes F-actin depolymerization
ipaB	Translocon pore	Invasion; lyses vacuole; controls T3S secretion; anti-secretion plug
ipaC	Binds β-catenin	Invasion; induces cytoskeletal reorganization; forms membrane pores with IpaB
ipaD	Needle tip	Forms plug to modulate secretion
ipgB1	RhoG mimic	Activates Rac1; stimulates membrane ruffling through the ELMO-Dock180 pathway
ipgB2	RhoA mimic	Triggers formation of actin stress fibers
ipgD	Phosphatidylinositol 4-phosphatase	Induces membrane blebbing and ruffling at entry site in the host cell
icsA	Autotransporter	Catalyzes actin polymerization and intra-, intercellular spread
virA	Activates calpain	Degrades p53 and inhibits p53/NF-κB pathway
ipaH7.8		Escape from vacuole
ipaH9.8	E3 ubiquitin ligase	Downregulates host inflammatory response
ospB		Downregulates host inflammatory response
ospC1		Downregulates host inflammatory response
ospG	Protein kinase	Downregulates host inflammatory response
ospF	Phosphothreonine lyase	Downregulates host inflammatory response
ospI	Glutamine deamidase	Downregulates host inflammatory response
ospZ	Methyltransferase	Downregulates host inflammatory response
ospD3/senA	Enterotoxin	Induces fluid secretion
ospE		Stabilizes focal adhesins and prevents detachment of infected cells
Chaperones		Partners

TABLE 7.1 *Shigella* virulence-associated loci encoded on the virulence plasmid—cont'd

Locus	Protein/activity	Role in virulence
ipgC		IpaB and IpaC
ipgA		IcsA
ipgE		IpgD
spa15		IpaA, IpgB1, OspC3
Structural elements of T3SS		
mxi/spa	~20 proteins	
Regulatory factors		
virF	Transcriptional activator	Temperature regulation of virB
virB	Transcriptional activator	Temperature regulation of *ipa*, *mxi*, and *spa* virulence operons

164 million (with 1.1 million deaths), the vast majority of cases occurring in developing countries (Kotloff et al., 1999). Over-crowding, poor sanitation, substandard hygiene and unsafe water supplies are conditions that contribute to outbreaks of shigellosis in these areas. The greatest frequency of illness due to *Shigella* occurs among children less than 6 years of age (Kotloff et al., 1999). In the US, outbreaks of shigellosis and other diarrheal diseases are increasing in day care centers as more single-parent and two parent working families turn to these facilities to care for their children (Levine and Levine, 1994; Arvelo et al., 2009). Inadequate personal hygiene habits coupled with typical toddler behavior such as oral exploration of the environment create conditions ideally suited to transmission of bacterial, protozoan, and viral pathogens that are spread by fecal contamination. Transmission of *Shigella* in this population is very efficient and the low infectious dose for causing disease increases the risk for shigellosis. Family contacts of day care attendees are also at increased risk. Shigellosis can be endemic in other institutional settings as well. Prisons, mental hospitals, and nursing homes, where crowding and/or insufficient hygienic conditions may exist, can provide an environment for direct fecal-oral contamination.

When natural or man-made disasters destroy a region's sanitary waste treatment and water purification infrastructure, developed countries take on the conditions of a developing country. These conditions place a population at risk for diarrheal diseases such as cholera and dysentery. Examples include famine and political upheaval in Somalia and the war in Bosnia (Levine and Levine, 1994).

In addition, massive population displacement caused by civil wars in Burundi and Rwanda in 1993–94 led to explosive epidemics of diarrheal disease caused by *Vibrio cholerae* and *S. dysenteriae* 1 (Kerneis et al., 2009).

Reservoirs and vehicles of infection

Shigella spp. are highly host-adapted pathogens. Humans are the natural reservoir of *Shigella* infections and non-human primates such as rhesus macaque monkeys are the only animals in which *Shigella* cause disease. Interestingly, several cases of transmission from monkeys to man have been reported. In one instance, three animal caretakers at a monkey house complained of having diarrhea. *S. flexneri* 1b was isolated from stool samples of these employees and further investigation showed that four monkeys were shedding the identical serotype. The disease was apparently spread by direct contact of the caretakers with excrement from the infected monkeys (Kennedy et al., 1993). Asymptomatic carriers of *Shigella* may contribute to the maintenance and spread of this pathogen in developing countries. Two studies, one in Bangladesh (Hossain et al., 1994) and the other in Mexico (Guerrero et al., 1994), showed that *Shigella* could be isolated from stool samples from asymptomatic children under the age of 5 years. *Shigella* were rarely found in infants under the age of 6 months.

Clinical features

The incubation period for shigellosis is 1 to 7 days, but the illness usually begins within 3 days. Signs and symptoms associated with dysentery include fever, severe abdominal cramps, tenesmus, and diarrhea composed of watery stools containing mucus and traces of blood. While nearly all patients with shigellosis experience abdominal pain and diarrhea, fever occurs in about one-third and gross blood in the stools in about 40% of the cases (DuPont, 2005). The feces also contain high numbers of PMNs and viable infectious *Shigella*.

The clinical picture of shigellosis ranges from a mild watery diarrhea to severe dysentery. The dysentery stage of the disease may or may not be preceded by watery diarrhea. This stage probably reflects the transient multiplication of the bacteria as they pass through the small bowel. Enterotoxins produced by *Shigella* and EIEC are believed to induce fluid secretion in the small intestine. In addition, as bacterial invasion and destruction of the colonic mucosa commences, jejunal secretions probably are not effectively reabsorbed in the colon due to transport abnormalities caused by tissue destruction (Kinsey et al., 1976). The dysentery stage of disease correlates with extensive bacterial colonization of the colonic mucosa. The bacteria invade the epithelial cells of the colon, spread from cell to cell but penetrate only as far as the lamina propria. Foci of individually infected cells produce microabscesses that coalesce, forming large abscesses and mucosal ulcerations. Dead cells of the mucosal surface slough off as the infection progresses, leading to the presence of blood, pus and mucus in the stools. Bacillary dysentery is also characterized by massive amounts of PMNs which migrate from the subepithelium into the intestinal lumen. The intense host inflammatory response is initiated by *Shigella* and sustained by release of pro-inflammatory mediators.

S. dysenteriae 1 causes the most severe disease while *S. sonnei* produces the mildest. *S. flexneri* and *S. boydii* infections can be either mild or severe. Volunteer studies and clinical reports show that EIEC produce dysentery with a clinical presentation indistinguishable from that produced by *Shigella* with symptoms ranging from mild to severe (DuPont et al., 1971; Tulloch et al., 1973). Despite the severity of the disease, shigellosis is self-limiting. If left untreated, clinical illness usually persists for 1 to 2 weeks (although it may be as long as a month) and the patient recovers.

Complications

Dysentery can be a very painful and incapacitating disease and is more likely to require hospitalization than other bacterial diarrheas. It is not usually life-threatening and mortality is rare except in malnourished children, immunocompromised individuals, and the elderly (Bennish et al., 1990). However, serious complications can arise from the disease and include severe dehydration, intestinal perforation, toxic megacolon, sepsis, seizures, HUS, and Reiter's syndrome (Bennish, 1991). HUS is a rare but potentially fatal complication associated with infection by *S. dysenteriae* 1 (Butler, 2012). The syndrome is characterized by hemolytic anemia, thrombocytopenia, and acute renal failure. Epidemiological studies suggest that Shiga toxin produced by *S. dysenteriae* 1 is the cause of HUS. This hypothesis is supported by the fact that HUS is also caused by strains of *E. coli*, especially enterohemorrhagic *E. coli*, that produce high levels of Shiga toxin (see Chapter 5) (Melton-Celsa, et al., 2012). Shiga toxin causes HUS by entering the bloodstream and damaging vascular endothelial cells such as those in the kidney (O'Loughlin and Robins-Browne, 2001). Reiter's syndrome, a form of reactive arthritis, is a post infection sequela to shigellosis that is strongly associated with individuals of the HLA-B27 histocompatibility group (Simon et al., 1981). The syndrome is comprised of urethritis, conjunctivitis, and arthritis with the latter being the most dominant symptom. Reactive arthritis is an ill-defined clinical syndrome that also results from infections by several other Gram-negative enteric pathogens (Townes, 2010). A PubMed search of the literature (November 2012) yielded no reports of reactive arthritis as a complication of infection with EIEC.

Treatment, control, and prevention

Stool fluid losses with dysentery are not as massive as with other bacterial diarrheas. However, the diarrhea associated with dysentery, combined with water loss due to fever and decreased water intake due to anorexia, may result in severe dehydration (Bennish et al., 1990). Although intravenous rehydration may be required in treatment of very young and elderly patients, oral intake can generally replace these fluid losses.

The antibiotic of choice for treatment of dysentery is trimethoprim-sulfamethoxazole (DuPont, 2005). However, there is some controversy regarding the use of antibiotics in treating dysentery. Since the infection is self-limited in normally healthy patients and full recovery occurs without the use of antibiotics, drug therapy is usually not indicated. In addition, multiple antibiotic resistance among isolates of *Shigella* is becoming more common. Clinical isolates resistant to ampicillin, trimethoprim-sulfamethoxazole, sulfamethoxazole-sulfisoxazole, tetracycline, nalidixic acid, chloramphenicol, streptomycin, and ciprofloxacin have been found and azithromycin is now being used against highly resistant strains (Shiferaw et al., 2012). However, plasmid-mediated resistance to azithromycin may limit the usefulness of azithromycin for treatment of shigellosis in the near future (Howie et al., 2010). Extensive use of antibiotics selects for drug resistant organisms and, therefore, many believe that antimicrobial therapy for shigellosis should be reserved only for the most severely ill patients. On the other hand, there are persuasive public health arguments for the use of antibiotics to manage shigellosis. Antibiotic treatment limits the duration of disease and shortens the period of fecal excretion of bacteria (Kabir et al., 1986). Since an infected person or asymptomatic carrier can be an index case for person-to-person and food and water borne spread, antibiotic treatment of these individuals can be an effective public health tool to contain the spread of dysentery. However, antibiotics are not a substitute for improved hygienic conditions to contain secondary spread of dysentery. The single most effective means of preventing secondary transmission is hand washing (Arvelo et al., 2009). It should be noted that, unlike with enterohemorrhagic *E. coli*, antibiotic treatment of infection with *S. dysenteriae* 1 is not contraindicated as the gene for Shiga toxin in this organism is not encoded on an inducible bacteriophage.

Despite many years of intensive effort, no licensed vaccine against shigellosis has yet been developed. Live-attenuated, inactivated whole-cell, subunit antigen, and conjugate oligosaccharide vaccines are currently being tested (Steele et al., 2012). An important consideration is whether any vaccine can afford cross protection against all species and serotypes of *Shigella* as well as against EIEC.

CONCLUSION

A century of research has yielded significant insights into the mechanisms used by *Shigella* and EIEC to interact with host tissues. These insights have revealed a family of pathogens that are master cell biologists that reprogram host cells and exploit the actions of host phagocytes to serve the needs of the parasite. Furthermore, the unique evolution of these bacteria has promoted understanding of pathogen evolution and driven discovery of novel evolutionary pathways.

Thus, *Shigella* and EIEC have served as models of several aspects of pathogen biology and provided paradigms that have been applied to other disease causing bacteria. Yet despite the advances in our knowledge of *Shigella* and EIEC pathogenesis, there remain many mysteries at the genetic and cell biology levels. Furthermore, in the absence of an effective vaccine against dysentery, these pathogens will continue to be a formidable global threat to public health.

ACKNOWLEDGMENTS

Thanks to Kym Bliven, Sabrina Joseph, Miranda Gray and Manon Rosselin for their insightful comments. Special thanks to Kym Bliven for drawing Figure 7.1. Research on the genetics of *Shigella* virulence in the Maurelli laboratory is supported by Public Health Service grant AI24656 from the National Institute of Allergy and Infectious Diseases. The opinions or assertions contained herein are the private ones of the author and are not to be construed as official or as reflecting the views of the Department of Defense or the Uniformed Services University.

REFERENCES

Adler, B., Sasakawa, C., Tobe, T., Makino, S., Komatsu, K., Yoshikawa, M., 1989. A dual transcriptional activation system for the 230 kb plasmid genes coding for virulence-associated antigens of *Shigella flexneri*. Mol. Microbiol. 3, 627–635.

Alexander, W.A., Hartman, A.B., Oaks, E.V., Venkatesan, M.M., 1996. Construction and characterization of *virG (icsA)*-deleted *Escherichia coli* K12-*Shigella flexneri* hybrid vaccine strains. Vaccine 14, 1053–1061 doi:0264410X96000023 [pii].

Alto, N.M., Shao, F., Lazar, C.S., et al., 2006. Identification of a bacterial type III effector family with G protein mimicry functions. Cell 124, 133–145.

Arvelo, W., Hinkle, C.J., Nguyen, T.A., et al., 2009. Transmission risk factors and treatment of pediatric shigellosis during a large daycare center-associated outbreak of multidrug resistant *Shigella sonnei*: implications for the management of shigellosis outbreaks children. Pediatr. Infect. Dis. J. 28, 976–980.

Ashida, H., Kim, M., Schmidt-Supprian, M., Ma, A., Ogawa, M., Sasakawa, C., 2010. A bacterial E3 ubiquitin ligase IpaH9.8 targets NEMO/IKKgamma to dampen the host NF-kappaB-mediated inflammatory response. Nat. Cell. Biol. 12, 66–73.

Barbagallo, M., Martino, M.L., Marcocci, L., et al., 2011. A new piece of the *Shigella* pathogenicity puzzle: spermidine accumulationby silencing of the *speG* gene. PLoS One 6, e27226.

Barnoy, S., Baqar, S., Kaminski, R.W., et al., 2011. *Shigella sonnei* vaccine candidates WRSs2 and WRSs3 are as immunogenic as WRSS1, a clinically tested vaccine candidate, in a primate model of infection. Vaccine 29, 6371–6378.

Beloin, C., Dorman, C.J., 2003. An extended role for the nucleoid structuring protein H-NS in the virulence gene regulatory cascade of *Shigella flexneri*. Mol. Microbiol. 47, 825–838.

Bennish, M.L., 1991. Potentially lethal complications of shigellosis. Rev. Infect. Dis. 13 (Suppl. 4), S319–S324.

Bennish, M.L., Harris, J.R., Wojtyniak, B.J., Struelens, M., 1990. Death in shigellosis: incidence and risk factors in hospitalized patients. J. Infect. Dis. 161, 500–506.

Bergounioux, J., Elisee, R., Prunier, A.L., et al., 2012. Calpain activation by the *Shigella flexneri* effector VirA regulates key steps in the formation and life of the bacterium's epithelial niche. Cell Host. Microbe. 11, 240–252.

Bernardini, M.L., Mounier, J., d'Hauteville, H., Coquis-Rondon, M., Sansonetti, P.J., 1989. Identification of *icsA*, a plasmid locus of *Shigella flexneri* that governs bacterial intra- and intercellular spread through interaction with F-actin. Proc. Natl. Acad. Sci. USA 86, 3867–3871.

Blocker, A., Jouihri, N., Larquet, E., et al., 2001. Structure and composition of the *Shigella flexneri* needle complex, a part of its type III secreton. Mol. Microbiol. 39, 652–663.

Bourdet-Sicard, R., Rudiger, M., Jockusch, B.M., Gounon, P., Sansonetti, P.J., Nhieu, G.T., 1999. Binding of the *Shigella* protein IpaA to vinculin induces F-actin depolymerization. EMBO J. 18, 5853–5862.

Brahmbhatt, H.N., Lindberg, A.A., Timmis, K.N., 1992. *Shigella* lipopolysaccharide: structure, genetics, and vaccine development. Curr. Top. Microbiol. Immunol. 180, 45–64.

Brenner, D.J., Fanning, G.R., Johnson, K.E., Citarella, R.V., Falkow, S., 1969. Polynucleotide sequence relationships among members of Enterobacteriaceae. J. Bacteriol. 98, 637–650.

Buchrieser, C., Glaser, P., Rusniok, C., et al., 2000. The virulence plasmid pWR100 and the repertoire of proteins secreted by the type III secretion apparatus of *Shigella flexneri*. Mol. Microbiol. 38, 760–771.

Butler, T., 2012. Haemolytic uraemic syndrome during shigellosis. Trans. R. Soc. Trop. Med. Hyg. 106, 395–399.

Casalino, M., Latella, M.C., Prosseda, G., Colonna, B., 2003. CadC is the preferential target of a convergent evolution driving enteroinvasive *Escherichia coli* toward a lysine decarboxylase-defective phenotype. Infect. Immun. 71, 5472–5479.

Cilia, V., Lafay, B., Christen, R., 1996. Sequence heterogeneities among 16S ribosomal RNA sequences, and their effect on phylogenetic analyses at the species level. Mol. Biol. Evol. 13, 451–461.

Day Jr., W.A., Fernandez, R.E., Maurelli, A.T., 2001. Pathoadaptive mutations that enhance virulence: genetic organization of the *cadA* regions of *Shigella* spp. Infect. Immun. 69, 7471–7480.

Demers, B., Sansonetti, P.J., Parsot, C., 1998. Induction of type III secretion in *Shigella flexneri* is associated with differential control of transcription of genes encoding secreted proteins. EMBO J. 17, 2894–2903.

Dorman, C.J., 2004. H-NS: a universal regulator for a dynamic genome. Nat. Rev. Microbiol. 2, 391–400.

DuPont, H.L., 2005. *Shigella* species (Bacillary dysentery). In: Mandell, G.L., Bennett, J.E., Dolin, R. (Eds.), Principles and Practice of Infectious Diseases, sixth ed., Churchill Livingstone Inc., New York, NY, pp. 2655–2661.

DuPont, H.L., Formal, S.B., Hornick, R.B., et al., 1971. Pathogenesis of *Escherichia coli* diarrhea. N. Engl. J. Med. 285, 1–9.

DuPont, H.L., Levine, M.M., Hornick, R.B., Formal, S.B., 1989. Inoculum size in shigellosis and implications for expected mode of transmission. J. Infect. Dis. 159, 1126–1128.

Egile, C., d'Hauteville, H., Parsot, C., Sansonetti, P.J., 1997. SopA, the outer membrane protease responsible for polar localization of IcsA in *Shigella flexneri*. Mol. Microbiol. 23, 1063–1073.

Egile, C., Loisel, T.P., Laurent, V., et al., 1999. Activation of the CDC42 effector N-WASP by the *Shigella flexneri* IcsA protein promotes actin nucleation by Arp2/3 complex and bacterial actin-based motility. J. Cell. Biol. 146, 1319–1332.

Epler, C.R., Dickenson, N.E., Olive, A.J., Picking, W.L., Picking, W.D., 2009. Liposomes recruit IpaC to the *Shigella flexneri* type III secretion apparatus needle as a final step in secretion induction. Infect. Immun. 77, 2754–2761.

Espina, M., Olive, A.J., Kenjale, R., et al., 2006. IpaD localizes to the tip of the type III secretion system needle of *Shigella flexneri*. Infect. Immun. 74, 4391–4400.

Ewing, E.H., 1986. Edwards and Ewing's Identification of Enterobacteriacae, fourth ed. Elsevier Science Publishing Co. Inc, New York.

Faherty, C.S., Maurelli, A.T., 2009. Spa15 of *Shigella flexneri* is secreted through the type III secretion system and prevents staurosporine-induced apoptosis. Infect. Immun. 77, 5281–5290.

Falkow, S., Schneider, H., Baron, L.S., Formal, S.B., 1963. Virulence of *Escherichia-Shigella* genetic hybrids for the guinea pig. J. Bacteriol. 86, 1251–1258.

Fasano, A., Noriega, F.R., Maneval Jr., D.R., et al., 1995. *Shigella* enterotoxin 1: an enterotoxin of *Shigella flexneri* 2a active in rabbit small intestine in vivo and in vitro. J. Clin. Invest. 95, 2853–2861.

Fernandez-Prada, C.M., Hoover, D.L., Tall, B.D., Hartman, A.B., Kopelowitz, J., Venkatesan, M.M., 2000. *Shigella flexneri* IpaH(7.8) facilitates escape of virulent bacteria from the endocytic vacuoles of mouse and human macrophages. Infect. Immun. 68, 3608–3619.

Formal, S.B., LaBrec, E.H., Kent, T.H., Falkow, S., 1965. Abortive intestinal infection with an *Escherichia coli-Shigella flexneri* hybrid strain. J. Bacteriol. 89, 1374–1382.

Galan, J.E., Wolf-Watz, H., 2006. Protein delivery into eukaryotic cells by type III secretion machines. Nature 444, 567–573.

Germane, K.L., Ohi, R., Goldberg, M.B., Spiller, B.W., 2008. Structural and functional studies indicate that *Shigella* VirA is not a protease and does not directly destabilize microtubules. Biochemistry 47, 10241–10243.

Goldberg, M.B., Barzu, O., Parsot, C., Sansonetti, P.J., 1993. Unipolar localization and ATPase activity of IcsA, a *Shigella flexneri* protein involved in intracellular movement. J. Bacteriol. 175, 2189–2196.

Gorden, J., Small, P.L.C., 1993. Acid resistance in enteric bacteria. Infect. Immun. 61, 364–367.

Guerrero, L., Calva, J.J., Morrow, A.L., et al., 1994. Asymptomatic *Shigella* infections in a cohort of Mexican children younger than two years of age. Pediatr. Infect. Dis. J. 13, 597–602.

Hale, T.L., Oaks, E.V., Formal, S.B., 1985. Identification and antigenic characterization of virulence-associated, plasmid-coded proteins of *Shigella* spp. and enteroinvasive *Escherichia coli*. Infect. Immun. 50, 620–629.

Handa, Y., Suzuki, M., Ohya, K., et al., 2007. *Shigella* IpgB1 promotes bacterial entry through the ELMO-Dock180 machinery. Nat. Cell. Biol. 9, 121–128.

Harris, J.R., Wachsmuth, I.K., Davis, B.R., Cohen, M.L., 1982. High-molecular-weight plasmid correlates with *Escherichia coli* enteroinvasiveness. Infect. Immun. 37, 1295–1298.

Hartman, A.B., Venkatesan, M., Oaks, E.V., Buysse, J.M., 1990. Sequence and molecular characterization of a multicopy invasion plasmid antigen gene, *ipaH*, of *Shigella flexneri*. J. Bacteriol. 172, 1905–1915.

Hayward, R.D., Cain, R.J., McGhie, E.J., Phillips, N., Garner, M.J., Koronakis, V., 2005. Cholesterol binding by the bacterial type III translocon is essential for virulence effector delivery into mammalian cells. Mol. Microbiol. 56, 590–603.

High, N., Mounier, J., Prevost, M.C., Sansonetti, P.J., 1992. IpaB of *Shigella flexneri* causes entry into epithelial cells and escape from the phagocytic vacuole. EMBO J. 11, 1991–1999.

Hossain, M.A., Hasan, K.Z., Albert, M.J., 1994. *Shigella* carriers among non-diarrhoeal children in an endemic area of shigellosis in Bangladesh. Trop. Geogr. Med. 46, 40–42.

Howie, R.L., Folster, J.P., Bowen, A., Barzilay, E.J., Whichard, J.M., 2010. Reduced azithromycin susceptibility in *Shigella sonnei*, United States. Microb. Drug Res. 16, 245–248.

Hromockyj, A.E., Maurelli, A.T., 1989. Identification of *Shigella* invasion genes by isolation of temperature-regulated *inv::lacZ* operon fusions. Infect. Immun. 57, 2963–2970.

Jennison, A.V., Verma, N.K., 2007. The acid-resistance pathways of *Shigella flexneri* 2457T. Microbiology 153, 2593–2602.

Jin, Q., Yuan, Z., Xu, J., et al., 2002. Genome sequence of *Shigella flexneri* 2a: insights into pathogenicity through comparison with genomes of *Escherichia coli* K12 and O157. Nucleic Acids Res. 30, 4432–4441.

Kabir, I., Butler, T., Khanam, A., 1986. Comparative efficacies of single intravenous doses of ceftriaxone and ampicillin for shigellosis in a placebo-controlled trial. Antimicrob. Agents. Chemother. 29, 645–648.

Kane, C.D., Schuch, R., Day Jr., W.A., Maurelli, A.T., 2002. MxiE regulates intracellular expression of factors secreted by the *Shigella flexneri* 2a type III secretion system. J. Bacteriol. 184, 4409–4419.

Kayath, C.A., Hussey, S., El hajjami, N., Nagra, K., Philpott, D., Allaoui, A., 2010. Escape of intracellular *Shigella* from autophagy requires binding to cholesterol through the type III effector, IcsB. Microbes. Infect. 12, 956–966.

Kennedy, F.M., Astbury, J., Needham, J.R., Cheasty, T., 1993. Shigellosis due to occupational contact with non-human primates. Epidemiol. Infect. 110, 247–251.

Kerneis, S., Guerin, P.J., von Seidlein, L., Legros, D., Grais, R.F., 2009. A look back at an ongoing problem: *Shigella dysenteriae* type 1 epidemics in refugee settings in Central Africa (1993–1995). PLoS One 4, e4494.

Kim, D.W., Lenzen, G., Page, A.L., Legrain, P., Sansonetti, P.J., Parsot, C., 2005. The *Shigella flexneri* effector OspG interferes with innate immune responses by targeting ubiquitin-conjugating enzymes. Proc. Natl. Acad. Sci. USA 102, 14046–14051.

Kim, M., Ogawa, M., Fujita, Y., et al., 2009. Bacteria hijack integrin-linked kinase to stabilize focal adhesions and block cell detachment. Nature 459, 578–582.

Kinsey, M.D., Formal, S.B., Dammin, G.J., Giannella, R.A., 1976. Fluid and electrolyte transport in rhesus monkeys challenged intracecally with *Shigella flexneri* 2a. Infect. Immun. 14, 368–371.

Kotloff, K.L., Winickoff, J.P., Ivanoff, B., et al., 1999. Global burden of *Shigella* infections: implications for vaccine development and implementation of control strategies. Bull. World Health Organ. 77, 651–666.

LaBrec, E.H., Schneider, H., Magnani, T.J., Formal, S.B., 1964. Epithelial cell penetration as an essential step in the pathogenesis of bacillary dysentery. J. Bacteriol. 88, 1503–1518.

Lan, R., Alles, M.C., Donohoe, K., Martinez, M.B., Reeves, P.R., 2004. Molecular evolutionary relationships of enteroinvasive *Escherichia coli* and *Shigella* spp. Infect. Immun. 72, 5080–5088.

Lan, R., Lumb, B., Ryan, D., Reeves, P.R., 2001. Molecular evolution of large virulence plasmid in *Shigella* clones and enteroinvasive *Escherichia coli*. Infect. Immun. 69, 6303–6309.

Le Gall, T., Mavris, M., Martino, M.C., Bernardini, M.L., Denamur, E., Parsot, C., 2005. Analysis of virulence plasmid gene expression defines three classes of effectors in the type III secretion system of *Shigella flexneri*. Microbiology 151, 951–962.

Levine, M.M., Levine, O.S., 1994. Changes in human ecology and behavior in relation to the emergence of diarrheal diseases, including cholera. Proc. Natl. Acad. Sci. USA 91, 2390–2394.

Magdalena, J., Hachani, A., Chamekh, M., et al., 2002. Spa32 regulates a switch in substrate specificity of the type III secreton of *Shigella flexneri* from needle components to Ipa proteins. J. Bacteriol. 184, 3433–3441.

Makino, S., Sasakawa, C., Kamata, K., Kurata, T., Yoshikawa, M., 1986. A genetic determinant required for continuous reinfection of adjacent cells on large plasmid in *S. flexneri* 2a. Cell 46, 551–555.

Maurelli, A.T., 2007. Black holes, antivirulence genes, and gene inactivation in the evolution of bacterial pathogens. FEMS Microbiol. Lett. 267, 1–8.

Maurelli, A.T., Baudry, B., d'Hauteville, H., Hale, T.L., Sansonetti, P.J., 1985. Cloning of plasmid DNA sequences involved in invasion of HeLa cells by *Shigella flexneri*. Infect. Immun. 49, 164–171.

Maurelli, A.T., Blackmon, B., Curtiss, R.I., 1984. Temperature-dependent expression of virulence genes in *Shigella* species. Infect. Immun. 43, 195–201.

Maurelli, A.T., Fernandez, R.E., Bloch, C.A., Rode, C.K., Fasano, A., 1998. Black holes and bacterial pathogenicity: a large genomic deletion that enhances the virulence of *Shigella* spp. and enteroinvasive *Escherichia coli*. Proc. Natl. Acad. Sci. USA 95, 3943–3948.

Maurelli, A.T., Sansonetti, P.J., 1988. Identification of a chromosomal gene controlling temperature-regulated expression of *Shigella* virulence. Proc. Natl. Acad. Sci. USA 85, 2820–2824.

Mavris, M., Page, A.L., Tournebize, R., Demers, B., Sansonetti, P., Parsot, C., 2002. Regulation of transcription by the activity of the *Shigella flexneri* type III secretion apparatus. Mol. Microbiol. 43, 1543–1553.

McCormick, B.A., Fernandez, M.I., Siber, A.M., Maurelli, A.T., 1999. Inhibition of *Shigella flexneri*-induced transepithelial migration of polymorphonuclear leucocytes by cadaverine. Cell Microbiol. 1, 143–155.

McCormick, B.A., Siber, A.M., Maurelli, A.T., 1998. Requirement of the *Shigella flexneri* virulence plasmid in the ability to induce trafficking of neutrophils across polarized monolayers of the intestinal epithelium. Infect. Immun. 66, 4237–4243.

Melton-Celsa, A., Mohawk, K., Teel, L., O'Brien, A., 2012. Pathogenesis of Shiga-toxin producing *Escherichia coli*. Curr. Top. Microbiol. Immunol. 357, 67–103.

Menard, R., Prevost, M.C., Gounon, P., Sansonetti, P., Dehio, C., 1996. The secreted Ipa complex of *Shigella flexneri* promotes entry into mammalian cells. Proc. Natl. Acad. Sci. USA 93, 1254–1258.

Menard, R., Sansonetti, P.J., Parsot, C., 1993. Nonpolar mutagenesis of the *ipa* genes defines IpaB, IpaC, and IpaD as effectors of *Shigella flexneri* entry into epithelial cells. J. Bacteriol. 175, 5899–5906.

Menard, R., Sansonetti, P., Parsot, C., 1994a. The secretion of the *Shigella flexneri* Ipa invasins is activated by epithelial cells and controlled by IpaB and IpaD. EMBO J. 13, 5293–5302.

Menard, R., Sansonetti, P., Parsot, C., Vasselon, T., 1994b. Extracellular association and cytoplasmic partitioning of the IpaB and IpaC invasins of *S. flexneri*. Cell 79, 515–525.

Miura, M., Terajima, J., Izumiya, H., Mitobe, J., Komano, T., Watanabe, H., 2006. OspE2 of *Shigella sonnei* is required for the maintenance of cell architecture of bacterium-infected cells. Infect. Immun. 74, 2587–2595.

Monack, D.M., Theriot, J.A., 2001. Actin-based motility is sufficient for bacterial membrane protrusion formation and host cell uptake. Cell Microbiol. 3, 633–647.

Moss, J.E., Cardozo, T.J., Zychlinsky, A., Groisman, E.A., 1999. The *selC*-associated SHI-2 pathogenicity island of *Shigella flexneri*. Mol. Microbiol. 33, 74–83.

Mumy, K.L., Bien, J.D., Pazos, M.A., Gronert, K., Hurley, B.P., McCormick, B.A., 2008. Distinct isoforms of phospholipase A2 mediate the ability of *Salmonella enterica* serotype typhimurium and *Shigella flexneri* to induce the transepithelial migration of neutrophils. Infect. Immun. 76, 3614–3627.

Nassif, X., Mazert, M.C., Mounier, J., Sansonetti, P.J., 1987. Evaluation with an *iuc*::Tn*10* mutant of the role of aerobactin production in the virulence of *Shigella flexneri*. Infect. Immun. 55, 1963–1969.

Nataro, J.P., Seriwatana, J., Fasano, A., et al., 1995. Identification and cloning of a novel plasmid-encoded enterotoxin of enteroinvasive *Escherichia coli* and *Shigella* strains. Infect. Immun. 63, 4721–4728.

Niebuhr, K., Giuriato, S., Pedron, T., et al., 2002. Conversion of PtdIns(4,5)P(2) into PtdIns(5)P by the *S. flexneri* effector IpgD reorganizes host cell morphology. EMBO J. 21, 5069–5078.

Nygren, B.L., Schilling, K.A., Blanton, E.M., Silk, B.J., Cole, D.J., Mintz, E.D., 2012. Foodborne outbreaks of shigellosis in the USA, 1998–2008. Epidemiol. Infect. 1–9.

O'Loughlin, E.V., Robins-Browne, R.M., 2001. Effect of Shiga toxin and Shiga-like toxins on eukaryotic cells. Microbes. Infect. 3, 493–507.

Oaks, E.V., Hale, T.L., Formal, S.B., 1986. Serum immune response to *Shigella* protein antigens in rhesus monkeys and humans infected with *Shigella* spp. Infect. Immun. 53, 57–63.

Ochman, H., Whittam, T.S., Caugant, D.A., Selander, R.K., 1983. Enzyme polymorphism and genetic population structure in *Escherichia coli* and *Shigella*. J. Gen. Microbiol. 129 (Pt 9), 2715–2726.

Ogawa, M., Sasakawa, C., 2006. Bacterial evasion of the autophagic defense system. Curr. Opin. Microbiol. 9, 62–68.

Ogawa, M., Yoshimori, T., Suzuki, T., Sagara, H., Mizushima, N., Sasakawa, C., 2005. Escape of intracellular *Shigella* from autophagy. Science 307, 727–731.

Ohya, K., Handa, Y., Ogawa, M., Suzuki, M., Sasakawa, C., 2005. IpgB1 is a novel *Shigella* effector protein involved in bacterial invasion of host cells. Its activity to promote membrane ruffling via Rac1 and Cdc42 activation. J. Biol. Chem. 280, 24022–24034.

Olive, A.J., Kenjale, R., Espina, M., Moore, D.S., Picking, W.L., Picking, W.D., 2007. Bile salts stimulate recruitment of IpaB to the *Shigella flexneri* surface, where it colocalizes with IpaD at the tip of the type III secretion needle. Infect. Immun. 75, 2626–2629.

Parsot, C., 2005. *Shigella* spp. and enteroinvasive *Escherichia coli* pathogenicity factors. FEMS Microbiol. Lett. 252, 11–18.

Perdomo, J.J., Gounon, P., Sansonetti, P.J., 1994a. Polymorphonuclear leukocyte transmigration promotes invasion of colonic epithelial monolayer by *Shigella flexneri*. J. Clin. Invest. 93, 633–643.

Perdomo, O.J., Cavaillon, J.M., Huerre, M., Ohayon, H., Gounon, P., Sansonetti, P.J., 1994b. Acute inflammation causes epithelial invasion and mucosal destruction in experimental shigellosis. J. Exp. Med. 180, 1307–1319.

Prunier, A.L., Schuch, R., Fernandez, R.E., et al., 2007. *nadA* and *nadB* of *Shigella flexneri* 5a are antivirulence loci responsible for the synthesis of quinolinate, a small molecule inhibitor of *Shigella* pathogenicity. Microbiology 153, 2363–2372.

Pupo, G.M., Karaolis, D.K., Lan, R., Reeves, P.R., 1997. Evolutionary relationships among pathogenic and nonpathogenic *Escherichia coli* strains inferred from multilocus enzyme electrophoresis and *mdh* sequence studies. Infect. Immun. 65, 2685–2692.

Pupo, G.M., Lan, R., Reeves, P.R., 2000. Multiple independent origins of *Shigella* clones of *Escherichia coli* and convergent evolution of many of their characteristics. Proc. Natl. Acad. Sci. USA 97, 10567–10572.

Purdy, G.E., Payne, S.M., 2001. The SHI-3 iron transport island of *Shigella boydii* 0-1392 carries the genes for aerobactin synthesis and transport. J. Bacteriol. 183, 4176–4182.

Rajakumar, K., Sasakawa, C., Adler, B., 1997. Use of a novel approach, termed island probing, identifies the *Shigella flexneri she* pathogenicity island which encodes a homolog of the immunoglobulin A protease-like family of proteins. Infect. Immun. 65, 4606–4614.

Ray, K., Bobard, A., Danckaert, A., et al., 2010. Tracking the dynamic interplay between bacterial and host factors during pathogen-induced vacuole rupture in real time. Cell Microbiol. 12, 545–556.

Reiterer, V., Grossniklaus, L., Tschon, T., Kasper, C.A., Sorg, I., Arrieumerlou, C., 2011. *Shigella flexneri* type III secreted effector OspF reveals new crosstalks of proinflammatory signaling pathways during bacterial infection. Cell Signal 23, 1188–1196.

Rohde, J.R., Breitkreutz, A., Chenal, A., Sansonetti, P.J., Parsot, C., 2007. Type III secretion effectors of the IpaH family are E3 ubiquitin ligases. Cell Host. Microbe. 1, 77–83.

Rolland, K., Lambert-Zechovsky, N., Picard, B., Denamur, E., 1998. *Shigella* and enteroinvasive *Escherichia coli* strains are derived from distinct ancestral strains of *E. coli*. Microbiology 144 (Pt 9), 2667–2672.

Rout, W.R., Formal, S.B., Giannella, R.A., Dammin, G.J., 1975. Pathophysiology of *Shigella* diarrhea in the rhesus monkey: intestinal transport, morphological, and bacteriological studies. Gastroenterology 68, 270–278.

Sanada, T., Kim, M., Mimuro, H., et al., 2012. The *Shigella flexneri* effector OspI deamidates UBC13 to dampen the inflammatory response. Nature 483, 623–626.

Sandlin, R.C., Goldberg, M.B., Maurelli, A.T., 1996. Effect of O side-chain length and composition on the virulence of *Shigella flexneri* 2a. Mol. Microbiol. 22, 63–73.

Sandlin, R.C., Maurelli, A.T., 1999. Establishment of unipolar localization of IcsA in *Shigella flexneri* 2a is not dependent on virulence plasmid determinants. Infect. Immun. 67, 350–356.

Sansonetti, P.J., Arondel, J., Fontaine, A., d'Hauteville, H., Bernardini, M.L., 1991. *ompB* (osmoregulation) and *icsA* (cell-to-cell spread) mutants of *Shigella flexneri*: vaccine candidates and probes to study the pathogenesis of shigellosis. Vaccine 9, 416–422.

Sansonetti, P.J., Arondel, J., Huerre, M., Harada, A., Matsushima, K., 1999. Interleukin-8 controls bacterial transepithelial translocation at the cost of epithelial destruction in experimental shigellosis. Infect. Immun. 67, 1471–1480.

Sansonetti, P.J., d'Hauteville, H., Formal, S.B., Toucas, M., 1982a. Plasmid-mediated invasiveness of *Shigella*-like *Escherichia coli*. Ann. Microbiol. (Paris). 133, 351–355.

Sansonetti, P.J., Hale, T.L., Oaks, E.V., 1985. Genetics of virulence in enteroinvasive *Escherichia coli*. In: Schlessinger, D. (Ed.), Microbiology-1985, American Society for Microbiology, Washington, D.C., pp. 74–77.

Sansonetti, P.J., Kopecko, D.J., Formal, S.B., 1981. *Shigella sonnei* plasmids: evidence that a large plasmid is necessary for virulence. Infect. Immun. 34, 75–83.

Sansonetti, P.J., Kopecko, D.J., Formal, S.B., 1982b. Involvement of a plasmid in the invasive ability of *Shigella flexneri*. Infect. Immun. 35, 852–860.

Schroeder, G.N., Hilbi, H., 2008. Molecular pathogenesis of *Shigella* spp.: controlling host cell signaling, invasion, and death by type III secretion. Clin. Microbiol. Rev. 21, 134–156.

Shere, K.D., Sallustio, S., Manessis, A., D'Aversa, T.G., Goldberg, M.B., 1997. Disruption of IcsP, the major *Shigella* protease that cleaves IcsA, accelerates actin-based motility. Mol. Microbiol. 25, 451–462.

Shiferaw, B., Solghan, S., Palmer, A., et al., 2012. Antimicrobial susceptibility patterns of *Shigella* isolates in Foodborne Diseases Active Surveillance Network (FoodNet) sites, 2000–2010. Clin. Infect. Dis. 54 (Suppl. 5), S458–S463.

Shu, S., Setianingrum, E., Zhao, L., et al., 2000. I-CeuI fragment analysis of the *Shigella* species: evidence for large-scale chromosome rearrangement in *S. dysenteriae* and *S. flexneri*. FEMS Microbiol. Lett. 182, 93–98.

Silva, R.M., Toledo, M.R., Trabulsi, L.R., 1980. Biochemical and cultural characteristics of invasive *Escherichia coli*. J. Clin. Microbiol. 11, 441–444.

Simon, D.G., Kaslow, R.A., Rosenbaum, J., Kaye, R.L., Calin, A., 1981. Reiter's syndrome following epidemic shigellosis. J. Rheumatol. 8, 969–973.

Sims, G.E., Kim, S.H., 2011. Whole-genome phylogeny of *Escherichia coli/Shigella* group by feature frequency profiles (FFPs). Proc. Natl. Acad. Sci. USA 108, 8329–8334.

Skoudy, A., Mounier, J., Aruffo, A., et al., 2000. CD44 binds to the *Shigella* IpaB protein and participates in bacterial invasion of epithelial cells. Cell Microbiol. 2, 19–33.

Small, P.L., Falkow, S., 1988. Identification of regions on a 230-kilobase plasmid from enteroinvasive *Escherichia coli* that are required for entry into HEp-2 cells. Infect. Immun. 56, 225–229.

Sokurenko, E.V., Hasty, D.L., Dykhuizen, D.E., 1999. Pathoadaptive mutations: gene loss and variation in bacterial pathogens. Trends. Microbiol. 7, 191–195.

Steele, D., Riddle, M., van de Verg, L., Bourgeois, L., 2012. Vaccines for enteric diseases: a meeting summary. Expert Rev. Vaccines 11, 407–409.

Strockbine, N.A., Maurelli, A.T., 2005. Shigella. In: Garrity, G.M. (Ed.), Bergey's Manual of Systematic Bacteriology, second ed. The Proteobacteria Part B: The Gammaproteobacteria, vol. 2. Springer, New York, pp. 811–823.

Suzuki, T., Mimuro, H., Suetsugu, S., Miki, H., Takenawa, T., Sasakawa, C., 2002. Neural Wiskott-Aldrich syndrome protein (N-WASP) is the specific ligand for *Shigella* VirG among the WASP family and determines the host cell type allowing actin-based spreading. Cell Microbiol. 4, 223–233.

Suzuki, T., Nakanishi, K., Tsutsui, H., et al., 2005. A novel caspase-1/toll-like receptor 4-independent pathway of cell death induced by cytosolic *Shigella* in infected macrophages. J. Biol. Chem. 280, 14042–14050.

Tamano, K., Aizawa, S., Katayama, E., et al., 2000. Supramolecular structure of the *Shigella* type III secretion machinery: the needle part is changeable in length and essential for delivery of effectors. EMBO J. 19, 3876–3887.

Tamano, K., Katayama, E., Toyotome, T., Sasakawa, C., 2002. *Shigella* Spa32 is an essential secretory protein for functional type III secretion machinery and uniformity of its needle length. J. Bacteriol. 184, 1244–1252.

Tobe, T., Nagai, S., Okada, N., Adler, B., Yoshikawa, M., Sasakawa, C., 1991. Temperature-regulated expression of invasion genes in *Shigella flexneri* is controlled through the transcriptional activation of the *virB* gene on the large plasmid. Mol. Microbiol. 5, 887–893.

Tobe, T., Yoshikawa, M., Mizuno, T., Sasakawa, C., 1993. Transcriptional control of the invasion regulatory gene *virB* of *Shigella flexneri*: activation by *virF* and repression by H-NS. J. Bacteriol. 175, 6142–6149.

Touchon, M., Hoede, C., Tenaillon, O., et al., 2009. Organised genome dynamics in the *Escherichia coli* species results in highly diverse adaptive paths. PLoS Genet. 5, e1000344.

Townes, J.M., 2010. Reactive arthritis after enteric infections in the United States: the problem of definition. Clin. Infect. Dis. 50, 247–254.

Tran, V.N., Ben Ze'ev, A., Sansonetti, P.J., 1997. Modulation of bacterial entry into epithelial cells by association between vinculin and the *Shigella* IpaA invasin. EMBO J. 16, 2717–2729.

Tran, V.N., Caron, E., Hall, A., Sansonetti, P.J., 1999. IpaC induces actin polymerization and filopodia formation during *Shigella* entry into epithelial cells. EMBO J. 18, 3249–3262.

Tulloch Jr., E.F., Ryan, K.J., Formal, S.B., Franklin, F.A., 1973. Invasive enteropathic *Escherichia coli* dysentery. An outbreak in 28 adults. Ann. Intern. Med. 79, 13–17.

Turner, E.C., Dorman, C.J., 2007. H-NS antagonism in *Shigella flexneri* by VirB, a virulence gene transcription regulator that is closely related to plasmid partition factors. J. Bacteriol. 189, 3403–3413.

Uchiya, K., Tobe, T., Komatsu, K., et al., 1995. Identification of a novel virulence gene, *virA*, on the large plasmid of *Shigella*, involved in invasion and intercellular spreading. Mol. Microbiol. 17, 241–250.

van den Beld, M.J., Reubsaet, F.A., 2012. Differentiation between *Shigella*, enteroinvasive *Escherichia coli* (EIEC) and noninvasive *Escherichia coli*. Eur. J. Clin. Microbiol. Infect. Dis. 31, 899–904.

Venkatesan, M.M., Buysse, J.M., Kopecko, D.J., 1989. Use of Shigella flexneri *ipaC* and *ipaH* gene sequences for the general identification of *Shigella* spp. and enteroinvasive *Escherichia coli*. J. Clin. Microbiol. 27, 2687–2691.

Venkatesan, M.M., Buysse, J.M., Oaks, E.V., 1992. Surface presentation of *Shigella flexneri* invasion plasmid antigens requires the products of the *spa* locus. J. Bacteriol. 174, 1990–2001.

Venkatesan, M.M., Goldberg, M.B., Rose, D.J., Grotbeck, E.J., Burland, V., Blattner, F.R., 2001. Complete DNA sequence and analysis of the large virulence plasmid of *Shigella flexneri*. Infect. Immun. 69, 3271–3285.

Vokes, S.A., Reeves, S.A., Torres, A.G., Payne, S.M., 1999. The aerobactin iron transport system genes in *Shigella flexneri* are present within a pathogenicity island. Mol. Microbiol. 33, 63–73.

Watarai, M., Funato, S., Sasakawa, C., 1996. Interaction of Ipa proteins of *Shigella flexneri* with alpha5beta1 integrin promotes entry of the bacteria into mammalian cells. J. Exp. Med. 183, 991–999.

Watarai, M., Tobe, T., Yoshikawa, M., Sasakawa, C., 1995. Contact of *Shigella* with host cells triggers release of Ipa invasins and is an essential function of invasiveness. EMBO J. 14, 2461–2470.

Yang, F., Yang, J., Zhang, X., et al., 2005. Genome dynamics and diversity of *Shigella* species, the etiologic agents of bacillary dysentery. Nucleic Acids Res. 33, 6445–6458.

Yoshida, S., Handa, Y., Suzuki, T., et al., 2006. Microtubule-severing activity of *Shigella* is pivotal for intercellular spreading. Science 314, 985–989.

Zhang, L., Ding, X., Cui, J., et al., 2012. Cysteine methylation disrupts ubiquitin-chain sensing in NF-kappaB activation. Nature 481, 204–208.

Zurawski, D.V., Mitsuhata, C., Mumy, K.L., McCormick, B.A., Maurelli, A.T., 2006. OspF and OspC1 are *Shigella flexneri* type III secretion system effectors that are required for postinvasion aspects of virulence. Infect. Immun. 74, 5964–5976.

Zurawski, D.V., Mumy, K.L., Badea, L., et al., 2008. The NleE/OspZ family of effector proteins is required for polymorphonuclear transepithelial migration, a characteristic shared by enteropathogenic *Escherichia coli* and *Shigella flexneri* infections. Infect. Immun. 76, 369–379.

Zurawski, D.V., Mumy, K.L., Faherty, C.S., McCormick, B.A., Maurelli, A.T., 2009. *Shigella flexneri* type III secretion system effectors OspB and OspF target the nucleus to downregulate the host inflammatory response via interactions with retinoblastoma protein. Mol. Microbiol. 71, 350–368.

Zychlinsky, A., Kenny, B., Menard, R., Prevost, M.C., Holland, I.B., Sansonetti, P.J., 1994. IpaB mediates macrophage apoptosis induced by *Shigella flexneri*. Mol. Microbiol. 11, 619–627.

Zychlinsky, A., Prevost, M.C., Sansonetti, P.J., 1992. *Shigella flexneri* induces apoptosis in infected macrophages. Nature 358, 167–169.

Zychlinsky, A., Sansonetti, P.J., 1997. Apoptosis as a proinflammatory event: what can we learn from bacteria-induced cell death? Trends. Microbiol. 5, 201–204.

Chapter 8

Enteroaggregative
Escherichia coli

Nadia Boisen[1], Karen A. Krogfelt[2], and James P. Nataro[1]

[1]*University of Virginia School of Medicine, Charlottesville, VA, USA,* [2]*Statens Serum Institut, Copenhagen, Denmark*

INTRODUCTION

Diarrhea is a significant cause of morbidity and mortality worldwide, particularly in children under 5 years of age. In addition to almost one million directly attributable deaths per year (Black et al., 2010), morbidity associated with repeated episodes during childhood diarrhea can be lifelong (Petri et al., 2008). Repeated bouts of diarrhea during infancy result in malabsorption of nutrients leading to developmental disabilities, which include growth shortfalls and impaired cognition.

Collectively, the diarrheagenic *Escherichia coli* (DEC) represent the most common bacterial pathogen worldwide (Farthing, 2000; Wanke, 2001; Ina et al., 2003). Included among the DEC is enteroaggregative *Escherichia coli* (EAEC), a pathogen of emerging significance. However, identification and isolation of pathogenic strains remain elusive, given an as yet imperfect definition of this pathotype. Thus, the global burden of diarrheal diseases resulting from EAEC may be vastly underestimated (Bryce et al., 2005). This underestimation is also true of the sequelae of persistent diarrhea for which EAEC is reportedly a prevalent cause (Fang et al., 1995; Lima et al., 2000). Hence, detailed understanding of EAEC heterogeneity and pathogenic mechanisms are important.

ENTEROAGGREGATIVE *ESCHERICHIA COLI* (EAEC) HISTORY

EAEC was first described by the senior author in 1983, while a student in the lab of Dr. James Kaper. During adhesion studies of *E. coli* to Hep-2 cells it was observed that three distinct adhesion patterns could be described as diffuse, localized and aggregative patterns. The role of these patterns in disease was yet to be elucidated. Thus, the first association of EAEC with diarrheal disease was published in 1987, as part of a prospective study of pediatric diarrhea in

Escherichia coli. http://dx.doi.org/10.1016/B978-0-12-397048-0.00008-5

Chile (Nataro et al., 1987). Shortly thereafter, EAEC was associated with persistent diarrhea among children in three studies (Bhan et al., 1989a,b; Cravioto et al., 1991).

EAEC is a pathotype of DEC defined as *E. coli* that do not secrete the heat-stable (ST) or heat-labile (LT) toxins of enterotoxigenic *Escherichia coli* (ETEC), and which manifest a characteristic aggregative or 'stacked brick' pattern (AA) of adherence to HEp2-cells in culture (Figure 8.1).

Rapid developments in molecular microbiology lead to novel definitions for EAEC, yet the gold standard is the adhesion pattern on cell monolayers.

Typical EAEC applies to EAEC strains possessing the AggR regulon (described below). Typical EAEC strains have been linked to acute diarrhea (Sarantuya et al., 2004; Huang et al., 2006). However, atypical EAEC lack the AggR regulon and are not reliably associated with diarrhea (Nataro, 2005). Since the original description, EAEC has emerged as an important pathogen in several clinical scenarios, including travelers' diarrhea (Adachi et al., 1999, 2001, 2002; Glandt et al., 1999; Tompkins et al., 1999), endemic pediatric diarrhea among children in developed countries (Tompkins et al., 1999), and developing countries (Okeke et al., 2000a), as well as persistent diarrhea amongst HIV-infected patients (Wanke et al., 1998a,b; Durrer et al., 2000; Mossoro et al., 2002; Gassama-Sow et al., 2004). A meta-analysis by Huang et al. (2006) showed that EAEC is a cause of acute diarrheal illness among different subpopulations in both developing and industrialized regions.

EPIDEMIOLOGY

EAEC epidemiology is poorly understood. There is no evidence for an animal reservoir of the bacterium (Huang et al., 2004a). Foodborne outbreaks have

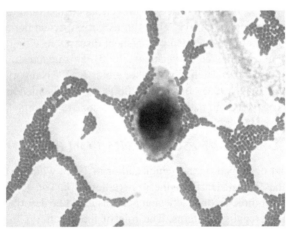

FIGURE 8.1   Characteristic aggregative or 'stacked brick' pattern of adherence (AA) to HEp2-cells in culture by EAEC prototype strain 042. Photograph by Nadia Boisen.

been described (Itoh et al., 1997). Risk factors for EAEC infection include travel to developing countries, ingestion of contaminated food and water, poor hygiene, host susceptibility, and possibly immunosuppression (Nataro et al., 1985; Okeke and Nataro, 2001; Huang and Dupont, 2004, 2006).

Definitive data demonstrating pathogenicity of individual EAEC strains is sparse. In volunteer studies, archetype strain 042 elicited clinical diarrhea in three of five subjects, and enteric symptoms in another (Nataro et al., 1995). In a prior study, strain JM221 elicited enteric symptoms in some patients. Interestingly, however, some other EAEC strains failed to cause diarrhea in volunteers. These results suggest heterogeneity of EAEC virulence, although the molecular basis of this variation remains elusive. Some studies suggest that children are more likely to be affected in the first month of life (Gonzalez et al., 1997; Gascon et al., 1998), while others have stressed that most cases arise in older children (Okeke et al., 2000a). These disparate epidemiologic observations could be reconciled by diverse virulence characteristics of the strains circulating at the respective sites, in combination with a variety of host immunity and host resistance.

Outbreaks

The first reported EAEC outbreak occurred in a Serbian nursery in 1995 (Cobeljic et al., 1996); of the 19 afflicted infants, three developed persistent diarrhea. There have been two EAEC outbreaks reported in Mexico City (Eslava et al., 1993). Itoh et al. (1997) described a massive outbreak of EAEC diarrhea among Japanese children, affecting nearly 2700 patients. Outbreaks have also been reported among adults in the UK (Smith et al., 1997).

Morabito et al. (1998) described an outbreak of hemolytic uremic syndrome (HUS) in France caused by a Shiga toxin-producing EAEC strain. Sporadic small outbreaks of hemolytic uremic syndrome (HUS) have been attributed to Stx-producing EAEC of the O111:H2, O111:H21, and O104:H4 serotypes over the last 15 years (Morabito et al., 1998; Frank et al., 2011; Scheutz et al., 2011); however, they were localized to small populations in France, Ireland, and the Republic of Georgia and were not disseminated throughout Europe.

From May through June 2011, Europe was struck by an outbreak of massive proportions, infecting 4137 individuals and resulting in 54 deaths (see Chapter 11). By epidemiological linking, contaminated sprouts were found to be the source of the outbreak, although not isolated from the source (Frank et al., 2011; Scheutz et al., 2011). There are troubling differences between the German outbreak and previous large outbreaks: (i) HUS represented 22% (845) of the ascertained cases, which is a much higher proportion than in other outbreaks (6–10%); (ii) it was predominately seen in healthy adults, which was highly uncommon with diarrhea-associated HUS, that occurs primarily in children; and (iii) the causative agent was a non-O157 Stx-producing EAEC strain of serotype O104:H4 (Frank et al., 2011).

Rasko et al. (2011) performed genetic characterization of the outbreak strain classifying it within the EAEC pathotype. EAEC of serotype O104:H4 strains are closely related and form a distinct clade among *E. coli* and EAEC strains. However, the genome of the German outbreak strain can be distinguished from those of other O104:H4 strains because it contains a prophage encoding Shiga toxin 2 and a distinct set of additional virulence (such as AAF/I, Pic, SepA, and SigA) and antibiotic-resistance genes. Stepwise horizontal genetic exchange events allowed for the emergence of the highly virulent Shiga-toxin-producing EAEC strain (Rasko et al., 2011). Comparing genomes of related strains isolated during the outbreak suggested that environmental selection enhanced the magnitude of the outbreak (Grad et al., 2012).

Endemic diarrhea in developing countries

EAEC is best known for its role in persistent diarrhea (>14 days) in infants and children in developing countries. Studies in Mongolia (Sarantuya et al., 2004), India (Dutta et al., 1999), Brazil (Piva et al., 2003; Zamboni et al., 2004), Nigeria (Okeke et al., 2000a,b), Israel (Shazberg et al., 2003), Venezuela (Gonzalez et al., 1997), Congo (Jalaluddin et al., 1998) and many other countries, have identified EAEC as a highly prevalent (often the most prevalent) *E. coli* pathotype in infants. The role of EAEC as a cause of persistent diarrhea and malnutrition in Brazil has been demonstrated repeatedly (Lima and Guerrant, 1992; Fang et al., 1995; Huang et al., 2006). In one study, 68% of those with persistent diarrhea shed EAEC in their stools (Fang et al., 1995). In Guinea-Bissau the most common bacteria isolated from feces of children <2 years of age with diarrhea was EAEC (Valentier-Branth et al., 2003).

Endemic diarrhea in developed countries

A large prospective study in England (Tompkins et al., 1999), in which over 3600 cases of diarrhea and controls were studied, found that typical EAEC was the second most common bacterial cause of gastroenteritis, following *Campylobacter*. EAEC was significantly associated with diarrhea in both prospective cohorts and in patients presenting to physicians' attention. Typical EAEC were found as major cause of bacterial diarrhea among infants in Cincinnati, OH, USA (Coohen et al., 2005). A large prospective study of diarrhea was conducted across all ages in Baltimore, MD, USA and New Haven, CT, USA and EAEC was significantly associated with diarrhea, being the most common bacterial cause of diarrhea at both sites (Nataro et al., 2006). This observation has recently been confirmed in a similar study performed in New Jersey (Cennimo et al., 2009). Moreover, a Scandinavian case-control study, found EAEC in significantly more diarrheal cases than controls (Bhatnagar et al., 1993). This is supported by studies from Germany (Huppertz et al., 1997) and

Austria (Presterl et al., 1999), which emphasizes the role of EAEC in developed countries.

EAEC as a cause of diarrhea in AIDS patients

The role of EAEC as an important pathogen in AIDS patients continues to develop, and EAEC now ranks among the most important enteric pathogens in this sub-population (Wanke et al., 1998a,b). EAEC was reported as the predominant cause of diarrhea among HIV-infected patients in the Central African Republic (Germani et al., 1998). The importance of EAEC in persistent diarrhea among African AIDS patients was re-emphasized during a study in Senegal (Gassame-Sow et al., 2004). The finding that HIV replication is enhanced by inflammation and activation of NF-κB raises the concern that EAEC-induced inflammation may add to the rapid downward decline of millions of African AIDS patients (Klein et al., 2000).

EAEC in travelers

As outbreaks suggest, EAEC is capable of causing diarrhea in adults. In a recent review of all published studies of traveler's diarrhea, EAEC was in aggregate second only to ETEC as the most common pathogen among patients with traveler's diarrhea (Shah et al., 2009). EAEC was a major cause of diarrhea among Spanish travelers going to the developing world, with an incidence identical to that of ETEC (Gascon et al., 1998). Studies in Mexico have shown that in contrast with ETEC, in which travelers are most susceptible during the first weeks of exposure, travelers remain susceptible to EAEC infection throughout their stay (Adachi et al., 2002), likely reflecting the particular ability of EAEC to evade the immune system and cause persistent diarrhea (Okhuysen et al., 2010). Travelers in Mexico showed that the rate of EAEC colonization increased proportionally with the length of the stay (Adachi et al., 2002).

EAEC and malnutrition

Investigators working in Fortaleza, Brazil, have repeatedly implicated EAEC as the predominant agent of persistent diarrhea (Wanke et al., 1991; Fang et al., 1995), which is associated with growth retardation. Interestingly, in this study population even asymptomatic patients infected with EAEC exhibit growth retardation (Steiner et al., 1998) compared with uninfected controls. In longitudinal studies of an infant cohort in Guinea-Bissau, EAEC infection was highly prevalent and was accompanied by growth retardation, although a direct link could not be established (Valentiner-Branth et al., 2001). Given the high rate of asymptomatic excretion of EAEC in much of the developing world (Fang et al., 1995), understanding its potential role in malnutrition and growth retardation is a high priority.

CLINICAL MANIFESTATIONS OF INFECTION

The clinical manifestations of EAEC infection involve complex host–pathogen interaction. At play are the heterogeneity of EAEC strains, number of EAEC ingested, and specific host susceptibility (Huang et al., 2004b). Jiang et al. (2003) identified a specific host genetic determinant that influences the clinical manifestations of EAEC infection. These investigators found that a single nucleotide polymorphism in the promoter of the −251 site of IL-8 conferred an increased risk of developing EAEC diarrhea as well as presence of elevated fecal IL-8 (Jiang et al., 2003).

Data on the clinical presentation of EAEC infection are derived primarily from volunteer studies, studies on traveler's diarrhea patients, and from outbreak investigations. The common clinical features of EAEC infection include watery diarrhea with or without passage of blood and mucus, abdominal pain, nausea, vomiting, borborygmi (abdominal gurgling), and fever (Nataro et al., 1995; Huppertz et al., 1997; Glandt et al., 1999; Bouckenooghe et al., 2000; Infante et al., 2004; Kahali et al., 2004; Regua-Mangia et al., 2004). As reviewed by Huang et al. (2006) the incubation period spans from 8 to 18 hours. Malnourished hosts, especially children living in developing countries, may be unable to repair mucosal damage and thus become prone to persistent (>14 days) diarrhea (Huang et al., 2006).

During the recent outbreak the disease manifestation was profound and severe since the EAEC strain acquired a Shiga-toxin-producing phage (Scheutz et al., 2011).

MICROBIAL PATHOGENESIS

EAEC as currently defined most likely encompasses both pathogenic and non-pathogenic *E. coli* strains (Kaper et al., 2004). This conundrum may be resolved upon further understanding of EAEC pathogenesis, which is becoming slowly elucidated. The EAEC genome is highly mosaic, with many putative virulence genes flanked by insertion sequences, and predictably, many *E. coli* strains harbor a few of the identified genes in various combinations (Boisen et al., 2012). Thus, a major obstacle in understanding EAEC is the inability to define a fundamental pathogenetic strategy. EAEC has unmistakably been associated with diarrhea in some individuals (e.g. in volunteers and outbreak patients), but subclinical colonization in endemic areas is common.

The essential differences between pathogenic and non-pathogenic strains are largely unknown, but pathogenesis studies suggest three general stages of infection: (i) adherence to the intestinal mucosa by virtue of aggregative adherence fimbriae (AAF) or other adherence factors (Tzipori et al., 1992; Hicks et al., 1996); (ii) stimulation of mucus production, forming a biofilm on the surface of the mucosa (Hicks et al., 1996); and (iii) toxicity to the mucosa, manifested by cytokine release, cell exfoliation, intestinal secretion, and induction of

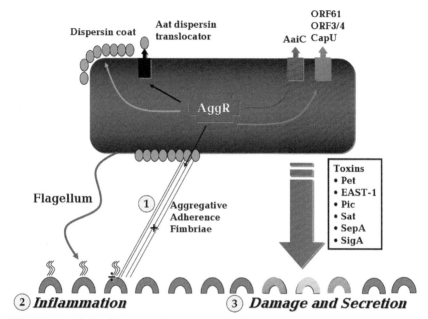

FIGURE 8.2   The basic features of EAEC pathogenesis. Stages 1–3 highlighted in yellow illustrate the three main steps in EAEC pathogenesis. See text for discussion. *Modified after Harrington et al. (2006).*

mucosal inflammation (Steiner et al., 1998, 2000; Bouckenooghe et al., 2000; Jiang et al., 2002; Harrington et al., 2005) (Figure 8.2).

Virulence determinants

Several putative virulence factors have been identified in EAEC (Table 8.1), including enterotoxins and cytotoxins, secreted proteins and many more. The virulence factors so far described are encoded either on the large virulence plasmid of EAEC called pAA, or on the chromosome. The clinical roles of these factors remain uncertain (Nataro, 2005).

The best-studied EAEC factor is AggR, putatively the master regulator of EAEC virulence. AggR is a member of the AraC/XylS family of transcriptional activators with its nearest neighbors being Rns of ETEC and RegA of *Citrobacter rodentium*, while ToxT of *Vibrio cholerae* is a distant relative.

Virulence factors regulated by AggR

AggR controls expression of adherence factors, a dispersin surface coat protein and a large cluster of genes encoded on the EAEC chromosome. Nataro (2005) have therefore suggested that a 'package' of plasmid-borne and chromosomal virulence factors for EAEC are required to execute pathogenesis, and that this

TABLE 8.1 Putative EAEC virulence genes and factors

Factor	Description	Unique to EAEC
Master regulator gene		
AggR	Master regulator of a package of EAEC plasmid virulence genes, including aggregative adherence factors, fimbriae AAF/I-AAF/IV, and a large cluster of genes inserted on a pathogenicity island at the PheU locus (Bernier et al., 2002)	Yes
AggR regulated genes		
AatA	Encodes ABC protein responsible for transporting the dispersin protein out of the outer membrane of EAEC (Nishi et al., 2003)	ND
Aap	Encodes a 10-kDa secreted protein named dispersin, and is responsible for 'dispersing' EAEC across the intestinal mucosa (Sheikh et al., 2002)	No
AggA	Encodes AAF/I, mediates adherence to colonic mucosa and hemagglutination of erythrocytes (Nataro et al., 1992)	Yes
AafA	Encodes AAF/II, mediates adherence to colonic mucosa and hemagglutination of erythrocytes (Czeczulin et al., 1997)	Yes
Agg3A	Encodes AAF/III hemagglutination of erythrocytes (Bernier et al., 2002)	Yes
Agg4A	Encodes AAF/IV, mediates adherence to colonic mucosa and hemagglutination of erythrocytes (Boisen et al., 2008)	Yes
AaiC	AaiC, secreted protein. Encoded on the chromosome. Mode of action	ND
ORF3/4[a]	Co-regulated two-gene cluster with homology to isoprenoid synthesis genes	ND
ORF61[a]	Expressed contact hemolysis of erythrocytes	ND
CapU	Hexosyltransferase homolog (Czeczulin et al., 1999)	ND
Toxin		
Pet	A 108-kDa autotransporter protein that functions as a heat-labile enterotoxin and cytotoxin (Navarro-Garcia et al., 1998)	Yes
SigA	IgA protease-like homolog (Rajakumar et al., 1997)	No

TABLE 8.1 Putative EAEC virulence genes and factors—cont'd

Factor	Description	Unique to EAEC
Pic	Pic protein has mucinase activity and is capable of causing hemagglutination of erythrocytes (Henderson et al., 1999a)	No
SepA	*Shigella* extracellular protein. May induce mucosal atrophy and tissue inflammation in *S. flexneri* (Benjelloun-Touimi et al., 1998)	No
Sat	Secreted autotransporter toxin. Might induce cytoskeletal perturbation in intestinal epithelium accompanied by rearrangement of tight junction proteins (Guyer et al., 2000)	No
EAST-1	Encodes the enteroaggregative heat-stable toxin (EAST-1), which has physical and mechanistic similarities to *E. coli* STa enterotoxin (Savarino et al., 1993)	No
Others		
EilA	*Salmonella* HilA homolog, activates the bacterial surface protein Air (Sheikh et al., 2006)	ND
Air	Possible aggregation and adherence (Sheikh et al., 2006)	ND
Pro. Fl.	Single protein (from strain 042) moiety that has a sequence similar to that of a flagelin from *Shigella dysenteriae* (Steiner et al., 2000)	No
Irp2	Encodes an iron-uptake system mediated by siderophore yersiniabactin that plays a role in iron transport and regulation (Schubert et al., 1998)	No
Lectin	A complex carbohydrate that shows cross-reactivity to the bindings subunit of cholera toxin, and induces morphological changes in HEp-2 cells and fluid accumulation in the rabbit ileal loop (Basu et al., 2004)	No

[a]Unpublished; ND, not determined ; Pro. Fl, proinflamatory flagelins.

set of factors is under the coordinate control of AggR. This hypothesis has not been rigorously tested.

The natural conditions that result in activation of AggR are unknown. However, activating conditions for the related RegA, Rns, and ToxT have been described: Rns (Grewal et al., 1997) and ToxT (Prouty et al., 2005) regulons are activated by bile, whereas sodium bicarbonate activates the RegA-dependent genes (Hart et al., 2008).

Once ingested the localization of EAEC in the gastrointestinal tract has not been well defined. Electron microscopy of infected small intestinal mucosa revealed bacteria in association with a thick mucus layer above an intact enterocyte brush border, which contained extruded cell fragments. In the colon EAEC induces cytotoxic effects (Hicks et al., 1996). All of these findings suggest that even though EAEC binds to many segments of the small intestinal tract, it is most pathogenic in the colonic epithelium (Huang et al., 2004a).

Aggregative adherence (AA) to the intestinal mucosa represents the first step in the pathogenesis of EAEC (Okeke et al., 2000a) (Figure 8.2). The EAEC-defining phenotype, AA, suggests that adhesins have an important role in pathogenesis, as they do for all enteric pathogens.

Aggregative adherence fimbriae (AAF) are the principal EAEC mucosal adhesins of which at least five variants are known (Nataro et al., 1992; Czeczulin et al., 1997; Bernier et al., 2002; Boisen et al., 2008). The AAFs are members of the chaperone-usher pili family (see Chapter 12) and display a high level of conservation of accessory genes, but with much greater divergence of the fimbrial (pilin) genes. Four structural subunits encoded by *aggA* (AAF/I), *aafA* (AAF/II) *agg3A* (AAF/III), and *agg4A* (AAF/IV) on the pAA plasmid have been described. AAF/I and AAF/IV confer hemagglutination to human erythrocytes and are responsible for the aggregative phenotype (Nataro et al., 1992; Boisen et al., 2008). Recently, a fifth AAF/V has been discovered (Protein ID number BAI44132.1) which shares ~35% identities of the mature protein with Agg3A from prototype strain 55989 (Bernier et al., 2002). AAF/II is implicated in intestinal adherence (Czeczulin et al., 1997). AAF/III also functions as a cellular adhesin, and the *agg3* biogenesis genes are closely related to those of the *agg* and *aaf* operons of AAF/I and AAF/II respectively (Bernier et al., 2002). The structural subunits of AAF/I and II are 25% identical and 47% similar. AAF/I is expressed by 31% of EAEC and AAF/II by 12% (Czeczulin et al., 1997, 1999; Elias et al., 1999).

The AAF family is related to the Dr-family of adhesins found in uropathogenic and diffuse adhering *E. coli* (DAEC) (see Chapters 9 and 11) (Savarino et al., 1994; Elias et al., 1999). The fimbriae themselves are composed of multimers of major and minor fimbrial subunits. The fimbriae of AAF/I, AAF/II, and AAF/III (AggA, AafA, and Agg3A, respectively) comprise a distinct phylogenetic cluster, as do the original Dr-family, comprising F1845 and the AFA adhesins (Boisen et al., 2008). Collectively, these four variants may account for AA in the large majority of EAEC strains (Boisen et al., 2008, 2009).

The genetic organization of AAF/I and AAF/IV comprises genes encoding a chaperone, usher, putative invasin, and major pilin protein in a single gene cluster (Savarino et al., 1994; Boisen et al., 2008).

The morphology of AAF/II from EAEC archetype strain 042 exhibits a semi-flexible bundle forming structure, which wraps around neighboring bacteria (Figure 8.3). In contrast to the bundle-forming morphology of AAF/I, II and AAF/IV, the AAF/III appear as peritrichous, long and flexible filaments. Transmission electron microscopy shows that fimbriae of AAF/II are thicker (5 nm diameter) compared to those of AAF/I, which are 2–3 nm in diameter (Czeczulin et al., 1997).

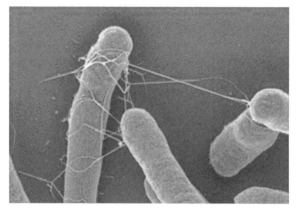

FIGURE 8.3   Scanning electron microscopy of EAEC wt strain 042. White arrow indicates AAF/II fimbriae. *From Sheikh et al. (2002).*

The Afa/Dr adhesins bind to several receptors on epithelial cells, including decay-accelerating factor (DAF) as well as extracellular matrix (ECM) proteins, in particular type IV collagen (reviewed by Servin, 2005). Recent studies have found that EAEC strain 042 expressing AAF/II binds the ECM fibronectin, laminin, and type IV collagen proteins (Farfan et al., 2008).

Studies performed with EAEC strain 042 showed that AAF/II appears to be important for biofilm (defined here as the association of bacteria into communities on abiotic surfaces) formation in EAEC strains producing these fimbriae (Sheikh et al., 2001). In addition, it was established that the pAA plasmid from EAEC strain 042 (pAA2) and EAEC strain C1010-00 (pAA4) was sufficient to confer biofilm formation on commensal *E. coli* strain HS (Harrington et al., 2005; Boisen et al., 2008). It was found that many EAEC strains not expressing known AAFs developed biofilm under specific conditions suggesting that more AAFs remain to be discovered (Sheikh et al., 2001).

Most (approximately 80%) EAEC strains secrete a 10-kDa protein called dispersin, encoded by the *aap* gene. Dispersin secretion is dependent upon a system called Aat (encoded by *aatPABCD*). Once outside the bacterium, dispersin remains non-covalently attached to the outer membrane. In dispersin mutants, the AAF fimbriae are formed, but collapse and adhere tightly to the surface of the bacterium (Sheikh et al., 2002). Consequently, dispersin serves to insulate the surface of the bacterium so that the positively charged AAF is free to extend away from the negatively charged cell surface.

A cluster of AggR-activated genes has been identified on the chromosome of strain 042; these genes comprise a type VI secretion system and are encoded within a chromosomal island called AAI which comprises a cluster a 16 predicted genes designated *aaiA-P* (Dudley et al., 2006). The contribution of the *aai* genes to EAEC pathogenesis is currently unknown, but these genes do not appear to play a role in adherence to abiotic surfaces or intestinal epithelial cells, as do previously described AggR-regulated genes (Harrington et al., 2006). Colony hybridization indicates that *aaiA* and *aaiC* can be found in 67% of strains from a library of worldwide EAEC isolates (Dudley et al., 2006). As expected, the *aai* genes are characteristic of typical EAEC strains (Jenkins et al., 2005).

The second stage (Figure 8.2) of pathogenesis of EAEC involves production of a mucus layer probably stemming from contributions of both bacteria and intestinal mucosa. EAEC survives within the mucus layer on the surface of enterocytes (Hicks et al., 1996).

Virulence factors not regulated by AggR

The third stage (Figure 8.2) of EAEC pathogenesis involves the release of toxins and inducers of an inflammatory response, mucosal toxicity, and intestinal secretion. The mechanism of EAEC-induced mucosal toxicity is not completely elucidated (Hicks et al., 1996). However, several EAEC toxins have been described and they are encoded either on the pAA plasmid or the chromosome (Henderson et al., 1999a, 2000; Henderson and Nataro, 2001; Okeke and Nataro, 2001; Boisen et al., 2009; Ruiz-Perez et al., 2009).

Infection of human intestinal explants suggests that most EAEC strains elicit obvious mucosal damage, accompanied by rounding and exfoliation of colonocytes. The autotransporter protease Pet is responsible for the induction of cytotoxic effects on human intestinal explants infected with EAEC in vitro (Henderson et al., 1999b). However, a paradox exists: only a small minority of EAEC strains carries the *pet* gene, though a much larger number of strains cause toxic effects to explants. This paradox occurs in the context of substantial heterogeneity of EAEC adhesins and other putative virulence factors, presenting a confusing clinical and epidemiologic scenario (Huang et al., 2004b). It has been shown (Huang et al., 2004b) that most EAEC strains express, if not Pet, some related cytotoxin, the most common of which is Sat, a cytoskeleton-cleaving protease (like Pet), that was initially described in uropathogenic *E. coli*.

Serine protease autotransporters of Enterobacteriaceae (SPATEs) comprise a large group of trypsin-like serine proteases which are secreted by *Shigella* spp., uropathogenic *E. coli*, and all of the DEC pathotypes (see also Chapter 16) (Benjelloun-Touimi et al., 1995; Stein et al., 1996; Brunder et al., 1997; Eslava et al., 1998; Henderson et al., 1999a; Al-Hasani et al., 2000; Guyer et al., 2000; Patel et al., 2004). The toxins are translocated across the outer membrane by the autotransporter pathway, in which translocation requires a dedicated C-terminal beta barrel moiety. The N-terminal, mature SPATE toxins are 104–110 kDa in size and feature a typical N-terminal serine protease catalytic domain, followed by a highly conserved beta-helix motif, which is present in nearly all autotransporters (Henderson et al., 2004). Notably, SPATEs have not been identified in a non-pathogenic organism (Henderson et al., 2001).

The SPATEs have been organized phylogenetically into two classes. Members of the class I SPATEs (which include Pet) are all cytotoxic to epithelial cells (Dutta et al., 2002). In addition to Pet, the class I SPATEs include among others EspP from enterohemorrhagic *E. coli*, EspC from enteropathogenic *E. coli*, SigA from *Shigella flexneri*, and Sat, from uropathogenic and diffusely adhering *E. coli* (Henderson and Nataro, 2001). Class II SPATEs are more diverse with regard to phenotype, though several are known to cleave mucin. This class comprises, among others, Pic and SepA from EAEC and *S. flexneri* (Benjelloun-Touimi et al., 1995; Henderson et al., 1999a, Harrington et al., 2009) and Tsh from avian pathogenic *E. coli* (Provence and Curtis 1994). Class II separates further into two subclasses where one comprises of several potential O-glycoproteases (Ruiz-Perez et al., unpublished).

Pet is a protease encoded on the pAA plasmid of strain 042 and other EAEC strains (Eslava et al., 1998). The prevalence of Pet among EAEC isolates varies between 18–44%, although the toxin is unique to EAEC (Vila et al., 2000; Yamazaki et al., 2000). The toxin enters the eukaryotic cell and moves through the vesicular system which appears to be required for the induction of cytopathic effects (Navarro-Garcia et al., 2001). It has been shown that Pet cleaves erythroid spectrin in vitro and Pet intoxication is accompanied by degradation of spectrin species (components of the cytoskeleton) and clumping of spectrin in intoxicated HEp2 cells (Villaseca et al., 2000; Sui et al., 2003). The mucosal intestinal toxicity of Pet results in dilation of crypt openings (Figure 8.4), rounding and exfoliation of colonic enterocytes, widening of intercrypt crevices and loss of apical mucus from goblet cells (Henderson and Nataro, 1999).

Sat is cytotoxic to urinary epithelial cells in vitro (Guyer et al., 2000), and Sat induces cytoskeletal perturbation in intestinal epithelium accompanied by rearrangement of tight junction proteins (Guignot et al., 2007). Though the fundamental mode of action of Sat is unknown, it was suggested that the protein enters epithelial cells and directly cleaves spectrin (Maroncle et al., 2006), an effect also seen in Pet (Canizalez-Roman et al., 2003).

The *sigA* gene is situated on the *she* pathogenicity island of *Shigella flexneri* 2a which also contains the *pic* gene. SigA is a IgA protease-like homolog lying 3.6 kb downstream and in a reverted orientation with respect to *pic* (Rajakumar et al., 1997). Mature SigA is 103 kDa, slightly smaller than the other class 1 SPATE proteins (Al-Hasani et al., 2000). Al-Hasani et al. (2000) showed that SigA is secreted as a temperature-regulated serine protease capable of degrading casein; these investigators also reported that SigA is cytopathic for HEp-2 cells, suggesting that it may be a cell-altering toxin with a role in the pathogenesis of *Shigella* infections. SigA was at least partly responsible for the ability of *S. flexneri* to stimulate fluid accumulation in ligated rabbit ileal loops.

Two EAEC toxins are encoded on the same chromosomal locus, embedded on opposite strands. The larger gene encodes Pic. The opposite strand encodes the oligomeric enterotoxin that is known as *Shigella* enterotoxin 1 (ShET1, encoded by *setAB* genes), owing to its presence in most *Shigella flexneri* of serotype 2a (Fasano et al., 1997; Henderson et al., 1999b). Hence ShET1 is encoded within the *pic* open-reading frame, on the antisense strand. It is thought

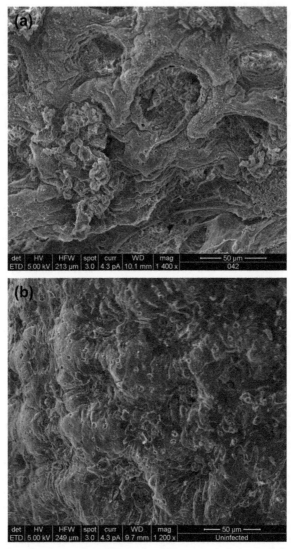

FIGURE 8.4   Scanning electron photomicrographs of *Macaca fascicularis* colon infected with EAEC strain 042 for 3 hours (a) showing biofilm formation and mucosal intestinal toxicity which results in dilation of crypt openings, rounding and exfoliation of colonic enterocytes, widening of intercrypt crevices and loss of apical mucus from goblet cells. (b) Uninfected. Bar 50 μm. Photograph by Nadia Boisen.

that the *pic/set* genes are expressed in the intestinal lumen, and further results suggest a novel mechanism of ShET1 regulation and the existence of pathogen-specific regulators of *pic* and *setAB*. Pic is a 109-kDa mucinase, which degrades intestinal mucin and may promote intestinal colonization via an unknown mechanism (Henderson et al., 1999a). Harrington et al., (2009) showed that the *pic* gene in strain 042 increased fitness for growth in mucus scraped from the surface of the mouse intestine. Moreover, *pic* mutants in EAEC are less adept at colonizing the mouse intestine, whereas a *Shigella pic* mutant elicited less intense intestinal inflammation in the rabbit ligated-loop model (Henderson and Nataro, 2001).

The ShET1 (55-kDa) mode of action has not been defined, but it does not appear to act via the traditional mechanisms of toxin-induced intestinal secretion, such as via cyclic AMP and cyclic GMP. ShET1 induces fluid secretion in mucosal tissue explants but does not induce cytotoxic effects (Fasano et al., 1997). Studies have suggested that most EAEC strains from patients with diarrhea express the ShET1 toxin (Czeczulin et al., 1999; Vila et al., 2000). ShET1 may contribute to the secretory diarrhea that accompanies EAEC and *Shigella* infections (Kaper et al., 2004).

SepA, a SPATE toxin, was originally described in *Shigella flexneri* 2a, where it is among the most abundant secreted proteins. The protease is encoded on the *Shigella* virulence plasmid, suggesting a contribution to pathogenicity (Benjelloun-Touimi et al., 1995). Using human colonic tissue, it was shown that purified SepA toxin elicits mucosal damage (Coron et al., 2009). SepA mutants of *S. flexneri* induced less mucosal inflammation in ligated rabbit ileal loops compared with the wild-type parent strain (Benjelloun-Touimi et al., 1998). The precise mode of action of SepA remains to be discovered. We have found SepA to be common and epidemiologically important among EAEC strains: among EAEC strains isolated from a case-control study of children's diarrhea in Mali, SepA was the only factor strongly associated with diarrheal illness (Boisen et al., 2012).

The first EAEC virulence factor that was implicated as a potential cause of diarrhea was the enteroaggregative heat-stable toxin, EAST1. Epidemiological studies demonstrate that EAST1 is not only associated with EAEC but is also present in a wide range of pathogenic *E. coli* such as ETEC, DAEC, enteropathogenic *E. coli* (EPEC), and enterohemorrhagic *E. coli*. Furthermore, *E. coli* strains harboring no known virulence factors other than EAST1 were found in the feces of humans with diarrhea (Paiva de Sousa and Dubreuil, 2001; Menard and Dubreuil, 2002). EAST1 is encoded by the *astA* gene which spans 117 bp. It is a 38-aminoacid peptide with homology to the heat-stable (ST) enterotoxin of ETEC (Savarino et al., 1991, 1993). The *astA* gene can be found on either plasmids or on the chromosome, and sometimes both, in one or several copies (Menard and Dubreuil, 2002). EAST1 is immunologically different from STa, as no cross-neutralization was observed with polyclonal anti-STa antibodies (Savarino et al., 1996). It is conceivable that EAST1 could contribute to watery diarrhea in EAST1-positive EAEC strains (41% of EAEC strains harbor the *astA* gene). However, the *astA* gene is also present in up to 38% of commensal *E. coli* strains (Savarino et al., 1996; Yamamoto and Echeverria, 1996; Zhou et al., 2002). EAST1 may exist as a series of allelic variants, some of which may be more virulent than others (Menard et al., 2004).

We have described in EAEC several putative pathogenicity islands that may encode additional virulence factors, as predicted from in silico homology to better characterized systems. Some EAEC strains encode a type 3 secretion system called ETT-2; putative effectors for this system were found elsewhere on the genome (Sheikh et al., 2006). Moreover, we have shown that these putative effectors are under transcriptional control of a factor called EilA, homologous to the HilA activator from *Salmonella* strains. EilA also activates the bacterial surface protein Air, which features predicted immunoglobulin-like repeats. This new putative virulence-related regulon in EAEC may include adherence and aggregation (Sheikh et al., 2006).

INFLAMMATION IN EAEC PATHOGENESIS

EAEC is an inflammatory pathogen, as demonstrated both in clinical (Greenberg et al., 2002) and laboratory (Steiner et al., 1998) reports. Clinical studies have shown that lactoferrin, IL-8, and IL-β can be detected in feces from cases of EAEC diarrhea at a higher level than in stools of patients infected with non-EAEC diarrhea (Jiang et al., 2002). The virulence factors of typical EAEC (including *aggA*, *aggR*, *aafA*, and *aap*) are associated with increased levels of fecal cytokines and inflammatory markers, and may be observed whether or not the patient manifests diarrhea. The inflammatory effect was linked to expression of a novel EAEC flagellin protein (Donnelly and Steiner, 2002), which is homologous to a flagellin encoded by *S. dysenteriae*. The EAEC flagellin induced IL-8 from intestinal epithelial cells (IECs) in culture (Steiner et al., 1997, 1998).

It was shown that flagellin was the major pro-inflammatory factor of EAEC on intestinal epithelial cells in culture (Okhuysen and Dupont, 2010). Infection of polarized monolayers of the human colonic intestinal cell line T84 with EAEC strain 042 caused both IL-8 release (Harrington et al., 2005) and a drop in trans-epithelial electric resistance (TEER) when compared with the uninfected control and with non-pathogenic *E. coli* HS (Strauman et al., 2010). It is now confirmed in vitro that the fimbriae mediate release of IL-8 and drop in TEER. Furthermore it was shown that AAF/II fimbriae are sufficient to induce transmigration of neutrophils across an epithelial layer in vitro (Qadri et al., 1994). Since a suitable animal model is not in use, a xenotransplant model was used and it was shown that the aggregative fimbriae are important in the development of inflammation in the human intestine. These data suggest that the AAF adhesins may be not only colonization factors, but may also be both necessary and sufficient for induction of mucosal inflammation (Boll et al., 2012).

STRAIN HETEROGENEITY

EAEC strains belong to a diverse range and combination of O:H serotypes (Vial et al., 1988; Yamamoto et al., 1992; Qadri et al., 1994; Huppertz et al.,

1997; Olesen et al., 2005; Boisen et al., 2012). Moreover, a high frequency of EAEC strains expresses untypable O antigens and H antigens or are non-motile (Nataro and Kaper, 1998; Uber et al., 2006; Regua-Mangia et al., 2009). Nevertheless, there exist commonly isolated EAEC serotypes (O44:H18, O111:H12, O125, and O126 strains). Studies show that some serotypes of the traditional enteropathogenic *E. coli* such as O55, O111, O86, O126, and O128 can be found in EAEC (Nataro et al., 1998; Suzart et al., 1999; Elias et al., 2002) although the most commonly found serogroups reported in EAEC are O86, O126, and O125 (Spencer et al., 1999; Sarantuya et al., 2004; Uber et al., 2006). It was suggested that the occurrence of EPEC O serogroups (O126, O128, and O158) along with EAEC markers influenced the positive association of *E. coli* strain with diarrhea (Pereira et al., 2007). Serotyping is often useful in the characterization of other pathogenic *E. coli*, however it is of little value in EAEC diagnostics (Okeke and Nataro, 2001). However, in a detailed study on genomic characterization by Boisen et al. (2012), targeting a collection of 121 EAEC strains isolated from children in Mali with or without moderate to severe diarrhea, it was found that strains expressing the H33 flagellar antigen were found significantly more often in cases than in controls. This association may signify the existence of a specific set of virulence genes in strains of this H type (Boisen et al., 2012). A phylogenetic framework was presented identifying three major clusters of DEC containing EAEC. Members of each group show conserved plasmids and chromosomal loci, which indicates that most EAEC, like EPEC, feature conserved linkage of virulence genes (Czeczulin et al., 1999). EAEC strains that did not fall into the above-mentioned clusters could be milder pathogens eliciting inflammation without diarrhea (Steiner et al., 1998). Notably, the pAA plasmid of EAEC is, however, heterogeneous with regard to fimbria and toxin expression. In volunteer studies, three strains expressing the AAF/I variant did not induce diarrhea, whereas the one strain expressing AAF/II caused diarrhea in the majority of infected subjects (Nataro et al., 1995).

IDENTIFICATION OF EAEC

The gold standard for identifying EAEC is the HEp2-cell adherence assay (Nataro and Kaper, 1998) in as much as the pathogen was initially defined by the presence of a characteristic stacked brick pattern, designated aggregative adherence (AA) in this assay (Nataro et al., 1987). The HEp2-cell adherence assay is unfortunately not designed to screen large numbers of colonies from stool samples as it is very time consuming and is further limited by the risk of contamination of the cell cultures. A DNA probe, CVD432 from the pAA plasmid of EAEC, has been reported to be specific for EAEC but varies in sensitivity (Baudry et al., 1990). As reviewed, the sensitivity variation was between 20% to 89% when compared to the HEp2-cell adherence assay (Okeke, 2009). The CVD432 probe has been shown to correspond to the *aatA* gene, which encodes a transporter for the dispersin protein (Aap), also regulated by AggR

(Nishi et al., 2003). Aap of EAEC is secreted by many EAEC strains and was suggested as a possible target for diagnosis (Sheikh et al., 2002). However, a recent report demonstrates that Aap is also produced by non-EAEC strain. A multiplex PCR assay was developed that detects the three AA plasmid-borne genes (*aatA, aggR,* and *aap*) and studies showed that these loci are commonly but not invariably linked (Opintan et al., 2010). The authors of this study found that 82% of the EAEC strains isolated from patients with diarrhea were positive for the three loci and that use of a multiplex assay increases both the sensitivity and the specificity of EAEC detection (Cerna et al., 2003). Several studies have applied PCR in detecting EAEC targeting genes *aggR* and/or *aatA* (Kahali et al., 2004; Sarantuya et al., 2004; Pereira et al., 2007; Regua-Mangia et al., 2009; Gomez-Duarte et al., 2010; Opintan et al., 2010; Rugeles et al., 2010). The variable results obtained using molecular diagnostics may be due to the heterogeneity in EAEC pathogenic mechanisms (Okeke and Nataro, 2001). PCR targeting both virulence factors on the pAA plasmid and the chromosomal EAEC loci, such as *aaiC* (Dudley et al., 2006) might prove to be the most advantageous approach in detecting EAEC.

REFERENCES

Adachi, J., Glandt, M., Jiang, Z.-D., et al., 1999. Enteroaggregative *Escherichia coli* as a major etiologic agent in traveler's diarrhea in three regions of the world. 37th Annual Meeting of the Infectious Diseases Society of America. 1999. Infectious Diseases Society of America, Philadelphia, p. 23.

Adachi, J.A., Ericsson, C.D., Jiang, Z.D., DuPont, M.W., Pallegar, S.R., DuPont, H.L., 2002. Natural history of enteroaggregative and enterotoxigenic *Escherichia coli* infection among US travelers to Guadalajara, Mexico. J. Infect. Dis. 185 (11), 1681–1683.

Adachi, J.A., Jiang, Z.D., Mathewson, J.J., et al., 2001. Enteroaggregative *Escherichia coli* as a major etiologic agent in traveler's diarrhea in 3 regions of the world. Clin. Infect. Dis. 32 (12), 1706–1709.

Al-Hasani, K., Henderson, I.R., Sakellaris, H., et al., 2000. The sigA gene which is borne on the she pathogenicity island of *Shigella flexneri* 2a encodes an exported cytopathic protease involved in intestinal fluid accumulation. Infect. Immun. 68 (5), 2457–2463.

Basu, S., Ghosh, S., Ganguly, N.K., Majumdar, S., 2004. A biologically active lectin of enteroaggregative *Escherichia coli*. Biochimie. 86 (9–10), 657–666.

Baudry, B., Savarino, S.J., Vial, P., Kaper, J.B., Levine, M.M., 1990. A sensitive and specific DNA probe to identify enteroaggregative *Escherichia coli*, a recently discovered diarrheal pathogen. J. Infect. Dis. 161 (6), 1249–1251.

Benjelloun-Touimi, Z., Sansonetti, P.J., Parsot, C., 1995. SepA, the major extracellular protein of *Shigella flexneri*: autonomous secretion and involvement in tissue invasion. Mol. Microbiol. 17 (1), 123–135.

Benjelloun-Touimi, Z., Si Tahar, M., Montecucco, C., Sansonetti, P.J., Parsot, C., 1998. SepA, the 110 kDa protein secreted by *Shigella flexneri*: two-domain structure and proteolytic activity. Microbiology 144 (Pt 7), 1815–1822.

Bernier, C., Gounon, P., Le Bouguenec, C., 2002. Identification of an aggregative adhesion fimbria (AAF) type III-encoding operon in enteroaggregative *Escherichia coli* as a sensitive probe for detecting the AAF-encoding operon family. Infect. Immun. 70 (8), 4302–4311.

Bhan, M.K., Khoshoo, V., Sommerfelt, H., Raj, P., Sazawal, S., Srivastava, R., 1989a. Enteroaggregative *Escherichia coli* and *Salmonella* associated with nondysenteric persistent diarrhea. Pediatr. Infect. Dis. J. 8 (8), 499–502.

Bhan, M.K., Raj, P., Levine, M.M., et al., 1989b. Enteroaggregative *Escherichia coli* associated with persistent diarrhea in a cohort of rural children in India. J. Infect. Dis. 159 (6), 1061–1064.

Bhatnagar, S., Bhan, M.K., Sommerfelt, H., Sazawal, S., Kumar, R., Saini, S., 1993. Enteroaggregative *Escherichia coli* may be a new pathogen causing acute and persistent diarrhea. Scand. J. Infect. Dis. 25 (5), 579–583.

Black, R.E., Cousens, S., Johnson, H.L., et al., 2010. Global, regional, and national causes of child mortality in 2008: a systematic analysis. Lancet 375 (9730) 1969–1987.

Boisen, N., Ruiz-Perez, F., Scheutz, F., Krogfelt, K.A., Nataro, J.P., 2009. Short report: high prevalence of serine protease autotransporter cytotoxins among strains of enteroaggregative *Escherichia coli*. Am. J. Trop. Med. Hyg. 80 (2), 294–301.

Boisen, N., Scheutz, F., Rasko, D.A., et al., 2012. Genomic characterization of enteroaggregative *Escherichia coli* from children in Mali. J. Infect. Dis. 205 (3), 431–444.

Boisen, N., Struve, C., Scheutz, F., Krogfelt, K.A., Nataro, J.P., 2008. New adhesin of enteroaggregative *Escherichia coli* related to the Afa/Dr/AAF family. Infect. Immun. 76 (7), 3281–3292.

Boll, E.J., Struve, C., Sander, A., Demma, Z., Krogfelt, K.A., McCormick, B.A., 2012. Enteroaggregative *Escherichia coli* promotes transepithelial migration of neutrophils through a conserved 12-lipoxygenase pathway. Cell Microbiol. 14 (1), 120–132.

Bouckenooghe, A.R., Dupont, H.L., Jiang, Z.D., et al., 2000. Markers of enteric inflammation in enteroaggregative *Escherichia coli* diarrhea in travelers. Am. J. Trop. Med. Hyg. 62 (6), 711–713.

Brunder, W., Schmidt, H., Karch, H., 1997. EspP, a novel extracellular serine protease of enterohaemorrhagic *Escherichia coli* O157:H7 cleaves human coagulation factor V. Mol. Microbiol. 24 (4), 767–778.

Bryce, J., Boschi-Pinto, C., Shibuya, K., Black, R.E., 2005. WHO estimates of the causes of death in children. Lancet 365 (9465), 1147–1152.

Canizalez-Roman, A., Navarro-Garcia, F., 2003. Fodrin CaM-binding domain cleavage by Pet from enteroaggregative *Escherichia coli* leads to actin cytoskeletal disruption. Mol. Microbiol. 48 (4), 947–958.

Cennimo, D., Abbas, A., Huang, D.B., Chiang, T., 2009. The prevalence and virulence characteristics of enteroaggregative *Escherichia coli* at an urgent-care clinic in the USA: a case-control study. J. Med. Microbiol. 58 (4), 403–407.

Cerna, J.F., Nataro, J.P., Estrada-Garcia, T., 2003. Multiplex PCR for detection of three plasmid-borne genes of enteroaggregative *Escherichia coli* strains. J. Clin. Microbiol. 41 (5), 2138–2140.

Cobeljic, M., Miljkovic-Selimovic, B., Paunovic-Todosijevic, D., et al., 1996. Enteroaggregative *Escherichia coli* associated with an outbreak of diarrhoea in a neonatal nursery ward. Epidemiol. Infect. 117 (1), 11–16.

Cohen, M.B., Nataro, J.P., Bernstein, D.I., Hawkins, J., Roberts, N., Staat, M.A., 2005. Prevalence of diarrheagenic *Escherichia coli* in acute childhood enteritis: a prospective controlled study. J. Pediatr. 146 (1), 54–61.

Coron, E., Flamant, M., Aubert, P., et al., 2009. Characterisation of early mucosal and neuronal lesions following *Shigella flexneri* infection in human colon. PLoS One 4 (3), e4713.

Cravioto, A., Tello, A., Navarro, A., et al., 1991. Association of *Escherichia coli* HEp-2 adherence patterns with type and duration of diarrhoea. Lancet 337 (8736), 262–264.

Czeczulin, J.R., Balepur, S., Hicks, S., et al., 1997. Aggregative adherence fimbria II, a second fimbrial antigen mediating aggregative adherence in enteroaggregative *Escherichia coli*. Infect. Immun. 65 (10), 4135–4145.

Czeczulin, J.R., Whittam, T.S., Henderson, I.R., Navarro-Garcia, F., Nataro, J.P., 1999. Phylogenetic analysis of enteroaggregative and diffusely adherent *Escherichia coli*. Infect. Immun. 67 (6), 2692–2699.

Donnelly, M.A., Steiner, T.S., 2002. Two nonadjacent regions in enteroaggregative *Escherichia coli* flagellin are required for activation of toll-like receptor 5. J. Biol. Chem. 277 (43), 40456–40461.

Dudley, E.G., Thomson, N.R., Parkhill, J., Morin, N.P., Nataro, J.P., 2006. Proteomic and microarray characterization of the AggR regulon identifies a pheU pathogenicity island in enteroaggregative *Escherichia coli*. Mol. Microbiol. 61 (5), 1267–1282.

Durrer, P., Zbinden, R., Fleisch, F., et al., 2000. Intestinal infection due to enteroaggregative *Escherichia coli* among human immunodeficiency virus-infected persons [In Process Citation]. J. Infect. Dis. 182 (5), 1540–1544.

Dutta, P.R., Cappello, R., Navarro-Garcia, F., Nataro, J.P., 2002. Functional comparison of serine protease autotransporters of enterobacteriaceae. Infect. Immun. 70 (12), 7105–7113.

Dutta, S., Pal, S., Chakrabarti, S., Dutta, P., Manna, B., 1999. Use of PCR to identify enteroaggregative *Escherichia coli* as an important cause of acute diarrhoea among children living in Calcutta, India. J. Med. Microbiol. 48 (11), 1011–1016.

Elias Jr., W.P., Czeczulin, J.R., Henderson, I.R., Trabulsi, L.R., Nataro, J.P., 1999. Organization of biogenesis genes for aggregative adherence fimbria II defines a virulence gene cluster in enteroaggregative *Escherichia coli*. J. Bacteriol. 181 (6), 1779–1785.

Elias, W.P., Barros, S.F., Moreira, C.G., Trabulsi, L.R., Gomes, T.A., 2002. Enteroaggregative *Escherichia coli* strains among classical enteropathogenic *Escherichia coli* O serogroups. J. Clin. Microbiol. 40 (9), 3540–3541.

Eslava, C., Navarro-Garcia, F., Czeczulin, J.R., Henderson, I.R., Cravioto, A., Nataro, J.P., 1998. Pet, an autotransporter enterotoxin from enteroaggregative *Escherichia coli*. Infect. Immun. 66 (7), 3155–3163.

Eslava, C.V.J., Morales, R., Navarro, A., Cravioto, A., 1993. Abstracts of the 93rd General Meeting of the American Society for Microbiology 1993. American Society for Microbiology, Washington, DC. Identification of a protein with toxigenic activity produced by enteroaggregative *Escherichia coli*, abstr. B105, p. 44.

Fang, G.D., Lima, A.A., Martins, C.V., Nataro, J.P., Guerrant, R.L., 1995. Etiology and epidemiology of persistent diarrhea in northeastern Brazil: a hospital-based, prospective, case-control study. J. Pediatr. Gastroenterol Nutr. 21 (2), 137–144.

Farfan, M.J., Inman, K.G., Nataro, J.P., 2008. The major pilin subunit of the AAF/II fimbriae from enteroaggregative *Escherichia coli* mediates binding to extracellular matrix proteins. Infect. Immun. 76 (10), 4378–4384.

Farthing, M.J., 2000. Diarrhoea: a significant worldwide problem. Int. J. Antimicrob. Agents 14 (1), 65–69.

Fasano, A., Noriega, F.R., Liao, F.M., Wang, W., Levine, M.M., 1997. Effect of shigella enterotoxin 1 (ShET1) on rabbit intestine in vitro and in vivo. Gut 40 (4), 505–511.

Frank, C., Werber, D., Cramer, J.P., et al., 2011. Epidemic profile of Shiga-toxin-producing *Escherichia coli* O104:H4 outbreak in Germany. N. Engl. J. Med. 365 (19), 1771–1780.

Gascon, J., Vargas, M., Quinto, L., Corachan, M., 1998. Jimenez de Anta MT, Vila J. Enteroaggregative *Escherichia coli* strains as a cause of traveler's diarrhea: a case-control study. J. Infect. Dis. 177 (5), 1409–1412.

Gassama-Sow, A., Sow, P.S., Gueye, M., et al., 2004. Characterization of pathogenic *Escherichia coli* in human immunodeficiency virus-related diarrhea in Senegal. J. Infect. Dis. 189 (1), 75–78.

Germani, Y., Minssart, P., Vohito, M., et al., 1998. Etiologies of acute, persistent, and dysenteric diarrheas in adults in Bangui, Central African Republic, in relation to human immunodeficiency virus serostatus. Am. J. Trop. Med. Hyg. 59 (6), 1008–1014.

Glandt, M., Adachi, J.A., Mathewson, J.J., et al., 1999. Enteroaggregative *Escherichia coli* as a cause of traveler's diarrhea: clinical response to ciprofloxacin. Clin. Infect. Dis. 29 (2), 335–338.

Gomez-Duarte, O.G., Arzuza, O., Urbina, D., et al., 2010. Detection of *Escherichia coli* entero-pathogens by multiplex polymerase chain reaction from children's diarrheal stools in two Caribbean-Colombian cities. Foodborne Pathog. Dis. 7 (2), 199–206.

Gonzalez, R., Diaz, C., Marino, M., Cloralt, R., Pequeneze, M., Perez-Schael, I., 1997. Age-specific prevalence of *Escherichia coli* with localized and aggregative adherence in Venezuelan infants with acute diarrhea. J. Clin. Microbiol. 35 (5), 1103–1107.

Grad, Y.H., Lipsitch, M., Feldgarden, M., et al., 2012. Genomic epidemiology of the *Escherichia coli* O104:H4 outbreaks in Europe, 2011. Proc. Natl. Acad. Sci. USA 109 (8), 3065–3070.

Greenberg, D.E., Jiang, Z.D., Steffen, R., Verenker, M.P., DuPont, H.L., 2002. Markers of inflam-mation in bacterial diarrhea among travelers, with a focus on enteroaggregative *Escherichia coli* pathogenicity. J. Infect. Dis. 185 (7), 944–949.

Grewal, H.M., Valvatne, H., Bhan, M.K., van Dijk, L., Gaastra, W., Sommerfelt, H., 1997. A new putative fimbrial colonization factor, CS19, of human enterotoxigenic *Escherichia coli*. Infect. Immun. 65 (2), 507–513.

Guignot, J., Chaplais, C., Coconnier-Polter, M.H., Servin, A.L., 2007. The secreted autotrans-porter toxin, Sat, functions as a virulence factor in Afa/Dr diffusely adhering *Escherichia coli* by promoting lesions in tight junction of polarized epithelial cells. Cell Microbiol. 9 (1), 204–221.

Guyer, D.M., Henderson, I.R., Nataro, J.P., Mobley, H.L., 2000. Identification of sat, an autotrans-porter toxin produced by uropathogenic *Escherichia coli*. Mol. Microbiol. 38 (1), 53–66.

Harrington, S.M., Dudley, E.G., Nataro, J.P., 2006. Pathogenesis of enteroaggregative *Escherichia coli* infection. FEMS Microbiol. Lett. 254 (1), 12–18.

Harrington, S.M., Sheikh, J., Henderson, I.R., Ruiz-Perez, F., Cohen, P.S., Nataro, J.P., 2009. The Pic protease of enteroaggregative *Escherichia coli* promotes intestinal colonization and growth in the presence of mucin. Infect. Immun. 77 (6), 2465–2473.

Harrington, S.M., Strauman, M.C., Abe, C.M., Nataro, J.P., 2005. Aggregative adherence fimbriae contribute to the inflammatory response of epithelial cells infected with enteroaggregative *Escherichia coli*. Cell Microbiol. 7 (11), 1565–1578.

Hart, E., Yang, J., Tauschek, M., et al., 2008. RegA, an AraC-like protein, is a global transcriptional regulator that controls virulence gene expression in. Citrobacter rodentium. Infect. Immun. 76 (11), 5247–5256.

Henderson, I.R., Cappello, R., Nataro, J.P., 2000. Autotransporter proteins, evolution and redefining protein secretion. Trends. Microbiol. 8 (12), 529–532.

Henderson, I.R., Czeczulin, J., Eslava, C., Noriega, F., Nataro, J.P., 1999a. Characterization of pic, a secreted protease of *Shigella flexneri* and enteroaggregative *Escherichia coli*. Infect. Immun. 67 (11), 5587–5596.

Henderson, I.R., Hicks, S., Navarro-Garcia, F., Elias, W.P., Philips, A.D., Nataro, J.P., 1999b. Involvement of the enteroaggregative *Escherichia coli* plasmid-encoded toxin in causing human intestinal damage. Infect. Immun. 67 (10), 5338–5344.

Henderson, I.R., Nataro, J.P., 1999. A conserved motif in the hexosyltransferases. Mol. Microbiol. 33 (1), 222.

Henderson, I.R., Nataro, J.P., 2001. Virulence functions of autotransporter proteins. Infect. Immun. 69 (3), 1231–1243.

Henderson, I.R., Navarro-Garcia, F., Desvaux, M., Fernandez, R.C., Ala'Aldeen, D., 2004. Type V protein secretion pathway: the autotransporter story. Microbiol. Mol. Biol. Rev. 68 (4), 692–744.

Hicks, S., Candy, D.C., Phillips, A.D., 1996. Adhesion of enteroaggregative *Escherichia coli* to pediatric intestinal mucosa in vitro. Infect. Immun. 64 (11), 4751–4760.

Huang, D.B., Dupont, H.L., 2004. Enteroaggregative *Escherichia coli*: an emerging pathogen in children. Semin. Pediatr. Infect. Dis. 15 (4), 266–271.

Huang, D.B., Koo, H., DuPont, H.L., 2004a. Enteroaggregative *Escherichia coli*: an emerging pathogen. Curr. Infect. Dis. Rep. 6 (2), 83–86.

Huang, D.B., Mohanty, A., DuPont, H.L., Okhuysen, P.C., Chiang, T., 2006. A review of an emerging enteric pathogen: enteroaggregative *Escherichia coli*. J. Med. Microbiol. 55 (Pt 10), 1303–1311.

Huang, D.B., Nataro, J.P., DuPont, H.L., et al., 2006. Enteroaggregative *Escherichia coli* is a cause of acute diarrheal illness: a meta-analysis. Clin. Infect. Dis. 43 (5), 556–563.

Huang, D.B., Okhuysen, P.C., Jiang, Z.D., DuPont, H.L., 2004b. Enteroaggregative *Escherichia coli*: an emerging enteric pathogen. Am. J. Gastroenterol. 99 (2), 383–389.

Huppertz, H.I., Rutkowski, S., Aleksic, S., Karch, H., 1997. Acute and chronic diarrhoea and abdominal colic associated with enteroaggregative *Escherichia coli* in young children living in western Europe. Lancet 349, 1660–1662.

Ina, K., Kusugami, K., Ohta, M., 2003. Bacterial hemorrhagic enterocolitis. J. Gastroenterol. 38 (2), 111–120.

Infante, R.M., Ericsson, C.D., Jiang, Z.D., et al., 2004. Enteroaggregative *Escherichia coli* diarrhea in travelers: response to rifaximin therapy. Clin. Gastroenterol. Hepatol. 2 (2), 135–138.

Itoh, Y., Nagano, I., Kunishima, M., Ezaki, T., 1997. Laboratory investigation of enteroaggregative *Escherichia coli* O untypeable:H10 associated with a massive outbreak of gastrointestinal illness. J. Clin. Microbiol. 35 (10), 2546–2550.

Jalaluddin, S., de Mol, P., Hemelhof, W., et al., 1998. Isolation and characterization of enteroaggregative *Escherichia coli* (EAggEC) by genotypic and phenotypic markers, isolated from diarrheal children in Congo. Clin. Microbiol. Infect. 4 (4), 213–219.

Jenkins, C., van Ijperen, C., Dudley, E.G., et al., 2005. Use of a microarray to assess the distribution of plasmid and chromosomal virulence genes in strains of enteroaggregative *Escherichia coli*. FEMS Microbiol. Lett. 253 (1), 119–124.

Jiang, Z.D., Greenberg, D., Nataro, J.P., Steffen, R., DuPont, H.L., 2002. Rate of occurrence and pathogenic effect of enteroaggregative *Escherichia coli* virulence factors in international travelers. J. Clin. Microbiol. 40 (11), 4185–4190.

Jiang, Z.D., Okhuysen, P.C., Guo, D.C., et al., 2003. Genetic susceptibility to enteroaggregative *Escherichia coli* diarrhea: polymorphism in the interleukin-8 promotor region. J. Infect. Dis. 188 (4), 506–511.

Kahali, S., Sarkar, B., Rajendran, K., et al., 2004. Virulence characteristics and molecular epidemiology of enteroaggregative *Escherichia coli* isolates from hospitalized diarrheal patients in Kolkata, India. J. Clin. Microbiol. 42 (9), 4111–4120.

Kaper, J.B., Nataro, J.P., Mobley, H.L., 2004. Pathogenic *Escherichia coli*. Nat. Rev. Microbiol. 2 (2), 123–140.

Klein, S.A., Klebba, C., Kauschat, D., et al., 2000. Interleukin-18 stimulates HIV-1 replication in a T-cell line. Eur. Cytokine. Netw. 11 (1), 47–52.

Lima, A.A., Guerrant, R.L., 1992. Persistent diarrhea in children: epidemiology, risk factors, pathophysiology, nutritional impact, and management. Epidemiol. Rev. 14, 222–242.

Lima, A.A., Moore, S.R., Barboza Jr., M.S., et al., 2000. Persistent diarrhea signals a critical period of increased diarrhea burdens and nutritional shortfalls: a prospective cohort study among children in northeastern Brazil. J. Infect. Dis. 181 (5), 1643–1651.

Maroncle, N.M., Sivick, K.E., Brady, R., Stokes, F.E., Mobley, H.L., 2006. Protease activity, secretion, cell entry, cytotoxicity, and cellular targets of secreted autotransporter toxin of uropathogenic *Escherichia coli*. Infect. Immun. 74 (11), 6124–6134.

Menard, L.P., Dubreuil, J.D., 2002. Enteroaggregative *Escherichia coli* heat-stable enterotoxin 1 (EAST1): a new toxin with an old twist. Crit. Rev. Microbiol. 28 (1), 43–60

Menard, L.P., Lussier, J.G., Lepine, F., Paiva de Sousa, C., Dubreuil, J.D., 2004. Expression, purification, and biochemical characterization of enteroaggregative *Escherichia coli* heat-stable enterotoxin 1. Protein Expr. Purif. 33 (2), 223–231.

Morabito, S., Karch, H., Mariani-Kurkdjian, P., et al., 1998. Enteroaggregative, Shiga toxin-producing *Escherichia coli* O111:H2 associated with an outbreak of hemolytic-uremic syndrome. J. Clin. Microbiol. 36 (3), 840–842.

Mossoro, C., Glaziou, P., Yassibanda, S., et al., 2002. Chronic diarrhea, hemorrhagic colitis, and hemolytic-uremic syndrome associated with HEp-2 adherent *Escherichia coli* in adults infected with human immunodeficiency virus in Bangui, Central African Republic. J. Clin. Microbiol. 40 (8), 3086–3088.

Nataro, J.P., 2005. Enteroaggregative *Escherichia coli* pathogenesis. Curr. Opin. Gastroenterol. 21 (1), 4–8.

Nataro, J.P., Baldini, M.M., Kaper, J.B., Black, R.E., Bravo, N., Levine, M.M., 1985. Detection of an adherence factor of enteropathogenic *Escherichia coli* with a DNA probe. J. Infect. Dis. 152 (3), 560–565.

Nataro, J.P., Deng, Y., Cookson, S., et al., 1995. Heterogeneity of enteroaggregative *Escherichia coli* virulence demonstrated in volunteers. J. Infect. Dis. 171 (2), 465–468.

Nataro, J.P., Deng, Y., Maneval, D.R., German, A.L., Martin, W.C., Levine, M.M., 1992. Aggregative adherence fimbriae I of enteroaggregative *Escherichia coli* mediate adherence to HEp-2 cells and hemagglutination of human erythrocytes. Infect. Immun. 60 (6), 2297–2304.

Nataro, J.P., Kaper, J.B., 1998. Diarrheagenic *Escherichia coli*. Clin. Microbiol. Rev. 11 (1), 142–201.

Nataro, J.P., Kaper, J.B., Robins-Browne, R., Prado, V., Vial, P., Levine, M.M., 1987. Patterns of adherence of diarrheagenic *Escherichia coli* to HEp-2 cells. Pediatr. Infect. Dis. J. 6 (9), 829–831.

Nataro, J.P., Steiner, T., Guerrant, R.L., 1998. Enteroaggregative *Escherichia coli*. Emerg. Infect. Dis. 4 (2), 251–261.

Nataro, J.P., Mai, V., Johnson, J., et al., 2006. Diarrheagenic *Escherichia coli* infection in Baltimore, Maryland, and New Haven, Connecticut. Clin. Infect. Dis. 43 (4), 402–407.

Navarro-Garcia, F., Canizalez-Roman, A., Luna, J., Sears, C., Nataro, J.P., 2001. Plasmid-encoded toxin of enteroaggregative *Escherichia coli* is internalized by epithelial cells. Infect. Immun. 69 (2), 1053–1060.

Navarro-Garcia, F., Eslava, C., Villaseca, J.M., et al., 1998. In vitro effects of a high-molecular-weight heat-labile enterotoxin from enteroaggregative *Escherichia coli*. Infect. Immun. 66 (7), 3149–3154.

Nishi, J., Sheikh, J., Mizuguchi, K., et al., 2003. The export of coat protein from enteroaggregative *Escherichia coli* by a specific ATP-binding cassette transporter system. J. Biol. Chem. 278 (46), 45680–45689.

Okeke, I.N., 2009. Diarrheagenic *Escherichia coli* in sub-Saharan Africa: status, uncertainties and necessities. J. Infect. Dev. Ctries. 3 (11), 817–842.

Okeke, I.N., Lamikanra, A., Czeczulin, J., Dubovsky, F., Kaper, J.B., Nataro, J.P., 2000a. Heterogeneous virulence of enteroaggregative *Escherichia coli* strains isolated from children in Southwest Nigeria. J. Infect. Dis. 181 (1), 252–260.

Okeke, I.N., Lamikanra, A., Steinruck, H., Kaper, J.B., 2000b. Characterization of *Escherichia coli* strains from cases of childhood diarrhea in provincial southwestern Nigeria. J. Clin. Microbiol. 38 (1), 7–12.

Okeke, I.N., Nataro, J.P., 2001. Enteroaggregative *Escherichia coli*. Lancet Infect. Dis. 1 (5), 304–313.

Okhuysen, P.C., Dupont, H.L., 2010. Enteroaggregative *Escherichia coli* (EAEC): A cause of acute and persistent diarrhea of worldwide importance. J. Infect. Dis. 202 (4), 503–505.

Olesen, B., Neimann, J., Bottiger, B., et al., 2005. Etiology of diarrhea in young children in Denmark: a case-control study. J. Clin. Microbiol. 43 (8), 3636–3641.

Opintan, J.A., Bishar, R.A., Newman, M.J., Okeke, I.N., 2010. Carriage of diarrhoeagenic *Escherichia coli* by older children and adults in Accra, Ghana. Trans. R. Soc. Trop. Med. Hyg. 104 (7), 504–506.

Paiva de Sousa, C., Dubreuil, J.D., 2001. Distribution and expression of the astA gene (EAST1 toxin) in *Escherichia coli* and *Salmonella*. Int. J. Med. Microbiol. 291 (1), 15–20.

Patel, S.K., Dotson, J., Allen, K.P., Fleckenstein, J.M., 2004. Identification and molecular characterization of EatA, an autotransporter protein of enterotoxigenic *Escherichia coli*. Infect. Immun. 72 (3), 1786–1794.

Pereira, A.L., Ferraz, L.R., Silva, R.S., Giugliano, L.G., 2007. Enteroaggregative *Escherichia coli* virulence markers: positive association with distinct clinical characteristics and segregation into 3 enteropathogenic *E. coli* serogroups. J. Infect. Dis. 195 (3), 366–374.

Petri Jr., W.A., Miller, M., Binder, H.J., Levine, M.M., Dillingham, R., Guerrant, R.L., 2008. Enteric infections, diarrhea, and their impact on function and development. J. Clin. Invest. 118 (4), 1277–1290.

Piva, I.C., Pereira, A.L., Ferraz, L.R., et al., 2003. Virulence markers of enteroaggregative *Escherichia coli* isolated from children and adults with diarrhea in Brasilia. Brazil. J. Clin. Microbiol. 41 (5), 1827–1832.

Presterl, E., Nadrchal, R., Wolf, D., Rotter, M., Hirschl, A.M., 1999. Enteroaggregative and enterotoxigenic *Escherichia coli* among isolates from patients with diarrhea in Austria. Eur. J. Clin. Microbiol. Infect. Dis. 18 (3), 209–212.

Prouty, M.G., Osorio, C.R., Klose, K.E., 2005. Characterization of functional domains of the *Vibrio cholerae* virulence regulator ToxT. Mol. Microbiol. 58 (4), 1143–1156.

Provence, D.L., Curtiss 3rd, R., 1994. Isolation and characterization of a gene involved in hemagglutination by an avian pathogenic *Escherichia coli* strain. Infect. Immun. 62 (4), 1369–1380.

Qadri, F., Haque, A., Faruque, S.M., Bettelheim, K.A., Robins-Browne, R., Albert, M.J., 1994. Hemagglutinating properties of enteroaggregative *Escherichia coli*. J. Clin. Microbiol. 32 (2), 510–514.

Rajakumar, K., Sasakawa, C., Adler, B., 1997. Use of a novel approach, termed island probing, identifies the *Shigella flexneri* she pathogenicity island which encodes a homolog of the immunoglobulin A protease-like family of proteins. Infect. Immun. 65 (11), 4606–4614.

Rasko, D.A., Webster, D.R., Sahl, J.W., et al., 2011. Origins of the *E. coli* strain causing an outbreak of hemolytic-uremic syndrome in Germany. N. Engl. J. Med. 365 (8), 709–717.

Regua-Mangia, A.H., Gomes, T.A., Vieira, M.A., Andraqde, J.R., Irino, K., Teixeira, L.M., 2004. Frequency and characteristics of diarrhoeagenic *Escherichia coli* strains isolated from children with and without diarrhoea in Rio de Janeiro. Brazil. J. Infect. 48 (2), 161–167.

Regua-Mangia, A.H., Gomes, T.A., Vieira, M.A., Irino, K., Teixeira, L.M., 2009. Molecular typing and virulence of enteroaggregative *Escherichia coli* strains isolated from children with and without diarrhoea in Rio de Janeiro city, Brazil. J. Med. Microbiol. 58 (Pt 4), 414–422.

Rugeles, L.C., Bai, J., Martinez, A.J., Vanegas, M.C., Gomez-Duarte, O.G., 2010. Molecular characterization of diarrheagenic *Escherichia coli* strains from stools samples and food products in Colombia. Int. J. Food Microbiol. 138 (3), 282–286.

Ruiz-Perez, F., Henderson, I.R., Leyton, D.L., Rossiter, A.E., Zhang, Y., Nataro, J.P., 2009. Roles of periplasmic chaperone proteins in the biogenesis of serine protease autotransporters of Enterobacteriaceae. J. Bacteriol. 191 (21), 6571–6583.

Sarantuya, J., Nishi, J., Wakimoto, N., et al., 2004. Typical enteroaggregative *Escherichia coli* is the most prevalent pathotype among *E. coli* strains causing diarrhea in Mongolian children. J. Clin. Microbiol. 42 (1), 133–139.

Savarino, S.J., Fasano, A., Robertson, D.C., Levine, M.M., 1991. Enteroaggregative *Escherichia coli* elaborate a heat-stable enterotoxin demonstrable in an in vitro rabbit intestinal model. J. Clin. Invest. 87 (4), 1450–1455.

Savarino, S.J., Fasano, A., Watson, J., et al., 1993. Enteroaggregative *Escherichia coli* heat-stable enterotoxin 1 represents another subfamily of *E. coli* heat-stable toxin. Proc. Natl. Acad. Sci. USA 90 (7), 3093–3097.

Savarino, S.J., Fox, P., Deng, Y., Nataro, J.P., 1994. Identification and characterization of a gene cluster mediating enteroaggregative *Escherichia coli* aggregative adherence fimbria I biogenesis. J. Bacteriol. 176 (16), 4949–4957.

Savarino, S.J., McVeigh, A., Watson, J., et al., 1996. Enteroaggregative *Escherichia coli* heat-stable enterotoxin is not restricted to enteroaggregative *E. coli*. J. Infect. Dis. 173 (4), 1019–1022.

Scheutz, F., Nielsen, E.M., Frimodt-Moller, J., et al., 2011. Characteristics of the enteroaggregative Shiga toxin/verotoxin-producing *Escherichia coli* O104:H4 strain causing the outbreak of haemolytic uraemic syndrome in Germany, May to June 2011. Euro. Surveill. 16 (24). pii=19889. Available online: http://www.eurosurveillance.org/ViewArticle.aspx?ArticleId=19889.

Schubert, S., Rakin, A., Karch, H., Carniel, E., Heesemann, J., 1998. Prevalence of the high-pathogenicity island of Yersinia species among *Escherichia coli* strains that are pathogenic to humans. Infect. Immun. 66 (2), 480–485.

Servin, A.L., 2005. Pathogenesis of Afa/Dr diffusely adhering *Escherichia coli*. Clin. Microbiol. Rev. 18 (2), 264–292.

Shah, N., DuPont, H.L., Ramsey, D.J., 2009. Global etiology of travelers' diarrhea: systematic review from 1973 to the present. Am. J. Trop. Med. Hyg. 80 (4), 609–614.

Shazberg, G., Wolk, M., Schmidt, H., Sechter, I., Gottesman, G., Miron, D., 2003. Enteroaggregative *Escherichia coli* serotype O126:H27, Israel. Emerg. Infect. Dis. 9 (9), 1170–1173.

Sheikh, J., Czeczulin, J.R., Harrington, S., et al., 2002. A novel dispersin protein in enteroaggregative *Escherichia coli*. J. Clin. Invest. 110 (9), 1329–1337.

Sheikh, J., Dudley, E.G., Sui, B., Tamboura, B., Suleman, A., Nataro, J.P., 2006. EilA, a HilA-like regulator in enteroaggregative *Escherichia coli*. Mol. Microbiol. 61 (2), 338–350.

Sheikh, J., Hicks, S., Dall'Agnol, M., Phillips, A.D., Nataro, J.P., 2001. Roles for Fis and YafK in biofilm formation by enteroaggregative *Escherichia coli*. Mol. Microbiol. 41 (5), 983–997.

Smith, H.R., Cheasty, T., Rowe, B., 1997. Enteroaggregative *Escherichia coli* and outbreaks of gastroenteritis in UK. Lancet 350, 814–815.

Spencer, J., Smith, H.R., Chart, H., 1999. Characterization of enteroaggregative *Escherichia coli* isolated from outbreaks of diarrhoeal disease in England. Epidemiol. Infect. 123 (3), 413–421.

Stein, M., Kenny, B., Stein, M.A., Finlay, B.B., 1996. Characterization of EspC, a 110-kilodalton protein secreted by enteropathogenic *Escherichia coli* which is homologous to members of the immunoglobulin A protease-like family of secreted proteins. J. Bacteriol. 178 (22), 6546–6554.

Steiner, T., Flores, C., Pizarro, T., Guerrant, R., 1997. Fecal lactoferrin, interleukin-1ß, and interleukin-8 are elevated in patients with severe *Clostridium difficile* colitis. Clin. Diag. Lab. Immunol. 4 (6), 179–722.

Steiner, T.S., Lima, A.A., Nataro, J.P., Guerrant, R.L., 1998. Enteroaggregative *Escherichia coli* produce intestinal inflammation and growth impairment and cause interleukin-8 release from intestinal epithelial cells. J. Infect. Dis. 177 (1), 88–96.

Steiner, T.S., Nataro, J.P., Poteet-Smith, C.E., Smith, J.A., Guerrant, R.L., 2000. Enteroaggregative *Escherichia coli* expresses a novel flagellin that causes IL-8 release from intestinal epithelial cells. J. Clin. Invest. 105 (12), 1769–1777.

Strauman, M.C., Harper, J.M., Harrington, S.M., Boll, E.J., Nataro, J.P., 2010. Enteroaggregative *Escherichia coli* disrupts epithelial cell tight junctions. Infect. Immun. 78 (11), 4958–4964.

Sui, B.Q., Dutta, P.R., Nataro, J.P., 2003. Intracellular expression of the plasmid-encoded toxin from enteroaggregative *Escherichia coli*. Infect. Immun. 71 (9), 5364–5370.

Suzart, S., Aparecida, T., Gomes, T., Guth, B.E., 1999. Characterization of serotypes and outer membrane protein profiles in enteroaggregative *Escherichia coli* strains. Microbiol. Immunol. 43 (3), 201–205.

Tompkins, D.S., Hudson, M.J., Smith, H.R., et al., 1999. A study of infectious intestinal disease in England: microbiological findings in cases and controls. Commun. Dis. Public Health 2 (2), 108–113.

Tzipori, S., Montanaro, J., Robins-Browne, R.M., Vial, P., Gibson, R., Levine, M.M., 1992. Studies with enteroaggregative *Escherichia coli* in the gnotobiotic piglet gastroenteritis model. Infect. Immun. 60 (12), 5302–5306.

Uber, A.P., Trabulsi, L.R., Irino, K., et al., 2006. Enteroaggregative *Escherichia coli* from humans and animals differ in major phenotypical traits and virulence genes. FEMS Microbiol. Lett. 256 (2), 251–257.

Valentiner-Branth, P., Steinsland, H., Fischer, T.K., et al., 2003. Cohort study of Guinean children: incidence, pathogenicity, conferred protection, and attributable risk for enteropathogens during the first 2 years of life. J. Clin. Microbiol. 41 (9), 4238–4245.

Valentiner-Branth, P., Steinsland, H., Santos, G., et al., 2001. Community-based controlled trial of dietary management of children with persistent diarrhea: sustained beneficial effect on ponderal and linear growth. Am. J. Clin. Nutr. 73 (5), 968–974.

Vial, P.A., Robins-Browne, R., Lior, H., et al., 1988. Characterization of enteroadherent-aggregative *Escherichia coli*, a putative agent of diarrheal disease. J. Infect. Dis. 158 (1), 70–79.

Vila, J., Vargas, M., Henderson, I.R., Gascon, J., Nataro, J.P., 2000. Enteroaggregative *Escherichia coli* virulence factors in traveler's diarrhea strains. J. Infect. Dis. 182 (6), 1780–1783.

Villaseca, J.M., Navarro-Garcia, F., Mendoza-Hernandez, G., Nataro, J.P., Cravioto, A., Eslava, C., 2000. Pet toxin from enteroaggregative *Escherichia coli* produces cellular damage associated with fodrin disruption. Infect. Immun. 68 (10), 5920–5927.

Wanke, C.A., 2001. To know *Escherichia coli* is to know bacterial diarrheal disease. Clin. Infect. Dis. 32 (12), 1710–1712.

Wanke, C.A., Gerrior, J., Blais, V., Mayer, H., Acheson, D., 1998a. Successful treatment of diarrheal disease associated with enteroaggregative *Escherichia coli* in adults infected with human immunodeficiency virus. J. Infect. Dis. 178 (5), 1369–1372.

Wanke, C.A., Mayer, H., Weber, R., Zbinden, R., Watson, D.A., Acheson, D., 1998b. Enteroaggregative *Escherichia coli* as a potential cause of diarrheal disease in adults infected with human immunodeficiency virus. J. Infect. Dis. 178 (1), 185–190.

Wanke, C.A., Schorling, J.B., Barrett, L.J., Desouza, M.A., Guerrant, R.L., 1991. Potential role of adherence traits of *Escherichia coli* in persistent diarrhea in an urban Brazilian slum. Pediatr. Infect. Dis. J. 10 (10), 746–751.

Yamamoto, T., Echeverria, P., 1996. Detection of the enteroaggregative *Escherichia coli* heat-stable enterotoxin 1 gene sequences in enterotoxigenic *E. coli* strains pathogenic for humans. Infect. Immun. 64 (4), 1441–1445.

Yamamoto, T., Echeverria, P., Yokota, T., 1992. Drug resistance and adherence to human intestines of enteroaggregative *Escherichia coli*. J. Infect. Dis. 165 (4), 744–749.

Yamazaki, M., Inuzuka, K., Matsui, H., et al., 2000. Plasmid encoded enterotoxin (Pet) gene in enteroaggregative *Escherichia coli* isolated from sporadic diarrhea cases. Jpn. J. Infect. Dis. 53 (6), 248–249.

Zamboni, A., Fabbricotti, S.H., Fagundes-Neto, U., Scaletsky, I.C., 2004. Enteroaggregative *Escherichia coli* virulence factors are found to be associated with infantile diarrhea in Brazil. J. Clin. Microbiol. 42 (3), 1058–1063.

Zhou, Z., Ogasawara, J., Nishikawa, Y., et al., 2002. An outbreak of gastroenteritis in Osaka, Japan due to *Escherichia coli* serogroup O166:H15 that had a coding gene for enteroaggregative *E. coli* heat-stable enterotoxin 1 (EAST1). Epidemiol. Infect. 128 (3), 363–371.

Chapter 9

Uropathogenic *Escherichia coli*

Rachel R. Spurbeck and Harry L.T. Mobley
University of Michigan Medical School, Ann Arbor, MI, USA

BACKGROUND

Classification and evolution of uropathogenic *E. coli*

Uropathogenic *Escherichia coli* (UPEC) are a heterogeneous group of strains within the broader classification of extraintestinal pathogenic *E. coli*. There is no core set of virulence factors shared by all UPEC. Most UPEC belong to B2 and D phylogroups (Chapter 1) and can be classified based on the O (lipopolysaccharide) serotype into uropathogenic clones, with 58% of UPEC isolates belonging to one of eight serogroups (O1, O2, O4, O6, O8, O9, O18, and O83) (Blanco et al., 1994, 1996). UPEC can be classified by the type or severity of infection from which the bacteria are isolated. Pyelonephritis or urosepsis isolates cause the most severe UTI, infecting the kidneys and allowing access to the bloodstream. Cystitis isolates cause bladder infections, and asymptomatic bacteriuria (ABU) strains stably colonize the urinary tract without causing disease symptoms in an almost commensal state. Comparative genomic hybridization comparing the genomic content of ten UPEC strains and four fecal/commensal strains revealed 10 pathogen-specific genomic islands comprising 13% of the genome of pyelonephritis isolate *E. coli* CFT073 (Lloyd et al., 2007). Three islands were known pathogenicity islands containing P fimbriae and other characterized virulence factors. Furthermore, 52% of the genome is shared among UPEC and commensal isolates, and only 131 genes of the 5379 genes present in *E. coli* CFT073 were UPEC-specific (Lloyd et al., 2007).

ABU isolates, like UPEC isolates that cause symptomatic disease, have expanded genomes when compared to fecal commensal *E. coli*. However, a study of 112 ABU isolates, in which bacteriuria lasted from 1–74 days with at least one urine culture of $\geq 10^5$ CFU/ml, suggests that ABU isolates have undergone reductive evolution within human hosts. Point mutations and deletions in genes encoding virulence factors reduce the virulence of the ABU isolates over time (Salvador et al., 2012). Therefore, UPEC acquired virulence through horizontal gene transfer events, such as acquisition of PAIs, but reductive evolution attenuates ABU strains into a more commensal-like state.

Escherichia coli. http://dx.doi.org/10.1016/B978-0-12-397048-0.00009-7

Epidemiology and global impact

Urinary tract infections (UTIs) are the second most common infection of humans, causing 8.1 million physician visits in the United States in 2007 (Schappert and Rechtsteiner, 2008), and an estimated 10 million cases in western Europe. UTIs are usually caused by bacteria from the intestinal tract contaminating the peri-urethral opening, followed by ascension of the urinary tract. Women are more susceptible to UTI due to the relative proximity of the periurethral space to the anus and to the fact that the urethra is much shorter than that of men, thus allowing easier access to the bladder by fecal bacteria. It is estimated that over 50% of all women will have a UTI in their lifetime, 25% will then experience a second UTI, and 3% will have a third UTI within 6 months of the initial infection (Foxman, 2003). *E. coli* is the primary cause of UTIs, responsible for an estimated 80% of all uncomplicated UTIs (Stamm and Hooton, 1993; Ronald, 2002) and 18–35% of long-term indwelling catheter-associated UTIs (Nicolle, 2005).

MOLECULAR PATHOGENESIS

Entry and ascension of the urinary tract

UTIs caused by *E. coli* are ascending infections in which bacterial contamination of the periurethral space allows UPEC access to the otherwise sterile urinary tract. Ascension of the urinary tract by UPEC is mediated by the action of flagella, which propel bacteria up the urethra to the bladder, where the bacteria utilize a multitude of fimbriae and non-fimbrial adhesins to colonize, iron acquisition systems to facilitate growth, and toxins to avoid the innate immune response (Table 9.1 and Figure 9.1). A subpopulation of *E. coli* can detach from the bladder and move up the ureters to the kidneys (mediated by flagella) (Lane et al., 2007a; Walters et al., 2012). Flagellar motility is essential for uropathogenesis, as non-motile mutants of UPEC that cannot express FliC, the main component of the flagellum, are attenuated in the mouse model of ascending UTI (Wright et al., 2005; Lane et al., 2007a). *E. coli* reciprocally regulates fimbrial and flagellar biosynthesis so that when fimbriae are expressed, flagella are down-regulated causing a shift between motile and sessile lifestyles (Lane et al., 2007b). Several transcription factors, including H-NS (Korea et al., 2010), Lrp (Simms and Mobley, 2008a), cAMP-CRP (Yokota and Gots, 1970; Muller et al., 2009), and LeuX (Ritter et al., 1995), as well as non-specific regulators, such as DNA topology (Dorman and Corcoran, 2009), can affect the switch between motility and sessility. Most of these transcription factors regulate the expression of the recombinases FimB and FimE, which affect the orientation of the promoter of the type 1 fimbrial operon as described below (Ritter et al., 1995; Simms and Mobley, 2008b; Dorman and Corcoran, 2009; Muller et al., 2009). Simultaneously, the same regulators affect the transcription of *flhDC*, which encode the master regulator of flagellar biosynthesis, in an opposite manner, so that when flagella are expressed, fimbrial biosynthesis is repressed.

TABLE 9.1 Evidence that virulence factors contribute to colonization and pathogenesis of urinary tract infection

Factor	Epidemiology[1]	Attenuated Mutant[2]	Complementation[3]	Volunteers[4]
Type 1 fimbriae	✓	✓	✓	
P fimbriae	✓	✓		✓
Ygi fimbriae	✓	✓	✓	
Yad fimbriae	✓	✓		
Auf fimbriae	✓			
F9 fimbriae	✓			
F1C fimbriae	✓			
S fimbriae	✓	✓[6]	✓	
Yfc fimbriae	✓			
Dr adhesins[5]	✓	✓		
Type IV pilus 2	✓			
TosA	✓	✓		
FdeC		✓		
Ag43a	✓	✓		
UpaB		✓		
UpaG	✓			
UpaH		✓		
TonB	✓	✓		
Hemolysin	✓	✓[7]		
CNF-1	✓	✓		
LPS O-antigen	✓	✓		
proP, guaA, argC		✓		
Sat	✓	✓[7]		
Pic	✓			
Tsh	✓			
ChuA	✓	✓		

Continued

TABLE 9.1 Evidence that virulence factors contribute to colonization and pathogenesis of urinary tract infection—cont'd

Factor	Epidemiology[1]	Attenuated Mutant[2]	Complementation[3]	Volunteers[4]
Hma	✓	✓		
IutA	✓	✓		
IroN	✓	✓		
FyuA	✓			
Iha	✓	✓		
IreA		✓		
TraT	✓			
CdtB	✓			
IbeA	✓			
OmpT	✓			
K1 kpsM	✓			
kpsM II	✓			
RfaH		✓		
TcpC				
DraD/AfaD				
Flagella				
DegS		✓	✓	
DegP		✓	✓	
PhoU		✓	✓	
PhoP		✓	✓	
SisA and SisB	✓ (SisA only)	✓	✓	

[1]Gene or expression more frequent in UPEC than control strains.
[2]Mutant in gene less able to cause UTI in animal model.
[3]Ability to cause UTI restored to mutant by re-introduction of gene.
[4]Evidence of role in UTI from experiments in humans.
[5]Attenuation and complementation confirmed in murine model of chronic pyelonephritis.
[6]Attenuated mutant also was hemolysin and serum resistance negative.
[7]Did not reduce colonization in mouse model, however, there was significantly less damage observed by histology.

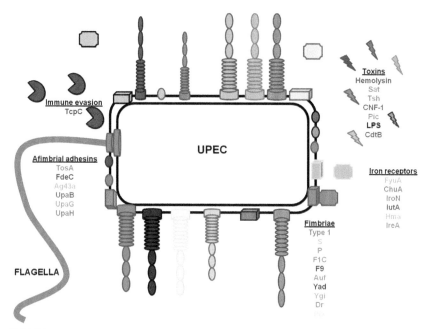

FIGURE 9.1   UPEC encode a variety of virulence factors. Depicted here are the fimbriae, iron receptors, afimbrial adhesins, flagella, toxins, and immune evasion proteins that have been tested in the murine model of ascending UTI.

Fimbrial operons can encode regulatory elements themselves that affect motility and flagellar expression. For example, a regulatory gene associated with the Sfa_{II} fimbrial operon, $sfaX_{II}$, when overproduced, reduces the amount of FliC, corresponding with lower motility. This same protein also negatively affects type 1 fimbrial expression (Sjostrom et al., 2009). PapX, a protein encoded by the P fimbrial operon that has 96% amino acid sequence identity to SfaX, also inhibits flagellar gene expression in UPEC (Simms and Mobley, 2008b). Although the mechanisms are unknown, the expression of type 1 and Yad fimbriae also inhibit flagellar expression (Lane et al., 2007b; Simms and Mobley, 2008a; Spurbeck et al., 2011), demonstrating that production of fimbriae can have indirect effects on motility.

Adherence

Colonization of the urinary tract is an essential step in the infection process. Without binding to the host epithelium, bacteria would be washed out of the urinary tract by the force of urination. UPEC have several different mechanisms for adherence to the uroepithelium, including: fimbriae, rod-like appendages that protrude from the bacterial cell surface culminating in a tip adhesin; fibrillae, flexible, extended conformations with adhesins present throughout the structure (i.e. not only at the tip); and afimbrial adhesins such as autotransporters. Extensive regulatory systems coordinate expression of these different adhesins (Snyder et al., 2005). Study of the regulatory networks coordinating expression of fimbrial adhesins is an active area of research, which could eventually lead to the development of novel therapeutic targets.

Fimbriae

The prototypical pyelonephritis strain *E. coli* CFT073 encodes 12 fimbrial operons in its genome (Welch et al., 2002) including 10 chaperone-usher fimbriae (Chapter 12) and two putative type IV pili (Chapter 13). Eight of the chaperone-usher fimbriae have been studied for effects on uropathogenesis: type 1, F1C, Auf, Ygi, Yad, F9, and two P fimbriae. The remaining four, Yeh; Yfc; and two type IV pili, remain to be studied. Pix fimbriae have been associated with uropathogenic *E. coli* and have been studied in the pyelonephitis isolate *E. coli* 536, however, molecular epidemiologic data demonstrate that this fimbria is rare (Spurbeck et al., 2011). Dr fimbriae are fibrillar structures that have been associated with uropathogenic strains of *E. coli* as well as diffuse adhering *E. coli* (DAEC), and will be discussed below.

Type 1 fimbriae

Type 1 fimbriae, encoded in the genomes of most (99%) *E. coli* strains (Vigil et al., 2011a, b), are necessary for colonization of the oropharynx as a precursor to intestinal colonization (Orndorff and Bloch, 1990). Type 1 fimbriae primarily consist of FimA, the main structural subunit, several minor subunits, and the adhesin, FimH, which is found at the tip of the fimbriae as well as intermittently throughout the shaft (Klemm et al., 1990; Krogfelt et al., 1990). These fimbriae enhance pathogenicity in the urinary tract by mediating adherence to and invasion of bladder epithelial cells by binding mannose moieties (Brinton, 1959) and uroplakin 1a with the tip adhesin FimH (Zhou et al., 2001), as well as by binding to muscle cells and vascular walls (Virkola et al., 1988). Type 1 fimbriae cause damage to the urinary tract by increasing inflammation during infection (Connell et al., 1996). The promoter for type 1 fimbriae resides within an invertible element (designated the *fim* switch or *fimS*), and thus type 1 fimbriae are subject to phase variation at the level of transcription. This phase variation is regulated by the activity of two recombinases, FimB and FimE, on the *fim* switch (see below in the section on the regulation hierarchy of fimbriae). Type 1 fimbriae are a virulence factor of UPEC in experimental models, as shown in the studies by Connell et al. (1996) that demonstrated deletion of *fimH* reduced colonization and inflammation in the mouse model of ascending UTI. Furthermore, colonization of the murine urinary tract and inflammatogenicity were restored in the *fimH* mutant by expression of *fimH* from a plasmid, fulfilling Koch's postulates (Connell et al., 1996).

P fimbriae

P fimbria was the first UPEC virulence factor characterized (Eden et al., 1976). The *pap* (pyelonephritis associated pili) operon encoding P fimbriae consists of 11 genes, with the main fimbrial subunit encoded by *papA* and the tip adhesin encoded by *papG* (Hull et al., 1981). The adherence mediated by P fimbriae is mannose-resistant, unlike that of type 1 fimbriae, and is specific to P blood group antigen, a glycosphingolipid, in which the key epitope is a digalactoside, α-D-Gal-(1-4)-β-D-Gal (Källenius and Möllby, 1979). There are three alleles of *papG*, each with substrate specificity to a different region of the Gal-Gal disaccharide-containing glycosphingolipids (Stromberg et al., 1991). P fimbriae are encoded in the genomes of 54–70% of UPEC isolates as compared to 20–25% of fecal *E. coli* strains, with the *papG* allelic distribution being *papG*1 (0–1% UPEC, 0–1% fecal), *papG*2 (36–46% UPEC, 16–18% fecal), *papG*3 (17–23% UPEC, 5–9% fecal) (Johanson et al., 1993; Johnson et al., 1998; Spurbeck et al., 2011). P fimbriae enhance the severity of UTI by promoting adherence strongly to vascular endothelium, the muscular layer, and weakly to bladder epithelial surfaces (Virkola et al., 1988), as well as increasing mucosal inflammation. Although there is a positive correlation between severity of infection and presence of P fimbriae (Spurbeck et al., 2011; Vigil et al., 2011b), and antibodies to P fimbriae are present in the serum of infected individuals (de Ree and van den Bosch, 1987), there is no defect in colonization of the mouse model of ascending UTI in the P fimbria isogenic mutant (Mobley et al., 1993). However, in a pyelonephritis model in cynomolgus monkeys wild-type *E. coli* DS17 colonized the urinary tract longer and caused a loss of kidney function, whereas *E. coli* DS17-8 (Δ*papG*) was cleared earlier and did not cause as extensive kidney damage (Roberts et al., 1994). While P fimbriae are thought to play only a subtle role in uropathogenesis, the presence of the *pap* operon is clearly a marker of highly virulent strains.

F1C fimbriae

In a recent study, the prevalence of F1C fimbriae was not significantly different between UPEC (16%) and fecal *E. coli* strains (10%) (Spurbeck et al., 2011), however, other studies have shown that the presence of F1C fimbriae correlates with pyelonephritis isolates (Johnson et al., 2005b). An operon of seven genes, *focAICDFGH*, encodes F1C fimbria (Riegman et al., 1990), with FocA representing the major subunit of the pilus and FocH the tip adhesin. F1C fimbriae bind receptors present in both the bladder and kidneys (Virkola et al., 1988). F1C fimbriae mediate adhesion to the endothelium and the muscular layer of the bladder, but not to the epithelium (Virkola et al., 1988). In the kidneys, F1C mediates adherence primarily to the distal tubules and collecting ducts, however, these fimbriae also bind in the glomeruli and to the vascular endothelium (Virkola et al., 1988). The receptor structure of F1C fimbriae was determined to be the GalNAcβ1-4Galβ sequence of glycolipids (Khan et al., 2000). Therefore, F1C fimbriae have the potential to be a virulence factor involved in colonization of the urinary tract. However, there has been no experimental infection conducted to demonstrate the importance of F1C fimbriae during a urinary tract infection.

S fimbriae

S fimbriae are encoded by 15% of UPEC isolates as compared to 5% fecal *E. coli* strains, and are highly associated with cystitis and ABU isolates (Spurbeck et al., 2011). Genetically identical to F1C fimbriae, with exception of the tip adhesin, S fimbriae are often grouped with F1C fimbriae in epidemiological studies (Ott et al., 1988). However, the difference in the tip adhesin between F1C and S fimbriae confers different binding specificities; thus, these similar fimbriae should be considered different virulence factors for UPEC (Marre et al., 1990). SfaS, the tip adhesin, binds sialic acid present on epithelial cells in the human kidney (Morschhauser et al., 1990), and can be inhibited by addition of the receptor analog sialyl(α2-3)lactose (Korhonen et al., 1986). Specifically, S fimbriae bind to vascular endothelium in the large vessels of kidney tissue and the capillary endothelium in the interstitium and glomerulus. S fimbriae also bind to the epithelium in the lumen of the proximal and distal tubules, the collecting ducts, and the glomerular epithelium (Korhonen et al., 1986). In the bladder, S fimbriae bind to the epithelial and muscular layers as well as connective tissue (Virkola et al., 1988). In a rat model of pyelonephritis, deletion of a region of the chromosome that encodes hemolysin, S fimbriae, and serum resistance from the pyelonephritis isolate *E. coli* 536 attenuated the mutant. When the mutant was complemented with a plasmid containing the *sfa* operon, the number of bacteria recovered from the kidneys of infected rats was increased 20-fold (Marre et al., 1986). Thus, S fimbriae fulfill the criteria of a virulence determinant in UPEC.

F9 fimbriae

F9 fimbriae (encoded by the *f9* operon) are encoded by 78% of UPEC isolates as compared to 56% fecal *E. coli* strains (Spurbeck et al., 2011). Although F9 fimbriae do not agglutinate red blood cells from human, dog, horse, or sheep, nor promote adherence to HeLa epithelial cells, this fimbria mediates biofilm formation in M9 minimal salts medium when expressed in a non-fimbriated strain of *E. coli* (Ulett et al., 2007a). However, when the *f9* operon was deleted from *E. coli* CFT073, no significant difference in biofilm formation was observed, even when type 1 fimbriae and F1C fimbriae, which have been implicated in biofilm formation, were also deleted from the chromosome (Ulett et al., 2007a). Thus, F9 fimbriae play, at most, a minor role in biofilm formation by UPEC on abiotic surfaces. F9 fimbriae are expressed when *E. coli* CFT073 is cultured in human urine, which may indicate that these fimbriae have some role in virulence (Spurbeck et al., 2011). However, the role of F9 fimbriae in vivo or on adherence to uroepithelial cells has not been assessed.

Ygi fimbriae

Ygi fimbriae are encoded by 61% of UPEC isolates as compared to 24% fecal *E. coli* strains, suggesting these fimbriae may be urovirulence factors (Spurbeck et al., 2011). Deletion of the *ygi* operon from the pyelonephritis strain *E. coli* CFT073 was sufficient to reduce adherence to the human kidney epithelial cell line, HEK 293, and reduce biofilm formation on abiotic surfaces (Spurbeck et al., 2011). These deficiencies were fully complemented by introduction of the *ygi* operon on a plasmid. Ygi fimbriae are expressed both in LB medium and in human urine. In the mouse model of ascending UTI, the *ygi* deletion mutant was outcompeted by wild-type *E. coli* CFT073 in the kidneys of mice and this phenotype was fully complementable, fulfilling molecular Koch's postulates, demonstrating that these fimbriae are involved in the pathogenesis of UTI (Spurbeck et al., 2011).

Yad fimbriae

Yad fimbriae are encoded by 46% of UPEC compared to 21% fecal *E. coli* strains (Spurbeck et al., 2011). Deletion of the *yad* operon from *E. coli* CFT073 resulted in diminished adherence to the human bladder epithelial cell line, UM-UC-3, biofilm formation on abiotic surfaces, and motility (Spurbeck et al., 2011). Furthermore, Yad fimbriae are expressed when the bacteria are cultured in human urine, and a double mutant, in which both the *yad* and *ygi* fimbrial operons are deleted from the chromosome, was outcompeted by wild-type *E. coli* CFT073 in bladder and kidneys of the mouse model of ascending UTI. Since the *ygi* deletion mutant alone is only outcompeted by wild-type *E. coli* CFT073 in the kidneys, this suggests that Yad fimbriae, along with Ygi fimbriae, are involved in colonization of the bladder (Spurbeck et al., 2011).

Auf fimbriae

Auf fimbriae (another upec fimbriae, encoded by *aufABCDEFG*) are encoded by 67% of UPEC compared to 27% fecal *E. coli* strains (Spurbeck et al., 2011). Auf fimbriae, when expressed in *E. coli* BL21-AI cells, did not agglutinate human, guinea pig, or sheep erythrocytes, nor mediate adherence to T24 bladder epithelial cells or HEp-2 laryngeal carcinoma cells (Buckles et al., 2004). Although the *auf* operon is transcribed by *E. coli* CFT073 and 10 other UPEC strains, no protein was detected by immunoblot in vitro. After mice were infected with either wild-type *E. coli* CFT073 or an *aufC* mutant, the mice infected with the wild-type strain elicited an immune response to AufA antigen as detected by ELISA, whereas only 1 out of 10 mice infected with the mutant strain had an immune response. This suggested that Auf fimbriae are synthesized in vivo (Buckles et al., 2004). However, results from the mouse model of ascending UTI demonstrate that Auf fimbriae do not play a significant role in colonization, as there was no significant difference in colonization between wild-type *E. coli*

CFT073 and the *auf* mutant and the mutant was not outcompeted by wild-type in competition studies (Buckles et al., 2004).

Pix fimbriae

Pix fimbriae are a relatively rare fimbrial type found in 8% of UPEC compared to 4% of fecal *E. coli* strains (Spurbeck et al., 2011). Pix fimbriae were initially discovered on a pathogenicity island found in the uropathogenic *E. coli* strain X2194 (Lugering et al., 2003). Pix fimbriae are related to P fimbriae, but are differentially regulated due to a truncation in the regulatory region associated with P fimbriae. Expression of Pix was found to be temperature-dependent during stationary phase. Pix fimbriae promote adherence to HeLa cells (Lugering et al., 2003), demonstrating that these fimbriae, although rare, can be an important virulence factor for some UPEC. The importance of Pix fimbriae has yet to be tested in the mouse model of ascending UTI.

Dr/Afa fimbriae

Dr and Afa adhesins are a family of *E. coli* adhesins that recognize the Dr blood group antigen (a cell membrane protein in erythrocytes also called the decay-accelerating factor) as their receptor (Nowicki et al., 1990). While some Dr family adhesins are afimbrial adhesins, such as AFA-I and AFA-III, others are fimbrial adhesins like F1845 and Dr fimbriae (Nowicki et al., 1990). The best-studied member of the Dr family of adhesins, the Dr fimbriae (aka O75X), are found in 7% of UPEC compared to 2% of fecal *E. coli* strains (Spurbeck et al., 2011) and bind type IV collagen in vitro (Westerlund et al., 1989). In vivo, Dr fimbriae bind to peritubular connective tissue in the human kidney (Nowicki et al., 1986), the connective tissue between muscle cell layers in the bladder, human neutrophils (Johnson et al., 1995), and weakly to epithelial cells (Virkola et al., 1988). Dr fimbriae also facilitate cell invasion (Goluszko et al., 1997; Fang et al., 2004; Das et al., 2005), allowing bacteria to avoid the humoral immune response by hiding within the intracellular space. Vaccination with purified Dr fimbriae reduced mortality associated with UTI in a mouse model of infection (Goluszko et al., 2005), however, since the majority of UPEC do not encode these fimbriae, vaccination against Dr fimbriae would prevent only a minor subset of UPEC infections.

Putative fimbriae (Yeh, type IV pili 1 and 2, Yfc)

The presence of two putative fimbriae encoded by *E. coli* CFT073, Yfc fimbriae (encoded by *yfcOPQRSUV*, present in 73% UPEC isolates v. 33% fecal isolates) and type IV pilus 2 (encoded by *c2394-c2395*, present in 51% UPEC v. 12% fecal isolates), are correlated with uropathogenic isolates (Spurbeck et al., 2011). However, Yeh fimbriae (encoded by *yehABCD*, present in 94% of *E. coli* isolates) and type IV pilus 1 (encoded by *ppdD*, *hofB*, and *hofC*, present in 99% of *E. coli* isolates) are found ubiquitously in the *E. coli* population, and are therefore, not markers of pathogenic isolates.

Regulatory hierarchy of fimbrial expression

Regulation of type 1 fimbriae has been the focus of several studies, since this is one of the major factors required for virulence in the murine model of ascending UTI. The promoter of the type 1 fimbrial operon is located on an invertible sequence (designated the *fim* switch) and causes type 1 fimbriae to be phase-variable. Phase variation is dependent on the activity of two recombinases, FimB and FimE. FimB can reverse the orientation of the *fim* switch from either the on-to-off or off-to-on direction, whereas FimE primarily works only in the on-to-off direction. Therefore, any regulation that changes the ratio of these recombinases or enhances or suppresses their activity, will affect the proportion of *E. coli* expressing type 1 fimbriae. P and S fimbriae can affect the expression of type 1 fimbriae via inhibition of FimB activity, which then forces the *fim* switch into the off position from the action of FimE (Xia et al., 2000; Holden et al., 2001, 2006). The negative regulation of type 1 fimbrial expression by F1C, S and P fimbriae is due to regulatory proteins, FocB, SfaB, or PapB, associated with the respective fimbrial operons (Xia et al., 2000; Holden et al., 2001; Lindberg et al., 2008). Furthermore, P fimbriae positively regulate the expression of S fimbriae (Morschhauser et al., 1994), suggesting that both fimbriae are utilized by the bacteria to colonize the same niche. When taken together with the negative regulation of type 1 fimbriae, the hierarchy of regulation begins to appear such that while type 1 fimbriae are needed for adherence in an earlier niche, such as in the bladder, they may hinder adherence later, such as in the kidneys, where P and S fimbriae are utilized. This type of hierarchical regulation could be essential for a pathogen like UPEC to ensure that only fimbriae necessary for adherence to a particular substrate are expressed at any given time, as well as to limit exposure of each antigen to the immune system, allowing the pathogen to avoid clearance brought on by an antibody response. How the uncharacterized fimbriae feed into this regulatory system remains to be elucidated.

Non-fimbrial adhesins

Factor adherence *E. coli*: FdeC

FdeC is an ExPEC adhesin that is found in 99% of ExPEC, 82.6% of commensal, and 93–100% of intestinal pathogenic *E. coli* strains (Nesta et al., 2012). FdeC is only expressed upon contact with host cells and has structural similarity to intimin (from enteropathogenic *E. coli*) and invasin (from *Yersinia pseudotuberculosis*), two bacterial virulence factors that mediate host cell interactions. Recombinant FdeC binds epithelial cells and collagens V and VI (both present in the human urinary tract) with high affinity and, in vivo, *E. coli* 536 wild-type significantly outcompetes a *fdeC* mutant in colonization of the bladder and kidneys (Nesta et al., 2012). Thus, FdeC is a non-fimbrial adhesin necessary for full urovirulence.

RTX-like adhesin: TosA

TosA is a repeat-in-toxin (RTX) protein highly associated with UPEC (Vigil et al., 2011b). Originally thought to encode a toxin due to sequence similarity with other RTX proteins such as hemolysin, *tosA* was of interest to researchers studying the pathogenesis of UPEC for two reasons. First, *tosA* is encoded on a pathogenicity island that, when deleted, rendered *E. coli* CFT073 attenuated in the mouse model of ascending UTI (Lloyd et al., 2009a).Upon deletion of *tosA* itself, the deletion strain was outcompeted in vivo, demonstrating that the attenuation of the PAI deletion strain was due, at least in part, to the deletion of *tosA* (Lloyd et al., 2009a). Second, *tosA* was hit in an IVIAT screen, a method for identification of antigenic proteins that are expressed only in vivo (Vigil et al., 2011a). TosA is outer membrane-associated in *E. coli*, but does not have a cytotoxic effect on mammalian cells (Vigil et al., 2012). Rather, when over-expressed, TosA increases adherence of *E. coli* CFT073 to kidney epithelial cell lines, suggesting that TosA acts as an afimbrial adhesin mediating intimate attachment to the upper urinary tract (Vigil et al., 2012).

Autotransporters

Autotransporters are proteins that undergo type V secretion (Chapter 16), and can either be secreted into the extracellular mileu or remain attached to the bacterial cell surface. Eleven autotransporter-encoding genes have been identified in the uropathogenic *E. coli* strain CFT073 (Parham et al., 2004; Allsopp et al., 2010). Some, like Sat, are secreted toxins, but most autotransporters in UPEC remain membrane-associated. Here we discuss the autotransporter proteins that enhance adherence to uroepithelial cells or biofilm formation.

UpaG

UpaG is associated with *E. coli* phylogroups B2 and D, which include the uropathogens (Valle et al., 2008; Totsika et al., 2012). UpaG mediates adherence to human bladder epithelial cells in vitro (Valle et al., 2008; Totsika et al., 2012) and promotes cell aggregation, biofilm formation, and binding to the human extracellular matrix proteins fibronectin and laminin (Valle et al., 2008). However, when tested in the mouse model of ascending UTI, a *upaG* deletion mutant had no fitness defect when compared to wild-type *E. coli* CFT073 (Valle et al., 2008), suggesting that the role of this protein during infection is too subtle to be detected as a fitness defect.

UpaH

UpaH is highly prevalent in all *E. coli* strains tested, as it is found in 86% of A, 85% of B1, 84% of B2, and 70% of phylogenetic group D (Allsopp et al., 2010). However, in *E. coli* K-12 the pseudogene encoding UpaH, *ydbA*, is truncated by insertion sequences (Allsopp et al., 2010). In wild-type *E. coli* CFT073, UpaH promotes biofilm formation on abiotic surfaces in M9 minimal medium. UpaH is also expressed in vivo during infection and the *upaH* deletion strain was significantly outcompeted by wild-type *E. coli* CFT073 in the mouse bladder, demonstrating that this autotransporter adhesin is involved in uropathogenesis

(Allsopp et al., 2010). However, in two other UPEC strains that express UpaH there was no significant defect in colonization when *upaH* was deleted. Thus, UpaH appears to be a strain-specific virulence factor (Allsopp et al., 2010).

UpaB

Although there is no significant difference in the prevalence of UpaB in UPEC (58%) versus fecal isolates (42%), UpaB is disrupted or absent in the genomes of all diarrheagenic *E. coli* genomes (Allsopp et al., 2012). UpaB is expressed by the prototypical pyelonephritis isolate *E. coli* CFT073 and is present on the surface of the bacterium. Demonstrating the potential for UpaB to contribute to uropathogenesis, UpaB from *E. coli* CFT073 expressed in *E. coli* K-12 was found to mediate adherence to the extracellular matrix proteins fibronectin, fibrinogen, and laminin, but not collagen types I–V, elastin, or heparin sulfate (Allsopp et al., 2012). Furthermore, when tested in vivo, wild-type *E. coli* CFT073 outcompeted Δ*upaB* in colonization of the murine bladder (Allsopp et al., 2012), demonstrating that this autotransporter protein is involved in uropathogenesis.

UpaC

UpaC, which was characterized alongside UpaB, is found at similar prevalence in UPEC strains (47%) as in commensal *E. coli* strains (34%) (Allsopp et al., 2012), however, UpaC is disrupted in the commensal isolates *E. coli* MG1655, DH1, and HS. While UpaC was undetectable by western blot from wild-type *E. coli* CFT073, deletion of *hns* enhanced UpaC expression, demonstrating that H-NS acts a repressor of UpaC expression. When over-expressed in *E. coli* K-12, UpaC promotes biofilm formation to abiotic surfaces. However, when assessed in co-challenge in the murine ascending UTI model, there was no significant difference in in vivo fitness between wild-type *E. coli* CFT073 and Δ*upaB* (Allsopp et al., 2012). Therefore, UpaC may play only a subtle role in vivo.

Ag43a

Antigen 43a (encoded by *fluA*, Ag43a, UpaF) is a surface-located adhesin that undergoes phase variation and is associated with urovirulence (Anderson et al., 2003). Ag43a mediates cell aggregation and biofilm formation on abiotic surfaces when expressed from a plasmid in *E. coli* K-12. However, in *E. coli* CFT073, cell aggregation and biofilm formation were not affected by Ag43a, as this is a short surface-associated structure and the longer surface structures such as flagella and fimbriae expressed simultaneously by *E. coli* CFT073 do not allow the close cell–cell contact necessary for these phenotypes to be observed (Ulett et al., 2007a). In the mouse model of ascending UTI, deletion of *fluA*, the gene encoding Ag43a in *E. coli* CFT073, reduced long-term persistence in the bladder (Ulett et al., 2007b). Therefore, Ag43a may have a role in persistence in vivo.

Invasion and intracellular bacterial communities

Invasion of epithelial cells by UPEC is a relatively rare event (Mysorekar and Hultgren, 2006). However, in the murine urinary tract, invasion occurs by a zipper mechanism in which type 1 fimbriae, or DraD/AfaD bind to receptors on the host cell that mediate actin cytoskeletal rearrangements (Zalewska et al., 2001; Wang et al., 2008). Actin filament rearrangement induces a phagocytic cup that envelops the bacterium, thus allowing entry into the host cell (Wang et al., 2008). In DraD/AfaD-mediated internalization, *E. coli* binds the host cell utilizing the Dr fimbrial adhesin, DraE/AfaE (Garcia et al., 2000). Adherent bacteria enter the epithelial cell through clathrin-coated vesicles, after a threshold level of the invasin, AfaD/DraD is achieved (Zalewska et al., 2001). Cytotoxic necrotizing factor-1 (CNF-1) toxin, also mediates invasion into host cells by stimulating rearrangement of the actin cytoskeleton (Falzano et al., 1993). Once in the cytosol of the host cell, CNF-1 deamidates glutamine 61 of Rac-1 and glutamine 63 of RhoA activating these Rho GTPases. The Toll-interacting protein (Tollip) interacts with the activated Rac-1, and binds Tom1, which recruits clathrin to the plasma membrane, causing internalization of the bacterium through the endocytic pathway (Visvikis et al., 2011). The active Rho GTPases are then ubiquitylated and degraded by the proteasome (Doye et al., 2002).

Type 1 fimbriae mediate invasion of cultured human and murine mannosylated superficial umbrella cells, leading to formation of intercellular bacterial communities (IBCs) (Martinez et al., 2000; Justice et al., 2004). IBCs develop in the cytosol from a single bacterium that replicates to form a biofilm-like community encased in a polysaccharide-rich matrix (Anderson et al., 2003). While the mechanistic basis for IBC formation is not fully understood, three factors are necessary for complete IBC maturation. First, SurA, a periplasmic prolylisomerase/chaperone that facilitates biogenesis and assembly of outer membrane proteins and fimbriae, is essential for both invasion and maturation of IBCs, demonstrating that fimbriae and/or other outer membrane proteins are involved in bacterial uptake and IBC development (Justice et al., 2006b). Second, OmpA, an outer membrane protein, promotes IBC maturation, as disruption of *ompA* does not affect adherence or invasion of bladder epithelial cells, but leads to significantly lower numbers of IBCs. This defect was complemented, demonstrating the decrease in mature IBCs was due to the loss of OmpA (Nicholson et al., 2009). Finally, the K capsule polysaccharide itself is necessary for IBC formation (Goller and Seed, 2010). After IBCs are fully matured, UPEC undergo a morphological change into a filamentous form. These filamentous bacteria then escape into the lumen of the bladder in a process called fluxing, which allows the infection to spread to other bladder cells or ascend the urinary tract (Mulvey et al., 2001). The phenomena of fluxing and IBC formation led to the hypothesis that recurrent UTIs are caused by UPEC that avoid killing by antibiotics by persisting in IBCs, followed by reemergence into the bladder lumen by fluxing. However, this has yet to be substantiated in humans.

IBCs, while mostly studied in the context of murine bladder infections, were found in 22% of exfoliated epithelial cell samples from human UTIs (Rosen et al., 2007). However, whether IBCs play a role in pathogenesis still needs to be elucidated.

Iron acquisition

To survive in the human urinary tract, UPEC have evolved mechanisms to scavenge iron from the host. The human body utilizes heme, lactoferrin, transferrin, and lipocalin-2 to sequester iron because this element, while essential for life, is also toxic in high quantities. UPEC, therefore, use iron-chelating siderophores to rip iron away from host iron-bound proteins, and also evolved mechanisms to use the iron in heme itself. UPEC strains can synthesize up to four siderophores: the hydoxamateaerobactin, two catecholates: enterobactin and salmochelin, and yersiniabactin, a mixed-type siderophore. Each iron-bound siderophore is recognized by a cognate iron receptor: IroN mediates salmochelin uptake (Russo et al., 2002), IutA mediates aerobactin uptake (Torres et al., 2001), Iha mediates enterobactin uptake (Johnson et al., 2005a, b), and FyuA mediates yersiniabactin uptake (Heesemann et al., 1993; Schubert et al., 2002). ChuA and Hma mediate direct heme uptake (Torres et al., 2001; Hagan and Mobley, 2009). All of the iron acquisition systems are TonB-dependent, and demonstrating the importance of iron scavenging in UTI, a *tonB* mutant is attenuated in the murine model of ascending UTI (Torres et al., 2001). Complementation of the mutation by expression of *tonB* from a vector restored virulence, thereby fulfilling Koch's molecular postulates (Torres et al., 2001). While iron acquisition is necessary for UPEC to cause infections, there is redundancy in the iron acquisition systems, as no one siderophore-receptor system is essential for UTI. However, there is a functional hierarchy during urinary tract colonization, in which enterobactin and salmochelin receptors contribute the least and the heme and noncatecholate siderophore receptors contribute the most to iron acquisition during infection (Garcia et al., 2011).

Host response to UPEC, pathogen avoidance of host responses

Mechanical host response

The luminal surface of the bladder is made up of superficial umbrella cells which are coated by uroplakins to which the type 1 fimbriae or UPEC bind (Wu et al., 2009). This triggers exfoliation of the bladder epithelium (Zhou et al., 2001). The shed uroepithelial cells, along with any adherent or intracellular bacteria, are then flushed out of the urinary tract by the flow of urine. Furthermore, a protein present in high amounts in human urine, known as uromodulin, binds type 1 fimbriae, thereby titrating UPEC away from receptors on the uroepithelium (Serafini-Cessi et al., 2003).

Innate immune response

LPS and fimbriae of UPEC stimulate the inflammatory response through the Toll-like receptor 4 (TLR4) pathway and bacterial adherence can initiate cell death through apoptosis (Bien et al., 2012). TLR4 causes infected cells to produce cytokines leading to an influx of PMNs and exfoliation of the infected uroepithelium (Bien et al., 2012). This leads to activation of NFκB, the secretion of antimicrobial peptides, and synthesis of the chemokines IL-6/IL-8 which further recruit neutrophils to kill UPEC (Bien et al., 2012). The inflammatory response also leads to the generation of reactive nitrogen and oxygen species as well as other antimicrobial compounds (Mulvey et al., 2000; Bower et al., 2005).

PhoP

PhoP is the response regulator of a conserved two component signal transduction system in both pathogenic and non-pathogenic genera of the Enterobacteriaceae, with PhoQ as the sensor kinase. In both pathogenic and commensal *E. coli* the PhoPQ system regulates a network of genes in response to low Mg^{2+} (Kato et al., 1999) and acidic pH (Eguchi et al., 2011). However, in uropathogenic *E. coli*, a deletion mutant of *phoP* is attenuated in the mouse model of ascending UTI, and when complemented with *phoPQ* on a plasmid, virulence is restored (Alteri et al., 2011). Microarray and phenotypic analysis demonstrates that the PhoP regulon includes genes that mediate motility, acid tolerance, and lipopolysaccharide modification. Furthermore, the *phoP* mutant was highly susceptible to antimicrobial peptides demonstrating that the PhoP regulon is involved in avoidance of the host innate immune response (Alteri et al., 2011). An energized membrane is essential for all of the phenotypes observed in the *phoP* mutant, suggesting that the PhoP regulon is involved in maintenance of membrane potential. This is important during pathogenesis, as the host responds to invading pathogens by producing antimicrobial peptides and an influx of neutrophils and macrophages that produce reactive oxygen species and decrease the pH in an effort to kill bacterial pathogens (Bower et al., 2005). The PhoP regulon includes several gene products such as SodC which detoxify the effects of superoxides, therefore the PhoP regulon represents a bacterial defense system against the host innate immune response to infection (Alteri et al., 2011).

TcpC

The host immune system recognizes bacterial pathogens using Toll-like receptor (TLR) signaling mechanisms that initiate the innate immune response, characterized by release of the cytokines and chemokines that recruit inflammatory cells to the site of infection. The toll/interleukin-1 receptor domain (TIR) is essential to the function of TLRs, as mutation of this domain abolishes LPS recognition. To avoid killing by the inflammatory response, UPEC has evolved a system to avoid this TLR-mediated response. UPEC express and secrete TcpC, a homolog of the TIR domain (Cirl et al., 2008). TcpC reduces secretion of the cytokines TNFα (produced by macrophages) and IL-6 (produced by uro-epithelial cells) by binding the MyD88 adaptor protein. Further study showed that TcpC also inhibits TRIF and IL-6/IL-1 signaling pathways, which activate NFκB and IRF3/7, unlike MyD88 (Yadav et al., 2010), further suppressing the innate immune response. The avoidance of the innate immune response directly affects the ability of UPEC to cause pyelonephritis. In the mouse model of infection, the *tcpC* mutant was significantly reduced in its ability to colonize the kidneys as compared to the wild-type strain. Furthermore, the reduction in colo-nization correlated with a decrease in the formation of kidney abscesses during infection. Both colonization and abscess formation in the kidneys were restored upon complementation of *tcpC* on a plasmid, fulfilling molecular Koch's pos-tulates (Cirl et al., 2008).

Repression of inflammation: SisA and SisB

SisA and SisB are cytosolic proteins expressed by 70% of uropathogenic *E.coli* in vivo (Lloyd et al., 2009b) and are encoded by 85% and 22% of pyelonephri-tis isolates respectively (Mao et al., 2012). While the mechanism of action is unknown, SisA and SisB suppress the host inflammatory response. In *E. coli* CFT073, a double deletion mutant Δ*sisA*Δ*sisB* was attenuated in both the blad-der and kidneys at 6 hours post infection and had significantly more inflam-matory foci in the kidneys. These foci were more severe than those observed during infection by wild-type *E. coli* CFT073. However, by 24 or 48 hours post infection, the double mutant was no longer attenuated, suggesting that SisA and SisB are involved in early colonization events in vivo. Furthermore, the double deletion mutant could be complemented by either SisA or SisB expressed from a plasmid, demonstrating that these proteins have redundant function (Lloyd et al., 2009b). More work is necessary to determine what host component SisA and SisB interact with to reduce the inflammatory response during the first 6 hours of infection.

Avoidance of detection: AmpG, WaaL, and Alr

The innate immune system continually samples the human body for pathogen-associated molecular patterns (PAMPs) by pattern recognition receptors, which include the TLRs. PAMPs produced by UPEC and commensal *E. coli* and rec-ognized by TLRs include LPS, flagella, type 1 and P fimbriae. Once recognized by a TLR, such as TLR4, a signal transduction cascade leads to the production of cytokines and chemokines including IL-6 and IL-8. Deletion of three genes, *ampG*, *waaL*, and *alr*, from the uropathogenic *E. coli* strain NU14 increased IL-8 secretion from urothelial cells demonstrating that these genes are involved in reduction of PAMP recognition. AmpG transports peptidoglycan fragments from the periplasm back into the cytoplasm, thus reducing the amount of pep-tidoglycan released into the extracellular milieu that can be recognized by the innate immune system. WaaL ligates O antigen onto the lipid A core subunit of LPS which reduces detection by TLR4 in the urinary tract, and Alr converts L-alanine to D-alanine for peptidoglycan synthesis (Billips et al., 2008). Deletion strains of *ampG*, *waaL*, or *alr* were outcompeted by wild-type *E. coli* NU14 in bladder colonization in the mouse model of infection, demonstrating that modification of PAMPs is important for uropathogenesis (Billips et al., 2008).

α-Hemolysin (HlyA)

Alpha-hemolysin is a prototypical member of the repeat in toxin (RTX) family (Welch, 1991). While deletion of *hlyA* did not attenuate UPEC in the mouse model of ascending UTI, significantly less sloughing of the uroepithelium and bladder hemorrhage was observed in comparison to wild-type *E. coli* (Smith et al., 2008). Epidemiological evidence associates α-hemolysin with highly virulent strains, as α-hemolysin is more prevalent in pyelonephritis and urosepsis isolates (Opal et al., 1990; Blanco et al., 1996; Johnson et al., 2005b). α-Hemolysin is secreted by UPEC by the type I mechanism (Chapter 16), and complexes with LPS which then exploits the CD14/LPS-binding protein to bring hemolysin into contact with the host cell (Mansson et al., 2007). HlyA then inserts into the cell membrane of uroepithelial cells, and forms pores, permeabilizing the cell membrane causing cell death at high concentrations (Felmlee and Welch, 1988; Welch et al., 1992; Stanley et al., 1998). At sublytic concentrations, α-hemolysin affects signal transduction pathways in the host cells by inducing an oscillatory change intracellular calcium concentration (Uhlen et al., 2000), which leads to suppression of cytokine release (Konig and Konig, 1993; Hilbert et al., 2012). Further study revealed that insertion of HlyA into the cell membrane of epithelial cells and macrophages induces activation of mesotrypsin, a serine protease (Dhakal and Mulvey, 2012). Mesotrypsin then degrades paxillin, a cytoskeletal scaffold protein and components of the pro-inflammatory NFκB signaling cascade, thus suppressing the inflammatory response and disabling macrophages (Dhakal and Mulvey, 2012). α-Hemolysin integration in the host cell membrane also activates caspases, leading to host cell death (Dhakal and Mulvey, 2012).

Cytotoxic necrotizing factor-1 (CNF-1)

Cytotoxic necrotizing factor-1 (CNF-1) has been epidemiologically linked to urovirulent strains of *E. coli*, with 61% of UTI isolates and only 10% of fecal isolates encoding *cnf-1* (Yamamoto et al., 1995). As mentioned above, CNF-1 affects host cells by deamidation of small GTPases leading to uroepithelial cell invasion (Falzano et al., 1993; Visvikis et al., 2011). Other effects of CNF-1 include reduction of complement receptor-3-dependent phagocytosis in monocytes (Capo et al., 1998), and decreased transmigration of PMNs (Hofman et al., 1998). CNF-1 activation of RhoA causes an increase in F-actin in PMNs resulting in actin polymerization and cell spreading. CNF-1 increases the adhesiveness of PMNs to epithelial cells, impairing the transmigration of PMNs and eliciting an oxidative burst that contributes to epithelial cell damage characteristic of UTI (Hofman et al., 2000). Furthermore, phagocytosis by PMNs is also inhibited by CNF-1, again impeding clearance of UTI by the innate immune system (Hofman et al., 2000). In a mouse model of ascending UTI CNF-1-positive strains induced higher levels of inflammation and colonized the urine and bladders better than *cnf-1* isogenic mutants (Rippere-Lampe et al., 2001). However, another study showed the opposite, which could be due to differences in the mouse strain studied, or the type of UPEC isolates characterized (Johnson et al., 2000). In the study that showed no defect in colonization by a *cnf-1* isogenic mutant, the *E. coli* strain was isolated from a patient with cystitis. The strains in which the *cnf-1* mutant had an effect were isolated from the blood of a patient with urosepsis. A third hypothesis for the discrepancy in the results is that the strain used in the study by Johnson et al. might not produce CNF-1. It was never tested for the ability to actually produce the toxin. Therefore, more data are necessary before a generalized conclusion can be made as to whether CNF-1 plays a role in vivo during UTI. However, the evidence supports the hypothesis that CNF-1, when produced, can effectively inhibit the innate immune system response to UPEC.

Serine protease autotransporters of Enterobacteriaceae

Uropathogenic *E. coli* encode three serine protease autotransporters (SPATES), secreted by the type V mechanism (Chapter 16), and designated Sat, Pic, and Tsh. All three SPATES are more prevalent in UPEC than fecal *E. coli* isolates: *sat* is present in 68% of pyelonephritis isolates compared to 14% of fecal isolates (Guyer et al., 2000), *pic* is present in 31% of pyelonephritis isolates compared to 7% of fecal *E. coli* isolates,and *tsh* is present in 63% of all UPEC strains and only 33% of fecal *E. coli* strains (Heimer et al., 2004). Sat is expressed during UTI in the mouse model of infection, and has a cytopathic effect on kidney and bladder epithelial cells (Guyer et al., 2000). While there is no significant difference in colonization between a Sat-deficient mutant and wild-type in the mouse model of ascending UTI, histological changes were apparent in the kidneys (Guyer et al., 2002). Specifically, vacuolation of proximal tubule cells and dissolution of the glomerular membrane was observed, indicating that Sat is a vacuolating cytotoxin (Guyer et al., 2002). Further study demonstrated Sat was taken up by host cells and interacts with the cytoskeleton, causing actin loss and cytoskeletal contraction within intoxicated cells (Maroncle et al., 2006).

In *E. coli* CFT073, both *pic* and *tsh* are temperature-regulated, with higher expression at 37°C. Bacteria from the urine of infected mice also express both SPATES (Heimer et al., 2004). Both Pic and Tsh cleave surface glycoproteins from leukocytes, including CD43, PSGL-1, CD44, CD45, CD93, and fractalkine/CX3CL1 (Ruiz-Perez et al., 2011). All targets of Pic and Tsh are involved in leukocyte attraction and migration (Ruiz-Perez et al., 2011). Thus, the neutrophil infiltration of the urinary tract in response to UPEC can be delayed by the secretion of SPATES.

Extracellular polysaccharides

In a signature tagged mutagenesis study of *E. coli* CFT073, mutation of genes involved in biosynthesis of group II capsule and the enterobacterial common antigen were attenuated in the mouse model of ascending UTI (Bahrani-Mougeot et al., 2002). Similarly, an isogenic mutant lacking the K2 capsule biosynthetic genes was outcompeted by wild-type *E. coli* CFT073 in the mouse model of ascending UTI (Buckles et al., 2009). Furthermore, the K2 capsule was found to be necessary for serum resistance (Buckles et al., 2009). Recently, it has been suggested that capsule polysaccharides are involved in the formation of intracellular biofilm-like communities by some UPEC strains, as deletion of biosynthetic K capsule genes reduced IBC formation in the mouse model of UTI due to infiltration of neutrophils (Anderson et al., 2010). Again, the mechanism by which K capsule promotes IBC formation has yet to be discovered. Several roles for capsular polysaccharides in the avoidance of the host response to infection have been suggested, but as of yet, none have been verified experimentally. Polysaccharides may reduce phagocytosis, opsonization, killing by antimicrobial peptides, and the formation of the membrane attack complex (Goller and Seed, 2010). Further studies are necessary to fully understand the role polysaccharides play in vivo during UTI.

SulA and filamentation

SulA is an inhibitor of cell division that is activated by the SOS response to DNA damage, such as that caused by the oxidative burst from PMNs (Huisman et al., 1980). The inhibition of cell division causes *E. coli* to take on a filamentous morphology, which is thought to confer resistance to killing by PMNs (Justice et al., 2004). In the cystitis strain *E. coli* UTI89, a *sulA* mutant is not able to form filaments in vivo, suggesting that SulA is responsible for the filamentous UPEC observed after fluxing from IBCs (Justice et al., 2006a). The *sulA* mutant was attenuated in bladder colonization in the mouse model of ascending UTI, but in *tlr4*[−/−] knockout mice, the *sulA* mutant was not significantly different in bladder colonization from the wild-type strain *E. coli* UTI89 suggesting that filamentation is involved in evasion of the innate immune response to cystitis (Justice et al., 2006a). The mechanism by which filamentation aids the bacteria to avoid the innate immune system has not been elucidated as of yet, but is hypothesized to be due to decreased phagocytosis by PMNs.

Other virulence factors

Mutant strains of DegS (Redford et al., 2003), DegP (Redford and Welch, 2006), and PhoU (Buckles et al., 2006), were attenuated in the mouse model of infection, and complementation restored virulence. However, the role that these proteins play in uropathogenesis is still unknown.

CLINICAL MANIFESTATIONS

Transmission

Host factors predisposing women to recurrent UTI and upper UTI

Several host-associated factors predispose women to recurrent UTI. Frequent sexual intercourse (>9 times/month) is the factor with the strongest correlation to recurrent UTI (Kodner and Thomas Gupton, 2010). Other factors that can predispose otherwise healthy women to UTI include choice of contraceptive, anatomical factors, local pH fluctuations and the antibody titers in the vaginal tract (Kodner and Thomas Gupton, 2010). Development of upper UTI, such as pyelonephritis, is associated with diabetes, immunosuppression, and anatomical urinary tract obstructions (Tseng et al., 2002).

Clinical features and diagnosis

Urinary tract infection clinical features

UTI is categorized by the site within the urinary tract to which the infection is localized and the severity of disease: bacteriuria (the urine, otherwise unspecified), cystitis (the bladder), pyelonephritis (the kidneys), and urosepsis (often, but not always blood). Cystitis, infection of the bladder, is characterized by inflammation of the lower urinary tract and presents with symptoms such as dysuria, increased frequency of urination, and suprapubic pain. Bacteriuria is present and a urinary dipstick test is often positive for leukocyte esterase and nitrates (Kodner and Thomas Gupton, 2010; Norinder et al., 2012). Diagnosis of cystitis is confirmed when $\geq 10^3$ bacteria/ml in a midstream clean-catch urine sample from a patient with symptoms of UTI is observed, although the diagnosis can be made without culture confirmation in women with typical features and a prior history. Acute pyelonephritis-associated features include flank pain, pyuria with casts, costovertebral angle tenderness, rigors, bacteriuria with or without diaphoresis, nausea, vomiting, abdominal or groin pain, and fever (Kodner and Thomas Gupton, 2010).

Complications

Intense inflammation caused by UPEC in the kidneys occurs in wedge-shaped areas from papillae to renal cortex, and tubules are filled with polymorphonuclear leukocytes (PMNs). Renal abscesses may form in areas of localized inflammation, and bacteria may travel to the bloodstream causing bacteremia. In fact, 12% of patients with acute pyelonephritis have bacteremia. This increases the severity of disease, and when untreated, can be fatal. Complications from pyelonephritis result in a 25% mortality rate with 36 000 deaths per annum in the USA (Stamm and Norrby, 2001). Pyelonephritis can also lead to scarring of the kidneys, reduced renal concentrating ability, and renal failure.

Treatment

UTIs are commonly treated with oral antibiotics based on the symptoms described by patients before the diagnosis is confirmed (Mishra et al., 2012). However, urine dipstick testing or wet mount microscopy to test for pyuria are quick confirmatory tests that can be conducted prior to antibiotic therapy to ensure that the diagnosis is correct. Antibiotics are selected based on local resistance profiles and those commonly prescribed include sulfamethoxazole-trimethoprim, nitrofurantoin, ciprofloxacin, norfloxacin, or ofloxacin (Kodner and Thomas Gupton, 2010). Symptoms usually resolve within a few days of treatment. Treatment lasts from three to ten days. Women who suffer from recurrent UTI are often treated with continuous or postcoital antibiotics to prophylactically prevent subsequent infections (Kodner and Thomas Gupton, 2010).

Control and prevention

Currently there is no effective vaccine to prevent UTI caused by *E. coli*. To prevent UTI, physicians recommend use of contraceptives that do not include spermicides. For recurrent infections, antibiotics taken continuously or prophylactically after sexual intercourse is a common preventative practice (Kodner and Thomas Gupton, 2010). A non-prescription preventative method is drinking cranberry juice, however, there is no clear correlation between dose or frequency of intake and a reduction of incidence of UTI (Sen, 2006).

CONCLUSIONS

Uropathogenic *E. coli* utilize a variety of virulence factors to ascend and colonize the urinary tract (Table 9.1 and Figure 9.1). Flagella are necessary to propel bacteria from the periurethral space to the bladder, where UPEC then use multiple fimbrial and afimbrial adhesins to colonize the uroepithelium. The host responds to UTI by exfoliation of bladder epithelial cells and stimulation of the inflammatory response. Further limitation of infection occurs by iron sequestration. To overcome iron limitation, UPEC produce several siderophores that chelate iron from the host proteins and heme receptors to take up iron from the host. To evade killing by neutrophils, UPEC produce TcpC to reduce the TLR4 response, as well as secrete toxins that reduce leukocyte motility. Flagella can also direct UPEC up the ureters to the kidneys, where damage from the robust inflammatory response can cause renal scarring and ultimately renal failure. Fimbrial and afimbrial adhesins again are used to bind to the kidney epithelial cells. While several virulence factors are now known for UPEC, the mechanism of action is still to be determined for some newly recognized factors. UPEC are highly versatile pathogens, able to commensally colonize the human gastrointestinal tract, while opportunistically causing disease in the urinary tract.

REFERENCES

Allsopp, L.P., Beloin, C., Ulett, G.C., et al., 2012. Molecular characterization of UpaB and UpaC, two new autotransporter proteins of uropathogenic *Escherichia coli* CFT073. Infect. Immun. 80 (1), 321–332.

Allsopp, L.P., Totsika, M., Tree, J.J., et al., 2010. UpaH is a newly identified autotransporter protein that contributes to biofilm formation and bladder colonization by uropathogenic *Escherichia coli* CFT073. Infect. Immun. 78 (4), 1659–1669.

Alteri, C.J., Lindner, J.R., Reiss, D.J., Smith, S.N., Mobley, H.L., 2011. The broadly conserved regulator PhoP links pathogen virulence and membrane potential in *Escherichia coli*. Mol. Microbiol. 82 (1), 145–163.

Anderson, G.G., Goller, C.C., Justice, S., Hultgren, S.J., Seed, P.C., 2010. Polysaccharide capsule and sialic acid-mediated regulation promote biofilm-like intracellular bacterial communities during cystitis. Infect. Immun. 78 (3), 963–975.

Anderson, G.G., Palermo, J.J., Schilling, J.D., Roth, R., Heuser, J., Hultgren, S.J., 2003. Intracellular bacterial biofilm-like pods in urinary tract infections. Science 301 (5629), 105–107.

Bahrani-Mougeot, F.K., Buckles, E.L., Lockatell, C.V., et al., 2002. Type 1 fimbriae and extracellular polysaccharides are preeminent uropathogenic *Escherichia coli* virulence determinants in the murine urinary tract. Mol. Microbiol. 45 (4), 1079–1093.

Bien, J., Sokolova, O., Bozko, P., 2012. Role of uropathogenic *Escherichia coli* virulence factors in development of urinary tract infection and kidney damage. Int. J. Nephrol. 2012, 681473.

Billips, B.K., Schaeffer, A.J., Klumpp, D.J., 2008. Molecular basis of uropathogenic *Escherichia coli* evasion of the innate immune response in the bladder. Infect. Immun. 76 (9), 3891–3900.

Blanco, M., Blanco, J.E., Alonso, M.P., Blanco, J., 1994. Virulence factors and O groups of *Escherichia coli* strains isolated from cultures of blood specimens from urosepsis and non-urosepsis patients. Microbiologia 10 (3), 249–256.

Blanco, M., Blanco, J.E., Alonso, M.P., Blanco, J., 1996. Virulence factors and O groups of *Escherichia coli* isolates from patients with acute pyelonephritis, cystitis and asymptomatic bacteriuria. Eur. J. Epidemiol. 12 (2), 191–198.

Bower, J.M., Eto, D.S., Mulvey, M.A., 2005. Covert operations of uropathogenic *Escherichia coli* within the urinary tract. Traffic 6 (1), 18–31.

Brinton Jr., C.C., 1959. Non-flagellar appendages of bacteria. Nature 183 (4664), 782–786.

Buckles, E.L., Bahrani-Mougeot, F.K., Molina, A., et al., 2004. Identification and characterization of a novel uropathogenic *Escherichia coli*-associated fimbrial gene cluster. Infect. Immun. 72 (7), 3890–3901.

Buckles, E.L., Wang, X., Lane, M.C., et al., 2009. Role of the K2 capsule in *Escherichia coli* urinary tract infection and serum resistance. J. Infect. Dis. 199 (11), 1689–1697.

Buckles, E.L., Wang, X., Lockatell, C.V., Johnson, D.E., Donnenberg, M.S., 2006. PhoU enhances the ability of extraintestinal pathogenic *Escherichia coli* strain CFT073 to colonize the murine urinary tract. Microbiology 152 (Pt 1), 153–160.

Capo, C., Meconi, S., Sanguedolce, M.V., et al., 1998. Effect of cytotoxic necrotizing factor-1 on actin cytoskeleton in human monocytes: role in the regulation of integrin-dependent phagocytosis. J. Immunol. 161 (8), 4301–4308.

Cirl, C., Wieser, A., Yadav, M., et al., 2008. Subversion of Toll-like receptor signaling by a unique family of bacterial Toll/interleukin-1 receptor domain-containing proteins. Nat. Med. 14 (4), 399–406.

Connell, I., Agace, W., Klemm, P., Schembri, M., Marild, S., Svanborg, C., 1996. Type 1 fimbrial expression enhances *Escherichia coli* virulence for the urinary tract. Proc. Natl. Acad. Sci. USA 93 (18), 9827–9832.

Das, M., Hart-Van Tassell, A., Urvil, P.T., et al., 2005. Hydrophilic domain II of *Escherichia coli* Dr fimbriae facilitates cell invasion. Infect. Immun. 73 (9), 6119–6126.

de Ree, J.M., van den Bosch, J.F., 1987. Serological response to the P fimbriae of uropathogenic *Escherichia coli* in pyelonephritis. Infect. Immun. 55 (9), 2204–2207.

Dhakal, B.K., Mulvey, M.A., 2012. The UPEC pore-forming toxin alpha-hemolysin triggers proteolysis of host proteins to disrupt cell adhesion, inflammatory, and survival pathways. Cell Host. Microbe. 11 (1), 58–69.

Dorman, C.J., Corcoran, C.P., 2009. Bacterial DNA topology and infectious disease. Nucleic Acids Res. 37 (3), 672–678.

Doye, A., Mettouchi, A., Bossis, G., et al., 2002. CNF1 exploits the ubiquitin-proteasome machinery to restrict Rho GTPase activation for bacterial host cell invasion. Cell 111 (4), 553–564.

Eden, C.S., Hanson, L.A., Jodal, U., Lindberg, U., Akerlund, A.S., 1976. Variable adherence to normal human urinary-tract epithelial cells of *Escherichia coli* strains associated with various forms of urinary-tract infection. Lancet 1 (7984), 490–492.

Eguchi, Y., Ishii, E., Hata, K., Utsumi, R., 2011. Regulation of acid resistance by connectors of two-component signal transduction systems in *Escherichia coli*. J. Bacteriol. 193 (5), 1222–1228.

Falzano, L., Fiorentini, C., Donelli, G., et al., 1993. Induction of phagocytic behaviour in human epithelial cells by *Escherichia coli* cytotoxic necrotizing factor type 1. Mol. Microbiol. 9 (6), 1247–1254.

Fang, L., Nowicki, B.J., Urvil, P., et al., 2004. Epithelial invasion by *Escherichia coli* bearing Dr fimbriae is controlled by nitric oxide-regulated expression of CD55. Infect. Immun. 72 (5), 2907–2914.

Felmlee, T., Welch, R.A., 1988. Alterations of amino acid repeats in the *Escherichia coli* hemolysin affect cytolytic activity and secretion. Proc. Natl. Acad. Sci. USA 85 (14), 5269–5273.

Foxman, B., 2003. Epidemiology of urinary tract infections: incidence, morbidity, and economic costs. Dis. Mon. 49 (2), 53–70.

Garcia, E.C., Brumbaugh, A.R., Mobley, H.L., 2011. Redundancy and specificity of *Escherichia coli* iron acquisition systems during urinary tract infection. Infect. Immun. 79 (3), 1225–1235.

Garcia, M.I., Jouve, M., Nataro, J.P., Gounon, P., Le Bouguenec, C., 2000. Characterization of the AfaD-like family of invasins encoded by pathogenic *Escherichia coli* associated with intestinal and extra-intestinal infections. FEBS Lett. 479 (3), 111–117.

Goller, C.C., Seed, P.C., 2010. Revisiting the *Escherichia coli* polysaccharide capsule as a virulence factor during urinary tract infection: contribution to intracellular biofilm development. Virulence 1 (4), 333–337.

Goluszko, P., Goluszko, E., Nowicki, B., Nowicki, S., Popov, V., Wang, H.Q., 2005. Vaccination with purified Dr fimbriae reduces mortality associated with chronic urinary tract infection due to *Escherichia coli* bearing Dr adhesin. Infect. Immun. 73 (1), 627–631.

Goluszko, P., Popov, V., Selvarangan, R., Nowicki, S., Pham, T., Nowicki, B.J., 1997. Dr fimbriae operon of uropathogenic *Escherichia coli* mediate microtubule-dependent invasion to the HeLa epithelial cell line. J. Infect. Dis. 176 (1), 158–167.

Guyer, D.M., Henderson, I.R., Nataro, J.P., Mobley, H.L., 2000. Identification of *sat*, an autotransporter toxin produced by uropathogenic *Escherichia coli*. Mol. Microbiol. 38 (1), 53–66.

Guyer, D.M., Radulovic, S., Jones, F.E., Mobley, H.L., 2002. Sat, the secreted autotransporter toxin of uropathogenic *Escherichia coli*, is a vacuolating cytotoxin for bladder and kidney epithelial cells. Infect. Immun. 70 (8), 4539–4546.

Hagan, E.C., Mobley, H.L., 2009. Haem acquisition is facilitated by a novel receptor Hma and required by uropathogenic *Escherichia coli* for kidney infection. Mol. Microbiol. 71 (1), 79–91.

Heesemann, J., Hantke, K., Vocke, T., et al., 1993. Virulence of *Yersinia enterocolitica* is closely associated with siderophore production, expression of an iron-repressible outer membrane polypeptide of 65,000 Da and pesticin sensitivity. Mol. Microbiol. 8 (2), 397–408.

Heimer, S.R., Rasko, D.A., Lockatell, C.V., Johnson, D.E., Mobley, H.L., 2004. Autotransporter genes *pic* and *tsh* are associated with *Escherichia coli* strains that cause acute pyelonephritis and are expressed during urinary tract infection. Infect. Immun. 72 (1), 593–597.

Hilbert, D.W., Paulish-Miller, T.E., Tan, C.K., et al., 2012. Clinical *Escherichia coli* isolates utilize alpha-hemolysin to inhibit *in vitro* epithelial cytokine production. Microbes. Infect. 14 (7-8), 628–638.

Hofman, P., Flatau, G., Selva, E., et al., 1998. *Escherichia coli* cytotoxic necrotizing factor 1 effaces microvilli and decreases transmigration of polymorphonuclear leukocytes in intestinal T84 epithelial cell monolayers. Infect. Immun. 66 (6), 2494–2500.

Hofman, P., Le Negrate, G., Mograbi, B., et al., 2000. *Escherichia coli* cytotoxic necrotizing factor-1 (CNF-1) increases the adherence to epithelia and the oxidative burst of human polymorphonuclear leukocytes but decreases bacteria phagocytosis. J. Leukoc. Biol. 68 (4), 522–528.

Holden, N.J., Totsika, M., Mahler, E., et al., 2006. Demonstration of regulatory cross-talk between P fimbriae and type 1 fimbriae in uropathogenic *Escherichia coli*. Microbiology 152 (Pt 4), 1143–1153.

Holden, N.J., Uhlin, B.E., Gally, D.L., 2001. PapB paralogues and their effect on the phase variation of type 1 fimbriae in *Escherichia coli*. Mol. Microbiol. 42 (2), 319–330.

Huisman, O., D'Ari, R., George, J., 1980. Further characterization of sfiA and sfiB mutations in *Escherichia coli*. J. Bacteriol. 144 (1), 185–191.

Hull, R.A., Gill, R.E., Hsu, P., Minshew, B.H., Falkow, S., 1981. Construction and expression of recombinant plasmids encoding type 1 or D-mannose-resistant pili from a urinary tract infection *Escherichia coli* isolate. Infect. Immun. 33 (3), 933–938.

Johanson, I.M., Plos, K., Marklund, B.I., Svanborg, C., 1993. *Pap*, *papG* and *prsG* DNA sequences in *Escherichia coli* from the fecal flora and the urinary tract. Microb. Pathog. 15 (2), 121–129.

Johnson, D.E., Drachenberg, C., Lockatell, C.V., Island, M.D., Warren, J.W., Donnenberg, M.S., 2000. The role of cytotoxic necrotizing factor-1 in colonization and tissue injury in a murine model of urinary tract infection. FEMS Immunol. Med. Microbiol. 28 (1), 37–41.

Johnson, J.R., Brown, J.J., Maslow, J.N., 1998. Clonal distribution of the three alleles of the Gal(alpha1-4)Gal-specific adhesin gene *papG* among *Escherichia coli* strains from patients with bacteremia. J. Infect. Dis. 177 (3), 651–661.

Johnson, J.R., Jelacic, S., Schoening, L.M., et al., 2005a. The IrgA homologue adhesin Iha is an *Escherichia coli* virulence factor in murine urinary tract infection. Infect. Immun. 73 (2), 965–971.

Johnson, J.R., Owens, K., Gajewski, A., Kuskowski, M.A., 2005b. Bacterial characteristics in relation to clinical source of *Escherichia coli* isolates from women with acute cystitis or pyelonephritis and uninfected women. J. Clin. Microbiol. 43 (12), 6064–6072.

Johnson, J.R., Skubitz, K.M., Nowicki, B.J., Jacques-Palaz, K., Rakita, R.M., 1995. Nonlethal adherence to human neutrophils mediated by Dr antigen-specific adhesins of *Escherichia coli*. Infect. Immun. 63 (1), 309–316.

Justice, S.S., Hung, C., Theriot, J.A., et al., 2004. Differentiation and developmental pathways of uropathogenic *Escherichia coli* in urinary tract pathogenesis. Proc. Natl. Acad. Sci. USA 101 (5), 1333–1338.

Justice, S.S., Hunstad, D.A., Seed, P.C., Hultgren, S.J., 2006a. Filamentation by *Escherichia coli* subverts innate defenses during urinary tract infection. Proc. Natl. Acad. Sci. USA 103 (52), 19884–19889.

Justice, S.S., Lauer, S.R., Hultgren, S.J., Hunstad, D.A., 2006b. Maturation of intracellular *Escherichia coli* communities requires SurA. Infect. Immun. 74 (8), 4793–4800.

Källenius, G., Möllby, R., 1979. Adhesion of *Escherichia coli* to human periurethral cells correlated to mannose-resistant agglutination of human erythrocytes. FEMS Microbiol. Lett. 5, 295–299.

Kato, A., Tanabe, H., Utsumi, R., 1999. Molecular characterization of the PhoP-PhoQ two-component system in *Escherichia coli* K-12: identification of extracellular Mg2+-responsive promoters. J. Bacteriol. 181 (17), 5516–5520.

Khan, A.S., Kniep, B., Oelschlaeger, T.A., Van Die, I., Korhonen, T., Hacker, J., 2000. Receptor structure for F1C fimbriae of uropathogenic *Escherichia coli*. Infect. Immun. 68 (6), 3541–3547.

Klemm, P., Krogfelt, K.A., Hedegaard, L., Christiansen, G., 1990. The major subunit of *Escherichia coli* type 1 fimbriae is not required for D-mannose-specific adhesion. Mol. Microbiol. 4 (4), 553–559.

Kodner, C.M., Thomas Gupton, E.K., 2010. Recurrent urinary tract infections in women: diagnosis and management. Am. Fam. Physician. 82 (6), 638–643.

Konig, B., Konig, W., 1993. Induction and suppression of cytokine release (tumour necrosis factor-alpha; interleukin-6, interleukin-1 beta) by *Escherichia coli* pathogenicity factors (adhesions, alpha-haemolysin). Immunology 78 (4), 526–533.

Korea, C.G., Badouraly, R., Prevost, M.C., Ghigo, J.M., Beloin, C., 2010. *Escherichia coli* K-12 possesses multiple cryptic but functional chaperone-usher fimbriae with distinct surface specificities. Environ. Microbiol. 12 (7), 1957–1977.

Korhonen, T.K., Parkkinen, J., Hacker, J., et al., 1986. Binding of *Escherichia coli* S fimbriae to human kidney epithelium. Infect. Immun. 54 (2), 322–327.

Krogfelt, K.A., Bergmans, H., Klemm, P., 1990. Direct evidence that the FimH protein is the mannose-specific adhesin of *Escherichia coli* type 1 fimbriae. Infect. Immun. 58 (6), 1995–1998.

Lane, M.C., Alteri, C.J., Smith, S.N., Mobley, H.L., 2007a. Expression of flagella is coincident with uropathogenic *Escherichia coli* ascension to the upper urinary tract. Proc. Natl. Acad. Sci. USA 104 (42), 16669–16674.

Lane, M.C., Simms, A.N., Mobley, H.L., 2007b. Complex interplay between type 1 fimbrial expression and flagellum-mediated motility of uropathogenic *Escherichia coli*. J. Bacteriol. 189 (15), 5523–5533.

Lindberg, S., Xia, Y., Sonden, B., Goransson, M., Hacker, J., Uhlin, B.E., 2008. Regulatory interactions among adhesin gene systems of uropathogenic *Escherichia coli*. Infect. Immun. 76 (2), 771–780.

Lloyd, A.L., Henderson, T.A., Vigil, P.D., Mobley, H.L., 2009a. Genomic islands of uropathogenic *Escherichia coli* contribute to virulence. J. Bacteriol. 191 (11), 3469–3481.

Lloyd, A.L., Rasko, D.A., Mobley, H.L., 2007. Defining genomic islands and uropathogen-specific genes in uropathogenic *Escherichia coli*. J. Bacteriol. 189 (9), 3532–3546.

Lloyd, A.L., Smith, S.N., Eaton, K.A., Mobley, H.L., 2009b. Uropathogenic *Escherichia coli* suppresses the host inflammatory response via pathogenicity island genes *sisA* and *sisB*. Infect. Immun. 77 (12), 5322–5333.

Lugering, A., Benz, I., Knochenhauer, S., Ruffing, M., Schmidt, M.A., 2003. The Pix pilus adhesin of the uropathogenic *Escherichia coli* strain X2194 (O2: K(-): H6) is related to Pap pili but exhibits a truncated regulatory region. Microbiology 149 (Pt 6), 1387–1397.

Mansson, L.E., Kjall, P., Pellett, S., et al., 2007. Role of the lipopolysaccharide-CD14 complex for the activity of hemolysin from uropathogenic *Escherichia coli*. Infect. Immun. 75 (2), 997–1004.

Mao, B.H., Chang, Y.F., Scaria, J., et al., 2012. Identification of *Escherichia coli* genes associated with urinary tract infections. J. Clin. Microbiol. 50 (2), 449–456.

Maroncle, N.M., Sivick, K.E., Brady, R., Stokes, F.E., Mobley, H.L., 2006. Protease activity, secretion, cell entry, cytotoxicity, and cellular targets of secreted autotransporter toxin of uropathogenic *Escherichia coli*. Infect. Immun. 74 (11), 6124–6134.

Marre, R., Hacker, J., Henkel, W., Goebel, W., 1986. Contribution of cloned virulence factors from uropathogenic *Escherichia coli* strains to nephropathogenicity in an experimental rat pyelonephritis model. Infect. Immun. 54 (3), 761–767.

Marre, R., Kreft, B., Hacker, J., 1990. Genetically engineered S and F1C fimbriae differ in their contribution to adherence of *Escherichia coli* to cultured renal tubular cells. Infect. Immun. 58 (10), 3434–3437.

Martinez, J.J., Mulvey, M.A., Schilling, J.D., Pinkner, J.S., Hultgren, S.J., 2000. Type 1 pilus-mediated bacterial invasion of bladder epithelial cells. EMBO J. 19 (12), 2803–2812.

Mishra, B., Srivastava, S., Singh, K., Pandey, A., Agarwal, J., 2012. Symptom-based diagnosis of urinary tract infection in women: are we over-prescribing antibiotics? Int. J. Clin. Pract. 66 (5), 493–498.

Mobley, H.L., Jarvis, K.G., Elwood, J.P., et al., 1993. Isogenic P-fimbrial deletion mutants of pyelonephritogenic *Escherichia coli*: the role of alpha Gal(1-4) beta Gal binding in virulence of a wild-type strain. Mol. Microbiol. 10 (1), 143–155.

Morschhauser, J., Hoschutzky, H., Jann, K., Hacker, J., 1990. Functional analysis of the sialic acid-binding adhesin SfaS of pathogenic *Escherichia coli* by site-specific mutagenesis. Infect. Immun. 58 (7), 2133–2138.

Morschhauser, J., Vetter, V., Emody, L., Hacker, J., 1994. Adhesin regulatory genes within large, unstable DNA regions of pathogenic *Escherichia coli*: cross-talk between different adhesin gene clusters. Mol. Microbiol. 11 (3), 555–566.

Muller, C.M., Aberg, A., Strasevinene, J., Emody, L., Uhlin, B.E., Balsalobre, C., 2009. Type 1 fimbriae, a colonization factor of uropathogenic *Escherichia coli*, are controlled by the metabolic sensor CRP-cAMP. PLoS Pathog. 5 (2), e1000303.

Mulvey, M.A., Schilling, J.D., Hultgren, S.J., 2001. Establishment of a persistent *Escherichia coli* reservoir during the acute phase of a bladder infection. Infect. Immun. 69 (7), 4572–4579.

Mulvey, M.A., Schilling, J.D., Martinez, J.J., Hultgren, S.J., 2000. Bad bugs and beleaguered bladders: interplay between uropathogenic *Escherichia coli* and innate host defenses. Proc. Natl. Acad. Sci. USA 97 (16), 8829–8835.

Mysorekar, I.U., Hultgren, S.J., 2006. Mechanisms of uropathogenic *Escherichia coli* persistence and eradication from the urinary tract. Proc. Natl. Acad. Sci. USA 103 (38), 14170–14175.

Nesta, B., Spraggon, G., Alteri, C., et al., 2012. FdeC, a novel broadly conserved *Escherichia coli* adhesin eliciting protection against urinary tract infections. MBio 3 (2), e00010–12. doi:10.1128/ mBio.00010-12.

Nicholson, T.F., Watts, K.M., Hunstad, D.A., 2009. OmpA of uropathogenic *Escherichia coli* promotes postinvasion pathogenesis of cystitis. Infect. Immun. 77 (12), 5245–5251.

Nicolle, L.E., 2005. Catheter-related urinary tract infection. Drugs Aging. 22 (8), 627–639.

Norinder, B.S., Koves, B., Yadav, M., Brauner, A., Svanborg, C., 2012. Do *Escherichia coli* strains causing acute cystitis have a distinct virulence repertoire? Microb. Pathog. 52 (1), 10–16.

Nowicki, B., Holthofer, H., Saraneva, T., Rhen, M., Vaisanen-Rhen, V., Korhonen, T.K., 1986. Location of adhesion sites for P-fimbriated and for O75X-positive *Escherichia coli* in the human kidney. Microb. Pathog. 1 (2), 169–180.

Nowicki, B., Labigne, A., Moseley, S., Hull, R., Hull, S., Moulds, J., 1990. The Dr hemagglutinin, afimbrial adhesins AFA-I and AFA-III, and F1845 fimbriae of uropathogenic and diarrhea-associated *Escherichia coli* belong to a family of hemagglutinins with Dr receptor recognition. Infect. Immun. 58 (1), 279–281.

Opal, S.M., Cross, A.S., Gemski, P., Lyhte, L.W., 1990. Aerobactin and alpha-hemolysin as virulence determinants in *Escherichia coli* isolated from human blood, urine, and stool. J. Infect. Dis. 161 (4), 794–796.

Orndorff, P.E., Bloch, C.A., 1990. The role of type 1 pili in the pathogenesis of *Escherichia coli* infections: a short review and some new ideas. Microb. Pathog. 9 (2), 75–79.

Ott, M., Hoschutzky, H., Jann, K., Van Die, I., Hacker, J., 1988. Gene clusters for S fimbrial adhesin (*sfa*) and F1C fimbriae (*foc*) of *Escherichia coli*: comparative aspects of structure and function. J. Bacteriol. 170 (9), 3983–3990.

Parham, N.J., Srinivasan, U., Desvaux, M., Foxman, B., Marrs, C.F., Henderson, I.R., 2004. PicU, a second serine protease autotransporter of uropathogenic *Escherichia coli*. FEMS Microbiol. Lett. 230 (1), 73–83.

Redford, P., Roesch, P.L., Welch, R.A., 2003. DegS is necessary for virulence and is among extraintestinal *Escherichia coli* genes induced in murine peritonitis. Infect. Immun. 71 (6), 3088–3096.

Redford, P., Welch, R.A., 2006. Role of sigma E-regulated genes in *Escherichia coli* uropathogenesis. Infect. Immun. 74 (7), 4030–4038.

Riegman, N., Kusters, R., Van Veggel, H., et al., 1990. F1C fimbriae of a uropathogenic *Escherichia coli* strain: genetic and functional organization of the *foc* gene cluster and identification of minor subunits. J. Bacteriol. 172 (2), 1114–1120.

Rippere-Lampe, K.E., O'Brien, A.D., Conran, R., Lockman, H.A., 2001. Mutation of the gene encoding cytotoxic necrotizing factor type 1 (cnf(1)) attenuates the virulence of uropathogenic *Escherichia coli*. Infect. Immun. 69 (6), 3954–3964.

Ritter, A., Blum, G., Emody, L., et al., 1995. tRNA genes and pathogenicity islands: influence on virulence and metabolic properties of uropathogenic *Escherichia coli*. Mol. Microbiol. 17 (1), 109–121.

Roberts, J.A., Marklund, B.I., Ilver, D., et al., 1994. The Gal(alpha 1-4)Gal-specific tip adhesin of *Escherichia coli* P-fimbriae is needed for pyelonephritis to occur in the normal urinary tract. Proc. Natl. Acad. Sci. USA 91 (25), 11889–11893.

Ronald, A., 2002. The etiology of urinary tract infection: traditional and emerging pathogens. Am. J. Med. 113 (Suppl. 1A), 14S–19S.

Rosen, D.A., Hooton, T.M., Stamm, W.E., Humphrey, P.A., Hultgren, S.J., 2007. Detection of intracellular bacterial communities in human urinary tract infection. PLoS Med. 4 (12), e329.

Ruiz-Perez, F., Wahid, R., Faherty, C.S., et al., 2011. Serine protease autotransporters from *Shigella flexneri* and pathogenic *Escherichia coli* target a broad range of leukocyte glycoproteins. Proc. Natl. Acad. Sci. USA 108 (31), 12881–12886.

Russo, T.A., McFadden, C.D., Carlino-MacDonald, U.B., Beanan, J.M., Barnard, T.J., Johnson, J.R., 2002. IroN functions as a siderophore receptor and is a urovirulence factor in an extraintestinal pathogenic isolate of *Escherichia coli*. Infect. Immun. 70 (12), 7156–7160.

Salvador, E., Wagenlehner, F., Kohler, C.D., et al., 2012. Comparison of asymptomatic bacteriuria *Escherichia coli* isolates from healthy individuals versus those from hospital patients shows that long-term bladder colonization selects for attenuated virulence phenotypes. Infect. Immun. 80 (2), 668–678.

Schappert, S.M., Rechtsteiner, E.A., 2008. Ambulatory medical care utilization estimates for 2006. Natl. Health Stat. Rep. (8), 1–29.

Schubert, S., Picard, B., Gouriou, S., Heesemann, J., Denamur, E., 2002. Yersinia high-pathogenicity island contributes to virulence in *Escherichia coli* causing extraintestinal infections. Infect. Immun. 70 (9), 5335–5337.

Sen, A., 2006. Recurrent cystitis in non-pregnant women. Clin. Evid. (15), 2558–2564.

Serafini-Cessi, F., Malagolini, N., Cavallone, D., 2003. Tamm-Horsfall glycoprotein: biology and clinical relevance. Am. J. Kidney. Dis. 42 (4), 658–676.

Simms, A.N., Mobley, H.L., 2008a. Multiple genes repress motility in uropathogenic *Escherichia coli* constitutively expressing type 1 fimbriae. J. Bacteriol. 190 (10), 3747–3756.

Simms, A.N., Mobley, H.L., 2008b. PapX, a P fimbrial operon-encoded inhibitor of motility in uropathogenic *Escherichia coli*. Infect. Immun. 76 (11), 4833–4841.

Sjostrom, A.E., Sonden, B., Muller, C., et al., 2009. Analysis of the sfaX(II) locus in the *Escherichia coli* meningitis isolate IHE3034 reveals two novel regulatory genes within the promoter-distal region of the main S fimbrial operon. Microb. Pathog. 46 (3), 150–158.

Smith, Y.C., Rasmussen, S.B., Grande, K.K., Conran, R.M., O'Brien, A.D., 2008. Hemolysin of uropathogenic *Escherichia coli* evokes extensive shedding of the uroepithelium and hemorrhage in bladder tissue within the first 24 hours after intraurethral inoculation of mice. Infect. Immun. 76 (7), 2978–2990.

Snyder, J.A., Haugen, B.J., Lockatell, C.V., et al., 2005. Coordinate expression of fimbriae in uropathogenic *Escherichia coli*. Infect. Immun. 73 (11), 7588–7596.

Spurbeck, R.R., Stapleton, A.E., Johnson, J.R., Walk, S.T., Hooton, T.M., Mobley, H.L., 2011. Fimbrial profiles predict virulence of uropathogenic *Escherichia coli* strains: contribution of *ygi* and *yad* fimbriae. Infect. Immun. 79 (12), 4753–4763.

Stamm, W.E., Hooton, T.M., 1993. Management of urinary tract infections in adults. N. Engl. J. Med. 329 (18), 1328–1334.

Stamm, W.E., Norrby, S.R., 2001. Urinary tract infections: disease panorama and challenges. J. Infect. Dis. 183 (Suppl. 1), S1–4.

Stanley, P., Koronakis, V., Hughes, C., 1998. Acylation of *Escherichia coli* hemolysin: a unique protein lipidation mechanism underlying toxin function. Microbiol. Mol. Biol. Rev. 62 (2), 309–333.

Stromberg, N., Nyholm, P.G., Pascher, I., Normark, S., 1991. Saccharide orientation at the cell surface affects glycolipid receptor function. Proc. Natl. Acad. Sci. USA 88 (20), 9340–9344.

Torres, A.G., Redford, P., Welch, R.A., Payne, S.M., 2001. TonB-dependent systems of uropathogenic *Escherichia coli*: aerobactin and heme transport and TonB are required for virulence in the mouse. Infect. Immun. 69 (10), 6179–6185.

Totsika, M., Wells, T.J., Beloin, C., et al., 2012. Molecular characterization of the EhaG and UpaG trimeric autotransporter proteins from pathogenic *Escherichia coli*. Appl. Environ. Microbiol. 78 (7), 2179–2189.

Tseng, C.C., Wu, J.J., Liu, H.L., Sung, J.M., Huang, J.J., 2002. Roles of host and bacterial virulence factors in the development of upper urinary tract infection caused by *Escherichia coli*. Am. J. Kidney Dis. 39 (4), 744–752.

Uhlen, P., Laestadius, A., Jahnukainen, T., et al., 2000. Alpha-haemolysin of uropathogenic *E. coli* induces Ca2+ oscillations in renal epithelial cells. Nature 405 (6787), 694–697.

Ulett, G.C., Mabbett, A.N., Fung, K.C., Webb, R.I., Schembri, M.A., 2007a. The role of F9 fimbriae of uropathogenic *Escherichia coli* in biofilm formation. Microbiology 153 (Pt 7), 2321–2331.

Ulett, G.C., Valle, J., Beloin, C., Sherlock, O., Ghigo, J.M., Schembri, M.A., 2007b. Functional analysis of antigen 43 in uropathogenic *Escherichia coli* reveals a role in long-term persistence in the urinary tract. Infect. Immun. 75 (7), 3233–3244.

Valle, J., Mabbett, A.N., Ulett, G.C., et al., 2008. UpaG, a new member of the trimeric autotransporter family of adhesins in uropathogenic *Escherichia coli*. J. Bacteriol. 190 (12), 4147–4161.

Vigil, P.D., Alteri, C.J., Mobley, H.L., 2011a. Identification of *in vivo*-induced antigens including an RTX family exoprotein required for uropathogenic *Escherichia coli* virulence. Infect. Immun. 79 (6), 2335–2344.

Vigil, P.D., Stapleton, A.E., Johnson, J.R., et al., 2011b. Presence of putative repeat-in-toxin gene *tosA* in *Escherichia coli* predicts successful colonization of the urinary tract. MBio 2 (3), e00066–e00011.

Vigil, P.D., Wiles, T.J., Engstrom, M.D., Prasov, L., Mulvey, M.A., Mobley, H.L., 2012. The repeat-in-toxin family member TosA mediates adherence of uropathogenic *Escherichia coli* and survival during bacteremia. Infect. Immun. 80 (2), 493–505.

Virkola, R., Westerlund, B., Holthofer, H., Parkkinen, J., Kekomaki, M., Korhonen, T.K., 1988. Binding characteristics of *Escherichia coli* adhesins in human urinary bladder. Infect. Immun. 56 (10), 2615–2622.

Visvikis, O., Boyer, L., Torrino, S., et al., 2011. *Escherichia coli* producing CNF1 toxin hijacks Tollip to trigger Rac1-dependent cell invasion. Traffic 12 (5), 579–590.

Walters, M.S., Lane, M.C., Vigil, P.D., Smith, S.N., Walk, S.T., Mobley, H.L., 2012. Kinetics of uropathogenic *Escherichia coli* metapopulation movement during urinary tract infection. MBio 3 (1).

Wang, H., Liang, F.X., Kong, X.P., 2008. Characteristics of the phagocytic cup induced by uropathogenic *Escherichia coli*. J. Histochem. Cytochem. 56 (6), 597–604.

Welch, R.A., 1991. Pore-forming cytolysins of gram-negative bacteria. Mol. Microbiol. 5 (3), 521–528.

Welch, R.A., Burland, V., Plunkett 3rd, G., et al., 2002. Extensive mosaic structure revealed by the complete genome sequence of uropathogenic *Escherichia coli*. Proc. Natl. Acad. Sci. USA 99 (26), 17020–17024.

Welch, R.A., Forestier, C., Lobo, A., Pellett, S., Thomas Jr., W., Rowe, G., 1992. The synthesis and function of the *Escherichia coli* hemolysin and related RTX exotoxins. FEMS Microbiol. Immunol. 5 (1-3), 29–36.

Westerlund, B., Kuusela, P., Risteli, J., et al., 1989. The O75X adhesin of uropathogenic *Escherichia coli* is a type IV collagen-binding protein. Mol. Microbiol. 3 (3), 329–337.

Wright, K.J., Seed, P.C., Hultgren, S.J., 2005. Uropathogenic *Escherichia coli* flagella aid in efficient urinary tract colonization. Infect. Immun. 73 (11), 7657–7668.

Wu, X.R., Kong, X.P., Pellicer, A., Kreibich, G., Sun, T.T., 2009. Uroplakins in urothelial biology, function, and disease. Kidney Int. 75 (11), 1153–1165.

Xia, Y., Gally, D., Forsman-Semb, K., Uhlin, B.E., 2000. Regulatory cross-talk between adhesin operons in *Escherichia coli*: inhibition of type 1 fimbriae expression by the PapB protein. EMBO J. 19 (7), 1450–1457.

Yadav, M., Zhang, J., Fischer, H., et al., 2010. Inhibition of TIR domain signaling by TcpC: MyD88-dependent and independent effects on *Escherichia coli* virulence. PLoS Pathog. 6 (9), e1001120.

Yamamoto, S., Tsukamoto, T., Terai, A., Kurazono, H., Takeda, Y., Yoshida, O., 1995. Distribution of virulence factors in *Escherichia coli* isolated from urine of cystitis patients. Microbiol. Immunol. 39 (6), 401–404.

Yokota, T., Gots, J.S., 1970. Requirement of adenosine 3′, 5′-cyclic phosphate for flagella formation in *Escherichia coli* and *Salmonella typhimurium*. J. Bacteriol. 103 (2), 513–516.

Zalewska, B., Piatek, R., Cieslinski, H., Nowicki, B., Kur, J., 2001. Cloning, expression, and purification of the uropathogenic *Escherichia coli* invasin DraD. Protein Expr. Purif. 23 (3), 476–482.

Zhou, G., Mo, W.J., Sebbel, P., et al., 2001. Uroplakin Ia is the urothelial receptor for uropathogenic *Escherichia coli*: evidence from *in vitro* FimH binding. J. Cell. Sci. 114 (Pt 22), 4095–4103.

Chapter 10

Meningitis-associated
Escherichia coli

Kwang Sik Kim

Johns Hopkins University School of Medicine, Baltimore, MD, USA

INTRODUCTION

Gram-negative bacillary meningitis continues to be an important cause of mortality and morbidity throughout the world. Case fatality rates have ranged between 15% and 40%, and approximately 50% of survivors sustain neurological sequelae (Gladstone et al., 1990; Unhanand et al., 1993; Dawson et al., 1999; Klinger et al., 2000; Stevens et al., 2003). A major contributing factor to such mortality and morbidity is our incomplete understanding of the pathogenesis of this disease (Kim, 2001, 2002, 2003, 2008, 2010, 2012). Both clinical and experimental data indicate limited efficacy with antimicrobial therapy alone for the treatment of Gram-negative bacillary meningitis (McCracken et al., 1984; Kim, 1985). *E. coli* is the most common Gram-negative organism that causes meningitis, particularly during the neonatal period.

An emergence of antibiotic resistance is an additional contributing factor to mortality and morbidity associated with *E. coli* meningitis. Recent reports of neonatal meningitis caused by *E. coli* strains producing CTX-M-type or TEM-type extended-spectrum β-lactamases are a particular concern (Blanco et al., 2011; Moissenet et al., 2011). Drug-resistant clonal group ST131 is prevalent among *E. coli* strains causing extraintestinal infection and some carry CTX-M-type extended-spectrum β-lactamases (Blanco et al., 2011). These findings indicate that a novel strategy is needed to identify new targets for prevention and therapy of *E. coli* meningitis.

Several lines of evidence from human cases of *E. coli* meningitis and animal models of experimental hematogenous *E. coli* meningitis indicate that *E. coli* invasion into the brain follows a high level of bacteremia and cerebral capillaries are the portal of entry into the brain (Berman and Banker, 1966; Kim et al., 1992), but how meningitis-causing *E. coli* strains invade the blood–brain barrier and penetrate into the brain remains incompletely understood.

Escherichia coli. http://dx.doi.org/10.1016/B978-0-12-397048-0.00010-3

Given the plethora of *E. coli* serotypes, it is striking that *E. coli* strains possessing the K1 capsular polysaccharide are predominant (approximately 80%) among isolates from neonatal *E. coli* meningitis (Robbins et al., 1974; Gross et al., 1982; Korhonen et al., 1985) and most of these K1 isolates are associated with a limited number of O serotypes (e.g. O18, O7, O16, O1, O45), belonging to phylogenetic group B2 and to a lesser extent, to group D (Sarff et al., 1975; Bingen et al., 1998; Johnson et al., 2001; Bonacorsi et al., 2003). The basis of this association of phylogenetic groups B2 and D with *E. coli* meningitis has not been elucidated.

The development of both in vitro and in vivo models of the blood–brain barrier has facilitated our investigations on the mechanisms of the microbial traversal of the blood–brain barrier, a key step required for the development of *E. coli* meningitis (Kim, 2001, 2002, 2003, 2008, 2010, 2012). The blood–brain barrier protects the brain from microbes circulating in the blood, but meningitis-causing pathogens have been shown to traverse the blood–brain barrier transcellularly, paracellularly and/or by means of infected phagocytic cells (Trojan-horse mechanism) (Kim, 2008). Recent studies have demonstrated that meningitis-causing *E. coli* K1 strains exhibit the ability to traverse the blood–brain barrier by transcellular penetration and cause central nervous system (CNS) inflammation, resulting in meningitis (Kim, 2001, 2002, 2003, 2008, 2010, 2012).

E. COLI TRAVERSAL OF THE BLOOD–BRAIN BARRIER

The blood–brain barrier is a structural and functional barrier that is formed by brain microvascular endothelial cells (BMEC), astrocytes, and pericytes (Rubin and Staddon, 1999). It regulates the passage of molecules into and out of the brain to maintain the neural microenvironment, and astrocytes and pericytes help maintain the barrier property of BMEC. The contributions of astrocytes and pericytes to *E. coli* traversal of the blood–brain barrier are, however, shown to be minimal.

The in vitro blood–brain barrier model has been developed with human brain microvascular endothelial cells (HBMEC). Upon cultivation on collagen-coated Transwell inserts these HBMEC exhibit morphologic and functional properties of tight junction formation and form a polarized monolayer. These properties are shown by the demonstrations of tight junction proteins (such as claudin 5 and ZO-1) and adherens junction proteins (such as VE-cadherin and β-catenin) and their spatial separation, as well as development of high transendothelial electrical resistance (Stins et al., 1997, 2001; Kim et al., 2004; Ruffer et al., 2004). Our previous studies with scanning and transmission electron microscopy documented the internalization of meningitis-causing *E. coli* K1 strains into HBMEC, as shown by the demonstration that internalized bacteria are found within membrane-bound vacuoles of HBMEC (Figure 10.1). *E. coli* K1 transmigrates the HBMEC monolayer through an

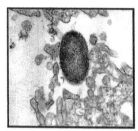 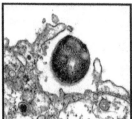 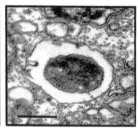

FIGURE 10.1   Transmission electron micrograph of human brain microvascular endothelial cell monolayer infected with meningitis-causing *E. coli* K1 strain RS218 (O18:K1, B2 group). Cellular protrusions surround internalizing *E. coli* (left) and internalized *E. coli* are found within membrane-bound vacuoles (right). Scale bar = 1 µm. *(Adapted from Kim, 2003, figure 3.)*

enclosed vacuole without intracellular multiplication and without any change in the integrity of the HBMEC monolayer (Nemani et al., 1999; Stins et al., 2001; Kim, 2003). No free bacteria are found in the cytoplasm of HBMEC or between adjacent HBMEC.

The in vivo blood–brain barrier model has been developed by inducing hematogenous meningitis in infant rats and mice. In this animal model, *E. coli* is administered orally, subcutaneously, intracardially, or intravenously, resulting in bacteremia and subsequent entry into the CNS (Kim et al., 1992; Huang et al., 1995, 1999; Wang et al., 1999; Hoffman et al., 2000; Khan et al., 2002; Wang and Kim, 2002; Wang et al., 2004; Zhu et al., 2010a,b). Studies in experimental hematogenous meningitis models indicate that the primary site of entry into the CNS for circulating *E. coli* is the cerebral vasculature, not the choroid plexus (Kim et al., 1992). *E. coli* entry into the CNS was documented without any change in the blood–brain barrier permeability as well as without concomitant presence of host inflammatory cells (e.g. PMNs, macrophages) (Kim et al., 1992, 1997), excluding the possibility that *E. coli* penetrates into the brain using the paracellular mechanism or the Trojan-horse mechanism via transmigration of *E. coli*-infected phagocytic cells. Taken together, these findings indicate a transcellular penetration of meningitis-causing *E. coli* K1 across the blood–brain barrier.

It is important to point out that *E. coli* is an enteric organism and there are no data supporting the concept that *E. coli* meningitis occurs following inhalation or intranasal spread (Mittal et al., 2010) and also there are no data indicating that an exposure to passive smoking (e.g. nicotine) increases the development of *E. coli* meningitis (Chi et al., 2011).

Studies with the above-mentioned in vitro and in vivo models of the blood–brain barrier have demonstrated that successful traversal of the blood–brain barrier by circulating *E. coli* requires (a) a high degree of bacteremia; (b) *E. coli* binding to and invasion of HBMEC; and (c) traversal of the blood–brain barrier as live bacteria (Kim et al., 1992; Hoffman et al., 1999; Kim 2001, 2002, 2003, 2008, 2010, 2012) (Table 10.1).

TABLE 10.1 Mechanisms involved in *E. coli* traversal of the blood–brain barrier

Mechanisms	*E. coli* factors	Host factors
1. A high-degree of bacteremia	K1, O-LPS, Nlpl	PMNs, complement
2. *E. coli* binding to HBMEC Nlpl	FimH, FliC, OmpA	CD48, gp96,
3. *E. coli* invasion of HBMEC	Ibe proteins, AslA, CNF1	37LRP
4. Traversal of the BBB as live bacteria	K1	Rab7

A threshold level of bacteremia required for *E. coli* penetration into the brain

Several studies in humans and experimental animals point to a relationship between the magnitude of bacteremia and the development of meningitis due to *E. coli*. For example, a significantly higher incidence of *E. coli* meningitis was noted in neonates who had bacterial counts in blood higher than 10^3 colony-forming units (CFU)/ml (6 out of 11 or 55%), compared to those with blood bacterial counts lower than 10^3 CFU/ml (1 out of 19 or 5%) (Dietzman et al., 1974). A high degree of bacteremia was also shown to be a primary determinant for penetration into the brain by circulating *E. coli* K1 in neonatal and adult animals with experimental hematogenous *E. coli* meningitis (Kim et al., 1992; Huang et al., 1995, 1999; Wang et al., 1999, 2004; Hoffman et al., 2000; Khan et al., 2002; Wang and Kim, 2002), but an approximately 10^6-fold greater inoculum of *E. coli* K1 is required to induce a similar high-level bacteremia in adult animals compared to neonatal animals (Kim et al., 1992). These findings suggest that the age dependency of *E. coli* meningitis is most likely due to the relative resistance of adults to high-level bacteremia, which precedes the development of meningitis, and less likely due to greater invasion of meningitis-causing *E. coli* in HBMEC derived from neonates compared to those from adults. This concept is shown by our demonstration that the abilities of meningitis-causing *E. coli* K1 to bind and invade BMEC are similar between BMEC derived from young and old rats as well as HBMEC derived from different ages (Stins et al., 1999). Thus, one of the reasons for the close association of meningitis-causing *E. coli* strains with neonatal meningitis is their ability to escape from host defenses and then to achieve a threshold level of bacteremia necessary for invasion of the meninges. Taken together, these findings indicate that the prevention of bacterial multiplication in the blood that is required for penetration into the brain would be one potential approach for prevention of *E. coli* meningitis.

Previous studies have identified that the expression of K1 capsular polysaccharide and O-lipopolysaccharide (LPS) are shown to be critical for induction of a high degree of bacteremia (Cross et al., 1986; Kim et al., 1986, 1988), but the feasibility of using the K1 capsule and O-LPS for the prevention of *E. coli* bacteremia has been shown to be limited (Soderstrom et al., 1984; Finne et al., 1987; Cross et al., 1994).

Recent functional *E. coli* genomic studies identified several *E. coli* factors that are shown to contribute to bacteremia (Kim, 2008; Xie et al., 2008; Moriel et al., 2010). For example, NlpI is a lipoprotein located in the outer membrane and has been shown to contribute to a high-level *E. coli* K1 bacteremeia. NlpI's evasion of serum-mediated killing is through regulation of complement regulator C4bp deposition on the bacterial surface (Tseng et al., 2012). Studies are in progress to determine the protective and broadly conserved antigens or to develop a multi-epitope subunit vaccine for the prevention of *E. coli* bacteremia and subsequent meningitis (Moriel et al., 2010; Wieser et al., 2012).

E. coli binding to and invasion of HBMEC

Subsequent studies have shown that a high degree of bacteremia is necessary, but not sufficient for *E. coli* penetration of the blood–brain barrier in vivo, and that *E. coli* binding to and invasion of HBMEC is a prerequisite for penetration into the brain (Huang et al., 1995, 1999; Wang et al., 1999; Hoffman et al., 2000; Khan et al., 2002; Wang and Kim, 2002), the essential step for the development of *E. coli* meningitis. This was shown by the demonstration in infant rats with experimental hematogenous meningitis that several isogenic mutants of meningitis-causing *E. coli* K1 strain RS218 deleted of factors contributing to HBMEC binding (e.g. OmpA) and invasion (e.g. Ibe proteins, CNF1) were significantly less able to induce meningitis than the parent strain despite similar levels of bacteremia (Table 10.2). These findings indicate that those *E. coli* factors contributing to HBMEC binding and invasion are necessary for crossing the blood–brain barrier in vivo.

E. coli *factors contributing to HBMEC binding*

Infections caused by pathogenic *E. coli* are often initiated by binding of the bacteria to the host cell surface, and this concept is likely to be important for circulating *E. coli* to withstand the blood flow in vivo and cross the blood–brain barrier. Several *E. coli* factors are identified to be involved in binding to HBMEC that subsequently affect invasion into HBMEC. These bacterial factors include type 1 fimbriae (FimH), flagella (FliC), outer membrane protein A (OmpA) and a lipoprotein (NlpI) (Khan et al., 2003, 2007; Shin et al., 2005; Teng et al., 2005, 2010; Parthasarathy et al., 2007). The roles of these *E. coli* factors in HBMEC binding were verified by deletion and complementation experiments. For example, isogenic deletion mutants were significantly less able to bind HBMEC and their binding ability was restored to the level of the parent strain by complementation with respective wild-type genes.

Fimbriae

Our experiment with an *E. coli* DNA microarray comparing the gene expression patterns of HBMEC-associated versus non-associated *E.coli* K1 revealed that type 1 fimbriae play an important role in *E. coli* K1 binding to HBMEC (Teng et al., 2005). The HBMEC-associated *E. coli* K1 showed significantly higher expression levels of the *fim* cluster genes compared to non-associated bacteria. Expression of type 1 fimbriae in wild-type *E. coli* is regulated by phase variation in which each bacterium can alternate between fimbriated and non-fimbriated states, so-called phase-ON and phase-OFF, respectively (see Chapter 12). We have shown that *E. coli* K1 associated with HBMEC are predominantly type 1 fimbriae phase-ON bacteria. We constructed type 1 fimbria locked-ON and locked-OFF mutants of *E. coli* K1 strain RS218, whose *fim* promoters are fixed in the ON and OFF orientations, respectively, and showed that *E. coli* K1 binding to HBMEC is significantly greater with the locked-ON mutant than the wild-type strain, while it is significantly less with the locked-OFF mutant (Teng et al., 2005). Decreased binding as the result of the *fimH* deletion or the locked-

TABLE 10.2 Development of bacteremia and meningitis (defined as positive CSF cultures) in newborn rats receiving meningitis-causing *E. coli* K1 strain RS218 or its deletion mutants

E. coli strain	No. of animals	Bacteremia (log CFU/ml blood)	No (% of animals with positive CSF)	References
RS218	19	7.18 ± 0.63	12 (63)	Wang and Kim, 2002
ΔompA	22	7.05 ± 0.49	6 (27)[a]	
RS218	24	7.51 ± 1.25	16 (67)	Huang et al., 1995
ΔibeA	25	6.97 ± 1.21	4 (16)[a]	
RS218	27	7.01 ± 1.17	15 (56)	Huang et al., 1999
ΔibeB	25	7.06 ± 1.29	4 (16)[a]	
RS218	24	7.53 ± 0.40	18 (75)	Wang et al., 1999
ΔibeC	24	7.80 ± 0.67	10 (42)[a]	
RS218	17	7.50 ± 0.32	14 (82)	Hoffman et al., 2000
ΔaslA	22	7.60 ± 0.49	7 (32)[a]	
RS218	26	7.64 ± 1.00	20 (77)	Khan et al., 2002
Δcnf1	27	7.24 ± 1.60	12 (44)[a]	

[a]Significantly less than RS218.
Modified with permission from Infection and Immunity (Kim, 2001).

OFF mutant resulted in decreased invasion into HBMEC (Teng et al., 2005; Khan et al., 2007).

From our *E. coli* DNA microarray experiments, we identified a novel site-specific recombinase, HbiF, which inverted the molecular switch *fimS* independent of the two known recombinases, FimB and FimE, that invert *fimS* and control the expression of the downstream *fim* operon (Xie et al., 2006a,b). Discovery of HbiF-mediated *fimS* switching provides a new opportunity for regulating type 1 fimbriae expression, which will help in developing a novel strategy for the prevention and therapy of *E. coli* bacteremia and meningitis.

We also identified that FimH interacts with a glycosylphosphatidylinositol-anchored receptor, CD48 on the surface of HBMEC, and FimH-CD48 interaction contributes to *E. coli* binding to HBMEC and increases in intracellular Ca^{2+} ($[Ca^{2+}]i$) in HBMEC (Khan et al., 2007; Kim et al., 2008). This concept is shown by the demonstration that CD48 antibody blocks FimH-mediated binding to HBMEC and FimH-induced $[Ca^{2+}]i$ changes in HBMEC (Table 10.3).

S fimbriae, which bind to terminal NeuAc α2,3-galactose sequences present on glycoproteins, have been implicated in *E. coli* binding to HBMEC. For example, purified S fimbriae or a recombinant *E. coli* strain HB101 expressing S fimbriae was shown to bind to the luminal surfaces of the brain vascular endothelium in neonatal rat brain tissues (Parkkinen et al., 1988). We have previously shown using S fimbriated transformants of *E. coli* strain HB101 that S fimbriae allowed this laboratory strain of *E. coli* to bind to HBMEC (Prasadarao et al., 1993; Stins et al., 1994), suggesting that S fimbriae contribute to *E. coli* binding to HBMEC. However, in-frame deletion of the S fimbriae operon in

TABLE 10.3 Blood–brain barrier receptors used by *E. coli* and other CNS-infecting microorganisms

Receptor	Organism and ligand	References
Gp96	*E. coli* OmpA	Khan et al., 2003
	L. monocytogenes Vip	Cabanes et al., 2005
37LRP	*E. coli* CNF1	Chung et al., 2003
	N. meningitidis PilQ/PorA	Orihuela et al., 2009
	S. pneumoniae CbpA	Orihuela et al., 2009
	H. influenza type b Omp2	Orihuela et al., 2009
	Prion protein	Gauczynski et al., 2001
	Viruses (sindbis, dengue, tick-borne encephalitis, Venezuelan equine encephalitis, adeno-associated)	Wang et al., 1992 Ludwig et al., 1996 Thepparit and Smith, 2004 Akache et al., 2006

meningitis-causing *E. coli* K1 did not significantly affect *E. coli* binding to and invasion of HBMEC, and also did not affect *E. coli* K1 penetration into the brain in the experimental hematogenous meningitis animal model (Wang et al., 2004). These findings suggest that S fimbriae are not critical to HBMEC binding in vitro and traversal of the blood–brain barrier in vivo by meningitis-causing *E. coli* K1.

Flagella

The external whip-like structures involved in bacterial motility have been shown to contribute to the virulence of many enteric bacteria, including *E. coli* (Parthasarathy et al., 2007). Our experiment with *E. coli* DNA microarray comparing the gene expression patterns of HBMEC-associated versus non-associated *E.coli* K1 also demonstrated that flagella play an important role in *E. coli* K1 binding to HBMEC (Parthasarathy et al., 2007; Xie et al., 2008). The flagellum is synthesized from three classes of genes. The class I region consists of the master operon *flhDC*, the class II region comprises the genes involved in the synthesis of the flagella basal body and hook, and the class III region consists of genes involved in flagella synthesis and motility. Using defined deletion mutants constructed for each class of genes, we showed that the mutants deleted of the flagella structure exhibited significant defects in HBMEC binding and invasion, while deletion of a gene affecting only motility had no defects in HBMEC binding and invasion (Parthasarathy et al., 2007). These findings indicate that bacterial flagellin plays an important role in meningitis-causing *E. coli* K1 binding to and invasion of HBMEC. This concept is supported by the demonstration that recombinant flagellin (FliC) binds directly to the surface of HBMEC and exogenous recombinant flagellin decreases *E. coli* K1 binding to HBMEC (Parthasarathy et al., 2007).

OmpA

OmpA is one of the major outer membrane proteins in *E. coli* and its N-terminal domain crosses the outer membrane eight times in antiparallel β-strands with four hydrophilic surface-exposed loops and short periplasmic turns. We have shown that the N-terminal portion of OmpA and its surface-exposed loops contribute to binding to HBMEC (Nemani et al., 1996a,b; Shin et al., 2005; Maruvada and Kim, 2011). We also showed that OmpA interacts with HBMEC through N-acetylglucosamine (GlcNAc) residues of gp 96 (Nemani et al., 1996b; Khan et al., 2003). The chitooligomers (GlcNAc β1, 4-GlcNAc oligomers) and chitohexose block meningitis-causing *E. coli* K1 invasion of HBMEC and traversal of the blood–brain barrier in the infant rat model of experimental hematogenous meningitis (Nemani et al., 1996b; Maruvada and Kim, 2011).

Our recent study with an *E. coli* DNA microarray comparing the *ompA* deletion mutant with its parent *E. coli* K1 strain RS218, however, revealed that the *ompA* deletion mutant exhibited significantly lower expression of the *fim* cluster genes, and lower expression of type 1 fimbriae on the bacterial surface (Teng et al., 2006). These findings suggest that decreased binding of the *ompA* deletion mutant may be related to its lower expression of type 1 fimbriae. The *ompA* deletion mutant was significantly less efficient in its penetration into the brain in vivo compared to the parent *E. coli* K1 strain (Wang and Kim, 2002). Additional studies are needed to determine whether these in vitro and in vivo defects of the *ompA* deletion mutant are in part related to its decreased expression of type 1 fimbriae and to also understand how the deletion of *ompA* affects type 1 fimbria expression.

NlpI

NlpI is shown to be an important factor of Crohn's disease-associated *E. coli* strain LF82 (083:H1) to interact with intestinal epithelial cells (Barnich et al., 2004). Deletion of *nlpI* in *E. coli* strain LF82 decreased expression of type 1 fimbriae and flagella (Barnich et al., 2004). We showed that NlpI is an outer membrane-anchored protein and contributes to meningitis-causing *E. coli* K1 binding to and invasion of HBMEC (Teng et al., 2010). Unlike strain LF82, deletion of *nlpI* in meningitis-causing *E. coli*, however, did not affect the expression of type 1 fimbriae, flagella and OmpA, indicating that the contribution of NlpI to HBMEC binding and invasion is independent of those bacterial factors in meningitis-causing *E. coli*. This concept is shown by the demonstration that mutants deleted of type 1 fimbriae, OmpA and NlpI exhibited significantly decreased HBMEC binding and invasion compared to mutants deleted of individual factors or a combination of the two factors (Teng et al., 2010). These findings suggest that type 1 fimbriae, OmpA and NlpI are likely to contribute to HBMEC binding and invasion independent of each other. It remains, however, incompletely understood how and why several bacterial factors of meningitis-causing *E. coli* are involved in HBMEC binding.

E. coli *structures contributing to invasion of HBMEC*

Previous studies using *TnphoA* mutagenesis, signature-tagged mutagenesis and differential fluorescence induction with screening of a *gfp* fusion library identified several *E. coli* determinants contributing to invasion of HBMEC, which include Ibe (named after invasion of brain endothelial cell) proteins and cytotoxic necrotizing factor 1 (CNF1) (Huang et al., 1995, 1999; Wang et al., 1999; Badger et al., 2000a,b; Hoffman et al., 2000; Khan et al., 2002). Isogenic deletion mutants were significantly less invasive in HBMEC and less able to penetrate into the brain in vivo (Table 10.2) and their invasive defects were restored to the levels of the parent strain by complementation with respective wild-type genes.

Ibe proteins

Ibe A, B, and C proteins were identified via *TnphoA* mutagenesis of meningitis-causing *E. coli* K1 strain RS218 and screening of the mutants for loss of invasion by use of both the in vitro and in vivo blood–brain barrier models (Huang et al., 1995, 1999; Wang et al., 1999).

Exogenous recombinant Ibe proteins were shown to inhibit meningitis-causing *E. coli* K1 invasion of HBMEC (Huang et al., 1995), suggesting that Ibe proteins contribute to HBMEC invasion by ligand–receptor interactions. This concept was supported by the demonstration of a HBMEC surface protein interactive with IbeA, and a polyclonal antibody raised against this receptor protein inhibited *E. coli* K1 invasion of HBMEC (Kim, 2001). The gene, *ibeA*, was shown to be prevalent in meningitis-causing *E. coli* K1 of phylogenetic B2 group (Johnson et al., 2002; Bonacorsi et al., 2003), while *ibeB* and *ibeC* were found to have homologs in non-meningitis isolates of *E. coli*. It remains unclear whether the *ibeB* and *ibeC* homologs from non-meningitis isolates will exhibit the HBMEC invasion phenotype similar to that of meningitis-causing *E. coli* K1 strains. The mechanisms involved with IbeA, IbeB, and IbeC in their contributions to *E. coli* K1 invasion of HBMEC, however, appear to be similar and their contributions to HBMEC invasion are likely to be redundant. This is shown by the demonstration that HBMEC invasion frequencies did not differ significantly between mutants deleted of *ibeA*, *ibeB*, and *ibeC* versus mutants deleted of single or double genes (Kim, 2002). Additional studies are needed to elucidate the mechanisms involved with IbeA, IbeB, and IbeC for their contribution to *E. coli* meningitis.

CNF1

CNF1 is a bacterial virulence factor associated with extraintestinal pathogenic *E. coli* strains causing urinary tract infection and meningitis (Bouquet, 2001). CNF1 is an AB-type toxin, composed of the N-terminal cell binding domain and the C-terminal catalytic domain possessing a deaminase activity through the site-specific deamination of a Gln residue to Glu (Flatau et al., 1997; Schmidt et al., 1997). CNF1 has been shown to activate RhoGTPases such as RhoA and induce uptake of latex beads, bacteria, and apoptotic bodies into non-professional phagocytes such as epithelial and endothelial cells by macropinocytosis (Fabbri et al., 2002).

CNF1 contributes to *E. coli* K1 invasion of HBMEC in vitro and traversal of the blood–brain barrier in vivo, and these in vitro and in vivo effects of CNF1 are dependent upon RhoA activation (Khan et al., 2002). These conclusions were shown by (a) significantly decreased invasion and RhoA activation with the CNF1 deletion mutant compared to the parent strain in HBMEC and (b) restoration of the CNF1 mutant's invasion frequency to the level of the parent strain in HBMEC expressing constitutively active RhoA.

CNF1 has been suggested to be internalized via receptor-mediated endocytosis upon binding to a cell surface receptor (Bouquet, 2001). We have identified the HBMEC receptor for CNF1 by yeast two-hybrid screening of the HBMEC cDNA library using the N-terminal cell-binding domain of CNF1 as bait (Chung et al., 2003). This receptor, 37-kDa laminin receptor precursor (LRP) interacted with the N-terminal CNF1 and full-length CNF1 but not with the C-terminal CNF1. CNF1-mediated RhoA activation and bacterial uptake were inhibited by LRP antisense oligodeoxynucleotides,whereas they were increased in LRP-overexpressing cells, demonstrating correlation between effects of CNF1 and levels of LRP expression in HBMEC (Chung et al., 2003). These findings indicate that CNF1 interaction with its receptor, LRP, is the initial step required for CNF1-mediated RhoA activation and bacterial uptake in eukaryotic cells. The 37-kDa LRP is a ribosome-associated cytoplasmic protein and shown to be a precursor of 67-kDa laminin receptor (67 LR). It is unclear how 67 LR is matured and synthesized from the LRP, but mature 67 LR is shown to be present on the cell surface and functions as a membrane receptor for the adhesive basement membrane protein laminin (Massia et al., 1993). We have shown that incubation of HBMEC with CNF1-expressing *E. coli* K1 up-regulates 67 LR expression and recruits 67 LR to the site of invading *E. coli* K1 in a CNF1-dependent manner (Kim et al., 2003), supporting the participation of 67 LR in CNF1-expressing *E. coli* K1 invasion of HBMEC.

Of interest, LRP is also shown to be a cellular target for other CNS-infecting microorganisms, including *S. pneumoniae*, *N. meningitidis*, *H. influenzae* type b, dengue virus, adeno-associated virus, Venezuelan equine encephalitis virus, and prion protein (Table 10.3). The mechanisms by which the same receptor is involved in CNS penetration by different organisms remain to be established.

E. coli *invasion of HBMEC involves host cell actin cytoskeleton rearrangements*

Pathogenic microbes exploit various strategies to penetrate into their host cells (Knodler et al., 2001; Cossart and Sansonetti, 2004; Kim, 2008). Microbial internalization into non-professional phagocytes such as endothelial and epithelial cells is shown to occur mainly via two different mechanisms involving host cell actin cytoskeleton rearrangements: a zipper mechanism involving the formation of cell protrusions in contact with the pathogens and a trigger mechanism involving the formation of membrane ruffling around the pathogens (Knodler et al., 2001; Cossart and Sansonetti, 2004).

Studies with scanning and transmission electron microscopy have shown that internalization of meningitis-causing *E. coli* K1 into HBMEC is associated with microvilli-like protrusions at the entry site on the surface of HBMEC (Nemani et al., 1999; Kim, 2003) (Figure 10.1), suggesting the involvement of host cell actin cytoskeleton rearrangement in *E. coli* K1 invasion of HBMEC. This concept is supported by the demonstrations that F-actin condensation occurs with invading bacteria and blockade of actin condensation with microfilament-disrupting agents such as cytochalasin D inhibits *E. coli* K1 invasion of HBMEC (Nemani et al., 1999).

Meningitis-causing E. coli *K1 strains exploit host cell signaling molecules for invasion of HBMEC*

Several host cell signal transduction pathways have been shown to be involved in *E. coli* K1 invasion of HBMEC. These include focal adhesion kinase (FAK), paxillin, phosphatidylinositol 3-kinase (PI3K), Src kinase, signal transducers and activators of transcription 3 (STAT3), Rho GTPases (RhoA and Rac1), cytosolic phospholipase A2 (cPLA2), 5-lipoxygenase and cysteinyl leukotrienes, vascular endothelial growth factor (VEGF) receptor-1, ezrin, radixin and moesin (ERM), calmodulin-dependent myosin light-chain kinase, and protein kinase C (PKC) (Reddy et al., 2000; Kim et al., 2008; Kim, 2008, 2010, 2012; Teng et al., 2010; Zhao et al., 2010; Zhu et al., 2010a,b; Maruvada and Kim, 2011, 2012). Despite this substantial list, the underlying host–microbial factors have been incompletely elucidated.

It is important to note that activations of the above-mentioned host cell signaling molecules occur in response to specific microbial factors of meningitis-causing *E. coli* K1 and their interactions with HBMEC, and participation of the same bacterial or host factors does not necessarily lead to activation of the same host cell signaling molecules. For example, FimH of meningitis-causing *E. coli* has been shown to induce RhoA activation, not FAK activation in HBMEC (Khan et al., 2007). In contrast, FimH of uropathogenic *E. coli* induces FAK activation in bladder epithelial cells (Marinez et al., 2000). A similar concept is shown with host factors, e.g. gp96 functions as the receptor for *E. coli* OmpA and *L. monocytogenes* Vip (Table 10.3). The OmpA–gp96 interaction resulted in FAK activation in HBMEC (Reddy et al., 2000), but no FAK activation occurred with the Vip–gp96 interaction in mouse fibroblasts (Cabanes et al., 2005).

Our current knowledge on the mechanisms involved with the above-mentioned host cell signaling molecules for their contribution to HBMEC invasion has been derived from the following two approaches, (a) identification of the *E. coli* factors participating in specific host cell signaling molecules and (b) examination of the interrelationship of the host cell signaling molecules involved in *E. coli* invasion of HBMEC.

Our findings so far demonstrate that OmpA and Ibe proteins of meningitis-causing *E. coli* K1 are involved in FAK and PI3K activations, OmpA and IbeA in STAT3 and Rac1 activations, FimH and CNF1 in RhoA activation, OmpA and NlpI in cPLA2 and PKC activations, and CNF1 in ERM activation (Figure 10.2). This information has been useful for elucidating how and why several bacterial factors contribute to *E. coli* binding to and invasion of HBMEC, and also whether or not their contributions are redundant. For example, Rac1 activation occurs in response to OmpA or IbeA, and RhoA activation occurs in response to CNF1 or FimH, while cPLA2 activation occurs in response to OmpA or NlpI (Khan et al., 2007; Teng et al., 2010; Maruvada and Kim, 2011, 2012). We showed that mutants deleted of OmpA and CNF1, OmpA and FimH, or FimH and NlpI exhibit significantly greater defects in invasion of HBMEC compared to individual deletion mutants. In contrast, mutants deleted of OmpA

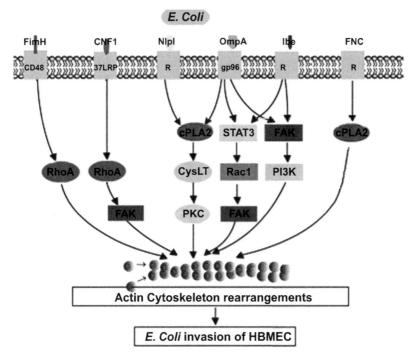

FIGURE 10.2 Host signaling molecules involved in actin cytoskeleton rearrangements and *E. coli* K1 invasion of the blood–brain barrier. *E. coli* K1 invasion of the blood–brain barrier requires specific bacterial factors (FimH, CNF1, NlpI, OmpA, Ibe proteins, and FliC) and their interactions with host factors (CD48, 37LRP, R, gp96) on human brain microvascular endothelial cells (HBMEC), involving Rho GTPases (RhoA and Rac1), focal adhesion kinase (FAK), cytosolic phospholipase A2α (cPLA2α), cysteinyl leukotrienes (CysLT), protein kinase C (PKC), signal transducers and activators of transcription 3 (STAT3), and phosphatidylinositol 3-kinase (PI3K); CNF1, cytotoxic necrotizing factor 1; 37LRP, 37 laminin receptor precursor; R, receptor; gp96, glycoprotein 96.

and Ibe are not shown to exhibit significantly greater defects in HBMEC invasion compared to individual deletion mutants. Thus, the reasons for the additive (non-redundant) versus indifferent (redundant) effects of different bacterial factors in *E. coli* binding to and invasion of HBMEC can be in part explained by their underlying host cell signaling mechanisms (e.g. involving different or same host cell signaling molecules, respectively).

We have also shown that meningitis-causing *E. coli* K1 strains exploit FAK and PI3K for invasion of HBMEC, as shown by significantly decreased invasion in HBMEC expressing dominant-negative FAK and PI3K and pharmacological inhibition of FAK and PI3K. In addition, we have shown that PAK is upstream of PI3K in *E. coli* K1 invasion of HBMEC. This is shown by the demonstration that PI3K activation was abolished in HBMEC expressing dominant-negative FAK (Reddy et al., 2000). Similarly, both STAT3 and Rac1 are involved in meningitis-causing *E. coli* K1 invasion of HBMEV, but STAT3 is upstream of

Rac1, as shown by blockade of Rac1 activation in HBMEC expressing dominant-negative STAT3 (Maruvada and Kim, 2012). Also, cPLA2 and PKC are involved in meningitis-causing *E. coli* K1 invasion of HBMEV, but cPLA2 is upstream of PKC, as shown by the demonstration that inhibition of cPLA2 prevents PKC activation in response to meningitis-causing E. *coli* K1 in HBMEC (Zhu et al., 2010a).

Despite the above information, it remains incompletely understood why several microbial factors contribute to HBMEC binding and invasion. It remains to be determined whether complete abolition of such phenotypes requires deletion of all the non-redundant bacterial factors contributing to HBMEC binding and invasion.

It should also be noted that host cell actin cytoskeleton rearrangements are required for HBMEC invasion by meningitis-causing bacteria such as *E. coli*, group B *Streptococcus* and *L. monocytogenes*, as shown by inhibition of their invasion into HBMEC by cytochalasin D (Nizet, 1997; Greiffenberg et al., 1998; Nemani et al., 1999), but the host cell signaling mechanisms involved in actin cytoskeleton rearrangements and HBMEC invasion differ between meningitis-causing bacteria. For example, *E. coli* K1 invasion of HBMEC depends on activations of FAK, Src, and cPLA2. In contrast, group B *Streptococcus* invasion of HBMEC is independent of Src and *L. monocytogenes* invasion of HBMEC is independent of FAK and cPLA2 activation (Das et al., 2001; Kim, 2001).

E. coli *traversal of the blood–brain barrier as live bacteria*

Another crucial factor for the development of meningitis is the ability of meningitis-causing bacteria to cross the blood–brain barrier as live bacteria. *E. coli* K1 has been shown to traverse the blood–brain barrier as live bacteria without altering the integrity of the HBMEC monolayer and without affecting the blood–brain barrier permeability as well as without accompanying host inflammatory cells (Kim et al., 1992, 1997; Stins et al., 2001).

As indicated above, internalized *E. coli* are located within membrane-bound vacuoles of HBMEC and transmigrate the HBMEC monolayer (Figure 10.1). We showed that HBMEC have the complete intracellular trafficking machinery required to deliver the microbe-containing vacuoles to cathepsin D-containing components (i.e. lysosomes) (Kim et al., 2003). For example, vacuoles containing *E. coli* K1 capsule deletion mutant interact sequentially with early endosomal marker proteins (e.g. early endosomal auto-antigen 1 and transferrin receptor) and late endosome and late endosome/lysosomal markers (e.g. Rab7 and lysosome-associated membrane proteins, respectively), and allow lysosomal fusion with subsequent degradation inside vacuoles. In contrast, vacuoles containing *E. coli* K1 encapsulated strain obtained early and late endosomes but without fusion with lysosomes, thereby allowing *E. coli* K1 to cross the blood–brain barrier as live bacteria (Kim et al., 2003), indicating that *E. coli* K1 capsule modulates intracellular trafficking to avoid lysosomal fusion in HBMEC (Figure 10.3).

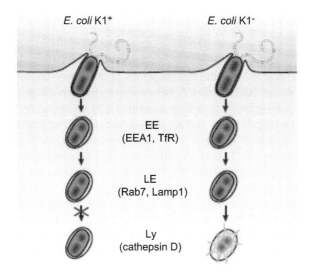

FIGURE 10.3   Intracellular trafficking of *E. coli* containing vacuoles in HBMEC monolayer. In the absence of the K1 capsule (K1⁻, right), the vacuoles mature to fuse with lysosomes and display sequential markers of early endosome (EE), late endosome (LE) and lysosomes (Ly). By contrast, the presence of the K1 capsule (K1⁺, left) interferes with the vacuole maturation to prevent fusion with lysosomes. This allows *E. coli* K1 to traverse HBMEC monolayer as live bacteria. EEA, early endosomal antigen; TfR, transferring receptor; LAMP, lysosome-associated membrane protein. *(Adapted from Kim, 2003, figure 3.)*

   E. coli K1 capsule is well recognized for its serum resistance and anti-phagocytic properties (Cross et al., 1986; Kim et al., 1986, 1988), which are the essence of inducing a high degree of bacteremia. Another novel property of the K1 capsule is, therefore, to modulate the maturation process of vacuoles containing *E. coli* K1 encapsulated strain and prevent their fusion with lysosomes, which is an event necessary for traversal of the blood–brain barrier as live bacteria. Additional studies are needed to understand how the K1 capsule is able to modulate intracellular trafficking of *E. coli* K1-containing vacuoles to avoid fusion with lysosomes in HBMEC and whether similar events occur with other meningitis-causing microorganisms.

IDENTIFICATION OF MICROBIAL FACTORS INVOLVED IN *E. COLI* MENINGITIS BY FUNCTIONAL GENOMIC APPROACHES

Genome sequencing of meningitis-causing microorganisms provides new tools for elucidating the pathogenesis of meningitis. For example, comparative genome analysis of the prototypic meningitis-causing *E. coli* K1 strain RS218 (O18:K1) versus laboratory K-12 strain MG1655 identified 22 RS 218-derived islands (RDIs) that are larger than 10 kb but are absent in strain MG1655 (Xie et al., 2006a,b). The total length of these RDIs is approximately 793 kb, which replaced approximately 80 kb of MG1655-specific sequences, resulting in approximately 713 kb larger chromosome size in RS218. This difference is 160 kb larger than the previously estimated genome size difference between RS218 and MG1655 (Rode et al., 1999). Previous studies using comparative macrorestriction mapping and subtractive hybridization of the chromosomes of meningitis-causing *E. coli* K1 (e.g. O18:K1 strains RS218 and C5, belonging to phylogenetic group B2) compared to non-pathogenic *E. coli* have identified 500 kb spread over at least 12 chromosome loci specific to *E. coli* K1 (Bloch et al., 1996; Bonacorsi et al., 2000). Mapping studies reveal that those *E. coli* loci are located at different regions of the *E. coli* chromosome. Twenty-two RDIs have also been shown to be located at different regions of the *E. coli* K1 RS218 chromosome (Xie et al., 2006a,b).

Using RDI deletion mutants constructed from strain RS218, eight RDIs have been shown to be involved in the pathogenesis of meningitis (i.e. induction of a high degree of bacteremia, HBMEC binding, and invasion). Two RDIs include a P4-family integrase and are directly adjacent to tRNAs (RDI 4-*serX* and RDI 21-*leuX*) and four RDIs have markedly lower GC percentages compared to the whole RS218 genome (Xie et al., 2006a,b), suggesting that these RDIs are acquired through horizontal gene transfer. Further identification and characterization of microbial factors from those RDIs that are involved in the pathogenesis of *E. coli* meningitis should help in elucidating microbial factors involved in meningitis.

At present, the microbial factors identified from prototypic meningitis-causing O18:K1 *E. coli* strains (e.g. strains RS218 and C5) have been used to understand the pathogenesis of *E. coli* meningitis (Kim, 2001, 2002, 2003, 2008; Xie et al., 2006a,b; Yao et al., 2006), but it is unclear whether the information derived from these *E. coli* K1 strains is relevant to other *E. coli* K1 strains isolated from CSF. We have conducted a comparative genomic hybridization (CGH) with an *E. coli* DNA microarray to examine the basis of meningitis caused by representative *E. coli* K1 strains isolated from CSF, belonging to phlyogenetic groups B2 and D (Yao et al., 2006). These strains include RS218 (O18:K1, B2 group), C5 (18:K1, B2 group), IHE3034 (O18:K1, B2 group), EC10 (O7:K1, D group), A90 (O1:K1, B2 group), RS168 (O1:K1, D group), RS167 (O16:K1, B2 group), S88 (O45:K1, B2 group), and S95 (O45:K1, Be group). Our hierarchical clustering revealed that these strains can be categorized into two groups. Group 1 includes strains RS218, C5, IHE3034, A90, RS167, S88, and S95, while strains EC10 and RS168 belong to group 2. All group 1 strains belong to the phylogenetic group B2, which is predominant in CSF isolates, and group 2 strains belong to less common phylogenetic groups A and D (Yao et al., 2006). Of interest, a type III secretion system was found to be present in group 2 strains of *E. coli* but not in group 1 strains, isolated from neonatal meningitis, and was shown to contribute to *E. coli* invasion of HBMEC and intracellular survival (Yao et al., 2009). The existence of a type III secretion system as well as its effectors and homologs was verified by genome sequencing of strain EC10 (Lu et al., 2011). This is the first demonstration of a type III secretion system in meningitis-causing *E. coli*, but its underlying mechanisms involved in *E. coli* meningitis remain to be established.

We also examined by CGH the distribution of the eight RDIs that are shown to be involved in the pathogenesis of *E. coli* meningitis among representative *E. coli* K1 strains (Xie et al., 2006a,b). RDI 16 harbors the K1 capsule biosynthesis gene cluster and, as expected, is present in all of these *E. coli* K1 strains. The other pathogenic RDIs are found to exist in strains belonging to group 1 and phylogenetic group B2. For example, RDI 1, 7, 13, 20, and 22 are widely distributed among this group of *E. coli* K1 strains. Previous studies using PCR, dot blot, and Southern blot suggest that PAI III$_{536}$-like, PAI II$_{J96}$-like, and GimA-like ectochromosomal DNA domains (ECDNA) are prevalent among O18:K1 strains, the most common serogroup in meningitis-causing *E. coli* (Bonacorsi et al., 2003). Based on their virulence signatures, those ECDNAs correspond to RDI 4, 21, and 22, respectively (Xie et al., 2006a,b). The distribution of these three islands among O18:K1 strains based on CGH is consistent with previous findings (Bonacorsi et al., 2003; Xie et al., 2006a,b). Our CGH analysis also revealed that type VI secretion system (T6SS)-like gene clusters, including the *icmF*-like component, *clpV*, *dotU*, and *hcp2*, are present in the RDI 1 (Xie et al., 2006a,b). Of interest, the T6SS clusters have two *hcp*-like genes located next to each other in the chromosome of strain RS218, and we showed that the two Hcp family proteins have different roles in meningitis-causing *E. coli* K1 infection and coordinately contribute to the pathogenicity of *E. coli* K1 interaction with HBMEC, e.g. *E. coli* binding to and invasion of HBMEC as well as release of IL-6 and IL-8 from HBMEC (Zhou et al., 2012).

In addition, microbial DNA microarrays offer new opportunities for exploring microbial gene expression profiling during microbial–host interactions. For example, using *E. coli* DNA microarray analysis with microarray-grade bacterial RNA isolated from *E. coli* K1 interacting with HBMEC, we showed that the expression of the type 1 fimbriae and flagella genes are significantly greater for *E. coli* associated with HBMEC compared to *E. coli* not associated with HBMEC (Teng et al., 2005; Parthasarathy et al., 2007; Xie et al., 2008). We subsequently showed that type 1 fimbriae and flagellin play an important role in *E. coli* K1 binding to and invasion of HBMEC (Teng et al., 2005; Parthasarathy et al., 2007), indicating that microbial DNA microarray analysis has a potential for elucidating microbial–host interactions that contribute to the pathogenesis of *E. coli* meningitis (Kim et al., 2005; Korczak et al., 2005).

PREVENTION OF *E. COLI* PENETRATION INTO THE BRAIN BY TARGETING THE MICROBIAL–HOST FACTORS CONTRIBUTING TO *E. COLI* INVASION OF HBMEC MONOLAYER

We have shown that meningitis-causing *E. coli* K1 penetration into the brain requires *E. coli* binding to and invasion of HBMEC, involving specific microbial–host interactions (so-called ligand–receptor interactions) and host cell signaling molecules (Kim, 2008, 2010, 2012). For example, CNF1-mediated *E. coli* uptake and RhoA activation in HBMEC were inhibited by LRP antisense oligodeoxynucleotides, whereas they were increased in LRP-overexpressing cells (Chung et al., 2003). These findings support the concept that the expression level of host cell receptor(s) dictates the fate of *E. coli* interaction with HBMEC in a bacterial ligand-dependent manner. In addition, pharmacological inhibition and gene deletion of host cell signaling molecules (e.g. cPLA2) involved in *E. coli* invasion of HBMEC was efficient in preventing *E. coli* penetration into the brain (Zhu et al., 2010a,b). These findings suggest that pharmacological inhibition of the HBMEC receptors that interact with *E. coli* factors and host cell signaling molecules contributing to *E. coli* invasion of HBMEC might be a novel strategy for prevention of *E. coli* meningitis.

The feasibility of this novel strategy was shown by our demonstration that pharmacological inhibition of the host cell receptor (e.g. LRP) and host cell signaling molecules (e.g. cysteinyl leukotrienes) involved in *E. coli* invasion of HBMEC was efficient in preventing *E. coli* penetration into the brain (Kim, 2010, 2012; Zhu et al., 2010a,b). Additional studies are needed to elucidate the microbial–host factors that contribute to *E. coli* invasion of HBMEC and also can serve as novel targets for prevention and therapy of *E. coli* meningitis.

An additional novel strategy for prevention and therapy of *E. coli* meningitis is to modulate the interaction with the blood–brain barrier of the microbial factors contributing to HBMEC binding and invasion by affecting their expression and/or secretion. For example, CNF1 is a key factor contributing to *E. coli* invasion of HBMEC and penetration into the brain via the interaction with its receptor (LRP) on HBMEC (Khan et al., 2002; Chung et al., 2003). CNF1, however, is a cytoplasmic protein and execution of its contribution to *E. coli* invasion of the blood–brain barrier requires its secretion from the bacterial cytoplasm. No signal peptide is found in the CNF1 sequence. CNF1 secretion is, therefore, a strategy utilized by meningitis-causing *E. coli* to invade the blood–brain barrier. It, however, remains unclear how CNF1 secretion occurs across the bacterial inner membrane and outer membrane as well as into outer membrane vesicles. We have shown that ferredoxin is involved in CNF1 secretion across the bacterial inner membrane and YgfZ is involved in CNF1 secretion into outer membrane vesicles (Yu and Kim, 2010, 2012). Elucidation of the mechanisms involved in CNF1 secretion will, therefore, enhance our knowledge on the pathogenesis of *E. coli* meningitis and also help in developing a novel strategy targeting CNF1 secretion in prevention of *E. coli* meningitis. Taken together, these findings suggest that modulation of bacterial secretion systems (CNF1 secretion, type III secretion, type VI secretion) is likely to represent a novel target for prevention of *E. coli* meningitis.

THE BASIS FOR NEUROTROPISM IN *E. COLI* MENINGITIS

E. coli strains possessing the K1 capsular polysaccharide are predominant (approximately 80%) among isolates from neonatal *E. coli* meningitis and most of these K1 isolates are associated with a limited number of O sero-types (e.g. O18, O7, O16, O1, O45), belonging to phylogenetic group B2 and to a lesser extent, to group D (Sarff et al., 1975; Korhonen et al., 1985; Bonacorsi et al., 2003; Yao et al., 2006). The basis of their association with *E. coli* meningitis, however, remains unclear. We have shown that arachidonic acid metabolism pathways contribute to meningitis-causing *E. coli* K1 strains of phlyogenetic groups B2 and D for their invasion of HBMEC and penetration specifically into the brain (Zhu et al., 2010a). This was shown by the demonstration that gene deletion and pharmacological inhibition of arachidonic acid metabolism pathways prevent *E. coli* penetration specifically into the brain without affecting the level of bacteremia and penetration into non-brain organs (e.g. kidney, spleen, and lung). These findings suggest that determination of the microbial–host factors contributing to arachidonic acid metabolism pathways leading to penetration of the blood–brain barrier is likely to elucidate the novel concept of neurotropism in *E. coli* meningitis and also provide the information on novel targets for prevention and therapy of *E. coli* meningitis.

THE MECHANISMS INVOLVED IN CNS INFLAMMATION IN RESPONSE TO BACTERIAL MENINGITIS

Bacterial meningitis is characterized by inflammation of the meninges that occurs in response to bacteria and bacterial products, resulting in release of cytokines and chemokines as well as pathophysiological alterations such as infiltration of leukocytes and blood–brain barrier dysfunction (Kim, 2003). Recent studies have shown that the mechanisms involved in microbial invasion of the blood–brain barrier differ from those involved in the release of cyto-kines and chemokines in response to meningitis-causing pathogens (Kim, 2008, 2010, 2012). For example, interleukin-8 secretion in response to *E. coli* strain K1 happens in HBMEC, but not in non-brain endothelial cells (e.g. human umbilical vein endothelial cells). However, *E. coli* factors involved in binding to and invasion of HBMEC did not affect the release of interleukin 8 from HBMEC (Galanakis et al., 2006). Similar findings were demonstrated for group B *Streptococcus* and *N. meningitidis* (Sokolova et al., 2004; Banerjee et al., 2011). These findings suggest that targets for prevention of bacterial penetration into the brain are likely to differ from those involved in CNS inflammation associated with bacterial meningitis.

NEURONAL INJURY FOLLOWING *E. COLI* MENINGITIS

Neurological sequelae are common in survivors of neonatal *E. coli* meningitis, but the underlying mechanisms remain incompletely understood (Kim, 2003). Neuronal injury stemming from cortical necrosis and hippocampal apopto-sis is considered a major contributing factor. Experimental animal studies demonstrate that adjunctive treatment with dexamethasone was detrimental in hippocampal apoptosis, while hypertonic saline was beneficial (Spreer et al., 2006; Liu et al., 2011). Recent studies suggest an endogenous potential of neural progenitor cell renewal in the hippocampal formation in bacterial meningitis (Gerber et al., 2009; Hofer et al., 2011) and studies are needed to determine whether stimulation of endogenous neurogenesis can be used as a potential option in survivors of bacterial meningitis.

CONCLUSIONS

A major limitation to advances in prevention and therapy of *E. coli* meningitis is our incomplete understanding of the pathogenesis and pathophysiology of this disease. Given the plethora of *E. coli* serotypes, it is unclear why K1 and limited numbers of O types of *E. coli* phylogenetic groups B2 and D account for most cases of meningitis. As indicated above, studies with the in vitro and in vivo blood–brain barrier models have shed light on the mechanisms of microbial translocation of the blood–brain barrier, a key step for the development of meningitis. We have shown that *E. coli* traverses the blood–brain barrier without altering the integrity of the HBMEC monolayer and without affecting blood–brain barrier permeability as well as without accompanying infiltrating phagocytes (Kim et al., 1992, 1997; Stins et al., 2001). *E. coli* K1 penetration into the brain requires a high degree of bacteremia, and *E. coli* binding to and invasion of HBMEC, as well as traversal of the blood–brain barrier as live bacteria. *E. coli* invasion of the blood–brain barrier occurs as the result of specific bacteria–host interactions involving specific host cell signal transduction pathways. Complete understanding of the microbial–host factors that are involved in *E. coli* penetration of the blood–brain barrier as well as *E. coli*-induced neuronal injury should help in developing new strategies for prevention and therapy of *E. coli* meningitis and associated neuronal injury.

ACKNOWLEDGMENTS

This work was supported by NIH grants R01 NS 26310 and AI 84984.

REFERENCES

Akache, B., Grimm, D., Pandey, K., Yant, S.R., Xu, H., Kay, M.A., 2006. The 37/67-kilodalton laminin receptor is a receptor for adeno-associated virus serotypes 8, 2, 3, and 9. J. Virol. 80, 9831–9836.

Badger, J.B., Wass, C.A., Kim, K.S., 2000a. Identification of *E. coli* KI genes contributing to human brain microvascular endothelia cell invasion by differential fluorescence induction. Mol. Micro. 36, 174–182.

Badger, J., Wass, C., Weissman, S., Kim, K.S., 2000b. Application of signature-tagged mutagenesis for the identification of *E. coli* K1 genes that contribute to invasion of the blood–brain barrier. Infect. Immun. 68, 5056–5061.

Banerjee, A., Kim, B.J., Carmona, E.M., et al., 2011. Bacterial pili exploit integrin machinery to promote immune actrivation and efficient blood–brain barrier penetration. Nat. Commun. 2, 139–151.

Barnich, N., Bringer, M.A., Claret, L., Darfeuille-Michaud, A., 2004. Involvement of lipoprotein NlpI in the virulence of adherent invasive *Escherichia coli* strain LF82 isolated from a patient with Crohn's disease. Infect. Immun. 72, 2484–2493.

Berman, P.H., Banker, B.Q., 1966. Neonatal meningitis. A clinical and pathologic study of 29 cases. Pediatrics 38, 6–24.

Bingen, E., Picard, B., Brahimi, N., et al., 1998. Phylogenetic analysis of *Escherichia coli* strains causing neonatal meningitis suggests horizontal gene transfer from a predominant pool of highly virulent B2 group strains. J. Infect. Dis. 177, 642–650.

Blanco, J., Mora, A., Mamani, R., et al., 2011. National survey of *Escherichia coli* causing extraintestinal infections reveal the spread of drug-resistant clonal groups O25b:H4-B2-ST131, O15:H5-D-ST393 and CGA-D-ST69 with high virulence gene content in Spain. J. Antimicrob. Chemother. 66, 2011–2012.

Bloch, C.A., Huang, S.H., Rode, C.K., Kim, K.S., 1996. Mapping of noninvasion TnphoA mutations on the *Escherichia coli* O18:K1:H7 chromosome. FEMS Microbiol. Lett. 144, 171–176.

Bonacorsi, S., Clermont, O., Houdouin, V., et al., 2003. Molecular analysis and experimental virulence of French and North American *Escherichia coli* neonatal meningitis isolates: identification of a new virulent clone. J. Infect. Dis. 187, 1895–1906.

Bonacorsi, S.P., Clermont, O., Tinsley, C., et al., 2000. Identification of regions of the *Escherichia coli* chromosome specific for neonatal meningitis-associated strains. Infect. Immun. 68, 2096–2101.

Bouquet, P., 2001. The cytotoxic necrotizing factor 1 (CNF1) from *Escherichia coli*. Toxicon 39, 1673–1680.

Cabanes, D., Sousa, S., Cebria, A., Lecuit, M., Portillo, F., Cossart, P., 2005. Gp96 is a receptor for a novel *Listeria monocytogenes* virulence factor, Vip, a surface protein. EMBO J. 25, 2827–2838.

Chi, F., Wang, L., Zheng, X., et al., 2011. Meningitic *Escherichia coli* K1 penetration and neutrophil transmigration across the blood–brain barrier are modulated by alpha7 nicotine receptor. PLoS One 6, e25016.

Chung, J.W., Hong, S.J., Kim, K.J., et al., 2003. 37 kDa laminin receptor precursor modulates cytotoxic necrotizing factor 1-mediated RhoA activation and bacterial uptake. J. Biol. Chem. 278, 16857–16862.

Cossart, P., Sansonetti, P.J., 2004. Bacterial invasion: the paradigms of enteroinvasive pathogens. Science 304, 242–248.

Cross, A., Artenstein, A., Que, J., et al., 1994. Safety and immunogenicity of a polyvalent *Escherichia coli* vaccine in human volunteers. J. Infect. Dis. 170, 834–840.

Cross, A.S., Kim, K.S., Wright, D.C., Sadoff, J.C., Gemski, P., 1986. Role of lipopolysaccharide and capsule in the serum resistance of bacteremic strains of *Escherichia coli*. J. Infect. Dis. 154, 497–503.

Das, A., Asatryan, L., Reddy, M.A., et al., 2001. Differential role of cytosolic phospholipase A2 in the invasion of brain microvascular endothelial cells by *Escherichia coli* and *Listeria monocytogenes*. J. Infect. Dis. 184, 732–737.

Dawson, K.G., Emerson, J.C., Burns, J.L., 1999. Fifteen years of experience with bacterial meningitis. Pediatr. Infect. Dis. J. 18, 816–822.

Dietzman, D.E., Fischer, G.W., Schoenknecht, F.D., 1974. Neonatal *Escherichia coli* septicemia – bacterial counts in blood. J. Pediatr. 85, 128–130.

Fabbri, A., Falzano, L., Travaglione, S., et al., 2002. Rho-activating *Escherichia coli* cytotoxic necrotizing factor 1: macropinocytosis of apoptotic bodies in human epithelial cells. Int. J. Med. Microbiol. 291, 551–554.

Finne, J., Bitterman-Suermann, D., Goridis, C., Finne, U., 1987. An IgG monoclonal antibody to group B meningococci cross-reacts with developmentally regulated polysialic acid units of glycoproteins in neural and extraneural tissues. J. Immunol. 138, 4402–4407.

Flatau, G., Lemichez, E., Gauthier, M., et al., 1997. Toxin-induced activation of the G protein p21 Rho by deamidation of glutamine. Nature 387, 729–733.

Galanakis, E., DiCello, F., Paul-Satyaseela, M., Kim, K.S., 2006. *Escherichia coli* induces IL-8 expression in human brain microvascular endothelial cells. Eur. Cytokine Netw. 17, 260–265.

Gauczynski, S., Peyrin, J.M., Haïk, S., et al., 2001. The 37-kDa/67-kDa laminin receptor acts as the cell-surface receptor for the cellular prion protein. EMBO J. 20, 5863–5875.

Gerber, J., Tauber, S.C., Armbrecht, I., Schmidt, H., Bruck, W., Nau, R., 2009. Increased neuronal proliferation in human bacterial meningitis. Neurology 73, 1026–1032.

Gladstone, I.M., Ehrenkranz, R.A., Edberg, S.C., Baltimore, R.S., 1990. A ten-year review of neonatal sepsis and comparison with the previous fifty-year experience. Pediatr. Infect. Dis. 9, 819–825.

Greiffenberg, L., Goebel, W., Kim, K.S., Kuhn, M., 1998. Interaction of *Listeria monocytogenes* with human brain microvascular endothelial cells: InlB-dependent invasion, long-term intracellular growth and spread from macrophages to endothelial cells. Infect. Immun. 66, 5260–5267.

Gross, R.J., Ward, L.R., Threlfall, E.J., Cheasty, T., Rowe, B., 1982. Drug resistance among *Escherichia coli* strains isolated from cerebrospinal fluid. J. Hyg. 90, 195–198.

Hofer, S., Grandgirard, D., Burri, D., Frohlich, T.K., Leib, S.L., 2011. Bacterial meningitis impairs hippocampal neurogenesis. J. Neuropathol. Exp. Neurol. 70, 890–899.

Hoffman, J., Wass, C., Stins, M.F., Huang, S.H., Kim, K.S., 1999. The capsule supports survival but not traversal of K1 *E. coli* across blood–brain barrier. Infect. Immun. 67, 3566–3570.

Hoffman, J.A., Badger, J.L., Zhang, Y., Huang, S.H., Kim, K.S., 2000. *E. coli* K1 aslA contributed to invasion of brain microvascular endothelial cells *in vitro* and *in vivo*. Infect. Immun. 68, 5062–5067.

Huang, S.H., Chen, Y.H., Fu, Q., et al., 1999. Identification and characterization of an *E.coli* invasion gene locus ibeB required for penetration of brain microvascular endothelial cells. Infect. Immun. 67, 2103–2109.

Huang, S.H., Wass, C.A., Fu, Q., Nemani, P.V., Stins, M., Kim, K.S., 1995. *E. coli* invasion of brain microvascular endothelial cells *in vitro* and *in vivo*: molecular cloning and characterization of *E. coli* invasion gene ibe10. Infect. Immun. 63, 4470–4475.

Johnson, J.R., Delavari, P., O'Brien, T.T., 2001. *Escherichia coli* 018:K1:H7 isolates from patients with acute cystitis and neonatal meningitis exhibit common phylogenetic origins and virulence factor profiles. J. Infect. Dis. 183, 425–434.

Johnson, J.R., Oswald, E., O'Bryan, T.T., Kuskowski, M.A., Spanjaard, L., 2002. Phylogenetic distribution of virulence-associated genes among *Escherichia coli* isolates associated with neonatal bacterial meningitis in the Netherlands. J. Infect. Dis. 185, 774–784.

Khan, N.A., Kim, Y., Shin, S., Kim, K.S., 2007. FimH-mediated *Escherichia coli* K1 invasion of human brain microvascular endothelial cells. Cell Microbiol. 9, 169–178.

Khan, N.A., Shin, S., Chung, J.W., et al., 2003. Outer membrane protein A and cytotoxic necrotizing factor-1 use diverse signaling mechanisms for *Escherichia coli* K1 invasion of human brain microvascular endothelial cells. Microb. Pathogenesis 35, 35–42.

Khan, N.A., Wang, Y., Kim, K.J., Chung, J.W., Wass, C.A., Kim, K.S., 2002. Cytotoxic necrotizing factor 1 contributes to *Escherichia coli* K1 invasion of the central nervous system. J. Biol. Chem. 277, 15607–15612.

Kim, K.J., Chung, J.W., Kim, K.S., 2005. 67-kDa Laminin receptor promotes internalization of cytotoxic necrotizing factor 1-expressing *Escherichia coli* K1 into human brain microvascular endothelial cells. J. Biol. Chem. 280, 1360–1368.

Kim, K.J., Elliott, S.A., DiCello, F., Stins, M.F., Kim, K.S., 2003. The K1 capsule modulates trafficking of *E. coli*-containing vacuoles and enhances intracellular bacterial survival in human brain microvascular endothelial cells. Cell Microbiol. 5, 245–252.

Kim, K.S., 1985. Comparison of cefotaxime, imipenem-cilastatin, ampicillin-gentamicin and ampicillin-chloramphenicol in the treatment of experimental *E.coli* bacteremia and meningitis. Antimicrobial. Agents Chemother. 28, 433–436.

Kim, K.S., 2001. *E. coli* translocation at the blood–brain barrier. Infect. Immun. 69, 5217–5222.

Kim, K.S., 2002. Strategy of *E. coli* for crossing the blood–brain barrier. J. Infect. Dis. 186, S220–S224.

Kim, K.S., 2003. Neurological diseases: pathogenesis of bacterial meningitis: from bacteremia to neuronal injury. Nat. Rev. Neurosci. 4, 376–385.

Kim, K.S., 2008. Mechanisms of microbial traversal of the blood–brain barrier. Nat. Rev. Microbiol. 6, 625–634.

Kim, K.S., 2010. Acute bacterial meningitis in infants and children. Lancet. Infect. Dis. 10, 32–42.

Kim, K.S., 2012. Current concepts on the pathogenesis of *E. coli* meningitis; implications for prevention and therapy. Curr. Opin. Infect. Dis. 25, 273–278.

Kim, K.S., Itabashi, H., Gemski, P., Sadoff, J., Warren, R.L., Cross, A.S., 1992. The K1 capsule is the critical determinant in the development of *Escherichia coli* meningitis in the rat. J. Clin. Invest. 90, 897–905.

Kim, K.S., Kang, J.H., Cross, A.S., 1986. The role of capsular antigens in serum resistance and in vivo virulence of *Escherichia coli*. FEMS Microbiol. Lett. 35, 275–278.

Kim, K.S., Kang, J.H., Cross, A.S., Kaufman, B., Zollinger, W., Sadoff, J., 1988. Functional activities of monoclonal antibodies to O-side chain of *Escherichia coli* lipopolysaccharides in vitro and in vivo. J. Infect. Dis. 157, 47–53.

Kim, K.S., Wass, C.A., Cross, A.S., 1997. Blood–brain barrier permeability during the development of experimental bacterial meningitis in the rat. Exp. Neurol. 145, 253–257.

Kim, Y., Pearce, D., Kim, K.S., 2008. Ca2+/Calmodulin-dependent invasion of the human brain microvascular endothelial cells by *Escherichia coli* K1. Cell Tissue Res. 332, 427–433.

Kim, Y.V., DiCello, F., Hillaire, C.S., Kim, K.S., 2004. Protease-activated receptors of human brain microvascular endothelial cells: expression and differential Ca^{2+} signaling induced by thrombin and protease-activated receptor-1 activating peptide. Am. J. Physiol. Cell Physiol. 286, C31–C42.

Klinger, G., Chin, C.-N., Beyene, J., Perlman, M., 2000. Predicting the outcome of neonatal bacterial meningitis. Pediatrics 106, 477–482.

Knodler, L.A., Celli, J., Finlay, B.B., 2001. Pathogenic trickery: deception of host cell processes. Nat. Rev. Mol. Cell Biol. 2, 578–588.

Korczak, B., Frey, J., Schrenzel, J., et al., 2005. Use of diagnostic microarrays for determination of virulence gene patterns of *Escherichia coli* K1, a major cause of neonatal meningitis. J. Clin. Microbiol. 43, 1024–1031.

Korhonen, T.K., Valtonen, M.V., Parkkinen, J., et al., 1985. Serotypes, hemolysin production, and receptor recognition of *Escherichia coli* strains associated with neonatal sepsis and meningitis. Infect. Immun. 48, 486–491.

Liu, S., Li, L., Luo, Z., et al., 2011. Superior effect of hypertonic saline over mannitol to attenuate cerebral edema in a rabbit bacterial meningitis model. Crit. Care. Med. 39, 1467–1473.

Lu, S., Zhang, X., Zhu, Y., Kim, K., Yang, J., JIN, Q., 2011. Complete genome sequence of the neonatal meningitis-associated *Escherichia coli* strain CE10. J. Bacteriol. 193, 7005.

Ludwig, G.V., Kondig, J.P., Smith, J.F., 1996. A putative receptor for Venezuelan equine encephalitis virus from mosquito cells. J. Virol. 70, 5592–5599.

Marinez, J.J., Mulvey, M.A., Schilling, J.D., Pinkner, J.S., Hultgren, S.J., 2000. Type 1 pilus-medicated bacterial invasion of bladder epithelial cells. EMBO J. 19, 2803–2812.

Maruvada, R., Kim, K.S., 2011. Extracellular loops of the *E. coli* outer membrane protein A contribute to the pathogenesis of meningitis. J. Infect. Dis. 203, 131–140.

Maruvada, R., Kim, K.S., 2012. IbeA and OmpA of *Escherichia coli* K1 exploit Rac1 activation for invasion of human brain microvascular endothelial cells. Infect. Immun. 80, 2035–2041.

Massia, S.P., Rao, S.S., Hubbell, J.A., 1993. Covalently immobilized laminin peptide Tyr-Ile-Gly-Ser-Arg (YIGSR) supports cell spreading and co-localization of the 67-kilodalton laminin receptor with alpha-actinin and vinculin. J. Biol. Chem. 268, 8053–8059.

McCracken Jr., G.H., Threlkeld, N., Mize, S., et al., 1984. Moxalactam therapy for neonatal meningitis due to gram-negative sepsis enteric bacilli. JAMA 252, 1427–1432.

Mittal, R., Gonzalez-Gomez, I., Panigrahy, A., Goth, K., Bonnet, R., Prasadarao, N.V., 2010. IL-10 administration reduces PGE-2 levels and promotes CR3-mediated clearance of *Escherichia coli* K1 by phagocytes in meningitis. J. Exp. Med. 207, 1307–1319.

Moissenet, D., Slauze, B., Clermont, O., et al., 2011. Meningitis caused by *Escherichia coli* producing TEM-52 extended-spectrum β-lactamase within an extensive outbreak in a neonatal ward: epidemiological investigation and characterization of the strain. J. Clin. Micro. 48, 2459–2463.

Moriel, D.G., Bertoldi, I., Spagnuolo, A., et al., 2010. Identification of protective and broadly conserved vaccine antigens from the genome of extraintestinal pathogenic *Escherichia coli*. PNAS 107, 9072–9077.

Nemani, P.V., Stins, M., Wass, C.A., Shimada, H., Kim, K.S., 1999. Outer membrane A promoted cytoskeletal rearrangement of brain microvascular endothelial cells is required for *E. coli* invasion. Infect. Immun. 67, 5775–5783.

Nemani, P.V., Wass, C.A., Kim, K.S., 1996b. Endothelial cell GlcNAcB1-4 GlcNAc epitopes for outer membrane protein A traversal of *E. coli* across the blood–brain barrier. Infect. Immun. 64, 154–160.

Nemani, P.V., Wass, C., Stins, M.F., Weiser, J., Huang, S.H., Kim, K.S., 1996a. Outer membrane protein A of *E. coli* contributes to invasion of brain microvascular endothelial cells. Infect. Immun. 64, 146–153.

Nizet, V., Kim, K.S., Stins, M., Jonas, M., Nguyen, D., Rubens, C.E., 1997. Invasion of brain microvascular endothelial cells by group *B* streptococci. Infect. Immun. 65, 5074–5081.

Orihuela, C.J., Mahdavi, J., Thornton, J., et al., 2009. Laminin receptor initiates bacterial contact with the blood brain barrier in experimental meningitis models. J. Clin. Invest. 119, 1638–1646.

Parkkinen, J., Korhonen, T.K., Pere, A., Hacker, J., Soinila, S., 1988. Binding sites of the rat brain for *Escherichia coli* S-fimbriae associated with neonatal meningitis. J. Clin. Invest. 81, 860–865.

Parthasarathy, G., Yao, Y., Kim, K.S., 2007. Flagella promote *Escherichia coli* K1 association with and invasion of human brain microvascular endothelial cells. Infect. Immun. 75, 2937–2945.

Prasadarao, N.V., Wass, C.A., Hacker, J., Jann, K., Kim, K.S., 1993. Adhesin of S-fimbriated *Escherichia coli* to brain glycolipids mediated by sfaA gene-encoded protein of S-fimbriae. J. Biol. Chem. 268, 10356–10363.

Reddy, M.A., Nemani, P.V., Wass, C.A., Kim, K.S., 2000. Phosphatidylinositol 3-kinase activation and interaction with focal adhesion kinase in *E. coli* K1 invasion of human brain microvascular endothelial cells. J. Biol. Chem. 275, 36769–36774.

Robbins, J.B., McCracken Jr., G.H., Gotschlich, E.C., Orskov, F., Orskov, I., Hanson, L.A., 1974. *Escherichia coli* K1 capsular polysaccharide associated with neonatal meningitis. N. Engl. J. Med. 290, 1216–1220.

Rode, C.K., Melkerson-Watson, L.J., Johnson, A.T., Bloch, C.A., 1999. Type-specific contributions to chromosome size differences in *Escherichia coli*. Infect. Immun. 67, 230–236.

Rubin, L.L., Staddon, J.M., 1999. The cell biology of the blood–brain barrier. Annu. Rev. Neurosci. 22, 11–28.

Ruffer, C., Strey, A., Janning, A., Kim, K.S., Gerke, V., 2004. Cell–cell junctions of dermal microvascular endothelial cells contain tight and adherens junction proteins in spatial proximity. Biochemistry 43, 5360–5369.

Sarff, L.C., Mccracken Jr., G.H., Schiffer, M.S., et al., 1975. Epidemiology of *Escherichia coli* in healthy and diseased newborns. Lancet 1, 1099–1104.

Schmidt, G., Sehr, P., Wilm, M., Selzer, J., Mann, M., Aktories, K., 1997. Gln 63 of Rho is deami dated by *Escherichia coli* cytotoxic necrotizing factor-I. Nature 387, 725–729.

Shin, S., Lu, G., Cai, M., Kim, K.S., 2005. *Escherichia coli* outer membrane protein A adheres to human brain microvascular endothelial cells. Biochem. Biophys. Res. Commun. 330, 1199–1204.

Soderstrom, T., Hansson, G., Larson, G., 1984. The *Escherichia coli* K1 capsule shares antigenic determinants with the human ganglioside GM3 and GD3. N. Eng. J. Med. 15, 726–727.

Sokolova, O., Heppel, N., Jägerhuber, R., et al., 2004. Interaction of *Neisseria meningitides* with human brain microvascular endothelial cells: role of MAP- and tyrosine kinases in invasion and inflammatory cytokine release. Cell Microbiol. 6, 1153–1166.

Spreer, A., Gerber, J., Hanssen, M., Nau, R., 2006. Dexamethasone increases hippocampal neuronal apoptosis in a rabbit model of *Escherichia coli* meningitis. Pediatr. Res. 60, 210–215.

Stevens, J.P., Eames, M., Kent, A., Halket, S., Holt, D., Harvey, D., 2003. Long term outcome of neonatal meningitis. Arch. Dis. Child Fetal Neonatal. Ed. 88, F179–F184.

Stins, M.F., Badger, J.L., Kim, K.S., 2001. Bacterial invasion and transcytosis in transfected human brain microvascular endothelial cells. Microb. Pathogenesis 30, 19–28.

Stins, M.F., Gilles, F., Kim, K.S., 1997. Selective expression of adhesion molecules on human brain microvascular endothelial cells. J. Neuroimmunol. 76, 81–90.

Stins, M.F., Nemani, P.V., Wass, C., Kim, K.S., 1999. *E. coli* binding to and invasion of brain microvascular endothelial cells derived from human and rats of different ages. Infect. Immun. 67, 5522–5525.

Stins, M.F., Prasadarao, N.V., Ibric, L., Wass, C.A., Luckett, P., Kim, K.S., 1994. Binding characteristics of S fimbriated *Escherichia coli* to isolated brain microvascular endothelial cells. Am. J. Pathol. 145, 1228–1236.

Teng, C., Tseng, Y., Pearce, D., Paul-Satyaseela, M., Xie, Y., Kim, K.S., 2010. NlpI contributes to *Escherichia coli* K1 strain RS218 interaction with human brain microvascular endothelial cells. Infect. Immun. 78, 3090–3096.

Teng, C.H., Cai, M., Shin, S., et al., 2005. *Escherichia coli* K1 RS218 interacts with human brain microvascular endothelial cells via type 1 fimbria phase-on bacteria. Infect. Immun. 73, 2923–2931.

Teng, C.H., Xie, Y., Shin, S., et al., 2006. Effects of ompA deletion on *Escherichia coli* K1 strain RS218's type 1 fimbria expression and association with human brain microvascular endothelial cells. Infect. Immun. 74, 5609–5616.

Thepparit, C., Smith, D.R., 2004. Serotype-specific entry of dengue virus into liver cells: identification of the 37-kilodalton/67-kilodalton high-affinity laminin receptor as a dengue virus serotype 1 receptor. J. Virol. 78, 12647–12656.

Tseng, Y.T., Wang, S.W., Kim, K.S., et al., 2012. NlpI facilitates deposition of C4bp on *Escherichia coli* by blocking classical complement-mediated killing, resulting in high level bacteremia. Infect. Immun. (in press).

Unhanand, M., Musatafa, M.M., McCracken, G.H., Nelson, J.D., 1993. Gram-negative enteric bacillary meningitis: a twenty-one year experience. J. Pediatr. 122, 15–21.

Wang, K.S., Kuhn, R.J., Strauss, E.G., Ou, S., Strauss, J.H., 1992. High affinity laminin receptor is a receptor for Sindbisvirus in mammalian cells. J. Virol. 66, 4992–5001.

Wang, Y., Kim, K.S., 2002. Role of OmpA and IbeB in *Escherichia coli* invasion of brain microvascular endothelial cells *in vitro* and *in vivo*. Ped. Res. 51, 559–563.

Wang, Y., Huang, S.H., Wass, C., Kim, K.S., 1999. The gene locus yijP contributes to *E. coli* K1 invasion of brain microvascular endothelial cells. Infect. Immun. 67, 4751–4756.

Wang, Y., Wen, Z.G., Kim, K.S., 2004. Role of S. fimbriae in *E. coli* K1 binding to brain microvascular endothelial cells in vitro and penetration into the central nervous system in vivo. Microb. Pathogenesis 37, 287–293.

Wieser, A., Magistro, G., Norenberg, D., Hoffmann, C., Schubert, S., 2012. First multi-epitope subunit vaccine against extraintestinal pathogenic *Escherichia coli* delivered by a bacterial type -3 secretion system (T3SS). Int. J. Med. Microbiol. 302, 10–18.

Xie, Y., Kolisnychenko, V., Paul-Satyassela, M., et al., 2006a. Identification and characterization of *E. coli* RS 218 – derived islands in the pathogenesis of *E. coli* meningitis. J. Infect. Dis. 194, 358–364.

Xie, Y., Parthasarathy, G., Di Cello, F., Teng, C.H., Paul-Satyaseela, M., Kim, K.S., 2008. Transcriptome of *E. coli* K1 bound to human brain microvascular endothelial cells. Biochem. Biophys. Res. Commun. 365, 201–206.

Xie, Y., Yao, Y., Kolisnychenko, V., Teng, C., Kim, K.S., 2006b. HbiF regulates type 1 fimbriation independent of FimB and FimE. Infect. Immun. 74, 4039–4047.

Yao, Y., Xie, Y., Kim, K.S., 2006. Genomic comparison of *E. coli* strains isolated from the cerebrospinal fluid of patients with meningitis. Infect. Immun. 74, 2196–2206.

Yao, Y., Xie, Y., Pearce, D., et al., 2009. The type III secretion system is involved in the invasion and intracellular survival of *Escherichia coli* K1 in human brain microvascular endothelial cells. FEMS Microbiol. Lett. 300, 18–24.

Yu, H., Kim, K.S., 2010. Ferredoxin is involved in secretion of cytotoxic necrotizing factor 1 across the cytoplasmic membrane in *Escherichia coli* K1. Infect. Immun. 78, 838–844.

Yu, H., Kim, K.S., 2012. YgfZ contributes to secretion of cytotoxic necrotizing factor 1 into outer membrane vesicles in *Escherichia coli*. Microbiology 158, 612–621.

Zhao, W.D., Liu, W., Fang, W.G., Kim, K.S., Chen, Y.H., 2010. Vascular endothelial growth factor receptor 1 contributes to *Escherichia coli* K1 invasion of human brain microvascular endothelial cells through PI3K/Akt signaling pathway. Infect. Immun. 78, 4809–4816.

Zhou, Y., Yao, J., Yu, H., et al., 2012. Hcp-family proteins secreted via type VI secretion system coordinately regulate *Escherichia coli* K1 interaction with human brain microvascular endothelial cells. Infect. Immun. 80, 1243–1251.

Zhu, L., Maruvada, R., Sapirstein, A., Malik, K.U., Peters-Golden, M., Kim, K.S., 2010a. Arachidonic acid metabolism regulates *Escherichia coli* penetration of the blood–brain barrier. Infect. Immun. 78, 4302–4310.

Zhu, L., Pearce, D., Kim, K.S., 2010b. Prevention of *E. coli* K1 penetration of the blood–brain barrier by counteracting host cell receptor and signaling molecule involved in *E. coli* invasion of human brain microvascular endothelial cells. Infect. Immun. 78, 3554–3559.

Chapter 11

Hybrid and potentially pathogenic *Escherichia coli* strains

Victor A. Garcia-Angulo[1], Mauricio J. Farfan[2], and Alfredo G. Torres[1]

[1]*University of Texas Medical Branch, Galveston, TX, USA,* [2]*Universidad de Chile, Santiago, Chile*

DIFFUSELY ADHERENT *E. COLI* (DAEC)

Background

Definition

Diffusely adherent *E. coli* (DAEC) is one of six classical pathotypes of diarrheagenic *E. coli* (DEC), with the ability to adhere over the entire surface of HEp-2/HeLa cells in a diffuse adherent (DA) pattern (Scaletsky et al., 1984). Analysis of the virulence determinants of DAEC strains indicated that they are a diverse group of strains with virulence genes homologous to those found in DEC strains (EAEC, ETEC, or EPEC) or in extraintestinal *E. coli* strains associated with urinary tract infections (UTIs) (Czeczulin et al., 1999; Servin, 2005).

History and evolution

To recognize DAEC strains from diarrheal stool samples of patients, the ability of the bacteria to adhere to HEp-2/HeLa cells has been used (Mathewson and Cravioto, 1989). This assay differentiated three adherence phenotypes: localized, diffuse, or aggregative adherence (LA, DA, AA, respectively; Figure 11.1). Analysis of the factors involved in the DA pattern led to the discovery of two adhesins, the F1845 fimbriae and AIDA-I autotransporter (Benz and Schmidt, 1989; Bilge et al., 1989). Since then, DNA probes and PCR specific for these virulence factors have been used for the identification of DAEC strains (Jallat et al., 1993; Vidal et al., 2005), and to design epidemiological studies associating DAEC with diarrhea cases. In recent years, several groups have focused on the study of F1845 fimbriae producing strains, since these fimbriae are related to the Afa/Dr adhesins, which are expressed in a variety of pathogenic genetic

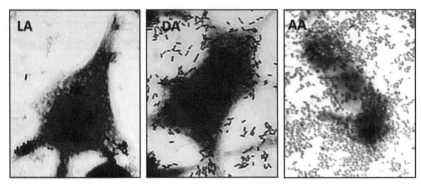

FIGURE 11.1   Adherence patterns of pathogenic *Escherichia coli* strains. Localized adherence (LA), diffuse adherence (DA), and aggregative adherence (AA). Magnification: ×100.

backgrounds (Escobar-Paramo et al., 2004), and have been associated with enteric and urinary tract infections.

Epidemiology and global impact

There are several studies that investigated the involvement of DAEC in diarrheal disease and, overall, they indicated that this association is still unsettled. Prevalence of DAEC isolates in stool samples of children with diarrhea is low compared to other DEC pathotypes (Gomez-Duarte et al., 2010; Rajendran et al., 2010). Some studies have associated DAEC strains with diarrheal disease in infants, children, and adults (Giron et al., 1991; Jallat et al., 1993). Further, an age-related incidence of DAEC strains expressing Afa/Dr adhesins has been described, implicating DAEC as a cause of diarrhea, particularly in children >12 months of age (Gunzburg et al., 1993; Levine et al., 1993). On the other hand, no differences in the frequency of DAEC strains were found in case-control studies and no relation with the type or duration of diarrhea in children was attributable to DAEC (Gomes et al., 1989; Cravioto et al., 1991). More strikingly, volunteer studies showed that DAEC strains orally administered failed to induce diarrhea in adults (Tacket et al., 1990).

A completely different situation is observed for DAEC strains associated with UTI infections. Afa/Dr DAEC strains are involved in 25–50% of cystitis cases in children, 30% of pyelonephritis cases in pregnant women, and one-third of recurrent UTIs in adults (Archambaud et al., 1988; Daigle et al., 1994). Moreover, a characterization of urinary *E. coli* isolates from 174 women with a first episode of UTI and assessment of the risk of a second UTI indicated that Afa/Dr adhesins were associated with a two-fold increased risk (Foxman et al., 1995). These observations have led to investigations about the function of Afa/Dr adhesins in DAEC pathogenicity, to understand the mechanism of DAEC interactions with the epithelia (Servin, 2005; Le Bouguenec and Servin, 2006).

Molecular pathogenesis

Mechanism of pathogenicity

Adherence to epithelial cells is probably the most important step in DAEC infection. This process includes highly specific interaction with the epithelium to: (i) recognize the infection site; (ii) compete and gain access to gastrointestinal and urinary tract regions that are occupied by the resident microflora; and (iii) activate signal transduction cascades that are necessary to induce diarrhea and/or inflammation (Benz and Schmidt, 1989; Bilge et al., 1989; Betis et al., 2003a,b). Many factors can contribute to DAEC adhesion to the host cells and some adhesins have been characterized (Betis et al., 2003a,b). Two main adhesins have been identified in DA strains, the adhesin involved in diffuse adherence (AIDA-1) and the F1845 fimbriae (Benz and Schmidt, 1989; Bilge et al., 1989). AIDA-1 is a plasmid-encoded 100 kDa autotransporter protein which was initially identified in an EPEC strain mediating adherence to HeLa cells and displaying the DA phenotype (Benz and Schmidt, 1989; Maurer et al., 1997). The nucleotide sequence of the F1845 fimbriae has significant homology with the Afa/Dr family of adhesins, which are virulence determinants in uropathogenic *E. coli* strains (Nowicki et al., 2001). Considering that approximately 75% of DAEC strains produce the F1845 or a related adhesin (Bilge et al., 1989), the role of Afa/Dr adhesins on the DAEC pathogenesis has been the focus of several studies.

The Afa/Dr adhesin family contains representatives of fimbrial and non-fimbrial structures, organized in operons consisting of at least five genes and belonging to the chaperone-usher family of adhesins. These operons are present in both diarrheal and uropathogenic *E. coli* strains and encode for the Afa/Dr adhesin, an invasin, and proteins required for the secretion and assembly of the adhesin at the bacterial cell surface (see Chapter 12) (Servin, 2005; Le Bouguenec and Servin, 2006). Afa/Dr adhesins are also classified on the basis of the host cell receptors that they bind, such as the Dr(a) blood-group antigen present on the decay-accelerating factor (DAF, CD55), the extracellular matrix protein collagen IV and/or carcinoembryonic antigen-related cellular adhesion molecules (CEACAMs) (Nowicki et al., 2001; Berger et al., 2004). All these receptors play a pivotal role in DEAC adherence to the gastrointestinal and urinary tract.

Adherence

Using polarized epithelial cultured cells it was established that Afa/Dr adhesins interact with DAF or CEACAM receptors (CEACAM6 for F1845) and, as a result, the receptors accumulate underneath the bacteria (Goluszko et al., 1999; Le Bouguenec et al., 2001). This interaction also triggers signal transduction cascades that control the growth of long finger-like cellular projections (lamellipodia) which wrap around the bacteria, promoting tight attachment of the bacteria to the cell surface (Cookson and Nataro, 1996).

In addition, DAEC adhesion is accompanied by brush border lesion, F-actin rearrangement and tight junction disruption (Bernet-Camard et al., 1996; Peiffer et al., 2000). As a result of the adherence process, the bacteria become firmly attached to the cell surface, initiating the colonization steps and activation of signal transduction pathways, resulting in the internalization of the bacteria and/or inflammatory responses.

Internalization

A small proportion of adherent Afa/Dr DAEC strains are able to invade cultured epithelial cells, a process that is also associated with the Afa/Dr adhesins (Guignot et al., 2009). During in vitro culture, Afa/Dr DAEC invades epithelial cells by a zipper-like mechanism, with the participation of microtubules, lipid rafts and $\alpha5\beta1$ integrins (Goluszko et al., 1999; Kansau et al., 2004; Guignot et al., 2009). Once in the cytoplasm, internalized bacteria reside in a large vacuole formed by the fusion of single bacterium-containing vacuoles originated early during the initial step of invasion (Plancon et al., 2003; Servin, 2005).

Inflammation

DAEC infection is characterized by the induction of an inflammatory response on epithelial cells. Afa/Dr DAEC infection of intestinal cultured cells induces the release of pro-inflammatory factors, such as IL-8, as a result of the activation of the mitogen-activated protein kinases pathway (Betis et al., 2003a; Arikawa et al., 2005). The inflammation induced by Afa/Dr DAEC promotes polymorphonuclear migration across the epithelial barrier, inducing the production of TNF-α and IL-1β, which stimulate the production of DAEC receptor DAF (Betis et al., 2003b). In addition, Afa/Dr DAEC increase the cell-surface expression of the major histocompatibility complex class I chain-like gene A, a key factor in the host innate immune response (Tieng et al., 2002). Overall, these data provide a mechanism by which Afa/Dr DAEC may participate in the development of inflammatory diarrhea in humans.

Clinical manifestations

Transmission and clinical features

Acquisition through food or water contaminated with human or animal feces comprises the main transmission route for this pathovar. Clinical presentation of diarrhea episodes caused by diarrhea-causing DAEC might include watery or bloody diarrhea, abdominal cramps, dehydration and fever (Gunzburg et al., 1993; Jallat et al., 1993); however, a unique clinical characteristic specific for DAEC intestinal infection has not been described. Interestingly, the finding that Afa/Dr DAEC adhesins are carried by *E. coli* causing UTIs supports the hypothesis of fecal–perineal–urethral transmission as the etiology for these infections (Daigle et al., 1994; Foxman et al., 1995).

Diagnosis

At first, the HEp-2/HeLa cells adherence assay was widely used to identify DAEC strains. However, the identification and characterization of F1845 and AIDA-1 led to the use of these adhesins as markers for DAEC strains. It is important to note that a significant number of strains with a DA phenotype are negative for both markers and therefore are not classified as DAEC (Jallat et al., 1993, 1994). Therefore, additional studies to identify novel, more reliable markers for the DAEC pathotype are warranted. Considering that the prevalence of Afa/Dr adhesins in DAEC is higher than that of AIDA-I, numerous assays to diagnose Afa/Dr DAEC have been developed using the F1845 or related adhesins. Among these assays, the molecular detection of genes located in the Afa/Dr operon has been explored. Different DNA probes for colony hybridization have been assayed, but this is a time-consuming technique that requires trained personnel, making difficult its implementation in a clinical laboratory. PCR-based techniques, such as multiplex PCR, to identify *E. coli* colonies obtained from cultures or stool samples offer a more practical, rapid and accurate diagnosis. Multiplex PCR assay to identify all six *E. coli* pathotypes has provided a valuable tool for epidemiological studies (Vidal et al., 2005).

Treatment

Diarrheal illness caused by DAEC, as well as other DEC pathotypes, might be ameliorated by specific antimicrobials, however clinical data demonstrating their efficacy are lacking. Moreover, DAEC strains are associated with a high frequency of antibiotic resistance. DAEC strains have presented high levels of resistance to antibiotics commonly used to treat enteric infections, compared to the rest of DEC strains (Ochoa et al., 2009). Moreover, the majority of the DAEC strains with resistance to multiple antibiotics are associated with conjugative plasmids, a situation that not only hinders the antibiotic treatment, but also facilitates the dissemination of the antibiotic resistance genes between enteric pathogens (Lopes et al., 2005). In many countries, the implementation of oral rehydration is the treatment of choice for DEC infections, including DAEC. This strategy has significantly reduced deaths caused by diarrhea, but in severe cases, an antibiotic treatment might be necessary. In this case, the knowledge of the profile of antibiotic susceptibility can guide the selection of an adequate antimicrobial management. UTIs caused by DAEC are treated according to guidelines for other *E. coli* UTIs (see Chapter 9).

Immune response

In order to investigate the role of DAEC in the development of diarrhea, volunteers were fed reference DAEC strains. None of them developed diarrhea and only of small number developed IgG or IgA antibodies to DAEC (Tacket et al., 1990). These results and the controversial role of DAEC as a diarrhea-causing pathogen have reduced the interest to analyze the immune response against this pathotype and the development of strategies to control DAEC infections. On the other hand, the data indicating that Afa/Dr DAEC strains induce an inflammatory response on polarized intestinal cells (Betis et al., 2003a), and the increasing number of studies associating the Afa/Dr adhesin with UTI, should lead to a renewed interest to understand the immune response of DAEC in humans.

Control and prevention

General guidelines for management of patients with diarrhea are also recommended for DAEC infections. Patients admitted to hospital with diarrhea should be barrier-nursed, preferably in a side-room, and the infection control department should be notified. To prevent UTIs, the most important recommendation is to practice good personal hygiene. Antibiotic prophylaxis remains unclear and contradictory (Greenfield, 2011).

ADHERENT AND INVASIVE *E. COLI* (AIEC)

Background

Definition and/or classification

Inflammatory bowel disease (IBD), comprises both Crohn's disease (CD) and ulcerative colitis (UC), and affects approximately 2.6 million people in the US and Europe (Loftus, 2004). CD is a chronic, granulomatous inflammatory condition characterized by a strong activation of the immune system. Although CD can affect any site of the intestinal tract, it is more common in the ileum. CD causes erosions in the inner surface of the intestine known as aphthous ulcers (Kaser et al., 2010), which in severe cases can lead to intestinal obstruction, among other complications. A complex interplay of genetic and environmental factors contributes to triggering abnormal immune responses in CD (Hugot et al., 2001; Ogura et al., 2001; Helbig et al., 2012). However, cumulative evidence suggests that intestinal bacteria play a major role in the onset and perpetuation of the disease (Sartor, 2005; Rolhion and Darfeuille-Michaud, 2007; Carvalho et al., 2009; Flanagan et al., 2011). A significant reduction in the gut microbiota diversity with an over-representation of *Enterobacteriaceae*, mainly *E. coli*, is observed in patients with CD (Kotlowski et al., 2007; Flanagan et al., 2011; Joossens et al., 2011). Although belonging to different serogroups, *E. coli* isolates from CD patients have highly related ribotype profiles and most of them belong to the B2 and D phylogroups (Kotlowski et al., 2007; Joossens et al., 2011). Unlike non-pathogenic commensal *E. coli*, isolates from CD lesions have been shown to efficiently adhere to, invade and survive inside epithelial cells and macrophages. As such, this group of isolates has been collectively named adherent-invasive *E. coli* (AIEC) (Darfeuille-Michaud et al., 2004). AIEC have been isolated from up to one-third of the ileal lesions of CD patients. Thus, AIEC comprise an *E. coli* category associated with persistence in these multifactorial CD. The criteria for inclusion in the AIEC category include: (i) ability to adhere to and invade intestinal epithelial cells with a macropinocytosis-like entry process; (ii) ability to survive and replicate within macrophages without triggering host cell death; and (iii) ability to induce the release of large amounts of TNF-α from infected macrophages (Darfeuille-Michaud, 2002; Darfeuille-Michaud et al., 2004).

History

In 1932, Dr. Burrill B. Crohn and colleagues presented a paper to the American Gastro-Enterological Association entitled 'Non-specific Granulomata of the Intestine', describing the features of what is now known as CD, a disease different from UC (Baron, 2000; Naser et al., 2012). It was speculated that an infectious agent could account for an environmental cause of the disease. In 1970, Rees and Mitchel showed that mice inoculated with CD tissue displayed focal granulomas (Mitchell and Rees, 1970). Subsequently, Cave and collaborators demonstrated that rabbits inoculated with human CD tissue homogenates developed granuloma (Cave et al., 1973) and this effect could be prevented by ampicillin pretreatment of the homogenates (Donnelly et al., 1977), suggesting the presence of a transmissible agent in the disease. Other studies showed that antibody titers against *E. coli* antigens were higher in CD patients than in controls (Brown and Lee, 1973; Tabaqchali et al., 1978) and that most CD tissue samples were positively labeled by anti-*E. coli* antibodies (Cartun et al., 1993; Liu et al., 1995). *E. coli* were recovered frequently from CD tissues (Darfeuille-Michaud et al., 1998), and these isolates were shown to be genetically related by their ribotype profiles (Masseret et al., 2001), and able to adhere to differentiated intestinal epithelial cells more frequently than isolates from healthy tissues (Darfeuille-Michaud et al., 1998; Masseret et al., 2001). Additionally, the reference AIEC LF82 strain was shown to invade and replicate inside intestinal epithelial cells, using a process dependent on a macropinocytosis-like mechanism (Darfeuille-Michaud, 2002). Based on those characteristics, the designation of adherent-invasive *E. coli* as a new group of *E. coli* associated with CD was proposed (Darfeuille-Michaud, 2002).

Evolution

The onset and perpetuation of CD has not been associated with any particular *E. coli* strain, but the use of culture-independent methods of identification suggested that most *E. coli* strains colonizing CD lesions are genetically related. Ribotyping of *E. coli* isolates from CD patients with early, recurrent, or chronic lesions, with or without endoscopic recurrence showed that, while compared to healthy controls, most *E. coli* from CD cases grouped in a single ribotype profile (Masseret et al., 2001). *E. coli* species comprises four phylogenetic groups: A, B1, B2, and D, and the CD-specific isolates were found to belong to B2 and D groups (Kotlowski et al., 2007). A recent study did not find any phylogenetic relationship among CD *E. coli* isolates and most of the phylogenetic groups were found equally distributed in CD patients and healthy individuals; however, the B2 phylogroup was more prevalent among AIEC-positive isolates (Martinez-Medina et al., 2009).

Phylogenetic analyses based on multilocus sequence typing (MLST) showed that AIEC isolates cluster with the uropathogenic *E. coli* (UPEC) strain CFT073 and the avian pathogenic *E. coli* (APEC) serotype O1:K1:H7 (Sepehri et al., 2009). The relationship between AIEC and either UPEC or APEC strains could be further validated because AIEC strains encode and express UPEC-related adhesins and autotransporter proteins (Darfeuille-Michaud et al., 1998; Kotlowski et al., 2007). Phylogenetic analyses and full genome sequence of prototype AIEC strains confirmed relatedness to these strains and expanded the spectrum of putative virulence factors, either AIEC-specific or shared with UPEC and APEC (Miquel et al., 2010a; Nash et al., 2010; Clarke et al., 2011; Krause et al., 2011). The genomic sequence of AIEC LF82 reveals that this strain shares a high percentage of common coding sequences with APEC-01 strain and that the putative virulence determinants include gene products promoting motility, serum resistance, iron uptake, capsule and LPS expression, as well as biofilm formation, adhesion to and invasion of epithelial cells (Miquel et al., 2010b). AIEC NRG857c strain also clusters in the B2 phylogroup with APEC-01 and the UPEC strains 536 and CFT073, and genomic analysis showed considerable sequence similarity and synteny with AIEC LF82 (Nash et al., 2010). Overall, AIEC strains seem to be evolutionarily and genetically related to extraintestinal *E. coli* strains, suggesting that both pathovars might use similar pathogenic strategies.

Epidemiology and global impact

CD affects mainly developed countries, with an annual incidence of up to 20.2 per 100 000 persons in North America, 12.7 in Europe and 5.0 in Asia and the Middle East (Molodecky et al., 2012). Systematic data reviews indicate that the incidence and prevalence of CD is increasing in both developed and developing nations, as the latter become more industrialized (Molodecky et al., 2012). Incidence rates are higher among the second to the fourth decade of life, which results in long-term costs to their productivity and the society (Yu et al., 2008; Molodecky et al., 2012). The extent of the contribution of AIEC to the onset and persistence of CD is still unclear and it has been proposed that CD is initiated by a dysregulated immunological response to normal intestinal micro-organisms, which creates an inflammatory milieu favoring expansion and invasion in a subset of patients (Strober, 2011). AIEC prevalence in CD patients varies in different studies from 22–52%, and a recent study found AIEC strains in 51.9% of ileal biopsies from CD patients, and only in 16.7% of healthy controls (Martinez-Medina et al., 2009). Another study found AIEC in 21.7% of chronic lesions, 36.4% in early lesions and 22.2% in healthy mucosa of CD patients, while only 6.2% of the mucosa was affected in controls (Darfeuille-Michaud et al., 2004). Thus, incidence rates suggest that AIEC strains could be contributing to CD in up to one-third of total cases of genetically susceptible individuals.

Molecular pathogenesis

Adherence and invasion

AIEC's ability to adhere to epithelial cells is mediated primarily by the type I pili, which interact with the host receptor CEACAM6. This receptor is abnormally over-expressed by the ileal epithelium in CD patients (Carvalho et al., 2009). AIEC-induced colitis in transgenic mice depends on both bacterial type I pili expression and human CEACAM6 expression in the intestine (Carvalho et al., 2009). The earliest lesions of CD are microscopic erosions of the follicle-associated epithelia lining the Peyer's patches (PPs). AIEC is able to adhere to the specialized M cells in PPs through the type I pili and the long polar fimbriae (Lpf) (Chassaing et al., 2011), favoring its translocation through M cells, which may explain the detection of AIEC in the lamina propria of CD patients. The invasion properties of AIEC are associated with the extrusion of microvillar extensions from the host cell membrane that engulf the bacteria, resembling macropinocytosis induced by other intracellular pathogenic bacteria (Darfeuille-Michaud, 2002). However, this mechanism for AIEC depends on both actin microfilaments and microtubules (Darfeuille-Michaud, 2002), while invasion for other pathogens depends exclusively on actin cytoskeleton rearrangements. Type I pili have been involved in the invasion of intestinal epithelial cells, and mutants in the *fimI* and *fimF* gene products, involved in the type I pili biogenesis, still adhere to IECs but their ability to induce membrane elongations at the site of contact is impaired (Boudeau et al., 2001). Further, outer membrane vesicles (OMV) containing biologically active proteins function as an efficient secretion pathway and are also important for AIEC invasion. For example, an *yfgL* mutant, showing reduced release of OMV, displays a diminished ability to invade intestinal cells (Rolhion et al., 2005). The endoplasmic reticulum stress protein Gp6 is also over-expressed in the ileum of CD patients; it co-localizes with CEACAM6 and functions as a receptor for the bacterial OmpA located on the surface of OMV (Rolhion et al., 2010). Although AIEC-specific content in the OMV is necessary for promoting invasion (Rolhion et al., 2010), the nature of the effector proteins delivered by OMV is still unknown.

Regulation

Compared to other *E. coli* pathotypes, little is known about the mechanisms regulating expression of virulence factors in AIEC. The expression of the *fim* operon is regulated by the orientation of an invertible DNA element that includes the *fim* promoter (the *fim* ON and OFF switch). The orientation of this invertible element is determined by the activity of the FimB and FimE recombinases (Klemm, 1986). In AIEC, *fim* is repressed in mutants defective in flagella assembly, generally as a result of a preferential switch towards the OFF state of the invertible element (Barnich et al., 2003, 2004; Claret et al., 2007). A reduction in the transcription of the *fimB* and *fimE* genes, together with a reduction in adherence and invasion, occurs in a strain defective in the NlpI lipoprotein (Barnich et al., 2004). The flagellar transcriptional regulator FlhD2C2 and the sigma factor FliA are required for the expression of type I pili, while the effect of FliA is partially mediated through YhjH, a phosphodiesterase involved in degradation of dimeric cGMP (Claret et al., 2007). While high osmolarity induces an increase in the ability of AIEC to interact with human cells and correlates with an increase in OmpC porin expression, an *ompC* null mutant expresses neither flagella nor type I pilli, and this effect is mediated by the RNA polymerase σ^E factor (Rolhion et al., 2007). In contrast, down-regulation of the histone-like protein Fis, occurring during AIEC infection of IECs, produces a repression of flagella expression and leads to the preferential ON status of the *fim* switch (Miquel et al., 2010a).

A *Caenorhabditis elegans*-based model for in vivo AIEC infection was developed and used to test whether Hfq RNA chaperone, which is involved in post-transcriptional regulation by small non-coding RNA, is required for full virulence. The *hfq* mutant is non-motile, less invasive and highly sensitive to chemical stress, indicating that post-transcriptional ribo-regulation mediates AIEC adaptation to this environmental niche (Simonsen et al., 2011). The transcription of the HtrA stress protein and the DsbA oxidoreductase, required for intramacrophage replication, are highly up-regulated in intramacrophagic bacteria and/or macrophage-mimicking stress culture conditions, but their activation is independent of the CpxRA two-component regulatory system (Bringer et al., 2005, 2007), which regulates HtrA and DsbA expression in non-pathogenic *E. coli* strains.

Avoidance of host responses

AIEC is able to translocate across in vitro cultured M cells monolayers (Chassaing et al., 2011). Enteric pathogens crossing the follicle-associated epithelia need to survive phagocytosis and bacterial killing by resident and recruited macrophages at the dome of the lymphoid follicle. Pathogenic bacteria counteract macrophage killing after phagocytosis, either by escaping phagosomes and inducing cell death, or by resisting the antimicrobial environment of the phagosome, which includes inhibition of its fusion with lysosomes. AIEC is able to survive and replicate within macrophages, in which they reside in a large vacuole without inducing cell death (Glasser et al., 2001). In contrast to other pathogens, AIEC-containing phagosomes traffic through the endocytic pathway and mature into active phagolysosomes where bacteria are exposed to low pH and the proteolytic activity of cathepsin D (Bringer et al., 2006). AIEC has developed mechanisms to resist phagosome stress conditions and mutations in *htrA* or *dsbA* increase the sensibility of the bacteria to acid and nutrient-limiting growth conditions and oxidative stress caused by hydrogen peroxide, which correlates with a defect in intramacrophagic replication (Bringer et al., 2005, 2007).

Putative genetic determinants predisposing humans to CD include polymorphisms of the gene encoding the intracellular receptor *NOD2/CARD15*

(Hugot et al., 2001; Ogura et al., 2001) and the genes *ATG16L1* and *IRGM*, involved in intracellular bacterial clearance through autophagy (Massey and Parkes, 2007; Rioux et al., 2007). NOD2/CARD15 stimulation by bacterial muramyl dipeptide activates various antibacterial responses through NF-κB and MAP kinase signaling pathways (Fritz et al., 2006). The autophagy constitutes an efficient host innate immune mechanism against intracellular replication of CD-associated AIEC (Lapaquette et al., 2010, 2012). AIEC may take advantage of a diminished surveillance function of monocytes with NOD2/CARD15 polymorphisms, since monocytes expressing CD-associated alleles showed a reduced early cytokine response to AIEC infection (Peeters et al., 2007). Similarly, reducing the expression of ATG16L1 or IRGM in IECs abrogates autophagy of intracellular AIEC. This effect could not be reversed by the expression of a CD-associated ATG16L1 variant (Lapaquette et al., 2010). In agreement, impairing expression of NOD2/CARD15, ATG16L1, or IRGM in macrophages produces an increase in intramacrophagic AIEC (Massey and Parkes, 2007). Thus, it is hypothesized that dysfunctional immunological responses toward intracellular bacteria in susceptible individuals harboring CD-associated polymorphisms contribute to the persistence of AIEC inside macrophages (Lapaquette et al., 2012).

Damage

As AIEC lacks typical virulence factors from other intestinal *E. coli* pathogens, the main damage elicited by AIEC colonization is restricted to its ability to sustain chronic, acute inflammation in CD patients. The reduction in the abundance and biodiversity of intestinal bacteria (dysbiosis) caused by *E. coli* over-representation displaces normal flora species known to possess anti-inflammatory properties (Sokol et al., 2008). Intramacrophagic AIEC bacteria can reside in a large vacuole without inducing cell death. Instead, the bacteria induce the aggregation and sometimes fusion of infected macrophages, to form multinucleated giant cells and subsequent recruitment of lymphocytes (Meconi et al., 2007). A hallmark of AIEC is the induction of pro-inflammatory TNF-α in infected macrophages (Glasser et al., 2001). AIEC also induces the production of inflammatory chemokines such as CCL20, as well as IL-8 (Eaves-Pyles et al., 2008) and IL-1β and IL-6 (Carvalho et al., 2008). Several lines of evidence support the role of AIEC in the inflammation of CD. TNF-α stimulates the expression of the AIEC receptor CEACAM6 in IECs (Carvalho et al., 2009). Moreover, AIEC, but not a non-pathogenic *E. coli* K-12 strain, induced colitis in mice with dextran sulfate sodium-injured colon (Carvalho et al., 2008) or in mice expressing human CEACAM6 (Carvalho et al., 2009). In addition, there is evidence that AIEC contribute to intestinal injury, disrupting the epithelial barrier by affecting tight junctions in CEACAM6-expressing mice (Denizot et al., 2012).

Clinical manifestations

Transmission

As AIEC strains are detected in a portion of the healthy population (Darfeuille-Michaud et al., 1998; Martinez-Medina et al., 2009), it is believed that in the absence of predisposing human host genetic factors, they are not able to trigger abnormal inflammatory responses and rather behave as commensal bacteria, not posing a risk for humans. However, truly adherent and invasive *E. coli* strains are found in dogs, cats, and swine; therefore, a putative zoonotic risk has been assigned in the infection with AIEC for the development of CD (Simpson et al., 2006; Martinez-Medina et al., 2011).

Complications

Although the exact role of AIEC strains needs to be further elucidated, the features of the CD lesions seem to be the result of a microbial process in the gut, including the onset of lymphoid aggregates, ulceration, micro-abscesses, fissures, fistulas, granulomas, and lymphangitis (Barnich and Darfeuille-Michaud, 2007). The exposure of the distal ileum to luminal contents after curative resection in patients with CD is associated with the onset of endoscopic lesions and increased inflammation (Rutgeerts et al., 1991). In severe CD cases, deeper and larger ulcers can develop, leading to bowel obstruction and holes in the intestinal wall, which could result in a disseminated infection (Barnich and Darfeuille-Michaud, 2007).

Diagnosis

Adhesion and invasion patterns of *E. coli* strains isolated from CD patients have been used to identify AIEC strains (Baumgart et al., 2007; Martinez-Medina et al., 2009). However, the identification of AIEC strains in a subset of healthy individuals and several AIEC-negative CD cases reflects the involvement of host genetic determinants as well as the possibility of different microbial priming in CD patients (Sartor, 2005). All these factors suggest that rendering solely on AIEC identification is an unreliable indicator of the disease status.

Treatment

The control of AIEC infection as a proposed method to treat CD has undergone two main directions, antibiotic treatment and feeding of probiotics as a maintenance therapy. Broad-spectrum antibiotics are clinically used to treat CD. Metronidazole administered alone or in combination with cotrimoxazole or ciprofloxacin has been shown to improve colonic but not ileal CD and to be effective at reducing post-operative recurrence (Rutgeerts et al., 1995; Gionchetti et al., 2006). Also, rifamycin derivatives induce remission in active CD (Khan et al., 2011). Although good results are obtained at maintaining UC remission with probiotics, more modest results are obtained in CD. Rifamixin administered with the probiotics preparation VSL#3 is effective at preventing post-operative recurrence of CD, while *Saccharomyces boulardii* improves the effect of mesalazine in the maintenance of remission (Gionchetti et al., 2006). Commensal *E. coli* strain Nissle improves maintenance of steroid-induced remission of colonic CD (Malchow, 1997) and is able to reduce the ability of AIEC to adhere, invade and induce pro-inflammatory cytokines (Huebner et al., 2011); however, it has no effect on the capacity of AIEC to adhere to biopsies from CD or healthy ileum (Jensen et al., 2011).

SHIGA TOXIN-PRODUCING *E. COLI* O104:H4

Background

Definition and/or classification

Escherichia coli O104:H4 is an infrequently isolated pathogenic *E. coli* serotype and the cause of the 2011 European diarrheal outbreak (Mellmann et al., 2011). Analysis of genome sequences obtained from several outbreak isolates showed that the *E. coli* O104:H4 strain is an enteroaggregative *E. coli* (EAEC; for further details, see Chapter 8) that has acquired the Shiga toxin genes, presumably by bacteriophage transduction (Brzuszkiewicz et al., 2011; Mellmann et al., 2011; Rasko et al., 2011; Rohde et al., 2011). Genome assembly confirmed two copies of the *stx2* prophage gene cluster; however, a set of additional virulence and antibiotic-resistance factors are also characteristic of this *E. coli* O104:H4 outbreak strain.

History

In May–July 2011, two outbreaks of bloody diarrhea and hemolytic uremic syndrome (HUS) occurred in Europe: one centered in Germany (around 4000 cases of bloody diarrhea, 850 cases of HUS and 50 deaths), and a much smaller outbreak in southwest France (15 cases of bloody diarrhea, 9 of which progressed to HUS) (Bielaszewska et al., 2011; Frank et al., 2011; Gault et al., 2011). Both outbreaks were caused by a Shiga toxin-producing *E. coli* (STEC) strain of serotype O104:H4 (Gault et al., 2011; Scheutz et al., 2011). This was not the first time this isolate was seen in Germany and the first isolates of *E. coli* O104:H4 carrying the Shiga toxin genes in this country date back to 2001, when the prototype strain known as HUSEC41 isolate 01-0991 was recovered from a HUS case (Mellmann et al., 2008; Bielaszewska et al., 2011; Kunne et al., 2012). *E. coli* O104:H4 had been also isolated in 2006, from a woman who contracted HUS in Korea and also from Italy in 2009 (Scavia et al., 2011). In the case of the HUS-producing isolate from South Korea (Bae et al., 2006), the strain appears not to be closely related to the German outbreak strain, differing in their toxin gene carriage, antibiotic resistance properties, and pulsed-field gel electrophoresis profiles (Kim et al., 2011).

Cumulative evidence by several groups confirmed that the two outbreaks in Germany and France were caused by a STEC O104:H4, with virulence features common to the EAEC pathogroup (Bielaszewska et al., 2011; Scheutz et al., 2011). This combination is very rare and was previously described in strains of serotype O111:H2 involved in a small outbreak of HUS in children in France (Morabito et al., 1998). The German outbreak strain, like the 55989 EAEC strain, was found to possess several virulence factors found in extraintestinal pathogenic *E. coli* as well as to have acquired resistance to numerous antibiotics, including third-generation cephalosporins, due to the presence of plasmid-encoded multi-antibiotic resistance cassettes (Denamur, 2011). The general consensus indicates that the *E. coli* O104:H4 outbreak strain is a recombinant of two pathogenic *E. coli* types, STEC and EAEC, and as a result, two models for *E. coli* O104:H4 evolution have been proposed. The first one indicates that *E. coli* O104:H4 and the African EAEC isolate 55989 (see below) evolved from a common Shiga toxin-producing *E. coli* O104:H4 progenitor strain by stepwise gain and loss of chromosomal and plasmid-encoded virulence factors (Mellmann et al., 2011). This process generated a highly pathogenic new hybrid strain, that has been proposed to be renamed 'Entero-Aggregative-Hemorrhagic *E. coli* (EAHEC)' (Brzuszkiewicz et al., 2011), but is also known as Shiga-toxin-producing *E. coli* serotype O104:H4. The second model is derived from comparative genomic analysis among the sequence of the German outbreak strain and the sequences of other O104:H4 EAEC from Africa and reference EAEC strains from serotypes different than O104:H4 (Rasko et al., 2011). This analysis indicated that the German *E. coli* O104:H4 strain belongs to the enteroaggregative clade, but could be distinguished from other *E. coli* O104:H4 strains due to the fact that it contains a recently acquired prophage encoding Shiga toxin 2 genes and a set of additional virulence and antibiotic-resistance factors. The second model supports the idea of horizontal genetic exchange, which allowed the emergence of the highly virulent Shiga-toxin-producing enteroaggregative *E. coli* O104:H4 strain (Rasko et al., 2011).

Evolution

Rapid identification of *E. coli* O104:H4 as the causative agent of the outbreak, followed by the application of high-throughput sequencing technologies, allowed draft genome sequences of several isolates (Brzuszkiewicz et al., 2011; Mellmann et al., 2011; Rasko et al., 2011; Rohde et al., 2011). A worldwide effort led to rapid phylogenetic analyses which indicated that a close relationship existed between the outbreak of *E. coli* O104:H4 isolates and EAEC 55989 (Rohde et al., 2011), a strain originally isolated in the Central African Republic from an HIV-infected individual who experienced persistent diarrhea (Bernier et al., 2002). Interestingly, Shiga-toxin-producing EAEC strains had been identified in the same population from which strain 55989 was isolated (Mossoro et al., 2002). Further comparisons of the outbreak isolate genome sequence with a number of EAEC O104:H4 strains from Africa confirmed the close relationship, suggesting a recent emergence of the outbreak strain (Rasko et al., 2011). The analysis highlighted changes associated with the emergence of the outbreak strain, including the acquisition of a prophage encoding Shiga toxin 2 genes and a plasmid encoding the extended beta-lactamase CTX-M-15 (Brzuszkiewicz et al., 2011; Mellmann et al., 2011; Rasko et al., 2011; Rohde et al., 2011).

Epidemiology

An epidemiological investigation of the outbreak involving all levels of European public-health and food-safety authorities was initiated to identify the vehicle of the outbreak infections and to prevent further cases of disease. The initial investigation suggested fresh vegetables were involved but the source of the outbreaks was eventually linked to contaminated sprouts (Buchholz et al., 2011). Epidemiological evidence indicates the outbreaks were connected to a fenugreek seed shipment from Egypt that arrived in Germany in 2009 (Grad et al., 2012). The seeds were distributed in North Germany, including the German farm that grew the contaminated sprouts. A second distribution chain occurred in the French outbreak, where the seeds were germinated at a children's community center, and the sprouts served on June 2011 (Gault et al., 2011).

Using a conventional molecular epidemiological approach (including virulence gene content assessment, serotyping, multilocus sequence typing, rep-PCR, pulsed-field gel electrophoresis, optical mapping, and antimicrobial susceptibility testing), the outbreak strains in Germany and France appear identical (Gault et al., 2011; Mariani-Kurkdjian et al., 2011). However, a recent molecular epidemiological study assessing single nucleotide polymorphisms in whole-genome sequencing data from multiple isolates showed remarkably little diversity among the German outbreak isolates (Grad et al., 2012). In contrast, a much greater diversity was found in isolates from seven individuals infected in the French outbreak. The differences in diversity observed between the German and French outbreak isolates suggested either restricted conditions preventing diversity in the German isolates, variation in mutation rates among the two *E. coli* outbreak populations, or uneven distribution of diversity in the seed populations that led to each outbreak (Grad et al., 2012).

Molecular pathogenesis

Virulence factors

No experimental evidence is currently available regarding the virulence and fitness factors of *E. coli* O104:H4, as well as the mechanisms mediating the interaction with the host cells and their targets are unknown. Therefore, the roles of Shiga toxins, biofilm-producing factors, as well as adhesins, iron uptake systems and other colonization factors encoded by these isolates need to be determined. Similarly, the resistance of *E. coli* O104:H4 to environmental stress and the factors that promote secretion or transport of its virulence factors need to be studied. Finally, appropriate infection models to study virulence and fitness under different conditions should be established. However, some clues regarding virulence can be deduced by the analysis of the genome sequence of *E. coli* O104:H4 and extrapolation of the role of similar virulence factors in other pathogenic *E. coli*.

Shiga toxin

Characterization of the epidemic strain showed that it carries the *stx2* genes but, in contrast to typical EHEC, it lacks the LEE pathogenicity island (genetic marker for Shiga-toxin-producing *E. coli* and other attaching and effacing *E. coli* strains; for further details, see Chapters 4 and 5) (Bielaszewska et al., 2011; Scheutz et al., 2011). *E. coli* O104:H4 acquired the bacteriophage encoding *Stx* and further subtyping showed that the gene corresponds to the *stx2a* variant (Scheutz et al., 2011). Sequence analysis of the *E. coli* O104:H4 *stx2a* gene showed 100% amino acid identity of the Stx2a protein found in the prototype STEC O157:H7 isolate EDL933, but differed by one nucleotide, making the nucleotide sequence identical to that found in sorbitol-fermenting STEC O157 strains from Germany (Scheutz et al., 2011). Therefore, the unusually high level of HUS development associated with the *E. coli* O104:H4 strain is not related to differences in toxin activity, but rather might be explained by the enteroaggregative adherence phenotype displayed by this strain, allowing the bacteria to efficiently colonize the intestinal mucosa, increasing the exposure of the tissue in the affected patient to the toxin.

Adherence

The *E. coli* O104:H4 strains isolated from these outbreaks were PCR-positive for different combinations of the *aggR, aatA, aaiC, aap, astA, sepA, pic, sigA, aggA* genes, commonly found in EAEC strains (for further details see Chapter 8) (Scheutz et al., 2011). The *aggA* gene encodes the major component of the aggregative adherence fimbriae (AAF/I), which is an adhesin associated with production of biofilms and hemagglutination of human erythrocytes (Harrington et al., 2006). In addition to the AAF/I fimbrial operon, *E. coli* O104:H4 also possesses the master virulence regulator gene *aggR*, typical of EAEC strains (Harrington et al., 2006). In general, limited evidence about the acquisition of AAF/I fimbrial operon by this serotype is available, because the closely related EAEC 55989 strain possesses instead the aggregative adherence fimbriae type III (AAF/III, encoded by the *agg3A* gene). It has been hypothesized that AAF-mediated enteroaggregative adherence allows *E. coli* O104:H4 colonization of the intestinal mucosa of infected patients (Scheutz et al., 2011). However, putatively inferred differences in the mechanism of adhesion in this isolate might explain why this strain is more likely to cause severe disease in adults rather than in children (i.e. they might differ in their susceptibility to the adherence and/or colonization properties of *E. coli* O104:H4).

Iron uptake systems

A wide variety of strategies for acquiring iron have been developed by EAEC and other pathogenic *E. coli*, the most common being the production of sidero-phores and the utilization of heme (Okeke et al., 2004). Interestingly, *E. coli* O104:H4 possesses the genes encoding for the siderophore aerobactin, an iron uptake system that is linked to pathogenesis in extraintestinal *E. coli* strains and commonly present in EAEC clinical isolates (Torres et al., 2001). A recent study in which a murine model mimicking enteropathogenicity of *E. coli* strains, in combination with bioluminescent imaging technology, was used to visualize the site of intestinal colonization as well as to define the role of the aerobactin transport system in the virulence of *E. coli* O104:H4 (Torres et al., 2012). The study found that the murine intestinal cecum was colonized by *E. coli* O104:H4, with bacteria persisting up to 7 days after intragastric inoculation. Further, the isogenic aerobactin receptor (*iutA*) mutant was out-competed by the wild-type *E. coli* O104:H4 during in vivo colonization, being unable to persist in the cecum and suggesting that this iron transport system is essential for intestinal fitness (Torres et al., 2012).

Antibiotic resistance

E. coli O104:H4 contains an array of antibiotic resistance genes conferring resistance to ampicillin, amoxicillin/clavulanic acid, piperacillin/sulbactam, piperacillin/tazobactam, cefuroxime, cefuroxime-axetil, cefoxitin, cefotaxime, cetfazidime, streptomycin, nalidixic acid, tetracycline, trimethoprim and sulfamethoxazol (Scheutz et al., 2011), exceeding the numerous resistance genes found in previous outbreak strains. The convergence of multiple antibiotic resistance genes not commonly found in EAEC strains suggests that this lethal bacterium might acquire the antibiotic markers by horizontal gene transfer of the multiresistant plasmid (Bezuidt et al., 2011). Despite the multiantibiotic resistance properties of *E. coli* O104:H4 and that antibiotic treatment of STEC infections is discouraged because it might increase HUS development, a number of patients in the German outbreak received prophylactic azithromycin treatment in combination with the C5 antibody eculizumab (see below) (Nitschke et al., 2012). Investigation of the patients receiving this treatment demonstrated that azithromycin administration was associated with a lower frequency of STEC O104:H4 shedding in several cases (Nitschke et al., 2012).

Clinical manifestations

Transmission

Shiga-toxin-producing *E. coli* serotype O104:H4 has been rarely associated with human cases in the past and has never been reported in animals (Auvray et al., 2012). Epidemiologic, microbiologic, and food trace-back and trace-forward investigations indicated that fenugreek sprouts were the vehicle of infection in this large outbreak of the HUS associated with *E. coli* O104:H4 (Buchholz et al., 2011). Because domestic ruminants, especially cattle, have been identified as natural reservoirs of Shiga-toxin-producing *E. coli*, a recent study investigated whether cattle could be a reservoir of *E. coli* O104:H4 and a potential source of transmission to humans (Auvray et al., 2012). A PCR approach targeting $stx2$, wzx_{O104}, $fliC_{H4}$ and $aggR$ genetic markers was used to evaluate cattle fecal carriage of the outbreak strain; however, the analysis was unable to detect this pathogen, suggesting that cattle is not a reservoir of the recently emerged *E. coli* O104:H4 pathotype (Auvray et al., 2012).

Clinical features

The clinical manifestations during the 2011 European outbreak were divided into a three-phase disease (German EHEC-HUS Registry, 2011). Most of the patients suffered from bloody diarrhea and a smaller portion presented with watery diarrhea. Some patients did not have stool abnormalities but presented with clinical or laboratory signs of thrombotic microangiopathy (TMA). Remarkably, the signs of TMA evolved over a period of 3–5 days in about a third of patients. Meanwhile, the majority of patients completely recovered from the infection. After an additional 3–10 days, almost 50% of the HUS patients needed dialysis and neurological symptoms developed in a majority of patients with TMA. Intriguingly, the neurological symptoms ranged from mild disorientation, qualitative and quantitative alterations of consciousness, double vision, dysphasia, hyperreflexia and apraxia to loss of reflexes or repeated epileptic seizures requiring intubation and mechanical ventilation (German EHEC-HUS Registry, 2011). Such neurological involvement had only been described in a less severe form of diarrhea-associated HUS and in very few patients (Nathanson et al., 2010).

Complications

The O104:H4 outbreak was characterized by a higher HUS incidence (25%) compared with other STEC outbreaks (Borgatta et al., 2012). While HUS is usually seen in children under 5 years of age suffering, i.e. of STEC O157:H7 infection (for further details, see Chapter 5), the *E. coli* O104:H4 outbreak affected mostly individuals above 20 years of age (87%). Also, there was a predominance of women in both *E. coli* O104:H4 infections (59%) and HUS (68%). As indicated above, about one-third of patients developed severe neurological complications and needed management in intensive care units. Patients with these neurological complications did not respond to therapeutic plasma exchange or eculizumab (see below) and because the complications arose 1 week after the onset of diarrhea, the use of an IgG immunoadsorption therapy was tested in a small number of patients (Greinacher et al., 2011). Although the study was not controlled, this rescue therapy appeared to show that IgG immunodepletion can improve neurological complications in patients with *E. coli* O104:H4-associated HUS and suggested that antibodies might be involved in the pathogenesis of severe neurological symptoms in these patients (Greinacher et al., 2011). The avoidance of ventilation or reduction of time on the ventilator in these patients might help to prevent respirator-associated complications.

Diagnosis

Because *E. coli* O104:H4 are considered emerging strains, a combination of molecular detection, culture and isolation of the pathogen was implemented to improve both outbreak detection and control and patient management. With regard to the microbiological detection and isolation, clinical samples should be streaked on extended-spectrum beta-lactamase (ESBL) plates for growth of the outbreak strain and inhibition of the majority of other *E. coli* strains (Scheutz et al., 2011). Also, growth is observed as light red colonies on cefiximetellurite sorbitol MacConkey (CT-SMAC) plates at 37°C. For rapid screening of clinical samples, slide agglutination with K9 antiserum (O104 O antigen is identical to the K9 capsular antigen) can also be used. Immediate positive reactions indicating the presence of *E. coli* O104:H4 need to be confirmed by conventional serotyping of O and H antigen, presence of the *stx2* gene and lack of the *eae* gene (to rule out other STEC strains) (Scheutz et al., 2011).

The *E. coli* O104:H4 strain can also be detected by a number of methods targeting the *stx2* gene such as PCR, RT-PCR, or commercial Stx detection kits. For example, PCR analysis indicated that *E. coli* O104:H4 lacked the LEE pathogenicity island, but are positive for a number of virulence genes typical of EAEC, including *pic*, *aggA*, *aggR*, *aap*, *set1*, and the virulence plasmid pAA (Bielaszewska et al., 2011). Colonies positive for *stx* genes were identified as O104:H4 through further testing for the O104 antigen-associated gene wzx_{O104} and the gene encoding the H4 flagellar antigen, $fliC_{H4}$ (Scheutz et al., 2011; Auvray et al., 2012).

In the case of food samples, standard methods (i.e. enrichment in buffered peptone water with incubation for 18–24 h at 37°C prior to DNA extraction) have been reported to detect and recover *E. coli* O104:H4 at a level of approximately 1 CFU/g of sprouts (Jinneman et al., 2012). The use of additional serotype-specific real-time PCR assays (the incorporation of the O104 *wzx* gene to the standard PCR method) and supplemental chromogenic media selection (i.e. Hardy CHROM ESBL agar or Rainbow O157 without tellurite or novobiocin) assisted for the detection and recovery of *E. coli* O104:H4.

In addition to the PCR methods, a pan-genomic analysis approach for the identification and validation of specific molecular targets for detection of *E. coli* O104:H4 outbreak strain has been recently reported (Ho et al., 2011). In this approach, unannotated genomic contigs of nine outbreak *E. coli* O104:H4 isolates were downloaded and used as target genomic data for in silico subtraction and thereby, to identify sequences specific to *E. coli* O104:H4. Primers specific to the selected targets were designed and successfully amplified the representative sequences from the outbreak strains with no false-positive results (Ho et al., 2011).

Treatment

Like other STEC strains, the use of antibiotics is not recommended because there is no evidence that antibiotics improve the course of disease, and it is thought that treatment with some antibiotics may precipitate kidney complications because of increased toxin expression. Because no causal therapy for HUS exists, and pooled analysis of previous data showed no benefit of antibiotics, alternative treatments have been utilized or developed. In this outbreak, novel strategies for treating patients with HUS were applied (i.e. antibody treatment with eculizumab, a monoclonal antibody inhibiting the terminal complement cascade, was associated with resolution of the severe neurological symptoms of three children with HUS) (Lapeyraque et al., 2011). Despite the fact that potential benefit in STEC-associated HUS has not been proven, therapeutic plasma exchange was used extensively on the basis of uncontrolled reports (Dundas et al., 1999).

Another approach for a prospective treatment was developed thanks to the public availability of the genome sequence of *E. coli* O104:H4, allowing the identification of the bacteriophage tail spike protein responsible for specific O-antigen binding encoded in the genome. The gene was fused to the tail fiber gene of an R-type pyocin, a phage tail-like bacteriocin. Upon binding to a receptor on the target bacterium, R-type pyocin are able to form a channel across the bacterial envelope and produce cell death. The novel bacteriocin containing the fusion to the phage tail fiber fusion from O104:H4 has bactericidal activity specifically against *E. coli* strains that produce the O104 lipopolysaccharide antigen (Scholl et al., 2012).

Immune response

No information is currently available regarding the protective host mechanisms, including immune response, and the role of commensal microbiota during *E. coli* O104:H4 infection. However, while searching for a therapeutic alternative treatment (especially for the neurological signs), the characteristic timespan between onset of gastroenteritis and the neurological complications (5–12 days) was a key clinical finding, because it indicated that additional auto-antibodies might be involved in the pathogenesis of the severe neurological complications (Lapeyraque et al., 2011). The intravascular reduction of these auto-antibodies by immunoabsorption should be considered one possible reason for the success of this therapy with regard to the neurological complications of HUS patients (Lapeyraque et al., 2011).

Control and prevention

Recommendations from the World Health Organization during the outbreak indicated that regular hand-washing, particularly before food preparation or consumption and after toilet contact, is highly recommended. These recommendations are particularly important for people who care for small children or are immunocompromised, as the bacterium could be passed from person to person, as well as through food and water.

CONCLUSIONS

The characterization of pathogenic *E. coli* with genomic and virulence properties that seems to be a mix of classic pathotypes, reflects the high genome plasticity of *E. coli* genomes. Events of horizontal transfer and divergent evolution are constantly occurring among bacteria that at some point, share a niche, and result in the development of emergent pathogenic types that acquire the capacity to colonize different niches in the host or to produce a variety of disease traits. In some cases, causality of disease by these strains remains to be undoubtedly proven, but in general, the characterization of their interactions with human cultured cell or animal models, such as the development of inflammatory responses or the requirement for a specific host genetic predisposition leads to the proposal of novel therapeutic approaches. In the same fashion, the studies about epidemic behavior and transmission routes, facilitated by culture-independent techniques of detection, have allowed public health officials to propose prevention and control measures that may impede further epidemic or endemic morbidity.

REFERENCES

Archambaud, M., Courcoux, P., Ouin, V., Chabanon, G., Labigne-Roussel, A., 1988. Phenotypic and genotypic assays for the detection and identification of adhesins from pyelonephritic *Escherichia coli*. Ann. Inst. Pasteur. Microbiol. 139, 557–573.

Arikawa, K., Meraz, I.M., Nishikawa, Y., Ogasawara, J., Hase, A., 2005. Interleukin-8 secretion by epithelial cells infected with diffusely adherent *Escherichia coli* possessing Afa adhesin-coding genes. Microbiol. Immunol. 49, 493–503.

Auvray, F., Dilasser, F., Bibbal, D., Kérourédan, M., Oswald, E., Brugère, H., 2012. French cattle is not a reservoir of the highly virulent enteroaggregative Shiga toxin-producing *Escherichia coli* of serotype O104:H4. Vet. Microbiol. Feb 28. [Epub ahead of print].

Bae, W.K., Lee, Y.K., Cho, M.S., et al., 2006. A case of hemolytic uremic syndrome caused by *Escherichia coli* O104:H4. Yonsei. Med. J. 47, 437–439.

Barnich, N., Boudeau, J., Claret, L., Darfeuille-Michaud, A., 2003. Regulatory and functional co-operation of flagella and type 1 pili in adhesive and invasive abilities of AIEC strain LF82 isolated from a patient with Crohn's disease. Mol. Microbiol. 48, 781–794.

Barnich, N., Bringer, M.A., Claret, L., Darfeuille-Michaud, A., 2004. Involvement of lipoprotein NlpI in the virulence of adherent invasive *Escherichia coli* strain LF82 isolated from a patient with Crohn's disease. Infect. Immun. 72, 2484–2493.

Barnich, N., Darfeuille-Michaud, A., 2007. Role of bacteria in the etiopathogenesis of inflammatory bowel disease. World J. Gastroenterol. 13, 5571–5576.

Baron, J.H., 2000. Inflammatory bowel disease up to 1932. Mt. Sinai J. Med. 67, 174–189.

Baumgart, M., Dogan, B., Rishniw, M., et al., 2007. Culture independent analysis of ileal mucosa reveals a selective increase in invasive *Escherichia coli* of novel phylogeny relative to depletion of Clostridiales in Crohn's disease involving the ileum. Isme. J. 1, 403–418.

Benz, I., Schmidt, M.A., 1989. Cloning and expression of an adhesin (AIDA-I) involved in diffuse adherence of enteropathogenic *Escherichia coli*. Infect. Immun. 57, 1506–1511.

Berger, C.N., Billker, O., Meyer, T.F., Servin, A.L., Kansau, I., 2004. Differential recognition of members of the carcinoembryonic antigen family by Afa/Dr adhesins of diffusely adhering *Escherichia coli* (Afa/Dr DAEC). Mol. Microbiol. 52, 963–983.

Bernet-Camard, M.F., Coconnier, M.H., Hudault, S., Servin, A.L., 1996. Pathogenicity of the diffusely adhering strain *Escherichia coli* C1845: F1845 adhesin-decay accelerating factor interaction, brush border microvillus injury, and actin disassembly in cultured human intestinal epithelial cells. Infect. Immun. 64, 1918–1928.

Bernier, C., Gounon, P., Le Bouguenec, C., 2002. Identification of an aggregative adhesion fimbria (AAF) type III-encoding operon in enteroaggregative *Escherichia coli* as a sensitive probe for detecting the AAF-encoding operon family. Infect. Immun. 70, 4302–4311.

Betis, F., Brest, P., Hofman, V., et al., 2003a. The Afa/Dr adhesins of diffusely adhering *Escherichia coli* stimulate interleukin-8 secretion, activate mitogen-activated protein kinases, and promote polymorphonuclear transepithelial migration in T84 polarized epithelial cells. Infect. Immun. 71, 1068–1074.

Betis, F., Brest, P., Hofman, V., et al., 2003b. Afa/Dr diffusely adhering *Escherichia coli* infection in T84 cell monolayers induces increased neutrophil transepithelial migration, which in turn promotes cytokine-dependent upregulation of decay-accelerating factor (CD55), the receptor for Afa/Dr adhesins. Infect. Immun. 71, 1774–1783.

Bezuidt, O., Pierneef, R., Mncube, K., Lima-Mendez, G., Reva, O.N., 2011. Mainstreams of horizontal gene exchange in enterobacteria: consideration of the outbreak of enterohemorrhagic *E. coli* O104:H4 in Germany in 2011. PLoS One 6, e25702.

Bielaszewska, M., Mellmann, A., Zhang, W., et al., 2011. Characterisation of the *Escherichia coli* strain associated with an outbreak of haemolytic uraemic syndrome in Germany, 2011: a microbiological study. Lancet Infect. Dis. 11, 671–676.

Bilge, S.S., Clausen, C.R., Lau, W., Moseley, S.L., 1989. Molecular characterization of a fimbrial adhesin, F1845, mediating diffuse adherence of diarrhea-associated *Escherichia coli* to HEp-2 cells. J. Bacteriol. 171, 4281–4289.

Borgatta, B., Kmet-Lunacek, N., Rello, J., 2012. *E. coli* O104:H4 outbreak and haemolytic-uraemic syndrome. Med. Intensiva. [Epub ahead of print].

Boudeau, J., Barnich, N., Darfeuille-Michaud, A., 2001. Type 1 pili-mediated adherence of *Escherichia coli* strain LF82 isolated from Crohn's disease is involved in bacterial invasion of intestinal epithelial cells. Mol. Microbiol. 39, 1272–1284.

Bringer, M.A., Barnich, N., Glasser, A.L., Bardot, O., Darfeuille-Michaud, A., 2005. HtrA stress protein is involved in intramacrophagic replication of adherent and invasive *Escherichia coli* strain LF82 isolated from a patient with Crohn's disease. Infect. Immun. 73, 712–721.

Bringer, M.A., Glasser, A.L., Tung, C.H., Meresse, S., Darfeuille-Michaud, A., 2006. The Crohn's disease-associated adherent-invasive *Escherichia coli* strain LF82 replicates in mature phagolysosomes within J774 macrophages. Cell Microbiol. 8, 471–484.

Bringer, M.A., Rolhion, N., Glasser, A.L., Darfeuille-Michaud, A., 2007. The oxidoreductase DsbA plays a key role in the ability of the Crohn's disease-associated adherent-invasive *Escherichia coli* strain LF82 to resist macrophage killing. J. Bacteriol. 189, 4860–4871.

Brown, W.R., Lee, E.M., 1973. Radioimmunologic measurements of naturally occurring bacterial antibodies. I. Human serum antibodies reactive with *Escherichia coli* in gastrointestinal and immunologic disorders. J. Lab. Clin. Med. 82, 125–136.

Brzuszkiewicz, E., Thürmer, A., Schuldes, J., et al., 2011. Genome sequence analyses of two isolates from the recent *Escherichia coli* outbreak in Germany reveal the emergence of a new pathotype: Entero-Aggregative-Haemorrhagic *Escherichia coli* (EAHEC). Arch. Microbiol. 193, 883–891.

Buchholz, U., Bernard, H., Werber, D., et al., 2011. German outbreak of *Escherichia coli* O104:H4 associated with sprouts. N. Engl. J. Med. 365, 1763–1770.

Cartun, R.W., Van Kruiningen, H.J., Pedersen, C.A., Berman, M.M., 1993. An immunocytochemical search for infectious agents in Crohn's disease. Mod. Pathol. 6, 212–219.

Carvalho, F.A., Barnich, N., Sauvanet, P., Darcha, C., Gelot, A., Darfeuille-Michaud, A., 2008. Crohn's disease-associated *Escherichia coli* LF82 aggravates colitis in injured mouse colon via signaling by flagellin. Inflamm. Bowel Dis. 14, 1051–1060.

Carvalho, F.A., Barnich, N., Sivignon, A., et al., 2009. Crohn's disease adherent-invasive *Escherichia coli* colonize and induce strong gut inflammation in transgenic mice expressing human CEACAM. J. Exp. Med. 206, 2179–2189.

Cave, D.R., Kane, S.P., Mitchell, D.N., Brooke, B.N., 1973. Further animal evidence of a transmissible agent in Crohn's disease. Lancet 308, 1120–1122.

Chassaing, B., Rolhion, N., de Vallee, A., et al., 2011. Crohn disease–associated adherent-invasive *E. coli* bacteria target mouse and human Peyer's patches via long polar fimbriae. J. Clin. Invest. 121, 966–975.

Claret, L., Miquel, S., Vieille, N., Ryjenkov, D.A., Gomelsky, M., Darfeuille-Michaud, A., 2007. The flagellar sigma factor FliA regulates adhesion and invasion of Crohn disease-associated *Escherichia coli* via a cyclic dimeric GMP-dependent pathway. J. Biol. Chem. 282, 33275–33283.

Clarke, D.J., Chaudhuri, R.R., Martin, H.M., et al., 2011. Complete genome sequence of the Crohn's disease-associated adherent-invasive *Escherichia coli* strain HM605. J. Bacteriol. 193, 4540.

Cookson, S.T., Nataro, J.P., 1996. Characterization of HEp-2 cell projection formation induced by diffusely adherent *Escherichia coli*. Microb. Pathog. 21, 421–434.

Cravioto, A., Tello, A., Navarro, A., et al., 1991. Association of *Escherichia coli* HEp-2 adherence patterns with type and duration of diarrhoea. Lancet 337, 262–264.

Czeczulin, J.R., Whittam, T.S., Henderson, I.R., Navarro-Garcia, F., Nataro, J.P., 1999. Phylogenetic analysis of enteroaggregative and diffusely adherent *Escherichia coli*. Infect. Immun. 67, 2692–2699.

Daigle, F., Harel, J., Fairbrother, J.M., Lebel, P., 1994. Expression and detection of pap-, sfa-, and afa-encoded fimbrial adhesin systems among uropathogenic *Escherichia coli*. Can. J. Microbiol. 40, 286–291.

Darfeuille-Michaud, A., 2002. Adherent-invasive *Escherichia coli*: a putative new *E. coli* pathotype associated with Crohn's disease. Int. J. Med. Microbiol. 292, 185–193.

Darfeuille-Michaud, A., Boudeau, J., Bulois, P., et al., 2004. High prevalence of adherent-invasive *Escherichia coli* associated with ileal mucosa in Crohn's disease. Gastroenterology 127, 412–421.

Darfeuille-Michaud, A., Neut, C., Barnich, N., et al., 1998. Presence of adherent strains in ileal mucosa of patients with Crohn's disease. Gastroenterology 115, 1405–1413.

Denamur, E., 2011. The 2011 Shiga toxin-producing *Escherichia coli* O104:H4 German outbreak: a lesson in genomic plasticity. Clin. Microbiol. Infect. 17, 1124–1125.

Denizot, J., Sivignon, A., Barreau, F., et al., 2012. Adherent-invasive *Escherichia coli* induce claudin-2 expression and barrier defect in CEABAC10 mice and Crohn's disease patients. Inflamm. Bowel Dis. 18, 294–304.

Donnelly, B.J., Delaney, P.V., Healy, T.M., 1977. Evidence for a transmissible factor in Crohn's disease. Gut 18, 360–363.

Dundas, S., Murphy, J., Soutar, R.L., Jones, G.A., Hutchinson, S.J., Todd, W.T., 1999. Effectiveness of therapeutic plasma exchange in the 1996 Lanarkshire *Escherichia coli* O157:H7 outbreak. Lancet Infect. Dis. 354, 1327–1330.

Eaves-Pyles, T., Allen, C.A., Taormina, J., et al., 2008. *Escherichia coli* isolated from a Crohn's disease patient adheres, invades, and induces inflammatory responses in polarized intestinal epithelial cells. Int. J. Med. Microbiol. 298, 397–409.

Escobar-Paramo, P., Clermont, O., Blanc-Potard, A.B., Bui, H., Le Bouguenec, C., Denamur, E., 2004. A specific genetic background is required for acquisition and expression of virulence factors in *Escherichia coli*. Mol. Biol. Evol. 21, 1085–1094.

Flanagan, P., Campbell, B.J., Rhodes, J.M., 2011. Bacteria in the pathogenesis of inflammatory bowel disease. Biochem. Soc. Trans. 39, 1067–1072.

Foxman, B., Zhang, L., Tallman, P., et al., 1995. Virulence characteristics of *Escherichia coli* causing first urinary tract infection predict risk of second infection. J. Infect. Dis. 172, 1536–1541.

Frank, C., Werber, D., Cramer, J.P., et al., 2011. Epidemic profile of Shiga-toxin-producing *Escherichia coli* O104:H4 outbreak in Germany. N. Engl. J. Med. 365, 1771–1780.

Fritz, J.H., Ferrero, R.L., Philpott, D.J., Girardin, S.E., 2006. Nod-like proteins in immunity, inflammation and disease. Nat. Immunol. 7, 1250–1257.

Gault, G., Weill, F.X., Mariani-Kurkdjian, P., et al., 2011. Outbreak of haemolytic uraemic syndrome and bloody diarrhoea due to *Escherichia coli* O104:H4, south-west France, June 2011. Euro. Surveill. 16, pii: 19905.

German EHEC-HUS Registry, 2011. The German 2011 epidemic of Shiga toxin-producing *E. coli*– the nephrological view. Nephrol. Dial. Trans. 26, 2723–2726.

Gionchetti, P., Rizzello, F., Lammers, K.M., et al., 2006. Antibiotics and probiotics in treatment of inflammatory bowel disease. World J. Gastroenterol. 12, 3306–3313.

Giron, J.A., Jones, T., Millan-Velasco, F., et al., 1991. Diffuse-adhering *Escherichia coli* (DAEC) as a putative cause of diarrhea in Mayan children in Mexico. J. Infect. Dis. 163, 507–513.

Glasser, A.L., Boudeau, J., Barnich, N., Perruchot, M.H., Colombel, J.F., Darfeuille-Michaud, A., 2001. Adherent invasive *Escherichia coli* strains from patients with Crohn's disease survive and replicate within macrophages without inducing host cell death. Infect. Immun. 69, 5529–5537.

Goluszko, P., Selvarangan, R., Popov, V., Pham, T., Wen, J.W., Singhal, J., 1999. Decay-accelerating factor and cytoskeleton redistribution pattern in HeLa cells infected with recombinant *Escherichia coli* strains expressing Dr family of adhesins. Infect. Immun. 67, 3989–3997.

Gomes, T.A., Blake, P.A., Trabulsi, L.R., 1989. Prevalence of *Escherichia coli* strains with localized, diffuse, and aggregative adherence to HeLa cells in infants with diarrhea and matched controls. J. Clin. Microbiol. 27, 266–269.

Gomez-Duarte, O.G., Arzuza, O., Urbina, D., et al., 2010. Detection of *Escherichia coli* enteropathogens by multiplex polymerase chain reaction from children's diarrheal stools in two Caribbean-Colombian cities. Foodborne Pathog. Dis. 7, 199–206.

Grad, Y.H., Lipsitch, M., Feldgarden, M., et al., 2012. Genomic epidemiology of the *Escherichia coli* O104:H4 outbreaks in Europe, 2011. Proc. Natl. Acad. Sci. USA 109, 3065–3070.

Greenfield, S.P., 2011. Antibiotic prophylaxis in pediatric urology: an update. Curr. Urol. Rep. 12, 126–131.

Greinacher, A., Friesecke, S., Abel, P., et al., 2011. Treatment of severe neurological deficits with IgG depletion through immunoadsorption in patients with *Escherichia coli* O104:H4-associated haemolytic uraemic syndrome: a prospective trial. Lancet 378, 1166–1173.

Guignot, J., Hudault, S., Kansau, I., Chau, I., Servin, A.L., 2009. Human decay-accelerating factor and CEACAM receptor-mediated internalization and intracellular lifestyle of Afa/Dr diffusely adhering *Escherichia coli* in epithelial cells. Infect. Immun. 77, 517–531.

Gunzburg, S.T., Chang, B.J., Elliott, S.J., Burke, V., Gracey, M., 1993. Diffuse and enteroaggregative patterns of adherence of enteric *Escherichia coli* isolated from aboriginal children from the Kimberley region of Western Australia. J. Infect. Dis. 167, 755–758.

Harrington, S.M., Dudley, E.G., Nataro, J.P., 2006. Pathogenesis of enteroaggregative *Escherichia coli* infection. FEMS Microbiol. Lett. 254, 12–18.

Helbig, K.L., Nothnagel, M., Hampe, J., et al., 2012. A case-only study of gene-environment interaction between genetic susceptibility variants in NOD2 and cigarette smoking in Crohn's disease aetiology. BMC Med. Genet. 13, 14.

Ho, C.C., Yuen, K.Y., Lau, S.K., Woo, P.C., 2011. Rapid identification and validation of specific molecular targets for detection of *Escherichia coli* O104:H4 outbreak strain by use of high-throughput sequencing data from nine genomes. J. Clin. Microbiol. 49, 3714–3716.

Huebner, C., Ding, Y., Petermann, I., Knapp, C., Ferguson, L.R., 2011. The probiotic *Escherichia coli* Nissle 1917 reduces pathogen invasion and modulates cytokine expression in Caco-2 cells infected with Crohn's disease-associated *E. coli* LF82. Appl. Environ. Microbiol. 77, 2541–2544.

Hugot, J.P., Chamaillard, M., Zouali, H., et al., 2001. Association of NOD2 leucine-rich repeat variants with susceptibility to Crohn's disease. Nature 411, 599–603.

Jallat, C., Darfeuille-Michaud, A., Rich, C., Joly, B., 1994. Survey of clinical isolates of diarrhoeogenic *Escherichia coli*: diffusely adhering *E. coli* strains with multiple adhesive factors. Res. Microbiol. 145, 621–632.

Jallat, C., Livrelli, V., Darfeuille-Michaud, A., Rich, C., Joly, B., 1993. *Escherichia coli* strains involved in diarrhea in France: high prevalence and heterogeneity of diffusely adhering strains. J. Clin. Microbiol. 31, 2031–2037.

Jensen, S.R., Fink, L.N., Nielsen, O.H., Brynskov, J., Brix, S., 2011. Ex vivo intestinal adhesion of *Escherichia coli* LF82 in Crohn's disease. Microb. Pathog. 51, 426–431.

Jinneman, K.C., Waite-Cusic, J.G., Yoshitomi, K.J., 2012. Evaluation of shiga toxin-producing *Escherichia coli* (STEC) method for the detection and identification of STEC O104 strains from sprouts. Food Microbiol. 30, 321–328.

Joossens, M., Huys, G., Cnockaert, M., et al., 2011. Dysbiosis of the faecal microbiota in patients with Crohn's disease and their unaffected relatives. Gut 60, 631–637.

Kansau, I., Berger, C., Hospital, M., et al., 2004. Zipper-like internalization of Dr-positive *Escherichia coli* by epithelial cells is preceded by an adhesin-induced mobilization of raft-associated molecules in the initial step of adhesion. Infect. Immun. 72, 3733–3742.

Kaser, A., Zeissig, S., Blumberg, R.S., 2010. Inflammatory bowel disease. Annu. Rev. Immunol. 28, 573–621.

Khan, K.J., Ullman, T.A., Ford, A.C., et al., 2011. Antibiotic therapy in inflammatory bowel disease: a systematic review and meta-analysis. Am. J. Gastroenterol. 106, 661–673.

Kim, J., Oh, K., Jeon, S., et al., 2011. *Escherichia coli* O104:H4 from 2011 European outbreak and strain from South Korea. Emerg. Infect. Dis. 17, 1755–1756.

Klemm, P., 1986. Two regulatory fim genes, *fimB* and *fimE*, control the phase variation of type 1 fimbriae in *Escherichia coli*. EMBO J. 5, 1389–1393.

Kotlowski, R., Bernstein, C.N., Sepehri, S., Krause, D.O., 2007. High prevalence of *Escherichia coli* belonging to the B2+D phylogenetic group in inflammatory bowel disease. Gut 56, 669–675.

Krause, D.O., Little, A.C., Dowd, S.E., Bernstein, C.N., 2011. Complete genome sequence of adherent invasive *Escherichia coli* UM146 isolated from Ileal Crohn's disease biopsy tissue. J. Bacteriol. 193, 583.

Kunne, C., Billion, A., Mshana, S.E., et al., 2012. Complete sequences of plasmids from the hemolytic-uremic syndrome-associated *Escherichia coli* strain HUSEC41. J. Bacteriol. 194, 532–533.

Lapaquette, P., Bringer, M.A., Darfeuille-Michaud, A., 2012. Defects in autophagy favour adherent-invasive *Escherichia coli* persistence within macrophages leading to increased proinflammatory response. Cell. Microbiol. 14, 791–807.

Lapaquette, P., Glasser, A.L., Huett, A., Xavier, R.J., Darfeuille-Michaud, A., 2010. Crohn's disease-associated adherent-invasive *E. coli* are selectively favoured by impaired autophagy to replicate intracellularly. Cell. Microbiol. 12, 99–113.

Lapeyraque, A.L., Malina, M., Fremeaux-Bacchi, V., et al., 2011. Eculizumab in severe Shiga-toxin-associated HUS. N. Engl. J. Med. 364, 2561–2563.

Le Bouguenec, C., Lalioui, L., du Merle, L., et al., 2001. Characterization of AfaE adhesins produced by extraintestinal and intestinal human *Escherichia coli* isolates: PCR assays for detection of Afa adhesins that do or do not recognize Dr blood group antigens. J. Clin. Microbiol. 39, 1738–1745.

Le Bouguenec, C., Servin, A.L., 2006. Diffusely adherent *Escherichia coli* strains expressing Afa/Dr adhesins (Afa/Dr DAEC): hitherto unrecognized pathogens. FEMS Microbiol. Lett. 256, 185–194.

Levine, M.M., Ferreccio, C., Prado, V., et al., 1993. Epidemiologic studies of *Escherichia coli* diarrheal infections in a low socioeconomic level peri-urban community in Santiago, Chile. Am. J. Epidemiol. 138, 849–869.

Liu, Y., van Kruiningen, H.J., West, A.B., Cartun, R.W., Cortot, A., Colombel, J.F., 1995. Immunocytochemical evidence of *Listeria, Escherichia coli*, and *Streptococcus* antigens in Crohn's disease. Gastroenterology 108, 1396–1404.

Loftus Jr., E.V., 2004. Clinical epidemiology of inflammatory bowel disease: incidence, prevalence, and environmental influences. Gastroenterology 126, 1504–1517.

Lopes, L.M., Fabbricotti, S.H., Ferreira, A.J., Kato, M.A., Michalski, J., Scaletsky, I.C., 2005. Heterogeneity among strains of diffusely adherent *Escherichia coli* isolated in Brazil. J. Clin. Microbiol. 43, 1968–1972.

Malchow, H.A., 1997. Crohn's disease and *Escherichia coli*. A new approach in therapy to maintain remission of colonic Crohn's disease? J. Clin. Gastroenterol. 25, 653–658.

Mariani-Kurkdjian, P., Bingen, E., Gault, G., Jourdan-Da Silva, N., Weill, F.X., 2011. *Escherichia coli* O104:H4 south-west France. Lancet Infect. Dis. 11, 732–733.

Martinez-Medina, M., Aldeguer, X., Lopez-Siles, M., et al., 2009. Molecular diversity of *Escherichia coli* in the human gut: new ecological evidence supporting the role of adherent-invasive *E. coli* (AIEC) in Crohn's disease. Inflamm. Bowel. Dis. 15, 872–882.

Martinez-Medina, M., Garcia-Gil, J., Barnich, N., Wieler, L.H., Ewers, C., 2011. Adherent-invasive *Escherichia coli* phenotype displayed by intestinal pathogenic *E. coli* strains from cats, dogs, and swine. Appl. Environ. Microbiol. 77, 5813–5817.

Masseret, E., Boudeau, J., Colombel, J.F., et al., 2001. Genetically related *Escherichia coli* strains associated with Crohn's disease. Gut 48, 320–325.

Massey, D.C., Parkes, M., 2007. Genome-wide association scanning highlights two autophagy genes, ATG16L1 and IRGM, as being significantly associated with Crohn's disease. Autophagy 3, 649–651.

Mathewson, J.J., Cravioto, A., 1989. HEp-2 cell adherence as an assay for virulence among diarrheagenic *Escherichia coli*. J. Infect. Dis. 159, 1057–1060.

Maurer, J., Jose, J., Meyer, T.F., 1997. Autodisplay: one-component system for efficient surface display and release of soluble recombinant proteins from *Escherichia coli*. J. Bacteriol. 179, 794–804.

Meconi, S., Vercellone, A., Levillain, F., et al., 2007. Adherent-invasive *Escherichia coli* isolated from Crohn's disease patients induce granulomas in vitro. Cell. Microbiol. 9, 1252–1261.

Mellmann, A., Bielaszewska, M., Kock, R., et al., 2008. Analysis of collection of hemolytic uremic syndrome-associated enterohemorrhagic *Escherichia coli*. Emerg. Infect. Dis. 14, 1287–1290.

Mellmann, A., Harmsen, D., Cummings, C.A., et al., 2011. Prospective genomic characterization of the German enterohemorrhagic *Escherichia coli* O104:H4 outbreak by rapid next generation sequencing technology. PLoS One 6, e22751.

Miquel, S., Claret, L., Bonnet, R., Dorboz, I., Barnich, N., Darfeuille-Michaud, A., 2010a. Role of decreased levels of Fis histone-like protein in Crohn's disease-associated adherent invasive *Escherichia coli* LF82 bacteria interacting with intestinal epithelial cells. J. Bacteriol. 192, 1832–1843.

Miquel, S., Peyretaillade, E., Claret, L., et al., 2010b. Complete genome sequence of Crohn's disease-associated adherent-invasive *E. coli* strain LF82. PLoS One 5.

Mitchell, D.N., Rees, R.J., 1970. Agent transmissible from Crohn's disease tissue. Lancet 2, 168–171.

Molodecky, N.A., Soon, I.S., Rabi, D.M., et al., 2012. Increasing incidence and prevalence of the inflammatory bowel diseases with time, based on systematic review. Gastroenterology 142, 46–54.

Morabito, S., Karch, H., Mariani-Kurkdjian, P., et al., 1998. Enteroaggregative, Shiga toxin-producing *Escherichia coli* O111:H2 associated with an outbreak of hemolytic uremic syndrome. J. Clin. Microbiol. 36, 840–842.

Mossoro, C., Glaziou, P., Yassibanda, S., et al., 2002. Chronic diarrhea, hemorrhagic colitis, and hemolytic-uremic syndrome associated with HEp-2 adherent *Escherichia coli* in adults infected with human immunodeficiency virus in Bangui, Central African Republic. J. Clin. Microbiol. 40, 3086–3088.

Naser, S.A., Arce, M., Khaja, A., et al., 2012. Role of ATG16L, NOD2 and IL23R in Crohn's disease pathogenesis. World J. Gastroenterol. 18, 412–424.

Nash, J.H., Villegas, A., Kropinski, A.M., et al., 2010. Genome sequence of adherent-invasive *Escherichia coli* and comparative genomic analysis with other *E. coli* pathotypes. BMC Genomics 11, 667.

Nathanson, S., Kwon, T., Elmaleh, M., et al., 2010. Acute neurological involvement in diarrhea-associated hemolytic uremic syndrome. Clin. J. Am. Soc. Nephrol. 5, 1218–1228.

Nitschke, M., Sayk, F., Härtel, C., et al., 2012. Association between azithromycin therapy and duration of bacterial shedding among patients with Shiga toxin-producing enteroaggregative *Escherichia coli* O104:H4. JAMA 307, 1046–1052.

Nowicki, B., Selvarangan, R., Nowicki, S., 2001. Family of *Escherichia coli* Dr adhesins: decay-accelerating factor receptor recognition and invasiveness. J. Infect. Dis. 183 (Suppl. 1), S24–27.

Ochoa, T.J., Ruiz, J., Molina, M., et al., 2009. High frequency of antimicrobial drug resistance of diarrheagenic *Escherichia coli* in infants in Peru. Am. J. Trop. Med. Hyg. 81, 296–301.

Ogura, Y., Bonen, D.K., Inohara, N., et al., 2001. A frameshift mutation in NOD2 associated with susceptibility to Crohn's disease. Nature 411, 603–606.

Okeke, I.N., Scaletsky, I.C., Soars, E.H., Macfarlane, L.R., Torres, A.G., 2004. Molecular epidemiology of the iron utilization genes of enteroaggregative *Escherichia coli*. J. Clin. Microbiol. 42, 36–44.

Peeters, H., Bogaert, S., Laukens, D., et al., 2007. CARD15 variants determine a disturbed early response of monocytes to adherent-invasive *Escherichia coli* strain LF82 in Crohn's disease. Int. J. Immunogenet. 34, 181–191.

Peiffer, I., Blanc-Potard, A.B., Bernet-Camard, M.F., Guignot, J., Barbat, A., Servin, A.L., 2000. Afa/Dr diffusely adhering *Escherichia coli* C1845 infection promotes selective injuries in the junctional domain of polarized human intestinal Caco-2/TC7 cells. Infect. Immun. 68, 3431–3442.

Plancon, L., Du Merle, L., Le Friec, S., et al., 2003. Recognition of the cellular beta1-chain integrin by the bacterial AfaD invasin is implicated in the internalization of afa-expressing pathogenic *Escherichia coli* strains. Cell Microbiol. 5, 681–693.

Rajendran, P., Ajjampur, S.S., Chidambaram, D., et al., 2010. Pathotypes of diarrheagenic *Escherichia coli* in children attending a tertiary care hospital in South India. Diagn. Microbiol. Infect. Dis. 68, 117–122.

Rasko, D.A., Webster, D.R., Sahl, J.W., et al., 2011. Origins of the *E. coli* strain causing an outbreak of hemolytic-uremic syndrome in Germany. N. Engl. J. Med. 365, 709–717.

Rioux, J.D., Xavier, R.J., Taylor, K.D., et al., 2007. Genome-wide association study identifies new susceptibility loci for Crohn disease and implicates autophagy in disease pathogenesis. Nat. Genet. 39, 596–604.

Rohde, H., Qin, J., Cui, Y., et al., 2011. Open-source genomic analysis of Shiga-toxin-producing *E. coli* O104:H4. N. Engl. J. Med. 365, 718–724.

Rolhion, N., Barnich, N., Bringer, M.A., et al., 2010. Abnormally expressed ER stress response chaperone Gp96 in CD favours adherent-invasive *Escherichia coli* invasion. Gut 59, 1355–1362.

Rolhion, N., Barnich, N., Claret, L., Darfeuille-Michaud, A., 2005. Strong decrease in invasive ability and outer membrane vesicle release in Crohn's disease-associated adherent-invasive *Escherichia coli* strain LF82 with the yfgL gene deleted. J. Bacteriol. 187, 2286–2296.

Rolhion, N., Carvalho, F.A., Darfeuille-Michaud, A., 2007. OmpC and the sigma(E) regulatory pathway are involved in adhesion and invasion of the Crohn's disease-associated *Escherichia coli* strain LF82. Mol. Microbiol. 63, 1684–1700.

Rolhion, N., Darfeuille-Michaud, A., 2007. Adherent-invasive *Escherichia coli* in inflammatory bowel disease. Inflamm. Bowel. Dis. 13, 1277–1283.

Rutgeerts, P., Goboes, K., Peeters, M., et al., 1991. Effect of faecal stream diversion on recurrence of Crohn's disease in the neoterminal ileum. Lancet 338, 771–774.

Rutgeerts, P., Hiele, M., Geboes, K., et al., 1995. Controlled trial of metronidazole treatment for prevention of Crohn's recurrence after ileal resection. Gastroenterology 108, 1617–1621.

Sartor, R.B., 2005. Does *Mycobacterium avium* subspecies *paratuberculosis* cause Crohn's disease? Gut 54, 896–898.

Scaletsky, I.C., Silva, M.L., Trabulsi, L.R., 1984. Distinctive patterns of adherence of enteropathogenic *Escherichia coli* to HeLa cells. Infect. Immun. 45, 534–536.

Scavia, G., Morabito, S., Tozzoli, R., et al., 2011. Similarity of Shiga toxin-producing *Escherichia coli* O104:H4 strains from Italy and Germany. Emerg. Infect. Dis. 17, 1957–1958.

Scheutz, F., Nielsen, E.M., Frimodt-Møller, J., et al., 2011. Characteristics of the enteroaggregative Shiga toxin/verotoxin-producing *Escherichia coli* O104:H4 strain causing the outbreak of haemolytic uraemic syndrome in Germany, May to June 2011. Euro. Surveill. 16, pii: 19889.

Scholl, D., Gebhart, D., Williams, S.R., Bates, A., Mandrell, R., 2012. Genome sequence of *E. coli* O104:H4 leads to rapid development of a targeted antimicrobial agent against this emerging pathogen. PLoS One 7, e33637.

Sepehri, S., Kotlowski, R., Bernstein, C.N., Krause, D.O., 2009. Phylogenetic analysis of inflammatory bowel disease associated *Escherichia coli* and the *fimH* virulence determinant. Inflamm. Bowel. Dis. 15, 1737–1745.

Servin, A.L., 2005. Pathogenesis of Afa/Dr diffusely adhering *Escherichia coli*. Clin. Microbiol. Rev. 18, 264–292.

Simonsen, K.T., Nielsen, G., Bjerrum, J.V., Kruse, T., Kallipolitis, B.H., Moller-Jensen, J., 2011. A role for the RNA chaperone Hfq in controlling adherent-invasive *Escherichia coli* colonization and virulence. PLoS One 6, e16387.

Simpson, K.W., Dogan, B., Rishniw, M., et al., 2006. Adherent and invasive *Escherichia coli* is associated with granulomatous colitis in boxer dogs. Infect. Immun. 74, 4778–4792.

Sokol, H., Pigneur, B., Watterlot, L., et al., 2008. *Faecalibacterium prausnitzii* is an anti-inflammatory commensal bacterium identified by gut microbiota analysis of Crohn disease patients. Proc. Natl. Acad. Sci. USA 105, 16731–16736.

Strober, W., 2011. Adherent-invasive *E. coli* in Crohn disease: bacterial agent provocateur. J. Clin. Invest. 121, 841–844.

Tabaqchali, S., Odonoghue, D.P., Bettelheim, K.A., 1978. *Escherichia coli* antibodies in patients with inflammatory bowel disease. Gut 19, 108–113.

Tacket, C.O., Moseley, S.L., Kay, B., Losonsky, G., Levine, M.M., 1990. Challenge studies in volunteers using *Escherichia coli* strains with diffuse adherence to HEp-2 cells. J. Infect. Dis. 162, 550–552.

Tieng, V., Le Bouguenec, C., du Merle, L., et al., 2002. Binding of *Escherichia coli* adhesin AfaE to CD55 triggers cell-surface expression of the MHC class I-related molecule MICA. Proc. Natl. Acad. Sci. USA 99, 2977–2982.

Torres, A.G., Cieza, R.J., Rojas-Lopez, M., et al., 2012. In vivo bioluminescence imaging of *E. coli* O104:H4 and characterization of virulence properties in a mouse model of infection. BMC Microbiol. 12, 112.

Torres, A.G., Redford, P., Welch, R.A., Payne, S.M., 2001. TonB-dependent systems of uropathogenic *Escherichia coli*: aerobactin and heme transport and TonB are required for virulence in the mouse. Infect. Immun. 69, 6179–6185.

Vidal, M., Kruger, E., Duran, C., et al., 2005. Single multiplex PCR assay to identify simultaneously the six categories of diarrheagenic *Escherichia coli* associated with enteric infections. J. Clin. Microbiol. 43, 5362–5365.

Yu, A.P., Cabanilla, L.A., Wu, E.Q., Mulani, P.M., Chao, J., 2008. The costs of Crohn's disease in the United States and other Western countries: a systematic review. Curr. Med. Res. Opin. 24, 319–328.

Escherichia coli virulence factors

Adhesive pili of the chaperone-usher family

Vasilios Kalas, Ender Volkan, Scott J. Hultgren

Washington University School of Medicine, St. Louis, MO, USA

INTRODUCTION

Bacterial surface appendages are commonly divided into two categories: flagellar and non-flagellar. Flagella are long, propeller-like structures that provide motility to bacteria, distinct from non-flagellar structures known as pili or fimbriae, which are thinner, hair-like structures involved in adherence, biofilm formation, and in the case of type IV pili, twitching motility (see Chapter 13). The first functional observation of adhesive pili may have been in 1908, when Guyot recorded the ability of certain strains of bacteria to hemagglutinate red blood cells (Guyot, 1908). Today, hemagglutination is a common assay for detecting and measuring various types of adhesive pili (Korhonen et al., 1984; Hultgren et al., 1986; Krasan et al., 2000). The first images of non-flagellar surface structures were made possible in the late 1940s–early 1950s by the invention of the electron microscope (Houwink and van Iterson, 1950; Ottow, 1975). Duguid was an early pioneer in characterizing molecular features of pili and used the word 'fimbriae' in 1955, meaning threads or fibers in Latin, to describe the surface structures of *Escherichia coli* involved in erythrocyte agglutination (Duguid et al., 1955). He went on to serologically distinguish these fibers using different agglutination assays with erythrocytes from various species as well as with yeast cells (Gillies and Duguid, 1958; Duguid et al., 1966). The term 'pili,' Latin for hair, was later used by Brinton in 1959 to describe non-flagellar *E. coli* surface appendages (Brinton, 1959). Soon thereafter, he reported the first X-ray diffraction patterns of pili (Brinton, 1965). This consequently led Salit and Gotschlich to establish pilus purification procedures, perform early biochemical and functional characterizations, and describe the unique phenomenon of streaming birefringence exhibited by pili (Salit and Gotschlich, 1977). In the 1980s, work by Stanley Falkow and Staffan Normark unraveled a genetic understanding of pili (Hull et al., 1981; Normark et al., 1983), and in subsequent

Escherichia coli. http://dx.doi.org/10.1016/B978-0-12-397048-0.00012-7

efforts, groups led by Stanley Falkow and Gordon Dougan were the first to clone pili gene clusters (Hull et al., 1981; Morrissey and Dougan, 1986). Meanwhile, afimbrial adhesins of *E. coli* were also discovered and cloned (Labigne-Roussel et al., 1985; Walz et al., 1985). These pioneering studies led to work focused on dissecting the biogenesis and structure of P pili of uropathogenic *E. coli* (UPEC), which provided a model and blueprint for the next three decades of work on these systems worldwide. The P pilus was discovered to be a multicomponent structure consisting of a stalk and an adhesive tip (Lindberg et al., 1987; Kuehn et al., 1992). At around the same time, PapD became the first pilus chaperone described in detail, a central factor in a molecular machine assembled by a process we would come to name the chaperone-usher (CU) pathway (Hultgren et al., 1989; Lindberg et al., 1989). The crystal structure of PapD described in 1989 was a landmark study that provided the first structural insights into CU pilus biogenesis (Holmgren et al., 1988; Holmgren and Branden, 1989). The surprising immunoglobulin (Ig)-like fold of PapD catapulted future structure–function analyses that led to the elucidation of fundamental principles of chaperone-assisted subunit folding and pilus biogenesis (Waksman and Hultgren, 2009). Thus, from the structural resolution of PapD and the discovery that it formed complexes with each of the pilus subunits, three decades of investigations were nucleated worldwide, which only recently culminated in the published crystal structure of an usher–chaperone–adhesin ternary complex (Phan et al., 2011). These historical findings served as the beginnings of a scientific era for understanding these virulence factors at the molecular level, which has now led to the development of therapeutics interfering with pilus assembly, pilus function, and hence, infection.

PILUS ARCHITECTURE

Pili of the CU system consist of multiple pilus subunits arranged into long, linear protein polymers. The morphology of CU pili varies across the six major clades – α, β, γ(1-4), κ, π, and σ – of the 189-membered CU pilus superfamily (as of 2007), ranging from thin, fibrillar structures to thick, helical rods topped by a fibrillar tip (Nuccio and Baumler, 2007). Paradigms for CU pilus architecture and assembly have been well established with the P pilus and type 1 pilus of UPEC, members of the π and γ1 clades, respectively. These two archetypal pili exhibit a bipartite organization, consisting of a long, helical rod connected to a thin tip fibrillum. The P pilus subunits PapG, PapF, PapE, PapK, PapA, and PapH arrange in order from fibrillar tip to rod base (Figure 12.1). PapG, the adhesin, lies at the distal end of the pilus. The tip adaptor PapF connects PapG to the main tip component PapE, which appears in 5–10 copies and has a width of ~2 nm. The adaptor PapK anchors PapE to the main rod component PapA, which appears in >1000 copies and gives rise to a right-handed, helical structure that displays a 6.8 nm width, 2.5 nm pitch, and 3.3 subunits per turn (Kuehn et al., 1992; Jacob-Dubuisson et al., 1993; Striker et al., 1994). Finally,

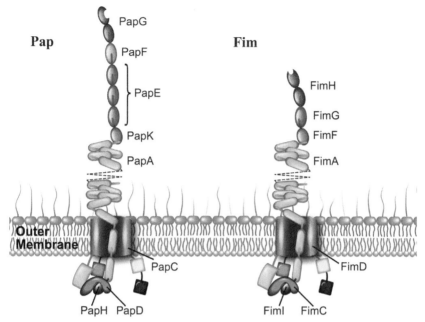

FIGURE 12.1 Architecture in P and type 1 pili. A graphic illustrating pili from the Pap (left) and type 1 (right) systems. See text for details.

PapH attaches at the base of the rod and terminates pilus biogenesis (Baga et al., 1987; Verger et al., 2006). Type 1 pili adopt a similar, yet condensed architecture (Hahn et al., 2002). The fibrillum consists of single copies of both the adhesin FimH and the tip subunit FimG, while the adaptor FimF links the fibrillum to the major rod component FimA, which appears in ~1000 copies (Figure 12.1). Preliminary studies suggest that FimI functions as the terminator subunit in type 1 pili (Valenski et al., 2003; Ignatov, 2009).

CHAPERONES

Structure and function

The chaperones of the CU pathway (~25–30 kDa) are all highly homologous in sequence and structure (Hung et al., 1996). They are composed of two complete Ig-like domains (Figure 12.2A). Chaperones function to transiently bind their cognate subunits, shielding their interactive surfaces to facilitate their proper folding and stability in the periplasm (Kuehn et al., 1991).

In PapD, the chaperone from the P pilus system, a conserved salt bridge stabilizes the two Ig-like domains to adopt an overall boomerang-like shape. These domains orient to form a cleft that is directly involved in subunit binding. In the absence of subunits, chaperones like PapD and SfaE form dimers and

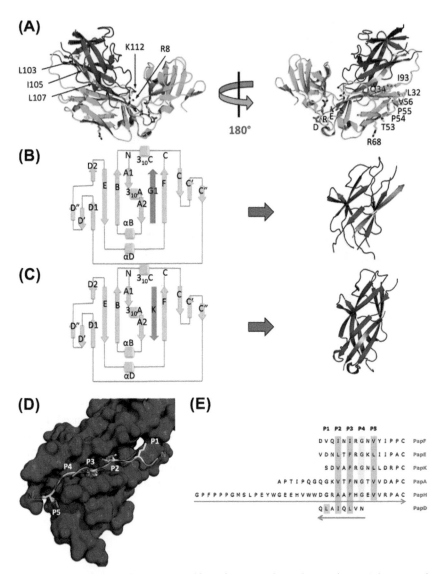

FIGURE 12.2 Conserved chaperone residues, donor strand complementation, and donor strand exchange. (A) A ribbon diagram of the PapD–PapE complex. PapE (red) binds the N-terminal domain of PapD (green). The conserved surfaces of PapD are shown in a ball-and-stick representation and highlighted in yellow, cyan, and orange. Their function is described in the text. (B) Donor strand complementation (DSC). The topology diagram on the left shows the secondary structure of donor strand complemented PapE (tan), indicating β-strands as arrows and α-helices as cylinders. PapD donates its G1 strand (green) parallel to the F strand of PapE. The ribbon diagram on the right also depicts DSC, in which the G1 strand of PapD (green) completes the fold of PapE (red). [PDB code: 1N0L]. (C) Donor strand exchange (DSE). The topology diagram on the left shows the secondary structure of donor strand exchanged PapE (tan), indicating β-strands as arrows and α-helices as cylinders. PapK donates its N-terminal extension (Nte, blue) antiparallel to the F strand of PapE.

Caf1M forms tetramers as a self-capping mechanism to prevent unfavorable interactions and proteolysis (Hung et al., 1999; Knight et al., 2002; Zavialov and Knight, 2007).

Interactive surfaces of the chaperone

Active site

Upon translocation of subunits across the SecYEG translocase to the periplasmic space, they are taken up by their cognate periplasmic chaperones, which use several interactive surfaces to provide stability to the subunits. The active site of the chaperone is comprised of residues R8 and K112, a conserved basic patch in the cleft formed between the two domains. Mutagenesis of these basic residues abolishes the ability of the chaperone to mediate pilus assembly (Slonim et al., 1992; Kuehn et al., 1993). Crystallography studies carried out on all of the P pili chaperone-subunit complexes (Sauer et al., 1999, 2002; Verger et al., 2006, 2007, 2008; Ford et al., 2012) further confirmed these interactions and the extended conformation by which the C-termini of subunits are anchored at the invariant R8 and K112 residues of the chaperone (Kuehn et al., 1993).

The G1 β-strand

The absence of the seventh β-strand in the subunit results in a deep groove on its surface that exposes its hydrophobic core. In a process termed donor strand complementation (DSC), the G1 strand of the chaperone is donated in trans in a non-canonical, parallel fashion to the exposed hydrophobic groove of the subunit to facilitate its proper folding (Sauer et al., 1999; Barnhart et al., 2000; Bann et al., 2004) (Figure 12.2B). The solvent-exposed set of alternating hydrophobic residues on the chaperone's G1 strand directly interact with subunit pockets

←———————————————————————————————————————

The ribbon diagram on the right also depicts DSE, in which the PapKNte (blue) completes the fold of PapE (red). [PDB code: 1N12]. (D) The P1–P5 pockets of PapE. The surface representation of PapE (red) emphasizes the pockets that allow the PapKNte (blue) and the PapD G1 strand (green) to mediate DSE and DSC, respectively. PapD residues L103, I105, and L107 (yellow, ball-and-stick representation) project into the P1–P3 pockets and are correspondingly referred to as P1–P3 residues. PapKNte residues (cyan, ball-and-stick representation) project into the P2–P5 pockets and are correspondingly referred to as P2–P5 residues. The N-terminal ends of the Nte and G1 strand are labeled in the corresponding color. Note the shallow nature of the P4 pocket, which only accommodates a glycine residue in the Nte sequence of Pap subunits, as shown in (E). (E) Sequence alignment of the PapD G1 strand and Nte's of all Pap subunits (except PapG, which lacks an Nte). Pap subunits have conserved P2–P5 residues at their Nte. P2, P3, and P5 residues are hydrophobic (blue), while the P4 residue is strictly glycine (yellow) to prevent steric clashes with the shallow P4 binding groove. PapD has three hydrophobic residues that correspond to the P1–P3 residues. The arrows indicate the N- to C-terminal direction for the sequences of Pap subunits (blue) and PapD (green).

(Sauer et al., 1999). In the case of PapD, residues L107, I105, and L103 are respectively in register with the subunits' hydrophobic P1–P3 pockets. Mutating residue 105 is particularly detrimental to pilus biogenesis (Hung et al., 1999). However, mutagenesis of residue L107 does not alter pilus assembly, which may be due to the plasticity of the P1 pocket (Ford et al., 2012). These residues are termed P1–P3 residues based on their corresponding acceptor sites, the subunit P1–P3 pockets (Figure 12.2D). Additionally, N101 forms hydrogen bonds above the shallow P4 subunit pocket. Thus, the interactions involved in DSC facilitate subunit folding and prevent subunit aggregation. The presence of unfolded subunits in the periplasmic space otherwise induces periplasmic stress responses targeting aggregated subunits for degradation by the DegP protease (Jones et al., 1997, 2002).

Once the subunit is folded, it remains bound to the chaperone until its incorporation into the pilus. The chaperone is exchanged for the N-terminal extension (Nte) of the incoming subunit via a 'zip-in, zip-out' mechanism termed donor strand exchange (DSE). DSE is initiated by the insertion of the Nte P5 residue into the open P5 pocket of the previously assembled subunit (Figure 12.2C–E). The chaperone G1 strand does not occupy the P5 pocket of subunits, which remains easily accessible in a chaperone–subunit complex to the P5 residue of the subunit Nte. Ultimately, insertion of Nte in an antiparallel fashion to the P2–P5 subunit pockets facilitates the removal of the chaperone's parallel-oriented G1 strand. The DSE interaction is more energetically favorable than the DSC interaction, which allows DSE and the stable docking of subunits with each other to occur (Sauer et al., 2002; Remaut et al., 2006).

Usher-binding surface

PapD-like chaperones have a set of conserved, solvent-exposed, hydrophobic residues (termed Set B: L32, Q34, T53, P54, P55, V56, R68, I93) situated at the N-terminal domain (Figure 12.2A). X-ray crystallography studies demonstrated that the FimC chaperone of the type 1 pilus system interacts with the N-terminal domain of the usher via these Set B residues (Nishiyama et al., 2003; Eidam et al., 2008), demonstrating that the Set B patch is a surface that interacts with the usher. Accordingly, point mutations in these residues negatively impact pilus biogenesis (Hung et al., 1999). In addition, small molecules called pilicides, designed to block pilus biogenesis (Pinkner et al., 2006; Chorell et al., 2010), were shown to bind Set B residues and prevent the targeting of chaperone–subunit complexes to the usher (Pinkner et al., 2006). These results implicate PapD's conserved hydrophobic patch Set B as the usher-targeting site. Other chaperone residues that may be involved in interactions with other domains of the usher are subjects of investigation.

Conserved cleft residues and the DRES motif

Another set of highly conserved residues (Set C: L78, P79, D81, R82, E83, S84) is located at the elbow region of PapD, which is not known to interact

with any protein partners. The stability of the chaperone depends on the formation of a buried salt bridge composed of D196, E83, and R116, which lies at
the interdomain region (Hung et al., 1999). E83 is part of the aspartate, arginine, glutamate, and serine (DRES) motif, which is highly conserved in the
chaperone superfamily (Hung et al., 1996). Situated at the end of the E1–F1
loop, part of the DRES motif packs against the hinge region connecting the two
domains (Holmgren and Branden, 1989; Hung et al., 1996), where the aspartate
and arginine side chains from the DRES motif point out into solution, seemingly suitable for protein–protein interactions (Figure 12.2A). It is plausible that
interactions with other proteins (subunits or usher domains) could cause conformational changes that would be transmitted to the interdomain region, possibly
disrupting the salt bridge and thus facilitating a reorientation of the domains and
causing allosteric conformational changes in the chaperone that play a role in
the assembly of the CU pilus.

SUBUNITS

Pilins

There are several major functional types of pilin that may comprise a CU pilus,
including adhesins, adaptors, tip fibrillar subunits, pilus rod subunits, and ter-
minators (Figure 12.1). Pilins, or pilus subunits, share an outstanding degree
of sequence and structural homology with one another, necessary for preserving common mechanisms of subunit–subunit, chaperone–subunit, and usher–
chaperone–subunit interactions needed for maturation of a functional pilus.
As described, the incomplete nature of the pilin fold insures either chaperone
binding (by DSC) or Nte binding (by DSE) to pilus subunits for their stability, proper folding, and ultimate incorporation into the growing pilus. Thus,
the Nte, chaperone G1 strand, and the incomplete Ig-like fold of pilin subunits,
particularly the P1–P5 pockets, serve as common recognition motifs in the CU
pathway for the use of pilins as building blocks in the construction of a pilus.

However, structural distinctions across homologous pilins permit their
unique positions and functions along the pilus chain. For example, differences
in Nte sequence and hydrophobic pocket characteristics dictate subunit ordering. DSE reactions performed by incubating all combinations of Nte peptides
(based on the five Nte-containing Pap subunits) with all chaperone–subunit
complexes showed a range of reactivities (Rose et al., 2008; Verger et al., 2008).
Reactions that occurred most rapidly were consistently those between cognate
groove–Nte partners. In addition to chaperone–subunit and subunit–subunit
interactions, subunits also contain specific surfaces that drive chaperone–
subunit–usher interactions. Differences in subunit structure and residue side
chains dictate selective trafficking to certain periplasmic domains of the usher
(see Ushers: Domain function and selectivity). Thus, adhesin and terminator
subunits, which occupy opposite ends of the pilus, adopt structures much more

varied from their pilin counterparts required for their essential functions in host adhesion and pilus assembly termination, respectively.

Adhesin

A mature CU pilus typically contains an adhesin at its distal end. Adhesins contain two distinctive domains: a lectin domain, which functions to bind specific receptors with stereochemical specificity and contribute to host and tissue tropisms, and an Ig-like pilin domain, which links the receptor-binding lectin domain to the fibrillar tip (Choudhury et al., 1999; Sauer et al., 1999). FimH and PapG, adhesins of the prototypical type 1 and P pilus systems, respectively, mediate host–pathogen interactions important in cystitis (Wu et al., 1996; Mulvey et al., 1998; Martinez et al., 2000; Bahrani-Mougeot et al., 2002) and pyelonephritis (Roberts et al., 1994). FimH mediates binding to α-D-mannose-containing receptors whereas PapG binds galabiose, a receptor that is present in the globo series glycolipids found in the kidney. Both lectin domains contain a β-barrel jelly-roll fold that is common to CU adhesin structures, but they contain different receptor-binding pockets. FimH contains a deep, negatively charged pocket at the tip of the lectin domain (Hung et al., 2002). In contrast, PapGII (class II of three PapG classes, each of which recognize different globo series of glycolipids) binds galabiose-containing receptors with a shallow binding pocket formed by three beta strands and a loop at the side of the lectin domain (Dodson et al., 2001). CU adhesins F17-G of ETEC (Buts et al., 2003) and GafD of UPEC (Merckel et al., 2003) bind N-acetyl-D-glucosamine residues of proteins at a shallow binding site on the side of the lectin domain at a location unrelated to the binding pocket of PapG (Figure 12.3). With the function of adhesion segregated to a distinct protein domain at the pilus tip, various bacterial strains can alter the structure and receptor specificity of the adhesin, thus allowing flexibility and selective advantage in establishing tropism.

Numerous studies suggest that the lectin domain of FimH does not operate independently of the pilin domain but instead interacts intimately with it to influence receptor binding. A crystal structure of the type 1 pilus tip fibrillum (Le Trong et al., 2010) indicates a markedly different conformation in FimH, compared to the conformation observed in FimH in the FimC–FimH complex (Choudhury et al., 1999) or in the mannose-bound FimC–FimH structure (Hung et al., 2002). In the tip structure, the pilin domain interacts with a compacted lectin domain in a stable, non-covalent manner, loosening the mannose-binding pocket at the distal end of the lectin domain and thereby inducing an inactive low-affinity state of the adhesin via a 'page-turning' allosteric mechanism (Le Trong et al., 2010). In contrast, FimH, when in complex with FimC and α-D-mannose, adopts an elongated, high-affinity state (Hung et al., 2002), which was later shown to be induced by application of shear force, presumably by destabilization of interdomain contacts. Indeed, cross-linking mutants along with positively selected residues in the pilin domain suggest that this

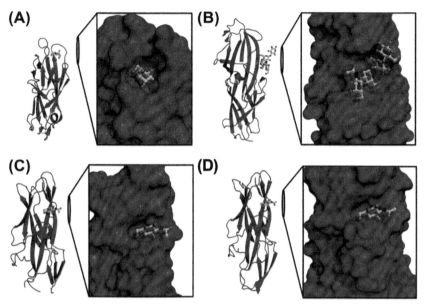

FIGURE 12.3   Co-crystal structures of CU adhesins with their receptors. (A) FimH in complex with D-mannose [PDB code:1KLF]. (B) PapG in complex with galabiose [PDB code: 1J8R]. (C) F17-G in complex with N-acetyl-D-glucosamine [PDB code: 1O9W]. (D) GafD in complex with N-acetyl-D-glucosamine [PDB code: 1OIO]. In every complex, the adhesin (magenta) is shown as a ribbon representation (left) and a magnified surface representation (right), with the bound ligand depicted as a ball-and-stick representation (green). The structures demonstrate that the binding pocket is in different locations of the adhesin fold and has different morphology, which dictates receptor specificity.

conformational equilibrium is important for mannose binding and infectivity in murine models of cystitis (Chen et al., 2009; Le Trong et al., 2010). Shear force does indeed enhance bacterial adhesion to target cells (Thomas et al., 2002), but whether this mechanism of tensile stress-induced binding enhancement applies to other pili systems or has a definitive role in pathogenesis remains to be seen.

Terminator

Studies in the Pap system suggest that pilus biogenesis ends with the incorporation of the PapH terminator at the base of the pilus rod. Deletions or mutations of PapH result in the formation of long pili that are shed from the extracellular surface (Baga et al., 1987). Overexpression of PapH results in short pili. Further, the structure of PapD–PapH indicates that PapH has an occluded P5 pocket, which prevents the initiation of DSE with further subunits, thereby terminating pilus biogenesis (Verger et al., 2006). Analogously, FimI may function as the terminator in type 1 pili (Valenski et al., 2003; Ignatov, 2009). Further studies will elucidate whether incorporation of a P5 pocket-lacking pilin is utilized by other CU pili for termination.

USHERS

Domain function and selectivity

The usher of the CU pathway catalyzes the translocation and assembly of the multisubunit pilus fiber across the outer membrane, while maintaining membrane integrity. The usher is an outer membrane protein comprised of five domains: a 24-stranded beta barrel channel, plug, N-terminal domain (NTD), and C-terminal domains (CTD1, CTD2) (Thanassi et al., 2002; Nishiyama et al., 2003; Capitani et al., 2006). In the apo state of the FimD usher, the plug resides in the lumen of the transmembrane channel, preventing flow of molecules across the outer membrane (Remaut et al., 2008; Huang et al., 2009); in the active form, the plug swings away from the channel, creating an unobstructed opening through which pilins translocate (Figure 12.4). This

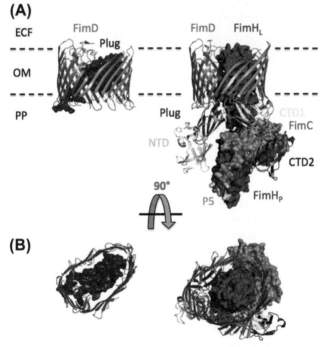

FIGURE 12.4   Structures of the gated and open FimDusher. (A) In the gated apo state, the plug (blue, surface representation) resides in the lumen of the usher pore (orange, ribbon diagram), preventing flow of molecules across the OM and maintaining membrane integrity. The remaining usher domains are not depicted since a FimD truncate was used to solve the apo structure [PDB code: 3OHN]. (B) In the FimD–FimC–FimH complex, the plug swings into the periplasm and binds NTD (light blue), while FimC (green) docks at CTD1 (yellow) and CTD2 (purple) with its bound FimH (magenta). The lectin domain of FimH now resides in the lumen of the usher. Note the unoccupied P5 pocket (light pink) in the FimH pilin domain, which is targeted by the Nte of the incoming subunit FimG for DSE [PDB code: 3RFZ]. Rotation of the usher by 90°, such that the usher is seen from the top, shows a kidney-shaped pore in the plug-gated usher but a nearly circular pore in the secreting usher. ECF, extracellular fluid; OM, outer membrane; PP, periplasm.

plug switch induces a conformational change in the β-barrel domain, shifting the shape of the channel pore from ovular (52 Å × 28 Å) to nearly circular (44 Å × 36 Å) (Phan et al., 2011). This in turn allows a range of subunit diameters (~20–25 Å) to pass through unobstructed. The plug then docks onto the NTD, as suggested in the crystal structure of the FimD usher (Phan et al., 2011) and from work in the Pap system (Volkan et al., 2012). The plug-NTD interaction results in the usher adopting an open state.

NTD and CTDs, which reside in the periplasmic space, serve as the workhorses of assembly within this molecular machine. Functional studies directly implicate the periplasmic domains of the usher in catalysis of pilus formation, as mutations in either NTD or CTD (Henderson et al., 2011; Thanassi et al., 2002) and deletions of the plug (Huang et al., 2009; Mapingire et al., 2009) abrogate assembly. To better dissect the role of each domain, affinities were measured for the interactions of chaperone–subunit complexes with purified NTD, CTD2, and plug domains from the PapC usher. NTD only binds the chaperone–adhesin complex with high affinity ($K_D = 1$ nM) and the chaperone–rod adaptor complex with much weaker affinity ($K_D = 1$ μM); plug and the plug–NTD complex bind the chaperone and all chaperone–subunit complexes with equal affinity ($K_D = 10$–100 nM); CTD2 binds chaperone and all chaperone–subunit complexes with equal affinity ($K_D = 1$ μM), but lacks affinity for the chaperone–terminator complex; and CTD2 dislocates the PapD–PapG complex from the NTD of the PapC usher (Volkan et al., 2012). These data suggest that (a) the tight-binding NTD serves as the initial anchoring site for the chaperone–adhesin complex, (b) plug or the NTD–plug complex serves as the docking site for all other chaperone–subunit complexes, and (c) CTD2 likely dissociates chaperone–subunits from NTD and plug, with the exception of the chaperone–terminator complex, which docks on plug or the NTD–plug complex and halts further pilus growth.

It is interesting to note that NTD can discriminate between subunits loaded onto the chaperone, despite the high structural homology that pilins share. Analogously to PapC NTD, the FimD NTD is selective; isothermal titration calorimetry (ITC) experiments show that FimD NTD binds the chaperone when the chaperone is loaded with FimH or FimF but not with FimG or FimA (Nishiyama et al., 2003). The molecular basis for this selectivity may be indirectly inferred from the crystal structures of the FimD NTD–FimC–FimH complex (Nishiyama et al., 2005) and FimD NTD–FimC–FimF complex (Eidam et al., 2008). In these crystal structures, FimD NTD consists of two flexible N- and C-terminal tails and a uniquely folded core, in which a strained, three-stranded, antiparallel β-sheet and a two-stranded, antiparallel β-sheet pack against one another and are bridged by a segment containing a 3_{10}-helix and two α-helices. These structures along with supporting NMR data reveal that the intrinsically disordered 24-residue N-terminal tail of NTD adopts an ordered conformation upon binding. As expected, deletion of the N-terminal tail of NTD completely abolishes pilus assembly in vivo (Nishiyama et al., 2005). In the ternary complex, the N-terminal tail of NTD contacts the chaperone and the subunit

(FimH or FimF), constituting 50–60% of the total contact surface area (1260 Å^2). While all components of these two ternary complexes adopt remarkably similar folds, there are subtle yet significant differences at the FimD NTD–subunit interface. First, the FimD NTD–FimH interface is 210 Å, larger than the FimD NTD–FimF interface of 120 Å. Second, the guanidinium moiety of FimD NTD R7 interacts with FimH via hydrogen bonding with the T200 side chain and hydrophobic contacts with the T212 side chain, but lacks any contact with FimF. Third, three hydrogen bonds exist at the FimD NTD–FimH interface – Y3/Q269, N5/T212, and R7/T200 – while only one hydrogen bond exists at the FimD NTD–FimF interface – N5/T212 (Eidam et al., 2008). Not surprisingly, FimD NTD has a higher affinity for FimC–FimH than it does for FimC–FimF. Presumably, the subunits FimG and FimA cannot bind NTD because they have an incomplete set or total lack of the hydrogen bond interactions with the N-terminal tail that were detailed above. A similar molecular logic likely holds true for the selectivity of the NTD domains of other CU ushers, although more work needs to be done to further bolster this argument.

Given their structural homology, it is not surprising that plug and CTD2 can recognize nearly all chaperone–subunit complexes. Both the CTD2 and plug domains of PapC adopt a small beta sandwich fold with a central Greek key motif – the only difference in fold being an additional β-strand, βG, at the C-terminus of CTD2 (Ford et al., 2010). In addition, many residues are conserved between the two usher domains. These conserved residues, mainly surface-exposed polar residues protruding from the βC, βB, βE, and βF side, may hint at a conserved function for this surface, like binding chaperone–subunit complexes. From the FimD–FimC–FimH ternary complex, the FimC–FimH complex primarily binds CTD1 and is seen in proximity to CTD2 (Phan et al., 2011). However, FimC–FimH does not approach CTD2 close enough to form van der Waals contacts or explain the moderate affinity interaction observed with the isolated domain (Volkan et al., 2012), suggesting the existence of another binding mode between CTD2 and the chaperone–subunit complexes. Likewise, there is no structural information regarding the contacts that plug forms with chaperone–subunit complexes, but given its ability to bind all complexes and chaperone alone, plug likely forms most of its contacts, if not all, with the chaperone itself. Whether the βC, βB, βE, and βF side of plug and of CTD2 is involved in binding chaperone–subunit complexes remains to be experimentally evaluated. Similarly, the basis for the terminator's ability to discriminate between plug and CTD2 remains unknown. More work is needed to elucidate the molecular basis of selectivity by usher domains for chaperone–subunit complexes to better understand the mechanism and subunit ordering of pilus assembly.

Monomer versus dimer

Initial attempts to probe the structure of the usher led to the discovery of a dimeric state (Thanassi et al., 1998). The first of these studies on the PapC usher employed cryo-EM and dynamic light scattering to reveal a twin-pore complex stable in solution and membrane-like lipid environments (Li et al., 2004). Each PapC molecule within the dimer adopts an ellipsoidal shape and has a pore of 2 nm in width. A PapC C-terminal deletion mutant still retained the ability to form dimers but by a different manner, as the mutant packed in length-wise fashion as opposed to the side-wise interaction observed in the wild-type crystals. However, the PapC truncate did not display in vivo activity consistent with the requirement of CTD to recruit chaperone–subunit complexes and/or allosterically relocate the chaperone–adhesin complex. The ability of FimD to interact with the PapC C-terminal truncate to rescue P pilus assembly (So and Thanassi, 2006), may relate more to donation of a related CTD than to a role for a dimer in the function and/or regulation of usher activity in vivo, but further experimentation is needed to resolve these complexities.

The protomers in the side-to-side dimer were only weakly associated (Remaut et al., 2008). This dimer only utilizes one channel within the dimer for secretion of the growing fiber, as suggested by electron density from cryo-EM. Further, the recent crystal structure of the FimD usher in complex with FimC–FimH, suggests that the usher monomer and its periplasmic domains should be sufficient for pilus assembly (Phan et al., 2011), without the need for side-to-side dimerization or head-to-head dimerization mediated by an interaction of exposed β4 strand edges from the β4–β5 hairpin that protrudes into the extracellular side of the β-barrel core (Huang et al., 2009).

Conformational dynamics

Molecular snapshots in the mechanism of pilus assembly provide signposts for the dynamic processes underlying the sequence of binding events at the usher that give rise to its catalytic activity. The first step in pilus biogenesis requires activation of the usher by displacement of plug from the channel. Early evidence from trypsin susceptibility assays (Saulino et al., 1998) and previously mentioned crystal structures do indeed indicate that the usher adopts at least two distinct conformations. It has been suggested that binding of the chaperone–adhesin complex to the periplasmic NTD primes the usher by inducing a conformational change in the usher that allows plug to swing away from the pore (Nishiyama et al., 2003; Phan et al., 2011). NMR studies on isolated FimD NTD and NTD bound to FimC–FimH give credence to this hypothesis. The substantial chemical shift changes observed in the linker region of FimD NTD upon binding FimC–FimH suggest a large movement in the linker (Nishiyama et al., 2005). This linker movement may tug at the β1 strand of the usher transmembrane domain, which may in turn induce changes in conformation of the β-barrel that facilitate plug exit from or plug re-entry into the pore. Simultaneously, the hinge movement may allow NTD to swing closer and unload its chaperone–adhesin cargo to CTDs. Because the usher functions independently of an external energy source (Jacob-Dubuisson et al., 1994), this large conformational rotation of plug and NTD must be powered from the energy of binding the chaperone–adhesin complex.

While the periplasmic domains NTD and CTD are freely mobile in the apo state (Remaut et al., 2008), once plug has swung into the periplasm and formed a complex with NTD (Phan et al., 2011; Volkan et al., 2012), the periplasmic domains must make deliberate and concerted movements for the efficient catalysis of pilus growth. Unfortunately, nothing is known about the dynamics involved in subsequent steps in pilus assembly. In our modeling efforts, a stationary in silico model of FimC–FimG docked at the NTD of the FimD–FimC–FimH complex shows steric clashes between the two chaperone–subunit complexes. The Nte of FimF, the incoming subunit, is almost but not quite properly oriented above the P5 pocket of FimH for DSE, suggesting an undiscovered usher conformation that partakes in the DSE reaction. Presumably, the chaperone–subunit complexes shuffle across these periplasmic domains so that the fiber can ascend through the channel and so that periplasmic domains are freed up for incoming chaperone–subunit complexes. Whether this shuffling occurs by rigid body movements of the NTD, CTDs, or both toward each other is unknown. Further work is needed to assess the dynamics and conformational changes within usher domains, which will greatly benefit our understanding of DSE and catalysis of pilus assembly.

Working model

The results and associated interpretations presented above suggest the following working model of CU pilus biogenesis (Figure 12.5):

1. The usher adopts a gated conformation in its inactive state, in which its plug domain lies in the center of the kidney-shaped β-barrel. NTD and CTD lie disordered in the periplasm, moving rapidly and randomly.
2. Once NTD binds the initiator of pilus assembly, the chaperone–adhesin complex, the binding energy induces movement in the NTD linker that thereby transduces conformational changes in the β-barrel domain, relaxing its shape from ovular to circular and preventing plug from re-entering the channel lumen once it has already entered the periplasm. Concurrently, NTD may swing to transfer its bound chaperone–adhesin complex to CTDs. Once NTD has unloaded its cargo, it binds the now periplasmic plug domain to form the NTD–plug complex, which can then bind and dock other incoming chaperone–subunit complexes.
3. Alternatively or concomitantly, the CTDs may swing over together to the NTD–plug complex to catalyze the dissociation of the chaperone–adhesin complex. A minor conformational species in the FimD–FimC–FimH complex as measured by electron paramagnetic resonance (EPR) (Phan et al., 2011) suggests the occurrence of an uncharacterized transient binding interaction between CTD2 and the chaperone–adhesin complex that may

Pilus Biogenesis via Chaperone-Usher System

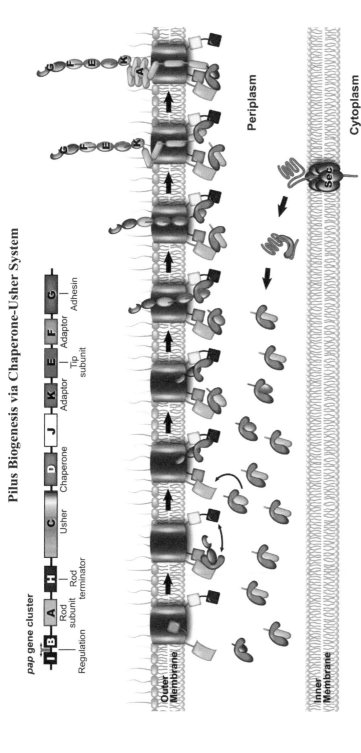

FIGURE 12.5   Model of pilus assembly. Upon translocation of subunits into the periplasm via the SEC machinery, chaperone proteins bind the subunits to maintain their proper fold and stability. At the outer membrane, the usher lies in a gated state, with the plug domain (blue) residing in the lumen of the usher pore (orange). Once the chaperone–adhesin complex binds the usher NTD (cyan), the plug domain swings away from the pore and docks onto the NTD, thus preparing the usher for pilus secretion. Soon thereafter, CTD1 (yellow) and CTD2 (purple) may swing over to bind and carry the chaperone–adhesin complex, freeing the NTD–plug complex in the process. The freed NTD–plug complex recruits the incoming chaperone–subunit complex, orienting it properly so that the Nte can participate in DSE with the CTD-docked chaperone–adhesin complex and the newly exchanged chaperone–subunit complex can transfer to the CTDs. The process repeats until the chaperone–terminator complex, bound at the NTD–plug complex, cannot transfer to the CTDs or engage in DSE, signaling the end of pilus assembly.

mediate this transfer. With the chaperone–adhesin complex bound primarily to CTD1, the lectin domain of the adhesin lies in the channel lumen, and the NTD–plug complex is available for the incoming chaperone–tip subunit complex.

4. Next, the NTD–plug complex recruits the upcoming chaperone–tip subunit complex, while the chaperone–adhesin complex is bound to the CTDs. A unique conformation in the NTD–plug complex brings the tip subunit into the ideal orientation with respect to the adhesin for DSE, allowing for the assembly of the growing pilus fiber. The reaction results in the displacement of the chaperone off of the CTD, allowing the newly incorporated chaperone–subunit complex to dock on the CTDs. The process repeats for all chaperone–subunit complexes in order of the subunit ordering observed in the mature pilus fiber.

5. Ultimately, the chaperone–terminator complex docks on the NTD–plug complex. Unable to bind CTD2 and unable to undergo DSE due to its lack of a P5 pocket, the chaperone–terminator complex signals the end of pilus growth. The plug may bind the chaperone–terminator complex with its βC, βB, βE, and βF side and swing into the β-barrel lumen with that same side facing the extracellular space, as seen in the apo state. The mature pilus fiber anchors to the outer membrane in this fashion and is ready to perform its adhesive function.

Overall, this model provides insight into how the usher assembles the CU pilus. With domain affinities, molecular snapshots of pilus biogenesis steps, and genetic and biochemical evidence, we can now understand how the usher efficiently transfers chaperone–subunit complexes using its periplasmic domains to perform its catalysis of fiber growth. Further work will be required to gain a deeper mechanistic understanding of this molecular machine to provide knowledge of targetable protein interfaces and dynamic processes for next generation small-molecule pilicides that can inhibit these virulent adhesive pili.

ROLE OF CU PILI IN INFECTIONS

UPEC introduced to the urinary tract from the fecal flora is the leading causative agent of urinary tract infections (UTIs), responsible for 85% of community-acquired UTIs (see Chapter 9) (Ronald et al., 2001). Genetic, biochemical, and imaging studies accompanied by murine models of UTIs revealed that CU pili are crucial factors for causing this disease.

For instance, type 1 pili, encoded by the *fim* operon, are required for bladder infection in a murine model of UTI (Hultgren et al., 1985; Connell et al., 1996; Mulvey et al., 1998; Anderson et al., 2003; Wright et al., 2007) and for biofilm formation in rat kidneys (Melican et al., 2011). The adhesin of type 1 pili, FimH, mediates binding to the mannosylated receptors on the surface of bladder urothelial cells for colonization and invasion of the superficial umbrella cells, which line the bladder lumen. Following invasion, UPEC grow inside the cytosol of host cells, giving rise to biofilm-like bacterial aggregates of 10^4–10^5 bacteria, known as intracellular bacterial communities (IBCs) (Anderson et al., 2003; Justice et al., 2004; Wright et al., 2007). Type 1 pili are required for the establishment and maintenance of IBCs, allowing proliferation of UPEC, which upon maturation can detach and spread to neighboring urothelial cells to continue the infection cascade. As IBCs have also been observed in human patients suffering from UTIs (Rosen et al., 2007), it is likely that involvement of type 1 pili and IBCs in human UTI is also common.

Another well-studied CU system that is important in urinary tract infections is P pili, encoded by the *pap* operon. P pilus adhesin PapG was shown to be important in pyelonephritis in primate models of infection (Roberts et al., 1994). In rat models of infection, P pili were shown to enhance early colonization of the kidney tubular epithelium (Melican et al., 2011). Most UPEC strains from human cases of pyelonephritis carry PapGII (Marschall et al., 2012; Otto et al., 2001), one of three classes of PapG, which binds to Galα1-4Gal receptor epitopes in the globoseries of glycolipids found in human renal tissue (Dodson et al., 2001).

In addition to P and type 1 pili's importance in UTIs, various other CU systems from different species are important in bacterial virulence. For instance, S pili, encoded by the *sfa* operon, is involved in meningitis (see Chapter 10) (Korhonen et al., 1984; Morschhauser et al., 1994), while class 5 ETEC fimbriae are important for the initiation of diarrheal disease (see Chapter 6) (Chattopadhyay et al., 2012). Considering the importance of CU pili in infection, understanding how these pili are assembled and the mechanisms by which they cause pathogenesis is critical in disease prevention and development of novel therapeutics.

CU PILI AS ANTIVIRULENCE TARGETS

Antibiotics have led to significant improvements in human health over decades, improving quality of life and human longevity. However, antibiotic resistance is escalating (Boucher et al., 2009) and multidrug-resistant uropathogens are spreading globally (Totsika et al., 2011). Accompanied by a lack of a significant effort to develop novel antibiotics and the outstanding risk that antibiotic usage can negatively impact gut microbiota and result in opportunistic infections, the search for antivirulence therapeutics is gaining attention. Due to the important role of CU pili in bacterial virulence, they have been important targets for vaccine development studies as well as antivirulence therapeutics. FimH was shown to be a successful vaccine option in both murine (Langermann et al., 1997) and primate (Langermann et al., 2000) models of infection. Keeping in mind the role of FimH in mediating adhesion with mannosylated uroplakin of uroepithelial cells, alky and phenyl-α-D-mannopyranosides, or mannosides, have been designed to competitively bind with FimH and interfere with adhesion on host

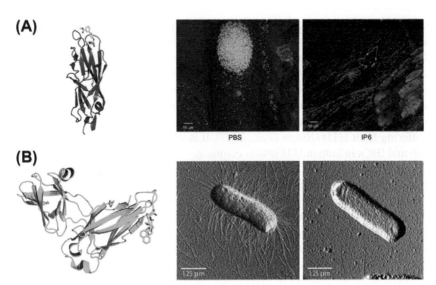

FIGURE 12.6 CU pili as antibacterial targets. (A) Crystal structure of a mannoside (cyan) in complex with the FimH lectin domain (magenta) [PDB code: 3MCY]. (B) Mannoside prevents IBC formation and can treat established infections (Cusumano et al., 2011). Confocal microscopy of mice bladders depicts an intracellular bacterial community (IBC, green) in the left panel and prevention of IBC formation with mannoside treatment in the right panel (white arrows indicate luminal bacteria). (C) Crystal structure of a pilicide (cyan) bound to PapD (green). The pilicide binds to the F1, C1, and D1" strands, a region thought to be the usher-targeting site of the protein, where many of the Set B residues coincide [PDB code: 2J7L]. (D) Pilicide inhibits type 1 pilus and P pilus assembly. Atomic force microscopy images show a piliated, untreated bacterium (left panel) and a naked, pilicide-treated bacterium (right panel), indicating that pilicides suppress pilus biogenesis. (B) and (D) are reproduced, with permission, from Cusumano et al. (2011) and Waksman and Hultgren (2009).

cells. These mannosides not only block adhesion but also counteract internalization and in vitro/in vivo biofilm formation on biotic and abiotic surfaces (Cusumano et al., 2011; Guiton et al., 2012) (Figure 12.6A,B). Furthermore, mannosides potentiate the efficacy of existing antibiotics in a murine model of UTI and have great oral bioavailability, increasing hopes that they can be used in drug development (Cusumano et al., 2011). Similar to FimH, PapG was also a target for antivirulence therapeutics design such that p-Methoxy-phenyl derivatives of galabiose inhibit PapG with low micromolar IC_{50} (half-maximal inhibitory concentration) values (Hultgren et al., 1989; George et al., 2001; Lin et al., 2012; Ohlsson et al., 2002).

In addition to targeting the receptor-binding site of the mature pilus, compounds that directly interfere with pilus biogenesis have also been developed. These rationally designed bicyclic 2-pyridone compounds, termed pilicides, inhibit UPEC hemagglutination of erythrocytes and biofilm formation by inhibiting both P and type 1 pilus biogenesis (Pinkner et al., 2006). X-ray crystallography and biochemical studies revealed an interaction of these compounds with PapD (Pinkner et al., 2006; Chorell et al., 2010). The pilicides were shown to bind with a conserved, hydrophobic, solvent-exposed patch at the N-terminal side of the chaperone (Figure 12.6C,D), the presumed usher-binding site, thus explaining the structural basis of their mechanism of action.

Translated to clinical practice, mannosides, pilicides, and vaccine options can be cost-effective ways to prevent and treat UTIs and other infections that require CU pili, while reducing antibiotic resistance. Using specific antivirulence therapeutics should have minimal impact on the composition of host microbiota, reducing the risk of opportunistic or recurrent infections.

CONCLUSION

Gram-negative bacteria use the CU pathway to assemble virulent surface appendages called pili. Structural biology combined with genetic and biochemical approaches has elucidated crucial protein–protein interactions made by the dedicated chaperone and ushers to facilitate the ordered assembly of pilins into the final pilus structure on the extracellular surface, as well as protein–carbohydrate interactions required for virulence. This work has generated new insights into protein folding and revealed novel mechanisms of macromolecular assembly. Chaperones stabilize the fold of each subunit in a chaperone–subunit complex, in which the subunit is held in a high-energy conformation primed to participate in pilus assembly at the usher. Ushers catalyze pilus assembly through interactions with each chaperone–subunit complex, coordinating the release of the chaperone and interactions of subunits with each other as they fold into their final condensed structures and translocate through the outermembrane usher pore. These multi-disciplinary approaches have revealed snapshots of a sophisticated protein assembly machinery, elucidated the virulence mechanisms of bacteria, and led to development of therapeutics that suppress infection in animal models. Ultimately, these efforts will lead to a better understanding of CU pilus-mediated infectious disease, giving rise to potent therapeutics that target acute, chronic, and recurrent infections for prevention and treatment of human disease.

REFERENCES

Anderson, G.G., Palermo, J.J., Schilling, J.D., Roth, R., Heuser, J., Hultgren, S.J., 2003. Intracellular bacterial biofilm-like pods in urinary tract infections. Science 301, 105–107.

Baga, M., Norgren, M., Normark, S., 1987. Biogenesis of E. coli Pap pili: papH, a minor pilin subunit involved in cell anchoring and length modulation. Cell 49, 241–251.

Bahrani-Mougeot, F.K., Buckles, E.L., Lockatell, C.V., et al., 2002. Type 1 fimbriae and extracellular polysaccharides are preeminent uropathogenic Escherichia coli virulence determinants in the murine urinary tract. Mol. Microbiol. 45, 1079–1093.

Bann, J.G., Pinkner, J.S., Frieden, C., Hultgren, S.J., 2004. Catalysis of protein folding by chaperones in pathogenic bacteria. Proc. Natl. Acad. Sci. USA 101, 17389–17393.

Barnhart, M.M., Pinkner, J.S., Soto, G.E., et al., 2000. PapD-like chaperones provide the missing information for folding of pilin proteins. Proc. Natl. Acad. Sci. USA 97, 7709–7714.

Boucher, H.W., Talbot, G.H., Bradley, J.S., et al., 2009. Bad bugs, no drugs: no ESKAPE! An update from the Infectious Diseases Society of America. Clin. Infect. Dis. 48, 1–12.

Brinton Jr., C.C., 1959. Non-flagellar appendages of bacteria. Nature 183, 782–786.

Brinton Jr., C.C., 1965. The structure, function, synthesis and genetic control of bacterial pili and a molecular model for DNA and RNA transport in gram negative bacteria. Trans. NY Acad. Sci. 27, 1003–1054.

Buts, L., Bouckaert, J., De Genst, E., et al., 2003. The fimbrial adhesin F17-G of enterotoxigenic *Escherichia coli* has an immunoglobulin-like lectin domain that binds N-acetylglucosamine. Mol. Microbiol. 49, 705–715.

Capitani, G., Eidam, O., Grutter, M.G., 2006. Evidence for a novel domain of bacterial outer membrane ushers. Proteins 65, 816–823.

Chattopadhyay, S., Tchesnokova, V., McVeigh, A., et al., 2012. Adaptive evolution of class 5 fimbrial genes in enterotoxigenic *Escherichia coli* and its functional consequences. J. Biol. Chem. 287, 6150–6158.

Chen, S.L., Hung, C.S., Pinkner, J.S., et al., 2009. Positive selection identifies an in vivo role for FimH during urinary tract infection in addition to mannose binding. Proc. Natl. Acad. Sci. USA 106, 22439–22444.

Chorell, E., Pinkner, J.S., Phan, G., et al., 2010. Design and synthesis of C-2 substituted thiazolo and dihydrothiazolo ring-fused 2-pyridones: pilicides with increased antivirulence activity. J. Med. Chem. 53, 5690–5695.

Choudhury, D., Thompson, A., Stojanoff, V., et al., 1999. X-ray structure of the FimC-FimH chaperone-adhesin complex from uropathogenic *Escherichia coli*. Science 285, 1061–1066.

Connell, I., Agace, W., Klemm, P., Schembri, M., Marild, S., Svanborg, C., 1996. Type 1 fimbrial expression enhances *Escherichia coli* virulence for the urinary tract. Proc. Natl. Acad. Sci. USA 93, 9827–9832.

Cusumano, C.K., Pinkner, J.S., Han, Z., et al., 2011. Treatment and prevention of urinary tract infection with orally active FimH inhibitors. Sci. Transl. Med. 3, 109–115.

Dodson, K.W., Pinkner, J.S., Rose, T., Magnusson, G., Hultgren, S.J., Waksman, G., 2001. Structural basis of the interaction of the pyelonephritic *E. coli* adhesin to its human kidney receptor. Cell 105, 733–743.

Duguid, J.P., Anderson, E.S., Campbell, I., 1966. Fimbriae and adhesive properties in Salmonellae. J. Pathol. Bacteriol. 92, 107–138.

Duguid, J.P., Smith, I.W., Dempster, G., Edmunds, P.N., 1955. Non-flagellar filamentous appendages (fimbriae) and haemagglutinating activity in *Bacterium coli*. J. Pathol. Bacteriol. 70, 335–348.

Eidam, O., Dworkowski, F.S., Glockshuber, R., Grutter, M.G., Capitani, G., 2008. Crystal structure of the ternary FimC-FimF(t)-FimD(N) complex indicates conserved pilus chaperone-subunit complex recognition by the usher FimD. FEBS Lett. 582, 651–655.

Ford, B., Rego, A.T., Ragan, T.J., et al., 2010. Structural homology between the C-terminal domain of the PapC usher and its plug. J. Bacteriol. 192, 1824–1831.

Ford, B., Verger, D., Dodson, K., et al., 2012. Structure of the PapD-PapGII pilin complex reveals an open and flexible P5 pocket. J. Bacteriol. 194, 6390–6397.

George, S.K., Schwientek, T., Holm, B., Reis, C.A., Clausen, H., Kihlberg, J., 2001. Chemoenzymatic synthesis of sialylated glycopeptides derived from mucins and T-cell stimulating peptides. J. Am. Chem. Soc. 123, 11117–11125.

Gillies, R.R., Duguid, J.P., 1958. The fimbrial antigens of Shigella flexneri. J. Hyg. 56, 303–318.

Guiton, P.S., Cusumano, C.K., Kline, K.A., et al., 2012. Combinatorial small-molecule therapy prevents uropathogenic *Escherichia coli* catheter-associated urinary tract infections in mice. Antimicrob. Agents. Chemother. 56, 4738–4745.

Guyot, G., 1908. Uber die bakterielle hamagglutination (bakterio-haemoagglutination). Zentralbl Bakteriol Parasitenkd Infektionskr Hyg. Abt. 1 Orig. 47, 640–653.

Hahn, E., Wild, P., Hermanns, U., et al., 2002. Exploring the 3D molecular architecture of *Escherichia coli* type 1 pili. J. Mol. Biol. 323, 845–857.

Henderson, N.S., Ng, T.W., Talukder, I., Thanassi, D.G., 2011. Function of the usher N-terminus in catalysing pilus assembly. Mol. Microbiol. 79, 954–967.

Holmgren, A., Branden, C.I., 1989. Crystal structure of chaperone protein PapD reveals an immunoglobulin fold. Nature 342, 248–251.

Holmgren, A., Branden, C.I., Lindberg, F., Tennent, J.M., 1988. Preliminary X-ray study of papD crystals from uropathogenic *Escherichia coli*. J. Mol. Biol. 203, 279–280.

Houwink, A.L., van Iterson, W., 1950. Electron microscopical observations on bacterial cytology; a study on flagellation. Biochimi. Biophys. Acta. 5, 10–44.

Huang, Y., Smith, B.S., Chen, L.X., Baxter, R.H., Deisenhofer, J., 2009. Insights into pilus assembly and secretion from the structure and functional characterization of usher PapC. Proc. Natl. Acad. Sci. USA 106, 7403–7407.

Hull, R.A., Gill, R.E., Hsu, P., Minshew, B.H., Falkow, S., 1981. Construction and expression of recombinant plasmids encoding type 1 or D-mannose-resistant pili from a urinary tract infection *Escherichia coli* isolate. Infect. Immun. 33, 933–938.

Hultgren, S.J., Lindberg, F., Magnusson, G., Kihlberg, J., Tennent, J.M., Normark, S., 1989. The PapG adhesin of uropathogenic *Escherichia coli* contains separate regions for receptor binding and for the incorporation into the pilus. Proc. Natl. Acad. Sci. USA 86, 4357–4361.

Hultgren, S.J., Porter, T.N., Schaeffer, A.J., Duncan, J.L., 1985. Role of type 1 pili and effects of phase variation on lower urinary tract infections produced by *Escherichia coli*. Infect. Immun. 50, 370–377.

Hultgren, S.J., Schwan, W.R., Schaeffer, A.J., Duncan, J.L., 1986. Regulation of production of type 1 pili among urinary tract isolates of *Escherichia coli*. Infect. Immun. 54, 613–620.

Hung, C.S., Bouckaert, J., Hung, D., et al., 2002. Structural basis of tropism of *Escherichia coli* to the bladder during urinary tract infection. Mol. Microbiol. 44, 903–915.

Hung, D.L., Knight, S.D., Woods, R.M., Pinkner, J.S., Hultgren, S.J., 1996. Molecular basis of two subfamilies of immunoglobulin-like chaperones. EMBO J. 15, 3792–3805.

Hung, D.L., Pinkner, J.S., Knight, S.D., Hultgren, S.J., 1999. Structural basis of chaperone self-capping in P pilus biogenesis. Proc. Natl. Acad. Sci. USA 96, 8178–8183.

Ignatov, O.V., 2009. The role of FimI protein in the assembly of type 1 pilus from *Escherichia coli*. Institute of Molecular Biology and Biophysics. ETH Zurich.

Jacob-Dubuisson, F., Heuser, J., Dodson, K., Normark, S., Hultgren, S., 1993. Initiation of assembly and association of the structural elements of a bacterial pilus depend on two specialized tip proteins. EMBO J. 12, 837–847.

Jacob-Dubuisson, F., Striker, R., Hultgren, S.J., 1994. Chaperone-assisted self-assembly of pili independent of cellular energy. J. Biol. Chem. 269, 12447–12455.

Jones, C.H., Danese, P.N., Pinkner, J.S., Silhavy, T.J., Hultgren, S.J., 1997. The chaperone-assisted membrane release and folding pathway is sensed by two signal transduction systems. EMBO J. 16, 6394–6406.

Jones, C.H., Dexter, P., Evans, A.K., Liu, C., Hultgren, S.J., Hruby, D.E., 2002. *Escherichia coli* DegP protease cleaves between paired hydrophobic residues in a natural substrate: the PapA pilin. J. Bacteriol. 184, 5762–5771.

Justice, S.S., Hung, C., Theriot, J.A., et al., 2004. Differentiation and developmental pathways of uropathogenic *Escherichia coli* in urinary tract pathogenesis. Proc. Natl. Acad. Sci. USA 101, 1333–1338.

Knight, S.D., Choudhury, D., Hultgren, S., Pinkner, J., Stojanoff, V., Thompson, A., 2002. Structure of the S pilus periplasmic chaperone SfaE at 2.2 A resolution. Acta. Crystallogr. D, Biol. Crystallogr. 58, 1016–1022.

Korhonen, T.K., Vaisanen-Rhen, V., Rhen, M., Pere, A., Parkkinen, J., Finne, J., 1984. *Escherichia coli* fimbriae recognizing sialyl galactosides. J. Bacteriol. 159, 762–766.

Krasan, G.P., Sauer, F.G., Cutter, D., et al., 2000. Evidence for donor strand complementation in the biogenesis of Haemophilus influenzae haemagglutinating pili. Mol. Microbiol. 35, 1335–1347.

Kuehn, M.J., Heuser, J., Normark, S., Hultgren, S.J., 1992. P pili in uropathogenic *E. coli* are composite fibres with distinct fibrillar adhesive tips. Nature 356, 252–255.

Kuehn, M.J., Normark, S., Hultgren, S.J., 1991. Immunoglobulin-like PapD chaperone caps and uncaps interactive surfaces of nascently translocated pilus subunits. Proc. Natl. Acad. Sci. USA 88, 10586–10590.

Kuehn, M.J., Ogg, D.J., Kihlberg, J., et al., 1993. Structural basis of pilus subunit recognition by the PapD chaperone. Science 262, 1234–1241.

Labigne-Roussel, A., Schmidt, M.A., Walz, W., Falkow, S., 1985. Genetic organization of the afimbrial adhesin operon and nucleotide sequence from a uropathogenic *Escherichia coli* gene encoding an afimbrial adhesin. J. Bacteriol. 162, 1285–1292.

Langermann, S., Mollby, R., Burlein, J.E., et al., 2000. Vaccination with FimH adhesin protects cynomolgus monkeys from colonization and infection by uropathogenic *Escherichia coli*. J. Infect. Dis. 181, 774–778.

Langermann, S., Palaszynski, S., Barnhart, M., et al., 1997. Prevention of mucosal *Escherichia coli* infection by FimH-adhesin-based systemic vaccination. Science 276, 607–611.

Le Trong, I., Aprikian, P., Kidd, B.A., et al., 2010. Structural basis for mechanical force regulation of the adhesin FimH via finger trap-like beta sheet twisting. Cell 141, 645–655.

Li, H., Qian, L., Chen, Z., et al., 2004. The outer membrane usher forms a twin-pore secretion complex. J. Mol. Biol. 344, 1397–1407.

Lin, L.Y., Tiemann, K.M., Li, Y., et al., 2012. Synthetic polymer nanoparticles conjugated with FimH(A) from *E. coli* pili to emulate the bacterial mode of epithelial internalization. J. Am. Chem. Soc. 134, 3938–3941.

Lindberg, F., Lund, B., Johansson, L., Normark, S., 1987. Localization of the receptor-binding protein adhesin at the tip of the bacterial pilus. Nature 328, 84–87.

Lindberg, F., Tennent, J.M., Hultgren, S.J., Lund, B., Normark, S., 1989. PapD, a periplasmic transport protein in P-pilus biogenesis. J. Bacteriol. 171, 6052–6058.

Mapingire, O.S., Henderson, N.S., Duret, G., Thanassi, D.G., Delcour, A.H., 2009. Modulating effects of the plug, helix, and N- and C-terminal domains on channel properties of the PapC usher. J. Biol. Chem. 284, 36324–36333.

Marschall, J., Zhang, L., Foxman, B., Warren, D.K., Henderson, J.P., 2012. Both host and pathogen factors predispose to *Escherichia coli* urinary-source bacteremia in hospitalized patients. Clin. Infect. Dis. 54, 1692–1698.

Martinez, J.J., Mulvey, M.A., Schilling, J.D., Pinkner, J.S., Hultgren, S.J., 2000. Type 1 pilus-mediated bacterial invasion of bladder epithelial cells. EMBO J. 19, 2803–2812.

Melican, K., Sandoval, R.M., Kader, A., et al., 2011. Uropathogenic *Escherichia coli* P and Type 1 fimbriae act in synergy in a living host to facilitate renal colonization leading to nephron obstruction. PLoS Pathogens 7, e1001298.

Merckel, M.C., Tanskanen, J., Edelman, S., Westerlund-Wikstrom, B., Korhonen, T.K., Goldman, A., 2003. The structural basis of receptor-binding by *Escherichia coli* associated with diarrhea and septicemia. J. Mol. Biol. 331, 897–905.

Morrissey, P.M., Dougan, G., 1986. Expression of a cloned 987P adhesion-antigen fimbrial determinant in *Escherichia coli* K-12 strain HB101. Gene 43, 79–84.

Morschhauser, J., Vetter, V., Emody, L., Hacker, J., 1994. Adhesin regulatory genes within large, unstable DNA regions of pathogenic *Escherichia coli*: cross-talk between different adhesin gene clusters. Mol. Microbiol. 11, 555–566.

Mulvey, M.A., Lopez-Boado, Y.S., Wilson, C.L., et al., 1998. Induction and evasion of host defenses by type 1-piliated uropathogenic *Escherichia coli*. Science 282, 1494–1497.

Nishiyama, M., Horst, R., Eidam, O., et al., 2005. Structural basis of chaperone-subunit complex recognition by the type 1 pilus assembly platform FimD. EMBO J. 24, 2075–2086.

Nishiyama, M., Vetsch, M., Puorger, C., Jelesarov, I., Glockshuber, R., 2003. Identification and characterization of the chaperone-subunit complex-binding domain from the type 1 pilus assembly platform FimD. J. Mol. Biol. 330, 513–525.

Normark, S., Lark, D., Hull, R., et al., 1983. Genetics of digalactoside-binding adhesin from a uropathogenic *Escherichia coli* strain. Infect. Immun. 41, 942–949.

Nuccio, S.P., Baumler, A.J., 2007. Evolution of the chaperone/usher assembly pathway: fimbrial classification goes Greek. Microbiol. Mol. Biol. Rev. 71, 551–575.

Ohlsson, J., Jass, J., Uhlin, B.E., Kihlberg, J., Nilsson, U.J., 2002. Discovery of potent inhibitors of PapG adhesins from uropathogenic *Escherichia coli* through synthesis and evaluation of galabiose derivatives. Chembiochem 3, 772–779.

Otto, G., Magnusson, M., Svensson, M., Braconier, J., Svanborg, C., 2001. pap genotype and P fimbrial expression in *Escherichia coli* causing bacteremic and nonbacteremic febrile urinary tract infection. Clin. Infect. Dis. 32, 1523–1531.

Ottow, J.C., 1975. Ecology, physiology, and genetics of fimbriae and pili. Annu. Rev. Microbiol. 29, 79–108.

Phan, G., Remaut, H., Wang, T., et al., 2011. Crystal structure of the FimD usher bound to its cognate FimC-FimH substrate. Nature 474, 49–53.

Pinkner, J.S., Remaut, H., Buelens, F., et al., 2006. Rationally designed small compounds inhibit pilus biogenesis in uropathogenic bacteria. Proc. Natl. Acad. Sci. USA 103, 17897–17902.

Remaut, H., Rose, R.J., Hannan, T.J., et al., 2006. Donor-strand exchange in chaperone-assisted pilus assembly proceeds through a concerted beta strand displacement mechanism. Mol. Cell 22, 831–842.

Remaut, H., Tang, C., Henderson, N.S., et al., 2008. Fiber formation across the bacterial outer membrane by the chaperone/usher pathway. Cell 133, 640–652.

Roberts, J.A., Marklund, B.I., Ilver, D., et al., 1994. The Gal(alpha 1-4)Gal-specific tip adhesin of *Escherichia coli* P-fimbriae is needed for pyelonephritis to occur in the normal urinary tract. Proc. Natl. Acad. Sci. USA 91, 11889–11893.

Ronald, A.R., Nicolle, L.E., Stamm, E., et al., 2001. Urinary tract infection in adults: research priorities and strategies. Int. J. Antimicrob. Agents. 17, 343–348.

Rose, R.J., Verger, D., Daviter, T., et al., 2008. Unraveling the molecular basis of subunit specificity in P pilus assembly by mass spectrometry. Proc. Natl. Acad. Sci. USA 105, 12873–12878.

Rosen, D.A., Hooton, T.M., Stamm, W.E., Humphrey, P.A., Hultgren, S.J., 2007. Detection of intracellular bacterial communities in human urinary tract infection. PLoS Med. 4, e329.

Salit, I.E., Gotschlich, E.C., 1977. Hemagglutination by purified type I *Escherichia coli* pili. J. Exp. Med. 146, 1169–1181.

Sauer, F.G., Futterer, K., Pinkner, J.S., Dodson, K.W., Hultgren, S.J., Waksman, G., 1999. Structural basis of chaperone function and pilus biogenesis. Science 285, 1058–1061.

Sauer, F.G., Pinkner, J.S., Waksman, G., Hultgren, S.J., 2002. Chaperone priming of pilus subunits facilitates a topological transition that drives fiber formation. Cell 111, 543–551.

Saulino, E.T., Thanassi, D.G., Pinkner, J.S., Hultgren, S.J., 1998. Ramifications of kinetic partitioning on usher-mediated pilus biogenesis. EMBO J. 17, 2177–2185.

Slonim, L.N., Pinkner, J.S., Branden, C.I., Hultgren, S.J., 1992. Interactive surface in the PapD chaperone cleft is conserved in pilus chaperone superfamily and essential in subunit recognition and assembly. EMBO J. 11, 4747–4756.

So, S.S., Thanassi, D.G., 2006. Analysis of the requirements for pilus biogenesis at the outer membrane usher and the function of the usher C-terminus. Mol. Microbiol. 60, 364–375.

Striker, R., Jacob-Dubuisson, F., Freiden, C., Hultgren, S.J., 1994. Stable fiber-forming and nonfiber-forming chaperone-subunit complexes in pilus biogenesis. J. Biol. Chem. 269, 12233–12239.

Thanassi, D.G., Saulino, E.T., Lombardo, M.J., Roth, R., Heuser, J., Hultgren, S.J., 1998. The PapC usher forms an oligomeric channel: implications for pilus biogenesis across the outer membrane. Proc. Natl. Acad. Sci. USA 95, 3146–3151.

Thanassi, D.G., Stathopoulos, C., Dodson, K., Geiger, D., Hultgren, S.J., 2002. Bacterial outer membrane ushers contain distinct targeting and assembly domains for pilus biogenesis. J. Bacteriol. 184, 6260–6269.

Thomas, W.E., Trintchina, E., Forero, M., Vogel, V., Sokurenko, E.V., 2002. Bacterial adhesion to target cells enhanced by shear force. Cell 109, 913–923.

Totsika, M., Beatson, S.A., Sarkar, S., et al., 2011. Insights into a multidrug resistant *Escherichia coli* pathogen of the globally disseminated ST131 lineage: genome analysis and virulence mechanisms. PloS One 6, e26578.

Valenski, M.L., Harris, S.L., Spears, P.A., Horton, J.R., Orndorff, P.E., 2003. The product of the fimI gene is necessary for *Escherichia coli* type 1 pilus biosynthesis. J. Bacteriol. 185, 5007–5011.

Verger, D., Bullitt, E., Hultgren, S.J., Waksman, G., 2007. Crystal structure of the P pilus rod subunit PapA. PLoS Pathog. 3, e73.

Verger, D., Miller, E., Remaut, H., Waksman, G., Hultgren, S., 2006. Molecular mechanism of P pilus termination in uropathogenic *Escherichia coli*. EMBO Rep. 7, 1228–1232.

Verger, D., Rose, R.J., Paci, E., et al., 2008. Structural determinants of polymerization reactivity of the P pilus adaptor subunit PapF. Structure 16, 1724–1731.

Volkan, E., Ford, B.A., Pinkner, J.S., et al., 2012. Domain activities of PapC usher reveal the mechanism of action of an *Escherichia coli* molecular machine. Proc. Natl. Acad. Sci. USA 109, 9563–9568.

Waksman, G., Hultgren, S.J., 2009. Structural biology of the chaperone-usher pathway of pilus biogenesis. Nat. Rev. Microbiol. 7, 765–774.

Walz, W., Schmidt, M.A., Labigne-Roussel, A.F., Falkow, S., Schoolnik, G., 1985. AFA-I, a cloned afimbrial X-type adhesin from a human pyelonephritic *Escherichia coli* strain. Purification and chemical, functional and serologic characterization. Eur. J. Biochem. 152, 315–321.

Wright, K.J., Seed, P.C., Hultgren, S.J., 2007. Development of intracellular bacterial communities of uropathogenic *Escherichia coli* depends on type 1 pili. Cell. Microbiol. 9, 2230–2241.

Wu, X.R., Sun, T.T., Medina, J.J., 1996. In vitro binding of type 1-fimbriated *Escherichia coli* to uroplakins Ia and Ib: relation to urinary tract infections. Proc. Natl. Acad. Sci. USA 93, 9630–9635.

Zavialov, A.V., Knight, S.D., 2007. A novel self-capping mechanism controls aggregation of periplasmic chaperone Caf1M. Mol. Microbiol. 64, 153–164.

Chapter 13

The type 2 secretion and type 4 pilus systems of *Escherichia coli*

Leon G. De Masi, Courntey D. Sturey, Joshua A. Lieberman,
Michael S. Donnenberg
University of Maryland School of Medicine, Baltimore, MD, USA

INTRODUCTION

Definition

The type 2 secretion (T2S) and the type 4 pilus (T4P) systems are two broadly conserved systems that play important roles in the virulence of pathogenic *Escherichia coli* and many other bacteria. These systems have been identified in enteropathogenic *E. coli* (EPEC), enterotoxigenic *E. coli* (ETEC), enterohemorrhagic *E. coli* (EHEC), and non-pathogenic *E. coli* strains. The T2S system exports exotoxins and important cell-surface anchored lipoprotein enzymes that digest components of the host cell, including proteins, lipids, and sugars (Economou et al., 2006). The T4P system produces pili that are involved in adherence, motility, and aggregation. Both systems consist of multicomponent machines that are broadly distributed but play distinct roles in bacterial pathogenesis. Components of the T2S and T4P machines are homologous in sequence, structure, and function. Only a few key differences account for the ability of T4P systems to anchor polymerized substrates in the inner membrane (IM) as opposed to the secretion of individual, folded substrates across the outer membrane (OM) by T2S systems. Under artificial conditions, T4P systems can assemble T2S pseudopilins into pili (Sauvonnet et al., 2000).

T2S represents one of at least six secretion systems in Gram-negative bacteria that function in the export of proteins to the extracellular environment (Economou et al., 2006). T2S has been referred to as the main terminal branch of the general secretory pathway (Gsp) and the second part of a two-step process. The first step is transport of the exoprotein across the IM by the Sec or Tat pathway (Cianciotto, 2005). The substrate precursor contains an N-terminal signal peptide that targets the protein to be transported through the IM. The signal peptide is cleaved off and the protein is released into the periplasm where it folds into its functional confirmation. The protein is then transported by the

T2S across the OM into the extracellular environment through a pore formed by the secretin protein (Genin and Boucher, 1994).

The T2S system in *E. coli* consists of 12 core proteins found in all T2S systems and two accessory proteins (Sandkvist, 2001). Mutations in genes encoding most of these proteins result in defects in the secretion process and accumulation of exoproteins in the periplasm (Filloux, 2004). There are chromosomal genes in the non-pathogenic *E. coli* K-12 and most other strains that encode a T2S system; however, under standard conditions secretion does not occur. The operon encoding this system is silenced by the nucleoid-structuring protein H-NS, and mutants deficient for H-NS can express the T2S system (Francetic et al., 2000). In *E. coli* pathogens, proteins secreted by accessory T2S systems play multiple roles, including adherence to eukaryotic cells and damage to host tissues. Examples include the EHEC StcE metalloprotease, which promotes adherence to host eukaryotic cells (Grys et al., 2005) and the ETEC heat-labile toxin (LT), a multimeric AB_5 enterotoxin functionally and structurally similar to cholera toxin of *Vibrio cholerae* (Sixma et al., 1993) that induces fluid secretion and promotes colonization of intestinal epithelia (Tauschek et al., 2002).

T4P are filamentous surface appendages required for adherence, motility, aggregation, and transformation in a wide array of bacterial and archaeal species. T4P were first identified in *Pseudomonas aeruginosa* and *Myxococcus xanthus* (Ottow, 1975). Further studies revealed that all T4P have a conserved set of core proteins (Pelicic, 2008). T4P are long, thin, flexible, polymeric, three-start helical filaments approximately 85 Å in diameter (Craig et al., 2004; Ramboarina et al., 2005). The pili are composed primarily of the major pilin protein (Craig et al., 2004), and are assembled by a complex biogenesis machine consisting of 10–18 proteins that spans both bacterial membranes (Sohel et al., 1996; Stone et al., 1996; Hwang et al., 2003; Peabody et al., 2003; Pelicic, 2008). Major pilins are first synthesized as precursor prepilins and then processed by a prepilin peptidase to create mature pilin proteins incorporated into the pilus (Strom et al., 1993).

T4P can be subdivided into two distinct classes, type 4 class A pili (T4aPs) and type 4 class B pili (T4bPs), based on differences in component structure, genetic organization, distribution, and function. T4bPs are characterized by larger major pilins compared to those of T4aPs that begin with an amino acid other than phenylalanine and are encoded in a contiguous operon often found on plasmids (Craig et al., 2004). As with the common T2S system, most strains of *E. coli* have the genes for a T4aP that has been called, somewhat confusingly, the hemorrhagic coli pilus (HCP) (Xicohtencatl-Cortes et al., 2007). In addition, some *E. coli* pathotypes express accessory T4bPs that play important roles in infection.

History

The T2S system was initially discovered in *Klebsiella oxytoca* and *Pseudomonas aeruginosa* and is widely conserved among Gram-negative bacteria (Filloux et al., 1990). Sequencing of plasmid pO157 from EHEC identified a T2S system conserved among all EHEC O157 and some non-O157 strains (Schmidt et al., 1997). A silenced T2S operon was also found in an *E. coli* K-12 strain (Blattner et al., 1997). This system can secrete endochitinase ChiA when the silencing protein is inactivated (Francetic et al., 2000). In ETEC a T2S system highly similar to that responsible for cholera toxin secretion in *Vibrio cholerae* was discovered to secrete LT (Tauschek et al., 2002). This gene cluster is absent from *E. coli* K12, however the K12 genome appears to have once contained this sequence.

Although discovered years earlier in other bacteria, T4P where first identified in *E. coli* with the discovery of the EPEC bundle-forming pilus (BFP) (Girón et al., 1991). Subsequent identification of the gene encoding the major structural subunit confirmed its similarity to other T4P pilin genes and suggested the presence of a large operon encoding genes required for pilus processing and biogenesis (Donnenberg et al., 1992). Additional *E. coli* T4P were discovered subsequently.

Distribution

Both T2S and T4P are produced by a diverse number of Gram-negative bacteria including many plant, animal, and human pathogens. Within *E. coli*, one T2S system and the HCP are ubiquitous, while additional T2S and T4P systems are found on specific subsets of pathotypes (Girón et al., 1993, 1994; Schmidt et al., 1997; Tauschek et al., 2002).

The components of T4P systems share significant sequence similarity and structural homology with components of T2S systems, DNA uptake systems (Averhoff and Friedrich, 2003; Peabody et al., 2003), and filamentous phage assembly systems (Linderoth et al., 1996; Russel et al., 1997). T4P assembly components also are orthologous to proteins involved in archaeal flagellum assembly (Peabody et al., 2003). The sequence and structural similarities across such a wide range of organisms strongly suggest an ancient and shared evolutionary history (Peabody et al., 2003; Pelicic, 2008). T4Ps have also been found in Gram-positive (Varga et al., 2006; Rodgers et al., 2011) and archaeal species (Herdendorf et al., 2002; Bardy and Jarrell, 2003).

Only a few T4bP systems have been identified in *E. coli*, where some are virulence factors used for adherence to eukaryotic cells, an essential first step for colonization. This is evident in the decreased virulence of EPEC strains deficient for functional pili (Bieber et al., 1998). The best-characterized system is the BFP of EPEC. Through the remainder of this chapter we will refer to the BFP system as the model of T4bP biogenesis, referring to other systems primarily to highlight differences when they occur. In addition to BFP, two related T4Ps have been identified in ETEC, the colonization factor antigen III (CFA/III) and Longus systems. Another system called the R64 pilus is encoded on a plasmid, and a T4aP called the hemorrhagic coli pilus or HCP has been described in

EHEC, although the genes encoding HCP are ubiquitous among *E. coli* strains (Xicohtencatl-Cortes et al., 2007).

BFP fibers are produced from an operon of 14 genes encoded on the large (~90 kb) adherence factor plasmid (pEAF) (Nataro et al., 1987; Stone et al., 1996). The BFP is a confirmed virulence factor (Bieber et al., 1998) that mediates the initial stages of adherence to the host intestinal epithelium (Cleary et al., 2004; Hyland et al., 2006a; Zahavi et al., 2011). CFA/III fibers are peritrichous pili that are 5–10 µm long and are often encoded on a large plasmid (Honda et al., 1984; Shinagawa et al., 1993). Some ETEC strains produce the Longus pilus, so named because it can grow over 20 µm in length, which is essentially allelic to CFA/III (Girón et al., 1994). The Longus T4P display a polar distribution on the cell surface (Girón et al., 1994). The N-termini of the major pilin proteins in each system, CofA and LngA, and of all accessory genes are nearly identical (Taniguchi et al., 1995; Gomez-Duarte et al., 2007). Despite their similarity, these pilus systems are expressed under different conditions and by different serogroups of ETEC (McConnell et al., 1989; Girón et al., 1995). The R64 pilus is encoded on a large conjugative plasmid. Unlike its counterparts in EPEC and ETEC, the R64 pilus has not been shown to have a role in adherence but rather is specialized for bacterial conjugation in liquid media (Yoshida et al., 1998). The HCP can grow to over 10 µm in length and is involved in adherence of EHEC to host cells (Xicohtencatl-Cortes et al., 2007).

GENETIC ORGANIZATION

The genetic organization of T2S systems is relatively well conserved. For all T2S systems there are 12 conserved genes, termed A–O, although in some cases additional genes may be required for T2S function (Peabody et al., 2003; Filloux, 2004). These genes are most often found in a single operon with slight variations depending on species (Sandkvist, 2001). The majority of T2S systems are located on the chromosome with the exception of a plasmid-encoded T2S system in *E. coli* O157 (Cianciotto, 2005).

Most T4P expressed in *E. coli* are T4bP and thus are found in a continuous operon on a large plasmid (Figure 13.1). The exception, the HCP pilus characterized in EHEC, is a T4aP and is thus encoded on several small operons in the *E. coli* genome, though little is known about the function of each gene in these operons outside of *hcpA* (also known as *ppdD*), the main pilus subunit (Xicohtencatl-Cortes et al., 2007). In EPEC, the *bfp* operon is found on pEAF (Sohel et al., 1996; Stone et al., 1996), and of the 14 genes, all except for *bfpH* must be expressed for BFP expression and function (Anantha et al., 2000). The operon begins with *bfpA*, which encodes the main pilus subunit precursor, prebundlin, and ends with *bfpL*. Expression of the *bfp* operon is controlled by *perA*, an AraC-family transcriptional activator (Gómez-Duarte and Kaper, 1995; Tobe et al., 1996).

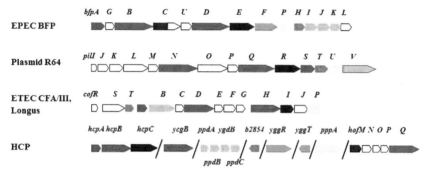

FIGURE 13.1   Comparison of *E. coli* T4P biogenesis genes. Arrows indicate genes encoding the following proteins: red (prepilin); magenta (secretin); dark green (extension ATPase); light green (retraction ATPase); dark blue (polytopic inner membrane protein); yellow (prepilin peptidase); light blue (lytic transglycosylase); orange (prepilin-like or minor pilin proteins); purple (bitopic inner membrane protein); no fill (no corresponding protein in other systems). The arrow depicting the *bfpC* gene is half-filled to indicate that the N-terminus of the protein has structural and functional homology to PilM. The slashes indicate that the HCP genes are not contiguous.

Like the *bfp* operon, the ETEC T4P systems Longus and CFA/III are encoded by a contiguous operon controlled by a *perA*-like promoter and found on a very large plasmid (Taniguchi et al., 1995, 2001; Gomez-Duarte et al., 2007). In both cases, the order of the genes differs substantially from that of the *bfp* operon, bearing more similarity to the *tcp* operon from *Vibrio cholerae* (Kirn et al., 2003). Another contiguous operon encodes the R64 conjugative pilus and is found on the plasmid of the same name (Kim and Komano, 1997; Sakai and Komano, 2002). Here, the genes encoding the putative tip adhesins are organized in a shufflon located at the end of the operon and the order of the genes encoding the assembly proteins differs markedly from both the BFP operon and the Longus/CFA/III operons. The shufflon is comprised of four DNA segments flanked by inverted repeat sequences. This allows for site-specific recombination between repeats and the formation of one of seven functional gene products, which have different LPS binding capabilities (Ishiwa and Komano, 2004).

REGULATION

T2S systems are regulated through specific quorum sensing and constitutive expression mechanisms depending upon the species. Regulation of the ETEC T2S system is under the control of the *yghJ* promoter, which is repressed by global regulatory proteins H-NS and StpA (Yang et al., 2007). It is expected that these proteins bind directly to the promoter region inhibiting open complex formation.

Multiple sensory and regulatory systems regulate expression of the BFP. The *pst* operon responds to inorganic phosphate starvation and positively regulates elaboration of the BFP machinery via the *per* operon (Ferreira and Spira, 2008).

The gene products of the *per* operon, PerA, PerB, and PerC, directly augment transcription of BFP genes (Tobe et al., 1999) and of a second transcriptional activator, Ler that in turn regulates additional EPEC genes involved in type 3 secretion (T3S) (Elliott et al., 2000) (see Chapter 4). Additionally, BFP receptor binding and retraction induce transient up-regulation of *bfp* transcription and down-regulates expression of Ler-induced virulence genes (Humphries et al., 2010). Retraction of BFP fibers is also required for efficient translocation of effector molecules secreted by this pathway (Zahavi et al., 2011).

Bacterial stress response pathways play an important role in regulating the expression of virulence factors in Gram-negative pathogens and the BFP is no exception (Nevesinjac and Raivio, 2005; MacRitchie et al., 2008; Vogt et al., 2010; Lieberman et al., 2012). In the case of the BFP system, full activation of the Cpx envelope stress response pathway represses transcription of BFP genes and inhibits pilus expression (Vogt et al., 2010). However, complete inactivation of the Cpx pathway in EPEC eliminates pilus biogenesis (Nevesinjac and Raivio, 2005) due to insufficient levels of periplasmic chaperone proteins such as DegP, CpxP, and DsbA (Vogt et al., 2010). A previous study revealed DsbA was required for stability of bundlin and BFP biogenesis (Zhang and Donnenberg, 1996). BFP can be expressed in K12 *E. coli* when the cloned *bfp* operon is placed under control of a strong promoter (Stone et al., 1996), but sufficient BFP production to allow associated phenotypes requires constitutive activation of Cpx (Price and Raivio, 2009).

STRUCTURAL COMPONENTS OF T2S AND T4P MACHINES

Introduction

The T2S and T4P machines are structurally and functionally homologous (Craig et al., 2004; Ayers et al., 2009; Korotkov et al., 2012). These complexes can be conceptualized as consisting of three distinct subcomplexes: the inner membrane (IM) subassembly associated with at least one cytoplasmic ATPase; the OM subassembly which consists of a transmembrane pore; and the translocated substrate. Remarkably few differences between the T2S and T4P machines account for their capacity to process very different substrates: the T2S extrudes fully folded proteins likely driven through the pore by a pseudopilus while the T4P system substrate is itself a pilus. Variations and accessory proteins do exist and are discussed after the conserved components. All components are listed using the General Secretory Pathway (Gsp) nomenclature used for ETEC (Table 13.1).

The IM subassembly

The IM subassemblies of both the T2S and T4P machines consist of a core set of homologous components, including several integral IM proteins and a hexameric ATPase that provides the energy driving secretion (Pelicic, 2008).

TABLE 13.1 Common and unique proteins of *E. coli* T2S and T4P systems

	System					
Protein description	Type 2 secretion	BFP	Longus	CFA/III	R64	HCP
Inner membrane protein, interacts with secretin	GspC	-	-	-	-	
Outer membrane secretin	GspD	BfpB	LngD	CofD	PilN	HofQ
Secretion ATPase	GspE	BfpD	LngH	CofH	PilQ	HcpB
Polytopic inner membrane platform protein	GspF	BfpE	LngI	CofI	PilR	HcpC
Major pilin or pseudopilin	GspG	BfpA	LngA	CofA	PilS	HcpA
Pilin-like protein, minor pilin or pseudopilin	GspH	BfpI	LngB	CofB	PilV	PpdC
Pilin-like protein, minor pilin or pseudopilin	GspI	BfpJ	-	-	-	YgdB
Pilin-like protein, minor pilin or pseudopilin	GspJ	BfpK	-	-	-	PpdB
Pilin-like protein, minor pilin or pseudopilin	GspK	-	-	-	-	PpdA
Bitopic inner-membrane protein	GspL	BfpC	LngE	CofE	-	
Bitopic inner-membrane protein	GspM	-	-	-	-	
Prepilin peptidase	GspO	BfpP	LngP	CofP	PilU	b2972
Pilotin	YghG	-	-	-	-	
Gene regulation		PerA	LngS	CofS	-	
Gene regulation		-	LngR	CofR	-	
Retraction ATPase	-	BfpF	-	-	-	YggR

Continued

TABLE 13.1 Common and unique proteins of *E. coli* T2S and T4P systems—cont'd

	System					
Protein description	Type 2 secretion	BFP	Longus	CFA/III	R64	HCP
Outer membrane protein, forms complex with secretin	-	BfpG	LngC	CofC	-	
Soluble protein essential for pilus biogenesis	-	BfpU	LngG	CofG	-	
Putative lytic transglycosylase	-	BfpH	LngT	CofT	PilT	b2854
Lipoprotein	-	-	-	-	PilL	
Bitopic inner-membrane protein	-	BfpL	-	-		
Protein of unknown function	-	-	LngF	CofF	-	
Protein of unknown function	-	-	LngJ	CofJ	-	
Inner membrane protein of unknown function	-	-	-	-	PilK	
Inner membrane protein of unknown function	-	-	-	-	PilM	
Inner membrane protein of unknown function	-	-	-	-	PilP	
Outer membrane protein of unknown function	-	-	-	-	PilO	

A polytopic IM protein, GspF, is not only well-conserved in the T4P and T2S machines, but also has homologs in the archaeal flagella system and Gram-positive competence systems (Peabody et al., 2003). A bitopic transmembrane protein, GspL, interacts in the IM with GspM, another bitopic protein that is predominantly periplasmic (Sandkvist et al., 1999, 2000; Abendroth et al., 2004a, 2009a). The cytoplasmic domain of GspL interacts with the hexameric

ATPase, GspE (Abendroth et al., 2005; Camberg and Sandkvist, 2005; Camberg et al., 2007). In addition, the T2S IM complex includes a membrane-anchored periplasmic protein, GspC, which interacts with the secretin of the system (Korotkov et al., 2006; Lybarger et al., 2009).

The IM complex of T4P biogenesis machines is very similar. Using the BFP of EPEC as a prototype, there is also a polytopic IM protein, BfpE (Blank and Donnenberg, 2001). The T4P system has a bitopic IM protein BfpC (Milgotina et al., 2011; Yamagata et al., 2012). These two proteins interact with each other as well as with BfpD, an ATPase that powers pilus extension (Milgotina et al., 2011). Remarkably, despite the absence of sequence similarity, the cytoplasmic domain of BfpC is a structural and functional homolog of GspL and of the T4aP component PilM (Yamagata et al., 2012). The Longus and CFA/III pili of ETEC have the same core IM proteins as BFP (Table 13.1). The same core components are also found in R64, however this system has several IM proteins with unknown function (Kim and Komano, 1997). BfpL is a bitopic IM protein essential for BFP biogenesis that interacts with BfpC, but homologs of BfpL have not yet been described in other systems (De Masi et al., 2012).

Although the core transmembrane proteins and ATPase are present in both T2S and T4P systems there are several important differences. At present, no homolog for GspC, which interacts with the secretin and is essential for stability of the IM complex, has been found in an *E. coli* T4P system. Also, unlike T2S systems, most T4P systems have the capacity to rapidly retract their pilus fibers, and BFP is no exception. This is accomplished through a dedicated retraction ATPase, BfpF (Anantha et al., 1998; Bieber et al., 1998). However, an equivalent in the T4P systems of ETEC has not been found.

The OM subassembly

The T2S and T4P systems encounter a common problem in exporting substrates: crossing the OM. Both machines solve this problem with pores formed by proteins in the secretin superfamily. Secretins form multimeric pores in the OM, most often consisting of 12 monomers, each of which is believed to be a fully folded beta barrel (Korotkov et al., 2011). Many of these proteins require a small lipoprotein, termed a pilotin, for stability and successful transport to the OM. In the T2S system, the OM pore is formed by a dodecamer of the secretin, GspD (Genin and Boucher, 1994; Reichow et al., 2010). This complex is stabilized by the pilotin, GspS (Hardie et al., 1996). The T4P OM complex also includes a secretin, BfpB (Ramer et al., 1996; Schmidt et al., 2001). However, the BFP secretin is itself a lipoprotein (Ramer et al., 1996) and can assemble and reach the OM when expressed alone (Daniel et al., 2006; Lieberman et al., 2012). To date no pilotin has been identified in the BFP system. Four secretins in both T2S and T4P systems have been experimentally validated as lipoproteins and none appear to have or require cognate pilotins (Hu et al., 1995; Ramer et al., 1996; Schmidt et al., 2001; Bose and Taylor, 2005; Viarre et al., 2009). Interestingly, the secretins of CFA/III, Longus, and R64 systems all contain canonical lipobox sequences (LxxC) in their N-termini and are likely liposecretins (Wu and Tokunaga, 1986; Viarre et al., 2009). Additionally, none of the other *E. coli* T4P operons appears to include pilotin genes, further suggesting that the secretins are all self-sorting lipoproteins.

Beyond the lack of a cognate pilotin, the T4P OM complex is noticeably different from the T2S system by virtue of two unique proteins. Both are soluble, periplasmic proteins that associate with the OM in the presence of the secretin (Daniel et al., 2006). BfpG is the product of the second gene in the BFP operon (Schmidt et al., 2001; Daniel et al., 2006). The second, BfpU, is found in both the cytoplasm and periplasm and has no apparent T2S or T4aP homologs (Schreiber et al., 2002). Both of these proteins are essential for pilus biogenesis and interact with BfpB to form the OM subassembly (Daniel et al., 2006). Interactions between the BfpU and BfpG have not been described.

Pseudopilus and pilus

According to generally accepted theory, both the T2S and T4P build filamentous appendages that move through a central channel formed by the IM and OM subassemblies. The T2S builds a pseudopilus that is of sufficient length to reach the vestibule formed by the secretin (Reichow et al., 2011). The pseudopilus is primarily composed of GspG, the major pseudopilin (Sauvonnet et al., 2000; Durand et al., 2003) and a set of two to four minor pseudopilins depending upon the system. In ETEC the four minor pseudopilins are GspH, I, J, and K (Yanez et al., 2008a,b). All pseudopilins share N-terminal sequence homology and are processed by a prepilin peptidase GspO (Bally et al., 1992; Dupuy et al., 1992).

The T4P forms a structure that protrudes through the secretin and serves as a fimbrial adhesin. The main pilus subunit precursor, pre-bundlin for BFP, is cleaved by the prepilin peptidase, BfpP (Zhang et al., 1994) to form the primary component of the pilus. The solution structure of bundlin lacking the N-terminal stretch of hydrophobic amino acids predicted to be buried in the pilus core has been solved. Core features common to all T4P pilins and T2S pseudopilins, including the N-terminal alpha helix, the variable alpha-beta region, and the antiparallel beta sheet, are apparent (Figure 13.2). However, the nature of the alpha-beta region, the topological arrangement of the beta strands in the sheet, and the enclosure of the sheet in alpha helices is unique. These features contrast, not only with those of T4aP pilin proteins, but even with T4bP pilin proteins such as CofA (Fukakusa et al., 2012; Kolappan et al., 2012), with which it shares only the most conserved pilin features (Figure 13.2).

T4P systems also contain analogs of the minor pseudopilins called pilinlike proteins: BfpI, BfpJ, and BfpK. These proteins, like the minor pilins in other T4P systems, contain the same peptidase cleavage site as bundlin and are processed by the prepilin peptidase (Ramer et al., 2002). Like the pseudopilins, the pilin-like proteins may form a complex with one another (Koomey, 1995; Ramer et al., 2002; Helaine et al., 2007). Immunoelectron microscopy has confirmed that BfpI is a true minor pilin.

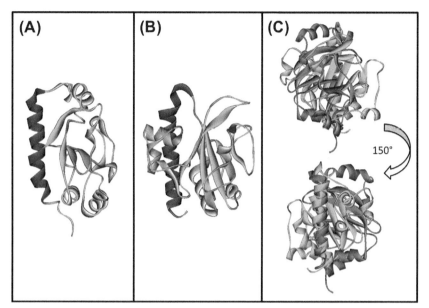

FIGURE 13.2   Structures of bundlin and CofA highlight similarities and differences among T4bP pilin proteins. Structures of soluble, N-terminally truncated versions of (A) bundlin (PDB 1zwt) and (B) CofA (PDB 3vor). The hydrophobic amino termini of both proteins, predicted to form contiguous alpha helices buried in the core of the pili, were omitted to facilitate purification and structure determination (Ramboarina et al., 2005; Fukakusa et al., 2012; Kolappan et al., 2012). Conserved features include the N-terminal alpha helix (color) and the central anti-parallel beta sheet (sheet). When overlaid (C), marked differences between bundlin (blue) and CofA (green) are apparent. Figures were drawn using PyMOL (The PyMOL Molecular Graphics System, Version 1.3, Schrödinger, LLC) by Kurt Piepenbrink.

There are several important differences between the T4P systems in *E. coli*. Unlike BFP, the CFA/III, Longus and R64 systems contain only one pilin-like protein (Gómez-Duarte et al., 1999; Yoshida et al., 1999; Taniguchi et al., 2001). Each of these is much larger in size than the pilin-like proteins in BFP. Second, the first amino acid after the prepilin peptidase cleavage site of BfpA is leucine and, as is the case with most T4P systems, is *N*-methylated (Donnenberg, 2012). However, in the R64 major pilin tryptophan is the first amino acid after the peptidase cleavage site and undergoes a distinct, but uncharacterized modification (Yoshida et al., 1998).

MECHANISM OF ACTION

Structure and function of type 2 secretion systems in *E. coli*

The T2S pseudopilus is believed to interact with substrate exoproteins in the periplasmic vestibule of the secretin and force the exoprotein into the extracellular space through an as yet undocumented piston mechanism (Reichow et al., 2011). The pseudopilus consists of the major pseudopilin, GspG, and minor pseudopilins GspH, I, J, and K, all of which are translocated across the IM via the Sec-dependent pathway (Francetic et al., 2007). A short, positively charged segment at the N-terminal region of each pseudopilin is cleaved off by the prepilin peptidase GspO (Bally et al., 1992; Dupuy et al., 1992). The peptidase is an aspartic acid protease with eight putative transmembrane helices and its catalytic residues in cytoplasmic loops (LaPointe and Taylor, 2000). The enzyme is bifunctional and also *N*-methylates the cleavage product (Strom et al., 1993). The pseudopilins GspI, J, and K appear to be incorporated as a trimer at the tip of the pseudopilus (Korotkov and Hol, 2008; Douzi et al., 2009) with GspK at the top and GspI/J forming the base. How monomers of GspG and the pseudopilin GspH are added to the pseudopilus remains unknown. All five pseudopilins have markedly different globular structures but share common features: an N-terminal alpha helix (the first half of which is hydrophobic) followed by a variable region and a conserved beta-sheet that varies in length with the various proteins.

T2S biogenesis requires the insertion of the IM proteins, GspC, F, L, and M, into the IM. These proteins create a complex together and appear to protect each other from proteolysis (Sandkvist et al., 1999). GspC contains a short cytoplasmic segment, a transmembrane helix, and two periplasmic domains: the homology region (HR) and PDZ domains (Bleves et al., 1999) although the PDZ domain is not universal across all T2S systems (Korotkov et al., 2006). The HR domain of GspC interacts with the periplasmic N-terminal regions of the secretin, GspD (Korotkov et al., 2011). The HR domain may also influence formation of the IM complex (Lybarger et al., 2009). GspF is the only polytopic protein in the complex, consisting of two cytoplasmic domains sharing some sequence homology and three transmembrane helices (Abendroth et al., 2009a,b). As discussed above, this polytopic transmembrane protein is one of the most highly conserved members of the T2S system and related machines (Peabody et al., 2003).

The GspL component recruits pseudopilins processed by GspO, the prepilin peptidase, to the IM complex (Sandkvist et al., 1995) and also recruits the ATPase, GspE, to the cytoplasmic face of the IM (Sandkvist et al., 2000; Abendroth et al., 2005). GspE is a member of a large ATPase family (Robien et al., 2003) providing the necessary energy for the secretion apparatus to function through ATP hydrolysis and, unlike most other proteins of its type contains zinc (Camberg and Sandkvist, 2005). The protein contains typical Walker boxes A and B (Planet et al., 2001) and is active as a hexamer (Satyshur et al., 2007). GspM contains a short cytoplasmic sequence, a transmembrane helix, and a ferrodoxin fold in its periplasmic domain that could serve as a binding site for another protein (Abendroth et al., 2004b). GspM interacts directly with GspL through contacts in the transmembrane domain and periplasm (Sandkvist et al., 1999; Py et al., 2001).

The secretin GspD is part of a large protein superfamily of multimeric, pore-forming OM proteins (Linderoth et al., 1996; Collins et al., 2004; Reichow et al., 2010). Although variations exist, some common themes of secretin structure have emerged. Like many of the secretin family members, GspD forms a dodecameric ring structure in the OM approximately 200 Å in height and 150–160 Å in width with a central pore through which exoprotein substrates pass (Korotkov et al., 2009; Reichow et al., 2010). The periplasmic side of the pore consists of a chamber that appears to serve as a docking site for secreted exoproteins (Reichow et al., 2011) with an opening of ~70 Å, which narrows to ~50 Å at a periplasmic gate (Korotkov et al., 2009; Reichow et al., 2010, 2011). The periplasmic chamber is formed by concentric rings composed of the N-termini of the monomers. How the secretin accommodates the passage of substrate proteins through the gates and narrow center of the channel, and how it prevents secretion of other molecules that are not substrates, remains unknown. The pore opens into a smaller chamber on the extracellular face that is capped. Since GspD is not a liposecretin, the pilotin GspS is required for secretin stability and transport to the OM via an interaction with the distal C-terminus of the secretin (Daefler et al., 1997; Nickerson et al., 2011). GspS is itself a lipoprotein that is acylated and transported to the OM via the Lol pathway (Collin et al., 2011).

Despite accumulating knowledge regarding the structures of the various T2S components, the actual mechanism of secreting a protein like LT of ETEC through the T2S apparatus has not been well established. However, a theoretical process has been proposed. Initially, the target proteins are synthesized in the cytoplasm and translocated via the Sec or Tat export pathways depending on whether the protein is folded in the periplasm or cytoplasm, respectively (Pugsley, 1993; Berks et al., 2005). Molecular modeling strongly suggests an interaction between the correctly folded substrate protein, the secretin, the pseudopilus tip complex of Gsp I/J/K, and GspC (Reichow et al., 2011), although interactions with additional machine components are possible. The binding of the exoprotein then stimulates the ATPase activity of GspE, leading to the addition of GspG monomers to the pseudopilus (Hobbs and Mattick, 1993; Shevchik et al., 1997). It is not known how the system chooses substrates from the plethora of periplasmic proteins available, or how the ATPase is specifically activated by substrate binding. The exoprotein is pushed upward through the secretin pore with the pseudopilus acting like a piston. However, it is not known how the pseudopilus is retracted. The trimer tip of the pseudopilus may interact with the secretin pore, inducing a conformational change destabilizing the GspG monomers below it, causing the pseudopilus to depolymerize (Korotkov et al., 2012). It notable that the degree of expression of the equivalent of GspK in the *Pseudomonas aeruginosa* T2S is inversely proportional to the length of the pseudopilus and an interaction between the GspK and GspG equivalent subunits destabilizes GspG (Durand et al., 2005).

Structure and function of type 4 pilus systems in *E. coli*

BFP biogenesis is believed to start with the Sec-dependent translocation of pre-bundlin and the pre-pilin-like proteins across the IM. The subunits are then processed by the BfpP prepilin peptidase (Zhang et al., 1994; Ramer et al., 2002), which cleaves pre-bundlin and the pilin-like proteins at a specific N-terminal cleavage site, removing a small leader sequence and adding a methyl group to the nascent N-terminal leucine residue. Simultaneously with pre-bundlin processing by BfpP at the cytoplasmic face of the IM, the globular head domain of bundlin is stabilized by DsbA, a periplasmic oxidoreductase that is not a component of the BFP system. DsbA catalyzes the formation of a disulfide bond in the C-terminus of bundlin critical for pilus assembly (Zhang and Donnenberg, 1996). The simultaneous and independent action of BfpP and DsbA on pre-bundlin indicates that prior to incorporation into the pili, pilins are integral transmembrane proteins.

T4P fibers are analogous to the pseudopilus complex of the T2S, except that they extend through the secretin into the extracellular space. T4P fibers are composed of polymerized pilin proteins and incorporated into a 3-start left-handed helix with the N-terminal alpha-helices of each subunit buried in the filament core and the C-terminal globular domains exposed on the surface of the pilus (Parge et al., 1995; Ramboarina et al., 2005). As with T2S, how each bundlin monomer is incorporated into the pilus is unknown. However, as BfpC is a homolog of GspL (Yamagata et al., 2012), it may recruit processed bundlin to the IM assembly complex. Like the minor pseudopilins of T2S, the pilin-like proteins BfpI, BfpJ, and BfpK may form a complex together (Ramer et al., 2002), as has been demonstrated for their T2S homologs. Pilin-like proteins have been found as integral components in T4P fibers of *Neisseria gonorrheae* (Helaine et al., 2007) and *Pseudomonas aeruginosa* (Giltner et al., 2010). As yet unpublished data indicate that BfpI is incorporated into BFP.

BfpE is a polytopic IM protein and the equivalent of GspF. However, in contrast to its counterpart, evidence suggests that BfpE may have four transmembrane segments, with two periplasmic domains, a small cytoplasmic domain, and both termini in the cytoplasm (Blank and Donnenberg, 2001). The extension ATPase BfpD provides the energy necessary to extract each pilin molecule and incorporate it into the pilus (Sohel et al., 1996; Stone et al., 1996; Milgotina et al., 2011). In a molecular model of the extraction of *P. aeruginosa* major pilin monomers from a membrane, the hydrolysis of as many as eight ATP molecules were required (Lemkul and Bevan, 2011). The ATPase activity of BfpD is slightly increased in the presence of the cytoplasmic N-termini of BfpE and BfpC (Yamagata et al., 2012). Another protein absolutely required for pilus biogenesis, BfpL, is also found in the IM (Ramer et al., 2002) and associates with the periplasmic face of BfpC (De Masi et al., 2012).

Once assembled, the pilus must be extruded through the membrane as more bundlin monomers are incorporated into the structure at its base. The secretin, BfpB, plays a crucial role in pilus biogenesis, as without a functional secretin multimer T4P biogenesis fails, a phenotype that manifests itself in EPEC as the loss of autoaggregation in *bfpB* and other mutants (Anantha et al., 2000; Durand et al., 2003). Here, secretins serve as the exit pore for pilus fibers as they extend and retract (Collins et al., 2005; Bayan et al., 2006; Korotkov et al., 2011). As with any secretin, conformational changes are required to permit transport, and significant changes would be required to accommodate the dynamic T4P fibers (Reichow et al., 2010, 2011). It has been suggested that the BFP machinery drives these conformational shifts, allowing passage of the assembled pilus (Hwang et al., 2003), though exactly how this occurs remains unknown.

Pilus retraction

While the process of pseudopilus retraction is unknown in the T2S, the phenotype is well characterized in BFP. Retraction is mediated by BfpF, a second hexameric ATPase which localizes to the cytoplasmic face of the IM. BfpF has homologs in some, but not all T4P systems and none in the T2S system (Anantha et al., 1998). Mutants deficient for *bfpF* do not disaggregate in culture (Bieber et al., 1998; Knutton et al., 1999) and may be hyperpiliated, though this observation has not been quantitatively examined. However, these phenotypic characteristics are consistent with a lack of pilus retraction in a *bfpF* mutant. Retraction of BFP facilitates host cell tight junction disruption and efficient delivery of bacterial host effectors during the course of EPEC pathogenesis (Zahavi et al., 2011). T4P fiber retraction is induced at least partly by BFP binding to *N*-acetyl lactosamine (Humphries et al., 2010).

T4P biogenesis machines can sustain forces of 100 pN during this rapid process of pilus retraction, during which T4P fibers undergo changes in quaternary structure (Biais et al., 2010). Such rapid changes in cell surface molecules likely exert considerable strain on the cell envelope. Similarly, simultaneously building many molecular machines that span both the IM and OM, such as those that elaborate T4P or the T2S, also may induce significant cell envelope stress. Indeed, the accumulation of very high levels of T4P subunits in the periplasm is cytotoxic in *Neisseria gonorrhoeae* (Wolfgang et al., 2000), although not in *N. meningitidis* (Carbonnelle et al., 2006).

It is thought that the extension and retraction ATPases in T4P act antagonistically during pilus biogenesis (Maier et al., 2004). It is not known how the activities of each ATPase are regulated to favor extension or retraction of the pilus in any system, however it has been demonstrated in *Myxococcus xanthus* that dynamic switching of the two ATPases across the cell causes extension or retraction, depending upon which ATPase is predominant at the polar regions (Bulyha et al., 2009). Further, since expression of the retraction ATPase is not required for pilus polymerization, a double mutant for the retraction ATPase and an accessory protein not absolutely necessary for pilus biogenesis will still express T4P on the surface even if pilus assembly is inefficient (Carbonnelle et al., 2006). This strategy can be utilized to determine which proteins are absolutely necessary for the polymerization process, for example T4P are expressed by *N. meningitidis* (Winther-Larsen et al., 2005) and *P. aeruginosa* (Giltner et al., 2010) in the absence of pilin-like proteins and the retraction ATPase, although their expression is required for normal T4P function (Brown et al., 2010). An ongoing analysis of *bfpF* double mutants indicates that BfpE, BfpB, BfpU, BfpG, and BfpL are absolutely required for pilus assembly, while BFP can be assembled in the absence of BfpI (unpublished data). However, how these findings apply to systems that lack a retraction ATPase, such as the T4P systems of ETEC, is not known.

Subcellular localization

The subcellular localization of T2S and T4P components has been the subject of some debate and may be of considerable functional importance. Studies with fluorescent fusion proteins have revealed the critical importance of gene dosage effects, while single molecule fluorescence and super-resolution microscopy techniques provide deeper insights into the actual distribution of machine components. In T2S systems, components fused to fluorescent proteins were observed at the poles when over-expressed, but these same fusion proteins produced peripheral foci of fluorescence when expressed at wild-type levels (Buddelmeijer et al., 2009; Lybarger et al., 2009). Thus, T2S are likely distributed around the cell and not exclusively at the cell poles.

The T4aPs of *Myxococcus xanthus* and *Pseudomonas aeruginosa* have been shown by transmission electron microscopy (TEM) and fluorescence microscopy to exit the cell primarily at a pole (Nudleman et al., 2006; Bulyha et al., 2009; Cowles and Gitai, 2010; Higashi et al., 2011), although non-polar fibers have been observed in *P. aeruginosa* (Cowles and Gitai, 2010). In these two systems, the T4P fibers mediate directional movement and their polar localization may be necessary for this function.

In EPEC, the BFP mediates initial attachment and aggregation into spherical aggregates. These interacting fibers create an extensive, overlapping, and complex meshwork and TEM has not yet demonstrated their subcellular origins. One recent study utilized photo-activated localization microscopy (PALM) to image single BfpB secretin molecules fused to photo-activated fluorophores and expressed in the context of the complete *bfp* operon (Lieberman et al., 2012). Such super-resolution techniques have made subdiffraction limit imaging possible and reveal unprecedented detail of T4P cell biology (Betzig et al., 2006; Hess et al., 2006). BfpB is localized to one or both of the bacterial cell poles in 20% of cells and is distributed around the cell envelope in the remainder; particularly when the cells are involved in aggregates, the secretin molecules form clusters (Lieberman et al., 2012). Furthermore, the secretin molecules may be distributed in a helix around the cell envelope (Lieberman et al., 2012), which may be a common phenotype for bacterial secretion systems (Aguilar et al., 2010, 2011; Lieberman et al., 2012). By distributing pili around the cell envelope, EPEC cells may be better able to form aggregates early in infection when BFP expression is critical (Zahavi et al., 2011) and thus confer a considerable advantage to the pathogen as it first colonizes the gastrointestinal tract, where

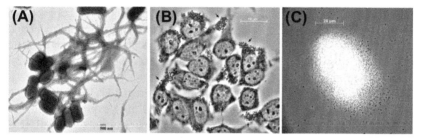

FIGURE 13.3 Reproduced with permission from Milgotina and Donnenberg, (2009). Phenotypes associated with bundle-forming pili expression in EPEC. (A) Transmission electron microscopy of EPEC bundle-forming pili. (B) Phase-contrast micrograph of EPEC displaying localized adherence to HeLa cells. Bacterial colonies indicated by arrows. (C) Phase-contrast image of auto-aggregation of wild-type bacteria in liquid culture upon induction of BFP expression. Scale bars are indicated.

it must compete with the commensal microbiota. The finding that BfpB is predominantly distributed around the cell periphery (Lieberman et al., 2012) contrasts with previous findings that BfpL and BfpF localize to the poles of cells expressing BFP (Ramer et al., 2002). The subcellular distribution of other T4P in *E. coli* has not been studied.

ROLE IN VIRULENCE

Many T4P play a crucial role in colonization and virulence by binding to cell surface receptors on the mucosal surface (Strom and Lory, 1993). BFP is a confirmed virulence factor (Bieber et al., 1998). BFP mediates the initial stages of adherence to the host intestinal epithelium (Cleary et al., 2004; Hyland et al., 2006a,b; Zahavi et al., 2011). Expression of BFPs is associated with a distinctive adherence pattern on the surface of host cells called localized adherence (LA) (Scaletsky et al., 1984), and the formation of aggregates in liquid culture (Anantha et al., 1998) (Figure 13.3).

Bundlin sequences vary (Blank et al., 2000). Those belonging to the alpha class bind to LacNac on the surface of host cells (Hyland et al., 2006a,b, 2008). A critical patch of surface-exposed residues determines this specificity (Humphries et al., 2009).

The BFP and Type 3 Secretion systems together orchestrate environmental sensing, machine assembly, substrate transport, and regulatory feedback networks that facilitate EPEC adherence to cells lining the intestinal epithelia. While BFPs are responsible for the initial attachment to host cells (Nougayrède et al., 2003), additional adhesins include the EspA filaments and intimin receptor of a T3S system. This system acts as a molecular syringe through which the bacteria inject numerous effector proteins into the host cell that modulate host cell processes (Kaper et al., 2004; Galán and Wolf-Watz, 2006; Dean and Kenny, 2009) (see Chapters 4, 14, and 15). The efficiency of intimate attachment depends upon initial adherence mediated by BFP, as has been demonstrated by analysis of a series of single, double, and triple mutant EPEC phenotypes (Cleary et al., 2004). While wild-type EPEC adhered robustly and exhibited LA, bacteria lacking BFP adhered poorly to host cells but could form intimate attachments, provided the EPEC strain still expressed both a functional T3S system and intimin.

In ETEC, the Longus pilus is important for aggregation. However, unlike pilus mutants in EPEC, ETEC mutants deficient for *lngA* expression can still adhere to human cell lines in vitro (Mazariego-Espinosa et al., 2010). ETEC strains expressing CFA/III can colonize infant mice and rabbits (Honda et al., 1984), human enterocytes (Knutton et al., 1989) and human carcinoma cells (Taniguchi et al., 2001). CFA/III expression is also associated with aggregation (Kolappan et al., 2012). However, CFA/III is one of numerous adhesins used by ETEC and its role in virulence is unknown.

IMMUNE RESPONSES

At present, very little has been published about the potential immune response to components in the inner and outer membrane assembly complexes from any bacterial T2S or T4P system or their potential as vaccine candidates. As the components of the IM assembly such as GspE or the prepilin peptidase are not surface expressed in bacteria they would likely make poor vaccination targets even if they were highly immunogenic. Mice immunized against the C-terminus of PilQ, the secretin of the T4P system of *Neisseria meningitidis,* generated serum bacteriocidal activity against homologous and heterologous strains (Haghi et al., 2012), but similar studies of *E. coli* secretins are lacking. Rather than focusing on the various components, most research focuses on the immunogenicity and vaccine efficacy of the proteins like LT secreted via the T2S pathway (Frech et al., 2008; Norton et al., 2011). The immunogenicity of each product differs according to the protein in question, but at least in the case of LT, immunizing volunteers with the exoprotein alone was enough to provide protection against travelers' diarrhea (Frech et al., 2008).

The main pilus subunit of EPEC, bundlin, is immunogenic (Loureiro et al., 1998; Fernandes et al., 2007) and antibodies against bundlin are produced in the course of an infection (Donnenberg et al., 1998; Parissi-Crivelli et al., 2000). There is evidence from volunteer studies of antibody responses against bundlin (Fernandes et al., 2007), but whether these responses are protective is unknown. Bundlin has been expressed in an attenuated *Salmonella enterica* serovar Typhimurium strain, but vaccine evaluation is difficult without a relevant animal model of infection (Schriefer et al., 1999). It is also possible that a bundlin-deficient mutant could be used as the basis for a live attenuated vaccine candidate as in the case of *V. cholerae* (Tacket et al., 1998). However, *bfpA* sequences vary. EPEC strains can be broken down into two broad categories of *bfpA* sequence similarity (Blank et al., 2000, 2003; Lacher et al., 2007). All alleles are at least 80% identical, with the alpha alleles having 97% sequence homology with one another, and the beta alleles are more divergent with 89% homology. The sequence diversity is concentrated at amino acids 137–155 in the C-terminal regions of bundlin, many of which are found on the surface of bundlin monomers polymerized into BFP (Ramboarina et al., 2005). While it is unknown whether anti-bundlin antibodies confer protective immunity, sera from volunteers rechallenged with an alpha bundlin strain of EPEC showed increases in response that were type-specific (Fernandes et al., 2007). Little research is available on the immune responses to other *E. coli* T4P.

THERAPEUTICS AND VACCINE PROSPECTS

As mentioned above, vaccine development for *E. coli* human pathogens is complicated by the lack of a valid model for study. Mouse models are commonly used for studying *E. coli* T3S homologs in *Citrobacter rodentium* (Frankel et al., 1996; Deng et al., 2003; Kelly et al., 2006), and a rabbit model for rabbit EPEC, has also been utilized (Peeters et al., 1988; Milon et al., 1992), as have porcine (Girard et al., 2005) and bovine in vitro cell models (Moxley and Francis, 1986). A mouse model for EPEC was previously proposed (Savkovic et al., 2005) but the adherence in both a mouse cell line and in vivo was very low, and the model has not been duplicated (Guttman et al., 2006). In the above instances, factors other than BFP are used by these pathogens for initial attachment to intestinal epithelial cells. These differences preclude using an existing animal model to study potential BFP vaccine candidates. Natural infections with BFP-expressing EPEC in dogs, birds and non-human primates hint of potential vaccine models yet to be developed (Beaudry et al., 1996; Schremmer et al., 1999; Carvalho et al., 2003).

ANTIVIRULENCE DRUGS

Most currently available antibiotics inhibit general bacterial growth or cause bacterial cell death, resulting in strong selective pressure for the emergence of antibiotic resistance. A focus on interrupting bacterial virulence may decrease this selective pressure. Due to the conservation of components and mechanism of action between T2S and T4P, drug design to target specific components of these systems could be advantageous. Both machines require tightly coordinated protein–protein interactions suggesting targets for chemotherapy. Most of the proteins in these systems have no human homologs, decreasing the likelihood of adverse effects (Peabody et al., 2003). Several studies have attempted to identify antivirulence drugs against T2S and T4P systems (Baron, 2010). Felise et al. identified a compound that inhibited both T2S and T3S in vitro and decreased secretion when cultured with eukaryotic cells (Felise et al., 2008). Further studies are needed to determine whether these advances will translate to clinical benefit.

CONCLUSIONS

The T2S and T4P systems of *E. coli* have long been known as important virulence factors. T2S systems secrete various toxins during pathogenesis depending upon the specific *E. coli* variant in question, while T4P are used by *E. coli* pathogens for initial adherence to host cells. While in most cases the T2S apparatus is encoded by genes on the bacterial chromosome, the T4P machinery is generally encoded on a contiguous operon located on a large plasmid. Both systems share a great deal of structural homology, as both contain a set of core proteins that form an IM subassembly complex, and an outer membrane secretin pore. Both systems polymerize a pilus or pilus-like structure as part of their function. While T2S systems are believed to use the pseudopilus to export exoproteins, such as toxins, into the extracellular space, the substrate of T4P systems is the pilus itself. Despite the amount of research published about both systems, the actual process by which the pseudopilin of the T2S or the pilin of the T4P is assembled remains poorly understood. As yet, little data have been published about the immunogenicity of components of either system and the prospects for vaccine development are hampered by the lack of suitable animal models. Despite these limitations, both systems hold promise as vaccine or drug targets given their importance in *E. coli* virulence.

REFERENCES

Abendroth, J., Bagdasarian, M., Sandkvist, M., Hol, W.G., 2004a. The structure of the cytoplasmic domain of EpsL, an inner membrane component of the type II secretion system of *Vibrio cholerae*: an unusual member of the actin-like ATPase superfamily. J. Mol. Biol. 344, 619–633.

Abendroth, J., Kreger, A.C., Hol, W.G., 2009a. The dimer formed by the periplasmic domain of EpsL from the type 2 secretion system of *Vibrio parahaemolyticus*. J. Struct. Biol. 168, 313–322.

Abendroth, J., Mitchell, D.D., Korotkov, K.V., et al., 2009b. The three-dimensional structure of the cytoplasmic domains of EpsF from the type 2 secretion system of *Vibrio cholerae*. J. Struct. Biol. 166, 303–315.

Abendroth, J., Murphy, P., Sandkvist, M., Bagdasarian, M., Hol, W.G., 2005. The X-ray structure of the type II secretion system complex formed by the N-terminal domain of EpsE and the cytoplasmic domain of EpsL of *Vibrio cholerae*. J. Mol. Biol. 348, 845–855.

Abendroth, J., Rice, A.E., McLuskey, K., Bagdasarian, M., Hol, W.G., 2004b. The crystal structure of the periplasmic domain of the type II secretion system protein EpsM from *Vibrio cholerae*: the simplest version of the ferredoxin fold. J. Mol. Biol. 338, 585–596.

Aguilar, J., Cameron, T.A., Zupan, J., Zambryski, P., 2011. Membrane and core periplasmic *Agrobacterium tumefaciens* virulence type IV secretion system components localize to multiple sites around the bacterial perimeter during lateral attachment to plant cells. MBio 2, e00218–11.

Aguilar, J., Zupan, J., Cameron, T.A., Zambryski, P.C., 2010. Agrobacterium type IV secretion system and its substrates form helical arrays around the circumference of virulence-induced cells. Proc. Natl. Acad. Sci. USA 107, 3758–3763.

Anantha, R.P., Stone, K.D., Donnenberg, M.S., 1998. The role of BfpF, a member of the PilT family of putative nucleotide-binding proteins, in type IV pilus biogenesis and in interactions between enteropathogenic *Escherichia coli* and host cells. Infect. Immun. 66, 122–131.

Anantha, R.P., Stone, K.D., Donnenberg, M.S., 2000. Effects of *bfp* mutations on biogenesis of functional enteropathogenic *Escherichia coli* type IV pili. J. Bacteriol. 182, 2498–2506.

Averhoff, B., Friedrich, A., 2003. Type IV pili-related natural transformation systems: DNA transport in mesophilic and thermophilic bacteria. Arch. Microbiol. 180, 385–393.

Ayers, M., Sampaleanu, L.M., Tammam, S., et al., 2009. PilM/N/O/P proteins form an inner membrane complex that affects the stability of the *Pseudomonas aeruginosa* type IV pilus secretin. J. Mol. Biol. 394, 128–142.

Bally, M., Filloux, A., Akrim, M., Ball, G., Lazdunski, A., Tommassen, J., 1992. Protein secretion in *Pseudomonas aeruginosa*: characterization of seven *xcp* genes and processing of secretory apparatus components by prepilin peptidase. Mol. Microbiol. 6, 1121–1131.

Bardy, S.L., Jarrell, K.F., 2003. Cleavage of preflagellins by an aspartic acid signal peptidase is essential for flagellation in the archaeon *Methanococcus voltae*. Mol. Microbiol. 50, 1339–1347.

Baron, C., 2010. Antivirulence drugs to target bacterial secretion systems. Curr. Opini. Microbiol. 13, 100–105.

Bayan, N., Guilvout, I., Pugsley, A.P., 2006. Secretins take shape. Mol. Microbiol. 60, 1–4.

Beaudry, M., Zhu, C., Fairbrother, J.M., Harel, J., 1996. Genotypic and phenotypic characterization of *Escherichia coli* isolates from dogs manifesting attaching and effacing lesions. J. Clin. Microbiol. 34, 144–148.

Berks, B.C., Palmer, T., Sargent, F., 2005. Protein targeting by the bacterial twin-arginine translocation (Tat) pathway. Curr. Opin. Microbiol. 8, 174–181.

Betzig, E., Patterson, G.H., Sougrat, R., et al., 2006. Imaging intracellular fluorescent proteins at nanometer resolution. Science 313, 1642–1645.

Biais, N., Higashi, D.L., Brujic, J., So, M., Sheetz, M.P., 2010. Force-dependent polymorphism in type IV pili reveals hidden epitopes. Proc. Natl. Acad. Sci. USA 107, 11358–11363.

Bieber, D., Ramer, S.W., Wu, C.Y., et al., 1998. Type IV pili, transient bacterial aggregates, and virulence of enteropathogenic *Escherichia coli*. Science 280, 2114–2118.

Blank, T.E., Donnenberg, M.S., 2001. Novel topology of BfpE, a cytoplasmic membrane protein required for type IV fimbrial biogenesis in enteropathogenic *Escherichia coli*. J. Bacteriol. 183, 4435–4450.

Blank, T.E., Lacher, D.W., Scaletsky, I.C.A., Zhong, H.L., Whittam, T.S., Donnenberg, M.S., 2003. Enteropathogenic *Escherichia coli* O157 strains from Brazil. Emer. Infect. Dis. 9, 113–115.

Blank, T.E., Zhong, H., Bell, A.L., Whittam, T.S., Donnenberg, M.S., 2000. Molecular variation among type IV pilin (*bfpA*) genes from diverse enteropathogenic *Escherichia coli* strains. Infect. Immun. 68, 7028–7038.

Blattner, F.R., Plunkett III, G., Bloch, C.A., et al., 1997. The complete genome sequence of *Escherichia coli* K-12. Science 277, 1453–1462.

Bleves, S., Gérard-Vincent, M., Lazdunski, A., Filloux, A., 1999. Structure-function analysis of XcpP, a component involved in general secretory pathway-dependent protein secretion in *Pseudomonas aeruginosa*. J. Bacteriol. 181, 4012–4019.

Bose, N., Taylor, R.K., 2005. Identification of a TcpC-TcpQ outer membrane complex involved in the biogenesis of the toxin-coregulated pilus of *Vibrio cholerae*. J. Bacteriol. 187, 2225–2232.

Brown, D.R., Helaine, S., Carbonnelle, E., Pelicic, V., 2010. Systematic functional analysis reveals that a set of seven genes is involved in fine-tuning of the multiple functions mediated by type IV pili in *Neisseria meningitidis*. Infect. Immun. 78, 3053–3063.

Buddelmeijer, N., Krehenbrink, M., Pecorari, F., Pugsley, A.P., 2009. Type II secretion system secretin PulD localizes in clusters in the *Escherichia coli* outer membrane. J. Bacteriol. 191, 161–168.

Bulyha, I., Schmidt, C., Lenz, P., et al., 2009. Regulation of the type IV pili molecular machine by dynamic localization of two motor proteins. Mol. Microbiol. 74, 691–706.

Camberg, J.L., Johnson, T.L., Patrick, M., Abendroth, J., Hol, W.G., Sandkvist, M., 2007. Synergistic stimulation of EpsE ATP hydrolysis by EpsL and acidic phospholipids. EMBO J. 26, 19–27.

Camberg, J.L., Sandkvist, M., 2005. Molecular analysis of the *Vibrio cholerae* type II secretion ATPase EpsE. J. Bacteriol. 187, 249–256.

Carbonnelle, E., Helaine, S., Nassif, X., Pelicic, V., 2006. A systematic genetic analysis in *Neisseria meningitidis* defines the Pil proteins required for assembly, functionality, stabilization and export of type IV pili. Mol. Microbiol. 61, 1510–1522.

Carvalho, V.M., Gyles, C.L., Ziebell, K., et al., 2003. Characterization of monkey enteropathogenic *Escherichia coli* (EPEC) and human typical and atypical EPEC serotype isolates from neotropical nonhuman primates. J. Clin. Microbiol. 41, 1225–1234.

Cianciotto, N.P., 2005. Type II secretion: a protein secretion system for all seasons. Trends Microbiol. 13, 581–588.

Cleary, J., Lai, L.-C., Donnenberg, M.S., Frankel, G., Knutton, S., 2004. Enteropathogenic *E. coli* (EPEC) adhesion to intestinal epithelial cells: role of bundle-forming pili (BFP), EspA filaments and intimin. Microbiology 150, 527–538.

Collin, S., Guilvout, I., Nickerson, N.N., Pugsley, A.P., 2011. Sorting of an integral outer membrane protein via the lipoprotein-specific Lol pathway and a dedicated lipoprotein pilotin. Mol. Microbiol. 80, 655–665.

Collins, R.F., Frye, S.A., Balasingham, S., Ford, R.C., Tonjum, T., Derrick, J.P., 2005. Interaction with type IV pili induces structural changes in the bacterial outer membrane secretin PilQ. J. Biol. Chem. 280, 18923–18930.

Collins, R.F., Frye, S.A., Kitmitto, A., Ford, R.C., Tønjum, T., Derrick, J.P., 2004. Structure of the *Neisseria meningitidis* outer membrane PilQ secretin complex at 12 A resolution. J. Biol. Chem. 279, 39750–39756.

Cowles, K.N., Gitai, Z., 2010. Surface association and the MreB cytoskeleton regulate pilus production, localization and function in *Pseudomonas aeruginosa*. Mol. Microbiol. 76, 1411–1426.

Craig, L., Pique, M.E., Tainer, J.A., 2004. Type IV pilus structure and bacterial pathogenicity. Nat. Rev. Microbiol. 2, 363–378.

Daefler, S., Guilvout, I., Hardie, K.R., Pugsley, A.P., Russel, M., 1997. The C-terminal domain of the secretin PulD contains the binding site for its cognate chaperone, PulS, and confers PulS dependence on pIVf1 function. Mol. Microbiol. 24, 465–475.

Daniel, A., Singh, A., Crowther, L.J., Fernandes, P.J., Schreiber, W., Donnenberg, M.S., 2006. Interaction and localization studies of enteropathogenic *Escherichia coli* type IV bundle-forming pilus outer membrane components. Microbiology 152, 2405–2420.

De Masi, L., Szmancinski, H., Schreiber, W., Donnenberg, M.S., 2012. BfpL is essential for type IV bundle-forming pilus biogenesis and interacts with the periplasmic face of BfpC. Microbiology 158, 2515–2526.

Dean, P., Kenny, B., 2009. The effector repertoire of enteropathogenic *E. coli*: ganging up on the host cell. Curr. Opin. Microbiol. 12, 101–109.

Deng, W., Vallance, B.A., Li, Y., Puente, J.L., Finlay, B.B., 2003. *Citrobacter rodentium* translocated intimin receptor (Tir) is an essential virulence factor needed for actin condensation, intestinal colonization and colonic hyperplasia in mice. Mol. Microbiol. 48, 95–115.

Donnenberg, M.S., 2012. The amino terminus of the type IV pilin bundlin is modified by N-methylation. Figshare. http://dx.doi.org/10.6084/m9.figshare.91923 Retrieved March 15, 2013 (GMT).

Donnenberg, M.S., Girón, J.A., Nataro, J.P., Kaper, J.B., 1992. A plasmid-encoded type IV fimbrial gene of enteropathogenic *Escherichia coli* associated with localized adherence. Mol. Microbiol. 6, 3427–3437.

Donnenberg, M.S., Tacket, C.O., Losonsky, G., et al., 1998. Effect of prior experimental human enteropathogenic *Escherichia coli* infection on illness following homologous and heterologous rechallenge. Infect. Immun. 66, 52–58.

Douzi, B., Durand, E., Bernard, C., et al., 2009. The XcpV/GspI pseudopilin has a central role in the assembly of a quaternary complex within the T2SS pseudopilus. J. Biol. Chem. 284, 34580–34589.

Dupuy, B., Taha, M.-K., Possot, O., Marchal, C., Pugsley, A.P., 1992. PulO, a component of the pullulanase secretion pathway of *Klebsiella oxytoca*, correctly and efficiently processes gonococcal type IV prepilin in *Escherichia coli*. Mol. Microbiol. 6, 1887–1894.

Durand, E., Bernadac, A., Ball, G., Lazdunski, A., Sturgis, J.N., Filloux, A., 2003. Type II protein secretion in *Pseudomonas aeruginosa*: the pseudopilus is a multifibrillar and adhesive structure. J. Bacteriol. 185, 2749–2758.

Durand, E., Michel, G., Voulhoux, R., Kurner, J., Bernadac, A., Filloux, A., 2005. XcpX controls biogenesis of the *Pseudomonas aeruginosa* XcpT-containing pseudopilus. J. Biol. Chem. 280, 31378–31389.

Economou, A., Christie, P.J., Fernandez, R.C., Palmer, T., Plano, G.V., Pugsley, A.P., 2006. Secretion by numbers: protein traffic in prokaryotes. Mol. Microbiol. 62, 308–319.

Elliott, S.J., Sperandio, V., Giron, J.A., et al., 2000. The locus of enterocyte effacement (LEE)-encoded regulator controls expression of both LEE- and non-LEE-encoded virulence factors in enteropathogenic and enterohemorrhagic *Escherichia coli*. Infect. Immun. 68, 6115–6126.

Felise, H.B., Nguyen, H.V., Pfuetzner, R.A., et al., 2008. An inhibitor of gram-negative bacterial virulence protein secretion. Cell Host. Microbe. 4, 325–336.

Fernandes, P.J., Guo, Q., Donnenberg, M.S., 2007. Functional consequences of sequence variation in bundlin, the enteropathogenic *Escherichia coli* type IV pilin protein. Infect. Immun. 75, 4687–4696.

Ferreira, G.M., Spira, B., 2008. The *pst* operon of enteropathogenic *Escherichia coli* enhances bacterial adherence to epithelial cells. Microbiology 154, 2025–2036.

Filloux, A., 2004. The underlying mechanisms of type II protein secretion. Biochim. Biophys. Acta. 1694, 163–179.

Filloux, A., Bally, M., Ball, G., Akrim, M., Tommassen, J., Lazdunski, A., 1990. Protein secretion in gram-negative bacteria: transport across the outer membrane involves common mechanisms in different bacteria. EMBO J. 9, 4323–4329.

Francetic, O., Badaut, C., Rimsky, S., Pugsley, A.P., 2000. The ChiA (YheB) protein of *Escherichia coli* K-12 is an endochitinase whose gene is negatively controlled by the nucleoid-structuring protein H-NS. Mol. Microbiol. 35, 1506–1517.

Francetic, O., Buddelmeijer, N., Lewenza, S., Kumamoto, C.A., Pugsley, A.P., 2007. Signal recognition particle-dependent inner membrane targeting of the PulG pseudopilin component of a type II secretion system. J. Bacteriol. 189, 1783–1793.

Frankel, G., Phillips, A.D., Hicks, S., Dougan, G., 1996. Enteropathogenic *Escherichia coli* - Mucosal infection models. Trans. R. Soc. Trop. Med. Hyg. 90, 347–352.

Frech, S.A., DuPont, H.L., Bourgeois, A.L., et al., 2008. Use of a patch containing heat-labile toxin from *Escherichia coli* against travellers' diarrhoea: a phase II, randomised, double-blind, placebo-controlled field trial. Lancet 371, 2019–2025.

Fukakusa, S., Kawahara, K., Nakamura, S., et al., 2012. Structure of the CFA/III major pilin subunit CofA from human enterotoxigenic *Escherichia coli* determined at 0.90 A resolution by sulfur-SAD phasing. Acta. Crystallogr. D. Biol. Crystallogr. 68, 1418–1429.

Galán, J.E., Wolf-Watz, H., 2006. Protein delivery into eukaryotic cells by type III secretion machines. Nature 444, 567–573.

Genin, S., Boucher, C.A., 1994. A superfamily of proteins involved in different secretion pathways in gram-negative bacteria: modular structure and specificity of the N-terminal domain. Mol. Gen. Genet. 243, 112–118.

Giltner, C.L., Habash, M., Burrows, L.L., 2010. *Pseudomonas aeruginosa* minor pilins are incorporated into type IV pili. J. Mol. Biol. 398, 444–461.

Girard, F., Batisson, I., Frankel, G.M., Harel, J., Fairbrother, J.M., 2005. Interaction of enteropathogenic and Shiga toxin-producing *Escherichia coli* and porcine intestinal mucosa: role of intimin and Tir in adherence. Infect. Immun. 73, 6005–6016.

Girón, J.A., Donnenberg, M.S., Martin, W.C., Jarvis, K.G., Kaper, J.B., 1993. Distribution of the bundle-forming pilus structural gene (*bfpA*) among enteropathogenic *Escherichia coli*. J. Infect. Dis. 168, 1037–1041.

Girón, J.A., Ho, A.S.Y., Schoolnik, G.K., 1991. An inducible bundle-forming pilus of enteropathogenic *Escherichia coli*. Science 254, 710–713.

Girón, J.A., Levine, M.M., Kaper, J.B., 1994. Longus: a long pilus ultrastructure produced by human enterotoxigenic *Escherichia coli*. Mol. Microbiol. 12, 71–82.

Girón, J.A., Viboud, G.I., Sperandio, V., et al., 1995. Prevalence and association of the longus pilus structural gene (*lngA*) with colonization factor antigens, enterotoxin types, and serotypes of enterotoxigenic *Escherichia coli*. Infect. Immun. 63, 4195–4198.

Gomez-Duarte, O.G., Chattopadhyay, S., Weissman, S.J., Giron, J.A., Kaper, J.B., Sokurenko, E.V., 2007. Genetic diversity of the gene cluster encoding longus, a type IV pilus of enterotoxigenic *Escherichia coli*. J. Bacteriol. 189, 9145–9149.

Gómez-Duarte, O.G., Kaper, J.B., 1995. A plasmid-encoded regulatory region activates chromosomal eaeA expression in enteropathogenic *Escherichia coli*. Infect. Immun. 63, 1767–1776.

Gómez-Duarte, O.G., Ruiz-Tagle, A., Gómez, D.C., et al., 1999. Identification of *lngA*, the structural gene of longus type IV pilus of enterotoxigenic *Escherichia coli*. Microbiology 145, 1809–1816.

Grys, T.E., Siegel, M.B., Lathem, W.W., Welch, R.A., 2005. The StcE protease contributes to intimate adherence of enterohemorrhagic *Escherichia coli* O157:H7 to host cells. Infect. Immun. 73, 1295–1303.

Guttman, J.A., Samji, F.N., Li, Y., Vogl, A.W., Finlay, B.B., 2006. Evidence that tight junctions are disrupted due to intimate bacterial contact and not inflammation during attaching and effacing pathogen infection in vivo. Infect. Immun. 74, 6075–6084.

Haghi, F., Peerayeh, S.N., Siadat, S.D., Zeighami, H., 2012. Recombinant outer membrane secretin PilQ(406-770) as a vaccine candidate for serogroup B *Neisseria meningitidis*. Vaccine 30, 1710–1714.

Hardie, K.R., Lory, S., Pugsley, A.P., 1996. Insertion of an outer membrane protein in *Escherichia coli* requires a chaperone-like protein. EMBO J. 15, 978–988.

Helaine, S., Dyer, D.H., Nassif, X., Pelicic, V., Forest, K.T., 2007. 3D structure/function analysis of PilX reveals how minor pilins can modulate the virulence properties of type IV pili. Proc. Natl. Acad. Sci. USA 104, 15888–15893.

Herdendorf, T.J., McCaslin, D.R., Forest, K.T., 2002. *Aquifex aeolicus* PilT, homologue of a surface motility protein, is a thermostable oligomeric NTPase. J. Bacteriol. 184, 6465–6471.

Hess, S.T., Girirajan, T.P., Mason, M.D., 2006. Ultra-high resolution imaging by fluorescence photoactivation localization microscopy. Biophys. J. 91, 4258–4272.

Higashi, D.L., Biais, N., Weyand, N.J., et al., 2011. *N. elongata* produces type IV pili that mediate interspecies gene transfer with *N. gonorrhoeae*. PLoS ONE 6, e21373.

Hobbs, M., Mattick, J.S., 1993. Common components in the assembly of type 4 fimbriae, DNA transfer systems, filamentous phage and protein-secretion apparatus: a general system for the formation of surface-associated protein complexes. Mol. Microbiol. 10, 233–243.

Honda, T., Arita, M., Miwatani, T., 1984. Characterization of new hydrophobic pili of human enterotoxigenic *Escherichia coli*: a possible new colonization factor. Infect. Immun. 43, 959–965.

Hu, N.T., Hung, M.N., Liao, C.T., Lin, M.H., 1995. Subcellular location of XpsD, a protein required for extracellular protein secretion by *Xanthomonas campestris* pv. *campestris*. Microbiology 141 (Pt 6), 1395–1406.

Humphries, R.M., Donnenberg, M.S., Strecker, J., et al., 2009. From alpha to beta: identification of amino acids required for the N-acetyllactosamine-specific lectin-like activity of bundlin. Mol. Microbiol. 72, 859–868.

Humphries, R.M., Griener, T.P., Vogt, S.L., , et al., 2010. N-acetyllactosamine-induced retraction of bundle-forming pili regulates virulence-associated gene expression in enteropathogenic *Escherichia coli*. Mol. Microbiol. 76, 1111–1126.

Hwang, J., Bieber, D., Ramer, S.W., Wu, C.Y., Schoolnik, G.K., 2003. Structural and topographical studies of the type IV bundle-forming pilus assembly complex of enteropathogenic *Escherichia coli*. J. Bacteriol. 185, 6695–6701.

Hyland, R.M., Beck, P., Mulvey, G.L., Kitov, P.I., Armstrong, G.D., 2006a. N-acetyllactosamine conjugated to gold nanoparticles inhibits enteropathogenic *Escherichia coli* colonization of the epithelium in human intestinal biopsy specimens. Infect. Immun. 74, 5419–5421.

Hyland, R.M., Griener, T.P., Mulvey, G.L., et al., 2006b. Basis for N-acetyllactosamine-mediated inhibition of enteropathogenic *Escherichia coli* localized adherence. J. Med. Microbiol. 55, 669–675.

Hyland, R.M., Sun, J., Griener, T.P., et al., 2008. The bundlin pilin protein of enteropathogenic *Escherichia coli* is an N-acetyllactosamine-specific lectin. Cell Microbiol. 10, 177–187.

Ishiwa, A., Komano, T., 2004. PilV adhesins of plasmid R64 thin pili specifically bind to the lipopolysaccharides of recipient cells. J. Mol. Biol. 343, 615–625.

Kaper, J.B., Nataro, J.P., Mobley, H.L., 2004. Pathogenic *Escherichia coli*. Nat. Rev. Microbiol. 2, 123–140.

Kelly, M., Hart, E., Mundy, R., et al., 2006. Essential role of the type III secretion system effector NleB in colonization of mice by *Citrobacter rodentium*. Infect. Immun. 74, 2328–2337.

Kim, S.R., Komano, T., 1997. The plasmid R64 thin pilus identified as a type IV pilus. J. Bacteriol. 179, 3594–3603.

Kirn, T.J., Bose, N., Taylor, R.K., 2003. Secretion of a soluble colonization factor by the TCP type 4 pilus biogenesis pathway in *Vibrio cholerae*. Mol. Microbiol. 49, 81–92.

Knutton, S., McConnell, M.M., Rowe, B., McNeish, A.S., 1989. Adhesion and ultrastructural properties of human enterotoxigenic *Escherichia coli* producing colonization factor antigens III and IV. Infect. Immun. 57, 3364–3371.

Knutton, S., Shaw, R.K., Anantha, R.P., Donnenberg, M.S., Zorgani, A.A., 1999. The type IV bundle-forming pilus of enteropathogenic *Escherichia coli* undergoes dramatic alterations in structure associated with bacterial adherence, aggregation and dispersal. Mol. Microbiol. 33, 499–509.

Kolappan, S., Roos, J., Yuen, A.S., Pierce, O.M., Craig, L., 2012. Structural characterization of CFA/III and Longus type IVb pili from enterotoxigenic *Escherichia coli*. J. Bacteriol. 194, 2725–2735.

Koomey, M., 1995. Prepilin-like molecules in type 4 pilus biogenesis: minor subunits, chaperones or mediators of organelle translocation? Trends. Microbiol. 3, 409–410 discussion 411–3.

Korotkov, K.V., Gonen, T., Hol, W.G., 2011. Secretins: dynamic channels for protein transport across membranes. Trends. Biochem. Sci. 36, 433–443.

Korotkov, K.V., Hol, W.G., 2008. Structure of the GspK-GspI-GspJ complex from the enterotoxigenic *Escherichia coli* type 2 secretion system. Nat. Struct. Mol. Biol. 15, 462–468.

Korotkov, K.V., Krumm, B., Bagdasarian, M., Hol, W.G., 2006. Structural and functional studies of EpsC, a crucial component of the type 2 secretion system from *Vibrio cholerae*. J. Mol. Biol. 363, 311–321.

Korotkov, K.V., Pardon, E., Steyaert, J., Hol, W.G., 2009. Crystal structure of the N-terminal domain of the secretin GspD from ETEC determined with the assistance of a nanobody. Structure 17, 255–265.

Korotkov, K.V., Sandkvist, M., Hol, W.G., 2012. The type II secretion system: biogenesis, molecular architecture and mechanism. Nat. Rev. Microbiol. 10, 336–351.

Lacher, D.W., Steinsland, H., Blank, T.E., Donnenberg, M.S., Whittam, T.S., 2007. Molecular evolution of typical enteropathogenic *Escherichia coli*: clonal analysis by multilocus sequence typing and virulence gene allelic profiling. J. Bacteriol. 189, 342–350.

LaPointe, C.F., Taylor, R.K., 2000. The type 4 prepilin peptidases comprise a novel family of aspartic acid proteases. J. Biol. Chem. 275, 1502–1510.

Lemkul, J.A., Bevan, D.R., 2011. Characterization of interactions between PilA from *Pseudomonas aeruginosa* strain K and a model membrane. J. Phys. Chem. B. 115, 8004–8008.

Lieberman, J.A., Frost, N.A., Hoppert, M., et al., 2012. Outer membrane targeting, ultrastructure and single molecule localization of the enteropathogenic *Escherichia coli* type IV pilus secretin BfpB. J. Bacteriol. 194, 1646–1658.

Linderoth, N.A., Model, P., Russel, M., 1996. Essential role of a sodium dodecyl sulfate-resistant protein IV multimer in assembly-export of filamentous phage. J. Bacteriol. 178, 1962–1970.

Loureiro, I., Frankel, G., Adu-Bobie, J., Dougan, G., Trabulsi, L.R., Carneiro-Sampaio, M.M., 1998. Human colostrum contains IgA antibodies reactive to enteropathogenic *Escherichia coli* virulence-associated proteins: intimin, BfpA, EspA, and EspB. J. Pediatr. Gastroenterol. Nutr. 27, 166–171.

Lybarger, S.R., Johnson, T.L., Gray, M.D., Sikora, A.E., Sandkvist, M., 2009. Docking and assembly of the type II secretion complex of *Vibrio cholerae*. J. Bacteriol. 191, 3149–3161.

MacRitchie, D.M., Ward, J.D., Nevesinjac, A.Z., Raivio, T.L., 2008. Activation of the Cpx envelope stress response down-regulates expression of several locus of enterocyte effacement-encoded genes in enteropathogenic *Escherichia coli*. Infect. Immun. 76, 1465–1475.

Maier, B., Koomey, M., Sheetz, M.P., 2004. A force-dependent switch reverses type IV pilus retraction. Proc. Natl. Acad. Sci. USA 101, 10961–10966.

Mazariego-Espinosa, K., Cruz, A., Ledesma, M.A., Ochoa, S.A., Xicohtencatl-Cortes, J., 2010. Longus, a type IV pilus of enterotoxigenic *Escherichia coli*, is involved in adherence to intestinal epithelial cells. J. Bacteriol. 192, 2791–2800.

McConnell, M.M., Chart, H., Scotland, S.M., Smith, H.R., Willshaw, G.A., Rowe, B., 1989. Properties of adherence factor plasmids of enteropathogenic *Escherichia coli* and the effect of host strain on expression of adherence to HEp-2 cells. J. Gen. Microbiol. 135, 1123–1134.

Milgotina, E., Donnenberg, M.S., 2009. The bundle-forming pilus and other type IVb pili. In: Jarrell, K. (Ed.), Pili and Flagella: Current Research and Future Trends, Caister Academic Press, Norfolk, UK, pp. 41–57.

Milgotina, E.I., Lieberman, J.A., Donnenberg, M.S., 2011. Corrigendum - The inner membrane subassembly of the enteropathogenic *Escherichia coli* bundle-forming pilus machine. Mol. Microbiol. 81, 1125–1127.

Milon, A., Esslinger, J., Camguilhem, R., 1992. Oral vaccination of weaned rabbits against enteropathogenic *Escherichia coli*-like *E. coli* O103 infection: use of heterologous strains harboring lipopolysaccharide or adhesin of pathogenic strains. Infect. Immun. 60, 2702–2709.

Moxley, R.A., Francis, D.H., 1986. Natural and experimental infection with an attaching and effacing strain of *Escherichia coli* in calves. Infect. Immun. 53, 339–346.

Nataro, J.P., Maher, K.O., Mackie, P., Kaper, J.B., 1987. Characterization of plasmids encoding the adherence factor of enteropathogenic *Escherichia coli*. Infect. Immun. 55, 2370–2377.

Nevesinjac, A.Z., Raivio, T.L., 2005. The Cpx envelope stress response affects expression of the type IV bundle-forming pili of enteropathogenic *Escherichia coli*. J. Bacteriol. 187, 672–686.

Nickerson, N.N., Tosi, T., Dessen, A., et al., 2011. Outer membrane targeting of secretin PulD relies on disordered domain recognition by a dedicated chaperone. J. Biol. Chem. 286 (45), 38833–38843.

Norton, E.B., Lawson, L.B., Freytag, L.C., Clements, J.D., 2011. Characterization of a mutant *Escherichia coli* heat-labile toxin, LT(R192G/L211A), as a safe and effective oral adjuvant. Clin. Vaccine Immunol. 18, 546–551.

Nougayrède, J.P., Fernandes, P.J., Donnenberg, M.S., 2003. Adhesion of enteropathogenic *Escherichia coli* to host cells. Cell Microbiol. 5, 359–372.

Nudleman, E., Wall, D., Kaiser, D., 2006. Polar assembly of the type IV pilus secretin in *Myxococcus xanthus*. Mol. Microbiol. 60, 16–29.

Ottow, J.C., 1975. Ecology, physiology, and genetics of fimbriae and pili. Annu. Rev. Microbiol. 29, 79–108.

Parge, H.E., Forest, K.T., Hickey, M.J., Christensen, D.A., Getzoff, E.D., Tainer, J.A., 1995. Structure of the fibre-forming protein pilin at 2.6 Å resolution. Nature 378, 32–38.

Parissi-Crivelli, A., Parissi-Crivelli, J.M., Girón, J.A., 2000. Recognition of enteropathogenic *Escherichia coli* virulence determinants by human colostrum and serum antibodies. J. Clin. Microbiol. 38, 2696–2700.

Peabody, C.R., Chung, Y.J., Yen, M.R., Vidal-Ingigliardi, D., Pugsley, A.P., Saier Jr., M.H., 2003. Type II protein secretion and its relationship to bacterial type IV pili and archaeal flagella. Microbiology 149, 3051–3072.

Peeters, J.E., Geeroms, R., Orskov, F., 1988. Biotype, serotype, and pathogenicity of attaching and effacing enteropathogenic *Escherichia coli* strains isolated from diarrheic commercial rabbits. Infect. Immun. 56, 1442–1448.

Pelicic, V., 2008. Type IV pili: e pluribus unum? Mol. Microbiol. 68, 827–837.

Planet, P.J., Kachlany, S.C., DeSalle, R., Figurski, D.H., 2001. Phylogeny of genes for secretion NTPases: identification of the widespread tadA subfamily and development of a diagnostic key for gene classification. Proc. Natl. Acad. Sci. USA 98, 2503–2508.

Price, N.L., Raivio, T.L., 2009. Characterization of the Cpx regulon in *Escherichia coli* strain MC4100. J. Bacteriol. 191, 1798–1815.

Pugsley, A.P., 1993. The complete general secretory pathway in gram-negative bacteria. Microbiol. Rev. 57, 50–108.

Py, B., Loiseau, L., Barras, F., 2001. An inner membrane platform in the type II secretion machinery of Gram-negative bacteria. EMBO Rep. 2, 244–248.

Ramboarina, S., Fernandes, P.J., Daniell, S., et al., 2005. Structure of the bundle-forming pilus from enteropathogenic *Escherichia coli*. J. Biol. Chem. 280, 40252–40260.

Ramer, S.W., Bieber, D., Schoolnik, G.K., 1996. BfpB, an outer membrane lipoprotein required for the biogenesis of bundle-forming pili in enteropathogenic *Escherichia coli*. J. Bacteriol. 178, 6555–6563.

Ramer, S.W., Schoolnik, G.K., Wu, C.Y., Hwang, J., Schmidt, S.A., Bieber, D., 2002. The type IV pilus assembly complex: biogenic interactions among the bundle-forming pilus proteins of enteropathogenic *Escherichia coli*. J. Bacteriol. 184, 3457–3465.

Reichow, S.L., Korotkov, K.V., Gonen, M., et al., 2011. The binding of cholera toxin to the periplasmic vestibule of the type II secretion channel. Channels (Austin.) 5, 215–218.

Reichow, S.L., Korotkov, K.V., Hol, W.G., Gonen, T., 2010. Structure of the cholera toxin secretion channel in its closed state. Nat. Struct. Mol. Biol. 17, 1226–1232.

Robien, M.A., Krumm, B.E., Sandkvist, M., Hol, W.G., 2003. Crystal structure of the extracellular protein secretion NTPase EpsE of *Vibrio cholerae*. J. Mol. Biol. 333, 657–674.

Rodgers, K., Arvidson, C.G., Melville, S., 2011. Expression of a *Clostridium perfringens* type IV pilin by *Neisseria gonorrhoeae* mediates adherence to muscle cells. Infect. Immun. 79, 3096–3105.

Russel, M., Linderoth, N.A., Šali, A., 1997. Filamentous phage assembly: variation on a protein export theme. Gene. 192, 23–32.

Sakai, D., Komano, T., 2002. Genes required for plasmid R64 thin-pilus biogenesis: identification and localization of products of the pilK, pilM, pilO, pilP, pilR, and pilT genes. J. Bacteriol. 184, 444–451.

Sandkvist, M., 2001. Biology of type II secretion. Mol. Microbiol. 40, 271–283.

Sandkvist, M., Bagdasarian, M., Howard, S.P., DiRita, V.J., 1995. Interaction between the autokinase EpsE and EpsL in the cytoplasmic membrane is required for extracellular secretion in *Vibrio cholerae*. EMBO J. 14, 1664–1673.

Sandkvist, M., Hough, L.P., Bagdasarian, M.M., Bagdasarian, M., 1999. Direct interaction of the EpsL and EpsM proteins of the general secretion apparatus in *Vibrio cholerae*. J. Bacteriol. 181, 3129–3135.

Sandkvist, M., Keith, J.M., Bagdasarian, M., Howard, S.P., 2000. Two regions of EpsL involved in species-specific protein-protein interactions with EpsE and EpsM of the general secretion pathway in *Vibrio cholerae*. J. Bacteriol. 182, 742–748.

Satyshur, K.A., Worzalla, G.A., Meyer, L.S., et al., 2007. Crystal structures of the pilus retraction motor PilT suggest large domain movements and subunit cooperation drive motility. Structure 15, 363–376.

Sauvonnet, N., Vignon, G., Pugsley, A.P., Gounon, P., 2000. Pilus formation and protein secretion by the same machinery in *Escherichia coli*. EMBO J. 19, 2221–2228.

Savkovic, S.D., Villanueva, J., Turner, J.R., Matkowskyj, K.A., Hecht, G., 2005. Mouse model of enteropathogenic *Escherichia coli* infection. Infect. Immun. 73, 1161–1170.

Scaletsky, I.C.A., Silva, M.L.M., Trabulsi, L.R., 1984. Distinctive patterns of adherence of enteropathogenic *Escherichia coli* to HeLa cells. Infect. Immun. 45, 534–536.

Schmidt, H., Henkel, B., Karch, H., 1997. A gene cluster closely related to type II secretion pathway operons of gram-negative bacteria is located on the large plasmid of enterohemorrhagic *Escherichia coli* O157 strains. FEMS Microbiolo. Lett. 148, 265–272.

Schmidt, S.A., Bieber, D., Ramer, S.W., Hwang, J., Wu, C.Y., Schoolnik, G., 2001. Structure-function analysis of BfpB, a secretin-like protein encoded by the bundle-forming-pilus operon of enteropathogenic *Escherichia coli*. J. Bacteriol. 183, 4848–4859.

Schreiber, W., Stone, K.D., Strong, M.A., DeTolla Jr., L.J., Hoppert, M., Donnenberg, M.S., 2002. BfpU, a soluble protein essential for type IV pilus biogenesis in enteropathogenic *Escherichia coli*. Microbiology 148, 2507–2518.

Schremmer, C., Lohr, J.E., Wastlhuber, U., et al., 1999. Enteropathogenic *Escherichia coli* in *Psittaciformes*. Avian. Pathology. 28, 349–354.

Schriefer, A., Maltez, J.R., Silva, N., Stoeckle, M.Y., Barral-Netto, M., Riley, L.W., 1999. Expression of a pilin subunit BfpA of the bundle-forming pilus of enteropathogenic *Escherichia coli* in an *aroA* live salmonella vaccine strain. Vaccine 17, 770–778.

Shevchik, V.E., Robert-Baudouy, J., Condemine, G., 1997. Specific interaction between OutD, an *Erwinia chrysanthemi* outer membrane protein of the general secretory pathway, and secreted proteins. EMBO J. 16, 3007–3016.

Shinagawa, H., Taniguchi, T., Yamaguchi, O., Yamamoto, K., Honda, T., 1993. Cloning of the genes that control formation of the fimbrial colonization factor antigen III (CFA/III) from an entero-toxigenic *Escherichia coli*. Microbiol. Immunol. 37, 689–694.

Sixma, T.K., Kalk, K.H., van Zanten, B.A., et al., 1993. Refined structure of *Escherichia coli* heat-labile enterotoxin, a close relative of cholera toxin. J. Mol. Biol. 230, 890–918.

Sohel, I., Puente, J.L., Ramer, S.W., Bieber, D., Wu, C.-Y., Schoolnik, G.K., 1996. Enteropatho-genic *Escherichia coli*: identification of a gene cluster coding for bundle-forming pilus mor-phogenesis. J. Bacteriol. 178, 2613–2628.

Stone, K.D., Zhang, H.-Z., Carlson, L.K., Donnenberg, M.S., 1996. A cluster of fourteen genes from enteropathogenic *Escherichia coli* is sufficient for biogenesis of a type IV pilus. Mol. Microbiol. 20, 325–337.

Strom, M.S., Lory, S., 1993. Structure-function and biogenesis of the type IV pili. Annu. Rev. Microbiol. 47, 565–596.

Strom, M.S., Nunn, D.N., Lory, S., 1993. A single bifunctional enzyme, PilD, catalyzes cleavage and N-methylation of proteins belonging to the type IV pilin family. Proc. Natl. Acad. Sci. USA 90, 2404–2408.

Tacket, C.O., Taylor, R.K., Losonsky, G., et al., 1998. Investigation of the roles of toxin-coregulated pili and mannose-sensitive hemagglutinin pili in the pathogenesis of *Vibrio cholerae* O139 infection. Infect. Immun. 66, 692–695.

Taniguchi, T., Akeda, Y., Haba, A., et al., 2001. Gene cluster for assembly of pilus colonization fac-tor antigen III of enterotoxigenic *Escherichia coli*. Infect. Immun. 69, 5864–5873.

Taniguchi, T., Fujino, Y., Yamamoto, K., Miwatani, T., Honda, T., 1995. Sequencing of the gene encoding the major pilin of pilus colonization factor antigen III (CFA/III) of human entero-toxigenic *Escherichia coli* and evidence that CFA/III is related to type IV pili. Infect. Immun. 63, 724–728.

Tauschek, M., Gorrell, R.J., Strugnell, R.A., Robins-Browne, R.M., 2002. Identification of a protein secretory pathway for the secretion of heat-labile enterotoxin by an enterotoxigenic strain of *Escherichia coli*. Proc. Natl. Acad. Sci. USA 99, 7066–7071.

Tobe, T., Schoolnik, G.K., Sohel, I., Bustamante, V.H., Puente, J.L., 1996. Cloning and character-ization of *bfpTVW*, genes required for the transcriptional activation of *bfpA* in enteropathogenic *Escherichia coli*. Mol. Microbiol. 21, 963–975.

Tobe, T., Tatsuno, I., Katayama, E., Wu, C.Y., Schoolnik, G.K., Sasakawa, C., 1999. A novel chro-mosomal locus of enteropathogenic *Escherichia coli* (EPEC), which encodes a *bfpT*-regulated chaperone-like protein, TrcA, involved in microcolony formation by EPEC. Mol. Microbiol. 33, 741–752.

Varga, J.J., Nguyen, V., O'Brien, D.K., Rodgers, K., Walker, R.A., Melville, S.B., 2006. Type IV pili-dependent gliding motility in the Gram-positive pathogen *Clostridium perfringens* and other Clostridia. Mol. Microbiol. 62, 680–694.

Viarre, V., Cascales, E., Ball, G., Michel, G.P., Filloux, A., Voulhoux, R., 2009. HxcQ liposecre-tin is self-piloted to the outer membrane by its N-terminal lipid anchor. J. Biol. Chem. 284, 33815–33823.

Vogt, S.L., Nevesinjac, A.Z., Humphries, R.M., Donnenberg, M.S., Armstrong, G.D., Raivio, T.L., 2010. The Cpx envelope stress response both facilitates and inhibits elaboration of the entero-pathogenic *Escherichia coli* bundle-forming pilus. Mol. Microbiol. 76, 1095–1110.

Winther-Larsen, H.C., Wolfgang, M., Dunham, S., et al., 2005. A conserved set of pilin-like molecules controls type IV pilus dynamics and organelle-associated functions in *Neisseria gonorrhoeae*. Mol. Microbiol. 56, 903–917.

Wolfgang, M., van Putten, J.P., Hayes, S.F., Dorward, D., Koomey, M., 2000. Components and dynamics of fiber formation define a ubiquitous biogenesis pathway for bacterial pili. EMBO J. 19, 6408–6418.

Wu, H.C., Tokunaga, M., 1986. Biogenesis of lipoproteins in bacteria. Curr. Top. Microbiol. Immunol. 125, 127–157.

Xicohtencatl-Cortes, J., Monteiro-Neto, V., Ledesma, M.A., et al., 2007. Intestinal adherence associated with type IV pili of enterohemorrhagic *Escherichia coli* O157:H7. J. Clin. Invest. 117, 3519–3529.

Yamagata, A., Milgotina, E., Scanlon, K., Craig, L., Tainer, J.A., Donnenberg, M.S., 2012. Structure of an essential type IV pilus biogenesis protein provides insights into pilus and type II secretion systems. J. Mol. Biol. 419, 110–124.

Yanez, M.E., Korotkov, K.V., Abendroth, J., Hol, W.G., 2008a. The crystal structure of a binary complex of two pseudopilins: EpsI and EpsJ from the type 2 secretion system of *Vibrio vulnificus*. J. Mol. Biol. 375, 471–486.

Yanez, M.E., Korotkov, K.V., Abendroth, J., Hol, W.G.J., 2008b. Structure of the minor pseudopilin EpsH from the type 2 secretion system of *Vibrio cholerae*. J. Mol. Biol. 377, 91–103.

Yang, J., Baldi, D.L., Tauschek, M., Strugnell, R.A., Robins-Browne, R.M., 2007. Transcriptional regulation of the *yghJ-pppA-yghG-gspCDEFGHIJKLM* cluster, encoding the type II secretion pathway in enterotoxigenic *Escherichia coli*. J. Bacteriol. 189, 142–150.

Yoshida, T., Furuya, N., Ishikura, M., et al., 1998. Purification and characterization of thin pili of IncI1 plasmids Collb-P9 and R64: formation of PilV-specific cell aggregates by type IV pili. J. Bacteriol. 180, 2842–2848.

Yoshida, T., Kim, S.R., Komano, T., 1999. Twelve *pil* genes are required for biogenesis of the R64 thin pilus. J. Bacteriol. 181, 2038–2043.

Zahavi, E.E., Lieberman, J.A., Donnenberg, M.S., et al., 2011. Bundle forming pilus retraction enhances enteropathogenic *Escherichia coli* infectivity. Mol. Biol. Cell. 22, 2436–2447.

Zhang, H.-Z., Donnenberg, M.S., 1996. DsbA is required for stability of the type IV pilin of enteropathogenic *Escherichia coli*. Mol. Microbiol. 21, 787–797.

Zhang, H.-Z., Lory, S., Donnenberg, M.S., 1994. A plasmid-encoded prepilin peptidase gene from enteropathogenic *Escherichia coli*. J. Bacteriol. 176, 6885–6891.

Chapter 14

Type 3 secretion systems

Liam J. Worrall, Julien R.C. Bergeron, Natalie C.J. Strynadka
University of British Columbia, Vancouver, BC, Canada

INTRODUCTION

In 1994, the group of Hans Wolf-Watz observed that the pathogenic Gram-negative bacterium *Yersinia pestis* could transfer toxin proteins (effectors) directly into the cytoplasm of infected mammalian cells. Since this process was independent from other secretion pathways known at the time, it was labeled type 3 secretion (T3S) (Salmond and Reeves, 1993; Rosqvist et al., 1994). A set of genes, organized in a complex operon structure, was shown to encode for the proteins necessary and sufficient for the translocation of effectors across both the bacterial membranes, as well as the mammalian cell membrane (Michiels et al., 1991). This set of proteins was therefore called the type 3 secretion system (T3SS).

Subsequently, homologous genes were identified in many human Gram-negative pathogens such as *Salmonella enterica* spp., *Burkholderia* spp., *Chlamidia* spp., *Pseudomonas aeruginosa*, *Vibrio parahaemolyticus*, *Shigella* spp., enterohemorrhagic *E. coli* (EHEC) and enteropathogenic *E. coli* (EPEC), as well as in the plant pathogens *Pseudomonas syringae*, *Ralstonia solanacearum* and *Erwinia chrysanthemi*. In most instances, the occurrence of T3S was confirmed, and has been shown to be an important factor for pathogenicity (for review, see Troisfontaines and Cornelis, 2005; Coburn et al., 2007; Tampakaki et al., 2010). The delivery of effector proteins into a host cell via T3S allows the bacterium to manipulate the bacterial–host cell interaction in its favor with effectors targeting a wide variety of cellular processes including the cytoskeleton, phagocytosis, apoptosis, and the inflammatory response (see Chapter 15).

The molecular basis for T3S was discovered in 1998, when Kubori et al. (1998) observed the presence of syringe-shaped structures on the surface of *Salmonella typhimurium*. These possessed a morphology similar to the hook-basal body organization of the bacterial flagellum, and were shown to be composed of proteins encoded by the Pathogenicity Island 1 (SPI-1), a major

Escherichia coli. http://dx.doi.org/10.1016/B978-0-12-397048-0.00014-0

pathogenicity component of *Salmonella typhimurium*. This assembly, referred to as the T3SS injectisome, was shown to be necessary for the translocation of effectors. Subsequently, the direct visualization of T3SS-encoded injectisomes have also been reported for *Yersinia pestis* (Journet et al., 2003), *Shigella flexneri* (Tamano et al., 2000) and EPEC (Sekiya et al., 2001). The similarity between the injectisome and the flagellum is not only morphological: confirming a phylogenetic link between the two systems, many flagellar proteins, specifically those localized at the inner membrane, are homologous to T3SS injectisome components, suggesting a common evolutionary origin (Ginocchio et al., 1994).

In this chapter, we will describe the various T3SSs present in *E. coli* and related *Shigella* strains (Chaudhuri and Henderson, 2012) (see also Chapter 7), and their role in pathogenicity. We will then review the current knowledge of the organization of the T3SS injectisome with special attention to the structural detail as well as current insights into the possible molecular mechanisms employed by these systems for efficient secretion of effector proteins (see also Chapter 15 for a review of T3SS effector function). Finally, an overview of the complex regulation implemented for T3SS assembly and secretion will be summarized. Our review will focus on the T3SS injectisome, but various proteins of orthologous function in the flagellar T3SS will also be discussed.

TYPE 3 SECRETION SYSTEMS IN *E. COLI*

LEE-encoded T3SS

The locus of enterocyte effacement (LEE) is a pathogenicity island originally identified in the EPEC strain O127:H6 (McDaniel et al., 1995), and later found in most EPEC, EHEC, and atypical enteropathogenic *E. coli* (aEPEC) strains. It is responsible for the distinctive formation of attaching and effacing (A/E) lesions by LEE-encoding strains, requiring first adhesion of the bacteria to the host intestinal epithelium followed by rapid recruitment of various host cytoskeletal elements that ultimately create the characteristic pedestal (lesion) beneath the adhered bacterium (Donnenberg and Kaper, 1992).

In most strains, the LEE is found inserted within tRNA-encoding sites (typically *selC*, *pheU*, or *pheV*) (Wieler et al., 1997; Muller et al., 2009). Sequence analysis shows a lower GC content within the LEE (38.4%) compared to the average for the overall *E. coli* genome (50.8%) as well as homology to transposons (Donnenberg et al., 1997) that suggests a potential mode of horizontal transfer between *E. coli* strains (Reid et al., 2000; Sandner et al., 2001).

The LEE is ~35 kb long, and contains 41 reading frames organized into five distinct operons (Figure 14.1A) (Elliott et al., 1998), which include genes encoding for several T3SS effectors and their chaperones, a T3SS injectisome and export apparatus, T3SS regulators, and an adhesin (Jarvis et al., 1995). Secretion of various T3SS effectors in EPEC-infected epithelial cells has been

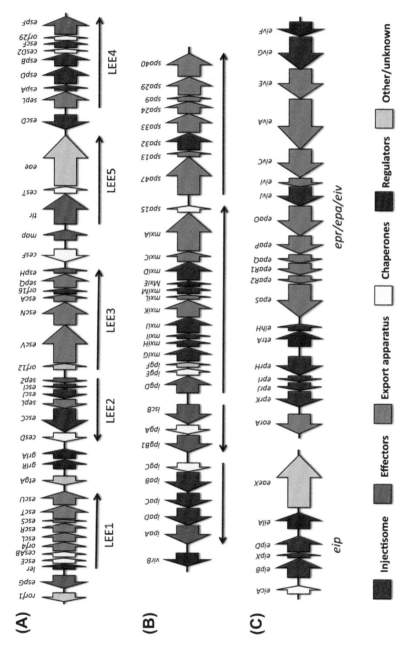

FIGURE 14.1 Genomic organization of the T3SSs found in *E. coli*. The gene arrangement for the EPEC LEE (A), *Shigella* (B) and EHEC ETT2 (C) secretion systems are shown, with known transcription operons shown in black arrows below. The color-coding for the role of various genes is indicated at the bottom.

observed (Kenny et al., 1996; Lai et al., 1997), directly confirming that the T3SS is functional. Additionally, deletion of individual LEE genes results in a lack of A/E lesion formation and reduced virulence (Rosenshine et al., 1992; Kenny and Finlay, 1995; Tacket et al., 2000).

Evolutionary, the T3SS encoded in the EPEC LEE is most closely related to the *Salmonella typhimurium* SPI-2 T3SS, as determined by sequence similarity (Hensel et al., 1998), which is active only in the *Salmonella*-containing vacuole. It remains to be seen if this phylogenetic proximity corresponds to similar features between these two systems, or if it simply reflects a common origin.

The Mxi/Spa system of *Shigella*

The main determinant of virulence in pathogenic *Shigella flexneri* is the presence of the large virulence plasmid pWR100 (Sansonetti et al., 1982, 1983), a 200 kb plasmid encoding for approximately 100 genes (Venkatesan et al., 2001). It is found in all pathogenic serotypes of *S. flexneri*, *S. dysenteriae* and *S. sonnei* (Yang et al., 2005), and is necessary for pathogenicity. Importantly, this plasmid can confer invasive properties to non-pathogenic *E. coli* strains (Sansonetti et al., 1983) and can be transmitted between bacteria (Sansonetti et al., 1982).

Within this plasmid, a 31 kb pathogenicity island was identified (Sasakawa et al., 1988), and sequence analysis showed that it contains 34 genes organized in two clusters transcribed in opposite directions (Figure 14.1B). Subsequent studies demonstrated that it forms a functional T3SS (Menard et al., 1994), labeled the Mxi/Spa system. It is one of the best-conserved regions of the virulence plasmid (Yang et al., 2005) and is necessary for epithelial cell invasion and macrophage killing (Sansonetti, 1991; Zychlinsky et al., 1992).

The Mxi/Spa system is closely related to the well-studied *inv* system, the *Salmonella typhimurium* SPI-1 encoded T3SS (Troisfontaines and Cornelis, 2005), which is responsible for the invasion into epithelial cells, suggesting a similar role for these two T3SSs.

The ETT2 system

More recently, sequence for a putative second T3SS was identified in the genome of the EHEC strain O157:H7 Sakai (Ohnishi et al., 2000), the source of a food-borne *E. coli* outbreak in Japan in 1999. This system, called ETT2 (*E. coli* Type Three secretion 2), was subsequently found in many EPEC and EHEC strains (Hartleib et al., 2003; Makino et al., 2003). Deletion of the ETT2 reduces pathogenicity (Ideses et al., 2005; Yao et al., 2009), but the molecular aspects of this effect remain unclear. In particular, secretion has not been observed to date for this system.

The ETT2 is formed by two separate operons, located in distinct loci of the genome (Ren et al., 2004): the *epr/epi/eiv* operon contains 19 genes, coding for putative elements of the injectisome, while the *eip* locus contains six genes including a putative translocon, effector and regulator (Figure 14.1C), related in sequence to the *inv/mxi/spa* systems including the *Salmonella typhimurium* SPI-1 and *Shigella* T3SSs. However, sequence comparison reveals that in most strains, many mutations and sequence deletions are present, leading to the proposition that this system is non-functional (Ren et al., 2004). Nonetheless, expression of several ETT2 genes, under the control of the putative regulator, has been reported (Sheikh et al., 2006).

STRUCTURE AND ORGANIZATION OF THE T3SS INJECTISOME

One of the most striking aspects of T3S is the encoded secretion system, a ~3.5 MDa macromolecular apparatus often referred to as the 'injectisome'. Since its first observation in *Salmonella typhimurium* (Kubori et al., 1998), our understanding of the T3SS structure and assembly has expanded through a combination of structural and genetic studies. It is composed of >20 different proteins, many of them in a highly oligomerized state. Electron microscopic (EM) analysis from various species shows a syringe-shaped structure (Figure 14.2) with a central hollow channel approximately 30 Å wide (Marlovits et al., 2004; Schraidt and Marlovits, 2011), constituting the proposed path through which T3S effectors are translocated.

For historical reasons, the nomenclature used to name the genes of the T3SS has not been standardized. As a consequence, the names of individual proteins vary between different T3SSs (see Table 14.1 for T3SS nomenclature from different systems). In order to simplify this chapter, the EPEC LEE T3SS nomenclature will be used for general description of individual proteins.

The basal body

The basal body of the injectisome refers to the large substructure spanning from the bacterial cytosol through to the inner and outer membranes of the bacterium (see Figure 14.2). The precise definition of what constitutes the basal body varies somewhat in the published literature; here, we will use this term to describe the structural components spanning both bacterial membranes and the periplasmic space, excluding the needle projecting outward as well as the inner-membrane export apparatus. High-resolution EM reconstruction of the basal body from the *Salmonella* SPI-1 T3SS (Marlovits et al., 2004, 2006; Schraidt and Marlovits, 2011), as well as that from *Shigella* (Blocker et al., 2001; Hodgkinson et al., 2009) have revealed they are composed of several concentric, symmetrical rings, resembling the basal body found in the flagellum.

The inner-membrane region of the basal body in the EPEC T3SS is composed of two proteins, EscD and EscJ, forming an intimately associated pair of 24-mer oligomeric rings (Yip et al., 2005b; Schraidt et al., 2010; Schraidt and Marlovits, 2011). By analogy with PrgK in the *Salmonella* SPI-1 system

(A)   **(B)**

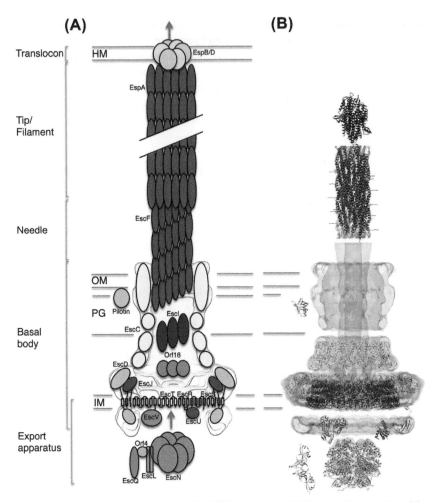

FIGURE 14.2   Structural organization of the T3SS injectisome. (A) Cartoon representation of the various proteins constituting the injectisome, in the EPEC T3SS. The location of the injectisome substructures (export apparatus, basal body, needle, filament, translocon) is indicated on the left. IM – inner membrane, PG - peptidoglycan, OM – outer membrane, HM – host membrane. (B) Known structures for individual injectisome subunits, in their putative localization relative to the EM map. Structures are from a variety of species and systems (see the text for detail).

(Marlovits et al., 2004; Sanowar et al., 2009; Spreter et al., 2009; Schraidt et al., 2010), EscJ is presumed to form the innermost ring component and, given the lipoprotein signal peptide at the N-terminus (Crepin et al., 2005), is localized to the inner membrane by a lipidation anchor (orthologs of EscJ in several T3SSs also possess an additional transmembrane helix at the C-terminus) (Kimbrough and Miller, 2000; Schuch and Maurelli, 2001). The EscJ structure consists of two independent small α/β domains of similar fold, joined by

TABLE 14.1 Nomenclature of the type 3 secretion system components

T3SS		EPEC/EHEC LEE	Shigella	EHEC ETT2	Salmonella SPI1	Yersinia sp.	Flagellar
Family		Esc	Inv/Mxi/Spa			Ysc	
Translocon	Small subunit	EspD	IpaB	EipB	SipB	YopB	
	Large subunit	EspB	IpaC	?	SipC	YopD	
	Tip	EspA	IpaD	EipD	SipD	LcrV	FliC
Needle		EscF	MxiH	EprI	PrgI	YscF	FligE
Basal body	Ruler	Orf16	Spa32	EivJ	InvJ	YscP	FliK
	Outer membrane	EscC	MxiD	EivG	InvG	YscC	
	Pilotin	N/A	MxiM	EivH	InvH	YscW (VirG)	
	Inner membrane	EscD	MxiG	EprH	PrgH	YscD	FliM
	Inner membrane	EscJ	MxiJ	EprK	PrgK	YscJ	FliF
	Rod	EscI	MxiI	EprJ	PrgJ	YscI	

Continued

TABLE 14.1 Nomenclature of the type 3 secretion system components—cont'd

T3SS		EPEC/EHEC LEE	Shigella	EHEC ETT2	Salmonella SPI1	Yersinia sp.	Flagellar
Export apparatus	Gatekeeper	SepD/SepL	MxiC	EivE	InvE	YopN	
	Central stalk	Orf15	Spa13	Eivl	InvI	YscO	FliJ
	ATPase	EscN	Spa47	EivC	InvC	YscN	FliI
	Gate	EscV	MxiA	EivA	InvA	LcrD (YscV)	FlhA
	Autoprotease	EscU	Spa40	EpaS	SpaS	YscU	FlhB
		EscR	Spa24	EpaP	SpaP	YscR	FliP
		EscT	Spa29	EpaR	SpaR	YscT	FliR
		EscS	Spa9	EpaQ	SpaQ	YscS	FliQ
	C-ring	SepQ (EscQ)	Spa33	EpaO	SpaO	YscQ	FliN
		Orf4	MxiK	EorA	OrgA	YscK	FliG?
	Peripheral stalk	EscL (Orf5)	MxiL	?	OrgB	YscL	FliH

a structured linker (Crepin et al., 2005; Yip et al., 2005b). Based on studies of the *Salmonella* SPI-1 ortholog of EscD, PrgH, this protein likely forms a 24-mer ring in intimate contact with the EscJ ring in the periplasm (Sanowar et al., 2009; Schraidt et al., 2010; Schraidt and Marlovits, 2011). PrgH contains a cytoplasmic domain and a periplasmic domain, which are separated by a transmembrane helix. The cytoplasmic domain possesses a Forkhead-domain associated (FHA) fold (McDowell et al., 2011; Barison et al., 2012; Lountos et al., 2012), and has been proposed to interact with elements of the export apparatus (Johnson et al., 2008). Whether this interaction is phosphorylation-dependent, as predicted from the FHA fold, remains unknown. The periplasmic region of PrgH is composed of three domains, each possessing a similar fold to the two domains of EscJ (Spreter et al., 2009). Collectively, the dual inner membrane ring system of the T3SS is proposed to act as a structural foundation for recruitment/stabilization and final assembly of the cytosolic ATPase and several multispan membrane proteins that create an energized export apparatus for chronological secretion of downstream T3SS components and effectors (see below for further details on the export apparatus components).

The outer-membrane ring is composed of a single protein, EscC, which belongs to the secretin family of outer-membrane channels (see Korotkov et al., 2011 for review). Its stoichiometry remains controversial, with 12-mers reported for the *Shigella* T3SS secretin MxiD (Hodgkinson et al., 2009), 13-mers for the *Yersinia* T3SS secretin YscC (Burghout et al., 2004b), and 15-mers for the *Salmonella* SPI-1 T3SS InvG (Schraidt and Marlovits, 2011). This discrepancy may reflect true system-to-system variation or may correspond to inherent plasticity in the oligomerization state of this component, particularly when isolated in vitro. High-resolution structural information on the transmembrane domain is lacking, but the structure of the periplasmic domain from the EPEC secretin EscC has been reported (Spreter et al., 2009); it is composed of two domains N0 and N1, also found in secretins from unrelated systems (Korotkov et al., 2009). The N1 domain possesses a similar fold to the domains of the periplasmic regions of EscJ and PrgH. Intriguingly, domains possessing this fold were also found in other oligomeric assemblies (Korotkov et al., 2009; Levdikov et al., 2012; Meisner et al., 2012), leading to the suggestion that this fold is a common 'ring-building motif' (Spreter et al., 2009).

In most T3SSs, a lipoprotein termed the pilotin promotes outer-membrane targeting, insertion and oligomerization of the secretin component (Crago and Koronakis, 1998; Burghout et al., 2004a). Pilotins are presumed to act through a concerted binding of the C-terminal end of the secretin and interaction with lipids in the membrane (Okon et al., 2008; Ross and Plano, 2011); however, the pilotins of different systems do not show any measurable sequence conservation, and even their overall structure is not conserved (Lario et al., 2005; Izore et al., 2011; Koo et al., 2012). Of note, a pilotin has not yet been identified in the EPEC T3SS, and the sequence of the corresponding secretin, EscC, appears to lack a C-terminal pilotin-binding domain, suggesting that this protein utilizes an alternative mechanism for its assembly in the outer membrane (Gauthier et al., 2003).

In addition to these structural components, analysis in the *Salmonella* SPI-1 system suggests the conserved but less well-characterized proteins PrgJ (EscI in EPEC) and InvJ (EscP in EPEC) are also essential components of the generalized T3SS basal body, with both proteins co-purifying with isolated injectisomes *in vitro* (Marlovits et al., 2004, 2006). These proteins are localized to the periplasmic region of the basal body, and are proposed to interact with the secretin and needle T3SS components, as well as with components of the inner membrane export apparatus (Kimbrough and Miller, 2000; Sukhan et al., 2003; Sal-Man et al., 2012). EscI (PrgJ in the SPI-I T3SS) is hypothesized (although not yet verified) to be a polymerized hollow rod that spans the T3SS basal body to form a continuous path with the downstream needle component (Marlovits et al., 2006; Sal-Man et al., 2012). This so-called 'inner-rod' has also been implicated in the transition from secretion of the needle protein to tip and translocon proteins (EspABD in EPEC), a process known as substrate switching (discussed below) (Marlovits et al., 2004; Sal-Man et al., 2012). NMR studies of a refolded, recombinant form of the EscI ortholog PrgJ shows that this protein largely lacks a defined structure (Zhong et al., 2012), although the relevance of this observation within the assembled injectisome remains to be determined. The biochemical role of InvJ/EscP is better characterized: its deletion creates T3SS variants with abnormal needle length (Tamano et al., 2002; Journet et al., 2003; Mota et al., 2005; Marlovits et al., 2006; Monjaras Feria et al., 2012), leading to the hypothesis that this protein acts as a 'molecular ruler' to regulate T3SS needle formation, although the mechanism by which such measurement is achieved remains unexplained.

The extracellular components: the needle, filament, and translocon

The T3SS needle is the portion of the apparatus extending away from the extracellular surface of the basal body; it is a ~70 Å wide superhelical structure, with a ~30 Å hollow channel in the middle (Figure 14.2), and varies in length depending on the system (~23 nm for the EPEC T3SS (Monjaras Feria et al., 2012), ~45 nm for the *Shigella* T3SS (Tamano et al., 2002)). The needle is composed of multiple copies of a single protein (EscF in EPEC), which is well conserved between species. EM helical reconstructions of the needle from the *Shigella* (MxiH) (Cordes et al., 2003; Fujii et al., 2012) and *Salmonella* SPI-1 (PrgI) T3SSs (Galkin et al., 2010) suggested that the needle helical arrangement possesses a pitch of 25 Å, and contains approximately 5.5 subunits per turn, which is similar to the flagellum filament.

Purified needle proteins spontaneously form extended needle-like oligomeric assemblies of several μm in length (Poyraz et al., 2010), which has impaired high-resolution X-ray crystallographic characterization in the context of the assembled needle; nonetheless, several structures of point and truncation mutants that inhibit oligomerization, or of needle monomers complexed with their specific chaperone (T3S encoded proteins that prevent premature polymerization of newly expressed needle components within the bacterial cytosol), have been reported (Deane et al., 2006; Zhang et al., 2006; Quinaud et al., 2007; Sun et al., 2008; Poyraz et al., 2010; for review, see Blocker et al., 2008). In all cases, a similar two-helix coiled-coil bundle, linked by a conserved turn defined by a PxxP sequence motif is observed. Recently, elegant studies of the *Salmonella typhimurium* needle using solid-state NMR (Poyraz et al., 2010; Loquet et al., 2012) has allowed for an accurate model of its assembled structure, providing atomic-resolution insights into the orientation of individual PrgI molecules within the needle, and their interaction with one another (Loquet et al., 2012).

Importantly, the needle not only forms a hollow conduit to passage effectors from the bacteria to the infected host, but has also been implicated in additional roles including mammalian cell sensing, and, along with the inner rod, in substrate switching during various stages of injectisome assembly and virulence effector secretion (Kenjale et al., 2005; Davis and Mecsas, 2007).

At the distal, extracellular end of the needle, multiple copies of a terminating tip protein have been reported in various species (EspA in EPEC, IpaD in *Shigella*); EM studies suggest the tip is pentameric, matching the cross-sectional symmetry of the needle (Deane et al., 2006; Broz et al., 2007; Epler et al., 2012). The proposed role of the tip is to sense the presence of mammalian cells, through direct interaction with membrane lipids (Olive et al., 2007; Veenendaal et al., 2007), as well as to serve as an adapter between the needle and the downstream translocon pore that inserts directly into the host cell membrane (Picking et al., 2005). The structures of the tip protein from several systems (Derewenda et al., 2004; Erskine et al., 2006; Johnson et al., 2007) show the presence of a core domain forming a two-helix coiled-coil similar to the needle protein. In addition, a small N-terminal helical domain regulates binding to the needle, requiring partial unfolding for the final assembly of the needle/tip complex (Wilharm et al., 2007; Lunelli et al., 2011). A third domain, found at the C-terminus, has been proposed to be involved in contacts with components of the translocon (Roehrich et al., 2010).

Of note, in EPEC and related species, an unusual ortholog of the T3SS tip protein, EspA, has been shown to be solely responsible for formation of the filament, a highly extended polymeric structure that protrudes from the bacteria, with lengths in excess of 600 nm (Knutton et al., 1998; Daniell et al., 2001; Sekiya et al., 2001). These extended translocation filaments in EPEC and EHEC are thought to help span the significant mucosal layer of infected epithelial cells in the host gut. EM helical reconstructions of sheared EspA filaments (Wang et al., 2006) shows a structure similar to the T3SS needle, with approximately 5.5 proteins per turn. Variations in helical parameters observed in these high-resolution EM maps suggest a potential conformational lability or 'breathing', proposed to help accommodate the significant cellular forces likely experienced at the bacterial/host interface and potentially to help propel virulence proteins internalized

within these hollow filaments along their considerable lengths to the translocation pore at the host cell membrane. Recombinant, purified EspA spontaneously
oligomerizes into filament-like structures, and the crystal structure of EspA in
complex with its T3SS encoded anti-polymerization chaperone CesA (Yip et al.,
2005a) revealed the canonical two-helix coiled-coil domain, suggesting an oligomerization mechanism similar to that of the needle, as well as an unstructured
region between these two domains that is likely surface-exposed.

The T3SS translocon is a pore-forming complex proposed to insert directly
into the eukaryotic host cell membrane, allowing T3S effector proteins entry
into the cytoplasm (for review see Mattei et al., 2011). The pore is a proposed
hetero-dimeric complex composed of two integral membrane proteins (EspB,
EspD in EPEC), which interact with the tip structure at the apical end of the
needle (Dorman, 2004). Like the needle and tip components, the translocon
proteins are exported by the injectisome itself, a process likely initiated upon
host-cell contact (Enninga et al., 2005). EspD is predicted to contain a two
transmembrane region and a coiled-coil domain (Dasanayake et al., 2011) while
EspB has one transmembrane spanning segment, a coiled-coil domain, and an
additional amphipathic helix (Luo and Donnenberg, 2011). Presumably the
coiled-coil domains in both pore components facilitate interactions with the
preceding needle tip/filament proteins.

At present there is limited structural information for the translocon, hindered
by poor expression of this pore-forming complex using standard recombinant
methods. Low-resolution EM and AFM analysis of the EPEC translocon in transfected sheep red blood cell membranes appears to suggest a non-symmetrical six
to eight subunit hetero-oligomeric stoichiometry (Ide et al., 2001) although it
remains to be confirmed if the structures observed in this study are indeed the
translocon. The translocon proteins have so far been recalcitrant to higher-reso-
lution structural study, however Barta et al. recently published the X-ray crystallographic structures of the protease-resistant domain of the small subunit from
Shigella (IpaB) and *Salmonella* SPI-1 (SipB) components (Barta et al., 2012).
The structures revealed a coiled-coil motif with observed conservation between
the various species (despite the relatively low sequence identity of ~ 22%). Further, structural comparisons revealed similarity to other bacterial pore-forming
proteins, notably colicin Ia, suggesting an evolutionary relationship.

Prior to secretion, the two hydrophobic translocon proteins are stabilized
within the bacterial cytoplasm by a common T3SS chaperone, CesD (classified
as a type II chaperone; for a recent review of T3SS chaperone architecture/func-
tion see Wilharm et al., 2007). The localized interaction interface of the translocon proteins with their cognate chaperone has recently been characterized,
revealing independent binding to a common surface on the chaperone. Although
these studies are beginning to illuminate our understanding of translocon function, many critical questions such as oligomerization state and stoichiometric
make-up of the pore and the consequent effect on translocon insertion into the
host membrane and effector secretion remain unanswered.

The inner-membrane export apparatus

The T3SS inner-membrane export apparatus consists of five integral membrane proteins (EscRSTUV), predicted to localize within the dual inner-membrane rings (EscJ/EscD) at the base of the basal body, along with several cytoplasmic membrane-associated proteins including the T3SS ATPase (EscN; see Figure 14.2). It should be noted that these proteins, which regulate secretion across the inner membrane, have been termed the export apparatus in the literature to distinguish them from the distinct and downstream translocating pore (the translocon) that passages substrate through the host membrane. The export apparatus proteins are amongst the most highly conserved proteins of the T3SS and have closely related homologs in the flagellar T3SS. EscRSTU are encoded on the same operon whilst the larger EscV is encoded separately (Figure 14.1A). Given their membrane-spanning nature, it is perhaps not surprising that little is currently understood at the molecular level regarding the coordinated function of the export apparatus, although it is generally believed to be central to recruitment and regulation of initial insertion of effectors into the injectisome (for review of the related flagellar export apparatus see Minamino et al., 2008b). Presumably, packing of EscRSTUV within the EscJ/D rings maintains sufficient room for passage of downstream needle/translocon and effector proteins through the inner membrane, as suggested by recent high-resolution EM studies in *Salmonella typhimurium* (Schraidt and Marlovits, 2011). The significant cytoplasmic domains of two of the export apparatus proteins, EscU and EscV, have been suggested to form a functional export gate controlling access to the inner channel (Saijo-Hamano et al., 2010). The soluble ATPase is thought to dock to this export gate (Minamino et al., 2012), where it uses the energy derived from ATP hydrolysis to dissociate effectors from their cognate chaperones prior to insertion into the inner-membrane channel.

Structural models for soluble domains of the export apparatus proteins have begun to emerge in recent years. EscV is the largest of the export apparatus proteins (~70 kDa) with eight predicted TM helices and a large C-terminal cytoplasmic domain (~40 kDa). X-ray crystallographic analysis of the analogous domain has recently been published for the *Salmonella* SPI-1 T3SS ortholog InvA (Lilic et al., 2010; Worrall et al., 2010) in addition to three structures of the flagellar homolog FlhA (Bange et al., 2010; Moore and Jia, 2010; Saijo-Hamano et al., 2010). These structures show a highly conserved overall fold consisting of four subdomains. Of interest, the structures capture the protein in different conformational states with variation in the degree of opening of a central cleft, although the functional relevance of this is at present unclear. The cytoplasmic domain of FlhA has been shown to bind subunits destined for the formation of the flagellar filament suggesting a role in substrate recognition and regulation of assembly (Bange et al., 2010; Diepold et al., 2011). Using a fluorescent tag, Cornelis and colleagues demonstrated that the *Yersinia* homolog YscV co-localizes with other T3SS proteins in multiple copies, unlike the other export apparatus proteins, suggesting it forms higher-order oligomers (Diepold et al., 2011).

EscU is predicted to contain two TM helices and a C-terminal ~20 kDa cytoplasmic domain. EscU orthologs have also been implicated in the temporal regulation of injectisome assembly, acting as part of a molecular 'switch' that regulates the chronological secretion of apparatus components and virulence effectors. It has been demonstrated that flagellar and non-flagellar homologs undergo a spontaneous autocleavage event in their C-terminal cytoplasmic domain leaving a cleaved fragment still associated with the C-terminal domain (Minamino and Macnab, 2000a; Lavander et al., 2002). The recent structures of the cytoplasmic domain of EscU along with orthologs from *Salmonella typhimurium* (SpaS), *Yersinia enterocolitica* (YscU) and *Shigella* (Spa40) show a small mixed alpha/beta domain (Deane et al., 2008a; Zarivach et al., 2008; Lountos et al., 2009; Wiesand et al., 2009). The structures revealed the molecular basis for an intein-like autocleavage reaction via asparagine cyclization on a highly conserved surface exposed loop, resulting in the remodeling of surface features and concomitant changes in electrostatics in the EscU cytosolic domain. Non-cleavable mutants show aberrant secretion of the tip or transloca-tor proteins suggesting cleavage is required for their recognition and a role in switching from translocon to effector secretion (Zarivach et al., 2008).

In addition to the membrane-embedded components of the export apparatus, several membrane-associated proteins are also involved in the regulation of secre-tion, notably the T3SS ATPase EscN. The identification of an ATPase associated with the flagellar T3SS (Vogler et al., 1991) and subsequently a homolog in the T3SS injectisome of *Yersinia* (Woestyn et al., 1994) led to the proposal that it was the energy source for protein translocation. More recently however, several studies have suggested that the proton-motive force (PMF), the electrochemical potential difference of protons across a membrane, provides the energy for protein unfold-ing and translocation (Minamino and Namba, 2008; Paul et al., 2008) supported by the observation that secretion can still occur, albeit less efficiently, in absence of the ATPase in both the flagellar T3SS (Minamino and Namba, 2008) and the T3SS injectisome (Wilharm et al., 2004). Based on the ability of the *Salmonella* SPI-1 T3SS ATPase to dissociate chaperone–effector complexes in an ATP-depen-dent manner (Akeda and Galan, 2005), it is currently believed that the ATPase is required for the effective targeting of T3S chaperone–effector complexes to the base of injectisome and subsequent effector release and initial unfolding prior to translocation. One intriguing feature of the ATPase is the similarity to the F/V/A-ATPases, rotary motors that couple transmembrane ion flow to ATP catalysis, first documented some 20 years ago (Vogler et al., 1991; Woestyn et al., 1994). Con-firming the close structural similarity, structures of the EPEC T3SS ATPase EscN (lacking the N-terminal domain) (Zarivach et al., 2007) and the flagellar ATPase FliI (Imada et al., 2007) were published several years ago. FliI crystallized as a homo-dimer conserved in nature to the hetero-dimer found in the F/V/A-ATPases and experimental evidence has confirmed that, like the F/V/A-ATPases, the type 3 ATPases likely function as hexamers in the physiological context (Kazetani et al., 2009). Initially reported for the flagellar T3SS, this evolutionary relationship has been extended to further soluble export apparatus proteins: firstly, the flagellar

T3SS FliH protein family, which was found to bind and regulate the activity of the ATPase (Minamino and Macnab, 2000b; Blaylock et al., 2006), and has significant sequence similarity to the peripheral stalk of the F/V/A-ATPases (Pallen et al., 2006), which acts as a stator to prevent rotation of the catalytic subunits. Secondly, structural similarity has also been observed for the small flagellar T3SS coiled-coil protein FliJ and the central stalk of the F/V/A-ATPases (Ibuki et al., 2011), which transduces the rotational force between the membrane-embedded and catalytic components. Both FliH and FliJ form complexes with the flagellar T3SS ATPase and have orthologs in the T3SS injectisome (EscL family and EscA family, respectively). The detected evolutionary relationship between the soluble export apparatus components of the T3SS and the F/V/A-ATPases, and the involvement of the PMF in T3SS transmembrane translocation, a pivotal feature of F/V/A-ATPases, supports a possible common evolutionary origin and related mechanism of these two transmembrane molecular machines.

MECHANISM OF SECRETION AND ASSEMBLY

Chaperones and effector recognition

The T3SS export apparatus must recognize and select a small number of effector proteins to export from the cellular milieu and these must be secreted in a temporally regulated manner. Effector recognition and secretion is regulated by a combination of a protein primary structure secretion signal, and by the binding of effector proteins to cytoplasmic chaperones.

The T3SS secretion signal is commonly harbored in the N-terminal 20–30 residues of the effector proteins (Sory et al., 1995; Schesser et al., 1996). It is remarkably variable in sequence, although it often shares similar physiochemical properties or sequence motifs (Lloyd et al., 2002; Arnold et al., 2009). One common feature of the signal sequence that has recently become evident is its intrinsically disordered nature and lack of tertiary structure (Lilic et al., 2006). Intrinsically disordered proteins have been implicated in protein–protein interactions where they can undergo disorder–order transition upon binding. Indeed, such a transition has been demonstrated for the interaction of the *Yersinia* effector YopE and its cognate chaperone SycE (Rodgers et al., 2008). In addition, secretion signals have also been documented in internal sequences and also at the C-terminus of effectors. For example, the EPEC translocon protein EspB has been shown to have internal and C-terminal motifs that determine secretion (Chiu et al., 2003) and the EPEC effector protein Tir (for translocated intimin receptor) has been shown to have a C-terminal signal sequence (Allen-Vercoe et al., 2005).

T3SS chaperones are usually small acidic, often dimeric, cytoplasmic proteins that bind to many effector proteins destined for secretion through the injectisome and exert their function in an ATP-independent manner. In addition to the secretion signal, some T3SS substrates require a cognate chaperone in order to be secreted. Chaperone proteins have been classified according to their binding partners: class IA chaperones bind one (or related) substrate protein while IB chaperones are more promiscuous and can assist several proteins, class II chaperones bind the translocon proteins and finally class III chaperones bind the substrates of the flagellar T3SS.

Numerous class IA and IB structures are now known (for review see Wilharm et al., 2007) including EPEC CesT (Luo et al., 2001) and *Shigella* Spa15 (van Eerde et al., 2004). Surprisingly, all appear as dimers with a conserved mixed alpha/beta fold despite low overall pairwise sequence identities of ~20%. It is proposed that this common fold represents a conserved interface for the binding of the chaperone-binding domains (CBD) of their cognate binding partners. Effector CBDs are usually contained within the N-terminal 50–100 residues following the secretion signal. Co-crystal structures of class I chaperone–substrate complexes show that the CBD of the effector is maintained in a conformation largely devoid of tertiary structure and it has been proposed that this keeps the effector in a state primed for unfolding prior to secretion and translocation (Stebbins and Galan, 2003). Interestingly, it has been shown that effector proteins often still maintain their enzymatic activity when bound by their chaperone (for example Luo et al., 2001) suggesting they are, other than the CBD, still folded. It has also been proposed that the shielding of the CBD by the chaperone acts to prevent non-productive interactions with other proteins (Woestyn et al., 1996) or even the membrane (Letzelter et al., 2006). Importantly, the chaperones play a role in targeting their substrates to the export machinery of the T3SS. Indeed, the EPEC chaperone CesT was shown to bind directly to the T3SS ATPase EscN independently of its effector and this interaction was important for the efficient secretion of the effector protein Tir (Gauthier et al., 2003). As mentioned, the T3SS ATPase has been further shown to dissociate chaperone–effector complexes in an ATP-dependent manner (Akeda and Galan, 2005) prior to unfolding and insertion into the transport channel.

More recently, structural information for the translocon-binding class II chaperones and their interaction with the two pore-forming translocon proteins has become available. Both components bind a common chaperone which, unlike the class I chaperones, shares considerable sequence identity even within distant species. Structures of *Shigella* IpgC (Lunelli et al., 2009), *Yersinia* SycD (Buttner et al., 2008) and *Pseudomonas* PcrH (Job et al., 2010) reveal a seven-helical tetratricopeptide repeat (TPR) fold, which binds the N-terminal sequences of both translocators in a concave surface formed by the TPR domain. Interestingly, this TPR-like fold has also been observed in structures of T3S chaperones for the needle subunits in *Pseudomonas* (Quinaud et al., 2007) and *Yersinia* (Sun et al., 2008) species suggesting this chaperone class mediates secretion of substrates involved in assembly of the injectisome, with the class I chaperones involved in secretion of the late effectors.

Injectisome assembly

Assembly of the injectisome occurs in a hierarchical and highly orchestrated manner (Figure 14.3). The process can be divided into two stages; firstly there is the formation of a secretion-competent but immature complex consisting of the

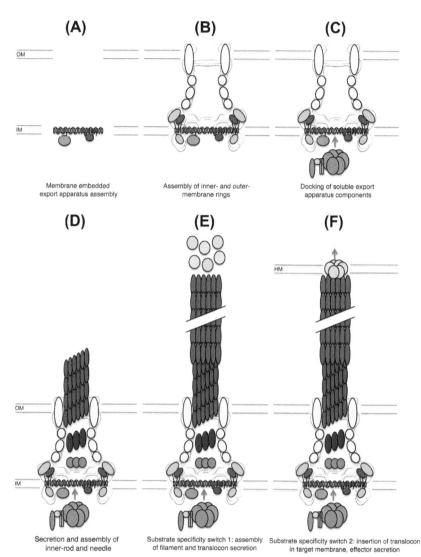

(A) Membrane embedded export apparatus assembly

(B) Assembly of inner- and outer-membrane rings

(C) Docking of soluble export apparatus components

(D) Secretion and assembly of inner-rod and needle

(E) Substrate specificity switch 1: assembly of filament and translocon secretion

(F) Substrate specificity switch 2: insertion of translocon in target membrane, effector secretion

FIGURE 14.3 Assembly of the T3SS injectisome. Cartoon representation of the major steps involved in the assembly of the T3SS injectisome. (A) Elements of the export apparatus assemble in the inner membrane. (B) This assembly serves as a scaffold for the arrangement of the basal body inner- and outer-membrane proteins. (C) The ATPase and related components assemble inside the cytoplasmic side, allowing secretion. (D) The ruler, rod, and needle proteins are secreted through the basal body, extending the needle. (E) A substrate specificity switch prevents further extension of the needle, which is capped by the tip/filament component. Translocon proteins are secreted. (F) A second substrate specificity switch, upon contact with mammalian cells, induces the formation of the translocation pore and the secretion of effector proteins. IM – inner membrane, OM – outer membrane, HM – host membrane.

basal body and export apparatus. This complex then secretes the parts necessary for completion of the injectisome (so-called early/mid substrates) including the inner-rod, needle, tip, and translocon.

Recent studies in *Salmonella typhimurium* (Wagner et al., 2010) and *Yersinia* (Diepold et al., 2011) have suggested a broadly conserved sequence of assembly in these two different systems. Based on observations of the *Salmonella* SPI-1 T3SS injectisome, assembly was proposed to proceed in an 'inside-out' manner, starting with the formation of the inner-membrane export apparatus with SpaP (EscR), SpaQ (EscS), and SpaR (EscT) forming a stable complex even in the absence of the needle complex (Figure 14.3A). This core complex acts as a nucleation point for the assembly of the remaining export apparatus proteins SpaS (EscU) and InvA (EscV) (the export apparatus proteins are presumed to be exported by the general secretory (sec) pathway) followed by the formation of the inner-membrane rings PrgK (EscJ) and PrgH (EscD), and outer-membrane secretin InvG (EscC) (the inner- and outer-rings are assembled via sec-mediated secretion). Independent studies into the assembly of the *Yersinia* injectisome revealed a similar pathway. Here, assembly is initiated by the oligomerization of the largest export apparatus protein YscV (EscV). The exact nature of this oligomer is unclear but, like in *Salmonella typhimurium*, several of the smaller export apparatus proteins are required for its assembly, namely YscRST (EscRST). This export apparatus complex is then capable of recruiting YscJ (EscJ), the smaller of the inner-membrane rings, which assembles into an oligomeric ring around YscV (EscV). In a separate process, the outer-membrane secretin YscC (EscC) assembles into a ring. These inner- and outer-membrane complexes are then connected via the larger of the inner-membrane rings, YscD (EscD), which envelops YscJ (EscJ) and interacts with the secretin (Figure 14.3B).

This nascent injectisome then recruits the ATPase and its associated proteins (Figure 14.3C). At this stage, it becomes secretion-competent, and the components required to complete assembly are secreted in a T3SS ATPase-dependent manner. Initially, early substrates including the inner-rod, which traverses the basal body, and the extracellular needle are secreted (Figure 14.3D). After termination of the needle, the 'mid' substrates are secreted; initially the tip protein is translocated and forms an oligomeric cap at the distal end of the needle, or in the case of EPEC EspA, a filamentous extension of the needle (see above). The two translocon proteins are then secreted (Figure 14.3E) and upon contact with a target cell, form a pore inside its membrane, attached to the needle tip, completing the assembly of the injectisome and forming a continuous channel from the bacterial cytoplasm to the host cell cytoplasm (Figure 14.3F) for the 'late' substrates, or effectors, to pass.

This hierarchical assembly requires a high degree of spatial and temporal regulation to ensure the correct formation of the injectisome substructures and order of component secretion. The first point of regulation is the substrate specificity switch between the secretion of the early and mid substrates, or between the secretion of the rod/needle proteins and the translocon proteins. This switch effectively defines the length of the needle, which must be long enough to bridge the extracellular gap between the bacterium and host and as such is tightly regulated. Needle length varies from species to species but within a species is highly controlled with needles typically in the 50–60 nm range (Kubori et al., 2000; Blocker et al., 2001; Journet et al., 2003).

The substrate specificity switch can be divided into two parts: the sensing of needle length and the switch from early to mid substrate export once the correct length has been achieved. Both of these functions are thought to be carried out by members of the T3S substrate specificity switch (T3S4) protein family, commonly referred to as ruler proteins (EscP in EPEC, Spa32 in *Shigella*, YscP in *Yersinia*, InvJ in *Salmonella typhimurium*), which are themselves secreted by the T3SS. Ruler knockouts tend to assemble longer needles and are impaired in the secretion of late substrates (Magdalena et al., 2002; Tamano et al., 2002; Journet et al., 2003). The ruler proteins can be divided into two domains, a poorly conserved, unstructured N-terminal domain, which is believed to sense needle length, and a small, globular C-terminal domain called the T3S4 domain (Mizuno et al., 2011), which interacts with the inner-membrane export apparatus protein EscU (Agrain et al., 2005). Once the correct needle length has been achieved, an interaction between the ruler T3S4 domain and EscU is thought to trigger the substrate specificity switch (Minamino and Macnab, 2000a; Botteaux et al., 2008). As discussed above, EscU contains a small cytoplasmic domain that undergoes an autocleavage event on a strictly conserved surface exposed loop. Non-cleavable mutants from the EPEC, *Salmonella* SPI-1 and *Yersinia* T3SSs result in aberrant secretion of the tip protein or translocators. It was initially thought that the cleavage event itself was linked to the switch although the elucidation of the intein-like cleavage mechanism from the recent crystal structures of EscU and SpaS would seem to imply that cleavage occurs immediately upon folding and thus some other property must be involved in the specificity switch itself (Zarivach et al., 2008).

The manner by which the ruler proteins sense needle length is still not clear and several theories have been proposed. The prevailing hypothesis is the molecular ruler/tape measure model, where a ruler protein is able to sense the length of the growing needle and trigger a change in substrate specificity once it reaches the required length. Studies of the *Yersinia* ruler protein YscP demonstrated that changes in the size of the N-terminal domain effected needle length accordingly (Journet et al., 2003) and it was thus proposed that YscP was acting as a physical ruler, which once extended to the correct length would interact with YscU and trigger the switch from secreting needle subunits to tip and translocon proteins. YscP has also been implicated in the assembly of the inner-rod and in its absence there is an over-secretion of the inner-rod protein YscI (Wood et al., 2008). The effect of the ruler on inner rod formation is also supported in the *Salmonella* SPI-1 T3SS, where InvJ ruler knockouts produce needle complexes with reduced amounts of PrgJ inner-rod and altered appearance at the base of the injectisome (Marlovits et al., 2006).

Once the needle is assembled, the tip and translocon components (mid effectors) must be secreted prior to secretion of the late effectors and a second substrate specificity switch is believed to control this transition. Several such negative regulatory controls have been described, some of which bind to the tip and physically block secretion while some act at the base of the injectisome. A family of proteins with shared but remote sequence similarity (SepD:SepL complex in EPEC, MxiC in *Shigella*, YopN:TyeA complex in *Yersinia*) have been shown to regulate translocator secretion (O'Connell et al., 2004; Deng et al., 2005; Hamad and Nilles, 2007; Deane et al., 2008b). It is thought that these proteins act as gatekeepers, allowing the secretion of translocon components but restricting the secretion of late effectors until an appropriate signal is detected and the secretion blockade relieved. The nature of this signal varies from species to species with environmental factors such as pH, temperature, small molecules and host cell contact demonstrated to trigger secretion in different systems.

Additionally, in a recent elegant study, Galan and colleagues described a sorting platform capable of ensuring secretion of the translocases prior to the late effectors (Lara-Tejero et al., 2011). This high-molecular-weight cytoplasmic complex was composed of three proteins – SpaO (the *Salmonella typhimurium* C-ring protein; SepQ/EscQ in EPEC), OrgA (Orf4 in EPEC) and OrgB (the ATPase peripheral stalk homolog; EscL in EPEC) – and was able to recognize the chaperones for both the mid and late effectors, delivering their complexes to the injectisome prior to secretion in an ATPase-dependent manner. It was found that the sorting platform had a higher affinity for the translocase chaperones ensuring the selective and ordered secretion. Importantly, similar complexes have been described in EPEC (Biemans-Oldehinkel et al., 2011) and *Shigella* (Johnson and Blocker, 2008) suggesting this may be a common means of regulation.

REGULATION OF TYPE 3 SECRETION

Type 3 secretion is a tightly regulated process; indeed, several pathogens (such as *Salmonella typhimurium* (Srikanth et al., 2011) and *Burkholderia* species (Sun and Gan, 2010)) possess several T3SSs, which are activated at different times during the course of infection. In practice, the secretion of a certain set of effectors is induced at one stage of the infection, and stopped at a later stage. Strikingly, it has been shown that induction of the *Salmonella* SPI-1 T3SS (Sturm et al., 2011) or *Yersinia* spp. T3SS (Brubaker and Surgalla, 1964; Wiley et al., 2007) leads to a heavy penalty in bacterial growth, illustrating the burden that unregulated type 3 secretion would cause on bacteria.

Environmental regulation

One common aspect of type 3 secretion regulation is the observation that it is induced upon contact with the cell of an infected organism (Cornelis, 2000). The molecular detail of eukaryotic cell sensing remains unknown to date; however, recent studies on the *Shigella* T3SS indicate that the translocon proteins IpaB and IpaD are key to host cell sensing (Veenendaal et al., 2007). This important observation indicates that at least in some systems, the injectisome is fully assembled prior to contact with the infected cell. This suggests that cell contact-dependent induction of secretion does not occur at the genetic level, but through an unknown signaling pathway, similar to the various substrate switching events observed during assembly.

Similarly, a number of environmental factors have been reported to influence the level of secretion, which varies in different systems. In EPEC, secretion of the LEE T3SS is maximal at 37°C, pH 7, and in the presence of sodium bicarbonate, calcium and $Fe(NO_3)_3$ (Kenny et al., 1997); these conditions are thought to correspond to the conditions in which T3S occurs *in vivo*. Similarly, the T3SS of *Shigella* is induced under anaerobic conditions, similar to the environment of cellular invasion during an infection (Marteyn et al., 2010). However, inducing conditions do not necessarily reflect the *in vivo* situation; for instance, in *Shigella* T3S can be induced *in vitro* by the addition of Congo red to the growth media (Bahrani et al., 1997).

Finally, quorum sensing has been shown to be an important factor in T3SS regulation. Quorum sensing, a process used by bacteria to communicate with one another (Bassler and Losick, 2006), is based on the secretion of particular small molecules (termed auto-inducers) into the media surrounding the bacteria, a chemical cross-talk which can be detected by others, and lead to appropriate changes in gene transcription. In EPEC and EHEC, it has been shown that the *luxS* gene, responsible for the production of the auto-inducers AI-2 and AI-3, controls the transcription of the LEE genes (Sperandio et al., 1999) through a complex cascade involving the QseBC and QseDEFG pathways (Antunes et al., 2010).

Genetic regulation of the LEE

All T3SSs encode regulatory genes that allow the synchronized transcription of the various genes in the system. However, these regulatory genes and their regulated transcription differ significantly depending on the system.

The LEE-encoded *ler* gene is a critical regulator of LEE transcription (Mellies et al., 1999), as it activates the transcription of the LEE2, LEE3, LEE4, and LEE5 operons (Figure 14.1A). Ler is homologous to the transcriptional repressor H-NS (a general housekeeping protein involved in many bacterial processes; see Dorman, 2004 for review). Ler is suggested to function by competing with and relieving the transcriptional repression of H-NS within the LEE (Bustamante et al., 2001). Importantly, *ler* transcription is itself regulated by a number of factors, including quorum sensing and the SOS response triggered by DNA damage (Sperandio et al., 2000). In addition, the PerC protein, a member of the AraC transcription factor family encoded in the EAF plasmid found in EPEC strains, has been shown to activate the transcription of *ler*, thus in turn activating the LEE (Gomez-Duarte and Kaper, 1995). Interestingly, the EAF plasmid is not present in EHEC strains, which still show LEE transcription; nonetheless, EHEC strains do possess PerC homologs in bacteriophage-acquired islands, several of which have been shown to induce LEE transcription (Iyoda and Watanabe, 2004; Porter et al., 2005).

In addition to *ler*, two other regulatory genes, *grlR* and *grlA*, have been identified in the LEE (Figure 14.1A). Both genes are activated by *ler*, but appear to have contradictory effects: GrlA is a LEE activator, forming a positive-feedback loop with ler, while GrlR binds to GrlA and prevents its function (Barba et al., 2005; Iyoda et al., 2006).

Of note, even though it is not yet established if the ETT2 system is functional, three regulatory genes have been identified for this system (Figure 14.1C). The *eilA* gene is homologous to the *Salmonella* SPI-1 T3SS regulator *hilA* (Ellermeier and Slauch, 2007), and its overexpression does induce transcription of ETT2 genes (Sheikh et al., 2006). In addition, the two regulatory genes *etrA* and *eivF* encoded by the ETT2 T3SS show a dramatic inhibitory effect on transcription of the LEE genes (Zhang et al., 2004), revealing a striking cross talk between T3SSs.

Regulation of the Mxi/Spa system

The *Shigella flexneri* T3SS possesses a regulatory gene, *virB* (Figure 14.1B), showing similarity to the LEE *ler*. Indeed, VirB was shown to function via a similar mechanism of displacing the H-NS repressor (Tobe et al., 1993). However, a second regulatory gene, *virF*, located 30 kb away from the T3SS locus, is required for the activation of *virB* (Adler et al., 1989). VirF belongs to the AraC family of transcription factors, but acts similarly to Ler or VirB, by displacing the H-NS repressor. It has been proposed that this dual regulatory pathway allows for a tighter control of transcriptional levels under changing conditions (Porter and Dorman, 1997).

A second regulatory gene, *mxiE*, has been identified in the T3SS locus (Kane et al., 2002), which promotes specifically the transcription of six effector proteins, but not of components of the secretion apparatus or injectisome. Interestingly, the protein MxiE also belongs to the AraC family of transcription factors, but acts through a different mechanism. Indeed, MxiE is not sufficient to activate transcription, but requires a direct binding to the effector chaperone IpgC (Mavris et al., 2002). It has been argued that the presence of free IpgC protein in the bacterial cytosol is used as a signal that the secretion apparatus is active, thus preventing the unnecessary transcription of effectors during apparatus assembly.

CONCLUSION

Since the discovery of type 3 secretion nearly 20 years ago, significant progress has been made towards understanding the detailed molecular mechanisms by which this process is performed, albeit with structural and mechanistic analysis of the full-length membrane-spanning components lagging well behind. Importantly, the T3SS has provided a new area of investigation for antibiotic development, as well as potential new antigens for vaccine therapies targeting the significant global pathogens that necessarily encode a T3SS. Nonetheless, several important questions remain unanswered:

- Firstly, the source of energy for T3S is still unclear. It has been shown that the role of the T3SS-associated ATPase likely involves dissociation of effector proteins from their cognate chaperones, facilitating effective secretion, while at the same time the proton-motive force has been shown to be integral to secretion. How these independent processes energize the system in a concerted manner so that effectors are secreted and translocated though the injectisome remains to be understood at the molecular level (Minamino et al., 2008a) although the suggestive evolutionary similarity to the F/V/A-ATPases hints at a related mechanism of transmembrane transport.
- Secondly, the details of injectisome assembly and substrate switching processes require further investigation. In particular, the precise roles of the rod, ruler, and various secretion apparatus components are subject to controversy (Erhardt et al., 2010). In many cases, it is very possible that different systems (as well as the flagellum) utilize different mechanisms, which would explain the difficulty to reconcile all the data into a single unified model.
- Finally, the translocation of the type 3 effector proteins across mammalian cell membrane remains very poorly characterized. In particular, recent evidence has challenged the dogma of a continuous channel from the bacterial cytoplasm to the mammalian cytoplasm. Instead, it has been proposed that effectors are secreted in the media, and that components of the translocon allow them to enter the mammalian cell independently of the needle complex (Edgren et al., 2012). These two models are not mutually exclusive, and could possibly reflect variations in effectors, species, and systems.

REFERENCES

Adler, B., Sasakawa, C., Tobe, T., Makino, S., Komatsu, K., Yoshikawa, M., 1989. A dual transcriptional activation system for the 230 kb plasmid genes coding for virulence-associated antigens of *Shigella flexneri*. Mol. Microbiol. 3 (5), 627–635.

Agrain, C., Callebaut, I., Journet, L., et al., 2005. Characterization of a type 3 secretion substrate specificity switch (T3S4) domain in YscP from *Yersinia enterocolitica*. Mol. Microbiol. 56 (1), 54–67.

Akeda, Y., Galan, J.E., 2005. Chaperone release and unfolding of substrates in type 3 secretion. Nature 437 (7060), 911–915.

Allen-Vercoe, E., Toh, M.C., Waddell, B., Ho, H., DeVinney, R., 2005. A carboxy-terminal domain of Tir from enterohemorrhagic *Escherichia coli* O157:H7 (EHEC O157:H7) required for efficient type 3 secretion. FEMS Microbiol. Lett. 243 (2), 355–364.

Antunes, L.C., Ferreira, R.B., Buckner, M.M., Finlay, B.B., 2010. Quorum sensing in bacterial virulence. Microbiology 156 (Pt 8), 2271–2282.

Arnold, R., Brandmaier, S., Kleine, F., et al., 2009. Sequence-based prediction of type 3 secreted proteins. PLoS Pathog. 5 (4), e1000376.

Bahrani, F.K., Sansonetti, P.J., Parsot, C., 1997. Secretion of Ipa proteins by *Shigella flexneri*: inducer molecules and kinetics of activation. Infect. Immun. 65 (10), 4005–4010.

Bange, G., Kummerer, N., Engel, C., Bozkurt, G., Wild, K., Sinning, I., 2010. FlhA provides the adaptor for coordinated delivery of late flagella building blocks to the type 3 secretion system. Proc. Natl. Acad. Sci. USA 107 (25), 11295–11300.

Barba, J., Bustamante, V.H., Flores-Valdez, M.A., Deng, W., Finlay, B.B., Puente, J.L., 2005. A positive regulatory loop controls expression of the locus of enterocyte effacement-encoded regulators Ler and GrlA. J. Bacteriol. 187 (23), 7918–7930.

Barison, N., Lambers, J., Hurwitz, R., Kolbe, M., 2012. Interaction of MxiG with the cytosolic complex of the type 3 secretion system controls *Shigella* virulence. FASEB J. 26 (4), 1717–1726.

Barta, M.L., Dickenson, N.E., Patil, M., et al., 2012. The structures of coiled-coil domains from type 3 secretion system translocators reveal homology to pore-forming toxins. J. Mol. Biol. 417 (5), 395–405.

Bassler, B.L., Losick, R., 2006. Bacterially speaking. Cell 125 (2), 237–246.

Biemans-Oldehinkel, E., Sal-Man, N., Deng, W., Foster, L.J., Finlay, B.B., 2011. Quantitative proteomic analysis reveals formation of an EscL-EscQ-EscN type 3 complex in enteropathogenic *Escherichia coli*. J. Bacteriol. 193 (19), 5514–5519.

Blaylock, B., Riordan, K.E., Missiakas, D.M., Schneewind, O., 2006. Characterization of the *Yersinia enterocolitica* type 3 secretion ATPase YscN and its regulator, YscL. J. Bacteriol. 188 (10), 3525–3534.

Blocker, A., Jouihri, N., Larquet, E., et al., 2001. Structure and composition of the *Shigella flexneri* "needle complex", a part of its type 3 secreton. Mol. Microbiol. 39 (3), 652–663.

Blocker, A.J., Deane, J.E., Veenendaal, A.K., et al., 2008. What's the point of the type 3 secretion system needle? Proc. Natl. Acad. Sci. USA 105 (18), 6507–6513.

Botteaux, A., Sani, M., Kayath, C.A., Boekema, E.J., Allaoui, A., 2008. Spa32 interaction with the inner-membrane Spa40 component of the type 3 secretion system of *Shigella flexneri* is required for the control of the needle length by a molecular tape measure mechanism. Mol. Microbiol. 70 (6), 1515–1528.

Broz, P., Mueller, C.A., Muller, S.A., et al., 2007. Function and molecular architecture of the *Yersinia* injectisome tip complex. Mol. Microbiol. 65 (5), 1311–1320.

Brubaker, R.R., Surgalla, M.J., 1964. The effect of Ca++ and Mg++ on lysis, growth, and production of virulence antigens by *Pasteurella pestis*. J. Infect. Dis. 114, 13–25.

Burghout, P., Beckers, F., de Wit, E., et al., 2004a. Role of the pilot protein YscW in the biogenesis of the YscC secretin in *Yersinia enterocolitica*. J. Bacteriol. 186 (16), 5366–5375.

Burghout, P., van Boxtel, R., Van Gelder, P., et al., 2004b. Structure and electrophysiological properties of the YscC secretin from the type 3 secretion system of *Yersinia enterocolitica*. J. Bacteriol. 186 (14), 4645–4654.

Bustamante, V.H., Santana, F.J., Calva, E., Puente, J.L., 2001. Transcriptional regulation of type 3 secretion genes in enteropathogenic *Escherichia coli*: Ler antagonizes H-NS-dependent repression. Mol. Microbiol. 39 (3), 664–678.

Buttner, C.R., Sorg, I., Cornelis, G.R., Heinz, D.W., Niemann, H.H., 2008. Structure of the *Yersinia enterocolitica* type 3 secretion translocator chaperone SycD. J. Mol. Biol. 375 (4), 997–1012.

Chaudhuri, R.R., Henderson, I.R., 2012. The evolution of the *Escherichia coli* phylogeny. Infect. Genet. Evol. 12 (2), 214–226.

Chiu, H.J., Lin, W.S., Syu, W.J., 2003. Type 3 secretion of EspB in enterohemorrhagic *Escherichia coli* O157:H7. Arch. Microbiol. 180 (3), 218–226.

Coburn, B., Sekirov, I., Finlay, B.B., 2007. Type 3 secretion systems and disease. Clin. Microbiol. Rev. 20 (4), 535–549.

Cordes, F.S., Komoriya, K., Larquet, E., et al., 2003. Helical structure of the needle of the type 3 secretion system of *Shigella flexneri*. J. Biol. Chem. 278 (19), 17103–17107.

Cornelis, G.R., 2000. Type 3 secretion: a bacterial device for close combat with cells of their eukaryotic host. Philos. Trans. R. Soc. Lond. B. Biol. Sci. 355 (1397), 681–693.

Crago, A.M., Koronakis, V., 1998. *Salmonella* InvG forms a ring-like multimer that requires the InvH lipoprotein for outer membrane localization. Mol. Microbiol. 30 (1), 47–56.

Crepin, V.F., Prasannan, S., Shaw, R.K., et al., 2005. Structural and functional studies of the enteropathogenic *Escherichia coli* type 3 needle complex protein EscJ. Mol. Microbiol. 55 (6), 1658–1670.

Daniell, S.J., Takahashi, N., Wilson, R., et al., 2001. The filamentous type 3 secretion translocon of enteropathogenic *Escherichia coli*. Cell Microbiol. 3 (12), 865–871.

Dasanayake, D., Richaud, M., Cyr, N., et al., 2011. The N-terminal amphipathic region of the *Escherichia coli* type 3 secretion system protein EspD is required for membrane insertion and function. Mol. Microbiol. 81 (3), 734–750.

Davis, A.J., Mecsas, J., 2007. Mutations in the *Yersinia pseudotuberculosis* type 3 secretion system needle protein, YscF, that specifically abrogate effector translocation into host cells. J. Bacteriol. 189 (1), 83–97.

Deane, J.E., Graham, S.C., Mitchell, E.P., Flot, D., Johnson, S., Lea, S.M., 2008a. Crystal structure of Spa40, the specificity switch for the *Shigella flexneri* type 3 secretion system. Mol. Microbiol. 69 (1), 267–276.

Deane, J.E., Roversi, P., Cordes, F.S., et al., 2006. Molecular model of a type 3 secretion system needle: implications for host-cell sensing. Proc. Natl. Acad. Sci. USA 103 (33), 12529–12533.

Deane, J.E., Roversi, P., King, C., Johnson, S., Lea, S.M., 2008b. Structures of the *Shigella flexneri* type 3 secretion system protein MxiC reveal conformational variability amongst homologues. J. Mol. Biol. 377 (4), 985–992.

Deng, W., Li, Y., Hardwidge, P.R., et al., 2005. Regulation of type 3 secretion hierarchy of translocators and effectors in attaching and effacing bacterial pathogens. Infect. Immun. 73 (4), 2135–2146.

Derewenda, U., Mateja, A., Devedjiev, Y., et al., 2004. The structure of *Yersinia pestis* V-antigen, an essential virulence factor and mediator of immunity against plague. Structure 12 (2), 301–306.

Diepold, A., Wiesand, U., Cornelis, G.R., 2011. The assembly of the export apparatus (YscR, S, T, U, V) of the *Yersinia* type 3 secretion apparatus occurs independently of other structural components and involves the formation of an YscV oligomer. Mol. Microbiol. 82 (2), 502–514.

Donnenberg, M.S., Kaper, J.B., 1992. Enteropathogenic *Escherichia coli*. Infect. Immun. 60 (10), 3953–3961.

Donnenberg, M.S., Lai, L.C., Taylor, K.A., 1997. The locus of enterocyte effacement pathogenicity island of enteropathogenic *Escherichia coli* encodes secretion functions and remnants of transposons at its extreme right end. Gene 184 (1), 107–114.

Dorman, C.J., 2004. H-NS: a universal regulator for a dynamic genome. Nat. Rev. Microbiol. 2 (5), 391–400.

Edgren, T., Forsberg, A., Rosqvist, R., Wolf-Watz, H., 2012. Type 3 secretion in *Yersinia*: injectisome or not? PLoS Pathog. 8 (5), e1002669.

Ellermeier, J.R., Slauch, J.M., 2007. Adaptation to the host environment: regulation of the SPI1 type 3 secretion system in *Salmonella enterica* serovar Typhimurium. Curr. Opin. Microbiol. 10 (1), 24–29.

Elliott, S.J., Wainwright, L.A., McDaniel, T.K., et al., 1998. The complete sequence of the locus of enterocyte effacement (LEE) from enteropathogenic *Escherichia coli* E2348/69. Mol. Microbiol. 28 (1), 1–4.

Enninga, J., Mounier, J., Sansonetti, P., Tran Van Nhieu, G., 2005. Secretion of type 3 effectors into host cells in real time. Nat. Methods. 2 (12), 959–965.

Epler, C.R., Dickenson, N.E., Bullitt, E., Picking, W.L., 2012. Ultrastructural analysis of IpaD at the tip of the nascent MxiH type 3 secretion apparatus of *Shigella flexneri*. J. Mol. Biol. 29 (420(1-2)), 29–39.

Erhardt, M., Hirano, T., Su, Y., et al., 2010. The role of the FliK molecular ruler in hook-length control in *Salmonella enterica*. Mol. Microbiol. 75 (5), 1272–1284.

Erskine, P.T., Knight, M.J., Ruaux, A., et al., 2006. High resolution structure of BipD: an invasion protein associated with the type 3 secretion system of *Burkholderia pseudomallei*. J. Mol. Biol. 363 (1), 125–136.

Fujii, T., Cheung, M., Blanco, A., Kato, T., Blocker, A.J., Namba, K., 2012. Structure of a type 3 secretion needle at 7-A resolution provides insights into its assembly and signaling mechanisms. Proc. Natl. Acad. Sci. USA 109 (12), 4461–4466.

Galkin, V.E., Schmied, W.H., Schraidt, O., Marlovits, T.C., Egelman, E.H., 2010. The structure of the *Salmonella typhimurium* type 3 secretion system needle shows divergence from the flagellar system. J. Mol. Biol. 396 (5), 1392–1397.

Gauthier, A., Puente, J.L., Finlay, B.B., 2003. Secretin of the enteropathogenic *Escherichia coli* type 3 secretion system requires components of the type 3 apparatus for assembly and localization. Infect. Immun. 71 (6), 3310–3319.

Ginocchio, C.C., Olmsted, S.B., Wells, C.L., Galan, J.E., 1994. Contact with epithelial cells induces the formation of surface appendages on *Salmonella typhimurium*. Cell 76 (4), 717–724.

Gomez-Duarte, O.G., Kaper, J.B., 1995. A plasmid-encoded regulatory region activates chromosomal eaeA expression in enteropathogenic *Escherichia coli*. Infect. Immun. 63 (5), 1767–1776.

Hamad, M.A., Nilles, M.L., 2007. Roles of YopN, LcrG and LcrV in controlling Yops secretion by *Yersinia pestis*. Adv. Exp. Med. Biol. 603, 225–234.

Hartleib, S., Prager, R., Hedenstrom, I., Lofdahl, S., Tschape, H., 2003. Prevalence of the new, SPI1-like, pathogenicity island ETT2 among *Escherichia coli*. Int. J. Med. Microbiol. 292 (7–8), 487–493.

Hensel, M., Shea, J.E., Waterman, S.R., et al., 1998. Genes encoding putative effector proteins of the type 3 secretion system of *Salmonella* pathogenicity island 2 are required for bacterial virulence and proliferation in macrophages. Mol. Microbiol. 30 (1), 163–174.

Hodgkinson, J.L., Horsley, A., Stabat, D., et al., 2009. Three-dimensional reconstruction of the *Shigella* T3SS transmembrane regions reveals 12-fold symmetry and novel features throughout. Nat. Struct. Mol. Biol. 16 (5), 477–485.

Ibuki, T., Imada, K., Minamino, T., Kato, T., Miyata, T., Namba, K., 2011. Common architecture of the flagellar type 3 protein export apparatus and F- and V-type ATPases. Nat. Struct. Mol. Biol. 18 (3), 277–282.

Ide, T., Laarmann, S., Greune, L., Schillers, H., Oberleithner, H., Schmidt, M.A., 2001. Characterization of translocation pores inserted into plasma membranes by type 3-secreted Esp proteins of enteropathogenic *Escherichia coli*. Cell Microbiol. 3 (10), 669–679.

Ideses, D., Gophna, U., Paitan, Y., Chaudhuri, R.R., Pallen, M.J., Ron, E.Z., 2005. A degenerate type 3 secretion system from septicemic *Escherichia coli* contributes to pathogenesis. J. Bacteriol. 187 (23), 8164–8171.

Imada, K., Minamino, T., Tahara, A., Namba, K., 2007. Structural similarity between the flagellar type 3 ATPase FliI and F1-ATPase subunits. Proc. Natl. Acad. Sci. USA 104 (2), 485–490.

Iyoda, S., Koizumi, N., Satou, H., et al., 2006. The GrlR-GrlA regulatory system coordinately controls the expression of flagellar and LEE-encoded type 3 protein secretion systems in enterohemorrhagic *Escherichia coli*. J. Bacteriol. 188 (16), 5682–5692.

Iyoda, S., Watanabe, H., 2004. Positive effects of multiple pch genes on expression of the locus of enterocyte effacement genes and adherence of enterohaemorrhagic *Escherichia coli* O157:H7 to HEp-2 cells. Microbiology 150 (Pt 7), 2357–2571.

Izore, T., Perdu, C., Job, V., Attree, I., Faudry, E., Dessen, A., 2011. Structural characterization and membrane localization of ExsB from the type 3 secretion system (T3SS) of *Pseudomonas aeruginosa*. J. Mol. Biol. 413 (1), 236–246.

Jarvis, K.G., Giron, J.A., Jerse, A.E., McDaniel, T.K., Donnenberg, M.S., Kaper, J.B., 1995. Enteropathogenic *Escherichia coli* contains a putative type 3 secretion system necessary for the export of proteins involved in attaching and effacing lesion formation. Proc. Natl. Acad. Sci. USA 92 (17), 7996–8000.

Job, V., Mattei, P.J., Lemaire, D., Attree, I., Dessen, A., 2010. Structural basis of chaperone recognition of type 3 secretion system minor translocator proteins. J. Biol. Chem. 285 (30), 23224–23232.

Johnson, D.L., Stone, C.B., Mahony, J.B., 2008. Interactions between CdsD, CdsQ, and CdsL, three putative *Chlamydophila pneumoniae* type 3 secretion proteins. J. Bacteriol. 190 (8), 2972–2980.

Johnson, S., Blocker, A., 2008. Characterization of soluble complexes of the *Shigella flexneri* type 3 secretion system ATPase. FEMS Microbiol. Lett. 286 (2), 274–278.

Johnson, S., Roversi, P., Espina, M., et al., 2007. Self-chaperoning of the type 3 secretion system needle tip proteins IpaD and BipD. J. Biol. Chem. 282 (6), 4035–4044.

Journet, L., Agrain, C., Broz, P., Cornelis, G.R., 2003. The needle length of bacterial injectisomes is determined by a molecular ruler. Science 302 (5651), 1757–1760.

Kane, C.D., Schuch, R., Day Jr., W.A., Maurelli, A.T., 2002. MxiE regulates intracellular expression of factors secreted by the *Shigella flexneri* 2a type 3 secretion system. J. Bacteriol. 184 (16), 4409–4419.

Kazetani, K., Minamino, T., Miyata, T., Kato, T., Namba, K., 2009. ATP-induced FliI hexamerization facilitates bacterial flagellar protein export. Biochem. Biophys. Res. Commun. 388 (2), 323–327.

Kenjale, R., Wilson, J., Zenk, S.F., et al., 2005. The needle component of the type 3 secreton of *Shigella* regulates the activity of the secretion apparatus. J. Biol. Chem. 280 (52), 42929–42937.

Kenny, B., Abe, A., Stein, M., Finlay, B.B., 1997. Enteropathogenic *Escherichia coli* protein secretion is induced in response to conditions similar to those in the gastrointestinal tract. Infect. Immun. 65 (7), 2606–2612.

Kenny, B., Finlay, B.B., 1995. Protein secretion by enteropathogenic *Escherichia coli* is essential for transducing signals to epithelial cells. Proc. Natl. Acad. Sci. USA 92 (17), 7991–7995.

Kenny, B., Lai, L.C., Finlay, B.B., Donnenberg, M.S., 1996. EspA, a protein secreted by enteropathogenic *Escherichia coli*, is required to induce signals in epithelial cells. Mol. Microbiol. 20 (2), 313–323.

Kimbrough, T.G., Miller, S.I., 2000. Contribution of *Salmonella typhimurium* type 3 secretion components to needle complex formation. Proc. Natl. Acad. Sci. USA 97 (20), 11008–11013.

Knutton, S., Rosenshine, I., Pallen, M.J., et al., 1998. A novel EspA-associated surface organelle of enteropathogenic *Escherichia coli* involved in protein translocation into epithelial cells. EMBO J. 17 (8), 2166–2176.

Koo, J., Burrows, L.L., Howell, P.L., 2012. Decoding the roles of pilotins and accessory proteins in secretin escort services. FEMS Microbiol. Lett. 328 (1), 1–12.

Korotkov, K.V., Gonen, T., Hol, W.G., 2011. Secretins: dynamic channels for protein transport across membranes. Trends. Biochem. Sci. 36 (8), 433–443.

Korotkov, K.V., Pardon, E., Steyaert, J., Hol, W.G., 2009. Crystal structure of the N-terminal domain of the secretin GspD from ETEC determined with the assistance of a nanobody. Structure 17 (2), 255–265.

Kubori, T., Matsushima, Y., Nakamura, D., et al., 1998. Supramolecular structure of the *Salmonella typhimurium* type 3 protein secretion system. Science 280 (5363), 602–605.

Kubori, T., Sukhan, A., Aizawa, S.I., Galan, J.E., 2000. Molecular characterization and assembly of the needle complex of the *Salmonella typhimurium* type 3 protein secretion system. Proc. Natl. Acad. Sci. USA 97 (18), 10225–10230.

Lai, L.C., Wainwright, L.A., Stone, K.D., Donnenberg, M.S., 1997. A third secreted protein that is encoded by the enteropathogenic *Escherichia coli* pathogenicity island is required for trans-duction of signals and for attaching and effacing activities in host cells. Infect. Immun. 65 (6), 2211–2217.

Lara-Tejero, M., Kato, J., Wagner, S., Liu, X., Galan, J.E., 2011. A sorting platform determines the order of protein secretion in bacterial type 3 systems. Science 331 (6021), 1188–1191.

Lario, P.I., Pfuetzner, R.A., Frey, E.A., et al., 2005. Structure and biochemical analysis of a secretin pilot protein. EMBO J. 24 (6), 1111–1121.

Lavander, M., Sundberg, L., Edqvist, P.J., Lloyd, S.A., Wolf-Watz, H., Forsberg, A., 2002. Proteolytic cleavage of the FlhB homologue YscU of *Yersinia pseudotuberculosis* is essential for bacterial survival but not for type 3 secretion. J. Bacteriol. 184 (16), 4500–4509.

Letzelter, M., Sorg, I., Mota, L.J., et al., 2006. The discovery of SycO highlights a new function for type 3 secretion effector chaperones. EMBO J. 25 (13), 3223–3233.

Levdikov, V.M., Blagova, E.V., McFeat, A., Fogg, M.J., Wilson, K.S., Wilkinson, A.J., 2012. Structure of components of an intercellular channel complex in sporulating *Bacillus subtilis*. Proc. Natl. Acad. Sci. USA. 109 (14), 5441–5445.

Lilic, M., Quezada, C.M., Stebbins, C.E., 2010. A conserved domain in type 3 secretion links the cytoplasmic domain of InvA to elements of the basal body. Acta. Crystallogr. D. Biol. Crystallogr. 66 (Pt 6), 709–713.

Lilic, M., Vujanac, M., Stebbins, C.E., 2006. A common structural motif in the binding of virulence factors to bacterial secretion chaperones. Mol. Cell. 21 (5), 653–664.

Lloyd, S.A., Sjostrom, M., Andersson, S., Wolf-Watz, H., 2002. Molecular characterization of type 3 secretion signals via analysis of synthetic N-terminal amino acid sequences. Mol. Microbiol. 43 (1), 51–59.

Loquet, A., Sgourakis, N.G., Gupta, R., et al., 2012. Atomic model of the type 3 secretion system needle. Nature 486 (7402), 276–279.

Lountos, G.T., Austin, B.P., Nallamsetty, S., Waugh, D.S., 2009. Atomic resolution structure of the cytoplasmic domain of *Yersinia pestis* YscU, a regulatory switch involved in type 3 secretion. Protein. Sci. 18 (2), 467–474.

Lountos, G.T., Tropea, J.E., Waugh, D.S., 2012. Structure of the cytoplasmic domain of *Yersinia pestis* YscD, an essential component of the type 3 secretion system. Acta. Crystallogr. D. Biol. Crystallogr. 68 (Pt 3), 201–209.

Lunelli, M., Hurwitz, R., Lambers, J., Kolbe, M., 2011. Crystal structure of PrgI-SipD: insight into a secretion competent state of the type three secretion system needle tip and its interaction with host ligands. PLoS Pathog. 7 (8), e1002163.

Lunelli, M., Lokareddy, R.K., Zychlinsky, A., Kolbe, M., 2009. IpaB-IpgC interaction defines binding motif for type 3 secretion translocator. Proc. Natl. Acad. Sci. USA 106 (24), 9661–9666.

Luo, W., Donnenberg, M.S., 2011. Interactions and predicted host membrane topology of the entero-pathogenic *Escherichia coli* translocator protein EspB. J. Bacteriol. 193 (12), 2972–2980.

Luo, Y., Bertero, M.G., Frey, E.A., et al., 2001. Structural and biochemical characterization of the type 3 secretion chaperones CesT and SigE. Nat. Struct. Biol. 8 (12), 1031–1036.

Magdalena, J., Hachani, A., Chamekh, M., et al., 2002. Spa32 regulates a switch in substrate speci-ficity of the type 3 secreton of *Shigella flexneri* from needle components to Ipa proteins. J. Bacteriol. 184 (13), 3433–3441.

Makino, S., Tobe, T., Asakura, H., et al., 2003. Distribution of the secondary type 3 secretion sys-tem locus found in enterohemorrhagic *Escherichia coli* O157:H7 isolates among Shiga toxin-producing *E. coli* strains. J. Clin. Microbiol. 41 (6), 2341–2347.

Marlovits, T.C., Kubori, T., Lara-Tejero, M., Thomas, D., Unger, V.M., Galan, J.E., 2006. Assem-bly of the inner rod determines needle length in the type 3 secretion injectisome. Nature 441 (7093), 637–640.

Marlovits, T.C., Kubori, T., Sukhan, A., Thomas, D.R., Galan, J.E., Unger, V.M., 2004. Structural insights into the assembly of the type 3 secretion needle complex. Science 306 (5698), 1040–1042.

Marteyn, B., West, N.P., Browning, D.F., et al., 2010. Modulation of *Shigella* virulence in response to available oxygen in vivo. Nature 465 (7296), 355–358.

Mattei, P.J., Faudry, E., Job, V., Izore, T., Attree, I., Dessen, A., 2011. Membrane targeting and pore formation by the type 3 secretion system translocon. FEBS J. 278 (3), 414–426.

Mavris, M., Page, A.L., Tournebize, R., Demers, B., Sansonetti, P., Parsot, C., 2002. Regulation of transcription by the activity of the *Shigella flexneri* type 3 secretion apparatus. Mol. Microbiol. 43 (6), 1543–1553.

McDaniel, T.K., Jarvis, K.G., Donnenberg, M.S., Kaper, J.B., 1995. A genetic locus of enterocyte effacement conserved among diverse enterobacterial pathogens. Proc. Natl. Acad. Sci. USA 92 (5), 1664–1668.

McDowell, M.A., Johnson, S., Deane, J.E., et al., 2011. Structural and functional studies on the N-terminal domain of the *Shigella* type 3 secretion protein MxiG. J. Biol. Chem. 286 (35), 30606–30614.

Meisner, J., Maehigashi, T., Andre, I., Dunham, C.M., Moran Jr., C.P., 2012. Structure of the basal components of a bacterial transporter. Proc. Natl. Acad. Sci. USA. 109 (14), 5446–5451.

Mellies, J.L., Elliott, S.J., Sperandio, V., Donnenberg, M.S., Kaper, J.B., 1999. The Per regulon of enteropathogenic *Escherichia coli*: identification of a regulatory cascade and a novel tran-scriptional activator, the locus of enterocyte effacement (LEE)-encoded regulator (Ler). Mol. Microbiol. 33 (2), 296–306.

Menard, R., Sansonetti, P., Parsot, C., 1994. The secretion of the *Shigella flexneri* Ipa invasins is activated by epithelial cells and controlled by IpaB and IpaD. EMBO J. 13 (22), 5293–5302.

Michiels, T., Vanooteghem, J.C., Lambert de Rouvroit, C., et al., 1991. Analysis of virC, an operon involved in the secretion of Yop proteins by *Yersinia enterocolitica*. J. Bacteriol. 173 (16), 4994–5009.

Minamino, T., Imada, K., Namba, K., 2008a. Mechanisms of type 3 protein export for bacterial flagellar assembly. Mol. Biosyst. 4 (11), 1105–1115.

Minamino, T., Imada, K., Namba, K., 2008b. Molecular motors of the bacterial flagella. Curr. Opin. Struct. Biol. 18 (6), 693–701.

Minamino, T., Kinoshita, M., Imada, K., Namba, K., 2012. Interaction between FliI ATPase and a flagellar chaperone FliT during bacterial flagellar protein export. Mol. Microbiol. 83 (1), 168–178.

Minamino, T., Macnab, R.M., 2000a. Domain structure of *Salmonella* FlhB, a flagellar export com-ponent responsible for substrate specificity switching. J. Bacteriol. 182 (17), 4906–4914.

Minamino, T., Macnab, R.M., 2000b. FliH, a soluble component of the type 3 flagellar export apparatus of *Salmonella*, forms a complex with FliI and inhibits its ATPase activity. Mol. Microbiol.
37 (6), 1494–1503.

Minamino, T., Namba, K., 2008. Distinct roles of the FliI ATPase and proton motive force in bacterial flagellar protein export. Nature 451 (7177), 485–488.

Mizuno, S., Amida, H., Kobayashi, N., Aizawa, S., Tate, S., 2011. The NMR structure of FliK, the
trigger for the switch of substrate specificity in the flagellar type 3 secretion apparatus. J. Mol.
Biol. 409 (4), 558–573.

Monjaras Feria, J., Garcia-Gomez, E., Espinosa, N., Minamino, T., Namba, K., Gonzalez-Pedrajo, B.,
2012. Role of EscP (Orf16) in injectisome biogenesis and regulation of type 3 protein secretion
in enteropathogenic *Escherichia coli*. J. Bacteriol. 194 (22), 6029–6045.

Moore, S.A., Jia, Y., 2010. Structure of the cytoplasmic domain of the flagellar secretion apparatus
component FlhA from *Helicobacter pylori*. J. Biol. Chem. 285 (27), 21060–21069.

Mota, L.J., Journet, L., Sorg, I., Agrain, C., Cornelis, G.R., 2005. Bacterial injectisomes: needle
length does matter. Science 307 (5713), 1278.

Muller, D., Benz, I., Liebchen, A., Gallitz, I., Karch, H., Schmidt, M.A., 2009. Comparative
analysis of the locus of enterocyte effacement and its flanking regions. Infect. Immun. 77 (8),
3501–3513.

O'Connell, C.B., Creasey, E.A., Knutton, S., et al., 2004. SepL, a protein required for enteropathogenic *Escherichia coli* type 3 translocation, interacts with secretion component SepD. Mol.
Microbiol. 52 (6), 1613–1625.

Ohnishi, M., Murata, T., Nakayama, K., et al., 2000. Comparative analysis of the whole set of rRNA
operons between an enterohemorrhagic *Escherichia coli* O157:H7 Sakai strain and an *Escherichia coli* K-12 strain MG1655. Syst. Appl. Microbiol. 23 (3), 315–324.

Okon, M., Moraes, T.F., Lario, P.I., et al., 2008. Structural characterization of the type-3 pilot-
secretin complex from *Shigella flexneri*. Structure 16 (10), 1544–1554.

Olive, A.J., Kenjale, R., Espina, M., Moore, D.S., Picking, W.L., Picking, W.D., 2007. Bile salts
stimulate recruitment of IpaB to the *Shigella flexneri* surface, where it colocalizes with IpaD at
the tip of the type 3 secretion needle. Infect. Immun. 75 (5), 2626–2629.

Pallen, M.J., Bailey, C.M., Beatson, S.A., 2006. Evolutionary links between FliH/YscL-like proteins from bacterial type 3 secretion systems and second-stalk components of the FoF1 and
vacuolar ATPases. Protein. Sci. 15 (4), 935–941.

Paul, K., Erhardt, M., Hirano, T., Blair, D.F., Hughes, K.T., 2008. Energy source of flagellar type 3
secretion. Nature 451 (7177), 489–492.

Picking, W.L., Nishioka, H., Hearn, P.D., et al., 2005. IpaD of *Shigella flexneri* is independently
required for regulation of Ipa protein secretion and efficient insertion of IpaB and IpaC into host
membranes. Infect. Immun. 73 (3), 1432–1440.

Porter, M.E., Dorman, C.J., 1997. Differential regulation of the plasmid-encoded genes in the
Shigella flexneri virulence regulon. Mol. Gen. Genet. 256 (2), 93–103.

Porter, M.E., Mitchell, P., Free, A., Smith, D.G., Gally, D.L., 2005. The LEE1 promoters from both
enteropathogenic and enterohemorrhagic *Escherichia coli* can be activated by PerC-like proteins from either organism. J. Bacteriol. 187 (2), 458–472.

Poyraz, O., Schmidt, H., Seidel, K., et al., 2010. Protein refolding is required for assembly of the
type three secretion needle. Nat. Struct. Mol. Biol. 17 (7), 788–792.

Quinaud, M., Ple, S., Job, V., et al., 2007. Structure of the heterotrimeric complex that regulates type
3 secretion needle formation. Proc. Natl. Acad. Sci. USA 104 (19), 7803–7808.

Reid, S.D., Herbelin, C.J., Bumbaugh, A.C., Selander, R.K., Whittam, T.S., 2000. Parallel evolution
of virulence in pathogenic *Escherichia coli*. Nature 406 (6791), 64–67.

Ren, C.P., Chaudhuri, R.R., Fivian, A., et al., 2004. The ETT2 gene cluster, encoding a second type 3 secretion system from *Escherichia coli*, is present in the majority of strains but has undergone widespread mutational attrition. J. Bacteriol. 186 (11), 3547–3560.

Rodgers, L., Gamez, A., Riek, R., Ghosh, P., 2008. The type 3 secretion chaperone SycE promotes a localized disorder-to-order transition in the natively unfolded effector YopE. J. Biol. Chem. 283 (30), 20857–20863.

Roehrich, A.D., Martinez-Argudo, I., Johnson, S., Blocker, A.J., Veenendaal, A.K., 2010. The extreme C terminus of *Shigella flexneri* IpaB is required for regulation of type 3 secretion, needle tip composition, and binding. Infect. Immun. 78 (4), 1682–1691.

Rosenshine, I., Donnenberg, M.S., Kaper, J.B., Finlay, B.B., 1992. Signal transduction between enteropathogenic *Escherichia coli* (EPEC) and epithelial cells: EPEC induces tyrosine phosphorylation of host cell proteins to initiate cytoskeletal rearrangement and bacterial uptake. EMBO J. 11 (10), 3551–3560.

Rosqvist, R., Magnusson, K.E., Wolf-Watz, H., 1994. Target cell contact triggers expression and polarized transfer of *Yersinia* YopE cytotoxin into mammalian cells. EMBO J. 13 (4), 964–972.

Ross, J.A., Plano, G.V., 2011. A C-terminal region of *Yersinia pestis* YscD binds the outer membrane secretin YscC. J. Bacteriol. 193 (9), 2276–2289.

Saijo-Hamano, Y., Imada, K., Minamino, T., et al., 2010. Structure of the cytoplasmic domain of FlhA and implication for flagellar type 3 protein export. Mol. Microbiol. 76 (1), 260–268.

Sal-Man, N., Deng, W., Finlay, B.B., 2012. EscI: a crucial component of the type 3 secretion system forms the inner rod structure in enteropathogenic *Escherichia coli*. Biochem. J. 442 (1), 119–125.

Salmond, G.P., Reeves, P.J., 1993. Membrane traffic wardens and protein secretion in Gram-negative bacteria. Trends. Biochem. Sci. 18 (1), 7–12.

Sandner, L., Eguiarte, L.E., Navarro, A., Cravioto, A., Souza, V., 2001. The elements of the locus of enterocyte effacement in human and wild mammal isolates of *Escherichia coli*: evolution by assemblage or disruption? Microbiology 147 (Pt 11), 3149–3158.

Sanowar, S., Singh, P., Pfuetzner, R.A., et al., 2009. Interactions of the transmembrane polymeric rings of the *Salmonella enterica* serovar Typhimurium type 3 secretion system. M. Bio. 1 (3), pii: e00158–10.

Sansonetti, P.J., 1991. Genetic and molecular basis of epithelial cell invasion by *Shigella* species. Rev. Infect. Dis. 13 (Suppl. 4). S285–S292.

Sansonetti, P.J., Hale, T.L., Dammin, G.J., Kapfer, C., Collins Jr., H.H., Formal, S.B., 1983. Alterations in the pathogenicity of *Escherichia coli* K-12 after transfer of plasmid and chromosomal genes from *Shigella flexneri*. Infect. Immun. 39 (3), 1392–1402.

Sansonetti, P.J., Kopecko, D.J., Formal, S.B., 1982. Involvement of a plasmid in the invasive ability of *Shigella flexneri*. Infect. Immun. 35 (3), 852–860.

Sasakawa, C., Kamata, K., Sakai, T., et al., 1988. Virulence-associated genetic regions comprising 31 kilobases of the 230-kilobase plasmid in *Shigella flexneri* 2a. J. Bacteriol. 170 (6), 2480–2484.

Schesser, K., Frithz-Lindsten, E., Wolf-Watz, H., 1996. Delineation and mutational analysis of the *Yersinia pseudotuberculosis* YopE domains which mediate translocation across bacterial and eukaryotic cellular membranes. J. Bacteriol. 178 (24), 7227–7233.

Schraidt, O., Lefebre, M.D., Brunner, M.J., et al., 2010. Topology and organization of the *Salmonella typhimurium* type 3 secretion needle complex components. PLoS Pathog. 6 (4), e1000824.

Schraidt, O., Marlovits, T.C., 2011. Three-dimensional model of *Salmonella*'s needle complex at subnanometer resolution. Science 331 (6021), 1192–1195.

Schuch, R., Maurelli, A.T., 2001. MxiM and MxiJ, base elements of the Mxi-Spa type 3 secretion system of *Shigella*, interact with and stabilize the MxiD secretin in the cell envelope. J. Bacteriol. 183 (24), 6991–6998.

Sekiya, K., Ohishi, M., Ogino, T., Tamano, K., Sasakawa, C., Abe, A., 2001. Supermolecular structure of the enteropathogenic *Escherichia coli* type 3 secretion system and its direct interaction with the EspA-sheath-like structure. Proc. Natl. Acad. Sci. USA 98 (20), 11638–11643.

Sheikh, J., Dudley, E.G., Sui, B., Tamboura, B., Suleman, A., Nataro, J.P., 2006. EilA, a HilA-like regulator in enteroaggregative *Escherichia coli*. Mol. Microbiol. 61 (2), 338–350.

Sory, M.P., Boland, A., Lambermont, I., Cornelis, G.R., 1995. Identification of the YopE and YopH domains required for secretion and internalization into the cytosol of macrophages, using the cyaA gene fusion approach. Proc. Natl. Acad. Sci. USA 92 (26), 11998–12002.

Sperandio, V., Mellies, J.L., Delahay, R.M., et al., 2000. Activation of enteropathogenic *Escherichia coli* (EPEC) LEE2 and LEE3 operons by Ler. Mol. Microbiol. 38 (4), 781–793.

Sperandio, V., Mellies, J.L., Nguyen, W., Shin, S., Kaper, J.B., 1999. Quorum sensing controls expression of the type 3 secretion gene transcription and protein secretion in enterohemorrhagic and enteropathogenic *Escherichia coli*. Proc. Natl. Acad. Sci. USA 96 (26), 15196–15201.

Spreter, T., Yip, C.K., Sanowar, S., et al., 2009. A conserved structural motif mediates formation of the periplasmic rings in the type 3 secretion system. Nat. Struct. Mol. Biol. 16 (5), 468–476.

Srikanth, C.V., Mercado-Lubo, R., Hallstrom, K., McCormick, B.A., 2011. *Salmonella* effector proteins and host-cell responses. Cell Mol. Life. Sci. 68 (22), 3687–3697.

Stebbins, C.E., Galan, J.E., 2003. Priming virulence factors for delivery into the host. Nat. Rev. Mol. Cell Biol. 4 (9), 738–743.

Sturm, A., Heinemann, M., Arnoldini, M., et al., 2011. The cost of virulence: retarded growth of *Salmonella* Typhimurium cells expressing type 3 secretion system 1. PLoS Pathog. 7 (7), e1002143.

Sukhan, A., Kubori, T., Galan, J.E., 2003. Synthesis and localization of the *Salmonella* SPI-1 type 3 secretion needle complex proteins PrgI and PrgJ. J. Bacteriol. 185 (11), 3480–3483.

Sun, G.W., Gan, Y.H., 2010. Unraveling type 3 secretion systems in the highly versatile *Burkholderia pseudomallei*. Trends. Microbiol. 18 (12), 561–568.

Sun, P., Tropea, J.E., Austin, B.P., Cherry, S., Waugh, D.S., 2008. Structural characterization of the *Yersinia pestis* type 3 secretion system needle protein YscF in complex with its heterodimeric chaperone YscE/YscG. J. Mol. Biol. 377 (3), 819–830.

Tacket, C.O., Sztein, M.B., Losonsky, G., et al., 2000. Role of EspB in experimental human enteropathogenic *Escherichia coli* infection. Infect. Immun. 68 (6), 3689–3695.

Tamano, K., Aizawa, S., Katayama, E., et al., 2000. Supramolecular structure of the *Shigella* type 3 secretion machinery: the needle part is changeable in length and essential for delivery of effectors. EMBO J. 19 (15), 3876–3887.

Tamano, K., Katayama, E., Toyotome, T., Sasakawa, C., 2002. *Shigella* Spa32 is an essential secretory protein for functional type 3 secretion machinery and uniformity of its needle length. J. Bacteriol. 184 (5), 1244–1252.

Tampakaki, A.P., Skandalis, N., Gazi, A.D., et al., 2010. Playing the Harp: evolution of our understanding of hrp/hrc genes. Annu. Rev. Phytopathol. 48, 347–370.

Tobe, T., Yoshikawa, M., Mizuno, T., Sasakawa, C., 1993. Transcriptional control of the invasion regulatory gene virB of *Shigella flexneri*: activation by virF and repression by H-NS. J. Bacteriol. 175 (19), 6142–6149.

Troisfontaines, P., Cornelis, G.R., 2005. Type 3 secretion: more systems than you think. Physiology (Bethesda) 20, 326–339.

van Eerde, A., Hamiaux, C., Perez, J., Parsot, C., Dijkstra, B.W., 2004. Structure of Spa15, a type 3 secretion chaperone from *Shigella flexneri* with broad specificity. EMBO Rep. 5 (5), 477–483.

Veenendaal, A.K., Hodgkinson, J.L., Schwarzer, L., Stabat, D., Zenk, S.F., Blocker, A.J., 2007. The type 3 secretion system needle tip complex mediates host cell sensing and translocon insertion. Mol. Microbiol. 63 (6), 1719–1730.

Venkatesan, M.M., Goldberg, M.B., Rose, D.J., Grotbeck, E.J., Burland, V., Blattner, F.R., 2001. Complete DNA sequence and analysis of the large virulence plasmid of *Shigella flexneri*. Infect. Immun. 69 (5), 3271–3285.

Vogler, A.P., Homma, M., Irikura, V.M., Macnab, R.M., 1991. *Salmonella typhimurium* mutants defective in flagellar filament regrowth and sequence similarity of FliI to F0F1, vacuolar, and archaebacterial ATPase subunits. J. Bacteriol. 173 (11), 3564–3572.

Wagner, S., Konigsmaier, L., Lara-Tejero, M., Lefebre, M., Marlovits, T.C., Galan, J.E., 2010. Organization and coordinated assembly of the type 3 secretion export apparatus. Proc. Natl. Acad. Sci. USA 107 (41), 17745–17750.

Wang, Y.A., Yu, X., Yip, C., Strynadka, N.C., Egelman, E.H., 2006. Structural polymorphism in bacterial EspA filaments revealed by cryo-EM and an improved approach to helical reconstruction. Structure 14 (7), 1189–1196.

Wieler, L.H., McDaniel, T.K., Whittam, T.S., Kaper, J.B., 1997. Insertion site of the locus of enterocyte effacement in enteropathogenic and enterohemorrhagic *Escherichia coli* differs in relation to the clonal phylogeny of the strains. FEMS Microbiol. Lett. 156 (1), 49–53.

Wiesand, U., Sorg, I., Amstutz, M., et al., 2009. Structure of the type 3 secretion recognition protein YscU from *Yersinia enterocolitica*. J. Mol. Biol. 385 (3), 854–866.

Wiley, D.J., Rosqvist, R., Schesser, K., 2007. Induction of the *Yersinia* type 3 secretion system as an all-or-none phenomenon. J. Mol. Biol. 373 (1), 27–37.

Wilharm, G., Dittmann, S., Schmid, A., Heesemann, J., 2007. On the role of specific chaperones, the specific ATPase, and the proton motive force in type 3 secretion. Int. J. Med. Microbiol. 297 (1), 27–36.

Wilharm, G., Lehmann, V., Krauss, K., et al., 2004. *Yersinia enterocolitica* type 3 secretion depends on the proton motive force but not on the flagellar motor components MotA and MotB. Infect. Immun. 72 (7), 4004–4009.

Woestyn, S., Allaoui, A., Wattiau, P., Cornelis, G.R., 1994. YscN, the putative energizer of the *Yersinia* Yop secretion machinery. J. Bacteriol. 176 (6), 1561–1569.

Woestyn, S., Sory, M.P., Boland, A., Lequenne, O., Cornelis, G.R., 1996. The cytosolic SycE and SycH chaperones of *Yersinia* protect the region of YopE and YopH involved in translocation across eukaryotic cell membranes. Mol. Microbiol. 20 (6), 1261–1271.

Wood, S.E., Jin, J., Lloyd, S.A., 2008. YscP and YscU switch the substrate specificity of the Yersinia type 3 secretion system by regulating export of the inner rod protein YscI. J. Bacteriol. 190 (12), 4252–4262.

Worrall, L.J., Vuckovic, M., Strynadka, N.C., 2010. Crystal structure of the C-terminal domain of the *Salmonella* type 3 secretion system export apparatus protein InvA. Protein. Sci. 19 (5), 1091–1096.

Yang, F., Yang, J., Zhang, X., et al., 2005. Genome dynamics and diversity of *Shigella* species, the etiologic agents of bacillary dysentery. Nucleic Acids Res. 33 (19), 6445–6458.

Yao, Y., Xie, Y., Perace, D., et al., 2009. The type 3 secretion system is involved in the invasion and intracellular survival of *Escherichia coli* K1 in human brain microvascular endothelial cells. FEMS Microbiol. Lett. 300 (1), 18–24.

Yip, C.K., Finlay, B.B., Strynadka, N.C., 2005a. Structural characterization of a type 3 secretion system filament protein in complex with its chaperone. Nat. Struct. Mol. Biol. 12 (1), 75–81.

Yip, C.K., Kimbrough, T.G., Felise, H.B., et al., 2005b. Structural characterization of the molecular platform for type 3 secretion system assembly. Nature 435 (7042), 702–707.

Zarivach, R., Deng, W., Vuckovic, M., et al., 2008. Structural analysis of the essential self-cleaving type 3 secretion proteins EscU and SpaS. Nature 453 (7191), 124–127.

Zarivach, R., Vuckovic, M., Deng, W., Finlay, B.B., Strynadka, N.C., 2007. Structural analysis of a prototypical ATPase from the type 3 secretion system. Nat. Struct. Mol. Biol. 14 (2), 131–137.

Zhang, L., Chaudhuri, R.R., Constantinidou, C., et al., 2004. Regulators encoded in the *Escherichia coli* type 3 secretion system 2 gene cluster influence expression of genes within the locus for enterocyte effacement in enterohemorrhagic *E. coli* O157:H7. Infect. Immun. 72 (12), 7282–7293.

Zhang, L., Wang, Y., Picking, W.L., Picking, W.D., De Guzman, R.N., 2006. Solution structure of monomeric BsaL, the type 3 secretion needle protein of *Burkholderia pseudomallei*. J. Mol. Biol. 359 (2), 322–330.

Zhong, D., Lefebre, M., Kaur, K., et al., 2012. The *Salmonella* type 3 secretion system inner rod protein PrgJ is partially folded. J. Biol. Chem. 287 (30), 25303–25311.

Zychlinsky, A., Prevost, M.C., Sansonetti, P.J., 1992. *Shigella flexneri* induces apoptosis in infected macrophages. Nature 358 (6382), 167–169.

Chapter 15

Type 3 secretion effectors

Abigail Clements, Cedric N. Berger, Mariella Lomma, Gad Frankel
Imperial College London, London, UK

INTRODUCTION

T3SS effector proteins of enterohemorrhagic *E. coli* (EHEC), enteropathogenic *E. coli* (EPEC) and *Shigella* affect diverse signaling pathways and physiological processes when translocated into the host cell. For EPEC and EHEC seven 'core' effectors are encoded with the T3S translocation machinery on the locus of enterocyte effacement (LEE) pathogenicity island (PI) (McDaniel et al., 1995), while other effectors are encoded within prophages and other integrative elements (Tobe et al., 2006). Although EPEC and EHEC show high levels of conservation between the LEE encoded effectors, there is significant diversity in their non-LEE effector (NLE) repertoire; EPEC strains E2348/69 and B171 encode at least 23 intact effector genes (Iguchi et al., 2009; Deng et al., 2012), whereas the EHEC O157 strain Sakai is estimated to have closer to 50 T3SS effectors (Table 15.1) (Tobe et al., 2006). The *Shigella* T3SS is encoded on a large virulence plasmid along with the majority of identified T3SS effectors (Table 15.2). Approximately 30 *Shigella* T3SS effectors are currently recognized (Parsot, 2009) although as more *Shigella* isolates of different species are sequenced this number may be revised. This chapter describes how these T3SS effectors modulate the host cytoskeleton, immune response, cell survival, and gut integrity.

CYTOSKELETON REMODELING

EPEC, EHEC and *Shigella* have acquired a subset of effectors to promote the colonization of the host by modulating the host cytoskeleton. The molecular mechanisms of pedestal formation, cell invasion, modulation of Rho GTPases and reorganization of microtubules and intermediate filaments will be described.

Intimate attachment and pedestal formation

EHEC and EPEC colonization is characterized by the T3SS-dependent formation of attaching and effacing (A/E) lesions on the apical surface of enterocytes. This presents as localized effacement of the microvilli (discussed further below),

Escherichia coli. **http://dx.doi.org/10.1016/B978-0-12-397048-0.00015-2**

TABLE 15.1 T3SS effectors of A/E pathogens

Effector	Function	Interacting proteins/ enzymatic targets	Homolog (*Shigella* homologs in bold)	EPEC1 E2348/69 O127:H6	EPEC2 B171 O11:NM	Atypical EPEC E110019 O11:H9
Cif	Rho GTPase modulation – prevents degradation of RhoA Induces cell cycle arrest through accumulation of p21 and p27	NEDD8, Ub (Cui et al., 2010)	YPK_1971, Cifbp	0(1) [b]	1	0
EspB	Inhibition of phagocytosis Microvilli effacement	Myosin 1c (Iizumi et al., 2007), α-catenin (Kodama et al., 2002), α_1-antitrypsin (Knappstein et al., 2004)	**IpaC**, YopD	1	1	1
EspF	Induces apoptosis Inhibition of phagocytosis TJ disruption Ion channel disruption Cytoskeletal integrity – intermediate filaments	ABCF2 (Nougayrede et al., 2007), SNX9 (Marches et al., 2006), CK18, 14-3-3zeta Viswanathan et al., 2004), ZO-1, ZO-2, Actin, Arp2/3, N-WASP, Profilin (Peralta-Ramirez et al., 2008)	TccP2/EspF$_U$	1	1	1
EspG	Cytoskeletal integrity - MT Disrupts cellular trafficking Ion channel disruption	Tubulin (Matsuzawa et al., 2004), Rab GTPases (Dong et al 2012) ARF 1/6, PAK1/2/3 (Selyunin et al., 2011), GM130, RACK1 (Clements et al., 2011)	**VirA**	2	1	1
EspH	Rho GTPase modulation – inactivation of DH-PH Rho GEFs Inhibition of phagocytosis (FCγR-mediated) Intimate adhesion and pedestal formation	DH-PH Rho GEFs (Dong et al., 2010)	-	1	1	1

Number of T3SS effectors present in each genome[a]

Rabbit EPEC E22 O103:H2	Atypical EPEC CB9615 O55:H7	EHEC EDL933 O157:H7	EHEC Sakai O157:H7	EHEC TW14359 O157:H7	EHEC EC4115 O157:H7	EHEC 11368 O26:H11	EHEC 11128 O111:H-	EHEC 12009 O103:H2	C. rodentium ICC168
1	0	0	0	0	0	0(1)	0(1)	0(1)	0
1	1	1	1	1	1	1	1	1	1
1	1	1	1	1	1	1	1	1	1
1	1	1	1	1	1	1	1	1	1
1	1	1	1	1	1	1	1	1	1

Continued

TABLE 15.1 T3SS effectors of A/E pathogens—cont'd

Effector	Function	Interacting proteins/ enzymatic targets	Homolog (*Shigella* homologs in bold)	EPEC1 E2348/69 O127:H6	EPEC2 B171 O11:NM	Atypical EPEC E110019 O11:H9
EspJ	Inhibition of phagocytosis (FCγR- and CR3-mediated)	-	SboC, HopF2	1	0	0
EspK	Unknown	-	GogB	0	0	1
EspL	Intimate adhesion and pedestal formation – increases F-actin bundling	Annexin 2 (Miyahara et al., 2009)	**OspD**	1(2)	1(2)	1
EspM	Rho GTPase modulation – RhoA GEF	RhoA (Arbeloa et al., 2008)	**IpgB1**, SopE	0	2 (1)	1
EspN	Unknown	-	SARI_01330&01464, SEHO0A_00314&01608, CNF	0	1	0
EspO	Cell integrity – maintains focal adhesions	Integrin-linked kinase (Kim et al., 2009)	**OspE**	0(1)	0	1(1)
EspR	Unknown	-	-	0	1	1
EspS	Unknown	-	**OspB**	0	1(1)	1
EspT	Rho GTPase modulation – Cdc42/ Rac1 GEF Immunomodulation- induces NF-κB, Erk1/2 and JNK pathways through Rac1	Cdc42, Rac1 (Bulgin et al., 2009)	**IpgB**, SopE	0	0	1
EspV	Actin remodeling – unknown mechanism	-	AvrA1, AvrBS1	0	0(1)	1 (1)
EspW	Unknown	-	HopPmaA	0	1	1
EspX	Unknown	-	SopA	0	0	0
EspY [c]	Unknown	-	SopD	0	0	0
EspZ	Inhibits cell cytotoxicity - maintains focal adhesions, reduces loss of mitochondrial membrane potential	CD98 (Shames et al., 2010), TIM17b (Shames et al., 2011)	-	1	1	1

Number of T3SS effectors present in each genome[a]

Rabbit EPEC E22 O103:H2	Atypical EPEC CB9615 O55:H7	EHEC EDL933 O157:H7	EHEC Sakai O157:H7	EHEC TW14359 O157:H7	EHEC EC4115 O157:H7	EHEC 11368 O26:H11	EHEC 11128 O111:H-	EHEC 12009 O103:H2	C. rodentium ICC168
1	1	1	1	0(1)	1	1	1	0	1
1	0	1	1	1	1	2	1	3	1
1	3	3	3(1)	3	3	1(2)	2(1)	2(1)	1
1	2	2	2	2	2	2	2	2	2(1)
1	0	1	1	1	1	1	1	1	1(2)
1	1(1)	2	2	2	2	2	2	2	2
1	3	3	3	3	3	1	1	1	0
1	0	0	0	1	1	1(1)	1	1 (1)	1
0	0	0	0	0	0	0	0	0	1
1	0(1)	0(1)	0(1)	0(1)	0(1)	0(1)	0(1)	0(2c)	1
1	0(2e)	1	1	1	1	1	1	1	0
1	0	0(1)	1	1	1	1	1	1	1
0	4	4	4	4	4	0	0	0	0
1	1	1	1	1	1	1	1	1	1

Continued

TABLE 15.1 T3SS effectors of A/E pathogens—cont'd

Effector	Function	Interacting proteins/ enzymatic targets	Homolog (*Shigella* homologs in bold)	EPEC1 E2348/69 O127:H6	EPEC2 B171 O11:NM	Atypical EPEC E110019 O11:H9
Map	Rho GTPase modulation – Cdc42 GEF Induces apoptosis – mitochondrial disruption TJ disruption Ion channel disruption	Cdc42 (Huang et al., 2009), EBP50 (NHERF1) (Simpson et al., 2006), NHERF2 (Martinez et al., 2010)	**IpgB2**	1	1	1
NleA/ EspI	TJ disruption – disrupts protein trafficking	Sec24 (Kim et al., 2007), Syntrophin, PDZK11, SNX27, MALS3, TCOF1, NHERF1/2, MAGI-3, SAP97, SAP102, PSD-95 (Lee et al., 2008)	-	1	1	1
NleB	Immunomodulation – inhibits TNFα induced NFκB activation	-	SARI_03503, LTSEUGA_5597,	2(1)	1(2)	1
NleC	Immunomodulation – cleaves p65, p50, c-Rel, IκB inhibiting NFκB activation	p65 (Baruch et al., 2011), p50, c-Rel (Pearson et al., 2011), IκB (Michail et al., 2003), p300 (Shames et al., 2011)	Aip56	1	0(2)	0
NleD	Immunomodulation – cleaves JNK and p38 to block AP-1 activation	JNK, p38 (Baruch et al., 2011)	SARI_02033, APM_0230,	1	0	0
NleE	Immunomodulation – methylates and inactivates TAB2 and 3 ubiquitin chain binding activity to disrupt NFκB signaling	TAB2,3 (Zhang et al., 2012)	**OspZ**	2	1	0
NleF	Unknown	-	PROVALCAL_00660, **SDY_P223**	1	1	1
NleG[d]	E3 ubiquitin ligase, unknown cellular function	UBE2D2 E2 ligase (Wu et al., 2010)	STY1076	1	5	0

Number of T3SS effectors present in each genome[a]

Rabbit EPEC E22 O103:H2	Atypical EPEC CB9615 O55:H7	EHEC EDL933 O157:H7	EHEC Sakai O157:H7	EHEC TW14359 O157:H7	EHEC EC4115 O157:H7	EHEC 11368 O26:H11	EHEC 11128 O111:H-	EHEC 12009 O103:H2	C. rodentium ICC168
1	1	1	1	1	1	1	1	1	1
1	1(1)	1	1	1	1	1(1)	1	1	1
2	2	2(1)	2(1)	2(1)	2(1)	1	2(1)	4	1(1)
1	1	1	1	1	1	1	2	0(2)	1
0	1	1	1	1	1	0	0	0	2
1	1	1	1	1	1	1	2	2	1
0	1	1	1	1	1	1	1	1	1
10	9	13	13	13	11	13	8	6	3(2)

Continued

TABLE 15.1 T3SS effectors of A/E pathogens—cont'd

Effector	Function	Interacting proteins/ enzymatic targets	Homolog (*Shigella* homologs in bold)	EPEC1 E2348/69 O127:H6	EPEC2 B171 O11:NM	Atypical EPEC E110019 O11:H9
NleH	Immunomodulation – prevents NFκB activation Inhibition of apoptosis	BI-1 (Hemrajani et al., 2010), NHERF2 (Martinez et al., 2010), RPS3 (Gao et al., 2009)	**OspG**	2(1)	1(1)	0(2)
TccP/ EspF$_U$	Intimate adhesion and pedestal formation	N-WASP (Garmendia et al., 2006), IRTKS (Vingadassalom et al., 2009), IRSp53 (Weiss et al., 2009), Cortactin (Mousnier et al., 2008)	EspF	0	1	1
Tir	Intimate adhesion and pedestal formation – receptor for intimin, induces actin polymerization	Nck (Gruenheid et al., 2001), IRTKS (Vingadassalom et al., 2009), IRSp53 (Weiss et al., 2009), PI3K (Sason et al., 2009), CK18 (Batchelor et al., 2004), Talin, Vinculin, α-actinin (Freeman et al., 2000), cortactin (Mousnier et al., 2008), 14-3-3tau (Patel et al., 2006)	-	1	1	1

[a]Effectors were identified by tBlastN searches using Integrated Microbial Genomes (DOE Joint Genome Institute).
[b]Numbers in brackets indicate pseudogenes present in genome which may occur through frameshifts, internal deletions, internal stop codons, N-terminal or C-terminal truncations, absent start codons, etc. Importantly, while some of these pseudogenes have been confirmed, others may be the result of sequencing errors (the effector may still be translocated and functional), and some mutations (e.g. internal or C-terminal deletions) may still result in a translocated and potentially functional effector.
[c]EspY proteins are grouped by the presence of an N-terminal WEX5F motif. Only homologs of the published EHEC Sakai EspY proteins (Tobe et al., 2006) are indicated and therefore these figures may be an underestimation of the total WEX5F containing effectors.
[d]NleG proteins were identified by the motif DUF1076.

Number of T3SS effectors present in each genome[a]

Rabbit EPEC E22 O103:H2	Atypical EPEC CB9615 O55:H7	EHEC EDL933 O157:H7	EHEC Sakai O157:H7	EHEC TW14359 O157:H7	EHEC EC4115 O157:H7	EHEC 11368 O26:H11	EHEC 11128 O111:H-	EHEC 12009 O103:H2	C. rodentium ICC168
1	2	2	2	2	2	2	2	1(1)	1
0	1	1(1)	1(1)	1(1)	1(1)	1	1	1	0
1	1	1	1	1	1	1	1	1	1

TABLE 15.2 T3SS effectors of *Shigella*

Effector	Function	Interacting proteins/ enzymatic targets	Homolog (EPEC/ EHEC homologs in bold)	Number of T3SS effectors present in each genome[a]			
				S. dysenteriae Sd197 300267	S. flexneri Sf301 198214	S. boydii Sb512 344609	S. sonnei Ss046 300269
IpaA	Invasion – induces actin filament depolymerization	Vinculin (Bourdet-Sicard et al., 1999)	SipA	1	1	1	1
IpaB	Invasion – interacts with NMEs to bring bacteria to host cell membrane. Induces apoptosis through IL-1β release. Induces cell cycle arrest	MAD2B (Iwai et al., 2007) Caspase 1 (Hilbi et al., 1998)	SipB, YopB	1	1	1	1
IpaC	Invasion – recruits Src and induces actin extensions around bacteria	-	SipC	1	1	1	1
IpaD	Invasion – interacts with NMEs to bring bacteria to host cell membrane	-	SipD	1	1	1	1
IpgB1	Rho GTPase modulation – Rac1/Cdc42 GEF	Rac1, Cdc42 (Huang et al., 2009), ELMO (Handa et al., 2007)	**EspM**, SifA, SifB	1	1	1	1
IpgB2	Rho GTPase modulation – RhoA GEF	RhoA (Klink et al., 2010)	**Map**, **EspT**, SopE	1	1	1	1

	Function	Target					
IpgD	Invasion – phosphoinositide signaling Inhibits apoptosis – degradation of p53	PtdIns(4,5)P(2) (Niebuhr et al., 2002)	SopB	1	1	1	1
VirA	Cytoskeletal remodeling – MT Inhibits apoptosis – degradation of p53	Tubulin (Yoshida et al., 2002)	**EspG**	1	1	1	1
IcsB	Immunomodulation – escape from autophagy	Cholesterol (Kayath et al., 2010)	BopA	1	1	1	1
OspB	Immunomodulation – inhibits polymorphonuclear (PMN) migration, MAPK signaling	Retinoblastoma (Rb) (Zurawski et al., 2009)	-	1	1	1	1
OspC	Immunomodulation – inhibits PMN migration	-	-	3(1)[b]	4	3(1)	3(1)
OspD	Unknown	-	**EspL**	3	3	3	3
OspE	Cell integrity – maintains focal adhesions	Integrin-linked kinase (Kim et al., 2009)	**EspO**	1	2	1	1(1)
OspF	Immunomodulation – dephosphorylates MAPK, reduces PMN migration	MAPK (ERK2) (Li et al., 2007), b-tubulin, Rb (Zurawski et al., 2009)	SpvC, HopA1	1	1	1	1
OspG	Immunomodulation – prevents IκB degradation and hence NFκB activation	UbcH5, UbcH7 (Kim et al., 2005)	**NleH**	1	1	1	1

Continued

TABLE 15.2 T3SS effectors of Shigella—cont'd

Effector	Function	Interacting proteins/enzymatic targets	Homolog (EPEC/EHEC homologs in bold)	Number of T3SS effectors present in each genome[a]			
				S. dysenteriae Sd197 300267	S. flexneri Sf301 198214	S. boydii Sb512 344609	S. sonnei Ss046 300269
OspI	Immunomodulation – reduces NFκB by deamidating UBC13	UBc13 (Sanada et al., 2012)	–	1	1	1	1
OspZ	Immunomodulation – disrupts NFκB signaling, reduces PMN migration	-	**NleE**	1	1	1	1
IpaH (plasmid)	Immunomodulation – blocks NFκB activation by E3 ubiquitin ligase activity	NEMO/IKKγ, ABIN-1 (Ashida et al., 2010) E2 Ub ligases (UBE2D1, UBE2D3 and UBE2D4) (Singer et al., 2008)	SspH	5	5	4	4
IpaH (chromosomal)	Potentially immunomodulation – E3 ubiquitin ligase	-	SspH	5(1)	4(3)	5	5

[a]Effectors were identified by tBlastN searches using Integrated Microbial Genomes (DOE Joint Genome Institute).
[b]Numbers in brackets indicate pseudogenes present in genome which may occur through frameshifts, internal deletions, internal stop codons, N-terminal or C-terminal truncations, absent start codons, etc. Importantly, while some of these pseudogenes have been confirmed, others may be the result of sequencing errors (the effector may still be translocated and functional), and some mutations (e.g. internal or C-terminal deletions) may still result in a translocated and potentially functional effector.

intimate attachment of the bacteria to the host membrane and actin polymerization underneath the bacterial attachment site. In cultured cells the actin accumulation is visualized as raised pedestal-like structures underneath adherent bacteria. EPEC and EHEC have derived an efficient way to ensure this important step in colonization occurs by expressing an outer-membrane adhesin, intimin, as well as the translocated intimin receptor (Tir), which the bacteria deliver into the host cell via the T3SS. Tir inserts into the host cell membrane in a hairpin loop topology with the extracellular loop interacting with intimin; the bacteria therefore provide both the adhesin and the 'host cell' receptor required for attachment (Figure 15.1A).

Intimin is encoded by the *eae* gene on the LEE PI and is secreted by the general secretory pathway to be inserted into the bacterial outer membrane (McDaniel et al., 1995; Touze et al., 2004). In addition to binding the T3SS effector Tir, intimin can also interact with endogenous host cell proteins, including β1-chain integrins (Isberg and Leong, 1990; Frankel et al., 1996) and nucleolin, which is up-regulated by Shiga toxin production (Robinson et al., 2006).

The T3SS effector Tir not only acts as the host cell receptor for intimin but, upon binding, intimin initiates Tir clustering which mediates protein signaling within epithelial cells. Diverse EPEC and EHEC strains have evolved different mechanisms of Tir signaling, all of which result in actin accumulation underneath the bacterial attachment site (Figure 15.1A). In some EPEC strains, Tir is phosphorylated at tyrosine 474 (Y474p) by host tyrosine kinases, (the specific kinases involved remain unclear) (Phillips et al., 2004; Swimm et al., 2004). Tir phosphorylation promotes its interaction with the SH2 domain of the adaptor protein Nck leading to the recruitment of neural Wiskott-Aldrich syndrome protein (N-WASP) via the Nck SH3 domain (Gruenheid et al., 2001; Phillips et al., 2004). Activation of N-WASP occurs by relieving its autoinhibition fold, allowing interaction with the actin-related protein 2/3 (Arp2/3) complex and initiation of actin polymerization (Gruenheid et al., 2001; Campellone et al., 2002). Proteins normally involved in endocytosis, clathrin (Veiga et al., 2007), CD-2-associated protein (CD2AP) (Guttman et al., 2010), and dynamin-2 (Unsworth et al., 2007), are also involved in pedestal formation via the Nck-dependent pathway, the significance of which requires further investigation. EHEC Tir lacks a tyrosine 474 equivalent and the process of actin polymerization is mediated via the non-LEE encoded T3SS translocated effector protein, TccP (Tir-cytoskeleton coupling protein) (Garmendia et al., 2004) also known as EspF$_U$ (*E. coli* secreted protein F in prophage U) (Campellone et al., 2004). TccP/EspF$_U$ interacts with the IRSp53/MIM proteins, IRTKS and IRSp53, which also bind Tir at an Asn-Pro-Tyr (NPY458) tripeptide in the Tir C-terminal domain thereby linking TccP/EspF$_U$ indirectly to Tir (Campellone and Leong, 2005; Brady et al., 2007; Vingadassalom et al., 2009). TccP/EspF$_U$ also interacts with and activates N-WASP to initiate Arp2/3 recruitment and actin polymerization (Cheng et al., 2008). In mammalian cells TccP/EspF$_U$ was also shown to induce Arp2/3 complex-dependent actin polymerization in the absence of N-WASP suggesting that alternative proteins can act in this pathway

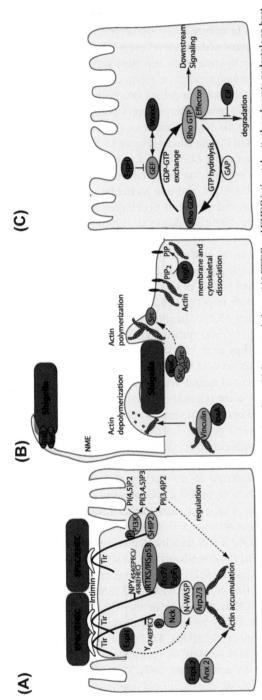

FIGURE 15.1 T3SS effectors from EPEC, EHEC, and *Shigella* interfere with host cytoskeleton. (A) EPEC and EHEC intimately attach and create pedestals on host cells. The bacterial outer-membrane protein, intimin, interacts with the T3SS effector Tir in the host cell membrane. In some EPEC strains Tir is phosphorylated at Y474 promoting interaction with the host adaptor protein Nck and recruitment of neural Wiskott-Aldrich syndrome protein (N-WASP). EHEC Tir lacks Y474 instead utilizing a NPY458 motif to promote actin polymerization. Tir NPY458 interacts with IRTKS/IRSp53, which binds another T3SS effector TccP/EspFU which in turn recruits N-WASP. In both cases N-WASP recruits the ARP2/3 complex to initiate actin polymerization. N-WASP independent activation of Arp2/3 by TccP/EspFU has also been described and EspH can recruit N-WASP and N-WASP interacting protein (WIP) to the pedestal independent of Tir residues Y474 and NPY458. In addition Tir can interact with phosphoinositide 3-kinase (PI3K) and the inositol-5-phosphatase SHIP2 to regulate actin accumulation in the pedestal through membrane phosphoinositide signaling. EspL2 interacts with Annexin2 (Anx2), increasing aggregation of Tir-induced actin. (B) Nanometer-thin micropodial extensions (NMEs) can interact with the *Shigella* T3SS effectors IpaB and IpaC and trigger NME retraction bringing the bacteria towards the cell. Invasion occurs through a combination of T3SS effectors; manipulation of Rho GTPases by IpgB1 and 2 (see Figure 15.1C), actin rearrangement through IpaC recruitment of Src, membrane and cytoskeleton dissociation by PIP2 to PIP conversion by IpgD and actin depolymerization by IpaA interaction with vincullin. (C) Rho GTPases cycle between inactive (GDP bound) and active (GTP bound) forms. Rho GEFs activate Rho GTPases by exchanging GDP for GTP and can also participate in binding effector proteins to influence downstream signaling pathways. T3SS effectors from EPEC/EHEC (Map, EspM, and EspT) and *Shigella* (IpgB1 and IpgB2) mimic Rho GEFs by binding to, and inducing a conformational change in, their respective Rho GTPases to allow GTP binding and activation. The T3SS effector Cif blocks RhoA degradation resulting in stress fiber formation while EspH inactivates one subset of mammalian Rho GEFs.

(Vingadassalom et al., 2010). The NPY motif is conserved in EPEC Tir (as NPY454) but in typical EPEC lineage 1 strains this pathway accounts only for low levels of actin polymerization as these strains do not have a TccP/EspF$_U$ homolog (Campellone and Leong, 2005). Both pathways appear to be utilized simultaneously in vitro for most non-O157 EHEC strains, EPEC O119:H6 (Whale et al., 2006) and EPEC lineage 2 strains which carry the homolog TccP2/EspF$_M$ (Ogura et al., 2007). The current conundrum is that neither pathway appears to be necessary for A/E lesion formation in vivo in EHEC animal models (infant rabbit and gnotobiotic piglet models (Ritchie et al., 2008) or the *C. rodentium* murine infection model), or in EPEC and EHEC infection of human intestinal in vitro organ cultures (IVOC) (Schuller et al., 2007; Crepin et al., 2010), indicating that the molecular mechanisms and function of A/E lesion formation during infection are far from understood.

EPEC Tir is also responsible for membrane phosphoinositide signaling as it binds host phosphoinositide 3-kinase (PI3K) at phosphorylated tyrosine 454 (Y454p) (Sason et al., 2009; Selbach et al., 2009) and the inositol-5-phosphatase SHIP2 (Smith et al., 2010). These two enzymes convert PI(4,5)P$_2$ to the predominant membrane phosphoinositide found in wild-type pedestals; PI(3,4)P$_2$. An intermediate, PI(3,4,5)P$_3$, which accumulates when SHIP2 recruitment is prevented, results in multiple elongated pedestals (Smith et al., 2010), suggesting a regulated process of initial PI(3,4,5)P$_3$-enhanced actin polymerization followed by signaling down-regulation by PI(3,4)P$_2$, may occur.

Other T3SS effectors implicated in this initial process of A/E lesions or pedestal formation are EspH, EspB, and EspL. EspH inhibits Rho GTPases activity and as such alters the actin cytoskeleton. In addition to modulating Rho GTPases activity, EspH can also promote N-WASP and WASP interacting protein (WIP) recruitment leading to Arp2/3-mediated actin accumulation, which is independent of Tir residues Y454 and Y474 (i.e. Tir:Nck and Tir:IRTKS/IRSp53 pathways) (Wong et al., 2012) revealing yet another Tir-mediated actin polymerization pathway that may contribute to A/E lesions in vivo.

EspB forms part of the translocation pore at the tip of the T3SS apparatus but in addition is translocated into the host cell where it has an effector function. Ectopic expression of EspB can redistribute host actin in the absence of any other effectors (Taylor et al., 1999) and can bind to host cell proteins including α-catenin (Kodama et al., 2002) and myosins (Iizumi et al., 2007), affecting pedestal formation and initiation of phagocytosis respectively. Both α-catenin and EspB can be seen in the pedestal of EHEC-infected tissue culture cells and are thought to participate in the rearrangement of actin molecules in the pedestal (Hamaguchi et al., 2008). Detailed understanding of EspB's role in pedestal formation has been hampered by its importance in T3SS pore formation, while the interaction of EspB and myosins has been separated from its pore-forming function, this has not occurred for the interaction of EspB and α-catenin.

EspL2 of EHEC can interact with annexin 2 and increase the ability of annexin 2 to aggregate the Tir-induced actin that accumulates underneath the bacterial attachment site (Miyahara et al., 2009). EspL2 also seems to have a Tir-independent function whereby it induces pseudopod-like protrusions of the host cell membrane to enhance bacterial adherence (Miyahara et al., 2009). EspL proteins are homologs of the *Shigella* OspD family, however the *Shigella* OspD proteins do not appear to be involved in actin aggregation (Nataro et al., 1995).

Like its distant homolog TccP/EspFu, EspF possesses N-WASP binding sites, which have been shown in vitro to be functional (Alto et al., 2007). EspF also binds to sorting nexin 9 (SNX9) and induces membrane remodeling and in vitro can nucleate a multiprotein complex of both N-WASP and SNX9 to potentially coordinate membrane remodeling and actin polymerization (Alto et al., 2007).

Invasion and intracellular spread

The uptake of *Shigella* by the host cell depends on a complex rearrangement of the host cell membrane and cytoskeleton, which is initially triggered by receptor binding but requires the T3SS effectors for complete uptake (Figure 15.1B). Recently it has been shown that bacteria can also be captured by filopodial extensions from the host cell, termed nanometer-thin micropodial extensions (NMEs), and this interaction occurs through the T3SS tip complex proteins IpaB and IpaD (Romero et al., 2011). Connexin-mediated signaling and extracellular ATP stimulates Erk1/2 activation, which controls actin retrograde flow in NMEs resulting in NME retraction, bringing the bacterium into contact with the cell body to allow invasion to occur.

Bacterial invasion begins with the formation of membrane ruffles, which require modulation of Rho GTPase activity by IpgB1 and 2. Rho GTPase modulation is discussed in detail elsewhere in this chapter. In addition, the carboxylterminal domain of IpaC induces actin polymerization responsible for the formation of cell extensions that engulf the bacterium (Tran Van Nhieu et al., 1999; Kueltzo et al., 2003). IpaC appears to do so by recruiting and activating Src tyrosine kinase (Mounier et al., 2009). IpaC also forms part of the T3SS pore and is thus necessary for delivery of T3SS effector proteins into the host cell. A *Shigella* strain mutated in the IpaC C-terminal domain was proficient in translocating T3SS effector proteins but was unable to recruit Src, uncoupling these two functions of IpaC (Mounier et al., 2009). Src appears to be initially recruited to the intimate contact site of the bacteria but quickly diffuses in an actin-dependent manner into extensions that surround the bacteria (Dumenil et al., 1998), the molecular details governing this process remain to be elucidated.

The T3SS effector IpgD dephosphorylates phosphatidylinositol-4,5-biphosphate [PI(4,5)P(2)] into phosphatidylinositol-5-phosphate [PI(5)P] (Niebuhr et al., 2002). This dephosphorylation event leads to a decrease in membrane tether force, which can be seen by increased membrane blebbing and actin filament remodeling upon ectopic expression of the protein (Niebuhr et al., 2002).

During infection this may permit localized dissociation between the membrane and cytoskeleton to allow membrane ruffling and filopodia extension through the manipulation of Rho GTPases by IpgB1 and IpgB2. The IpaA protein binds to the focal adhesion protein vinculin (Tran Van Nhieu et al., 1997), increasing the association of vinculin with F-actin and inducing actin filament depolymerization (Bourdet-Sicard et al., 1999), which may be required for completion of the entry process or for reappropriating actin.

In *Shigella*, actin-based motility occurs through the action of IcsA, an autotransporter rather than a T3SS effector, which recruits and activates N-WASP, Arp2/3, polymerizing actin.

Rho GTPase modulation

Many T3SS effectors subvert host cell actin dynamics by disrupting Rho GTPase signaling (Bulgin et al., 2010). The Rho family GTPases are crucial in the regulation of key cellular functions and the best characterized members are Cdc42, Rac1, and RhoA, which trigger filopodia, lamelipodia/ruffles, and stress fibers respectively (Hall, 1998). Rho GTPases cycle between an active, GTP-bound state (predominantly membrane associated) and an inactive GDP-bound form (predominantly cytoplasmic). As shown in Figure 15.1C, the exchange of GDP for GTP is stimulated by guanine exchange factors (GEFs) to activate Rho GTPases. Some RhoGEFs can also mediate interaction with effector molecules to determine downstream signaling pathways. GTP hydrolysis is enhanced by binding of GTPase-activating proteins (GAPs) resulting in inactive GDP-bound Rho GTPase, again some GAPs can scaffold protein complexes. In addition, GDP release is blocked by guanine nucleotide-dissociation inhibitors (GDIs) maintaining the inactive state of Rho GTPases in the cytosol. Therefore these regulators of Rho GTPases not only define the Rho GTPase activity of a molecule but also influence the formation of multiprotein complexes and hence downstream signaling events.

A number of bacterial effector proteins have been described which modulate Rho GTPases and many were grouped on the basis of a WxxxE motif and suggested to act as Rho GTPase mimics (Alto et al., 2006). Structural information helped establish that representatives of this family of effectors actually act as RhoGEFs by binding to their respective Rho GTPases and inducing a conformational change to allow GTP binding and hence activation (Buchwald et al., 2002; Ohlson et al., 2008). These include the EPEC/EHEC effectors Map (Cdc42 GEF[53]), EspM (RhoAGEF[54]), and EspT[55] (likely Rac1 and Cdc42 GEF, although direct binding has not been demonstrated) (Bulgin et al., 2009) and the *Shigella* effectors IpgB1 (a Rac1 and Cdc42 GEF) and IpgB2 (a RhoA GEF) (Huang et al., 2009; Klink et al., 2010).

Activation of Cdc42 by Map results in localized transient filopodia formation (Kenny et al., 2002; Berger et al., 2009). Map can polarize Cdc42 at the cell membrane to form actin-rich protrusions in the absence of external stimuli. This reaction is reliant on the interaction of Map with EBP50 (NHERF1), which interacts with Ezrin to link Map to the actin cytoskeleton (Orchard et al., 2012). During infection it is hypothesized that localized actin rearrangement at the bacterial attachment site is recognized by the Map-EBP50-ezrin complex, which recruits Cdc42 to the attachment site leading to further actin polymerization. EspM activation of RhoA results in stress fiber formation in a ROCK-dependent manner (Arbeloa et al., 2008), while EspT induces formation of membrane ruffles (through the Rac1 effector Wave2), lamellipodia, and bacterial internalization (Bulgin et al., 2009).

The *Shigella* Rho GEF IpgB1 associates with the plasma membrane and was previously shown to recruit the ELMO-Dock180 complex to the membrane to activate Rac1 (Handa et al., 2007), whether this occurs in addition to its Rho-GEF activity remains to be confirmed. IpgB1, but not IpgB2, is necessary for efficient invasion of HeLa cells (Ohya et al., 2005; Hachani et al., 2008). However in polarized cells only a double IpgB1/B2 mutant showed significantly reduced invasion (Hachani et al., 2008) and therefore the interplay between these effectors in triggering host cell invasion requires further investigation.

No T3SS GAPs or GDIs have been identified to date in EPEC/EHEC or *Shigella*, however other T3SS effectors can modulate Rho GTPases in different ways. The EPEC effector Cif stabilizes RhoA (Cui et al., 2010) resulting in stress fiber formation (Oswald et al., 1994). Cif deamidates the ubiquitin-like protein NEDD8, which in turn deactivates neddylated Cullin-RING ubiquitin ligases, a substrate of which is RhoA. RhoA is therefore not ubiquitinated and degraded when Cif is present (Cui et al., 2010). The EPEC/EHEC effector EspH subverts actin dynamics affecting filopodia and pedestal formation (Tu et al., 2003). EspH inactivates mammalian RhoGEFs which contain the Dbl-homology and pleckstrin-homology (DH-PH) domains, but does not inactivate the bacterial RhoGEFs (Wong et al., 2012). This poses an interesting scenario where EspH may potentially reduce a subset of the endogenous mammalian RhoGEFs in order for the bacterial RhoGEFs to take over cell signaling for the benefit of the bacteria.

T3SS effectors can also modulate Rho GTPase signaling pathways by binding directly to the Rho GTPase effector, bypassing the need for Rho GTPase activation. EspG and TccP/EspF$_U$ are examples of such effectors (Selyunin and Alto, 2011). Rho GTPase effectors such as N-WASP and p21 activated kinases (PAKs) have autoinhibitory Rho GTPase binding domains (GBD) which normally inhibit the activity-bearing domain (AD). For WASP the AD is a VCA domain (verprolin homology, central hydrophobic and acidic regions) which recruits the Arp2/3 complex while the PAK AD is a kinase domain. TccP/EspF$_U$ binds to the N-WASP GBD releasing the autoinhibition of N-WASP allowing it to recruit Arp2/3 and nucleate actin (Cheng et al., 2008), as described above. EspG binds to PAK releasing the kinase domain from autoinhibition, although to date this has only been demonstrated in vitro (Selyunin et al., 2011). The method of release and activation of the autoinhibited AD by these bacterial effectors is different to the mechanism by which endogenous Rho GTPases activate their effectors, indicating the bacterial effectors have developed a novel way to manipulate these host proteins.

Other cytoskeletal components (MT and IF)

The T3SS effector VirA was initially identified as being required for the efficient entry of bacteria into epithelial cells (Uchiya et al., 1995) and was subsequently demonstrated to be able to destabilize microtubules producing membrane ruffles and Rac1 activation (Yoshida et al., 2002). Further investigation suggested VirA had protease activity and could selectively degrade α-tubulin allowing intracellular and intercellular spreading (Yoshida et al., 2006). Recently VirA was shown not to directly degrade tubulin or microtubules (Davis et al., 2008; Germane et al., 2008) and the mechanism of microtubule disruption remains unclear. The VirA homolog EspG from EPEC and EHEC was also shown to induce microtubule disruption, releasing GEF-H1 from the cytoskeleton and activating RhoA resulting in actin stress fiber formation (Matsuzawa et al., 2004). New evidence suggests EspG acts as a Rab GTPase activating protein (RabGAP) trapping Rab GTPases in their inactive GDP bound form (Dong et al. 2012), in addition to binding PAKs and ARF GTPases (Selyunin et al., 2011), all of which may have downstream effects that result in cytoskeletal rearrangements.

EspF interacts with cytokeratin 18 (CK18), a protein that forms part of the intermediate filament network in epithelial cells and can be seen recruited to the site of bacterial attachment (Batchelor et al., 2004). CK18 can also bind Tir (Batchelor et al., 2004) and the adapter protein 14-3-3 (Patel et al., 2006), which EspF and Tir can also bind (Viswanathan et al., 2004). These interactions potentially allow Tir and EspF to coordinate collapse of the intermediate filament network with actin redistribution and polymerization.

MANIPULATION OF HOST IMMUNE RESPONSES

Manipulation of the host immune system is a common theme in bacterial infections and a requirement for successful colonization and dissemination. Bacteria must evade phagocytosis, modulate cell-intrinsic innate immunity and avoid autophagy in order to survive and proliferate.

Inhibition of phagocytosis

The first line of host immune defenses involves professional cells such as macrophages, neutrophils, and dendritic cells, which internalize and destroy bacteria and other invaders. Phagocytosis is a process by which phagocytic cells internalize particulate material and is therefore distinct from other forms of endocytosis such as the vesicular uptake of fluids. Phagocytosis is a multistep process triggered by the recognition of specific ligands by surface receptors and local remodeling of the actin cytoskeleton. Well-characterized phagocytic pathways involve Fc gamma receptor (FcγR) and complement receptor 3 (CR3) that bind IgG and C3bi respectively (Caron and Hall, 1998). Importantly EPEC and EHEC colonize gut epithelium but remain predominantly extracellular. Indeed EPEC is able not only to block its own uptake by professional phagocytes (*cis*-phagocytosis), but is able also to inhibit the phagocytosis of IgG-opsonized particles (*trans*-phagocytosis) (Goosney et al., 1999; Celli et al., 2001; Quitard et al., 2006). Inhibition of both *cis*- and *trans*-phagocytosis is T3SS dependent.

The first T3SS effector to be identified as responsible for inhibition of phagocytosis was EspF (Quitard et al., 2006; Marches et al., 2008). EPEC *espF* mutants are phagocytosed by mouse-derived macrophages to the same extent as EPEC lacking a functional T3SS (Quitard et al., 2006; Marches et al., 2008). The role of EspF in inhibition of phagocytosis seems to be correlated with its ability to inhibit PI3K (Quitard et al., 2006). The antiphagocytic activity of EspF is due to the N-terminal region that also contains binding sites for actin, profilin, and SNX9 (Quitard et al., 2006; Alto et al., 2007; Peralta-Ramirez et al., 2008). These partner proteins have not been shown to be involved in antiphagocytosis, but as a specific mechanism for EspF activity still remains to be elucidated, their involvement cannot be discounted. EspF was shown to block only *cis*- and FcγR-dependent phagocytosis (Marches et al., 2008).

EspB binds the actin-binding domain of multiple myosin family members (Iizumi et al., 2007). Myosins (myosin-1c, -2, -5, -6, and -10) are involved in phagocytosis as they localize at the phagocytic cup and their inhibition suppresses phagocytocis (Swanson et al., 1999). Furthermore, early reports have shown that tropomysin is recruited at the site of EPEC adherence (Goosney et al., 2001). Through its interaction with myosins EspB can compete with actin to bind myosins at the actin-binding domain and may then interfere with myosin-induced phagosome constriction. An internal region of EspB (amino acids 159–218) was identified as responsible for this interaction (Iizumi et al., 2007) and deletion of this region reduced the ability of EPEC to inhibit phagocytosis but was not required for bacterial adherence or delivery of T3SS effectors pointing to a specific role in antiphagocytosis (Iizumi et al., 2007).

A recently identified effector involved in inhibition of phagocytosis is EspH, which blocks the activation of Rho GTPases by binding DH-PH Rho GEFs (Dong et al., 2010). The impact of EspH on actin cytoskeleton dynamics is reflected at the level of phagocytosis as Rho GTPases control cytoskeleton remodeling during FcγR-dependent phagocytosis (Caron and Hall, 1998) and EPEC *espH* mutants have a reduced ability to inhibit FcγR-dependent phagocytosis (Dong et al., 2010).

Uniquely, EspJ alone can inhibit both FcγR- and CR3-dependent phagocytosis (Marches et al., 2008). EspJ expression appears to only block phagocytosis of opsonized particles (*trans*-phagocytosis) but not self-uptake (*cis*-phagocytosis). Similarly to EspF, translocated EspJ localizes to mitochondria (Kurushima et al., 2010) but this does not preclude a role for EspJ at the phagocytic cup. The mechanism by which EspJ blocks phagocytosis is currently unknown.

Subversion of innate immunity

In mammals a complex set of signaling networks initiate both innate and adaptive immune responses against bacterial pathogens. Pathogens have therefore evolved sophisticated mechanisms to subvert these signaling networks in order to circumvent host innate immune responses.

EPEC infection in vivo results in intestinal tissue damage, neutrophil infiltration of the infected mucosa and damage of the gut epithelium (Savkovic et al., 1996; Chakravortty and Kumar, 1999; Michail et al., 2003). This pathology has been linked to the inflammatory responses mounted by the cells of the infected tissues. EPEC and EHEC interaction with the gut mucosa results in the stimulation of pattern recognition receptors (PRRs), such as toll-like receptors (TLRs), that recognize specific pathogen-associated molecular patterns (PAMPs), including flagellin and lipopolysaccharide (LPS) (Khan et al., 2006; Schuller et al., 2009). TLRs initiate signaling pathways downstream of these receptors ultimately converging on a set of transcriptional activators including interferon-regulatory factors (IRFs), nuclear factor-κB (NF-κB), and mitogen-activated protein kinases (MAPKs) resulting in the expression and secretion of pro-inflammatory chemokines and cytokines such as interleukin 8 (IL-8), IL-6, IL-12, and tumor necrosis factor α (TNFα).

The pathway best described in the case of EPEC infection is the NF-κB pathway. The NF-κB family comprises five proteins that can form homo- and/ or heterodimeric complexes: RelA (p65), RelB, c-Rel, p50 (NF-κB1), and p52 (NF-κB2), but the most abundant form in mammalian tissues is the dimer p50/ p65 (Li and Verma, 2002). Under non-stimulating conditions, NF-κB is kept inactive in the cytoplasm through its association with inhibitory proteins (IκBs), this interaction masks the nuclear localization signal (NLS) of p65 thus preventing nuclear translocation. Molecular mechanisms that lead to NF-κB activation have been studied intensively. Briefly, ligand-bound TLRs recruit the adaptor protein MyD88, which in turn recruits the IL-1R-associated kinases 1 (IRAK1) and 4 (IRAK4), leading to their sequential autophosphorylation and activation (Silverman and Maniatis, 2001). Phosphorylated IRAK1 then associates with TNF receptor-associated factor 6 (TRAF6). This association activates the E3 ubiquitin ligase function of TRAF6, which, with the UBC13/UEV1 E2 ubiquitin-conjugating complex, catalyzes the synthesis of Lys-63-linked polyubiquitin chains (Deng et al., 2000). These ubiquitin chains serve as a scaffold to recruit both the transforming-growth-factor-β-activated kinase 1 (TAK1) and IκB kinase complexes (IKK) through their respective ubiquitin-binding subunits, TAK-binding protein 2 and 3 (TAB2/3) and NF-κB essential modulator (NEMO) (Kanayama et al., 2004; Ea et al., 2006; Wu et al., 2006). As a result of their proximity, TAK1 can phosphorylate the IKK-β subunit of IKK, which then phosphorylates IκBα. Release of NF-κB dimers occurs after phosphorylation of IκB which allows the ubiquitination by SCFβTrCP complex and subsequent proteasomal degradation of IκBα, allowing NF-κB translocation into the nucleus. Here, NF-κB binds specific DNA consensus sequences and activates expression of pro-inflammatory cytokines.

EPEC initially activates the NF-κB signaling pathway through a T3SS-independent mechanism, and subsequently utilizes a T3SS-dependent mechanism to inhibit NF-κB activation and production of pro-inflammatory cytokines (Hauf and Chakraborty, 2003; Maresca et al., 2005) (Figure 15.2A). In recent years an increasing number of studies have demonstrated that EPEC and EHEC encode and translocate into cells a set of protein effectors that target specific regulators of the NF-κB pathway to disarm the host inflammatory response (Nadler et al., 2010; Newton et al., 2010; Vossenkamper et al., 2010; Yen et al., 2010; Baruch et al., 2011; Muhlen et al., 2011; Pearson et al., 2011; Zhang et al., 2012).

NleE from EPEC/EHEC and OspZ from *Shigella* (Zurawski et al., 2008) belong to the same family and play a key role in modulating the innate immune response during infection by blocking the translocation of activated NF-κB (p65 and c-Rel but not p50) to the nucleus, thereby decreasing the expression and production of IL-8 (Nadler et al., 2010; Newton et al., 2010; Zhang et al., 2012). This effect is seen in both epithelial cells and dendritic cells (Vossenkamper et al., 2010). NleE prevents IκB degradation in response to TNFα or IL-1β suggesting that the effector targets components shared by these inflammatory signaling pathways.

NleE has a unique *S*-adenosyl-L-methionine-dependent methyltransferase (SAM) activity that specifically modifies a zinc-coordinating cysteine in the Npl4 zinc finger (NZF) domains in TAB2 and TAB3. This cysteine methylation disrupts ubiquitin chain binding of TAB2/3 and blocks TRAF6-induced activation of TAK1 (Zhang et al., 2012). A 6-aa motif IDSYMK$_{209-214}$ has been shown to be critical for the immunosuppressive function of NleE (Newton et al., 2010). Indeed, the crystal structure of NleE revealed a SAM-binding pocket, which involves Y212 and R107. Mutation of either residue abolished NleE methylation of TAB3-NZF (Zhang et al., 2012). Given the abundance of biological processes relying on the activation of zinc-finger motifs by cysteine-zinc binding, specific methylation of zinc-coordinating cysteine residues might regulate other eukaryotic pathways in addition to NF-κB signaling.

Another effector involved in the stabilization of IκB is the highly conserved effector NleB. NleB1 from EPEC inhibits NF-κB activation as effectively as NleE in response to TNFα but not IL-1β suggesting that NleB1 may act at a different point of the NF-κB signaling pathway and independently from MyD88/IRAK activation of TRAF6 (Kelly et al., 2006; Petty et al., 2010). The mechanism, however, is still unknown.

The NF-κB pathway is also targeted by the effector NleC. This effector contains a consensus zinc metalloprotease motif $_{183}$HEIIH$_{187}$ that is responsible for its proteolytic activity. Indeed NleC was recently described as a potent protease of EPEC that cleaves and inactivates the NF-κB p65 subunit (Petty et al., 2010; Yen et al., 2010; Baruch et al., 2011; Muhlen et al., 2011; Pearson et al., 2011). Site-directed mutagenesis of the histidines within the consensus sequence is

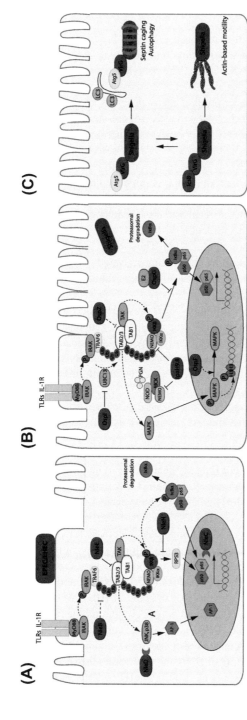

FIGURE 15.2 T3SS effectors from EPEC, EHEC, and *Shigella* interfere with host inflammatory pathways. (A) Infection of EPEC and EHEC results in the stimulation of TLRs and subsequent activation of inflammatory pathways. NleE, NleB, NleH, NleC and NleD T3SS effectors systematically dampen NF-kB and MAPK activation by targeting key regulatory proteins such as TAB2/3, IkBα, JNK, p38, and p65. (B) *Shigella* activates inflammatory pathways by stimulation of both NLRs and TLRs. OspZ, OspG, OspI, IpaH9.8, and OspF are involved in the subversion of NF-kB and MAPK pathways by targeting for example UBC13 UbcH5, NEMO/ ABIN-1, and histone H3. (C) *Shigella* escape autophagy by secreting IscB T3SS effector: Depending on the availability of IcsA, which is recognized by both IscB or Atg5, *Shigella* can be either trapped in septin cages or spread cell-to-cell via actin-based motility.

sufficient to abrogate NleC proteolytic activity (Petty et al., 2010; Yen et al., 2010; Baruch et al., 2011; Muhlen et al., 2011; Pearson et al., 2011). Furthermore, NleC cleaves p65 within its conserved DNA-binding domain, suggesting that other NF-κB family members containing the same sequence could be substrates of NleC (Baruch et al., 2011). The acetyltransferase p300 was recently identified as an additional target of NleC (Shames et al., 2011). p300 functions as a co-activator in the transcription of many genes by acetylating p65 (among other transcription factors), thereby enhancing expression of genes downstream of κB-containing enhancers (Chen et al., 2001) such as IL-8. Overexpression of p300 can indeed antagonize repression of IL-8 secretion induced by EPEC infection (Shames et al., 2011). While infection with EPEC *nleC* mutant does not result in increased IL-8 production, presumably due to functional redundancy with NleE and NleB, infection with an EPEC *nleEC* double mutant or an EPEC *nleBEC* triple mutant results in significantly higher IL-8 production by host cells compared with infection with EPEC *nleE* or EPEC *nleBE* alone (Baruch et al., 2011; Pearson et al., 2011). Thus, NleC appears to complement the activities of NleE and NleB by eliminating activated p65/p50 and/or its acetyltransferase p300 that would positively influence transcription and expression of pro-inflammatory cytokines.

Together with NleC, the effector NleD was identified as zinc-dependent metalloprotease as it also harbors the consensus sequence ($_{142}$HELLH$_{146}$) (Marches et al., 2005). NleD specifically targets MAP kinases JNK and p38 but not ERK, thereby blocking nuclear translocation of the AP-1 transcription factor, which regulates multiple cell processes including inflammation, differentiation, proliferation, and apoptosis (Baruch et al., 2011). Infection with an EPEC *nleBECD* mutant results in an increase of expression and secretion of IL-8 compared to the *nleBEC* mutant, indicating an anti-inflammatory role for NleD (Baruch et al., 2011).

The NleH family of proteins is also conserved amongst EPEC and EHEC strains and the C-termini of NleH1 and NleH2 proteins display significant amino acid sequence similarity with the *S. flexneri* T3SS anti-inflammatory effector OspG (Kim et al., 2005). NleH1 and NleH2 have critical roles in inhibiting host cell apoptosis; however, these effectors have also been implicated in the suppression of NF-κB activation. NleH1 and NleH2 decrease IκKβ-induced NF-κB activity by decreasing TNFα-induced degradation by preventing ubiquitination of phospho-IκBα. Although the mechanism is not clear yet, stabilization of IκBα is dependent on conserved lysine residues K159 in NleH1 and K169 in NleH2 from EPEC E2348/69 respectively (Royan et al., 2010). It was recently shown that the N-terminus of both NleH1 and NleH2 could bind the ribosomal protein S3 (RPS3) (Gao et al., 2009). RPS3 is a non-Rel subunit of NF-κB complexes (Wan et al., 2007). RPS3 is considered a 'specifier' subunit of NF-κB, because it facilitates high-affinity binding of DNA and determines the specificity of NF-κB for selected target genes (Wan et al., 2007). NleH1 of EHEC selectively blocks the transcription of NF-κB target genes by attenuating nuclear translocation of

RPS3 without affecting p65 localization (Gao et al., 2009). Nuclear transloca-tion of RPS3 is triggered by IκKβ kinase-dependent phosphorylation at S209, which enhances RPS3 association with nuclear importin-α. NleH1 specifically inhibits IκKβ-dependent phosphorylation of S209 thus blocking RPS3 function (Wan et al., 2011). Interestingly, another crucial target of IκKβ phosphorylation in the NF-κB pathway is IκBα but whether NleH can block IκKβ-dependent phosphorylation of IκBα remains to be elucidated as well as the mechanism employed by NleH to block such post-translational modification.

By translocating into the host cell cytoplasm effectors such as NleE, NleB, NleC, NleD, and NleH, EPEC and EHEC employ a highly coordinated strat-egy to subvert inflammatory signaling pathways. This coordinated attack allows the bacteria to establish infection and avoid rapid clearance by the host innate immune response.

Similar to EPEC and EHEC *Shigella* delivers a subset of effectors via its T3SS to subvert cellular and immune functions to promote infection (Figure 15.2B). During invasion and multiplication within the cells of the intestinal epi-thelium *Shigella* peptidoglycan stimulates the intracellular PRR receptor Nod-like receptor 1 (NOD1), which interacts with the serine-threonine kinase RICK (also called Ripk2 or RIP2) to induce NF-κB and MAPK p38, ERK, and JNK signaling (Kanneganti et al., 2007). Both MAPK and NF-κB pathways compo-nents are targeted by *Shigella* T3SS-delivered effectors (Ashida et al., 2011).

As mentioned above, among its repertoire of T3SS effectors *Shigella* trans-locates OspZ and OspG, homologs of NleE and NleH respectively. Similar to NleE, OspZ blocks the translocation of NF-κB to the nucleus but whether OspZ from *Shigella* shares the same cysteine methylase activity of EPEC NleE is not currently known (Zurawski et al., 2008; Newton et al., 2010).

A yeast two-hybrid screen showed that OspG is able to bind several ubiqui-tinated E2 Ub-conjugating enzymes, including UbcH5 (Kim et al., 2005). The interaction with UbcH5 seems to inhibit E2 activity, which is indispensable for the proper activation of the E3 Ub ligase SCFβ^{TrCP}, the specific SCF com-plex that controls IκBα degradation. Therefore, OspG negatively regulates the NF-κB inflammatory response by interfering with the proteasome-dependent degradation of IκBα. OspG also displays kinase activity and induces its auto-phosphorylation. This activity appears to be necessary for the function of OspG, but does not seem to be involved in the phosphorylation of IκBα before ubiquitination and proteasome-mediated degradation. Consistent with the OspG inhibiting SCFβ^{TrCP}activity, epithelial cells infected with *S. flexneri* wild-type strain, but not an *ospG* mutant, accumulate phosphorylated IκBα and the ospG mutant induced a much stronger inflammatory response during infection in vivo than the wild-type strain (Kim et al., 2005).

IpaH9.8 is another effector delivered by *Shigella* into host cells that inter-feres with the innate immune response. Once injected into the host cell IpaH9.8 translocates to the nucleus (Toyotome et al., 2001) and has been shown to be an E3 Ub ligase (Rohde et al., 2007). In vitro, IpaH9.8 displayed ubiquitin ligase activity specific for the yeast MAPK Ste7. Indeed, in yeast cells IpaH9.8 inter-rupts pheromone response signaling by promoting the proteasome-dependent destruction of Ste7 (Rohde et al., 2007) and replacement of an invariant Cys residue abolished the ubiquitin ligase activity of IpaH9.8 (Rohde et al., 2007). More recently IpaH9.8 has been shown to dampen the host NF-κB-mediated inflammatory response in mammalian cells (Ashida et al., 2010). IpaH9.8 inter-acts directly with NEMO/IKKγ and ABIN-1, an ubiquitin adaptor protein, pro-moting ABIN-1-dependent polyubiquitination of NEMO on two specific Lys residues (K309 and K321). Consequently, polyubiquitinated NEMO undergoes proteasome-dependent degradation and blocks NF-κB activation (Ashida et al., 2010).

OspF, another T3SS effector of *Shigella*, is translocated into the host cell nucleus, where it targets components of the MAPK pathway (Arbibe et al., 2007). OspF has a phosphothreonine lyase activity, which mediates irrevers-ible dephosphorylation of MAPKs thereby inactivating them. This irreversible modification inhibits the downstream phosphorylation of histone H3 at Ser10 and ultimately results in the inhibition of transcriptional activation of NF-κB-regulated genes, including IL-8 (Arbibe et al., 2007; Kramer et al., 2007; Li et al., 2007; Zurawski et al., 2009).

OspI was recently identified as a new T3SS effector of *Shigella* that selec-tively deamidates a glutamine residue to glutamic acid in the E2 enzyme UBC13. This modification abolishes UBC13 E2 ubiquitin-conjugating activity required for the activation and polyubiquitylation of TRAF6. The authors also demonstrated that through OspI activity, *Shigella* can block NF-κB-mediated inflammatory response at an early stage of epithelial invasion (Sanada et al., 2012). Indeed, OspI encoding gene (orf169b) was initially identified because its mutation led to a dramatic increase in the expression of pro-inflammatory chemokines and cytokines in HeLa-infected cells at early stages of infection. Although the impact of OspI activity on NF-κB pathway was clearly demon-strated, it still remains to be elucidated whether by abolishing TRAF6 activation OspI can also influence MAPKs mediated signaling pathways.

Escape from autophagy

Autophagy is an evolutionarily conserved catabolic pathway that allows eukaryotes to degrade proteins and organelles by sequestering them in special-ized double-membrane vesicles named autophagosomes (Levine et al., 2008; Yang et al., 2010). Although autophagosomes can sequester cytosolic material non-specifically, selective autophagy mediates degradation of various cellular structures, including protein aggregates, mitochondria, and invading microbes tagged by ubiquitin molecules (Kirkin et al., 2009; Kraft et al., 2010; Johansen and Lamark, 2011). Autophagic receptors (p62 and NDP52) can simultaneously bind ubiquitin and the autophagosome-associated ubiquitin-like proteins (i.e. LC3/GABARAP proteins) thus mediating docking of ubiquitinated cargo to the autophagosome. *Shigella* surface protein VirG/IcsA, which is required for actin-based motility, is targeted by the autophagic component Atg5, but this detrimental interaction is inhibited by a second bacterial protein, the T3SS effector IcsB (Ogawa et al., 2005). *icsB* mutants are indeed targeted to and degraded by the lysosome. Interestingly, septin cage-like structures are formed around intracellular *Shigella* and are essential for the targeting of the bacterium to the autophagy compartment (Mostowy et al., 2010). Septins are GTP-binding proteins that assemble in filaments and accumulate at the site of bacterial entry and VirG-induced actin polymerization, thus impeding bacterial mobility. Stalling of the bacteria in the cytosol allows autophagic recognition and subsequent bacterial clearance (Mostowy et al., 2010). Strikingly, wild-type *Shigella* can evade autophagy by delivering IcsB, which competitively binds to VirG to block Atg5 binding and enables the bacteria to evade autophagic recognition (Ogawa et al., 2005) (Figure 15.2C). Therefore, although *Shigella* induces autophagy upon cell invasion, they have also developed mechanisms to evade autophagic recognition (Ashida et al., 2011).

CELL DEATH AND SURVIVAL

EPEC, EHEC, and *Shigella* are in contact with a large variety of cells during infection and must prolong cell survival long enough to maintain and promote the infection or alternatively destroy the cells that are potentially harmful for them. This cell death must be regulated to avoid any host response and to achieve this, these bacteria have developed a large panel of effectors able to modulate the cell cycle, induce or inhibit apoptosis, and promote cell adhesion.

Altering cell cycle

The cell cycle can be divided into four phases: mitosis (M) when cells are dividing (further subdivided into prophase, metaphase, anaphase, and telophase); post-mitotic phase (G1) when cells grow and have a high rate of protein synthesis and metabolism; synthesis phase (S) characterized by DNA replication; and pre-mitotic phase (G2), when cells are preparing to undergo mitosis. Interphase comprises G1, S, and G2. Cell cycle progression is under the control of cyclin-dependent kinases (CDKs) where a Cdk1–Cyclin B complex triggers mitosis while Cdk2–Cyclin A/E or Cdk4–Cyclin D1 complexes trigger G1 to S phase transition.

Cell cycle progression is further regulated by cyclin-dependent kinase inhibitors (CKI). One family of CKIs is the CDK interacting protein/Kinase inhibitory protein family (CIP/KIP) that is composed of three proteins $p21^{Cpi1/WAF}$, $p27^{Kip1}$, and $p57^{Kip2}$. p21 binds and directly inhibits Cdk1–Cyclin B and Cdk2–Cyclin E complexes, blocking cells in G2/M and G1/S phases respectively (reviewed in Domingo-Sananes et al., 2011), while p27 binds Cdk2–Cyclin A/E and Cdk4–Cyclin D1, blocking the cells in G1/S phase. CIP/Kip family members are themselves regulated by different protein complexes which ubiquitinate them, promoting proteasomal degradation (Figure 15.3A and C). One of them is the Skp1-Cullin1-F-box protein (SCF) complexes, a subgroup of the larger Cullin Ring Ubiquitin ligase (CRL) family (Starostina and Kipreos, 2012). The SCF ubiquitination activity is stimulated by the binding of NEDD8 (neddylation) to cullins and by preventing the interaction between cullin and its inhibitor CAND1 (Duda et al., 2008).

A subset of EPEC and EHEC strains are able to block cells in G1/S and G2/M phases (Marches et al., 2003; Morikawa et al., 2010). The cell cycle arrest has been linked to the effector Cif which adopts a papain-like hydrolytic fold with a Cys-His-Asp/Asn/Glu/Gln catalytic triad (Hsu et al., 2008). During infection, Cif is targeted to the nuclear and perinuclear regions where it interacts with NEDD8, deamidating NEDD8 at Gln40 (Cui et al., 2010; Morikawa et al., 2010). Neddylation of SCF with deamidated NEDD8 impairs the ubiquitin ligase activity of the neddylated complex. SCF substrates (p21, p27, Nrf-2, HIF-α, Cyclin D1, Cdt1) (Cui et al., 2010; Jubelin et al., 2010; Morikawa et al., 2010), accumulate in cells, blocking the G1/S and G2/M phase transitions (Figure 15.3B and D). Importantly, the role of Cif has not been yet determined during in vivo infection.

The Anaphase Promoting Complex/Cyclosome ubiquitin ligases (APC/C) control the progression and the exit of mitosis. Like the SCF, the APC/C are part of the CRL family and control a large number of substrates through ubiquitination, such as Cyclin B which is destroyed after anaphase. Among the APC/C, two complexes are involved at different stages of mitosis. Binding of Cdc20 to APC/C (APC/C^{cdc20}) allows the initiation and progression of mitosis while binding of Cdh1 to APC/C (APC/C^{Cdh1}) activates the APC/C and ubiquitinates Cdc20 to inactivate APC/C^{cdc20} allowing the exit of mitosis. During G1 phases, Cdh1 is then inactivated. APC/C activation like the SCF is tightly controlled to avoid unscheduled degradation of substrate. The Early mitotic inhibitor 1 (Emi1) inhibits APC/C^{cdc20} by binding the newly synthesized Cdc20. During the prophase, Emi1 is phophorylated by Plk1 allowing the activation of APC/Cdc20 complex. Like Emi1, MAD2 (at spindle assembly checkpoint) and MAD2B (during G2/M phase) inhibit mainly APC/C^{cdc20} and APC/C^{Cdh1} respectively (Pfleger et al., 2001) (Figure 15.3E).

The translocon protein IpaB of *Shigella* can interact during infection with MAD2B but not with MAD2 and Emi1 (Iwai et al., 2007). Interestingly, MAD2B shares a region of similar sequence similarity with IpgC (aa 61–70), the IpaB chaperone. Interaction between IpaB and MAD2B decreases the binding of MAD2B to APC/C^{Cdh1} and induces an unscheduled activation of APC/C^{Cdh1}. This activation leads to the degradation of Cdc20, Plk1, and Cyclin B1, all protein substrates of the APC/C, and blocks the cell at G2/M phase (Iwai et al., 2007) (Figure 15.3F). Cell cycle arrest induced by IpaB during *Shigella* infection of rabbit intestinal ileal loops has been shown to play an important role in prolonging *Shigella* colonization of the tissues (Iwai et al., 2007). Importantly, cell cycle arrest induces cell death.

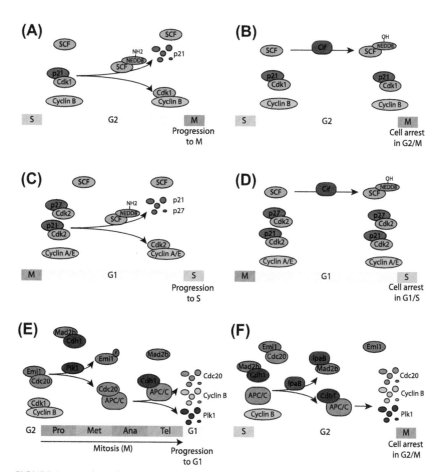

FIGURE 15.3   T3SS effectors from EPEC (Cif) and *Shigella* (IpaB) interfere with the host cell cycle. (A) At the cell cycle checkpoint between G2 and M phases, p21 is sent for degradation by neddylated SCF allowing the activation of Cdk1–cyclin B complex. (B) T3SS effector Cif methylates NEDD8 and inhibits SCF, blocking the formation of the Cdk1–cyclin B complex and inducing cell cycle arrest in G2/M phases. (C) At the cell cycle checkpoint between G1 and S phases, p21 and p27 are sent for degradation by the neddylated SCF allowing the activation of Cdk2–cyclin A/E complex. (D) The T3SS effector Cif methylates NEDD8 and inhibits SCF blocking the formation of the Cdk2–cyclin A/E complex and inducing cell cycle arrest in G1/S phases. (E) During the early phase of mitosis the inhibitor of Cdc20, Emi, is phophorylated by Plk1 releasing Cdc20 and activating the complex APC/Ccdc20. During anaphase, Cdh1 is released from its inhibitor MAD2B to activate the complex APC/Ccdh1 sending Cdc20, Plk1, and cyclin B for degradation. (F) The T3SS effector IpaB from *Shigella* activates during the G2 phase Cdh1, by binding its inhibitor MAD2B inducing the unscheduled destruction of Cdc20, Plk1, and cyclin B and blocking cells in G2/M phases.

Manipulating host cell death

Cell death participates in the renewal of the epithelium, cell maturation and response to infections. Controlled cell death (apoptosis) allows the organism to eliminate damaged or infected cells with few consequences for the organs, whereas uncontrolled cell death (necrosis) leads to massive inflammation and a strong immune response. Apoptosis can be instigated by extracellular signals (extrinsic pathway), such as cytokines (TNF-α or FAS), or by intracellular signals (intrinsic pathway), such as ER stress or organelle damage (mitochondria). The execution of apoptosis is under the control of caspases, cysteine proteases which cleave a large number of proteins and lead to the formation of apoptotic blebs vesiculating from the dying cells that are phagocytosed by macrophages. Inactive pro-caspases are auto-proteolysed or cleaved by other caspases to become active. One of these, caspase-3 (an executional caspase) is commonly used as a marker of apoptosis.

Some bacterial effectors have been shown to be able to induce apoptosis. Five minutes post-translocation, EspF is found in the mitochondria (Nougayrede and Donnenberg, 2004; Nagai et al., 2005), dependent on a mitochondrial targeting signal localized in the N-terminal 70 aa (Nougayrede and Donnenberg, 2004). Translocation of EspF to the mitochondria induces a loss of mitochondrial membrane potential ($\Delta\Psi_M$) and the release of cytochrome c, activation of the caspases (including caspase-3) leading to cell death (Nougayrede and Donnenberg, 2004; Nagai et al., 2005). EspF has been reported to bind and degrade Abcf2, which can increase caspase-3 activation and cell death, however the mechanism involved is unknown (Nougayrede et al., 2007).

At later stages of infection (after 48 h in IEC-6 intestinal epithelial cells) EPEC expressing Cif induces caspase-3 activation and apoptosis (Samba-Louaka et al., 2009). The catalytic domain of Cif is necessary for induction of apoptosis and therefore apoptosis induction may be linked to cell cycle arrest induced by Cif, although this has not been directly shown.

Shigella induces a form of apoptosis termed pyroptosis in macrophages (Raqib et al., 2002) which is dependent on the translocator IpaB (Zychlinsky et al., 1994). Following *Shigella* cell invasion, IpaB localizes to the cytoplasm (Chen et al., 1996) where it interacts with and then activates caspase-1 but not caspase-2 or caspase-3 (Hilbi et al., 1998). Activation of caspase-1 induces the release of IL-1β (Chen et al., 1996), an apoptotic inducer, leading to cell death and bacterial release. This process appears to be restricted in epithelial cells.

Non-phagocytic cells infected with EPEC, EHEC, or *Shigella* show limited signs of apoptosis early during infection despite delivery of these pro-apoptotic effectors (Mantis et al., 1996), suggesting other mechanisms exist to inhibit or slow down cell death. Indeed, cells or animals infected with an *nleH1* and *nleH2* double mutant present a high rate of apoptosis. By transfection, it has been shown that both proteins can inhibit apoptosis induced by a global apoptotic inducer or by ER stress (Hemrajani et al., 2010; Robinson et al., 2010). NleH is a serine threonine kinase which localizes at the membrane of the cell and at the ER. Interaction with the anti-apoptotic protein Bax inhibitor-1 at the surface of the ER is essential to the anti-apoptotic activity of NleH, however its kinase activity appears unnecessary (Hemrajani et al., 2010). NleH also has a c-terminal PDZ binding domain which can interact with the sodium–hydrogen exchanger regulatory factor 2 NHERF2, which is localized at the plasma membrane. Over-expression of NHERF2 leads to a decrease of the anti-apoptotic activity of NleH potentially by retaining NleH at the membrane of the infected cells (Martinez et al., 2010). EspH can also induce apoptosis which is partially inhibited by the bacterial Rho GEFs EspM2 and EspT (but not Map) (Wong et al., 2012). This inhibition may be achieved by modulation of MAPK and NF-κB pathways (Raymond et al., 2011) (which can regulate apoptosis), as activation of Rac1 and RhoA has been reported to be involved in these pathways (Hall, 2005).

Shigella induces DNA damage shortly after invasion of the cells, which promotes the DNA damage response via phosphorylation of ataxia telangiectasia kinase (ATM) and histone H2Ax, both of which are observed during *Shigella* infection. ATM activation leads to the phosphorylation of pro-apoptotic transcription factor p53 and prevents its interaction and degradation by the ubiquitin ligase Mdm2, stabilizing p53 and inducing cell death. To prevent p53-induced apoptosis, *Shigella* uses two different effectors. At the beginning of the infection, IpgD induces the phosphorylation of Mdm2, increasing its activity and inducing the degradation of p53 (Bergounioux et al., 2012). *Shigella* also promotes the activation of calpain, a protease that degrades p53, preventing the activation of the pro-apoptotic p53/NFκB signaling pathway, instead resulting in necrotic cell death. This activation is under the control of VirA, which induces the proteolysis of the calpain inhibitor, calpastatin (Bergounioux et al., 2012). The EPEC/EHEC homologs of VirA, EspG, and EspG2, also induce activation of calpain and promote necrotic cell death and cell detachment. EPEC and EHEC overcome this problem via another effector, Tir, which controls the activation of calpain during the infection (Dean et al., 2010).

Maintaining cell adhesion

Adhesion of the cell to the extracellular matrix (ECM) plays an important role in many biological processes. At the cell–matrix contact point are dynamic and tightly regulated structures known as focal adhesions (FA), an important component of which are integrins (e.g. β1-integrin). Upon binding to the ECM and activation, integrins induce the phosphorylation of focal adhesion kinase (FAK) on tyrosine 397. Phosphorylation of FAK will then promote the recruitment of the Src kinases which in turn phosphorylate tyrosines 576 and 577 of FAK. This active form of FAK will then promote the formation of the FA. Moreover, FAK also induces the phosphorylation of paxilin. This phosphorylation will then induce an increased turnover of the FA leading to their disassembly.

One of many protein complexes that regulate FA turnover is the IPP complex formed by integrin-linked kinase (ILK), and the adaptor proteins PINCH and parvin. After activation of β1-integrin, this complex promotes clustering of FAK with integrin and FAK phosphorylation (Legate et al., 2006). Other regulators of FA include the membrane protein CD98 which has been shown to increase FA formation and FAK phosphorylation by interacting with β1-integrin and affecting its function (Cai et al., 2005; Yan et al., 2008). FA turnover is also controlled by the Rho GTPases RhoA and Rac1 via FAK (reviewed in Tomar and Schlaepfer, 2009). Indeed, it has been shown that a number of Rho GEFs or Rho GAPs can interact with FAK to specifically activate or inhibit these Rho GTPases to promote or slowdown FA turnover and enhance cell migration (Tomar and Schlaepfer, 2009).

The OspE effector family (OspE1/2 from *Shigella* and EspO1/2 from EHEC) is able to modulate FA by interacting with ILK. Interaction between OspE and ILK does not change ILK activity or stability or alter the formation of the IPP complex, but it does lead to a decrease in phosphorylation of FAK (Y397) and paxilin, an increase in the active form of β1-integrin and a decreased turnover of β1-integrin. This mechanism serves to increase the membrane localization of ILK and stabilizes and strengthens FA by inhibiting FA disassembly and decreasing turnover. Infection of cells with an *ospE1* or *ospE2* mutant therefore results in cell rounding (Kim et al., 2009) and cell detachment of polarized cells (MDCK). Inhibition of the turnover of the FA and β1-integrin also leads to an inhibition of the cell motility. In animals an *ospE* mutant displays decreased colonization, inflammation, internal hemorrhaging, and diarrhea compared to the wild-type strain suggesting that the inhibition of cell motility plays a critical role during infection with *Shigella*.

Similarly, EPEC and EHEC have been shown to modulate FAK phosphorylation and FA turnover (Shifrin et al., 2002). One T3SS effector involved in this process is EspZ, which has been reported to bind CD98 and induce phosphorylation of FAK (Tyr 576/577). Like OspE, EspZ inhibits cell detachment in vitro (Shames et al., 2010) and plays a critical role during animal infection, with an *espZ* mutant showing strong attenuation (Deng et al., 2004). The action of EspZ is counteracted by another EPEC/EHEC effector, EspH which induces cell rounding and detachment (Dong et al., 2010; Wong et al., 2012) by progressively inducing disassembly of FA via a modulation of FAK concomitantly with actin disruption (Wong et al., 2012). This function may be achieved through Rho GTPase modulation, specifically through inhibition of Rac1 and partially RhoA pathways, but not Cdc42 as cell rounding or detachment induced by EspH can be counteracted by bacterial Rho GEFs SopE (a Rac1 RhoGEF from *Salmonella*), EspT or dominant positive of Rac1 and partially by EspM (Wong et al., 2012).

DISRUPTING GUT INTEGRITY: DIARRHEAGENIC MECHANISM

The intestinal epithelium is a selectively permeable barrier which must allow absorption of nutrients, electrolytes, and water while maintaining an effective barrier to noxious substances and pathogens. Diarrhea is the result of an imbalance in the absorption and secretion of ions and solute across the gut epithelium and subsequent movement of water into the intestine to address the imbalance.

Changing the absorptive surface area of the gut, modulating the junctional complexes, or altering selective channels/porins all effect the integrity of the intestinal epithelial layer. However, in addition the gut epithelium undergoes continuous self-renewal to maintain tissue homeostasis and eliminate damaged cells and changes in apoptosis will alter the turnover of gut epithelium and contribute to the disruption of gut integrity. Host inflammatory responses, such as migration of polymorphonuclear cells towards the gut lumen, also strongly disrupt the integrity of the epithelium layer and are thought to be the major contributors to *Shigella*- and EHEC-mediated diarrhea.

Microvilli effacement

Below the epithelial monolayer lies the lamina propria and submucosa. The undulating ridges of the intestinal submucosa, the folds of the lamina propria forming the villi and crypts, and the finger-like microvilli on epithelial cells maximize the absorptive surface of the small intestine. The effacement of the microvilli, a defining feature of the A/E lesion results in a reduced surface area for normal absorptive processes. Effectors involved in cytoskeleton remodeling and adherent junction manipulation contribute generally to the effacement of the microvilli, however specific interactions targeting microvilli effacement are less well understood. A single microvillus contains a core F-actin bundle which is internally stabilized by villin and fimbrin and laterally stabilized by myosin 1a:calmodulin bridges to the plasma membrane. The microvillus actin core is tethered into the cellular actin filament at the terminal web, which contains amongst other proteins, myosin 2, spectrin, tropomyosin, and α-actinin (Figure 15.4A).

Effacement has been described as a two-step process, requiring the cooperative action of three injected effectors (Map, EspF, and Tir) as well as intimin. It has been suggested that Tir and intimin predominate in the first step of effacement where bacteria sink into the brush border, while EspF and Map predominate in the ensuing microvillus destruction (Dean et al., 2006). Interestingly F-actin is not seen in the detached microvillar material but is thought to be contracted into the cell (Dean et al., 2006). As previously explained EspB can interact with the myosin superfamily, specifically with the C-terminal region of the motor domain of myosins, which normally interacts with actin filaments, competitively inhibiting the myosin:actin interaction (Iizumi et al., 2007). Myosins 1a and 2 (Figure 15.4A), as well as 1c, 5, and 6 localize to and stabilize the microvilli of enterocytes, and the interaction of EspB with myosins therefore directly contributes to the effacement of the microvilli.

Altering ion exchange and water transport

Increased secretion or decreased absorption of Cl⁻ increases the luminal Cl⁻ concentration and leads to increased water in the lumen. EPEC reduces

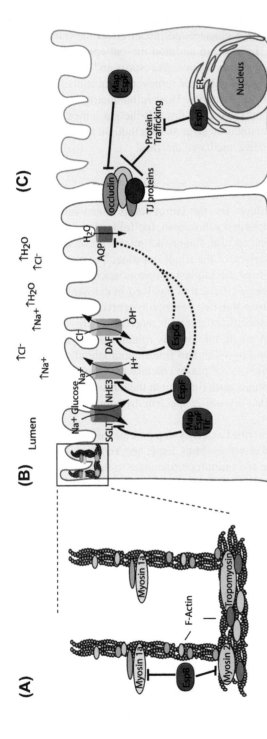

FIGURE 15.4 (A) Within a single microvillus a core of F-actin is stabilized by myosin 1a:calmodulin bridges to the plasma membrane. The F-actin core is tethered into the cellular actin filaments which contain amongst other proteins, myosin 2. EspB can interrupt myosin:actin binding causing microvilli retraction and effacement. (B) Map and EspF are involved in relocalizing the Na$^+$/glucose co-transporter SGLT1 to reduce its function. The Na$^+$/H$^+$ exchanger 2 (NHE2) activity is decreased by EspF, while EspG and EspG2 induce the internalization and hence reduce the activity of the down-regulated in adenoma (DRA) Cl$^-$/OH$^-$ exchanger. EspG and EspF, either directly or indirectly, contribute to the mis-localization of aquaporins (AQP). These changes result in increased Na$^+$, Cl$^-$ ions, and water in the lumen during infection manifesting as watery diarrhea. (C) Tight junction proteins regulate the passage of ions and electrolytes across the gut epithelium. Map and EspF redistribute the tight junction protein occludin, while EspI/NleA disrupts protein trafficking and hence replenishment of the tight junction proteins which are highly dynamic requiring constant renewal.

Cl⁻ absorption by targeting the down-regulated in adenoma (DRA) Cl⁻/OH⁻ exchanger (Gill et al., 2007). EspG and EspG2 are responsible for the internalization of DRA into a subapical vesicle pool, thus reducing the amount of surface-localized and functional DRA (Gill et al., 2007). While EspG was originally described to degrade microtubules it has recently been shown to act as a RabGAP trapping Rab GTPases in their inactive GDP bound form and thus altering cellular signalling pathways (Dong et al 2012). Additionally EspG can also bind the Golgi matrix protein GM130, p21-activated kinases (PAKs) and ADP-ribosylation factors (ARFs) (Clements et al., 2011; Selyunin et al., 2011) acting as a molecular scaffold to regulate host signaling cascades. EspG can interact with ARFs and Rabs simultaneously and by interacting with activated, membrane-embedded ARFs, EspG potentially transforms ARF activation into Rab inactivation at appropriate endomembranes. EspG can also bind PAKs and ARFs simultaneously; the EspG–ARF interaction prevents GTP hydrolysis by ARF–GAPs and EspG–PAK interaction prevents PAK autoinhibition, thus locking both ARF and PAK in their active states (Selyunin et al., 2011). EspG overexpression disrupts the structure of the Golgi apparatus and decreases protein secretion and receptor trafficking (Clements et al., 2011; Selyunin et al., 2011), which may contribute to the aberrant DRA localization during infection.

Diarrhea can also be caused by attenuation of Na⁺ absorption, which is primarily mediated by Na⁺/H⁺ exchangers (NHEs). EPEC differentially regulates two apical NHEs, increasing NHE2 activity and decreasing NHE3 activity dependent on the T3SS (Hecht et al., 2004). EspF appears responsible for decreasing NHE3 activity in EPEC-infected cells, although the mechanism is not yet known. The NHE regulatory factors NHERF1 and NHERF2 appear to also contribute to EPEC-induced down-regulation of NHE3, but whether this occurs through EspF is unclear (Hodges et al., 2008). Interestingly, while NHE3 activity is typically coupled to the DRA exchanger through interaction with NHERF proteins (Lamprecht et al., 2002), EPEC appears to use two different effector proteins, EspF and EspG, to individually target and disrupt the function of these Na⁺/Cl⁻ exchangers.

EPEC alters another route of Na⁺ uptake: the Na⁺/glucose cotransporter SGLT1 which is active only in the presence of glucose. Wild-type EPEC, but not an EPEC *map/tir/espF/intimin* quadruple mutant, rapidly inactivates SGLT-1 within 30 minutes of infection by altering its location from the apical surface to become mainly intracellular (Dean et al., 2006). Map and EspF appear to play major but redundant roles in this inactivation of SGLT, although the molecular mechanism of relocalization is unknown.

Aquaporins (AQP) are water and water/glycerol channels that are responsible for the rapid transport of water across membranes. In infected colonocytes AQP2 and AQP3 appeared to relocalize from the membrane to the cytoplasm and this relocalization was dependent on EspF and EspG (Guttman et al., 2007). The redistribution of aquaporins by EspF and EspG may be indirect as the host attempts to correct the altered ion levels induced by these effectors by reducing the intake of water, leading to increased water in the lumen, resulting in watery diarrhea.

This combined decrease in the activity of SGLT1, DRA, and NHE3 impairs the luminal NaCl absorption, leading to decreased water absorption potentially through incorrectly localized AQPs, producing the pathophysiology of EPEC-induced early diarrhea.

Altering junctional complexes

Intestinal epithelial cells are polarized so the apical side faces the lumen and contains microvilli and the basal side interacts with the underlying lamina. Junctional complexes (adherens junctions, densosomes, and tight junctions) form on the apical side of the epithelial cells to connect adjacent cells (adherens junctions and desmosomes) and to act as a regulatable barrier for passage of ions and electrolytes, resulting in cell polarity (tight junctions). Tight junctions are dynamic multi-protein complexes which contain occludin and members of the claudin family, cytoplasmic scaffolding proteins, including the zonula occludens family (ZO1, ZO2, and ZO3), and signaling molecules. EPEC can dephosphorylate and remove occludin from TJs in a T3SS-dependent manner (Simonovic et al., 2000) and in vivo EPEC was shown to redistribute occludin in the ileum and colon (Shifflett et al., 2005). The T3SS effectors Map and EspF are both able to effect occludin distribution in polarized Caco-2 monolayers (Dean and Kenny, 2004) although the mechanisms are unknown.

EspI (NleA) also disrupts intestinal tight junctions potentially through its ability to disrupt protein trafficking (Thanabalasuriar et al., 2010). EspI disrupts protein trafficking by binding to a protein in the COPII coat, SEC24, inhibiting COPII vesicle function (Kim et al., 2007). COPII vesicles mediate anterograde trafficking of proteins from their sites of synthesis in the endoplasmic reticulum to other membrane compartments in the cell and are a crucial first step in the cell secretion pathway. COPII cargo includes membrane and luminal proteins and Sec24 has an important role in sorting and selecting cargo for inclusion in COPII vesicles. It is therefore proposed that EspI disruption of COPII vesicles hampers the replenishment of TJ proteins, causing a sequential loss of functional TJ complexes (Thanabalasuriar et al., 2010).

A number of tight junction proteins including claudin-1, ZO-1, ZO-2, and occludin are altered by the presence of *Shigella*; however whether a T3SS effector is responsible for these changes is unknown (Sakaguchi et al., 2002).

CONCLUSION

EPEC/EHEC and *Shigella* have thus developed a diverse and sophisticated array of T3SS effectors to reprogram the host for the advantage of the pathogen. While some effectors are multifunctional, others appear to have overlapping or redundant functions. There are still a number of effectors whose function are unknown and many others that have not been fully explored. Importantly the effect in vivo of many of these effectors remains unresolved and determining the spatial-temporal activity of individual effectors in the overall context of infection has not been addressed.

REFERENCES

Alto, N.M., Shao, F., Lazar, C.S., et al., 2006. Identification of a bacterial type III effector family with G protein mimicry functions. Cell 124 (1), 133–145.

Alto, N.M., Weflen, A.W., Rardin, M.J., et al., 2007. The type III effector EspF coordinates membrane trafficking by the spatiotemporal activation of two eukaryotic signaling pathways. J. Cell Biol. 178 (7), 1265–1278.

Arbeloa, A., Bulgin, R.R., MacKenzie, G., et al., 2008. Subversion of actin dynamics by EspM effectors of attaching and effacing bacterial pathogens. Cell Microbiol. 10 (7), 1429–1441.

Arbibe, L., Kim, D.W., Batsche, E., et al., 2007. An injected bacterial effector targets chromatin access for transcription factor NF-kappaB to alter transcription of host genes involved in immune responses. Nat. Immun. 8 (1), 47–56.

Ashida, H., Kim, M., Schmidt-Supprian, M., Ma, A., Ogawa, M., Sasakawa, C., 2010. A bacterial E3 ubiquitin ligase IpaH9.8 targets NEMO/IKKgamma to dampen the host NF-kappaB-mediated inflammatory response. Nat. Cell Biol. 12 (1), 66–73, sup pp. 61–69.

Ashida, H., Ogawa, M., Mimuro, H., Kobayashi, T., Sanada, T., Sasakawa, C., 2011. Shigella are versatile mucosal pathogens that circumvent the host innate immune system. Curr. Opin. Immunol. 23 (4), 448–455.

Baruch, K., Gur-Arie, L., Nadler, C., et al., 2011. Metalloprotease type III effectors that specifically cleave JNK and NF-kappaB. EMBO J. 30 (1), 221–231.

Batchelor, M., Guignot, J., Patel, A., et al., 2004. Involvement of the intermediate filament protein cytokeratin-18 in actin pedestal formation during EPEC infection. EMBO reports 5 (1), 104–110.

Berger, C.N., Crepin, V.F., Jepson, M.A., Arbeloa, A., Frankel, G., 2009. The mechanisms used by enteropathogenic *Escherichia coli* to control filopodia dynamics. Cell Microbiol. 11 (2), 309–322.

Bergounioux, J., Elisee, R., Prunier, A.L., et al., 2012. Calpain activation by the *Shigella flexneri* effector VirA regulates key steps in the formation and life of the bacterium's epithelial niche. Cell Host Microbe 11 (3), 240–252.

Bourdet-Sicard, R., Rudiger, M., Jockusch, B.M., Gounon, P., Sansonetti, P.J., Nhieu, G.T., 1999. Binding of the *Shigella* protein IpaA to vinculin induces F-actin depolymerization. EMBO J. 18 (21), 5853–5862.

Brady, M.J., Campellone, K.G., Ghildiyal, M., Leong, J.M., 2007. Enterohaemorrhagic and enteropathogenic *Escherichia coli* Tir proteins trigger a common Nck-independent actin assembly pathway. Cell Microbiol. 9 (9), 2242–2253.

Buchwald, G., Friebel, A., Galan, J.E., Hardt, W.D., Wittinghofer, A., Scheffzek, K., 2002. Structural basis for the reversible activation of a Rho protein by the bacterial toxin SopE. EMBO J. 21 (13), 3286–3295.

Bulgin, R., Raymond, B., Garnett, J.A., et al., 2010. Bacterial guanine nucleotide exchange factors SopE-like and WxxxE effectors. Infect. Immun. 78 (4), 1417–1425.

Bulgin, R.R., Arbeloa, A., Chung, J.C., Frankel, G., 2009. EspT triggers formation of lamellipodia and membrane ruffles through activation of Rac-1 and Cdc42. Cell. Microbiol. 11 (2), 217–229.

Cai, S., Bulus, N., Fonseca-Siesser, P.M., et al., 2005. CD98 modulates integrin beta1 function in polarized epithelial cells. J. Cell Sci. 118 (Pt 5), 889–899.

Campellone, K.G., Giese, A., Tipper, D.J., Leong, J.M., 2002. A tyrosine-phosphorylated 12-amino-acid sequence of enteropathogenic *Escherichia coli* Tir binds the host adaptor protein Nck and is required for Nck localization to actin pedestals. Mol. Microbiol. 43 (5), 1227–1241.

Campellone, K.G., Leong, J.M., 2005. Nck-independent actin assembly is mediated by two phosphorylated tyrosines within enteropathogenic *Escherichia coli* Tir. Mol. Microbiol. 56 (2), 416–432.

Campellone, K.G., Robbins, D., Leong, J.M., 2004. EspFU is a translocated EHEC effector that interacts with Tir and N-WASP and promotes Nck-independent actin assembly. Dev. Cell 7 (2), 217–228.

Caron, E., Hall, A., 1998. Identification of two distinct mechanisms of phagocytosis controlled by different Rho GTPases. Science 282 (5394), 1717–1721.

Celli, J., Olivier, M., Finlay, B.B., 2001. Enteropathogenic *Escherichia coli* mediates antiphagocytosis through the inhibition of PI 3-kinase-dependent pathways. EMBO J. 20 (6), 1245–1258.

Chakravortty, D., Kumar, K.S., 1999. Interaction of lipopolysaccharide with human small intestinal lamina propria fibroblasts favors neutrophil migration and peripheral blood mononuclear cell adhesion by the production of proinflammatory mediators and adhesion molecules. Biochim. Biophys. Acta 1453 (2), 261–272.

Chen, L., Fischle, W., Verdin, E., Greene, W.C., 2001. Duration of nuclear NF-kappaB action regulated by reversible acetylation. Science 293 (5535), 1653–1657.

Chen, Y., Smith, M.R., Thirumalai, K., Zychlinsky, A., 1996. A bacterial invasin induces macrophage apoptosis by binding directly to ICE. EMBO J. 15 (15), 3853–3860.

Cheng, H.C., Skehan, B.M., Campellone, K.G., Leong, J.M., Rosen, M.K., 2008. Structural mechanism of WASP activation by the enterohaemorrhagic *E. coli* effector EspF(U). Nature 454 (7207), 1009–1013.

Clements, A., Smollett, K., Lee, S.F., Hartland, E.L., Lowe, M., Frankel, G., 2011. EspG of enteropathogenic and enterohemorrhagic *E. coli* binds the Golgi matrix protein GM130 and disrupts the Golgi structure and function. Cell. Microbiol. 13 (9), 1429–1439.

Crepin, V.F., Girard, F., Schuller, S., Phillips, A.D., Mousnier, A., Frankel, G., 2010. Dissecting the role of the Tir:Nck and Tir:IRTKS/IRSp53 signalling pathways in vivo. Mol. Microbiol. 75 (2), 308–323.

Cui, J., Yao, Q., Li, S., et al., 2010. Glutamine deamidation and dysfunction of ubiquitin/NEDD8 induced by a bacterial effector family. Science 329 (5996), 1215–1218.

Davis, J., Wang, J., Tropea, J.E., et al., 2008. Novel fold of VirA, a type III secretion system effector protein from *Shigella flexneri*. Protein Sci. 17 (12), 2167–2173.

Dean, P., Kenny, B., 2004. Intestinal barrier dysfunction by enteropathogenic *Escherichia coli* is mediated by two effector molecules and a bacterial surface protein. Mol. Microbiol. 54 (3), 665–675.

Dean, P., Maresca, M., Schuller, S., Phillips, A.D., Kenny, B., 2006. Potent diarrheagenic mechanism mediated by the cooperative action of three enteropathogenic *Escherichia coli*-injected effector proteins. Proc. Natl. Acad. Sci. USA 103 (6), 1876–1881.

Dean, P., Muhlen, S., Quitard, S., Kenny, B., 2010. The bacterial effectors EspG and EspG2 induce a destructive calpain activity that is kept in check by the co-delivered Tir effector. Cell. Microbiol. 12 (9), 1308–1321.

Deng, L., Wang, C., Spencer, E., et al., 2000. Activation of the IkappaB kinase complex by TRAF6 requires a dimeric ubiquitin-conjugating enzyme complex and a unique polyubiquitin chain. Cell 103 (2), 351–361.

Deng, W., Puente, J.L., Gruenheid, S., et al., 2004. Dissecting virulence: systematic and functional analyses of a pathogenicity island. Proc. Natl. Acad. Sci. USA 101 (10), 3597–3602.

Deng, W., Yu, H.B., de Hoog, C.L., et al., 2012. Quantitative proteomic analysis of type III secretome of enteropathogenic *Escherichia coli* reveals an expanded effector repertoire for attaching/effacing bacterial pathogens. Mol. Cell. Proteomics MCP 11 (9), 692–709.

Domingo-Sananes, M.R., Kapuy, O., Hunt, T., Novak, B., 2011. Switches and latches: a biochemical tug-of-war between the kinases and phosphatases that control mitosis. Philosophical transactions of the Royal Society of London. Ser. B. Biol. Sci. 366 (1584), 3584–3594.

Dong, N., Liu, L., Shao, F., 2010. A bacterial effector targets host DH-PH domain RhoGEFs and antagonizes macrophage phagocytosis. EMBO J. 29 (9), 1363–1376.

Dong, N., Zhu, Y., Lu, Q., Hu, L., Zheng, Y., Shao, F., 2012. Structurally distinct bacterial TBC-like GAPs link Arf GTPase to Rab1 inactivation to counteract host defenses. Cell 150 (5), 1029–41.

Duda, D.M., Borg, L.A., Scott, D.C., Hunt, H.W., Hammel, M., Schulman, B.A., 2008. Structural insights into NEDD8 activation of cullin-RING ligases: conformational control of conjugation. Cell 134 (6), 995–1006.

Dumenil, G., Olivo, J.C., Pellegrini, S., Fellous, M., Sansonetti, P.J., Nhieu, G.T., 1998. Interferon alpha inhibits a Src-mediated pathway necessary for *Shigella*-induced cytoskeletal rearrangements in epithelial cells. J. Cell Biol. 143 (4), 1003–1012.

Ea, C.K., Deng, L., Xia, Z.P., Pineda, G., Chen, Z.J., 2006. Activation of IKK by TNFalpha requires site-specific ubiquitination of RIP1 and polyubiquitin binding by NEMO. Mol. Cell 22 (2), 245–257.

Frankel, G., Lider, O., Hershkoviz, R., et al., 1996. The cell-binding domain of intimin from enteropathogenic *Escherichia coli* binds to beta1 integrins. J. Biol. Chem. 271 (34), 20359–20364.

Freeman, N.L., Zurawski, D.V., Chowrashi, P., et al., 2000. Interaction of the enteropathogenic *Escherichia coli* protein, translocated intimin receptor (Tir), with focal adhesion proteins. Cell Motil. Cytoskeleton 47 (4), 307–318.

Gao, X., Wan, F., Mateo, K., et al., 2009. Bacterial effector binding to ribosomal protein s3 subverts NF-kappaB function. PLoS Pathog. 5 (12), e1000708.

Garmendia, J., Carlier, M.F., Egile, C., Didry, D., Frankel, G., 2006. Characterization of TccP-mediated N-WASP activation during enterohaemorrhagic *Escherichia coli* infection. Cell. Microbiol. 8 (9), 1444–1455.

Garmendia, J., Phillips, A.D., Carlier, M.F., et al., 2004. TccP is an enterohaemorrhagic *Escherichia coli* O157:H7 type III effector protein that couples Tir to the actin-cytoskeleton. Cell. Microbiol. 6 (12), 1167–1183.

Germane, K.L., Ohi, R., Goldberg, M.B., Spiller, B.W., 2008. Structural and functional studies indicate that *Shigella* VirA is not a protease and does not directly destabilize microtubules. Biochemistry 47 (39), 10241–10243.

Gill, R.K., Borthakur, A., Hodges, K., et al., 2007. Mechanism underlying inhibition of intestinal apical Cl/OH exchange following infection with enteropathogenic *E. coli*. J. Clin. Invest. 117 (2), 428–437.

Goosney, D.L., Celli, J., Kenny, B., Finlay, B.B., 1999. Enteropathogenic *Escherichia coli* inhibits phagocytosis. Infect. Immun. 67 (2), 490–495.

Goosney, D.L., DeVinney, R., Finlay, B.B., 2001. Recruitment of cytoskeletal and signaling proteins to enteropathogenic and enterohemorrhagic *Escherichia coli* pedestals. Infect. Immun. 69 (5), 3315–3322.

Gruenheid, S., DeVinney, R., Bladt, F., et al., 2001. Enteropathogenic *E. coli* Tir binds Nck to initiate actin pedestal formation in host cells. Nat. Cell Biol. 3 (9), 856–859.

Guttman, J.A., Lin, A.E., Veiga, E., Cossart, P., Finlay, B.B., 2010. Role for CD2AP and other endocytosis-associated proteins in enteropathogenic *Escherichia coli* pedestal formation. Infect. Immun. 78 (8), 3316–3322.

Guttman, J.A., Samji, F.N., Li, Y., Deng, W., Lin, A., Finlay, B.B., 2007. Aquaporins contribute to diarrhoea caused by attaching and effacing bacterial pathogens. Cell. Microbiol. 9 (1), 131–141.

Hachani, A., Biskri, L., Rossi, G., et al., 2008. IpgB1 and IpgB2, two homologous effectors secreted via the Mxi-Spa type III secretion apparatus, cooperate to mediate polarized cell invasion and inflammatory potential of *Shigella flexneri*. Microbes Infect. 10 (3), 260–268.

Hall, A., 1998. Rho GTPases and the actin cytoskeleton. Science 279 (5350), 509–514.

Hall, A., 2005. Rho GTPases and the control of cell behaviour. Biochem. Soc. Trans. 33 (Pt 5), 891–895.

Hamaguchi, M., Hamada, D., Suzuki, K.N., Sakata, I., Yanagihara, I., 2008. Molecular basis of actin reorganization promoted by binding of enterohaemorrhagic *Escherichia coli* EspB to alpha-catenin. FEBS J. 275 (24), 6260–6267.

Handa, Y., Suzuki, M., Ohya, K., et al., 2007. *Shigella* IpgB1 promotes bacterial entry through the ELMO-Dock180 machinery. Nat. Cell Biol. 9 (1), 121–128.

Hauf, N., Chakraborty, T., 2003. Suppression of NF-kappa B activation and proinflammatory cytokine expression by Shiga toxin-producing *Escherichia coli*. J. Immunol. 170 (4), 2074–2082.

Hecht, G., Hodges, K., Gill, R.K., et al., 2004. Differential regulation of Na+/H+ exchange isoform activities by enteropathogenic *E. coli* in human intestinal epithelial cells. Am. J. Physiol. Gastrointest. Liver Physiol. 287 (2), G370–G378.

Hemrajani, C., Berger, C.N., Robinson, K.S., Marches, O., Mousnier, A., Frankel, G., 2010. NleH effectors interact with Bax inhibitor-1 to block apoptosis during enteropathogenic *Escherichia coli* infection. Proc. Natl. Acad. Sci. USA 107 (7), 3129–3134.

Hilbi, H., Moss, J.E., Hersh, D., et al., 1998. *Shigella*-induced apoptosis is dependent on caspase-1 which binds to IpaB. J. Biol. Chem. 273 (49), 32895–32900.

Hodges, K., Alto, N.M., Ramaswamy, K., Dudeja, P.K., Hecht, G., 2008. The enteropathogenic *Escherichia coli* effector protein EspF decreases sodium hydrogen exchanger 3 activity. Cell. Microbiol. 10 (8), 1735–1745.

Hsu, Y., Jubelin, G., Taieb, F., Nougayrede, J.P., Oswald, E., Stebbins, C.E., 2008. Structure of the cyclomodulin Cif from pathogenic *Escherichia coli*. J. Mol. Biol. 384 (2), 465–477.

Huang, Z., Sutton, S.E., Wallenfang, A.J., et al., 2009. Structural insights into host GTPase isoform selection by a family of bacterial GEF mimics. Nat. Struct. Mol. Biol. 16 (8), 853–860.

Iguchi, A., Thomson, N.R., Ogura, Y., et al., 2009. Complete genome sequence and comparative genome analysis of enteropathogenic *Escherichia coli* O127:H6 strain E2348/69. J. Bacteriol. 191 (1), 347–354.

Iizumi, Y., Sagara, H., Kabe, Y., et al., 2007. The enteropathogenic *E. coli* effector EspB facilitates microvillus effacing and antiphagocytosis by inhibiting myosin function. Cell Host Microbe 2 (6), 383–392.

Isberg, R.R., Leong, J.M., 1990. Multiple beta 1 chain integrins are receptors for invasin, a protein that promotes bacterial penetration into mammalian cells. Cell 60 (5), 861–871.

Iwai, H., Kim, M., Yoshikawa, Y., et al., 2007. A bacterial effector targets Mad2L2, an APC inhibitor, to modulate host cell cycling. Cell 130 (4), 611–623.

Johansen, T., Lamark, T., 2011. Selective autophagy mediated by autophagic adapter proteins. Autophagy 7 (3), 279–296.

Jubelin, G., Taieb, F., Duda, D.M., et al., 2010. Pathogenic bacteria target NEDD8-conjugated cullins to hijack host-cell signaling pathways. PLoS Pathog. 6 (9), e1001128.

Kanayama, A., Seth, R.B., Sun, L., et al., 2004. TAB2 and TAB3 activate the NF-kappaB pathway through binding to polyubiquitin chains. Mol. Cell 15 (4), 535–548.

Kanneganti, T.D., Lamkanfi, M., Nunez, G., 2007. Intracellular NOD-like receptors in host defense and disease. Immunity 27 (4), 549–559.

Kayath, C.A., Hussey, S., El hajjami, N., Nagra, K., Philpott, D., Allaoui, A., 2010. Escape of intracellular *Shigella* from autophagy requires binding to cholesterol through the type III effector, IcsB. Microbes. Infect. Institut. Pasteur. 12 (12–13), 956–966.

Kelly, M., Hart, E., Mundy, R., et al., 2006. Essential role of the type III secretion system effector NleB in colonization of mice by Citrobacter rodentium. Infect. Immun. 74 (4), 2328–2337.

Kenny, B., Ellis, S., Leard, A.D., Warawa, J., Mellor, H., Jepson, M.A., 2002. Co-ordinate regulation of distinct host cell signalling pathways by multifunctional enteropathogenic *Escherichia coli* effector molecules. Mol. Microbiol. 44 (4), 1095–1107.

Khan, M.A., Ma, C., Knodler, L.A., et al., 2006. Toll-like receptor 4 contributes to colitis development but not to host defense during Citrobacter rodentium infection in mice. Infect. Immun. 74 (5), 2522–2536.

Kim, D.W., Lenzen, G., Page, A.L., Legrain, P., Sansonetti, P.J., Parsot, C., 2005. The *Shigella flexneri* effector OspG interferes with innate immune responses by targeting ubiquitin-conjugating enzymes. Proc. Natl. Acad. Sci. USA 102 (39), 14046–14051.

Kim, J., Thanabalasuriar, A., Chaworth-Musters, T., et al., 2007. The bacterial virulence factor NleA inhibits cellular protein secretion by disrupting mammalian COPII function. Cell Host Microbe 2 (3), 160–171.

Kim, M., Ogawa, M., Fujita, Y., et al., 2009. Bacteria hijack integrin-linked kinase to stabilize focal adhesions and block cell detachment. Nature 459 (7246), 578–582.

Kirkin, V., McEwan, D.G., Novak, I., Dikic, I., 2009. A role for ubiquitin in selective autophagy. Mol. Cell 34 (3), 259–269.

Klink, B.U., Barden, S., Heidler, T.V., et al., 2010. Structure of *Shigella* IpgB2 in complex with human RhoA: implications for the mechanism of bacterial guanine nucleotide exchange factor mimicry. J. Biol. Chem. 285 (22), 17197–17208.

Knappstein, S., Ide, T., Schmidt, M.A., Heusipp, G., 2004. Alpha 1-antitrypsin binds to and interferes with functionality of EspB from atypical and typical enteropathogenic *Escherichia coli* strains. Infect. Immun. 72 (8), 4344–4350.

Kodama, T., Akeda, Y., Kono, G., Takahashi, A., Imura, K., Iida, T., Honda, T., 2002. The EspB protein of enterohaemorrhagic *Escherichia coli* interacts directly with alpha-catenin. Cell. Microbiol. 4 (4), 213–222.

Kraft, C., Peter, M., Hofmann, K., 2010. Selective autophagy: ubiquitin-mediated recognition and beyond. Nat. Cell Biol. 12 (9), 836–841.

Kramer, R.W., Slagowski, N.L., Eze, N.A., et al., 2007. Yeast functional genomic screens lead to identification of a role for a bacterial effector in innate immunity regulation. PLoS Pathog. 3 (2), e21.

Kueltzo, L.A., Osiecki, J., Barker, J., et al., 2003. Structure-function analysis of invasion plasmid antigen C (IpaC) from *Shigella flexneri*. J. Biol. Chem. 278 (5), 2792–2798.

Kurushima, J., Nagai, T., Nagamatsu, K., Abe, A., 2010. EspJ effector in enterohemorrhagic *E. coli* translocates into host mitochondria via an atypical mitochondrial targeting signal. Microbiol. Immunol. 54 (7), 371–379.

Lamprecht, G., Heil, A., Baisch, S., et al., 2002. The down regulated in adenoma (dra) gene product binds to the second PDZ domain of the NHE3 kinase A regulatory protein (E3KARP), potentially linking intestinal Cl-/HCO3- exchange to Na+/H+ exchange. Biochemistry 41 (41), 12336–12342.

Lee, S.F., Kelly, M., McAlister, A., et al., 2008. A C-terminal class I PDZ binding motif of EspI/NleA modulates the virulence of attaching and effacing *Escherichia coli* and Citrobacter rodentium. Cell. Microbiol. 10 (2), 499–513.

Legate, K.R., Montanez, E., Kudlacek, O., Fassler, R., 2006. ILK, PINCH and parvin: the tIPP of integrin signalling. Nat. Rev. Mol. Cell Biol. 7 (1), 20–31.

Levine, B. and Kroemer, G. (2008). Autophagy in the pathogenesis of disease. Cell 132, 27–42.

Li, H., Xu, H., Zhou, Y., et al., 2007. The phosphothreonine lyase activity of a bacterial type III effector family. Science 315 (5814), 1000–1003.

Li, Q., Verma, I.M., 2002. NF-kappaB regulation in the immune system. Nat. Rev. Immunol. 2 (10), 725–734.

Mantis, N., Prevost, M.C., Sansonetti, P., 1996. Analysis of epithelial cell stress response during infection by *Shigella flexneri*. Infect. Immun. 64 (7), 2474–2482.

Marches, O., Batchelor, M., Shaw, R.K., et al., 2006. EspF of enteropathogenic *Escherichia coli* binds sorting nexin 9. J. Bacteriol. 188 (8), 3110–3115.

Marches, O., Covarelli, V., Dahan, S., et al., 2008. EspJ of enteropathogenic and enterohaemorrhagic *Escherichia coli* inhibits opsono-phagocytosis. Cell. Microbiol. 10 (5), 1104–1115.

Marches, O., Ledger, T.N., Boury, M., et al., 2003. Enteropathogenic and enterohaemorrhagic *Escherichia coli* deliver a novel effector called Cif, which blocks cell cycle G2/M transition. Mol. Microbiol. 50 (5), 1553–1567.

Marches, O., Wiles, S., Dziva, F., et al., 2005. Characterization of two non-locus of enterocyte effacement-encoded type III-translocated effectors, NleC and NleD, in attaching and effacing pathogens. Infect. Immun. 73 (12), 8411–8417.

Maresca, M., Miller, D., Quitard, S., Dean, P., Kenny, B., 2005. Enteropathogenic *Escherichia coli* (EPEC) effector-mediated suppression of antimicrobial nitric oxide production in a small intestinal epithelial model system. Cell Microbiol. 7 (12), 1749–1762.

Martinez, E., Schroeder, G.N., Berger, C.N., et al., 2010. Binding to Na(+) /H(+) exchanger regulatory factor 2 (NHERF2) affects trafficking and function of the enteropathogenic *Escherichia coli* type III secretion system effectors Map, EspI and NleH. Cell. Microbiol. 12 (12), 1718–1731.

Matsuzawa, T., Kuwae, A., Yoshida, S., Sasakawa, C., Abe, A., 2004. Enteropathogenic *Escherichia coli* activates the RhoA signaling pathway via the stimulation of GEF-H1. EMBO J. 23 (17), 3570–3582.

McDaniel, T.K., Jarvis, K.G., Donnenberg, M.S., Kaper, J.B., 1995. A genetic locus of enterocyte effacement conserved among diverse enterobacterial pathogens. Proc. Natl. Acad. Sci. USA 92 (5), 1664–1668.

Michail, S.K., Halm, D.R., Abernathy, F., 2003. Enteropathogenic *Escherichia coli*: stimulating neutrophil migration across a cultured intestinal epithelium without altering transepithelial conductance. J. Pediatr. Gastroenterol. Nutr. 36 (2), 253–260.

Miyahara, A., Nakanishi, N., Ooka, T., Hayashi, T., Sugimoto, N., Tobe, T., 2009. Enterohemorrhagic *Escherichia coli* effector EspL2 induces actin microfilament aggregation through annexin 2 activation. Cell. Microbiol. 11 (2), 337–350.

Morikawa, H., Kim, M., Mimuro, H., et al., 2010. The bacterial effector Cif interferes with SCF ubiquitin ligase function by inhibiting deneddylation of Cullin1. Biochemi. Biophys. Res. Commun. 401 (2), 268–274.

Mostowy, S., Bonazzi, M., Hamon, M.A., et al., 2010. Entrapment of intracytosolic bacteria by septin cage-like structures. Cell Host Microbe 8 (5), 433–444.

Mounier, J., Popoff, M.R., Enninga, J., Frame, M.C., Sansonetti, P.J., Van Nhieu, G.T., 2009. The IpaC carboxyterminal effector domain mediates Src-dependent actin polymerization during *Shigella* invasion of epithelial cells. PLoS Pathog. 5 (1), e1000271.

Mousnier, A., Whale, A.D., Schuller, S., Leong, J.M., Phillips, A.D., Frankel, G., 2008. Cortactin recruitment by enterohemorrhagic *Escherichia coli* O157:H7 during infection in vitro and ex vivo. Infect. Immun. 76 (10), 4669–4676.

Muhlen, S., Ruchaud-Sparagano, M.H., Kenny, B., 2011. Proteasome-independent degradation of canonical NFkappaB complex components by the NleC protein of pathogenic *Escherichia coli*. J. Biol. Chem. 286 (7), 5100–5107.

Nadler, C., Baruch, K., Kobi, S., et al., 2010. The type III secretion effector NleE inhibits NF-kappaB activation. PLoS Pathog. 6 (1), e1000743.

Nagai, T., Abe, A., Sasakawa, C., 2005. Targeting of enteropathogenic *Escherichia coli* EspF to host mitochondria is essential for bacterial pathogenesis: critical role of the 16th leucine residue in EspF. J. Biol. Chem. 280 (4), 2998–3011.

Nataro, J.P., Seriwatana, J., Fasano, A., et al., 1995. Identification and cloning of a novel plasmid-encoded enterotoxin of enteroinvasive *Escherichia coli* and *Shigella* strains. Infect. Immun. 63 (12), 4721–4728.

Newton, H.J., Pearson, J.S., Badea, L., et al., 2010. The type III effectors NleE and NleB from enteropathogenic *E. coli* and OspZ from *Shigella* block nuclear translocation of NF-kappaB p65. PLoS Pathog. 6 (5), e1000898.

Niebuhr, K., Giuriato, S., Pedron, T., et al., 2002. Conversion of PtdIns(4,5)P(2) into PtdIns(5)P by the *S.flexneri* effector IpgD reorganizes host cell morphology. EMBO J. 21 (19), 5069–5078.

Nougayrede, J.P., Donnenberg, M.S., 2004. Enteropathogenic *Escherichia coli* EspF is targeted to mitochondria and is required to initiate the mitochondrial death pathway. Cell. Microbiol. 6 (11), 1097–1111.

Nougayrede, J.P., Foster, G.H., Donnenberg, M.S., 2007. Enteropathogenic *Escherichia coli* effector EspF interacts with host protein Abcf2. Cell. Microbiol. 9 (3), 680–693.

Ogawa, M., Yoshimori, T., Suzuki, T., Sagara, H., Mizushima, N., Sasakawa, C., 2005. Escape of intracellular *Shigella* from autophagy. Science 307 (5710), 727–731.

Ogura, Y., Ooka, T., Whale, A., et al., 2007. TccP2 of O157:H7 and non-O157 enterohemorrhagic *Escherichia coli* (EHEC): challenging the dogma of EHEC-induced actin polymerization. Infect. Immun. 75 (2), 604–612.

Ohlson, M.B., Huang, Z., Alto, N.M., et al., 2008. Structure and function of Salmonella SifA indicate that its interactions with SKIP, SseJ, and RhoA family GTPases induce endosomal tubulation. Cell Host Microbe 4 (5), 434–446.

Ohya, K., Handa, Y., Ogawa, M., Suzuki, M., Sasakawa, C., 2005. IpgB1 is a novel *Shigella* effector protein involved in bacterial invasion of host cells. Its activity to promote membrane ruffling via Rac1 and Cdc42 activation. J. Biol. Chem. 280 (25), 24022–24034.

Orchard, R.C., Kittisopikul, M., Altschuler, S.J., Wu, L.F., Suel, G.M., Alto, N.M., 2012. Identification of F-actin as the dynamic hub in a microbial-induced GTPase polarity circuit. Cell 148 (4), 803–815.

Oswald, E., Sugai, M., Labigne, A., et al., 1994. Cytotoxic necrotizing factor type 2 produced by virulent *Escherichia coli* modifies the small GTP-binding proteins Rho involved in assembly of actin stress fibers. Proc. Natl. Acad. Sci. USA 91 (9), 3814–3818.

Parsot, C., 2009. *Shigella* type III secretion effectors: how, where, when, for what purposes? Curr. Opin. Microbiol. 12 (1), 110–116.

Patel, A., Cummings, N., Batchelor, M., et al., 2006. Host protein interactions with enteropathogenic *Escherichia coli* (EPEC): 14-3-3tau binds Tir and has a role in EPEC-induced actin polymerization. Cell. Microbiol. 8 (1), 55–71.

Pearson, J.S., Riedmaier, P., Marches, O., Frankel, G., Hartland, E.L., 2011. A type III effector protease NleC from enteropathogenic *Escherichia coli* targets NF-kappaB for degradation. Mol. Microbiol. 80 (1), 219–230.

Peralta-Ramirez, J., Hernandez, J.M., Manning-Cela, R., et al., 2008. EspF Interacts with nucleation-promoting factors to recruit junctional proteins into pedestals for pedestal maturation and disruption of paracellular permeability. Infect. Immun. 76 (9), 3854–3868.

Petty, N.K., Bulgin, R., Crepin, V.F., et al., 2010. The Citrobacter rodentium genome sequence reveals convergent evolution with human pathogenic *Escherichia coli*. J. Bacteriol. 192 (2), 525–538.

Pfleger, C.M., Salic, A., Lee, E., Kirschner, M.W., 2001. Inhibition of Cdh1-APC by the MAD2-related protein MAD2L2: a novel mechanism for regulating Cdh1. Genes Deve. 15 (14), 1759–1764.

Phillips, N., Hayward, R.D., Koronakis, V., 2004. Phosphorylation of the enteropathogenic *E. coli* receptor by the Src-family kinase c-Fyn triggers actin pedestal formation. Nat. Cell Biol. 6 (7), 618–625.

Quitard, S., Dean, P., Maresca, M., Kenny, B., 2006. The enteropathogenic *Escherichia coli* EspF effector molecule inhibits PI-3 kinase-mediated uptake independently of mitochondrial targeting. Cell. Microbiol. 8 (6), 972–981.

Raqib, R., Ekberg, C., Sharkar, P., et al., 2002. Apoptosis in acute shigellosis is associated with increased production of Fas/Fas ligand, perforin, caspase-1, and caspase-3 but reduced production of Bcl-2 and interleukin-2. Infect. Immun. 70 (6), 3199–3207.

Raymond, B., Crepin, V.F., Collins, J.W., Frankel, G., 2011. The WxxxE effector EspT triggers expression of immune mediators in an Erk/JNK and NF-kappaB-dependent manner. Cell. Microbiol. 13 (12), 1881–1893.

Ritchie, J.M., Brady, M.J., Riley, K.N., et al., 2008. EspFU, a type III-translocated effector of actin assembly, fosters epithelial association and late-stage intestinal colonization by *E. coli* O157:H7. Cell. Microbiol. 10 (4), 836–847.

Robinson, C.M., Sinclair, J.F., Smith, M.J., O'Brien, A.D., 2006. Shiga toxin of enterohemorrhagic *Escherichia coli* type O157:H7 promotes intestinal colonization. Proc. Natl. Acad. Sci. USA 103 (25), 9667–9672.

Robinson, K.S., Mousnier, A., Hemrajani, C., Fairweather, N., Berger, C.N., Frankel, G., 2010. The enteropathogenic *Escherichia coli* effector NleH inhibits apoptosis induced by Clostridium difficile toxin B. Microbiology 156 (Pt 6), 1815–1823.

Rohde, J.R., Breitkreutz, A., Chenal, A., Sansonetti, P.J., Parsot, C., 2007. Type III secretion effectors of the IpaH family are E3 ubiquitin ligases. Cell Host Microbe 1 (1), 77–83.

Romero, S., Grompone, G., Carayol, N., et al., 2011. ATP-mediated Erk1/2 activation stimulates bacterial capture by filopodia, which precedes *Shigella* invasion of epithelial cells. Cell Host Microbe 9 (6), 508–519.

Royan, S.V., Jones, R.M., Koutsouris, A., et al., 2010. Enteropathogenic *E. coli* non-LEE encoded effectors NleH1 and NleH2 attenuate NF-kappaB activation. Mol. Microbiol. 78 (5), 1232–1245.

Sakaguchi, T., Kohler, H., Gu, X., McCormick, B.A., Reinecker, H.C., 2002. *Shigella flexneri* regulates tight junction-associated proteins in human intestinal epithelial cells. Cell. Microbiol. 4 (6), 367–381.

Samba-Louaka, A., Nougayrede, J.P., Watrin, C., Oswald, E., Taieb, F., 2009. The enteropathogenic *Escherichia coli* effector Cif induces delayed apoptosis in epithelial cells. Infect. Immun. 77 (12), 5471–5477.

Sanada, T., Kim, M., Mimuro, H., et al., 2012. The *Shigella flexneri* effector OspI deamidates UBC13 to dampen the inflammatory response. Nature 483 (7391), 623–626.

Sason, H., Milgrom, M., Weiss, A.M., et al., 2009. Enteropathogenic *Escherichia coli* subverts phosphatidylinositol 4,5-bisphosphate and phosphatidylinositol 3,4,5-trisphosphate upon epithelial cell infection. Mol. Biol. Cell 20 (1), 544–555.

Savkovic, S.D., Koutsouris, A., Hecht, G., 1996. Attachment of a noninvasive enteric pathogen, enteropathogenic *Escherichia coli*, to cultured human intestinal epithelial monolayers induces transmigration of neutrophils. Infect. Immun. 64 (11), 4480–4487.

Schuller, S., Chong, Y., Lewin, J., Kenny, B., Frankel, G., Phillips, A.D., 2007. Tir phosphorylation and Nck/N-WASP recruitment by enteropathogenic and enterohaemorrhagic *Escherichia coli* during ex vivo colonization of human intestinal mucosa is different to cell culture models. Cell. Microbiol. 9 (5), 1352–1364.

Schuller, S., Lucas, M., Kaper, J.B., Giron, J.A., Phillips, A.D., 2009. The ex vivo response of human intestinal mucosa to enteropathogenic *Escherichia coli* infection. Cell Microbiol 11 (3), 521–530.

Selbach, M., Paul, F.E., Brandt, S., et al., 2009. Host cell interactome of tyrosine-phosphorylated bacterial proteins. Cell Host Microbe 5 (4), 397–403.

Selyunin, A.S., Alto, N.M., 2011. Activation of PAK by a bacterial type III effector EspG reveals alternative mechanisms of GTPase pathway regulation. Small GTPases. 2 (4), 217–221.

Selyunin, A.S., Sutton, S.E., Weigele, B.A., et al., 2011. The assembly of a GTPase-kinase signalling complex by a bacterial catalytic scaffold. Nature 469 (7328), 107–111.

Sham, H.P., Shames, S.R., Croxen, M.A., et al., 2011. Attaching and effacing bacterial effector NleC suppresses epithelial inflammatory responses by inhibiting NF-kappaB and p38 mitogen-activated protein kinase activation. Infect. Immun. 79 (9), 3552–3562.

Shames, S.R., Bhavsar, A.P., Croxen, M.A., et al., 2011. The pathogenic *Escherichia coli* type III secreted protease NleC degrades the host acetyltransferase p300. Cell. Microbiol. 13 (10), 1542–1557.

Shames, S.R., Croxen, M.A., Deng, W., Finlay, B.B., 2011. The type III system-secreted effector EspZ localizes to host mitochondria and interacts with the translocase of inner mitochondrial membrane 17b. Infect. Immun. 79 (12), 4784–4790.

Shames, S.R., Deng, W., Guttman, J.A., et al., 2010. The pathogenic *E. coli* type III effector EspZ interacts with host CD98 and facilitates host cell prosurvival signalling. Cell. Microbiol. 12 (9), 1322–1339.

Shifflett, D.E., Clayburgh, D.R., Koutsouris, A., Turner, J.R., Hecht, G.A., 2005. Enteropathogenic *E. coli* disrupts tight junction barrier function and structure in vivo. Lab. Invest. 85 (10), 1308–1324.

Shifrin, Y., Kirschner, J., Geiger, B., Rosenshine, I., 2002. Enteropathogenic *Escherichia coli* induces modification of the focal adhesions of infected host cells. Cell. Microbiol. 4 (4), 235–243.

Silverman, N., Maniatis, T., 2001. NF-kappaB signaling pathways in mammalian and insect innate immunity. Genes Dev. 15 (18), 2321–2342.

Simonovic, I., Rosenberg, J., Koutsouris, A., Hecht, G., 2000. Enteropathogenic *Escherichia coli* dephosphorylates and dissociates occludin from intestinal epithelial tight junctions. Cell. Microbiol. 2 (4), 305–315.

Simpson, N., Shaw, R., Crepin, V.F., et al., 2006. The enteropathogenic *Escherichia coli* type III secretion system effector Map binds EBP50/NHERF1: implication for cell signalling and diarrhoea. Mol. Microbiol. 60 (2), 349–363.

Singer, A.U., Rohde, J.R., Lam, R., et al., 2008. Structure of the *Shigella* T3SS effector IpaH defines a new class of E3 ubiquitin ligases. Nat. Struct. Mol. Biolo. 15 (12), 1293–1301.

Smith, K., Humphreys, D., Hume, P.J., Koronakis, V., 2010. Enteropathogenic *Escherichia coli* recruits the cellular inositol phosphatase SHIP2 to regulate actin-pedestal formation. Cell Host Microbe 7 (1), 13–24.

Starostina, N.G., Kipreos, E.T., 2012. Multiple degradation pathways regulate versatile CIP/KIP CDK inhibitors. Trends Cell Biol. 22 (1), 33–41.

Swanson, J.A., Johnson, M.T., Beningo, K., Post, P., Mooseker, M., Araki, N., 1999. A contractile activity that closes phagosomes in macrophages. J. Cell Sci. 112 (Pt 3), 307–316.

Swimm, A., Bommarius, B., Li, Y., et al., 2004. Enteropathogenic *Escherichia coli* use redundant tyrosine kinases to form actin pedestals. Mol. Biol. Cell. 15 (8), 3520–3529.

Taylor, K.A., Luther, P.W., Donnenberg, M.S., 1999. Expression of the EspB protein of enteropathogenic *Escherichia coli* within HeLa cells affects stress fibers and cellular morphology. Infect. Immun. 67 (1), 120–125.

Thanabalasuriar, A., Koutsouris, A., Weflen, A., Mimee, M., Hecht, G., Gruenheid, S., 2010. The bacterial virulence factor NleA is required for the disruption of intestinal tight junctions by enteropathogenic *Escherichia coli*. Cell. Microbiol. 12 (1), 31–41.

Tobe, T., Beatson, S.A., Taniguchi, H., et al., 2006. An extensive repertoire of type III secretion effectors in *Escherichia coli* O157 and the role of lambdoid phages in their dissemination. Proc. Natl. Acad. Sci. USA 103 (40), 14941–14946.

Tomar, A., Schlaepfer, D.D., 2009. Focal adhesion kinase: switching between GAPs and GEFs in the regulation of cell motility. Curr. Opin. Cell Biol. 21 (5), 676–683.

Touze, T., Hayward, R.D., Eswaran, J., Leong, J.M., Koronakis, V., 2004. Self-association of EPEC intimin mediated by the beta-barrel-containing anchor domain: a role in clustering of the Tir receptor. Mol. Microbiol. 51 (1), 73–87.

Toyotome, T., Suzuki, T., Kuwae, A., et al., 2001. *Shigella* protein IpaH(9.8) is secreted from bacteria within mammalian cells and transported to the nucleus. J. Biol. Chem. 276 (34), 32071–32079.

Tran Van Nhieu, G., Ben-Ze'ev, A., Sansonetti, P.J., 1997. Modulation of bacterial entry into epithelial cells by association between vinculin and the *Shigella* IpaA invasin. EMBO J. 16 (10), 2717–2729.

Tran Van Nhieu, G., Caron, E., Hall, A., Sansonetti, P.J., 1999. IpaC induces actin polymerization and filopodia formation during *Shigella* entry into epithelial cells. EMBO J. 18 (12), 3249–3262.

Tu, X., Nisan, I., Yona, C., Hanski, E., Rosenshine, I., 2003. EspH, a new cytoskeleton-modulating effector of enterohaemorrhagic and enteropathogenic *Escherichia coli*. Mol. Microbiol. 47 (3), 595–606.

Uchiya, K., Tobe, T., Komatsu, K., et al., 1995. Identification of a novel virulence gene, virA, on the large plasmid of *Shigella*, involved in invasion and intercellular spreading. Mol. Microbiol. 17 (2), 241–250.

Unsworth, K.E., Mazurkiewicz, P., Senf, F., et al., 2007. Dynamin is required for F-actin assembly and pedestal formation by enteropathogenic *Escherichia coli* (EPEC). Cell. Microbiol. 9 (2), 438–449.

Veiga, E., Guttman, J.A., Bonazzi, M., et al., 2007. Invasive and adherent bacterial pathogens co-opt host clathrin for infection. Cell Host Microbe 2 (5), 340–351.

Vingadassalom, D., Campellone, K.G., Brady, M.J., et al., 2010. Enterohemorrhagic *E. coli* requires N-WASP for efficient type III translocation but not for EspFU-mediated actin pedestal formation. PLoS Pathog. 6 (8), e1001056.

Vingadassalom, D., Kazlauskas, A., Skehan, B., et al., 2009. Insulin receptor tyrosine kinase substrate links the *E. coli* O157:H7 actin assembly effectors Tir and EspF(U) during pedestal formation. Proc. Natl. Acad. Sci. USA 106 (16), 6754–6759.

Viswanathan, V.K., Lukic, S., Koutsouris, A., Miao, R., Muza, M.M., Hecht, G., 2004. Cytokeratin 18 interacts with the enteropathogenic *Escherichia coli* secreted protein F (EspF) and is redistributed after infection. Cell. Microbiol. 6 (10), 987–997.

Vossenkamper, A., Marches, O., Fairclough, P.D., et al., 2010. Inhibition of NF-kappaB signaling in human dendritic cells by the enteropathogenic *Escherichia coli* effector protein NleE. J. Immunol. 185 (7), 4118–4127.

Wan, F., Anderson, D.E., Barnitz, R.A., et al., 2007. Ribosomal protein S3: a KH domain subunit in NF-kappaB complexes that mediates selective gene regulation. Cell 131 (5), 927–939.

Wan, F., Weaver, A., Gao, X., Bern, M., Hardwidge, P.R., Lenardo, M.J., 2011. IKKbeta phosphorylation regulates RPS3 nuclear translocation and NF-kappaB function during infection with *Escherichia coli* strain O157:H7. Nat. Immun. 12 (4), 335–343.

Weiss, S.M., Ladwein, M., Schmidt, D., et al., 2009. IRSp53 links the enterohemorrhagic *E. coli* effectors Tir and EspFU for actin pedestal formation. Cell Host Microbe 5 (3), 244–258.

Whale, A.D., Garmendia, J., Gomes, T.A., Frankel, G., 2006. A novel category of enteropathogenic *Escherichia coli* simultaneously utilizes the Nck and TccP pathways to induce actin remodelling. Cell. Microbiol. 8 (6), 999–1008.

Wong, A.R., Clements, A., Raymond, B., Crepin, V.F., Frankel, G., 2012a. The interplay between the *Escherichia coli* Rho guanine nucleotide exchange factor effectors and the mammalian RhoGEF inhibitor EspH. mBio 3 (1), e00250-11.

Wong, A.R., Raymond, B., Collins, J.W., Crepin, V.F., Frankel, G., 2012b. The enteropathogenic *E. coli* effector EspH promotes actin pedestal formation and elongation via WASP-interacting protein (WIP). Cell. Microbiol. 14 (7), 1051–1070.

Wu, B., Skarina, T., Yee, A., et al., 2010. NleG Type 3 effectors from enterohaemorrhagic *Escherichia coli* are U-Box E3 ubiquitin ligases. PLoS Pathog. 6 (6), e1000960.

Wu, C.J., Conze, D.B., Li, T., Srinivasula, S.M., Ashwell, J.D., 2006. Sensing of Lys 63-linked polyubiquitination by NEMO is a key event in NF-kappaB activation [corrected]. Nat. Cell Biol. 8 (4), 398–406.

Yan, Y., Vasudevan, S., Nguyen, H.T., Merlin, D., 2008. Intestinal epithelial CD98: an oligomeric and multifunctional protein. Biochim. Biophys. Acta. 1780 (10), 1087–1092.

Yang, Z. and Klionsky, D.J. (2010). Eaten alive: a history of macroautophagy. Nat. Cell. Biol. 12, 814–822.

Yen, H., Ooka, T., Iguchi, A., Hayashi, T., Sugimoto, N., Tobe, T., 2010. NleC, a type III secretion protease, compromises NF-kappaB activation by targeting p65/RelA. PLoS Pathog. 6 (12), e1001231.

Yoshida, S., Handa, Y., Suzuki, T., et al., 2006. Microtubule-severing activity of *Shigella* is pivotal for intercellular spreading. Science 314 (5801), 985–989.

Yoshida, S., Katayama, E., Kuwae, A., Mimuro, H., Suzuki, T., Sasakawa, C., 2002. *Shigella* deliver an effector protein to trigger host microtubule destabilization, which promotes Rac1 activity and efficient bacterial internalization. EMBO J. 21 (12), 2923–2935.

Zhang, L., Ding, X.J., Cui, J., et al., 2012. Cysteine methylation disrupts ubiquitin-chain sensing in NF-kappa B activation. Nature 481 (7380), 204–208.

Zurawski, D.V., Mumy, K.L., Badea, L., et al., 2008. The NleE/OspZ family of effector proteins is required for polymorphonuclear transepithelial migration, a characteristic shared by enteropathogenic *Escherichia coli* and *Shigella flexneri* infections. Infect. Immun. 76 (1), 369–379.

Zurawski, D.V., Mumy, K.L., Faherty, C.S., McCormick, B.A., Maurelli, A.T., 2009. *Shigella flexneri* type III secretion system effectors OspB and OspF target the nucleus to downregulate the host inflammatory response via interactions with retinoblastoma protein. Mol. Microbiol. 71 (2), 350–368.

Zychlinsky, A., Kenny, B., Menard, R., Prevost, M.C., Holland, I.B., Sansonetti, P.J., 1994. IpaB mediates macrophage apoptosis induced by *Shigella flexneri*. Mol. Microbiol. 11 (4), 619–627.

Chapter 16

Type 1 and 5 secretion systems and associated toxins

Timothy J. Wells and Ian R. Henderson

University of Birmingham, Birmingham, UK

INTRODUCTION

Pathogenic *Escherichia coli* can cause disease in a variety of locations in the human body. The site of infection is dependent on the repertoire of virulence factors such as adhesins and toxins expressed by the pathogen. Most proteins involved in virulence need to traverse both the inner and outer membrane of the bacterium and in some cases the membrane of the host cell to perform their function. Given the diversity of function, multitude of targets and variety of structures of these secreted proteins, it is unsurprising to find that *E. coli* has multiple different secretion systems to effect translocation of proteins to the exterior of the cell.

Two of the simplest secretion systems that *E. coli* utilizes are the type 1 secretion system (T1SS) and the type 5 secretion system (T5SS). These secretion systems are widely distributed in other Gram-negative bacteria. In *E. coli* the best-known T1SS is that used to secrete hemolysin, a pore-forming toxin found in both uropathogenic (UPEC) and enterohemorrhagic (EHEC) strains of *E. coli*. The T5SS is widespread in *E. coli* with many of the virulence factors secreted by this mechanism specific for a particular pathotype. This chapter will focus on the T1SS and T5SS; the mechanism of biogenesis, the function of the secreted proteins and current attempts to use these systems in biotechnological applications. Other *E. coli* secretion systems are discussed in detail in other chapters.

THE TYPE 1 SECRETION SYSTEM

The T1SS of Gram-negative bacteria allows the secretion of proteins from the cytoplasm to the extracellular environment in a single step, with no periplasmic intermediate (Figure 16.1). Proteins secreted by the T1SS greatly vary in size, from 82 amino acids (aa) to over 8000 residues (Delepelaire, 2004).

Escherichia coli. http://dx.doi.org/10.1016/B978-0-12-397048-0.00016-4

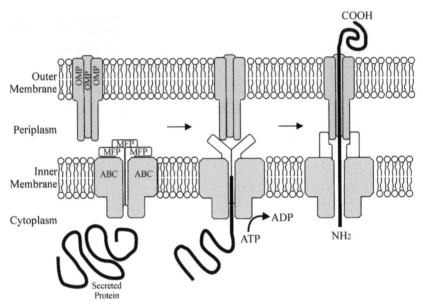

FIGURE 16.1　Model of the type 1 secretion system. Upon recognition of a C-terminal secretion signal on the secreted protein, the inner membrane complex formed by an energy-providing ABC cassette and MFP protein interacts with the trimeric OMP. A sealed channel assembly spanning across the two membranes of the Gram-negative cell envelope is formed, through which the substrate is secreted. The secreted protein is translocated in a single step, with no periplasmic intermediate.

The secreted proteins are diverse in function and include toxins, proteases, lipases, S-layer proteins, hemophores, bacteriocins, and others with as yet unknown functions (Delepelaire, 2004). Although divergent in sequence, the majority of proteins secreted by the T1SS have a C-terminal located secretion signal, which is recognized by the translocation machinery; the exception is dispersin of EAEC which is secreted to the periplasm by the Sec system before secretion via the T1SS to the extracellular milieu (Koronakis et al., 1989; Ghigo and Wandersman, 1994; Sheikh et al., 2002).

The translocating machinery of a T1SS is comprised of three proteins that span the cell envelope, all of which are essential for secretion (Letoffe et al., 1996). Two of the translocation proteins span the inner membrane (the ATP-binding cassette [ABC] and the membrane fusion/adapter protein [MFP]), while the final member of the translocator is an outer-membrane protein (OMP) (Figure 16.1). The ABC protein possesses a nucleotide-binding domain (NBD) fused to a transmembrane domain (TMD), and recognizes the C-terminal secretion signal of the substrate molecule; as such this protein is responsible for the specificity of the secretion machinery for the substrate molecule (Delepelaire, 2004). The MFP protein consists of a short cytoplasmic domain at the N-terminus, followed by a membrane anchor and a large periplasmic domain

(Dinh et al., 1994; Thanabalu et al., 1998). The OMP forms a trimeric, water-filled channel through the outer membrane, open to the extracellular milieu but constricted at the periplasmic end (Koronakis et al., 2000). The interaction of the secreted effector molecule with the ABC protein triggers the sequential assembly of the secretion complex by generating interactions between the ABC, MFP, and OMP (Thanabalu et al., 1998). When the MFP protein binds the OMP, it triggers the opening of the periplasmic end thus creating a complete channel from the cytoplasm to the extracellular medium (Thanabalu et al., 1998). Hydrolysis of ATP by the ABC protein provides the energy for the secretion of the substrate protein through the channel (Figure 16.1).

Proteins secreted via the T1SS can have diverse functions, however virtually all T1SS proteins found in *E. coli* belong to the repeat-in-toxin (RTX) family; the exception is dispersin of EAEC (Linhartova et al., 2010). RTX proteins are so named for the distinctive glycine-rich repeats (GGXGXDXXX) that specifically bind calcium (Welch, 2001). All RTX toxins are made as an inactive precursor and must undergo a common post-translational maturation step before export by the T1SS (Linhartova, et al., 2010). RTX family members can have a variety of functions, however the prominent and historically the first described group consists of toxins exhibiting a cytotoxic pore-forming ability, often first detected as a hemolytic halo surrounding bacterial colonies grown on blood agar plates (Goebel and Hedgpeth, 1982; Muller et al., 1983; Felmlee et al., 1985; Welch, 1991). RTX proteins found in *E. coli* include TosA from UPEC strains needed for adherence and colonization of the urinary tract (Parham et al., 2005; Iguchi et al., 2009; Lloyd et al., 2009; Vigil et al., 2011a,b, 2012), EhxA from EHEC strains (Bauer and Welch, 1996) and the prototypical member of the T1SS, hemolysin (HlyA), from uropathogenic *E. coli*.

Hemolysin

Regulation and maturation

Hemolysin of UPEC was the first described T1SS protein. Hemolysin is an important virulence factor in *E. coli* infections (see Chapter 9) owing to its cytolytic and cytotoxic activity against a wide range of mammalian cell types (Welch et al., 1981). The synthesis, activation and secretion of hemolysin is determined by the *hlyCABD* operon (Hess et al., 1986). In *E. coli* this operon is located either on chromosome-bound pathogenicity islands or on transmissible plasmids, suggesting that the T1SS has been acquired by horizontal gene transfer. The operon encodes the toxin activation protein (HlyC), the hemolysin itself (HlyA), the ABC transporter (HlyB), and the MFP protein (HlyD). The outer-membrane component, the multifunctional protein TolC, is found in all *E. coli* strains and is encoded elsewhere on the chromosome.

Full transcription of the *hlyCABD*operon is reliant on two factors, the 18-kDa protein RfaH (Bailey et al., 1992) and a motif 2-kb upstream from the operon termed the *ops* (operon polarity suppressor) (Bailey et al., 1996; Nieto et al., 1996; Leeds and Welch, 1997). RfaH is a bacterial elongation factor that increases expression of distal genes in several long, horizontally acquired operons, including those encoding lipopolysaccharide (LPS) and F (fertility) pilus biogenesis (Bailey et al., 1997). Loss of RfaH or the *ops* element shows a distinctive effect on *hlyCABD* transcription, modestly decreasing transcription of the first two genes, *hlyC* and *A,* but virtually abolishing transcription of *hlyB* and *D* (Bailey et al., 1992, 1996; Nieto et al., 1996). The *ops* element is required for RfaH recruitment to RNA polymerase (RNAP) (Bailey et al., 1996; Artsimovitch and Landick, 2002; Belogurov et al., 2009); it induces isomerization of the transcription elongation complex (TEC) into a distinct state necessary for the sequence-specific recruitment of RfaH to the non-template DNA strand. Following recruitment, RfaH remains bound to RNAP and acts as an antiterminator by reducing RNAP pausing and termination at some factor-independent and Rho-dependent signals (Bailey et al., 1996).

After translation of HlyA, the protein is inactive until matured intracellularly by the cotranslated HlyC, a fatty acid acyltransferase (Issartel et al., 1991). HlyC forms a homodimer that uses acyl–acyl carrier protein (ACP) as the fatty acid donor to acylate two internal lysine residues, K564 (K1) and K690 (KII) of HlyA. Although both lysines are acylated independently, both are required for in vivo toxin activity. However, acylation is only required for the hemolytic activity; the secretion of HlyA is independent of HlyC (Koronakis and Hughes, 1996).

After acylation HlyA is secreted via a T1SS formed by the proteins HlyB, HlyD, and TolC. HlyB, the ABC component of secretion, couples ATP hydrolysis to the secretion of HlyA (Koronakis et al., 1995). The topology of HlyB (707 aa) has been determined, suggesting that the protein is inserted in the inner membrane by eight hydrophobic, α-helical transmembrane domains (TMDs) extending from the amino acid positions 38–432 (Wang et al., 1991). The cytoplasmic loops of HlyB are large with many positively charged amino acids, whereas in contrast the periplasmic loops are quite small. HlyD is a prototypical and well-characterized member of the MFP family (Dinh et al., 1994). HlyD is anchored in the cytoplasmic membrane by a single TMD and possesses a large 100 aa C-terminal periplasmic domain (Schulein et al., 1992). This periplasmic domain is highly conserved not only within the MFP family but also within the superfamily of periplasmic efflux proteins (PEP) (Schulein et al., 1992). Contact between the HlyB:HlyD inner membrane complex and the final component, TolC in the outer membrane, is primarily mediated by HlyD in response to engagement by the HlyA substrate (Thanabalu et al., 1998). The fully active complex contains the substrate and all three export proteins, all of which undergo conformational change. The complex is transient however; once the substrate leaves the cell the inner-membrane and outer-membrane components disengage (Thanabalu et al., 1998). As TolC is involved with at least four different export systems, the mechanism behind forming the physical bridge across the periplasm has been well studied.

Secretion of HlyA via the TolC channel

TolC (55 kDa) is a multifunctional OMP in *E. coli* that can serve as an outer membrane component for several processes, including T1SS and drug export via RND (resistance, nodulation, division) systems (Wandersman and Delepelaire, 1990; Zgurskaya and Nikaido, 2000). Crystal structures of TolC reveal that it is a homotrimer in the outer membrane of *E. coli* where it forms a trans-periplasmic channel-tunnel. The trimer is about 140 Å in length, comprising a 40 Å β-barrel (channel domain) and a 100 Å long α-helical barrel that projects across the periplasmic space (tunnel domain) (Koronakis et al., 2000). The molecule forms a tapering tube that is almost closed at the periplasm end and wide open at the outer-membrane surface. The internal diameter of the channel is around 35 Å and consists of a large water-filled cavity open to the extracellular medium (Koronakis et al., 2000).

It is proposed that TolC opens at the proximal end by an iris-like mechanism, by unwinding the coiled-coiled helices so as to move the inner ring to the exterior. This would open the periplasmic end of the channel going from 5 Å in diameter to a maximal opening of around 20 Å (Figure 16.2) (Koronakis et al., 2000; Bavro et al., 2008). Crystal structures of TolC mutants either stuck in closed or semi-open states support this hypothesis (Bavro et al., 2008). Opening of TolC is triggered by members of the MFP class (HlyD in the case of hemolysin) (Janganan et al., 2011). Thus, HlyA binds HlyB, triggering a change in conformation in the HlyB/HlyD complex. HlyD then binds to TolC, triggering the opening of the periplasmic end of the TolC channel and making a complete pore from cytoplasm to extracellular medium. The HlyA protein is then secreted through the pore to the extracellular medium.

Mechanisms of action and role in virulence

Once secreted, *E. coli* HlyA acts as a pore-forming RTX cytotoxin, with a broad range of cytocidal activity on a wide spectrum of cells from a variety of species, including erythrocytes, leukocytes, granulocytes, monocytes, endothelial cells, and renal epithelial cells (Gadeberg and Orskov, 1984; Keane et al., 1987; Bhakdi et al., 1989, 1990; Mobley et al., 1990; Suttorp et al., 1990). HlyA alteration of membrane permeability causes lysis and death, which may provide iron and prevent phagocytosis. Once inserted, HlyA behaves as an integral membrane protein and causes target cell lysis by forming transmembrane pores that are cation-selective, pH-dependent, and apparently asymmetric (Bhakdi et al., 1986; Menestrina et al., 1995). Membrane insertion of HlyA is thought to be through a monomolecular mechanism. It has been estimated that only 1–3 HlyA molecules form the pore, with oligimerization occurring by subsequent addition of monomers within the membrane (Bhakdi et al., 1986; Benz et al., 1989). A conserved region toward the N-terminus of HlyA is essential for lysis and is predicted to be involved in pore formation, as it spans the only pronounced hydrophobic sequences in the otherwise hydrophilic HlyA protein (Stanley et al., 1998). The C-terminal glycine-rich repeat domain that binds Ca^{2+} is required for hemolysis but not for pore formation, and it seems likely that Ca^{2+}-induced conformational change may promote a subsequent insertion of the toxin into cell membranes (Bakas et al., 1998; Stanley et al., 1998).

In addition to pore formation, at very low, sublytic concentrations, HlyA is a potent trigger of G-protein-dependent generation of inositol triphosphate and diacylglycerol in granulocytes and endothelial cells, stimulating the respiratory burst and the secretion of vesicular constituents (Bhakdi and Martin, 1991). It has also been shown to stimulate the release of cytokines, including interleukin 1β and tumor necrosis factor (TNF) from a variety of human cells (Suttorp et al.,

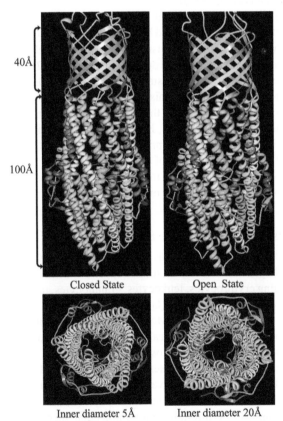

FIGURE 16.2   The structure of trimeric TolC modeled in the closed and open positions. The trimeric protein is made of a 40 Å channel domain in the outer membrane and a 100 Å tunnel domain in the periplasm. Binding of the MFP protein initiates opening of the TolC periplasmic domain by an iris-like movement enlarging the tunnel diameter from 5 Å to 20 Å. *Figure adapted from Protein Data Bank (Bavro et al., 2008).*

1993). All of these phenotypes suggest that HlyA would be a potent virulence factor in UPEC infections, however defining the actual role that HlyA may play in pathogenicity has been elusive (Linggood and Ingram, 1982; Mobley et al., 1990; Yamamoto et al., 1995; Smith et al., 2006). Recent studies however have shown that HlyA, although not contributing to colonization, does play a major role in evoking damage in the uroepithelium and in inducing hemorrhage in the bladder during the early stages of *E. coli*-mediated cystitis in the mouse (Smith et al., 2008).

THE TYPE 5 SECRETION SYSTEM

The T5SS is the simplest and most widespread protein secretion mechanism in Gram-negative bacteria. The secretion system is currently divided into five subclasses (types 5a to 5e) based on differences in gene organization and protein structure. Type 5a are better known as the classical autotransporters (ATs), type 5b as the two-partner secretion pathway (TPS), and type 5c as the trimeric autotransporter adhesins (TAA). Type 5d have only recently been described (Figure 16.3). The 5e subclass has only been added to the T5SS family recently, but includes the functionally well-characterized Intimin of *E. coli* and Invasin of *Yersinia*.

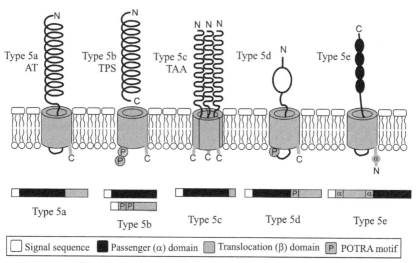

FIGURE 16.3   Model of the type 5 secretion pathway. The system is divided into five different pathways, the classical autotransporter (type 5a or ATs), the two-partner secretion (type 5b or TPS), the trimeric autotransporter adhesin (type 5c or TAA) and the recently described type 5d and type 5e pathways. All proteins have an N-terminal signal sequence, a translocation β-barrel and a passenger domain that is secreted to the surface. However they differ in gene number, organization and barrel similarities.

MOLECULAR ORGANIZATION

Autotransporters (ATs) are so named based on the hypothesis that all the functional elements required for secretion are contained in a single protein. Structurally, AT proteins are characterized by the presence of three distinct domains: (i) an N-terminal signal sequence; (ii) a secreted passenger/α-domain; and (iii) a C-terminal domain termed the translocation or β-domain (Jose et al., 1995). The signal-sequence mediates translocation across the inner membrane in a Sec-dependent manner. Once in the periplasm the translocation domain forms a β-barrel composed of 12 β-strands that inserts into the outer membrane (Henderson et al., 1998; Henderson and Nataro, 2001). Finally the β-barrel mediates translocation of the passenger domain to the surface where the secreted polypeptide is either cleaved or remains attached to the bacterium (Henderson et al., 2004). While the translocation domain is highly homologous among AT proteins (conserving the mechanism of transport), the secreted passenger domain demonstrates considerable amino acid sequence variation. ATs can also vary greatly in length, from as low as 500 aa to above 3000 aa.

Two-partner secretion (TPS) is similar to the AT mechanism, requiring a passenger domain and a β-barrel translocation domain. However, instead of being encoded on a single polypeptide, the passenger and translocation domains are produced as two separate proteins, each containing a Sec-dependent N-terminal signal sequence to mediate inner-membrane translocation to the periplasm (Jacob-Dubuisson et al., 2001). These proteins are referred to as TpsA (passenger) and TpsB (translocation) family members (Jacob-Dubuisson et al., 2001) and are frequently, but not always, encoded in an operon. Akin to the AT mechanism, once in the periplasm the translocation β-barrel mediates secretion of the passenger domain (Henderson et al., 2000). Whilst TpsA proteins can have widely differing functions, they share common features; they are large (many >3000 aa), they share significant sequence similarity within their N-terminal 300 aa (a region called the TPS domain) (Jacob-Dubuisson et al., 2001, 2004) and like the AT passenger domain adopt a β-helical conformation. The β-barrel of TpsB proteins is different to AT proteins being composed of 16 amphipathic β-strands instead of the usual 12 found in ATs (Guedin et al., 2000) and containing two polypeptide-transport-associated (POTRA) domains that descend into the periplasm and are thought to recognize the TpsA protein via the TPS domain. The POTRA domains are predicted to recruit nascent β-strands in a process called β-augmentation; in TPS proteins it is thought that the POTRA motif recognizes the TPS domain of TpsA proteins (Knowles et al., 2008; Jacob-Dubuisson et al., 2009; Delattre et al., 2011).

The third subgroup, the trimeric autotransporter adhesins (TAA) are produced as a single polypeptide similar to ATs, however they differ significantly in the size of the translocation domain. Instead of being comprised of ~300 aa forming 12 β-strands, a translocation domain of 70–100 aa is sufficient to mediate secretion of the passenger domain (Roggenkamp et al., 2003; Surana et al.,

2004). This shorter translocation domain has been shown to form four β-strands which only become a functional 12-stranded β-barrel by forming a homotrimer in the outer membrane (Cotter et al., 2005; Linke et al., 2006). Trimerization of the passenger domain is vital for stability and functionality of the protein (Cotter et al., 2006). As the name suggests all TAA proteins identified to date have a role in adhesion.

Recently, a patatin-like protein named PlpD was identified in an environmental *Pseudomonas* strain and described to belong to a fourth sub-group of T5SS (Salacha et al., 2010). Like the classical ATs, the protein is produced as a single polypeptide with an N-terminal signal sequence, a passenger domain and a β-barrel translocation unit, however the β-barrel domain in type 5d is composed of 16 β-strands and more closely related to the TpsB proteins of two-partner secretion. There is also a POTRA motif between the passenger and translocation domains, suggesting a gene fusion event combining the two components of a TPS system. No members of the type 5d have been identified in *E. coli* and these proteins will not be considered further here.

Finally, Intimin of *E. coli* (see Chapters 4 and 5) and Invasin of *Yersinia*, have long been studied for their role in adhesion but due to their vastly different genetic organization they have only recently been identified as belonging to T5SS. The proteins possess an N-terminal signal sequence, but in contrast to classical ATs, the order of the passenger and translocation domains are reversed (Tsai et al., 2010). In addition to this reversal, between the translocation β-barrel and signal peptide is an α-helical region and a LysM domain thought to bind to peptidoglycan (Fairman et al., 2012). The β-barrel, similar to classical ATs, is composed of 12 β-strands (Fairman et al., 2012). The inverted domain structure results in the C-terminus, not the N-terminus being secreted to the cell surface where it extends from the outer membrane to contact the host cell.

SECRETION

It was initially hypothesized that T5SS proteins were unique in that all elements required for secretion were found in either one or two proteins. However, recent studies reveal that the reality is more complicated, with periplasmic chaperones and machinery in the inner and outer membranes also required for secretion. The majority of work on T5SS secretion has been focused on classical ATs, so that little is yet known about the requirements of types 5b–e secreted proteins. However as the subgroups have many similarities in structure and domain organization it would be unsurprising if they shared many of the requirements of classical ATs.

All five subgroups of T5SS once translated are targeted to the Sec machinery, which catalyzes their energy-driven transport to the periplasm. Most T5SS proteins are synthesized with characteristic Sec-dependent signal peptides, typically 20–30 aa in length. These signal peptides typically consist of a positively charged N-domain, followed by a hydrophobic H domain and finally a C domain that contains a recognition site for a signal peptidase. On entering the periplasm the signal peptidase recognizes the uncharged residues at −1 and −3 relative to the cleavage site (Hegde and Bernstein, 2006). A significant subset of T5SS proteins (including members of types 5a, 5b, 5c, and 5e) possess unusually long signal peptides, typically 50–60 residues long. The purpose of the longer signal peptide has been studied for multiple proteins with differing results; however the consensus view is that the extended peptide has a subtle role, slowing translocation of the proteins across the inner membrane to prevent accumulation of misfolded species in the periplasm. The precise mechanism governing the decrease in translocation has yet to be elucidated (Leyton et al., 2010, 2012).

On entering the periplasm, all further steps in the secretion pathway must be independent of any energy input. Many OMPs have been shown to interact with periplasmic chaperones such as SurA, Skp, and the bifuncational periplasmic serine protease DegP. Many AT proteins also have been shown to interact with these proteins (Sklar et al., 2007; Baud et al., 2009; Knowles et al., 2009) suggesting that T5SS biogenesis follows the same rules as normal OMPs, whereby these chaperones mediate transit through the periplasm. Chaperones have also been shown to be involved for the other T5SS subtypes, with DegP interacting with the TPS protein FHA (Baud et al., 2009), and Intimin shown to interact with SurA, Skp, and DebP (Bodelon et al., 2009). The role of chaperones is currently unknown for TPS and TAA proteins. Thus, it seems that periplasmic chaperones act specifically to keep the passenger domain in a translocation-competent form and prevent aggregation and mis-folding (Lazar and Kolter, 1996; Leyton et al., 2012).

In addition to chaperones, the β-barrel assembly machinery (Bam) complex has been shown to be required for the biogenesis of many ATs (Ieva and Bernstein, 2009; Ruiz-Perez et al., 2009, 2010). The Bam complex is comprised of the integral β-barrel protein, BamA, and four peripheral lipoproteins, BamB–E (Knowles et al., 2009). BamA and BamD have been shown to be essential for the correct secretion of AT passenger domains across the outer membrane, whereas BamB, C, and E are not required (Rossiter et al., 2011b). However, redundant roles for BamB, C, and E have not been ruled out. BamA consists of a large C-terminal β-barrel in the outer membrane and five N-terminal POTRA domains that descend into the periplasm. POTRA 5 makes direct contact with the lipoprotein BamD (Kim et al., 2007), while POTRA 1 and 2 have been shown to bind peptides of OMPs (Knowles et al., 2008) and the chaperone SurA (Bennion et al., 2010). As SurA interacts with BamA, it is conceivable that ATs are delivered to the Bam complex by periplasmic chaperones. BamA has also been shown to be important for the secretion of proteins belonging to other subgroups of T5SS, including the TPS CdiA/B (Aoki et al., 2008), the TAA YadA (Lehr et al., 2010), and the type 5e secreted Intimin (Bodelon et al., 2009). The role of BamD in TPS, TAA, and type 5e is yet to be elucidated.

Finally, the translocation and assembly module (TAM) was recently described to be required for efficient secretion of ATs in proteobacteria. The TAM consists of two proteins, TamA in the outer membrane and TamB in the inner membrane that together span the entire cell envelope. TamA, like BamA and TpsB proteins discussed above, belongs to the Omp85 super-family of outer-membrane proteins. It consists of a C-terminal β-barrel in the outer membrane and three N-terminal POTRA domains thought to inter-act with TamB (Selkrig et al., 2012). Mutants lacking the TAM complex in *E.coli* failed to secrete the type 5a adhesins Antigen 43 and EhaA (Selkrig et al., 2012). It is as yet unknown if the TAM complex is required for effi-cient secretion of type 5b–e proteins. A summary of AT secretion is found in Figure 16.4.

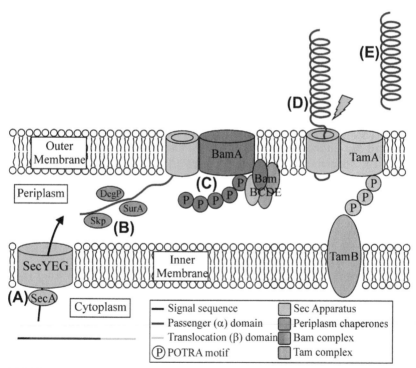

FIGURE 16.4   General model of autotransporter secretion. (A) In all proteins the signal sequence mediates inner-membrane translocation in a Sec-dependent manner. (B) On entering the periplasm many AT proteins have been shown to interact with periplasmic chaperones which keep the passen-ger domain in a translocation-competent form. (C) The β-barrel of the AT is inserted into the outer membrane via interactions with the BAM complex. (D) The translocation domain then mediates secretion of the passenger domain to the surface. Recently the Tam complex was shown to be critical for passenger domain secretion for AT proteins. (E) Once on the surface the passenger domain can either remain attached or be cleaved and secreted.

STRUCTURE OF T5SS DOMAINS

The structure of many AT proteins has been solved, including EspP and Hbp from *E. coli* strains. These structures have revealed a domain architecture comprising a signal sequence, the passenger domain, an autochaperone domain, an α-helix linker followed by the translocation domain. The translocation domain of AT proteins is highly conserved and consists of β-pleated sheets in the form of a β-barrel akin to most other integral outer-membrane proteins (Loveless and Saier, 1997; Barnard et al., 2007; Tajima et al., 2010). Although diverse in sequence, all solved AT β-barrels contain 12 antiparallel strands connected by extracellular loops and periplasmic turns of varying length, with a narrow (~1 × 1.25 nm) hydrophilic pore. Translocation domains also share a consensus amino acid motif at the C-terminus (Struyve et al., 1991; Jose et al., 1995; Loveless and Saier, 1997). The C-terminal 9 aa are generally alternating aromatic/hydrophobic and charged/hydrophilic with the last residue a tryptophan or phenylalanine. This sequence (in particular the C-terminal three residues) is predicted to play a role in outer-membrane localization and/or stability of outer-membrane proteins.

Encoded just before the translocation domain, AT proteins have a single α-helix linker, found to reside within the barrel lumen. The length of the linker is determined by whether the passenger domain remains attached (long), is cleaved extracellularly (long) or cleaved within the barrel (short) (Barnard et al., 2007; Tajima et al., 2010). The α-helix also ensures the pore is blocked post-translocation preserving the integrity of the outer membrane. In addition to this α-helix, in the two solved *E. coli* ATs, a long extracellular loop of the translocation domain also folds into the barrel, closing the pore from the outside (Barnard et al., 2007).

Passenger domains of AT proteins are diverse in sequence and function and thus structures are different between proteins, however there are some commonalities. Above 97% of AT passenger domains are predicted to form a right-handed β-helical structure despite large diversity in sequence, length, and function. The right-handed β-helical structure was first solved for the T5SS protein, Pertactin (Emsley et al., 1994, 1996). β-Helical structures typically contain three β sheets separated by three turns giving the protein a 'V' shape in cross-section (Jenkins and Pickersgill, 2001). A complete turn of the β-helix is known as a coil, with different proteins having varying numbers of coils. The structure displays extensive 'stacking' across its coils, whereby similar aliphatic residues occupy equivalent positions in neighboring β sheets leading to ridges of aliphatic residues across the coils (Jenkins and Pickersgill, 2001). This right-handed β-helical structure is thought to be conserved in AT proteins because the passenger domain adopts a predominantly unfolded conformation during its passage through the outer membrane, a process that occurs independent of ATP and proton gradients (Junker et al., 2006). The conserved β-helical structure may contribute to protein folding after transport through the translocation domain and may also play a role in presenting an adhesive functional tip away from the cell surface (Klemm et al., 2004; Junker et al., 2006).

The β-helical structure of the passenger domain also allows functional groups to be inserted within turns that protrude from the β helix without disturbing its structural integrity (Emsley et al., 1996). These functional groups can be small, such as the tripeptide Arg-Gly-Asp (RGD) adhesion motif of the *E. coli* adhesin Antigen 43 (Fernandez and Weiss, 1994; Klemm et al., 2004) or large globular domains with various activities. The serine protease ATs of *Enterobacteriacae* (SPATEs) are the best examples of this globular structure. The crystal structures of three *E. coli* SPATE proteins, Hbp, EspP, and Tsh, have been solved with all including an extended right-handed β-helical structure that forms the spine and a globular subdomain reminiscent of the chymotrypsin family of serine proteases, which carries the proteolytic function of the passenger. In Hbp and Tsh there is also a second subdomain (termed domain 2) that adopts a chitinase b-like fold but for which a functional role has not yet been identified (Figure 16.5).

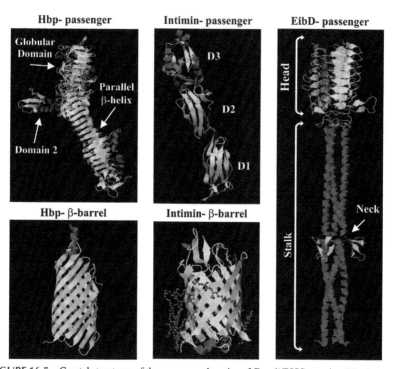

FIGURE 16.5   Crystal structures of the passenger domains of *E. coli* T5SS proteins. Hbp is a type 5a secreted SPATE with the typical stacked β-helices and globular domain. The β-barrel is composed of 12 strands with an α-helix linker. Intimin is a type 5e secreted protein with three Ig-like domains and a lectin-like domain which binds Tir. The β-barrel is composed of 12 strands with a periplasmic LysM domain. EibD is a type 5c secreted protein. Trimeric in structure it has clear head, stalk and neck regions. *Figures adapted from protein databank (Otto et al., 1998; Luo et al., 2000; Tajima et al., 2010; Leo et al., 2011; Fairman et al., 2012).*

Finally, at the extreme C-terminal of most AT passenger domains there is a β-strand hairpin structure that forms the autochaperone (AC) domain. This AC region is essential for folding of β-helical passenger domains, acting as a template as the passenger domain is secreted (Dutta et al., 2003; Oliver et al., 2003). In addition, residues in the AC domain of Hbp are essential for the initiation of passenger domain translocation (Soprova et al., 2010).

The structure of the passenger domain of TAA proteins has been extensively studied. The crystal structure of the *E. coli* TAA, EibD has been solved along with TAAs from other species (Leo et al., 2011). The structure of all TAAs have a clear organization of an N-terminal head, a connector/neck region, a stalk which is highly variable in length and a C-terminal anchor domain (Figure 16.5) (Linke et al., 2006). The stalk is a fibrous, highly repetitive structure rich in coiled coils and extremely variable in length across the TAAs. They function as spacers to project the head domain away from the bacterial surface; however they also can convey protection against host defenses such as serum resistance and antibody binding (Roggenkamp et al., 2003). The head domain of TAA proteins consists of β-sheets forming a coiled left-handed parallel β-roll (LPBR) (Nummelin et al., 2004). Interestingly these are similar to the β-helix of AT and TPS proteins suggesting that this structure may be important for either secretion or folding. EibD was the first TAA structure to have both a head domain and the entire coiled-coil stalk solved. The stalk begins as a right-handed superhelix, but switches handedness halfway down. Large cavities were found in the EibD structure that may explain how TAAs bend to bind their ligands (Leo et al., 2011).

The structure of the passenger domain of the type 5e protein, Intimin from *E. coli,* has also been solved (Luo et al., 2000). In contrast to all other passenger domains of T5SS, intimin is composed of 16 β-sheets together with four α-helices, three of which form Ig-like domains D0, D1, and D2 and the final forming a lectin-like domain, D3, which binds the translocated intimin receptor (Tir) (Figure 16.5). Recently the translocation domain of intimin was solved, revealing a 12-stranded β-barrel reminiscent of the ATs. Indeed, the barrels of the AT protein EspP and intimin superimpose closely (Fairman et al., 2012).

No structures have been solved in *E. coli* for members of the TPS pathway, however the structure has been solved for the prototypical TPS protein, filamentous hemagglutinin (FHA) from *Bordetella pertussis.* As expected the TpsB protein forms a 16-stranded β-barrel with two periplasmic POTRA domains and a large loop harboring a functional important motif (Jacob-Dubuisson et al., 2009). The TPS domain of the TpsA FHA was also solved and shows a β-helix, with three extrahelical motifs, a β-hairpin, a four-stranded β-sheet, and an N-terminal capping, mostly formed by the non-conserved regions of the TPS domain (Clantin et al., 2004). The structure explains why the TPS domain is able to initiate folding of the β-helical motifs that form the central domain of the adhesin, because it is itself a β-helical scaffold. It is likely that the TPS domains of other TpsA proteins form a similar fold. A β-helical structure has also been predicted for the majority of TPS-secreted TpsA proteins. The large size, the abundance of repetitive sequences and the predominance of β-structure thus seem to be common characteristics of both AT and TpsA proteins (reviewed extensively in Kajava and Steven, 2006). However TpsA proteins vary greatly in size, sequence, and function and thus other structures may exist.

PROCESSING OF T5SS PASSENGER DOMAINS

The fate of the passenger domain of T5SS proteins is dependent on the extent of processing that occurs at the bacterial cell surface. For AT proteins, some passenger domains remain intact (Wells et al., 2008, 2009), whereas others are cleaved from the β-domains and secreted (Eslava et al., 1998; Yen et al., 2008). Intriguingly, some AT passenger domains are cleaved but remain strongly associated with the translocation domain (Benz and Schmidt, 1989; Sherlock et al., 2005a). The mechanisms of passenger domain cleavage have been elucidated for a few proteins, however they already seem to be diverse, including processing by exogenous host proteases (Shere et al., 1997), processing by other AT proteins (van Ulsen et al., 2003) and autocatalytic reactions (Charbonneau et al., 2009).

The passenger domains of TpsA proteins can remain stably, but noncovalently, attached to the cell surface after translocation, whereas others are released into the extracellular milieu. Many TpsA proteins are synthesized with large pro-domains of unknown function, either at the N-terminus or the C-terminus, and these are removed at some point during the translocation process (Mazar and Cotter, 2007). In contrast to AT and TAA proteins, the passenger domains of all TAA proteins remain attached to the translocation domain. Finally, Intimin remains covalently attached to the cell surface after secretion.

DISTRIBUTION, FUNCTION, AND REGULATION

The T5SS is widely used by *E. coli*, with strains usually carrying up to 12 different T5SS encoding genes (Wells et al., 2010). These proteins are found across all *E. coli* pathotypes and evolutionary lineages (Chaudhuri et al., 2010) and are often associated with virulence. Although *E. coli* strains usually have a large complement of T5SS proteins only a few of these are conserved across the majority of strains. The T5SS proteins with specific virulence properties tend to be restricted to a single pathotype. Due to the variable nature of the passenger domain, T5SS proteins have a wide variety of functions, however some phenotypes such as autoaggregation, biofilm formation, and adhesion are common to many T5SS proteins. A summary of the functions of all known *E. coli* T5SS proteins is found in Table 16.1. As T5SS proteins are widespread in *E. coli* many different methods of regulation of the genes have been reported. A more detailed examination of function and regulation of four *E. coli* T5SS proteins from representative groups is presented below.

TABLE 16.1 Type 5 secreted proteins of *E. coli*

Protein	Pathotypes predominantly associated with the protein	Characterized virulence properties	Reference
AT			
SPATES			
EatA	ETEC	Serine protease activity	(Patel et al., 2004)
EpeA	EAEC	Protease and mucinase activity	(Leyton et al., 2003)
EspC	EPEC	Enterotoxin	(Stein et al., 1996; Mellies et al., 2001)
EspP	EHEC	Proteolytic toxin	(Brunder et al., 1997; Djafari et al., 1997)
Pet	EAEC	Proteolytic toxin	(Eslava et al., 1998)
PicU	UPEC	Mucinase	(Parham et al., 2004)
RpeA	EPEC	Colonization factor	(Leyton et al., 2007)
Sat	UPEC	Proteolytic toxin	(Guyer et al., 2000, 2002)
Tsh	APEC (Avian pathogenic *E. coli*)	Hemagglutinin/ hemoglobin binding	(Provence & Curtiss, 1994; Otto et al., 1998)
AIDA-I type			
AatA	APEC	Adhesion to chicken embryo fibroblast cells, contribution to APEC virulence	(Li et al., 2009)
AIDA-I	EAEC	Adhesion, autoaggregation, biofilm formation, self-recognition	(Benz and Schmidt, 1989; Sherlock et al., 2004)
Ag43	Widespread	Biofilm formation, autoaggregation, self-recognition, phase variable	(Kjaergaard et al., 2000b; Torres et al., 2002; Klemm et al., 2004; Ulett et al., 2006)

TABLE 16.1 Type 5 secreted proteins of *E. coli*—cont'd

Protein	Pathotypes predominantly associated with the protein	Characterized virulence properties	Reference
EhaA	EHEC	Adhesion to primary bovine cell lines. Autoaggregation, biofilm formation	(Wells et al., 2008)
EhaB/ UpaC	Widespread	Biofilm formation, Adhesion to laminin and collagen I	(Parham et al., 2004; Wells et al., 2008)
EhaD	EHEC	Biofilm formation	(Wells et al., 2008)
EhaJ	EHEC	Biofilm formation, adhesion to ECM molecules	(Easton et al., 2011)
UpaB	UPEC	Not yet characterized	(Parham et al., 2004)
UpaE	UPEC	Not yet characterized	(Parham et al., 2004)
UpaH	ETEC	Biofilm formation	(Allsopp et al., 2010)
TibA	UPEC	Adhesion/invasion, autoaggregation, biofilms formation, self-recognition	(Lindenthal and Elsinghorst, 2001; Sherlock et al., 2005b)
YcgV	Commensal	Biofilm formation	(Roux et al., 2005)
YdeK	Commensal	Pseudo-gene in K-12. Not characterized	
YejO	Commensal	Not yet characterized	(Roux et al., 2005)
YfaL/ EhaC	Commensal/ EHEC	YfaL increases biofilm formation	(Parham et al., 2004; Roux et al., 2005; Wells et al., 2008)
TPS			
CdiA, CdiB	UPEC	Contact-dependent bacterial growth inhibition	(Aoki et al., 2005)

Continued

TABLE 16.1 Type 5 secreted proteins of *E. coli*—cont'd

Protein	Pathotypes predominantly associated with the protein	Characterized virulence properties	Reference
EtpA, EtpB	ETEC	Adhesion to host epithelial cells	(Fleckenstein et al., 2006)
OtpA, OtpB	EHEC		(Choi et al., 2007)
TAA			
EibA-G	UPEC	Immunoglobulin binding and serum resistance	(Sandt and Hill, 2000, 2001)
Saa	EHEC	Adhesion to Hep-2 cells, autoaggregation	(Paton et al., 2001)
UpaG	UPEC	Adhesion to T24 cells, biofilm formation, autoaggregation, adhesion to ECM	(Valle et al., 2008a)
5e			
Intimin	EPEC, EHEC	Adhesion to eukaryotic cells by binding to Tir receptor	(Jerse et al., 1990)

Antigen 43

The prototypical *E. coli* AT is Antigen 43 (Ag43). Found in nearly all *E. coli* strains, Ag43 has the classic three-domain AT structure and functional properties common to many *E. coli* AT proteins; aggregation and biofilm formation (Henderson et al., 1997; Hasman et al., 1999; Danese et al., 2000; Klemm et al., 2004). The expression of Ag43 is phase variable, with switching rates of $\sim 10^{-3}$ per cell per generation due to the concerted actions of Dam-methylase (positive regulation) and OxyR (negative regulation) (Henderson and Owen, 1999; Waldron et al., 2002; Schembri et al., 2003a; Lim and van Oudenaarden, 2007). Ag43 can be expressed on the *E. coli* cell surface in very high numbers (up to 50 000 copies per cell), resulting in some strains in a characteristic frizzy colony morphology (Henderson and Owen, 1999; Hasman et al., 2000).

The most distinctive phenotype of Ag43 is the protein's ability to mediate aggregation which can be visualized macroscopically as flocculation and settling of cells in static liquid suspensions. Ag43-mediated aggregation is a self-recognition process mediated by receptor–ligand reactions between Ag43 molecules on adjacent cells (Hasman et al., 1999; Kjaergaard et al., 2000a). The autoaggregation property of Ag43 is also characteristic of many other autotransporter proteins including AIDA-I and TibA (Sherlock et al., 2004, 2005b).

Finally, Antigen 43 dramatically enhances biofilm formation on abiotic surfaces (Danese et al., 2000; Kjaergaard et al., 2000b; Torres et al., 2002; Reisner et al., 2003) and is specifically correlated with the biofilm mode of growth (Schembri et al., 2003b). Ag43 expression is correlated with biofilm formation by UPEC during infection of bladder cells (Anderson et al., 2003) and in enteropathogenic *E. coli* (Torres et al., 2002). Global gene expression profiling of *E. coli* during biofilm growth demonstrated that *agn43* expression is up-regulated when compared with both exponential and stationary phase planktonic cultures (Schembri et al., 2003b). Further, a comparison of the biofilm-forming capacity of Ag43 variants from different pathogenic *E. coli* strains demonstrated a conserved ability of Ag43 to enhance biofilm growth, albeit with different efficiency (Klemm et al., 2004). Ag43-mediated aggregation also protects bacteria against hydrogen peroxide, a phenomenon that may enhance the resistance of biofilm cells to antimicrobial agents (Schembri et al., 2003a).

Plasmid encoded toxin (Pet)

SPATEs are a subgroup of AT proteins that possess a consensus serine protease domain (Henderson et al., 1998). They have been identified in many pathogenic *E. coli* and other species such as *Shigella, Citrobacter*, and *Salmonella*. SPATE proteins possess several common features: (1) SPATE members do not cleave IgA1; (2) the serine protease domain is not involved in autoprocessing; (3) most SPATE proteins are the predominant secreted molecule of the strain; (4) SPATE proteins are highly associated with pathogenic strains; and (5) SPATEs are highly immunogenic. Interestingly, despite having the same serine protease domain, the SPATE proteins demonstrate distinct substrate specificities. SPATE proteins are among the best-characterized AT proteins of *E. coli* and possess a range of functional properties including cytotoxicity (Brunder et al., 1997; Eslava et al., 1998; Mellies et al., 2001; Guyer et al., 2002), mucinase activity (Parham et al., 2004), hemoglobin degradation (Provence and Curtiss, 1994; Otto et al., 1998) and promotion of intestinal colonization (Leyton et al., 2007). Some SPATE proteins including Pet and EspP are located on plasmids, while others are encoded on chromosomal pathogenicity islands.

One of the best-characterized SPATEs is the Plasmid-Encoded Toxin (Pet). Pet is encoded on pAA2, the large virulence plasmid of enteroaggregative *E. coli* (EAEC) strain 042. The expression of Pet was recently shown to be co-dependent on the two global regulators CRP and Fis (Rossiter et al., 2011a). This regulation is via a novel co-activation mechanism whereby CRP is placed at a non-optimal position for transcription initiation, creating dependence on Fis for full activation of *pet*. Other SPATEs were found to have similar promoters suggesting similar co-dependent regulation.

Pet is required for inducing dilation of crypt openings and rounding and extrusion of enterocytes in human tissue explants (Henderson et al., 1999). The serine-protease motif has been found to be essential for both the cytotoxicity and enterotoxicity of Pet. Once secreted and cleaved by an autocatalytic reaction, Pet enters eukaryotic cells, with internalization essential for cytotoxic activity. Pet is internalized by clathrin-coated vesicles and is trafficked to the Golgi apparatus and endoplasmic reticulum. Once internalized Pet binds and cleaves its substrate target, α-fodrin via the activity of the serine protease causing disruption of the cytoskeleton, cell rounding and finally cell death (Henderson et al., 1999; Betancourt-Sanchez and Navarro-Garcia, 2009).

EtpA

The plasmid locus *etpBAC* encodes proteins that form a TPS that are required for the glycosylation (EtpC) and secretion (EtpB) of the 170-kDa adhesin EtpA (Fleckenstein et al., 2006). EtpA was first identified as mediating adhesion of the enterotoxigenic *E. coli* (ETEC; see Chapter 6) strain H10407 to intestinal epithelial cells and is conserved across the ETEC pathotype (Sahl et al., 2011). EtpA is required for optimal intestinal colonization in a murine model, suggesting it is an important virulence factor of enterotoxigenic *E. coli* (Roy et al., 2008).

As EtpA is secreted, the protein theoretically must maintain contact with ETEC to promote adhesion. Recent experiments discovered that EtpA mimics and interacts with highly conserved regions of flagellin, the major subunit of the flagellum (Roy et al., 2009b). The flagellin molecules capture and hold the EtpA protein for presentation to eukaryotic receptors mediating indirect adhesion between flagella and host cells (Roy et al., 2009b). EtpA also shares similarities to proteins in other motile bacterial pathogens, suggesting a common mode of bacterial adhesion.

UpaG

Compared to ATs, only a few TAAs have been described in *E. coli*. The Saa and Eib TAA proteins range between 392 and 535 aa and have only been identified in two strains so far. In contrast, a large (~1800 aa) TAA is found in nearly all pathogenic *E. coli* strains where it is known as UpaG (from UPEC) or EhaG (from EHEC). UpaG was found to mediate adhesion to human bladder epithelial cells (Valle et al., 2008b). UpaG also promotes cell aggregation and biofilm formation on abiotic surfaces by CFT073 and various other UPEC strains as well as binding to the extracellular matrix (ECM) proteins fibronectin and laminin (Valle et al., 2008b). Prevalence studies indicated that *upaG* is frequently associated with extraintestinal *E. coli* (ExPEC) strains (Valle et al., 2008b). UpaG has also been identified as a potential protective antigen in ExPEC (Durant et al., 2007).

EhaG is the positional ortholog of UpaG in EHEC O157:H7 but contains significant sequence divergence within the passenger-encoding domain.

Cloning and expression of the *ehaG* gene revealed that the EhaG TAA although still mediating aggregation and biofilm formation, had different adhesion properties than those of UpaG. In contrast to UpaG, EhaG did not mediate adherence to human bladder cells, but did promote binding to colorectal cell lines (Totsika et al., 2012). These functional properties correlated with the distinct tissue tropism of EHEC and UPEC pathogens (Totsika et al., 2012). Interestingly, the UpaG/EhaG positional ortholog in *Salmonella enterica*, SadA, is also shown to mediate adhesion to human intestinal cell lines. The SadA protein was also found to elicit an IgG reaction in mice, and provide limited protection from *S. enterica* challenge (Raghunathan et al., 2011). Finally, the expression of many T5SS genes, including *upaG* and *ehaG*, have been shown to be dependent on the global regulator HNS (Totsika et al., 2012).

The role of the type 5e protein, Intimin, in pathogenesis has been covered extensively in Chapters 4 and 5 and thus has not been re-iterated here.

TYPE 1 AND 5 SECRETED PROTEINS AS PROSPECTS FOR VACCINES

The T1SS is being used primarily for two types of biotechnological applications; mass production of extracellular chimeric polypeptides and live vaccine delivery. For mass production, it was found that by fusing proteins to the C-terminus of HlyA, the whole protein was still recognized and secreted by *E. coli* via the type 1 secretion pathway (Blight and Holland, 1994; Gentschev et al., 1996; Tzschaschel et al., 1996). This mechanism allows an easy method to express and secrete fusion protein directly into the extracellular medium. There seem to be very few limitations concerning the size or origin of the heterologous regions of the fusion proteins, with sizes ranging from 20 aa to above 1000 aa and sequences derived from prokaryotes as well as eukaryotes (Catic et al., 1999; Orr et al., 1999).

A highly important application of the T1SS however, is the presentation of heterologous antigens in attenuated Gram-negative live vaccines. A bacterial carrier strain can be used to secrete antigens for virtually any bacterial, viral, and parasitic pathogens. A variety of antigens has been expressed and secreted in attenuated bacterial carrier strains for use as vaccines against pathogens including *Listeria monocytogenes* (Hess et al., 1996), *Shigella* (Su et al., 1992), *Clostridium difficile* (Ryan et al., 1997), and the measles virus (Spreng et al., 2000). Protection against some of these pathogens has been observed after vaccination with the recombinant live vaccines (Hess et al., 1996; Ryan et al., 1997). The T1SS has many advantages as a vehicle for antigen display including: (1) the lack of size limitations for the protein; (2) its suitability for the secretion of several antigens in a single carrier bacterium, allowing the constructions of multivalent vaccines (Catic et al., 1999); and finally (3) its function as a delivery system for immunocontraceptive vaccines and for co-expression and co-delivery of active cytokines (Hahn et al., 1998). Therefore, the use of a T1SS for delivery of antigens has been shown to induce protection and has many advantages over other live-vaccine methods.

T5S proteins present a great opportunity to develop novel vaccines against *E. coli*-mediated disease and T5SS proteins are already being used in current vaccine strategies against other Gram-negative pathogens. The primary example of T5SS protein vaccine is the acellular *B. pertussis* vaccine. For the last 50 years, efficacious whole-cell vaccines against whooping cough caused by *B. pertussis* have been available. Acellular pertussis vaccines have also been introduced that comprise from one to five proteins derived from *B. pertussis*, two of which are T5SS proteins. These proteins are the AT pertactin and the TPS protein filamentous hemagglutinin (FHA). Of these proteins, high levels of antibody to pertactin have the strongest correlation to a decreased likelihood of acquiring pertussis (Cherry et al., 1998; Hewlett and Halperin, 1998; Storsaeter et al., 1998).

A similar strategy against ETEC is currently being explored where the vaccine would be comprised of protein subunits from multiple ETEC antigens. To this end both AT and TPS proteins from ETEC strains are being tested for immunogenicity and ability to protect from infection. The ETEC TPS protein EtpA has been found to be both immunogenic and protective in a murine model (Roy et al., 2008, 2009a) while immune-proteomic studies identified not only EtpA but AT proteins TibA, EatA, and Antigen 43 suggesting that these proteins are expressed during infection both in mice and humans (Roy et al., 2010). Finally, immunization of mice with the passenger domains of two AT proteins protected against subsequent intestinal colonization by ETEC (Harris et al., 2011).With many other T5SS proteins from a variety of Gram-negative pathogens being found to be immunogenic (Turner et al., 2002, 2006; Cainelli Gebara et al., 2007; Litwin et al., 2007; Daigneault and Lo, 2009) and protective (Marr et al., 2008; Alamuri et al., 2009; Winter and Barenkamp, 2009, 2010; Chan et al., 2011) it is highly likely that more T5SS-based vaccines will be developed in the near future.

A further way that T5SS proteins are involved in vaccine delivery is exemplified by the live attenuated-vaccine strategy for *Shigella*. This application involves deleting the T5SS virulence protein IcsA (crucial for intracellular spread of *Shigella*) to create a live-attenuated vaccine strain (Venkatesan et al., 2002; Venkatesan and Ranallo, 2006).

The T5SS can also be utilized for vaccine development by using the system to display on the surface of bacterium, or secrete from the bacterium, chimeric antigens. By replacing the passenger domain with a protein of choice, the protein can be successfully secreted to the surface for display (Ruiz-Perez et al., 2002; Ruiz-Olvera et al., 2003) or into the extracellular milieu (Sevastsyanovich et al., 2012). This strategy provides a simple way to mass produce protein, or to develop live attenuated vaccines (Zhu et al., 2006; Chen and Schifferli, 2007).

CONCLUSIONS

The T1SS and T5SS of *E. coli* are relatively simple pathways for the bacterium to secrete a diverse range of virulence factors. Both pathways are capable of secreting very large proteins (>3000 aa) with a wide range of virulence functions including toxin, protease and lipase activity, adhesion to host cells and the formation of biofilms. As both pathways require only a few genes, there has been extensive horizontal gene transfer of virulence factors secreted via the T1SS and T5SS, many of which reside on pathotype-specific plasmids. This horizontal spread has allowed *E. coli* strains to quickly acquire and disperse an arsenal of virulence factors.

As many of the virulence factors secreted by these two pathways are only found in specific pathotypes of *E. coli*, they are ideal targets for vaccine development. In addition, both pathways are currently being used to secrete or display chimeric proteins and antigens for either mass production or attenuated bacterial live vaccines.

REFERENCES

Alamuri, P., Eaton, K.A., Himpsl, S.D., Smith, S.N., Mobley, H.L., 2009. Vaccination with proteus toxic agglutinin, a hemolysin-independent cytotoxin in vivo, protects against *Proteus mirabilis* urinary tract infection. Infect. Immun. 77, 632–641.

Allsopp, L.P., Totsika, M., Tree, J.J., et al., 2010. UpaH is a newly identified autotransporter protein that contributes to biofilm formation and bladder colonization by uropathogenic *Escherichia coli* CFT073. Infect. Immun. 78, 1659–1669.

Anderson, G.G., Palermo, J.J., Schilling, J.D., Roth, R., Heuser, J., Hultgren, S.J., 2003. Intracellular bacterial biofilm-like pods in urinary tract infections. Science 301, 105–107.

Aoki, S.K., Malinverni, J.C., Jacoby, K., et al., 2008. Contact-dependent growth inhibition requires the essential outer membrane protein BamA (YaeT) as the receptor and the inner membrane transport protein AcrB. Mol. Microbiol. 70, 323–340.

Aoki, S.K., Pamma, R., Hernday, A.D., Bickham, J.E., Braaten, B.A., Low, D.A., 2005. Contact-dependent inhibition of growth in *Escherichia coli*. Science 309, 1245–1248.

Artsimovitch, I., Landick, R., 2002. The transcriptional regulator RfaH stimulates RNA chain synthesis after recruitment to elongation complexes by the exposed nontemplate DNA strand. Cell 109, 193–203.

Bailey, M.J., Hughes, C., Koronakis, V., 1996. Increased distal gene transcription by the elongation factor RfaH, a specialized homologue of NusG. Mol. Microbiol. 22, 729–737.

Bailey, M.J., Hughes, C., Koronakis, V., 1997. RfaH and the ops element, components of a novel system controlling bacterial transcription elongation. Mol. Microbiol. 26, 845–851.

Bailey, M.J., Koronakis, V., Schmoll, T., Hughes, C., 1992. *Escherichia coli* HlyT protein, a transcriptional activator of haemolysin synthesis and secretion, is encoded by the rfaH (sfrB) locus required for expression of sex factor and lipopolysaccharide genes. Mol. Microbiol. 6, 1003–1012.

Bakas, L., Veiga, M.P., Soloaga, A., Ostolaza, H., Goni, F.M., 1998. Calcium-dependent conformation of *E. coli* alpha-haemolysin. Implications for the mechanism of membrane insertion and lysis. Biochim. Biophys. Acta. 1368, 225–234.

Barnard, T.J., Dautin, N., Lukacik, P., Bernstein, H.D., Buchanan, S.K., 2007. Autotransporter structure reveals intra-barrel cleavage followed by conformational changes. Nat. Struct. Mol. Biol. 14, 1214–1220.

Baud, C., Hodak, H., Willery, E., et al., 2009. Role of DegP for two-partner secretion in *Bordetella*. Mol. Microbiol. 74, 315–329.

Bauer, M.E., Welch, R.A., 1996. Characterization of an RTX toxin from enterohemorrhagic *Escherichia coli* O157:H7. Infect. Immun. 64, 167–175.

Bavro, V.N., Pietras, Z., Furnham, N., et al., 2008. Assembly and channel opening in a bacterial drug efflux machine. Mol. Cell. 30, 114–121.

Belogurov, G.A., Mooney, R.A., Svetlov, V., Landick, R., Artsimovitch, I., 2009. Functional specialization of transcription elongation factors. EMBO J. 28, 112–122.

Bennion, D., Charlson, E.S., Coon, E., Misra, R., 2010. Dissection of beta-barrel outer membrane protein assembly pathways through characterizing BamA POTRA 1 mutants of *Escherichia coli*. Mol. Microbiol. 77, 1153–1171.

Benz, I., Schmidt, M.A., 1989. Cloning and expression of an adhesin (*Aida-I*) involved in diffuse adherence of enteropathogenic *Escherichia coli*. Infect. Immun. 57, 1506–1511.

Benz, R., Schmid, A., Wagner, W., Goebel, W., 1989. Pore formation by the *Escherichia coli* hemolysin: evidence for an association-dissociation equilibrium of the pore-forming aggregates. Infect. Immun. 57, 887–895.

Betancourt-Sanchez, M., Navarro-Garcia, F., 2009. Pet secretion, internalization and induction of cell death during infection of epithelial cells by enteroaggregative *Escherichia coli*. Microbiology 155, 2895–2906.

Bhakdi, S., Greulich, S., Muhly, M., et al., 1989. Potent leukocidal action of *Escherichia coli* hemolysin mediated by permeabilization of target cell membranes. J. Exp. Med. 169, 737–754.

Bhakdi, S., Mackman, N., Nicaud, J.M., Holland, I.B., 1986. *Escherichia coli* hemolysin may damage target cell membranes by generating transmembrane pores. Infect. Immun. 52, 63–69.

Bhakdi, S., Martin, E., 1991. Superoxide generation by human neutrophils induced by low doses of *Escherichia coli* hemolysin. Infect. Immun. 59, 2955–2962.

Bhakdi, S., Muhly, M., Korom, S., Schmidt, G., 1990. Effects of *Escherichia coli* hemolysin on human monocytes. Cytocidal action and stimulation of interleukin 1 release. J. Clin. Invest. 85, 1746–1753.

Blight, M.A., Holland, I.B., 1994. Heterologous protein secretion and the versatile *Escherichia coli* haemolysin translocator. Trends Biotechnol. 12, 450–455.

Bodelon, G., Marin, E., Fernandez, L.A., 2009. Role of periplasmic chaperones and BamA (YaeT/Omp85) in folding and secretion of intimin from enteropathogenic *Escherichia coli* strains. J. Bacteriol. 191, 5169–5179.

Brunder, W., Schmidt, H., Karch, H., 1997. EspP, a novel extracellular serine protease of enterohaemorrhagic *Escherichia coli* O157:H7 cleaves human coagulation factor V. Mol. Microbiol. 24, 767–778.

Cainelli Gebara, V.C., Risoleo, L., Lopes, A.P., et al., 2007. Adjuvant and immunogenic activities of the 73kDa N-terminal alpha-domain of BrkA autotransporter and Cpn60/60kDa chaperonin of *Bordetella pertussis*. Vaccine 25, 621–629.

Catic, A., Dietrich, G., Gentschev, I., Goebel, W., Kaufmann, S.H., Hess, J., 1999. Introduction of protein or DNA delivered via recombinant *Salmonella typhimurium* into the major histocompatibility complex class I presentation pathway of macrophages. Microbes. Infect. 1, 113–121.

Chan, Y.G., Riley, S.P., Chen, E., Martinez, J.J., 2011. Molecular basis of immunity to rickettsial infection conferred through outer membrane protein B. Infect. Immun. 79, 2303–2313.

Charbonneau, M.E., Janvore, J., Mourez, M., 2009. Autoprocessing of the *Escherichia coli* AIDA-I autotransporter: a new mechanism involving acidic residues in the junction region. J. Biol. Chem. 284, 17340–17351.

Chaudhuri, R.R., Sebaihia, M., Hobman, J.L., et al., 2010. Complete genome sequence and comparative metabolic profiling of the prototypical enteroaggregative *Escherichia coli* strain 042. PLoS One 5, e8801.

Chen, H., Schifferli, D.M., 2007. Comparison of a fimbrial versus an autotransporter display system for viral epitopes on an attenuated *Salmonella* vaccine vector. Vaccine 25, 1626–1633.

Cherry, J.D., Gornbein, J., Heininger, U., Stehr, K., 1998. A search for serologic correlates of immunity to *Bordetella pertussis* cough illnesses. Vaccine 16, 1901–1906.

Choi, P.S., Dawson, A.J., Bernstein, H.D., 2007. Characterization of a novel two-partner secretion system in *Escherichia coli* O157:H7. J. Bacteriol. 189, 3452–3461.

Clantin, B., Hodak, H., Willery, E., Locht, C., Jacob-Dubuisson, F., Villeret, V., 2004. The crystal structure of filamentous hemagglutinin secretion domain and its implications for the two-partner secretion pathway. Proc. Natl. Acad. Sci. USA 101, 6194–6199.

Cotter, S.E., Surana, N.K., Grass, S., St. Geme III, J.W., 2006. Trimeric autotransporters require trimerization of the passenger domain for stability and adhesive activity. J. Bacteriol. 188, 5400–5407.

Cotter, S.E., Surana, N.K., St Geme, J.W., 2005. Trimeric autotransporters: a distinct subfamily of autotransporter proteins. Trends Microbiol. 13, 199–205.

Daigneault, M.C., Lo, R.Y., 2009. Analysis of a collagen-binding trimeric autotransporter adhesin from *Mannheimia haemolytica* A1. FEMS Microbiol. Lett. 300, 242–248.

Danese, P.N., Pratt, L.A., Dove, S.L., Kolter, R., 2000. The outer membrane protein, antigen 43, mediates cell-to-cell interactions within *Escherichia coli* biofilms. Mol. Microbiol. 37, 424–432.

Delattre, A.S., Saint, N., Clantin, B., et al., 2011. Substrate recognition by the POTRA domains of TpsB transporter FhaC. Mol. Microbiol. 81, 99–112.

Delepelaire, P., 2004. Type 1 secretion in gram-negative bacteria. Biochim. Biophys. Acta 1694, 149–161.

Dinh, T., Paulsen, I.T., Saier Jr., M.H., 1994. A family of extracytoplasmic proteins that allow transport of large molecules across the outer membranes of gram-negative bacteria. J. Bacteriol. 176, 3825–3831.

Djafari, S., Ebel, F., Deibel, C., Kramer, S., Hudel, M., Chakraborty, T., 1997. Characterization of an exported protease from Shiga toxin-producing *Escherichia coli*. Mol. Microbiol. 25, 771–784.

Durant, L., Metais, A., Soulama-Mouze, C., Genevard, J.M., Nassif, X., Escaich, S., 2007. Identification of candidates for a subunit vaccine against extraintestinal pathogenic *Escherichia coli*. Infect. Immun. 75, 1916–1925.

Dutta, P.R., Sui, B.Q., Nataro, J.P., 2003. Structure-function analysis of the enteroaggregative *Escherichia coli* plasmid-encoded toxin autotransporter using scanning linker mutagenesis. J. Biol. Chem. 278, 39912–39920.

Easton, D.M., Totsika, M., Allsopp, L.P., et al., 2011. Characterization of EhaJ, a new autotransporter protein from enterohemorrhagic and enteropathogenic *Escherichia coli*. Front. Microbiol. 2, 120.

Emsley, P., Charles, I.G., Fairweather, N.F., Isaacs, N.W., 1996. Structure of *Bordetella pertussis* virulence factor P.69 pertactin. Nature 381, 90–92.

Emsley, P., McDermott, G., Charles, I.G., Fairweather, N.F., Isaacs, N.W., 1994. Crystallographic characterization of Pertactin, a membrane-associated protein from *Bordetella pertussis*. J. Mol. Biol. 235, 772–773.

Eslava, C., Navarro-Garcia, F., Czeczulin, J.R., Henderson, I.R., Cravioto, A., Nataro, J.P., 1998. Pet, an autotransporter enterotoxin from enteroaggregative *Escherichia coli*. Infect. Immun. 66, 3155–3163.

Fairman, J.W., Dautin, N., Wojtowicz, D., et al., 2012. Crystal structures of the outer membrane domain of Intimin and Invasin from enterohemorrhagic *E. coli* and enteropathogenic. *Y. pseudotuberculosis*. Structure 20, 1233–1243.

Felmlee, T., Pellett, S., Welch, R.A., 1985. Nucleotide sequence of an *Escherichia coli* chromosomal hemolysin. J. Bacteriol. 163, 94–105.

Fernandez, R.C., Weiss, A.A., 1994. Cloning and sequencing of a *Bordetella pertussis* serum resistance locus. Infect. Immun. 62, 4727–4738.

Fleckenstein, J.M., Roy, K., Fischer, J.F., Burkitt, M., 2006. Identification of a two-partner secretion locus of enterotoxigenic *Escherichia coli*. Infect. Immun. 74, 2245–2258.

Gadeberg, O.V., Orskov, I., 1984. In vitro cytotoxic effect of alpha-hemolytic *Escherichia coli* on human blood granulocytes. Infect. Immun. 45, 255–260.

Gentschev, I., Mollenkopf, H., Sokolovic, Z., Hess, J., Kaufmann, S.H., Goebel, W., 1996. Development of antigen-delivery systems, based on the *Escherichia coli* hemolysin secretion pathway. Gene 179, 133–140.

Ghigo, J.M., Wandersman, C., 1994. A carboxyl-terminal four-amino acid motif is required for secretion of the metalloprotease PrtG through the *Erwinia chrysanthemi* protease secretion pathway. J. Biol. Chem. 269, 8979–8985.

Goebel, W., Hedgpeth, J., 1982. Cloning and functional characterization of the plasmid-encoded hemolysin determinant of *Escherichia coli*. J. Bacteriol. 151, 1290–1298.

Guedin, S., Willery, E., Tommassen, J., et al., 2000. Novel topological features of FhaC, the outer membrane transporter involved in the secretion of the *Bordetella pertussis* filamentous hemagglutinin. J. Biol. Chem. 275, 30202–30210.

Guyer, D.M., Henderson, I.R., Nataro, J.P., Mobley, H.L.T., 2000. Identification of Sat, an autotransporter toxin produced by uropathogenic *Escherichia coli*. Mol. Microbiol. 38, 53–66.

Guyer, D.M., Radulovic, S., Jones, F.E., Mobley, H.L.T., 2002. Sat, the secreted autotransporter toxin of uropathogenic *Escherichia coli*, is a vacuolating cytotoxin for bladder and kidney epithelial cells. Infect. Immun. 70, 4539–4546.

Hahn, H.P., Hess, C., Gabelsberger, J., Domdey, H., von Specht, B.U., 1998. A Salmonella typhimurium strain genetically engineered to secrete effectively a bioactive human interleukin (hIL)-6 via the *Escherichia coli* hemolysin secretion apparatus. FEMS Immunol. Med. Microbiol. 20, 111–119.

Harris, J.A., Roy, K., Woo-Rasberry, V., et al., 2011. Directed evaluation of enterotoxigenic *Escherichia coli* autotransporter proteins as putative vaccine candidates. PLoS Negl. Trop. Dis. 5, e1428.

Hasman, H., Chakraborty, T., Klemm, P., 1999. Antigen-43-mediated autoaggregation of *Escherichia coli* is blocked by fimbriation. J. Bacteriol. 181, 4834–4841.

Hasman, H., Schembri, M.A., Klemm, P., 2000. Antigen 43 and type 1 fimbriae determine colony morphology of *Escherichia coli* K-12. J. Bacteriol. 182, 1089–1095.

Hegde, R.S., Bernstein, H.D., 2006. The surprising complexity of signal sequences. Trends Biochem. Sci. 31, 563–571.

Henderson, I.R., Cappello, R., Nataro, J.P., 2000. Autotransporter proteins, evolution and redefining protein secretion. Trends Microbiol. 8, 529–532.

Henderson, I.R., Hicks, S., Navarro-Garcia, F., Elias, W.P., Philips, A.D., Nataro, J.P., 1999. Involvement of the enteroaggregative *Escherichia coli* plasmid-encoded toxin in causing human intestinal damage. Infect. Immun. 67, 5338–5344.

Henderson, I.R., Meehan, M., Owen, P., 1997. Antigen 43, a phase-variable bipartite outer membrane protein, determines colony morphology and autoaggregation in *Escherichia coli* K-12. FEMS Microbiol. Lett. 149, 115–120.

Henderson, I.R., Nataro, J.P., 2001. Virulence functions of autotransporter proteins. Infect. Immun. 69, 1231–1243.

Henderson, I.R., Navarro-Garcia, F., Desvaux, M., Fernandez, R.C., Ala'Aldeen, D., 2004. Type 5 protein secretion pathway: the autotransporter story. Microbiol. Mol. Biol. Rev. 68, 692–744.

Henderson, I.R., Navarro-Garcia, F., Nataro, J.P., 1998. The great escape: structure and function of the autotransporter proteins. Trends Microbiol. 6, 370–378.

Henderson, I.R., Owen, P., 1999. The major phase-variable outer membrane protein of *Escherichia coli* structurally resembles the immunoglobulin A1 protease class of exported protein and is regulated by a novel mechanism involving Dam and OxyR. J. Bacteriol. 181, 2132–2141.

Hess, J., Gentschev, I., Miko, D., et al., 1996. Superior efficacy of secreted over somatic antigen display in recombinant *Salmonella* vaccine induced protection against listeriosis. Proc. Natl. Acad. Sci. USA 93, 1458–1463.

Hess, J.L., Pyper, J.M., Clements, J.E., 1986. Nucleotide sequence and transcriptional activity of the caprine arthritis-encephalitis virus long terminal repeat. J. Virol. 60, 385–393.

Hewlett, E.L., Halperin, S.A., 1998. Serological correlates of immunity to *Bordetella pertussis*. Vaccine 16, 1899–1900.

Ieva, R., Bernstein, H.D., 2009. Interaction of an autotransporter passenger domain with BamA during its translocation across the bacterial outer membrane. Proc. Natl. Acad. Sci. 106, 19120–19125.

Iguchi, A., Thomson, N.R., Ogura, Y., et al., 2009. Complete genome sequence and comparative genome analysis of enteropathogenic *Escherichia coli* O127:H6 strain E2348/69. J. Bacteriol. 191, 347–354.

Issartel, J.P., Koronakis, V., Hughes, C., 1991. Activation of *Escherichia coli* prohaemolysin to the mature toxin by acyl carrier protein-dependent fatty acylation. Nature 351, 759–761.

Jacob-Dubuisson, F., Fernandez, R., Coutte, L., 2004. Protein secretion through autotransporter and two-partner pathways. Biochim. Biophys. Acta 1694, 235–257.

Jacob-Dubuisson, F., Locht, C., Antoine, R., 2001. Two-partner secretion in Gram-negative bacteria: a thrifty, specific pathway for large virulence proteins. Mol. Microbiol. 40, 306–313.

Jacob-Dubuisson, F., Villeret, V., Clantin, B., Delattre, A.S., Saint, N., 2009. First structural insights into the TpsB/Omp85 superfamily. Biol. Chem. 390, 675–684.

Janganan, T.K., Zhang, L., Bavro, V.N., et al., 2011. Opening of the outer membrane protein channel in tripartite efflux pumps is induced by interaction with the membrane fusion partner. J. Biol. Chem. 286, 5484–5493.

Jenkins, J., Pickersgill, R., 2001. The architecture of parallel beta-helices and related folds. Prog. Biophys. Mol. Biol. 77, 111–175.

Jerse, A.E., Yu, J., Tall, B.D., Kaper, J.B., 1990. A genetic-locus of enteropathogenic *Escherichia coli* necessary for the production of attaching and effacing lesions on tissue-culture cells. Proc. Natl. Acad. Sci. USA 87, 7839–7843.

Jose, J., Jahnig, F., Meyer, T.F., 1995. Common structural features of Iga1 protease-like outer-membrane protein autotransporters. Mol. Microbiol. 18, 378–380.

Junker, M., Schuster, C.C., McDonnell, A.V., et al., 2006. Pertactin beta-helix folding mechanism suggests common themes for the secretion and folding of autotransporter proteins. PNAS 103, 4918–4923.

Kajava, A.V., Steven, A.C., 2006. The turn of the screw: variations of the abundant beta-solenoid motif in passenger domains of type 5 secretory proteins. J. Struct. Biol. 155, 306–315.

Keane, W.F., Welch, R., Gekker, G., Peterson, P.K., 1987. Mechanism of *Escherichia coli* alpha-hemolysin-induced injury to isolated renal tubular cells. Am. J. Pathol. 126, 350–357.

Kim, S., Malinverni, J.C., Sliz, P., Silhavy, T.J., Harrison, S.C., Kahne, D., 2007. Structure and function of an essential component of the outer membrane protein assembly machine. Science 317, 961–964.

Kjaergaard, K., Schembri, M.A., Hasman, H., Klemm, P., 2000a. Antigen 43 from *Escherichia coli* induces inter- and intraspecies cell aggregation and changes in colony morphology of *Pseudomonas fluorescens*. J. Bacteriol. 182, 4789–4796.

Kjaergaard, K., Schembri, M.A., Ramos, C., Molin, S., Klemm, P., 2000b. Antigen 43 facilitates formation of multispecies biofilms. Environ. Microbiol. 2, 695–702.

Klemm, P., Hjerrild, L., Gjermansen, M., Schembri, M.A., 2004. Structure-function analysis of the self-recognizing Antigen 43 autotransporter protein from *Escherichia coli*. Mol. Microbiol. 51, 283–296.

Knowles, T.J., Jeeves, M., Bobat, S., et al., 2008. Fold and function of polypeptide transport-associated domains responsible for delivering unfolded proteins to membranes. Mol. Microbiol. 68, 1216–1227.

Knowles, T.J., Scott-Tucker, A., Overduin, M., Henderson, I.R., 2009. Membrane protein architects: the role of the BAM complex in outer membrane protein assembly. Nat. Rev. Microbiol. 7, 206–214.

Koronakis, E., Hughes, C., Milisav, I., Koronakis, V., 1995. Protein exporter function and in vitro ATPase activity are correlated in ABC-domain mutants of HlyB. Mol. Microbiol. 16, 87–96.

Koronakis, V., Hughes, C., 1996. Synthesis, maturation and export of the *E. coli* hemolysin. Med. Microbiol. Immunol. 185, 65–71.

Koronakis, V., Koronakis, E., Hughes, C., 1989. Isolation and analysis of the C-terminal signal directing export of *Escherichia coli* hemolysin protein across both bacterial membranes. EMBO J. 8, 595–605.

Koronakis, V., Sharff, A., Koronakis, E., Luisi, B., Hughes, C., 2000. Crystal structure of the bacterial membrane protein TolC central to multidrug efflux and protein export. Nature 405, 914–919.

Lazar, S.W., Kolter, R., 1996. SurA assists the folding of *Escherichia coli* outer membrane proteins. J. Bacteriol. 178, 1770–1773.

Leeds, J.A., Welch, R.A., 1997. Enhancing transcription through the *Escherichia coli* hemolysin operon, hlyCABD: RfaH and upstream JUMPStart DNA sequences function together via a postinitiation mechanism. J. Bacteriol. 179, 3519–3527.

Lehr, U., Schütz, M., Oberhettinger, P., et al., 2010. C-terminal amino acid residues of the trimeric autotransporter adhesin YadA of *Yersinia enterocolitica* are decisive for its recognition and assembly by BamA. Mol. Microbiol. no-no.

Leo, J.C., Lyskowski, A., Hattula, K., et al., 2011. The structure of *E. coli* IgG-binding protein D suggests a general model for bending and binding in trimeric autotransporter adhesins. Structure 19, 1021–1030.

Letoffe, S., Delepelaire, P., Wandersman, C., 1996. Protein secretion in gram-negative bacteria: assembly of the three components of ABC protein-mediated exporters is ordered and promoted by substrate binding. EMBO J. 15, 5804–5811.

Leyton, D.L., Adams, L.M., Kelly, M., et al., 2007. Contribution of a novel gene, rpeA, encoding a putative autotransporter adhesin to intestinal colonization by rabbit-specific enteropathogenic *Escherichia coli*. Infect. Immun. 75, 4664–4669.

Leyton, D.L., Sloan, J., Hill, R.E., Doughty, S., Hartland, E.L., 2003. Transfer region of pO113 from enterohemorrhagic *Escherichia coli*: similarity with R64 and identification of a novel plasmid-encoded autotransporter, EpeA. Infect. Immun. 71, 6307–6319.

Leyton, D.L., De Luna, M., d. G., Sevastsyanovich, Y.R., et al., 2010. The unusual extended signal peptide region is not required for secretion and function of an *Escherichia coli* autotransporter. FEMS Microbiol. Lett. 2, 133–139.

Leyton, D.L., Rossiter, A.E., Henderson, I.R., 2012. From self sufficiency to dependence: mechanisms and factors important for autotransporter biogenesis. Nat. Rev. Microbiol. 10, 213–225.

Li, G., Feng, Y., Kariyawasam, S., et al., 2009. AatA is a novel autotransporter and virulence factor of avian pathogenic *Escherichia coli*. Infect. Immun. IAI.00513–00509.

Lim, H.N., van Oudenaarden, A., 2007. A multistep epigenetic switch enables the stable inheritance of DNA methylation states. Nat. Genet. 39, 269–275.

Lindenthal, C., Elsinghorst, E.A., 2001. Enterotoxigenic *Escherichia coli* TibA glycoprotein adheres to human intestine epithelial cells. Infect. Immun. 69, 52–57.

Linggood, M.A., Ingram, P.L., 1982. The role of alpha haemolysin in the virulence of *Escherichia coli* for mice. J. Med. Microbiol. 15, 23–30.

Linhartova, I., Bumba, L., Masin, J., et al., 2010. RTX proteins: a highly diverse family secreted by a common mechanism. FEMS Microbiol. Rev. 34, 1076–1112.

Linke, D., Riess, T., Autenrieth, I.B., Lupas, A., Kempf, V.A.J., 2006. Trimeric autotransporter adhesins: variable structure, common function. Trends Microbiol. 14, 264–270.

Litwin, C.M., Rawlins, M.L., Swenson, E.M., 2007. Characterization of an immunogenic outer membrane autotransporter protein, Arp, of Bartonella henselae. Infect. Immun. 75, 5255–5263.

Lloyd, A.L., Henderson, T.A., Vigil, P.D., Mobley, H.L., 2009. Genomic islands of uropathogenic *Escherichia coli* contribute to virulence. J. Bacteriol. 191, 3469–3481.

Loveless, B.J., Saier, M.H., 1997. A novel family of channel-forming, autotransporting, bacterial virulence factors. Mol. Membr. Biol. 14, 113–123.

Luo, Y., Frey, E.A., Pfuetzner, R.A., et al., 2000. Crystal structure of enteropathogenic *Escherichia coli* intimin-receptor complex. Nature 405, 1073–1077.

Marr, N., Oliver, D.C., Laurent, V., Poolman, J., Denoel, P., Fernandez, R.C., 2008. Protective activity of the *Bordetella pertussis* BrkA autotransporter in the murine lung colonization model. Vaccine 26, 4306–4311.

Mazar, J., Cotter, P.A., 2007. New insight into the molecular mechanisms of two-partner secretion. Trends Microbiol. 15, 508–515.

Mellies, J.L., Navarro-Garcia, F., Okeke, I., Frederickson, J., Nataro, J.P., Kaper, J.B., 2001. espC pathogenicity island of enteropathogenic *Escherichia coli* encodes an enterotoxin. Infect. Immun. 69, 315–324.

Menestrina, G., Ropele, M., Dalla Serra, M., et al., 1995. Binding of antibodies to functional epitopes on the pore formed by *Escherichia coli* hemolysin in cells and model membranes. Biochim. Biophys. Acta. 1238, 72–80.

Mobley, H.L., Green, D.M., Trifillis, A.L., et al., 1990. Pyelonephritogenic *Escherichia coli* and killing of cultured human renal proximal tubular epithelial cells: role of hemolysin in some strains. Infect. Immun. 58, 1281–1289.

Muller, D., Hughes, C., Goebel, W., 1983. Relationship between plasmid and chromosomal hemolysin determinants of *Escherichia coli*. J. Bacteriol. 153, 846–851.

Nieto, J.M., Bailey, M.J., Hughes, C., Koronakis, V., 1996. Suppression of transcription polarity in the *Escherichia coli* haemolysin operon by a short upstream element shared by polysaccharide and DNA transfer determinants. Mol. Microbiol. 19, 705–713.

Nummelin, H., Merckel, M.C., Leo, J.C., Lankinen, H., Skurnik, M., Goldman, A., 2004. The Yersinia adhesin YadA collagen-binding domain structure is a novel left-handed parallel beta-roll. Embo. J. 23, 701–711.

Oliver, D.C., Huang, G., Fernandez, R.C., 2003. Identification of secretion determinants of the *Bordetella pertussis* BrkA autotransporter. J. Bacteriol. 185, 489–495.

Orr, N., Galen, J.E., Levine, M.M., 1999. Expression and immunogenicity of a mutant diphtheria toxin molecule, CRM(197), and its fragments in *Salmonella typhi* vaccine strain CVD 908-htrA. Infect. Immun. 67, 4290–4294.

Otto, B.R., van Dooren, S.J.M., Nuijens, J.H., Luirink, J., Oudega, B., 1998. Characterization of a hemoglobin protease secreted by the pathogenic *Escherichia coli* strain EB1. J. Exp. Med. 188, 1091–1103.

Parham, N.J., Pollard, S.J., Chaudhuri, R.R., et al., 2005. Prevalence of pathogenicity island IICFT073 genes among extraintestinal clinical isolates of *Escherichia coli*. J. Clin. Microbiol. 43, 2425–2434.

Parham, N.J., Srinivasan, U., Desvaux, M., Foxman, B., Marrs, C.F., Henderson, I.R., 2004. PicU, a second serine protease autotransporter of uropathogenic *Escherichia coli*. FEMS Microbiol. Lett. 230, 73–83.

Patel, S.K., Dotson, J., Allen, K.P., Fleckenstein, J.M., 2004. Identification and molecular characterization of EatA, an autotransporter protein of enterotoxigenic *Escherichia coli*. Infect. Immun. 72, 1786–1794.

Paton, A.W., Srimanote, P., Woodrow, M.C., Paton, J.C., 2001. Characterization of Saa, a novel autoagglutinating adhesin produced by locus of enterocyte effacement-negative Shiga-toxigenic *Escherichia coli* strains that are virulent for humans. Infect. Immun. 69, 6999–7009.

Provence, D.L., Curtiss, R., 1994. Isolation and characterization of a gene involved in hemagglutination by an avian pathogenic *Escherichia coli* strain. Infect. Immun. 62, 1369–1380.

Raghunathan, D., Wells, T.J., Morris, F.C., et al., 2011. SadA, a trimeric autotransporter from *Salmonella enterica* serovar Typhimurium, can promote biofilm formation and provides limited protection against infection. Infect. Immun. 79, 4342–4352.

Reisner, A., Haagensen, J.A., Schembri, M.A., Zechner, E.L., Molin, S., 2003. Development and maturation of *Escherichia coli* K-12 biofilms. Mol. Microbiol. 48, 933–946.

Roggenkamp, A., Ackermann, N., Jacobi, C.A., Truelzsch, K., Hoffmann, H., Heesemann, H., 2003. Molecular analysis of transport and oligomerization of the *Yersinia enterocolitica* adhesin YadA. J. Bacteriol. 185, 3735–3744.

Rossiter, A.E., Browning, D.F., Leyton, D.L., et al., 2011a. Transcription of the plasmid-encoded toxin gene from enteroaggregative *Escherichia coli* is regulated by a novel co-activation mechanism involving CRP and Fis. Mol. Microbiol. 81, 179–191.

Rossiter, A.E., Leyton, D.L., Tveen-Jensen, K., et al., 2011b. The essential beta-barrel assembly machinery complex components BamD and BamA are required for autotransporter biogenesis. J. Bacteriol. 193, 4250–4253.

Roux, A., Beloin, C., Ghigo, J.M., 2005. Combined inactivation and expression strategy to study gene function under physiological conditions: application to identification of new *Escherichia coli* adhesins. J. Bacteriol. 187, 1001–1013.

Roy, K., Bartels, S., Qadri, F., Fleckenstein, J.M., 2010. Enterotoxigenic *Escherichia coli* elicits immune responses to multiple surface proteins. Infect. Immun. 78, 3027–3035.

Roy, K., Hamilton, D., Allen, K.P., Randolph, M.P., Fleckenstein, J.M., 2008. The EtpA exoprotein of enterotoxigenic *Escherichia coli* promotes intestinal colonization and is a protective antigen in an experimental model of murine infection. Infect. Immun. 76, 2106–2112.

Roy, K., Hamilton, D., Ostmann, M.M., Fleckenstein, J.M., 2009a. Vaccination with EtpA glycoprotein or flagellin protects against colonization with enterotoxigenic *Escherichia coli* in a murine model. Vaccine 27, 4601–4608.

Roy, K., Hilliard, G.M., Hamilton, D.J., Luo, J., Ostmann, M.M., Fleckenstein, J.M., 2009b. Enterotoxigenic *Escherichia coli* EtpA mediates adhesion between flagella and host cells. Nature 457, 594–598.

Ruiz-Olvera, P., Ruiz-Perez, F., Sepulveda, N.V., et al., 2003. Display and release of the *Plasmodium falciparum* circumsporozoite protein using the autotransporter MisL of Salmonella enterica. Plasmid. 50, 12–27.

Ruiz-Perez, F., Henderson, I.R., Leyton, D.L., Rossiter, A.E., Zhang, Y., Nataro, J.P., 2009. Roles of periplasmic chaperone proteins in the biogenesis of serine protease autotransporters of Enterobacteriaceae. J. Bacteriol. 191, 6571–6583.

Ruiz-Perez, F., Henderson, I.R., Nataro, J.P., 2010. Interaction of FkpA, a peptidyl-prolyl cis/trans isomerase with EspP autotransporter protein. Gut. Microbes. 1, 339–344.

Ruiz-Perez, F., Leon-Kempis, R., Santiago-Machuca, A., et al., 2002. Expression of the *Plasmodium falciparum* immunodominant epitope (NANP)(4) on the surface of *Salmonella enterica* using the autotransporter MisL. Infect. Immun. 70, 3611–3620.

Ryan, E.T., Butterton, J.R., Smith, R.N., Caroll, P.A., Crean, T.I., Calderwood, S.B., 1997. Protective immunity against *Clostridium difficile* toxin A induced by oral immunization with a live, attenuated *Vibrio cholerae* vector strain. Infect. Immun. 65, 2941–2949.

Sahl, J.W., Steinsland, H., Redman, J.C., et al., 2011. A comparative genomic analysis of diverse clonal types of enterotoxigenic *Escherichia coli* reveals pathovar-specific conservation. Infect. Immun. 79, 950–960.

Salacha, R., Kovacic, F., Brochier-Armanet, C., et al., 2010. The *Pseudomonas aeruginosa* patatin-like protein PlpD is the archetype of a novel Type 5 secretion system. Environ. Microbiol. 12, 1498–1512.

Sandt, C.H., Hill, C.W., 2000. Four different genes responsible for nonimmune immunoglobulin-binding activities within a single strain of *Escherichia coli*. Infect. Immun. 68, 2205–2214.

Sandt, C.H., Hill, C.W., 2001. Nonimmune binding of human immunoglobulin A (IgA) and IgG Fc by distinct sequence segments of the EibF cell surface protein of *Escherichia coli*. Infect. Immun. 69, 7293–7303.

Schembri, M.A., Hjerrild, L., Gjermansen, M., Klemm, P., 2003a. Differential expression of the *Escherichia coli* autoaggregation factor antigen 43. J. Bacteriol. 185, 2236–2242.

Schembri, M.A., Kjaergaard, K., Klemm, P., 2003b. Global gene expression in *Escherichia coli* biofilms. Mol. Microbiol. 48, 253–267.

Schulein, R., Gentschev, I., Mollenkopf, H.J., Goebel, W., 1992. A topological model for the haemolysin translocator protein HlyD. Mol. Gen. Genet 234, 155–163.

Selkrig, J., Mosbahi, K., Webb, C.T., et al., 2012. Discovery of an archetypal protein transport system in bacterial outer membranes. Nat. Struct. Mol. Biol.

Sevastsyanovich, Y.R., Leyton, D.L., Wells, T.J., et al., 2012. A generalised module for the selective extracellular accumulation of recombinant proteins. Microb. Cell Fact. 11, 69.

Sheikh, J., Czeczulin, J.R., Harrington, S., et al., 2002. A novel dispersin protein in enteroaggregative *Escherichia coli*. J. Clin. Invest. 110, 1329–1337.

Shere, K.D., Sallustio, S., Manessis, A., D'Aversa, T.G., Goldberg, M.B., 1997. Disruption of IcsP, the major *Shigella* protease that cleaves IcsA, accelerates actin-based motility. Mol. Microbiol. 25, 451–462.

Sherlock, O., Munk-Vejborg, R., Klemm, P., 2005a. Self-associating autotransporters, Ag43, AIDA, and TibA: functional and structural similarities. Microbiology In press.

Sherlock, O., Schembri, M.A., Reisner, A., Klemm, P., 2004. Novel roles for the AIDA adhesin from diarrheagenic *Escherichia coli*: cell aggregation and biofilm formation. J. Bacteriol. 186, 8058–8065.

Sherlock, O., Vejborg, R.M., Klemm, P., 2005b. The TibA adhesin/invasin from enterotoxigenic *Escherichia coli* is self recognizing and induces bacterial aggregation and biofilm formation. Infect. Immun. 73, 1954–1963.

Sklar, J.G., Wu, T., Kahne, D., Silhavy, T.J., 2007. Defining the roles of the periplasmic chaperones SurA, Skp, and DegP in *Escherichia coli*. Genes Dev. 21, 2473–2484.

Smith, Y.C., Grande, K.K., Rasmussen, S.B., O'Brien, A.D., 2006. Novel three-dimensional organoid model for evaluation of the interaction of uropathogenic *Escherichia coli* with terminally differentiated human urothelial cells. Infect. Immun. 74, 750–757.

Smith, Y.C., Rasmussen, S.B., Grande, K.K., Conran, R.M., O'Brien, A.D., 2008. Hemolysin of uropathogenic *Escherichia coli* evokes extensive shedding of the uroepithelium and hemorrhage in bladder tissue within the first 24 hours after intraurethral inoculation of mice. Infect. Immun. 76, 2978–2990.

Soprova, Z., Sauri, A., van Ulsen, P., et al., 2010. A conserved aromatic residue in the autochaperone domain of the autotransporter Hbp is critical for initiation of outer membrane translocation. J. Biol. Chem. 285, 38224–38233.

Spreng, S., Gentschev, I., Goebel, W., Weidinger, G., ter Meulen, V., Niewiesk, S., 2000. *Salmonella* vaccines secreting measles virus epitopes induce protective immune responses against measles virus encephalitis. Microbes. Infect. 2, 1687–1692.

Stanley, P., Koronakis, V., Hughes, C., 1998. Acylation of *Escherichia coli* hemolysin: a unique protein lipidation mechanism underlying toxin function. Microbiol. Mol. Biol. Rev. 62, 309–333.

Stein, M., Kenny, B., Stein, M.A., Finlay, B.B., 1996. Characterization of EspC, a 110-kilodalton protein secreted by enteropathogenic *Escherichia coli* which is homologous to members of the immunoglobulin A protease-like family of secreted proteins. J. Bacteriol. 178, 6546–6554.

Storsaeter, J., Hallander, H.O., Gustafsson, L., Olin, P., 1998. Levels of anti-pertussis antibodies related to protection after household exposure to *Bordetella pertussis*. Vaccine 16, 1907–1916.

Struyve, M., Moons, M., Tommassen, J., 1991. Carboxy-terminal phenylalanine is essential for the correct assembly of a bacterial outer-membrane protein. J. Mol. Biol. 218, 141–148.

Su, G.F., Brahmbhatt, H.N., de Lorenzo, V., Wehland, J., Timmis, K.N., 1992. Extracellular export of Shiga toxin B-subunit/haemolysin A (C-terminus) fusion protein expressed in *Salmonella typhimurium* aroA-mutant and stimulation of B-subunit specific antibody responses in mice. Microb. Pathog. 13, 465–476.

Surana, N.K., Cutter, D., Barenkamp, S.J., St Geme, J.W., 2004. The *Haemophilus influenzae* Hia autotransporter contains an unusually short trimeric translocator domain. J. Biol. Chem. 279, 14679–14685.

Suttorp, N., Floer, B., Schnittler, H., Seeger, W., Bhakdi, S., 1990. Effects of *Escherichia coli* hemolysin on endothelial cell function. Infect. Immun. 58, 3796–3801.

Suttorp, N., Fuhrmann, M., Tannert-Otto, S., Grimminger, F., Bhadki, S., 1993. Pore-forming bacterial toxins potently induce release of nitric oxide in porcine endothelial cells. J. Exp. Med. 178, 337–341.

Tajima, N., Kawai, F., Park, S.Y., Tame, J.R., 2010. A novel intein-like autoproteolytic mechanism in autotransporter proteins. J. Mol. Biol. 402, 645–656.

Thanabalu, T., Koronakis, E., Hughes, C., Koronakis, V., 1998. Substrate-induced assembly of a contiguous channel for protein export from *E. coli*: reversible bridging of an inner-membrane translocase to an outer membrane exit pore. EMBO J. 17, 6487–6496.

Torres, A.G., Perna, N.T., Burland, V., Ruknudin, A., Blattner, F.R., Kaper, J.B., 2002. Characterization of Cah, a calcium-binding and heat-extractable autotransporter protein of enterohaemorrhagic *Escherichia coli*. Mol. Microbiol. 45, 951–966.

Totsika, M., Wells, T.J., Beloin, C., et al., 2012. Molecular characterisation of the EhaG and UpaG trimeric autotransporter proteins from pathogenic *Escherichia coli*. Appl. Environ. Microbiol. 78, 2179–2189.

Tsai, J.C., Yen, M.R., Castillo, R., Leyton, D.L., Henderson, I.R., Saier Jr., M.H., 2010. The bacterial intimins and invasins: a large and novel family of secreted proteins. PLoS One 5, e14403.

Turner, D.P., Wooldridge, K.G., Ala'Aldeen, D.A., 2002. Autotransported serine protease A of *Neisseria meningitidis*: an immunogenic, surface-exposed outer membrane, and secreted protein. Infect. Immun. 70, 4447–4461.

Turner, D.P., Marietou, A.G., Johnston, L., et al., 2006. Characterization of MspA, an immunogenic autotransporter protein that mediates adhesion to epithelial and endothelial cells in *Neisseria meningitidis*. Infect. Immun. 74, 2957–2964.

Tzschaschel, B.D., Guzman, C.A., Timmis, K.N., de Lorenzo, V., 1996. An *Escherichia coli* hemolysin transport system-based vector for the export of polypeptides: export of Shiga-like toxin IIeB subunit by *Salmonella typhimurium* aroA. Nat. Biotechnol. 14, 765–769.

Ulett, G.C., Webb, R.I., Schembri, M.A., 2006. Antigen-43-mediated autoaggregation impairs motility in *Escherichia coli*. Microbiology 152, 2101–2110.

Valle, J., Mabbett, A.N., Ulett, G.C., et al., 2008a. UpaG, a new member of the trimeric autotransporter family of adhesins in uropathogenic *Escherichia coli*. J. Bacteriol. JB.00122–00108.

Valle, J., Mabbett, A.N., Ulett, G.C., et al., 2008b. UpaG, a new member of the trimeric autotransporter family of adhesins in uropathogenic *Escherichia coli*. J. Bacteriol. 190, 4147–4161.

van Ulsen, P., van Alphen, L., ten Hove, J., Fransen, F., van der Ley, P., Tommassen, J., 2003. A Neisserial autotransporter NalP modulating the processing of other autotransporters. Mol. Microbiol. 50, 1017–1030.

Venkatesan, M.M., Hartman, A.B., Newland, J.W., et al., 2002. Construction, characterization, and animal testing of WRSd1, a *Shigella dysenteriae* 1 vaccine. Infect. Immun. 70, 2950–2958.

Venkatesan, M.M., Ranallo, R.T., 2006. Live-attenuated *Shigella* vaccines. Expert. Rev. Vaccines 5, 669–686.

Vigil, P.D., Alteri, C.J., Mobley, H.L., 2011a. Identification of in vivo-induced antigens including an RTX family exoprotein required for uropathogenic *Escherichia coli* virulence. Infect. Immun. 79, 2335–2344.

Vigil, P.D., Stapleton, A.E., Johnson, J.R., et al., 2011b. Presence of putative repeat-in-toxin gene tosA in *Escherichia coli* predicts successful colonization of the urinary tract. MBio 2, e00066–11.

Vigil, P.D., Wiles, T.J., Engstrom, M.D., Prasov, L., Mulvey, M.A., Mobley, H.L., 2012. The repeat-in-toxin family member TosA mediates adherence of uropathogenic *Escherichia coli* and survival during bacteremia. Infect. Immun. 80, 493–505.

Waldron, D.E., Owen, P., Dorman, C.J., 2002. Competitive interaction of the OxyR DNA-binding protein and the Dam methylase at the *antigen 43* gene regulatory region in *Escherichia coli*. Mol. Microbiol. 44, 509–520.

Wandersman, C., Delepelaire, P., 1990. TolC, an *Escherichia coli* outer membrane protein required for hemolysin secretion. Proc. Natl. Acad. Sci. USA 87, 4776–4780.

Wang, R.C., Seror, S.J., Blight, M., Pratt, J.M., Broome-Smith, J.K., Holland, I.B., 1991. Analysis of the membrane organization of an *Escherichia coli* protein translocator, HlyB, a member of a large family of prokaryote and eukaryote surface transport proteins. J. Mol. Biol. 217, 441–454.

Welch, R.A., 1991. Pore-forming cytolysins of gram-negative bacteria. Mol. Microbiol. 5, 521–528.

Welch, R.A., 2001. RTX toxin structure and function: a story of numerous anomalies and few analogies in toxin biology. Curr. Top. Microbiol. Immunol. 257, 85–111.

Welch, R.A., Dellinger, E.P., Minshew, B., Falkow, S., 1981. Haemolysin contributes to virulence of extra-intestinal *E. coli* infections. Nature 294, 665–667.

Wells, T.J., McNeilly, T.N., Totsika, M., Mahajan, A., Gally, D.L., Schembri, M.A., 2009. The *Escherichia coli* O157:H7 EhaB autotransporter protein binds to laminin and collagen I and induces a serum IgA response in O157:H7 challenged cattle. Environ. Microbiol. 11, 1803–1814.

Wells, T.J., Sherlock, O., Rivas, L., et al., 2008. EhaA is a novel autotransporter protein of entero-hemorrhagic *Escherichia coli* O157:H7 that contributes to adhesion and biofilm formation. Environ. Microbiol. 10, 589–604.

Wells, T.J., Totsika, M., Schembri, M.A., 2010. Autotransporters of *Escherichia coli*: a sequence-based characterization. Microbiology 156, 2459–2469.

Winter, L.E., Barenkamp, S.J., 2009. Antibodies specific for the Hia adhesion proteins of nontype-able *Haemophilus influenzae* mediate opsonophagocytic activity. Clin. Vaccine Immunol. 16, 1040–1046.

Winter, L.E., Barenkamp, S.J., 2010. Construction and immunogenicity of recombinant adenovirus vaccines expressing the HMW1, HMW2, or Hia adhesion protein of nontypeable *Haemophilus influenzae*. Clin. Vaccine Immunol. 17, 1567–1575.

Yamamoto, S., Tsukamoto, T., Terai, A., Kurazono, H., Takeda, Y., Yoshida, O., 1995. Distribution of virulence factors in *Escherichia coli* isolated from urine of cystitis patients. Microbiol. Immunol. 39, 401–404.

Yen, Y.T., Kostakioti, M., Henderson, I.R., Stathopoulos, C., 2008. Common themes and variations in serine protease autotransporters. Trends Microbiol. 16, 370–379.

Zgurskaya, H.I., Nikaido, H., 2000. Multidrug resistance mechanisms: drug efflux across two membranes. Mol. Microbiol. 37, 219–225.

Zhu, C., Ruiz-Perez, F., Yang, Z., et al., 2006. Delivery of heterologous protein antigens via hemolysin or autotransporter systems by an attenuated ler mutant of rabbit enteropathogenic *Escherichia coli*. Vaccine 24, 3821–3831.

Chapter 17

Capsule and lipopolysaccharide

Lisa M. Willis and Chris Whitfield

University of Guelph, Guelph, ON, Canada

INTRODUCTION

Pathogenic *E. coli* require many different virulence factors which allow them to invade the host, evade host immune defenses and colonize specific niches in the host where they can cause disease. The first interactions between *E. coli* and its host occur at the outer membrane and are mediated by proteins and carbo-hydrate-containing macromolecules (glycoconjugates) on the bacterial and host cell surfaces (Figure 17.1). Bacterial glycoconjugates provide crucial defenses against different elements of the innate and acquired immune system, as well as generating tremendous diversity in surface antigenicity. In most Gram-negative bacteria, the outer membrane is composed predominantly of the glycolipid known as lipopolysaccharide (LPS). In *E. coli*, this complex molecule is com-posed of three structurally distinct regions; the hydrophobic anchor called lipid A, a core oligosaccharide, and the long-chain polysaccharide called O antigen (or O-polysaccharide; O-PS). The differences in *E. coli* O-PS structures give rise to more than 180 distinct O antigens and these have been exploited in sero-typing classification methods (Orskov et al., 1977). Many *E. coli* isolates also produce another long-chain polysaccharide, known as capsular polysaccharide (CPS). CPS provides another major surface antigen, called the K-antigen, named after the German term 'kapsel'. There are more than 80 different K antigens in *E. coli*. The means of attachment of CPS to the cell surface is not known in all cases but these polymers create a coherent structural entity (the capsule) that is visible by light microscopy and extends 50–100 nm from the cell surface. As a result, the capsule often masks underlying O antigens in serotyping studies that exploit specific antisera and whole-cell agglutination methods. A single *E. coli* isolate can produce one O- and one K-antigen.

Together, LPS and CPS represent major virulence factors, which have been the target of numerous vaccines and therapies, and they provide the focus of this chapter. However, they are not the only glycoconjugates produced by *E. coli*. All isolates produce a polysaccharide called enterobacterial common antigen that provides resistance to organic acids in *E. coli* (Barua et al., 2002) and to bile salts and detergents in *Salmonella* (Ramos-Morales et al., 2003). Under certain stress conditions (e.g. perturbations of cell envelope integrity), some

Escherichia coli. http://dx.doi.org/10.1016/B978-0-12-397048-0.00017-6

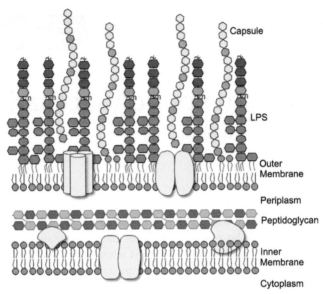

FIGURE 17.1   Schematic of the *E. coli* cell envelope. Phospholipids are purple and proteins are yellow. The diglucosamine backbone and inner core of LPS are colored blue while the O antigen is depicted as green and red. Frequently the first sugar in the O antigen is either GlcNAc or GalNAc but the polymer can be homopolymeric or heteropolymeric. The capsule, which is green and yellow, can also be a homo- or heteropolymer. The group 2 and 3 CPSs are known to be anchored in the membrane by a phospholipid, but it is not known whether the group 1 and 4 CPSs are covalently anchored to the membrane or just associated with it.

E. coli isolates also produce an exopolysaccharide called colanic acid which has a biosynthetic pathway that is part of the Rcs-regulon (Majdalani and Gottesman, 2005). Unlike CPS, colanic acid is poorly retained at the cell surface. Like several other bacterial species, the biofilm mode of growth in *E. coli* is supported by formation of bacterial cellulose and a by-polymer of *N*-acetylglucosamine (PNAG) (Cerca and Jefferson, 2008; Saldana et al., 2009). However, some commensal *E. coli* isolates also produce a polysaccharide of unknown structure that impairs biofilm formation by other bacteria including *Staphylococcus aureus*, potentially affecting community dynamics (Rendueles et al., 2011). Similar anti-adhesion properties have been reported for certain *E. coli* CPSs (Valle et al., 2006). In summary, the cell surfaces of *E. coli* isolates are rich in complex carbohydrates and these molecules play diverse, niche-dependent roles in the physiology of *E. coli*. In this chapter, we will focus only on LPS and CPS.

STRUCTURE AND BIOSYNTHESIS OF *E. COLI* LPS

The LPS molecule contains three regions, the lipid A, core, and O-PS. Lipid A is the most highly conserved portion of the molecule. The typical structure contains two phosphorylated glucosamine residues with six acyl chains (Figure 17.2A),

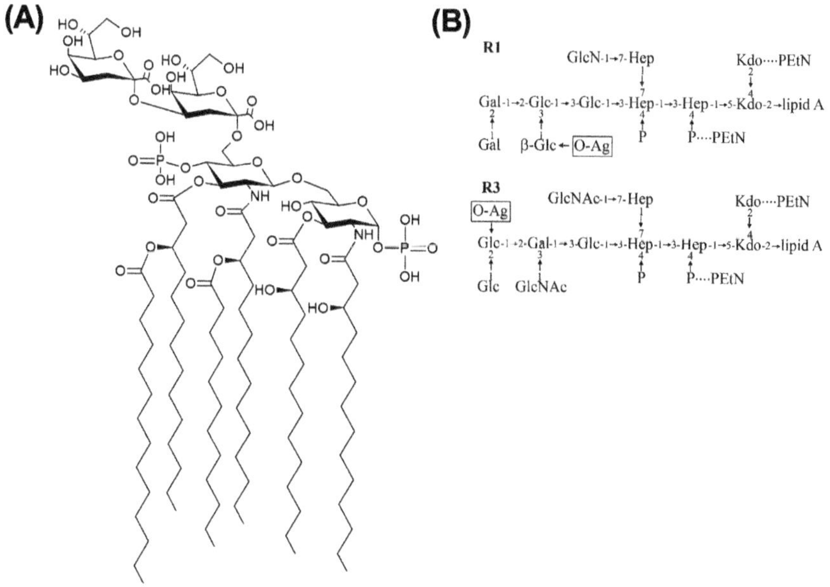

(A)

(B)

FIGURE 17.2   Structure of lipid A-core. (A) Structure of Kdo$_2$-lipid A. (B) Structure of the R1 and R3 core types. All linkages are in the alpha anomeric configuration unless otherwise noted. Dotted lines represent non-stochiometric modifications. The site of O antigen attachment is indicated.

though this can be modified in several places to alter the biological properties of the molecule (reviewed in Raetz et al., 2007) (see below). The core oligosaccharide can be divided into two regions. The inner core consists of two residues of 3-deoxy-D-*manno*-oct-2-ulosonic acid (Kdo) and three residues of L-glycero-D-*manno*-heptose (Hep) but it may also be modified with a number of other groups such as phosphate, pyrophosphorylethanolamine, or additional sugars (Figure 17.2B). This region is highly conserved in *E. coli* and *Salmonella* and is important for membrane stability (reviewed in Heinrichs et al., 1998). For example, loss of the phosphate moieties on Hep residues compromises outer membrane integrity and, in *Salmonella*, attenuates virulence (Yethon et al., 2000). The outer core serves as the point of attachment for O-PS and contains sugars such as glucose (Glc), galactose (Gal), glucosamine (GlcN), and N-acetylglucosamine (GlcNAc). Despite the potential for great diversity, there are only five known core types in *E. coli*, K-12, R1, R2, R3, and R4 (Heinrichs et al., 1998; Kaniuk et al., 2004). The core type R1 is most commonly found in *E. coli* causing extraintestinal infection while the verotoxin-producing *E. coli* isolates are predominantly R3 (Figure 17.2B).

Lipid A-core is synthesized as one unit in a step-wise manner by the Lpx and Waa enzymes, starting with the acylation of UDP-GlcNAc on the cytoplasmic face of the inner membrane (Figure 17.3) (reviewed in Raetz and Whitfield, 2002; Raetz et al., 2007). Most of the steps in this process were

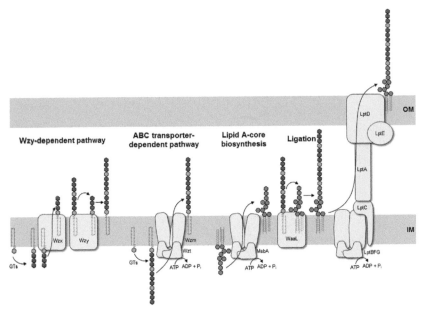

FIGURE 17.3   Biosynthesis of LPS. IM and OM are the inner and outer membranes respectively. O antigen is synthesized separately from lipid A-core through either the Wzy- or ABC transporter-dependent pathway. Lipid A-core is synthesized on the cytoplasmic face of the inner membrane, then flipped to the periplasm by MsbA where it can be ligated with the O antigen. The completed LPS molecule is then exported to the cell surface by the Lpt proteins where it makes up the majority of the outer membrane.

established by the efforts of Christian Raetz and his research group, leading to what is often known as the '*Raetz pathway*'. In this conserved pathway, formation of tetraacyl lipid A (called lipid IV_A) is followed by addition of the two inner core Kdo residues before the secondary acyl chains can be added to complete the hexaacyl lipid A. Once the lipid A-core is complete, it is flipped to the periplasmic side of the inner membrane by MsbA, an essential protein belonging to the ATP-binding cassette (ABC) transporter superfamily. MsbA plays a crucial role in quality control of lipid A structure since it has a preference for hexaacyl lipid A over earlier intermediates (such as lipid IV_A) as a substrate (Doerrler and Raetz, 2002). As a result, the minimal LPS structure for export and cell viability is normally considered to be Kdo_2-lipid A. However, suppressor mutants lacking Kdo (and therefore also devoid of secondary acyl chains) are viable, albeit with a compromised outer membrane barrier, leading to LPS molecules comprising lipid IV_A (Meredith et al., 2006). Alternatively, the absence of Kdo can also be overcome by overexpression of the secondary acyltransferases to generate pentaacyl and hexaacyl free lipid A (Reynolds and Raetz, 2009). After export of lipid A-core to the periplasmic face of the inner membrane, the ligase enzyme (WaaL) covalently links it to the O antigen, which is synthesized separately.

The O antigen is highly variable and typically consists of repeat units of 2–5 sugars in a polymer that can be more than 100 sugars long. The range of repeat unit numbers is specific for a particular strain and is controlled by different strategies, depending on which of the two biosynthetic pathways is used for O-antigen biosynthesis: the Wzy-dependent pathway or the ABC transporter-dependent pathway (Figure 17.3) (Raetz and Whitfield, 2002). In all cases, biosynthesis of O antigens begins at the cytoplasmic face of the inner membrane using the lipid carrier undecaprenyl phosphate. In *E. coli,* the enzyme WecA initiates O-PS biosynthesis by transferring GlcNAc-1-phosphate from UDP-GlcNAc to undecaprenyl phosphate to form undecaprenyl pyrophosphoryl-GlcNAc. In some cases, this is epimerized to undecaprenyl pyrophosphoryl-GalNAc (Rush et al., 2010). The GlcNAc/GalNAc residues occur in each repeat unit in O-PSs formed by the Wzy-dependent pathway, or just once at the reducing terminus in the ABC transporter-dependent process.

In the Wzy-dependent pathway, individual repeat units are synthesized on undecaprenyl pyrophosphoryl-GlcNAc/GalNAc using nucleoside phosphate sugar donors by enzymes encoded in the locus. These repeat units are then flipped to the periplasmic side of the inner membrane by the flippase Wzx and polymerized into the complete O antigen by the polymerase Wzy. Wzy transfers the growing glycan from one undecaprenyl pyrophosphate carrier to the incoming lipid-linked repeat unit. Insight into this pathway has benefitted from model systems, including the O7 and O86 antigens (Table 17.1), studied in the laboratories of Miguel Valvano and George Peng Wang, respectively. In the ABC transporter-dependent pathway, the O antigen is synthesized entirely at the cytoplasmic face of the inner membrane through the sequential action of glycosyltransferases that add sugars to undecaprenyl pyrophosphoryl-GlcNAc, before the completed chain is exported via an ABC transporter. This process is less common than the Wzy-dependent pathway and the resulting repeat-unit structures tend to be less elaborate. The serotype O8/O9/O9a antigens provide influential prototypes for this type of assembly process and many of the steps have been resolved by the research group of Klaus Jann and the Whitfield lab (Table 17.1). The *E. coli* WaaL ligase can operate effectively with nascent undecaprenyl-linked O-PSs from either pathway.

The LPS species extracted from a bacterial cell shows heterogeneity best illustrated by profiles separated by SDS-PAGE (Hitchcock and Brown, 1983). The most obvious variations occur in the chain-lengths of O-PS, evident as a ladder of high-molecular-weight molecules. The length of the O-PS has important functional consequences (see below) and is established by different mechanisms, depending on the assembly pathway. In the Wzy-dependent pathway, the length of the O antigen is controlled by the protein Wzz, a transmembrane protein with a large periplasmic domain that belongs to the polysaccharide co-polymerase (PCP) family (reviewed in Cuthbertson et al., 2009; Morona et al., 2009). In the ABC transporter-dependent pathway, chain length of some O-PS molecules is controlled by the addition of novel residues (ones not found in

TABLE 17.1 Representative O- and K-antigen structures

Serotype	Repeat unit structure	Reference

O7	-3)-β-Qui4NAc-(1-2)-α-Man-(1-4)-β-Gal-(1-3)-α-GlcNAc-(1-
	                     3
	      α-Rha-(1⌐
		L'vov et al., 1984

O8	-2)-α-Man-(1-2)-α-Man-(1-3)-β-Man-(1-	Jansson et al., 1985
O9a	-2)-α-Man-(1-2)-α-Man-(1-3)-α-Man-(1-3)-α-Man-(1-	Parolis et al., 1986
O9	-2)-[α-Man-(1-2)]₂-α-Man-(1-3)-α-Man-(1-3)-α-Man-(1-	Prehm et al., 1976

O86	-4)-α-Fuc-(1-2)-β-Gal-(1-3)-α-GalNAc-(1-3)-β-GalNAc-(1-
	                    3
	      α-Gal-(1⌐
		Andersson et al., 1989

O111	α-Col-(1⌐
	              3
	-3)-β-GlcNAc-(1-4)-α-Glc-(1,4)-α-Gal-(1-
	                 6
	        α-Col-(1⌐
		Kenne et al., 1983

O157	-2)-α-PerNAc-(1-3)-α-Fuc-(1-4)-β-Glc-(1-3)-α-GalNAc-(1-	Perry et al., 1986
K1	-8)-α-Neu5Ac-(2-	McGuire et al., 1964
K4	-4)-β-GlcA-(1-3)-β-GlcNAc-(1- 3 β-Fru-(1⌐	Rodriguez et al., 1988
K5	-4)-β-GlcA-(1-4)-α-GlcNAc-(1-	Vann et al., 1981

the repeat-unit domain) to the non-reducing end of the glycan (reviewed in Cuthbertson et al., 2010). These residues may be alternative sugars or non-glycose moieties and they not only prevent further elongation of the glycan, but are also required to engage the ABC transporter for export across the inner membrane. However, other O antigens in this pathway engage the ABC transporter in the absence of any identifiable terminating group and coordination of the activities of the O-PS elongating enzymes and the ABC transporter determine O-PS chain length (Cuthbertson et al., 2010).

The Lpt proteins, which span the inner and outer membranes as well as the periplasm, are responsible for transport of the mature LPS. Given the heterogeneity in LPS species, this system can effectively handle molecules with diverse structures and it is well conserved across bacterial species though it has been identified only recently. In the current working model, LptBFG form an ABC protein complex to extract nascent LPS out of the inner membrane. The molecule is passed via LptC to a filament of LptA molecules spanning the periplasm and then to LptDE located in the outer membrane, which completes translocation to the cell surface (reviewed in Ruiz et al., 2009; Sperandeo et al., 2009; Chng et al., 2010).

Most LPS modifications occur at the periplasmic face of either the inner or outer membrane and, as such, can serve as markers indicating the stage of transport. These include modification of the phosphate groups with 4-amino-4-deoxy-L-arabinose (L-Ara4N) or ethanolamine by enzymes found in the inner membrane, which increases bacterial resistance to innate immune defenses (see below). The outer membrane protein PagP adds a secondary palmitate in an acyloxyacyl linkage to the hydroxy-myristate located at the 2-position. The resulting heptaacyl species is a much less potent activator of cytokine induction (Raetz et al., 2007). PagP is normally latent in *E. coli* but is activated by membrane perturbations and by defects in acylation of the $3'$ position of the diglucosamine backbone with a myristol residue. However, this modification can alter host–pathogen interactions in other ways, because PagP activity in *E. coli* O157 has an indirect effect on the completion of the LPS core oligosaccharide, leading to loss of O antigen and serum sensitivity (Smith et al., 2008).

STRUCTURE AND BIOSYNTHESIS OF *E. COLI* CPSs

The CPSs of *E. coli* have been subdivided into four groups based on structural and genetic criteria (reviewed in Whitfield, 2006). Group 1 and 4 capsules share the same mode of synthesis, a Wzy-dependent pathway, and mainly differ in the chromosomal locations of key genes. Similarly, groups 2 and 3 capsules are both produced by ABC transporter-dependent processes but the genes are organized differently within the locus.

Group 1 (and 4) CPS are found in *E. coli* isolates causing intestinal infections. They are heteropolymers of repeating sugar units, as in many O antigens. Biosynthesis uses a process identical to the Wzy-dependent O antigens but the pathways diverge once the polymerized glycan is formed at the periplasmic face of the inner membrane (Figure 17.4). O antigens enter the LPS assembly pathway by ligation to lipid A-core and are transferred to the surface via Lpt proteins. Capsular K antigens have their own surface assembly process that involves four key components: (i) Wzc, an inner membrane PCP protein belonging to the PCP-2a subfamily; (ii) Wzb, a cytoplasmic protein tyrosine phosphatase; (iii) Wza, an outer membrane lipoprotein belonging to the OPX (outer membrane polysaccharide export) protein family (Cuthbertson et al., 2009); and (iv) an accessory outer membrane protein called Wzi. Understanding the group 1 CPS translocation processes has come from studies with the serotype K30 prototype in the Whitfield laboratory. In O-antigen biosynthesis, the corresponding PCP protein

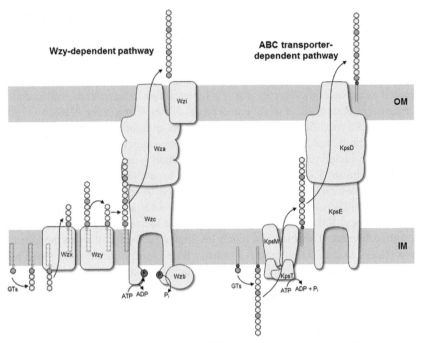

FIGURE 17.4 Biosynthesis of capsule. IM and OM are the inner and outer membranes respectively. Capsular polysaccharides can by synthesized by either the Wzy- or ABC transporter-dependent pathway.

belongs to sub-family 1 and its action appears to be confined to regulating the polymerization activity, although the exact mechanism is still unknown. PCP-2a proteins are more complex. Wzc certainly affects polymerization but this activity (and capsule formation) is dependent on the activity of a C-terminal tyrosine autokinase domain, as well as the dephosphorylation of these residues by the Wzb phosphatase (Wugeditsch et al., 2001). The autokinase domain is absent in PCP-1 family members. The additional role played by Wzc is mediated by its periplasmic domain, which is larger in PCP-2a proteins, and interacts with the extensive periplasmic region of Wza (Collins et al., 2007). The crystal structure of Wza reveals an octomer which forms a large periplasmic barrel connected to an outer membrane channel (Dong et al., 2006). The outer-membrane pore is formed by one α-helix contributed by each protamer and was the first example of an outer-membrane channel which is not a β-barrel (Dong et al., 2006). It is not known how the Wzc/Wza complex transports the capsule to the outer membrane or what the nature of the reducing end of the exported polymer is; i.e. is the CPS attached to an anchoring protein or lipid. In O-antigen biosynthesis, the ligase (WaaL) releases the nascent glycan from the undecaprenyl pyrophosphate carrier and links it to lipid A-core. There is no corresponding activity in the biosynthesis of group 1 and 4 K antigens. The CPS could retain its linkage to undecaprenyl pyrophosphate, or it could potentially be released by a leaky polymerization step, where the transfer of the growing chain to the incoming repeat unit may be incomplete. Wzi is important for surface retention of group 1 CPS although its precise function is not yet known and it is not present in group 4 systems (Rahn et al., 2003). The Wza-Wzb-Wzc proteins are highly conserved in all isolates possessing a group 1 or 4 capsule, indicating that they do not recognize any particular glycan repeat-unit structure. Group 1 K antigens are typically co-expressed with an O antigen synthesized by the ABC transporter pathway (e.g. O8/O9/O9a) and the corresponding gene clusters are both located near *his*. The group 1 gene cluster is allelic with genes for colanic acid biosynthesis, so expression of a capsule and colanic acid are mutually exclusive. However, group 4 K antigens really emphasize the parallels between O- and K-antigen biosynthesis. In these cases, the structures of the repeat units are identical and are produced by the same chromosomal locus. Some undecaprenyl pyrophosphate-linked molecules are diverted by WaaL into LPS, while others enter a CPS translocation pathway by Wza-Wzb-Wzc proteins that are encoded by a separated locus elsewhere on the chromosome (Peleg et al., 2005). Serotype O111 provided the first example, and the use of the same repeat structure led to the early description of '*O-antigen capsules*' (now group 4) (Goldman et al., 1982). It is now known that serotypes O127 and O157 also have group 4 capsules (Table 17.1) (Peleg et al., 2005; Shifrin et al., 2008). Isolates that can produce group 4 CPSs (unlike their group 1 counterparts) retain the genes for colanic acid production.

K antigens from extraintestinal pathogenic *E. coli* mostly fall into groups 2 and 3 (Whitfield, 2006). These include capsules from isolates causing urinary tract infections and meningitis. The production of many of these K antigens is temperature-regulated, unlike group 1 and 4 K antigens, with expression being 'on' at 37°C but 'off' at temperatures below 20°C. Some of the CPS structures resemble eukaryotic glycans and this is thought to aid virulence by preventing an effective immune response. For example, the K4 glycan is fructosylated chondroitin and the K5 glycan is heparosan, both similar to human glycosaminoglycans (Table 17.1). As well, the meningitis-causing K1 isolates possess a CPS containing α2,8-linked polysialic acid, which is identical in structure to the glycan found on neural cell adhesion molecule (NCAM) in the human brain. The K1 and K5 systems have been influential models for understanding the mechanisms of biosynthesis through studies in the laboratories of Eric Vimr, Willie Vann, and Ian Roberts. ABC transporter-dependent capsules are synthesized entirely on the cytoplasmic face of the inner membrane, similar to ABC transporter-dependent O antigens (Figure 17.4) (Whitfield, 2006). However, one critical difference is that undecaprenyl phosphate is apparently not involved (Finke et al., 1991). It has been shown that the mature capsular polysaccharide is attached at the reducing end to a phospholipid although the precise chemical structure has not been resolved (Gotschlich et al., 1981). In the essentially identical process in *N. meningitidis*, lipidation occurs before transport of polysialic acid CPS to the cell surface, but it is not known whether the glycan is synthesized directly on this phospholipid or if it is synthesized on another molecule before transfer to the phospholipid (Tzeng et al., 2005). Once the capsule is synthesized, it is transported through the inner membrane by the ABC transporter (KpsMT) (reviewed in Vimr and Steenbergen, 2009). The final steps of glycan translocation are mediated by KpsE, a representative of the PCP-3 subfamily, and KpsD, an OPX protein (Cuthbertson et al., 2009). A multiprotein export complex comprising KpsMTED is thought to span the cell envelope (Rigg et al., 1998; McNulty et al., 2006). Like the group 1 and 4 translocation export proteins, they are conserved in all isolates possessing a group 2 or 3 capsule, indicating that they also do not recognize any particular glycan repeat-unit structure. The genes encoding group 2 and 3 K antigens are located on the chromosome near *serA* and these glycans can be found together with a wide range of O serotypes, many formed by a Wzy-dependent system.

EVASION OF HOST CELL DEFENSES

Mammals have evolved with the pressure of bacteria, viruses, and fungi and so have developed ways of dealing with microbial infections; the most extensive defense machinery is the immune system. The human immune system consists of two branches, innate and adaptive immunity, between which there is extensive crosstalk. Innate immunity has developed as a first line of defense against challenges not necessarily seen before to protect the body from pathogens. Toll-like receptors (TLR) are the major effectors of innate immunity which recognize conserved structures in pathogens called pathogen-associated molecular patterns (PAMPs), somewhat of a misnomer, as they are found in non-pathogenic as well as pathogenic microbes; in Gram-negative bacteria, these include LPS, CPS, flagella, and nucleic acids. Binding of TLRs to their respective ligands activates a signaling cascade, resulting in recruitment and activation of immune cells, which can then eliminate the threat. In addition to the cellular receptors, there are also soluble complement proteins found in serum, which can bind directly to bacterial cell surfaces leading to opsonization, phagocytosis, and formation of the membrane attack complex leading to lysis. LPS and CPS are important parts of the arsenal of virulence factors used by bacteria to circumvent these processes.

Lipid A structure influences susceptibility to polycationic peptides

Cationic antimicrobial peptides (CAMPs) are short peptides secreted by immune and epithelial cells in response to bacterial products, like LPS, and other inflammatory signals (reviewed in Brown and Hancock, 2006). It is well established that the structure of lipid A has a profound effect on the susceptibility of bacteria to CAMPs and polycationic drugs such as polymyxin B. These compounds typically exploit negatively charged phosphate residues on the diglucosamine backbone to bind to the cell surface and then insert into the membrane, causing disruption of the permeability barrier and lysis of the cell (Brown and Hancock, 2006). Many bacteria can overcome susceptibility to these compounds by modifying the 1 and 4' phosphates with L-Ara4N or phosphorylethanolamine (PEtN) by the activity of ArnT and EptA respectively, following transfer of the nascent lipid A-core to the periplasm (Raetz et al., 2007). The enzymes involved in these processes in *E. coli* are activated under specific growth conditions (e.g. growth at low pH or in the presence of cationic peptides) and are under the regulation of the PmrAB and PhoPQ two-component systems (Guo et al., 1997; Gunn, 2008).

An additional modification involves phosphorylation of the 1-phosphate on the lipid A backbone to make pyrophosphate and is mediated by LpxT using undecaprenyl pyrophosphate as a donor (Touze et al., 2008). This activity helps recycle the essential lipid carrier for peptidoglycan, O antigen, group 1 K antigens, etc., but other cellular proteins can also fulfill this need (El Ghachi et al., 2005). The activity of LpxT is inhibited in PmrA-activated cells (Herrera et al., 2010).

LPS activates TLR4

During infections by Gram-negative bacteria, LPS molecules elicit an innate immune response resulting in an inflammatory response, a property that underlies LPS's description as 'endotoxin'. In this intended protective process, LPS activates macrophages, as well as some non-immune cells to secrete inflammatory mediators, such as cytokines. However, in certain situations, the response becomes dysregulated, generating a cascade of effector secretion. This can culminate in sepsis (or septic shock), involving hypotension and organ failure, although the details of the underlying physiological processes are not fully understood. The process begins with recognition of LPS molecules and leads through complex signaling pathways to cytokine induction. However, the outcome is highly dependent on the structure of lipid A, the critical determinant of endotoxic properties. In general, hexaacyl lipid A containing C_{12} or C_{14} acyl chains is a potent activator (or agonist) of the inflammatory pathways. In contrast, molecules lacking one or two acyl chains (e.g. lipid IV_A), or ones affected in phosphorylation of the diglucosamine backbone, typically have significantly diminished potency. These molecules are considered antagonists of the signaling pathway, but their status can vary depending on the macrophage source. Unlike *E. coli*, some bacterial species exploit the precise structural requirements of LPS-recognition proteins and produce naturally 'non-endotoxic' LPS (Munford, 2008). Differential recognition of lipid A structures has been exploited in the development of the universal antagonist called eritoran, a tetraacyl derivative, as a therapeutic strategy to block deleterious activation of signaling pathways by natural agonist LPS species (reviewed in Bryant et al., 2010).

LPS is an example of a PAMP and its recognition is mediated by a sensory complex comprising two critical and highly conserved proteins, Toll-like receptor 4 (TLR4) and myeloid-differentiation factor 2 (MD2) (reviewed in Munford, 2008; Bryant et al., 2010). TLR4 was the first member of the Toll-like receptor family of pattern recognition receptors to be described (Bryant et al., 2010). It is a type 1 transmembrane protein found on the surface of cells of the immune system, including monocytes and macrophages, as well as non-immune cells, such as epithelial and endothelial cells. The extracellular domain of TLR4 contains multiple leucine-rich repeats and these motifs confer the characteristic curved shape found in TLR4 and other related proteins (Figure 17.5). The cytoplasmic Toll-interleukin-1 receptor (TIR) domain of TLR4 activates downstream signaling pathways in the context of LPS binding. In order to perform this function, TLR4 forms a complex with MD2. MD2 is a member of a family of lipid-binding proteins and provides a soluble co-receptor for LPS (Inohara and Nunez, 2002). In mice, both TLR4 and MD2 are required to raise an immune response to LPS and mice lacking either protein are resistant to endotoxic shock.

The crystal structure of TLR4-MD2 in complex with an LPS molecule was invaluable in elucidating the details of LPS binding and TLR4 activation (Figure 17.5) (Park et al., 2009). Previous crystal structures of MD2 and TLR4-MD2 in complex with lipid IV_A and eritoran, respectively, revealed an LPS binding pocket in the β-sandwich fold of MD2 (Kim et al., 2007; Ohto et al., 2007). However, these antagonist structures only have four acyl chains and do not result in an activating conformational change in the receptor. In contrast, hexaacyl LPS promotes dimerization of TLR4-MD2 in an activated complex. The MD2 LPS-binding pocket cannot accommodate six acyl chains, leaving one (at C2) exposed, where it contributes to a hydrophobic region that forms a dimerization interface with TLR4. This repositions the diglucosamine backbone, allowing the phosphates to interact with each of the TLR4 molecules in the complex and the inner core Kdo and heptose

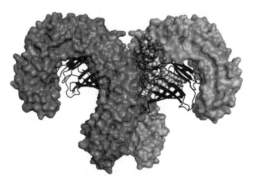

FIGURE 17.5 Crystal structure of the TLR4-MD2-LPS complex from PDB 3FXI (Park et al., 2009). The two TLR4 monomers are in green and magenta, the MD2 proteins are in blue and the LPS molecule is shown as spheres with phosphates colored orange. Five of the six acyl chains are bound in the MD2 pocket while the sixth forms an interaction with both MD2 and TLR4. Only the extracellular domain of TLR4 was crystallized; in the full-length protein the C-terminal transmembrane and TIR effector domains would be located at the base of the structure.

residues bind at the interface of the two TLR4 molecules, further strengthening the interaction. The conformational change in the complex is thought to promote the recruitment of effector proteins to the intracellular TIR domain to activate signaling.

Two signaling pathways lead to induction of immunomodulatory cytokines (reviewed in Kawai and Akira, 2010; Szabo et al., 2010). On binding LPS, TLR4 activates the MAL-MyD88 pathway, leading to a downstream signaling cascade resulting in activation of NF-κB and transcription of genes encoding proinflammatory cytokines like TNF-α and IL-1β. TLR4 also activates the TRAM-TRIF-dependent pathway, promoting sustained NF-κB activation and transcription of genes encoding immunomodulatory cytokines like IL-6 and RANTES.

LPS can bind directly to the TLR4-MD2 complex, or the interaction can be mediated by lipopolysaccharide binding protein (LBP) and CD14. LBP was discovered 25 years ago as a serum protein that binds LPS, although we now know it also binds other microbial components (Schumann, 2011). Experiments in mice have shown that LBP is essential for combating infections. LBP is primarily secreted by the liver, but can also be secreted by epithelial cells in the lungs, gastrointestinal tract, kidneys, and urinary tract. Upon binding to LPS in the bloodstream, LBP forms a stable complex with CD14, a receptor that can be anchored in the membrane of leukocytes or secreted into the blood in a soluble form (Thomas et al., 2002). The complex then mediates transfer of the LPS molecule to TLR4-MD2, starting a signaling cascade. The formation of the LBP-CD14-LPS complex allows the body to respond to LPS concentrations as low as 10 pg/mL (Thomas et al., 2002).

In addition to stimulation of the immune system through induction of proinflammatory cytokines, LPS also causes an increase in the permeability of epithelial and endothelial cells through a number of mechanisms which have yet to be fully characterized (Moriez et al., 2005; Sheth et al., 2007; Vandenbroucke et al., 2008; He et al., 2011). The permeability barrier is maintained through the action of tight junctions, multiprotein complexes that bridge adjacent cells. In response to LPS, many of the predominant tight junction proteins are redistributed, leading to the breakdown of tight junctions and increased permeability (Sheth et al., 2007; He et al., 2011). This is dependent on TLR4 signaling and phosphorylation of a number of signaling proteins, like c-Src and myosin light chain kinase (Moriez et al., 2005; Sheth et al., 2007). The breakdown of tight junctions is enhanced by rearrangement of actin filaments through the action of RhoA and NF-κB (Vandenbroucke et al., 2008; He et al., 2011). This causes contraction of the cell and further permeability of the barrier. Disruption of the barrier allows cells of the immune system to migrate to areas of infection, but also provides a portal for bacterial translocation, contributing to sepsis.

Dendritic cells (DCs) are potent antigen-presenting cells which link innate and adaptive immunity. DCs which are positive for DC-specific intracellular adhesion molecule 3 (ICAM-3)-grabbing non-integrin (DC-SIGN; CD209), which is a C-type lectin, bind primarily to mannose- and fucose-containing glycans, but may also bind GlcNAc-containing glycans as well (Zhang et al., 2006). DC-SIGN promotes phagocytosis and MHC class II antigen presentation of bacteria, viruses, fungi, and parasites, including *Helicobacter pylori* and *Mycobacterium tuberculosis*. In response to LPS from *E. coli* O55 (which contains GlcNAc, Gal, GalNAc, and colitose (Stenutz et al., 2006)), a subpopulation of monocytes differentiate into DC-SIGN+ dendritic cells in a TLR4/CD14-dependent manner (Cheong et al., 2010). These cells then migrate to lymph nodes where they can interact with T cells, involved in adaptive immunity.

Serum sensitivity and phagocytosis

In contrast to the membrane-bound TLRs, the complement system is a soluble network of proteins which respond to the presence of bacteria in the bloodstream. There are three main pathways for complement activation: the alternative pathway, the classical pathway, and the lectin pathway (Figure 17.6). The alternative pathway is activated by carbohydrates, lipids, and proteins on foreign surfaces. The classical pathway is activated by IgG or IgM binding to pathogens and is similar to the lectin pathway, which is activated by mannose-binding lectin (MBL) and MBL-associated proteins (MASPs). The three pathways follow different sequences that converge at the deposition of C3 protein on the bacterial surface and lead to opsonization, formation of membrane attack complexes, and formation of anaphylatoxins (reviewed in Sarma and Ward, 2011). When C3 is cleaved, it forms C3a and C3b. C3a is an anaphylatoxin, a small protein that causes a number of inflammatory responses, including acting as a chemoattractant for neutrophils and monocytes (phagocytes). C3b is an opsonin, which helps with phagocytosis. C3b also binds to either of the two C3 convertases to form a C5 convertase, C4bC2aC3b or C3bBbC3b. The C5 convertases cleave C5 into C5a and C5b. C5a is another anaphylatoxin while C5b interacts with C6, C7, C8, and multiple copies of C9 to form the membrane attack complex (MAC). The MAC inserts itself into the bacterial membrane, forming a pore and resulting in cell lysis.

Pathogens have evolved ways of circumventing the complement cascade in order to survive in the bloodstream. O antigens and capsules contribute to the virulence of pathogens by preventing phagocytosis and complement-mediated killing (Weiss et al., 1982; Johnson, 1991; Russo et al., 1993, 1994; Burns and Hull, 1999; Schneider et al., 2004; Buckles et al., 2009). One method for evasion of complement involves preventing complement activation. At the simplest level, this may be achieved by masking underlying antigens that would otherwise activate complement. Alternatively these surface glycans can prevent killing by complement, through decreasing attachment and internalization of the bacteria by PMNs, or by preventing formation of the membrane attack complex. However, in many cases the exact mechanism by which this is achieved is not clear and it may vary depending on the particular O or K antigen (or combination thereof) as well as the underlying cell surface (Cross et al., 1986; Falkenhagen et al., 1991). K5

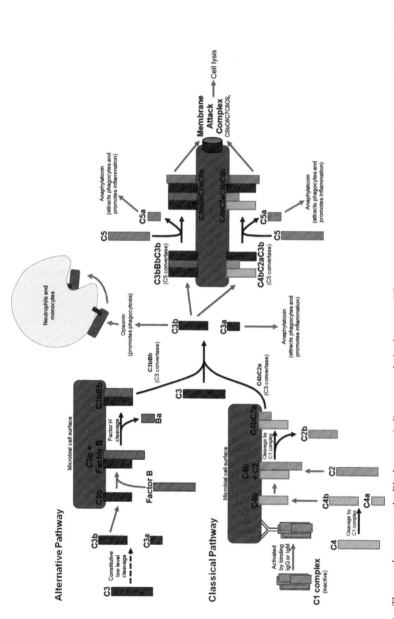

FIGURE 17.6 The complement cascade. Black arrows indicate proteolytic cleavage while gray arrows represent a movement (the next step). The lectin pathway is the same as the classical pathway in terms of C4 and C2 cleavage, except it is activated by MBL binding to carbohydrates on the cell surface and the associated MASPs cleave C4 and C2. All pathways converge at the deposition of C3b on the bacterial cell surface.

capsule has been shown to confer serum resistance in the absence of phagocytes and also decreases the attachment and internalization of bacteria by PMNs and monocytes (Burns and Hull, 1999). The K1, K2, and K54 capsule antigens have also been shown to be important for serum resistance in the absence of phagocytes (Russo et al., 1993; Buckles et al., 2009). In the case of the K54 capsule, the same amount of C3 was deposited on wild-type and capsule-minus strains, suggesting that the capsule may prevent killing by complement rather than activation of complement (Russo et al., 1993).

O antigen is also important for serum resistance. Isogenic mutants lacking the O antigen are extremely susceptible to killing by human serum containing PMNs (Burns and Hull, 1998). However, simple possession of the O antigen is not always sufficient as the chain length of the O antigen is also important. Studies with *E. coli* O-antigen biosynthesis mutants have shown that cells with successively shorter O antigens are more susceptible to serum mediated killing (Burns and Hull, 1998). *E. coli* Nissle 1917 is a well-known probiotic strain which is not pathogenic due to a mutation in the *wzy* gene (Grosdanov et al., 2002). This mutation results in a lipid A-core which is capped with only one O antigen repeat unit, rendering the cell sensitive to serum despite the presence of a K5 capsule. In all cases, loss of both capsule and O antigen has a synergistic effect on serum sensitivity (Burns and Hull, 1998; Russo et al., 1995; Schneider et al., 2004).

OTHER ROLES IN VIRULENCE

LT binds *E. coli* LPS

Heat-labile enterotoxin (LT) is a member of the AB_5 family of toxins and is found in many ETEC isolates (for more details, see Chapter 6). All proteins from this family, including cholera toxin (CT) and Shiga-like toxin, have the same basic structure (Beddoe et al., 2010). The A domain is an ADP-ribosylating enzyme which ultimately causes the efflux of water and electrolytes from epithelial cells into the lumen of the intestine. The B domain consists of a pentameric ring where each subunit binds a receptor on the epithelial cell surface. Ganglioside GM_1 is the receptor for both LT and CT and the binding sites for the ligand are found on one face of the ring, allowing for high avidity binding to the epithelial cell surface (Merritt et al., 1994). Both LT and CT are secreted from the bacterial cell, but over 95% of LT remains associated with outer-membrane vesicles through an interaction with LPS (Horstman and Kuehn, 2002). This association is mediated by a binding site on the B domain that is located on the opposite side of the ring to the GM_1-binding pocket (Mudrak et al., 2009). While LT may bind with higher affinity to full-length LPS, the minimum part of the LPS required for binding has been shown to be the core Kdo residues (Horstman et al., 2004). Interestingly, CT also binds to *E. coli* LPS but neither CT nor LT bind to *Vibrio* LPS, perhaps because the *Vibrio* LPS core contains a single phosphorylated Kdo residue (Horstman et al., 2004). It has been proposed that ETEC is a milder disease than cholera, in part due to the fact that large amounts of CT can reach host cells while LT is prevented from doing so because of its association with the outer-membrane vesicles.

LPS and hemolytic uremic syndrome

Hemolytic uremic syndrome (HUS) is a disease characterized by hemolytic anemia, acute kidney failure, and low platelet count (thrombocytopenia) and most often occurs following an infection with Shiga-toxin-producing *E. coli* such as strains belonging to serotype O157:H7. Activated platelets express P-selectin on their cell surfaces which causes platelet aggregation and leukocyte binding due to the fact that leukocytes constitutively express the P-selectin ligand, PSGL-1. Normally, this allows leukocytes to migrate to areas of inflammation but patients with HUS have increased levels of platelet–leukocyte aggregates in the blood (Stahl et al., 2009). These microparticles also express tissue factor (TF), which contributes to the prothrombotic state where the continued activation and subsequent aggregation of platelets leads to low platelet counts. In vitro, LPS is able to induce platelet and leukocyte activation and to increase the amount of platelet–leukocyte aggregates in blood by causing cells to release proinflammatory cytokines (Guessous et al., 2005; Harrison et al., 2005; Stahl et al., 2009). Treatment of human microvascular endothelial cells with O55-derived LPS causes an increase in the secretion of the neutrophil-recruiting cytokine IL-8, followed by secretion of leukocyte-activating cytokines, SDF-1α, and SDF-1β (Guessous et al., 2005). Macroarray and real-time PCR analysis of monocyte cell lines has shown that O157 LPS causes an up-regulation of proinflammatory cytokines, such as macrophage inflammatory protein 1α (MIP-1α), MIP-1β, TNF-α, IL-1β, IL-8, and growth-related oncogene β (GRO-2) (Harrison et al., 2005). This is followed by secretion of monocyte chemoattractants MIP-1α and MIP-1β, as well as the neutrophil chemoattractant GRO-2 (Harrison et al., 2005). LPS induces TF expression on monocytes and to a lesser degree on neutrophils, likely a result of the increase in platelet–leukocyte aggregates, which leads to an increase in the release of TF into the plasma (Stahl et al., 2009). O157 is a more potent inducer of platelet–leukocyte aggregation than other enterohemorrhagic *E. coli*-associated LPS antigens and has been detected using anti-O157 antibodies on platelets, monocytes, and neutrophils from several HUS patients (Stahl et al., 2009). LPS-induced aggregation is enhanced by high shear forces, such as those found in the renal cortex, which could contribute to the formation of microthrobi in the kidneys (Stahl et al., 2009).

Capsules, O antigen, and UPEC

The predominant cause of urinary tract infections (UTIs) is uropathogenic *E. coli* (UPEC). Acquiring a UTI involves migration of the bacteria through the urethra to the bladder, attachment to the bladder wall, invasion and intracellular bacterial community (IBC) formation, followed in more invasive cases by possible translocation to the blood to cause sepsis, or migration through the ureters to the kidneys (see Chapter 9). Most of the UPEC isolates produce group 2 K antigens (Schneider et al., 2004) and a number of different group 2 capsules have been shown to be important for UTI pathogenesis at different points in the infection (Johnson, 1991; Schneider et al., 2004; Snyder et al., 2004; Buckles et al., 2009; Anderson et al., 2010). For example, the K1 capsule has also been shown to contribute to the decreased IBC formation and resistance of those IBCs to neutrophil penetration (Anderson et al., 2010). There are more effectors of complement in the kidneys compared to the bladder, and certain capsules and O antigens confer protection to UPEC strains from complement, especially in upper UTI infections (Buckles et al., 2009).

Polysialic acid CPSs and meningitis

Meningitis is characterized by extreme inflammation of the meninges and subarachnoid space caused by release of proinflammatory cytokines in response to bacteria in the cerebrospinal fluid. This causes increased permeability of the blood–brain barrier (BBB) and influx of fluid and leukocytes which promote the inflammatory state. After group B *Streptococcus*, *E. coli* K1 is the most common cause of neonatal meningitis in the developed world (Gaschignard et al., 2011; Stoll et al., 2011). The *E. coli* K1 capsule consists of polysialic acid and the same glycan is also found in the human brain, where it contributes to the anti-adhesive properties of the neural cell adhesion molecule (NCAM). Because it is normally found in human tissues, it is considered non-immunogenic.

The transit of *E. coli* K1 to the cerebrospinal fluid is a multistep process (see Chapter 10) involving translocation of the bacteria from the gut to the blood, survival and multiplication in the bloodstream, followed by translocation of the bacteria through the BBB, all of which require the presence of the polysialic acid capsule (Kim et al., 1992; Mushtaq et al., 2004, 2005; Xie et al., 2004). In the bloodstream, the K1 capsule prevents phagocytosis and complement-mediated killing, as described above, allowing bacteremia to occur. High levels of bacteremia are correlated with an increase in development of meningitis (Xie et al., 2004). Once a threshold is reached, the bacteria can bind to brain microvascular endothelial cells (BMEC) where they are internalized into endosomes which traverse the cell and release the bacteria into the choroid plexus of the brain, where cerebrospinal fluid is produced. The polysialic acid capsule has actually been shown to decrease the ability of the bacteria to bind BMECs but once inside the BMECs, endosomes containing unencapsulated *E. coli* fuse with lysosomes. Only endosomes containing encapsulated bacteria traverse BMECs and are released into the cerebrospinal fluid (Kim et al., 2003). Once the bacteria are in the choroid plexus, polysialic acid expression is significantly reduced, and by the time they reach the meninges it is absent entirely (Zelmer et al., 2008). This is not due to a selection for capsule-deficient strains, since isolating and culturing these bacteria leads to restoration of the encapsulated phenotype (Zelmer et al., 2008). While many regulatory mechanisms for the inducement of capsule synthesis upon switching from 20°C to 37°C have been described, it is not known if any of these mechanisms are responsible for the loss of capsule when the bacteria colonize the brain. The endo-neuraminidase is an enzyme that hydrolyzes polysialic acid. Transcriptome analysis of *E. coli* K1 shows that the gene encoding the endo-neuraminidase (found near the polysialic acid O-acetyltransferase gene, *neuO*, on the CUS-3 bacteriophage insertion) is significantly up-regulated upon binding to human microvascular endothelial cells (Xie et al., 2008). However, more research is needed to determine whether this protein is responsible for the observed loss of capsule upon reaching the cerebrospinal fluid.

CONCLUSIONS

In summary, LPS and CPS molecules fulfill diverse roles in the biology and pathogenesis of *E. coli*. In some cases a function may be related to a precise structural motif or feature, whereas in other cases the presence of a hydrated layer provides protection regardless of the exact glycan structure. While significant progress has been made in understanding the structures and biosynthesis of LPS and CPS, there are still significant gaps in the details of their interactions with elements of the innate and acquired immune system.

REFERENCES

Anderson, G.G., Goller, C.C., Justice, S., Hultgren, S.J., Seed, P.C., 2010. Polysaccharide capsule and sialic acid-mediated regulation promote biofilm-like intracellular bacterial communities during cystitis. Infect. Immun. 78 (3), 963–975.

Andersson, M., Carlin, N., Leontein, K., Lindquist, U., Slettengren, K., 1989. Structural studies of the O-antigenic polysaccharide of *Escherichia coli* O86, which possesses blood-group B activity. Carbohydr. Res. 185, 211–223.

Barua, S., Yamashino, T., Hasegawa, T., Yokoyama, K., Torii, K., Ohta, M., 2002. Involvement of surface polysaccharides in the organic acid resistance of shiga toxin-producing *Escherichia coli* O157:H7. Mol. Microbiol. 43 (3), 629–640.

Beddoe, T., Paton, A.W., Le Nours, J., Rossjohn, J., Paton, J.C., 2010. Structure, biological functions and application of the AB5 toxins. Trends. Biochem. Sci. 35 (7), 411–418.

Brown, K.L., Hancock, R.E.W., 2006. Cationic host defense (antimicrobial) peptides. Curr. Opin. Immun. 18, 24–30.

Bryant, C.E., Spring, D.R., Gangloff, M., Gay, N.J., 2010. The molecular basis of the host response to lipopolysaccharide. Nat. Rev. Microbiol. 8, 8–14.

Buckles, E.L., Wang, X., Lane, M.C., et al., 2009. Role of the K2 capsule in *Escherichia coli* urinary tract infection and serum resistance. J. Infect. Dis. 199, 1689–1697.

Burns, S.M., Hull, S.I., 1998. Comparison of loss of serum resistance by defined lipopolysaccharide mutants and an acapsular mutant of uropathogenic *Escherichia coli* O75:K5. Infect. Immun. 66 (9), 4244–4253.

Burns, S.M., Hull, S.I., 1999. Loss of resistance to ingestion and phagocytic killing by O$^-$ and K$^-$ mutants of a uropathogenic *Escherichia coli* O75:K5 strain. Infect. Immun. 67 (8), 3757–3762.

Cerca, N., Jefferson, K.K., 2008. Effect of growth conditions on poly-N-acetylglucosamine expression and biofilm formation in *Escherichia coli*. FEMS Microbiol. Lett. 283 (1), 36–41.

Cheong, C., Matos, I., Choi, J.-H., et al., 2010. Microbial stimulation fully differentiates monocytes to DC-SIGN/CD209$^+$ dendritic cells for immune T cell areas. Cell 143, 416–429.

Chng, S.S., Gronenberg, L.S., Kahne, D., 2010. Proteins required for lipopolysaccharide assembly in *Escherichia coli* form a transenvelope complex. Biochemistry 49 (22), 4565–4567.

Collins, R.F., Beis, K., Dong, C., et al., 2007. The 3D structure of a periplasm-spanning platform required for assembly of group 1 capsular polysaccharides in *Escherichia coli*. Proc. Natl. Acad. Sci. 194 (7), 2390–2395.

Cross, A.S., Kim, K.S., Wright, D.C., Sadoff, J.C., Gemski, P., 1986. Role of lipopolysaccharide and capsule in the serum resistance of bacteremic strains of *Escherichia coli*. J. Infect. Dis. 154 (3), 497–503.

Cuthbertson, L., Kos, V., Whitfield, C., 2010. ABC transporters involved in export of cell surface glycoconjugates. Microbiol. Mol. Biol. Rev. 74 (3), 341–362.

Cuthbertson, L., Mainprize, I.L., Naismith, J.H., Whitfield, C., 2009. Pivitol roles of the outer membrane polysaccharide export and polysaccharide copolymerase protein families in export of extracellular polysaccharides in gram-negative bacteria. Microbiol. Mol. Biol. Rev. 73 (1), 155–177.

Doerrler, W.T., Raetz, C.R., 2002. ATPase activity of the MsbA lipid flippase of *Escherichia coli*. J. Biol. Chem. 277 (39), 36697–36705.

Dong, C., Beis, K., Nesper, J., et al., 2006. Wza, the translocon for *E. coli* capsular polysaccharides defines a new class of membrane protein. Nature 444, 226–229.

El Ghachi, M., Derbise, A., Bouhss, A., Mengin-Lecruilx, D., 2005. Identification of multiple genes encoding membrane proteins with undecaprenyl pyrophosphate phosphatase (UppP) activity in *Escherichia coli*. J. Biol. Chem. 280 (19), 18689–18695.

Falkenhagen, U., Zingler, G., Naumann, G., 1991. Serum resistance in different serotypes of *Escherichia coli*. Zentralblatt. fur. Bakteriol. 275 (2), 216–222.

Finke, A., Bronner, D., Nikolaev, A.V., Jann, B., Jann, K., 1991. Biosynthesis of the *Escherichia coli* K5 polysaccharide, a representative of group II capsular polysaccharides: polymerization in vitro and characterization of the product. J. Bacteriol. 173 (13), 4088–4094.

Gaschignard, J., Levy, C., Romain, O., et al., 2011. Neonatal bacterial meningitis: 444 cases in 7 years. Pediatr. Infect. Dis. J. 30 (3), 212–217.

Goldman, R.C., White, D., Orskov, F., et al., 1982. A surface polysaccharide of *Escherichia coli* O111 contains O-antigen and inhibits agglutination of cells by O-antiserum. J. Bacteriol. 151 (3), 1210–1221.

Gotschlich, E.C., Fraser, B.A., Nishimura, O., Robbins, J.B., Liu, T.-Y., 1981. Lipid on capsular polysaccharides of Gram-negative bacteria. J. Biol. Chem. 256 (17), 8915–8921.

Grosdanov, L., Zahringer, U., Blum-Oehler, G., et al., 2002. A single nucleotide exchange in the wzy gene is responsible for the semirough O6 lipopolysaccharide phenotype and serum sensitivity of *Escherichia coli* strain Nissle 1917. J. Bacteriol. 184 (21), 5912–5925.

Guessous, F., Marcinkiewicz, M., Polanowska-Grabowska, R., et al., 2005. Shiga toxin 2 and lipopolysaccharide induce human microvascular endothelial cells to release chemokines and factors that stimulate platelet function. Infect. Immun. 73 (12), 8306–8316.

Gunn, J.S., 2008. The *Salmonella* PmrAB regulon: lipopolysaccharide modifications, antimicrobial peptide resistance and more. Trends Microbiol. 16 (6), 284–290.

Guo, L., Lim, K.B., Gunn, J.S., et al., 1997. Regulation of lipid A modifications by *Salmonella typhimurium* virulence genes phoP-phoQ. Science 276 (5310), 250–253.

Harrison, L.M., van den Hoogen, C., van Haaften, W.C.E., Tesh, V.L., 2005. Chemokine expression in the monocytic cell line THP-1 in response to purified shiga toxin 1 and/or lipopolysaccharides. Infect. Immun. 73 (1), 403–412.

He, F., Peng, J., Deng, X.L., et al., 2011. RhoA and NF-κB are involved in lipopolysaccharide-induced brain microvascular cell line hyperpermeability. Neuroscience 188, 35–47.

Heinrichs, D.E., Yethon, J.A., Whitfield, C., 1998. Molecular basis for structural diversity in the core regions of the lipopolysaccharides of *Escherichia coli* and *Salmonella enterica*. Mol. Microbiol. 30 (2), 221–232.

Herrera, C.M., Hankins, J.V., Trent, M.S., 2010. Activation of PmrA inhibits LpxT-dependent phosphorylation of lipid A promoting resistance to antimicrobial peptides. Mol. Microbiol. 76 (6), 1444–1460.

Hitchcock, P.J., Brown, T.M., 1983. Morphological heterogeneity among *Salmonella* lipopolysaccharide chemotypes in silver-stained polyacrylamide gels. J. Bacteriol. 154 (1), 269–277.

Horstman, A.L., Bauman, S.J., Kuehn, M.J., 2004. Lipopolysaccharide 3-deoxy-D-manno-octulosonic acid (Kdo) core determines bacterial association of secreted toxins. J. Biol. Chem. 279 (9), 8070–8075.

Horstman, A.L., Kuehn, M.J., 2002. Bacterial surface association of heat-labile enterotoxin through lipopolysaccharide after secretion via the general secretory pathway. J. Biol. Chem. 277, 32538–32545.

Inohara, N., Nunez, G., 2002. ML – a conserved domain involved in innate immunity and lipid metabolism. Trends Biochem. Sci. 27 (5), 219–221.

Jansson, P.E., Lonngren, J., Widmalm, G., et al., 1985. Structural studies of the O-antigen polysaccharides of *Klebsiella* O5 and *Escherichia coli* O8. Carbohydr. Res. 145, 59–66.

Johnson, J.R., 1991. Virulence factors in *Escherichia coli* urinary tract infection. Clin. Microbiol. Rev. 4 (1), 80–128.

Kaniuk, N.A., Vinogradov, E., Li, J., Monteiro, M.A., Whitfield, C., 2004. Chromosomal and plasmid-encoded enzymes are required for assembly of the R3-type core oligosaccharide in the lipopolysaccharide of *Escherichia coli* O157:H7. J. Biol. Chem. 279 (30), 31237–31250.

Kawai, T., Akira, S., 2010. The role of pattern-recognition receptors in innate immunity: update on Toll-like receptors. Nat. Immun. 11, 373—384.

Kenne, L., Lindberg, B., Soderstrom, E., Bundle, D.R., Griffith, D.W., 1983. Structural studies of the O-antigens from *Salmonella greenside* and *Salmonella adelaide*. Carbohydr. Res. 111 (2), 289–296.

Kim, H.M., Park, B.S., Kim, J.I., et al., 2007. Crystal structure of the TLR4-MD2 complex with bound endotoxin antagonist Eritoran. Cell 130 (5), 906–917.

Kim, K.J., Elliott, J., Di Cello, F., Stins, M.F., Kim, K.S., 2003. The K1 capsule modulates trafficking of *E. coli*-containing vacuoles and enhances intracellular bacterial survival in human brain microvascular endothelial cells. Cell. Microbiol. 5 (4), 245–252.

Kim, K.S., Itabashi, H., Gemski, P., Sadoff, J., Warren, R.L., Cross, A.S., 1992. The K1 capsule is the critical determinant in the development of *Escherichia coli* meningitis in the rat. J. Clin. Invest. 90, 897–905.

L'vov, V.L., Shashkov, A.S., Dmitriev, B.A., Kochetkov, N.K., Jann, B., Jann, K., 1984. Structural studies of the O-specific side chain of the lipopolysaccharide from *Escherichia coli* O:7. Carbohydr. Res. 126, 249–259.

Majdalani, N., Gottesman, S., 2005. The Rcs phosphorelay: a complex signal transduction system. Annu. Rev. Microbiol. 59, 379–405.

McGuire, E.J., Binkley, S.B., 1964. The structure and chemistry of colominic acid. Biochemistry 3, 247–251.

McNulty, C., Thompson, J., Barrett, B., Lord, L., Andersen, C., Roberts, I.S., 2006. The cell surface expression of group 2 capsular polysaccharides in *Escherichia coli*: the role of KpsD, RhsA, and a multi-protein complex at the pole of the cell. Mol. Microbiol. 59 (3), 907–922.

Meredith, T.C., Aggarwal, P., Mamat, U., Lindner, B., Woodard, R.W., 2006. Redifining the requisite lipopolysaccharide structure in *Escherichia coli*. ACS Chem. Biol. 17 (1), 33–42.

Merritt, E.A., Sixma, T.K., Kalk, K.H., van Zanten, B.A., Hol, W.G., 1994. Galactose-binding site in *Escherichia coli* heat-labile enterotoxin (LT) and cholera toxin (CT). Mol. Microbiol. 13, 745–753.

Moriez, R., Salvador-Cartier, C., Theodorou, V., Fioramonti, J., Eutamene, H., Bueno, L., 2005. Myosin light chain kinase is involved in lipopolysaccharide-induced disruption of colonic epithelial barrier and bacterial translocation in rats. Am. J. Pathol. 167 (4), 1071–1079.

Morona, R., Purins, L., Tocilj, A., Matte, A., Cygler, M., 2009. Sequence-structure relationships in polysaccharide co-polymerase (PCP) proteins. Trends Biochem. Sci. 34 (2), 78–84.

Mudrak, B., Rodriguez, D.L., Kuehn, M.J., 2009. Residues of heat-labile enterotoxin involved in bacterial cell surface binding. J. Bacteriol. 191 (9), 2917–2925.

Munford, R.S., 2008. Sensing gram-negative bacterial lipopolysaccharides: a human disease determinant? Infect. Immun. 76 (2), 454–465.

Mushtaq, N., Redpath, M.B., Luzio, J.P., Taylor, P.W., 2004. Prevention and cure of systemic *Escherichia coli* K1 infection by modification of the bacterial phenotype. J. Antimicrob. Chemother. 48, 1503–1508.

Mushtaq, N., Redpath, M.B., Luzio, J.P., Taylor, P.W., 2005. Treatment of experimental *Escherichia coli* infection with recombinant bacteriophage-derived capsule depolymerase. J. Antimicrob. Chemother. 56, 160–165.

Ohto, U., Fukase, K., Miyake, K., Satow, Y., 2007. Crystal structures of human MD-2 and its complex with antiendotoxic lipid IV$_A$. Science 316 (5831), 1632–1634.

Orskov, I., Orskov, F., Jann, B., Jann, K., 1977. Serology, chemistry, and genetics of O and K antigens of *Escherichia coli*. Microbiol. Mol. Biol. Rev. 41 (3), 667–710.

Park, B.S., Song, D.H., Kim, H.M., Choi, B., Lee, H., Lee, J., 2009. The structural basis of lipopolysaccharide recognition by the TLR4-MD-2 complex. Nature 458, 1191–1195.

Peleg, A., Shifrin, Y., Ilan, O., et al., 2005. Identification of an *Escherichia coli* operon required for formation of the O-antigen capsule. J. Bacteriol. 187 (15), 5259–5266.

Perolis, L.A., Parolis, H., Dutton, G.G., 1986. Structural studies of the O-antigen polysaccharide of *Escherichia coli* O9a. Carbohydr. Res. 155, 272–276.

Perry, M.B., MacLean, L., Griffith, D.W., 1986. Structure of the O-chain polysaccharide of the phenol-phase soluble lipopolysaccharide of *Escherichia coli* O157:H7. Biochem. Cell Biol. 64 (1), 21–28.

Prehm, P., Jann, B., Jann, K., 1976. The O9 antigen of *Escherichia coli*. Structure of the polysaccharide chain. Eur. J. Biochem. 67, 53–56.

Raetz, C.R.H., Whitfield, C., 2002. Lipopolysaccharide endotoxins. Ann. Rev. Biochem. 71, 635–700.

Raetz, C.R., Reynolds, C.M., Trent, M.S., Bishop, R.E., 2007. Lipid A modification systems in gram-negative bacteria. Annu. Rev. Biochem. 76, 295–329.

Rahn, A., Beis, K., Naismith, J.H., Whitfield, C., 2003. A novel outer membrane protein, Wzi, is involved in surface assembly of the *Escherichia coli* K30 group 1 capsule J. Bacteriol. 185 (19), 5882–5890.

Ramos-Morales, F., Prieto, A.I., Beuzon, C.R., Holden, D.W., Casadesus, J., 2003. Role for *Salmonella enterica* enterobacterial common antigen in bile resistance and virulence. J. Bacteriol. 185 (17), 5328–5332.

Rendueles, O., Travier, L., Latour-Lambert, P., et al., 2011. Screening of *Escherichia coli* species biodiversity reveals new biofilm-associated antiadnesion polysaccharides. MBio 2 (3), e00043–11.

Reynolds, C.M., Raetz, C.R., 2009. Replacement of lipopolysaccharide with free lipid A molecules in *Escherichia coli* mutants lacking all core sugars. Biochemistry 48 (40), 9627–9640.

Rigg, G.P., Barrett, B., Roberts, I.S., 1998. The localization of KpsC, S and T, and KfiA, C and D proteins involved in the biosynthesis of the *Escherichia coli* K5 capsular polysaccharide: evidence for a membrane-bound complex. Microbiology 144, 2905–2914.

Rodriguez, M.L., Jann, B., Jann, K., 1988. Structure and serological characteristics of the capsular K4 antigen of *Escherichia coli* O5:K4:H4, a fructose-containing polysaccharide with a chondroitin backbone. Eur. J. Biochem. 177 (1), 117–124.

Ruiz, N., Kahne, D., Silhavy, T.J., 2009. Transport of lipopolysaccharide across the cell envelope: the long road of discovery. Nat. Rev. Microbiol. 7 (9), 677–683.

Rush, J.S., Alaimo, C., Robbiani, R., Wacker, M., Waechter, C.J., 2010. A novel epimerase that converts GlcNAc-P-P-undecaprenol to GalNAc-P-P-undecaprenol in *Escherichia coli* O157. J. Biol. Chem. 285 (3), 1671–1680.

Russo, T.A., Liang, Y., Cross, A.S., 1994. The presence of K54 capsular polysaccharide increases the pathogenicity of *Escherichia coli in vivo*. J. Infect. Dis. 169 (1), 112–118.

Russo, T.A., Moffitt, M.C., Hammer, C.H., Frank, M.M., 1993. Tn*phoA*-mediated disruption of K54 capsular polysaccharide genes in *Escherichia coli* confers serum specificity. Infect. Immun. 61 (8), 3578–3582.

Russo, T.A., Sharma, G., Brown, C.R., Campagnari, A.A., 1995. Loss of the O4 antigen moiety from the lipopolysaccharide of an extraintestinal isolate of *Escherichia coli* has only minor effects on serum sensitivity and virulence in vivo. Infect. Immun. 63 (4), 1263–1269.

Saldana, Z., Xicohtencatl-Cortes, J., Avelino, F., et al., 2009. Synergistic role of curli and cellulose in cell adherence and biofilm formation of attaching and effacing *Escherichia coli* and identification of Fis as a negative regulator of curli. Environ. Microbiol. 11 (4), 992–1006.

Sarma, J.V., Ward, P.A., 2011. The complement system. Cell Tissue Res. 343, 227–235.

Schneider, G., Dobrindt, U., Bruggemann, H., et al., 2004. The pathogenicity island-associated K15 capsule determinant exhibits a novel genetic structure and correlates with virulence in uropathogenic *Escherichia coli* strain 536. Infect. Immun. 72 (10), 5993–6001.

Schumann, R.R., 2011. Old and new findings on lipopolysaccharide-binding protein: a soluble pattern-recognition molecule. Biochemi. Soc. Trans. 39, 989–993.

Sheth, P., Delos Santos, N., Seth, A., LaRusso, N.F., Rao, R.K., 2007. Lipopolysaccharide disrupts tight junctions in cholangiocyte monolayers by a c-Src-, TLR4-, and LBP-dependent mechanism. Am. J. Physiol. Gastrointest. Liver Physiol. 293, G308–G318.

Shifrin, Y., Peleg, A., Ilan, O., et al., 2008. Transient shielding of intimin and the type III secretion system of enterohemorrhagic and enteropathogenic *Escherichia coli* by a group 4 capsule. J. Bacteriol. 190 (14), 5063–5074.

Smith, A.E., Kim, S.-H., Liu, F., et al., 2008. PagP activation in the outer membrane triggers R3 core oligosaccharide truncation in the cytoplasm of *Escherichia coli* O157:H7. J. Biol. Chem. 283 (7), 4332–4343.

Snyder, J.A., Haugen, B.J., Buckles, E.L., et al., 2004. Transcriptome of uropathogenic *Escherichia coli* during urinary tract infection. Infect. Immun. 72 (11), 6373–6381.

Sperandeo, P., Deho, G., Polissi, A., 2009. The lipopolysaccharide transport system of Gram-negative bacteria. Biochemi. Biophys. Acta 1791, 594–602.

Stahl, A., Sartz, L., Nelsson, A., Bekassy, Z.D., Karpman, D., 2009. Shiga toxin and lipopolysaccharide induce platelet-leukocyte aggregates and tissue factor release, a thrombotic mechanism in hemolytic uremic syndrome. PLOS One 4 (9), 1–12.

Stenutz, R., Weintraub, A., Widmalm, G., 2006. The structures of *Escherichia coli* O-polysaccharide antigens. FEMS Micriobiol. Rev. 30 (3), 382–403.

Stoll, B.J., Hansen, N.I., Sanchez, P.J., et al., 2011. Early onset neonatal sepsis: the burden of group B streptococcal and *E. coli* disease continues. Pediatrics 127 (5), 817–826.

Szabo, G., Bala, S., Petrasek, J., Gattu, A., 2010. Gut-liver axis and sensing microbes. Dig. Dis. 28, 737–744.

Thomas, C.J., Kapoor, M., Sharma, S., et al., 2002. Evidence of a trimolecular complex involving LPS, LPS binding protein and soluble CD14 as an effector of LPS response. FEBS Lett. 531, 184–188.

Touze, T., Tran, A.X., Hankins, J.V., Mengin-Lecreulx, D., Trent, M.S., 2008. Periplasmic phosphorylation of lipid A is linked to the synthesis of undecaprenyl phosphate. Mol. Microbiol. 67 (2), 264–277.

Tzeng, Y.-L., Datta, A.K., Strole, C.A., Lobritz, M.A., Carlson, R.W., Stephens, D.S., 2005. Translocation and surface expression of lipidated serogroup B capsular polysaccharide in *Neisseria meningitidis*. Infect. Immun. 73 (3), 1491–1505.

Valle, J., Da Re, S., Henry, N., et al., 2006. Broad-spectrum biofilm inhibition by a secreted bacterial polysaccharide. Proc. Nat. Acad. Sci. 103 (33), 12558–12563.

Vandenbroucke, E., Mehta, D., Minshall, R., Malik, A.B., 2008. Regulation of endothelial junctional permeability. Annals NY Acad. Sci. 1123, 134–145.

Vann, W.F., Schmidt, M.A., Jann, B., Jann, K., 1981. The structure of the capsular polysaccharide (K5 antigen) of urinary-tract-infective *Escherichia coli* O10:K5:H4. A polymer similar to desulfo-heparin. Eur. J. Biochem. 116 (2), 359–364.

Vimr, E.R., Steenbergen, S.M., 2009. Early molecular-recognition events in the synthesis and export of group 2 capsular polysaccharides. Microbiology 155 (Pt 1), 9–15.

Weiss, J., Victor, M., Cross, A.S., Elsbach, P., 1982. Sensitivity of K1-encapsulated *Escherichia coli* to killing by the bactericidal/permeability-increasing protein of rabbit and human neutrophils. Infect. Immun. 38 (3), 1149–1153.

Whitfield, C., 2006. Biosynthesis and assembly of capsular polysaccharides in *Escherichia coli*. Annu. Rev. Biochem. 75, 39–68.

Wugeditsch, T., Paiment, A., Hocking, J., Drummelsmith, J., Forrester, C., Whitfield, C., 2001. Phosphorylation of Wzc, a tyrosine autokinase, is essential for assembly of group 1 capsular polysaccharides I *Escherichia coli*. J. Biol. Chem. 276 (4), 2361–2371.

Xie, Y., Kim, K.J., Kim, K.S., 2004. Current concepts on *Escherichia coli* K1 translocation of the blood-brain barrier. FEMS Immunol. Med. Microbiol. 42, 271–279.

Xie, Y., Parthasarathy, G., Di Cello, F., Teng, C.H., Paul-Satyaseela, M., Kim, K.S., 2008. Transcriptome of *E. coli* K1 bound to human brain microvascular endothelial cells. Biochem. Biophys. Res. Commun. 365 (1), 201–206.

Yethon, J.A., Gunn, J.S., Ernst, R.K., et al., 2000. *Salmonella enterica* serovar typhimurium waaP mutants show increased susceptibility to polymyxin and loss of virulence *in vivo*. Infect. Immun. 68 (8), 4485–4491.

Zelmer, A., Bowen, M., Jokilammi, A., Finne, J., Luzio, J.P., Taylor, P.W., 2008. Differential expression of the polysialyl capsule during blood-to-brain transit of neuropathogenic *Escherichia coli* K1. Microbiology 154, 2522–2532.

Zhang, P., Snyder, S., Feng, P., et al., 2006. Role of N-acetylglucosamine within core lipopolysaccharide of several species of Gram-negative bacteria in targeting the DC-SIGN (CD209). J. Immun. 177, 4002–4011.

Note: Page numbers followed by "f" or "t" indicate figure or table respectively

Printed and bound by CPI Group (UK) Ltd, Croydon, CR0 4YY

16/10/2024

01774872-0002